XVII International Conference on the Use of Computers in Radiation Therapy

AA008670

(ICCR 2013)

Journal of Physics: Conference Series Volume 489

Melbourne, Australia
6-9 May 2013

Editors:

Annette Haworth **Tomas Kron**

ISBN: 978-1-63266-714-4
ISSN: 1742-6588

Printed from e-media with permission by:

Curran Associates, Inc.
57 Morehouse Lane
Red Hook, NY 12571

Some format issues inherent in the e-media version may also appear in this print version.

Printed by Curran Associates, Inc. (2014)

Institute of Physics
Dirac House, Temple Back
Bristol BS1 6BE UK

Phone: 44 1 17 929 7481
Fax: 44 1 17 920 0979

techtracking@iop.org

Additional copies of this publication are available from:

Curran Associates, Inc.
57 Morehouse Lane
Red Hook, NY 12571 USA
Phone: 845-758-0400
Fax: 845-758-2634
Email: curran@proceedings.com
Web: www.proceedings.com

TABLE OF CONTENTS

DOSE CALCULATION

012001 Common Platform of Monte Carlo Dose Calculation on Universal Grid Interface with Geant4 Based Particle Therapy Simulation Framework 1
G Iwai, W Takase, T Aso, Y Watase, T Sasaki, T Akagi, T Yamashita, Y Maeda, T Nishio

012002 Analytical Probabilistic Proton Dose Calculation and Range Uncertainties 6
M Bangert, P Hennig, U Oelfke

012003 Improvements in the Stopping Power Library LibdEDx and Release of the Web GUI dedx.au.dk 12
J Toftegaard, A Lühr, N Sobolevsky, N Bassler

012004 SHIELD-HIT12A - A Monte Carlo Particle Transport Program for Ion Therapy Research 17
N Bassler, D C Hansen, A Lühr, B Thomsen, J B Petersen, N Sobolevsky

012005 Monte Carlo Simulation of the Transit Dosimetric Response of an a-Si Electronic Portal Imaging Device 22
S J Blake, A L McNamara, P Vial, L Holloway, P B Greer, Z Kuncic

012006 The Influence of Monte Carlo Source Parameters on Detector Design and Dose Perturbation in Small Field Dosimetry 28
P H Charles, S B Crowe, T Kairn, R Knight, B Hill, J Kenny, C M Langton, J V Trapp

012007 Recommendations for Dose Calculations of Lung Cancer Treatment Plans Treated with Stereotactic Ablative Body Radiotherapy (SABR) 32
S Devpura, M S Siddiqui, D Chen, D Liu, H Li, S Kumar, J Gordon, M Ajlouni, B Movsas, I J Chetty

012008 in Vivo and Phantom Measurements Versus Eclipse TPS Prediction of Near Surface Dose for SBRT Treatments 36
G A Cho, A Ralston, M M Tin, D Martin, S Pickard, J-H Kim, R Tse

012009 Automated Segmentation and Dose-volume Analysis with DICOMautomaton 42
H Clark, S Thomas, V Moiseenko, R Lee, B Gill, C Duzenli, J Wu

012010 GMC ['gimik]: A One-variable Monte Carlo Dose Algorithm for Proton Therapy 46
N Depauw, B Clasie, T Madden, A Rosenfeld, H Kooy

012011 In Silico Radiobiology: Have We Reached the Limit of Monte Carlo Simulations? 51
Y Gholami, M Toghyani, C Champion, Z Kuncic

012012 an Improved Monte-carlo Model of the Varian EPID Separating Support Arm and Rear-housing Backscatter 56
M E Monville, Z Kuncic, P B Greer

012013 A Monte Carlo Tool for Raster-scanning Particle Therapy Dose Computation 61
U Jelen, M Radon, A Santiago, A Wittig, F Ammazzalorso

012014 Using Narrow Beam Profiles to Quantify Focal Spot Size, for Accurate Monte Carlo Simulations of SRS/SRT Systems 66
T Kairn, S B Crowe, P H Charles, J V Trapp

012015 In Vivo Verification of Radiation Dose Delivered to Healthy Tissue During Radiotherapy for Breast Cancer 70
P Lonski, M L Taylor, W Hackworth, A Phipps, R D Franich, T Kron

012016 Fast in Vivo Volume Dose Reconstruction Via Reference Dose Perturbation 75
W Lu, M Chen, X Mo, D Parnell, G Olivera, D Galmarini

012017 Open-Source Medical Devices (OSMD) Design of a Small Animal Radiotherapy System 80
S Prajapati, T R Mackie, R Jeraj

012018 Revealing the Underlying Mechanism of Microbeam Radiation Therapy with Low Energy Monte Carlo Simulations 86
A L McNamara, U Oelfke, Z Kuncic

012019 An Improved Method of Heterogeneity Compensation for the Convolution / Superposition Algorithm 90
R Jacques, T McNutt

012020 IMRT Treatment Monitor Unit Verification Using Absolute Calibrated BEAMnrc and Geant4 Monte Carlo Simulations 95
B M Oborn, M Williams, M Bailey, M G Carolan

012021 Formal Quality Control for a Proton Monte Carlo System in Radiation Therapy 100
J Perl

012022 Accounting for the Effects of Carbon Fibre Treatment Couches in the XiO Treatment Planning System 106
M Roche, S M Mahou-Lago, T Crabtree

012023 Calculation of the Characteristics of Clinical High-energy Photon Beams with EGS5-MPI 111
M Shimizu, Y Morishita, M Kato, T Kurosawa, T Tanaka, N Takata, N Saito

IMAGING FOR TREATMENT PLANNING AND MOTION

012024 Motion Artifact Detection in Four-dimensional Computed Tomography Images 115
G Bouilhol, M Ayadi, R Pinho, S Rit, D Sarrut

012025 a Comparison of Rigid Registration Methods for Prostate Localization on CBCT and the Dependence on Rectum Distension 121
C Boydev, D Pasquier, F Derraz, L Peyrodie, A Taleb-Ahmed, J P Thiran

012026 In-room Breathing Motion Estimation from Limited Projection Views Using a Sliding Deformation Model 127
V Delmon, J Vandemeulebroucke, R Pinho, M V Oliva, D Sarrut, S Rit

012027 Fast Cine-Magnetic Resonance Imaging Point Tracking for Prostate Cancer Radiation Therapy Planning 132
J Dowling, K Dang, C D Fox, S Chandra, S Gill, T Kron, D Pham, F Foroudi

012028 Contribution of FDOPA PET to Radiotherapy Planning for Advanced Glioma 138
N Dowson, M Fay, P Thomas, R Jeffree, R McDowall, C Winter, A Coulthard, J Smith, Y Gal, P Bourgeat, O Salvado, S Crozier, S Rose

012029 Fully Automated Shape Model Positioning for Bone Segmentation in Whole-body CT Scans 143
A Fränzle, M Sumkauskaite, J Hillengass, T Bäuerle, R Bendl

012030 Investigation of the Accuracy of Breast Tissue Segmentation Methods for the Purpose of Developing Breast Deformation Models for Use in Adaptive Radiotherapy 147
P Juneja, E J Harris, P M Evans

012031 Graded-threshold Parametric Response Maps: Towards a Strategy for Adaptive Dose Painting 151
A Lausch, N Jensen, J Chen, T Y Lee, M Lock, E Wong

012032 Effects of Registration Error on Parametric Response Map Analysis: A Simulation Study Using Liver CT-perfusion Images 156
A Lausch, N K G Jensen, J Chen, T Y Lee, M Lock, E Wong

012033 Audiovisual Biofeedback Improves Image Quality and Reduces Scan Time for Respiratory-gated 3D MRA 161
D Lee, P B Greer, J Arm, P Keall, T Kim

012034 Fully-Deformable Patient Motion Models from Cone-beam CT for Radiotherapy Applications 165
J Martin, J McClelland, C Yip, C Thomas, C Hartill, S Ahmad, I Meir, D Landau, D Hawkes

012035 Optimized Dynamic Contrast-enhanced Cone-beam CT for Target Visualization During Liver SBRT 169
B L Jones, C Altunbas, B Kavanagh, T Schefter, M Miften

012036 Construction of a Patient Observation System Using KINECT™ 173
K Miyaura, Y Kumazaki, C Fukushima, S Kato, H Saitoh

012037 2D-3D Registration for Brain Radiation Therapy Using a 3D CBCT and a Single Limited Field-of-View 2D kV Radiograph 177
R Munbodh, D J Moseley

012038 A Novel Marker Enhancement Filter (MEF) for Fluoroscopic Images 181
O Peshko, T N Davidson, J Modersitzki, T Terlaky, D J Moseley

012039 Inverse-consistent Rigid Registration of CT and MR for MR-based Planning and Adaptive Prostate Radiation Therapy 187
D Rivest-Hénault, N Dowson, P Greer, J Dowling

012040 Tissue Motion Tracking at the Edges of a Radiation Treatment Field Using Local Optical Flow Analysis 193
P T Teo, S Pistorius

012041 Establishing Tumour Tracking Accuracy in Free-breathing Respiratory Gated SBRT of Lung Cancer 197
C-D Wen, C Wong, T Ackerly, J Ruben, J Millar

012042 Computational Model to Simulate the Interplay Effect in dynamic IMRT delivery 202
S A Yoganathan, K J M Das, S Kumar

TREATMENT PLANNING AND OPTIMISATION

012043 GPU-accelerated Automatic Identification of Robust Beam Setups for Proton and Carbon-ion Radiotherapy 208
F Ammazzalorso, T Bednarz, U Jelen

012044 Ultra-fast Fluence Optimization for Beam Angle Selection Algorithms 212
M Bangert, P Ziegenhein, U Oelfke

012045 PyTRiP - A Toolbox and GUI for the Proton/Ion Therapy Planning System TRiP 217
J Toftegaard, J B Petersen, N Bassler

012046 A Case Study of IMRT Planning (Plan B) Subsequent to a Previously Treated Imrt Plan (Plan A) 221
F Cao, C Leong, J Schroeder, B Lee

012047 Radiation Treatment Dose Optimisation Using Poisson Tumour Control Probability Parameters 226
G A Cho, M A Ebert, L Holloway, Z Kuncic, C Baldock, D I Thwaites

012048 Automatic Atlas Based Electron Density and Structure Contouring for MRI-based Prostate Radiation Therapy on the Cloud 232
J A Dowling, N Burdett, P B Greer, J Sun, J Parker, P Pichler, P Stanwell, S Chandra, D Rivest-Hénault, S Ghose, O Salvado, J Fripp

012049 Integration of Second Cancer Risk Calculations in a Radiotherapy Treatment Planning System 238
M Hartmann, U Schneider

012050 GPU-Accelerated Ray-tracing for Real-time Treatment Planning 243
H Heinrich, P Ziegenhein, C P Kamerling, H Froening, U Oelfke

012051 Predicting the Likelihood of QA Failure Using Treatment Plan Accuracy Metrics 249
T Kairn, S B Crowe, J Kenny, R T Knight, J V Trapp

012052 A 3D Isodose Manipulation Tool for Interactive Dose Shaping 254
C P Kamerling, P Ziegenhein, H Heinrich, U Oelfke

012053 Bayesian Decision Support for Adaptive Lung Treatments 260
D McShan, Y Luo, M Schipper, R TenHaken

012054 Automatic Treatment Planning Implementation Using a Database of Previously Treated Patients 264
J A Moore, K Evans, W Yang, J Herman, T McNutt

012055 Clinical Implementation of Dose-volume Histogram Predictions for Organs-at-risk in IMRT Planning 269
K L Moore, L M Appenzoller, J Tan, J M Michalski, W L Thorstad, S Mutic

012056 Dose Distribution Transfer from Cyberknife to Varian Treatment Planning System 273
W Osewski, K Slosarek, B Karaszewska

012057 A Comparison of Coordinate Systems for Use in Determining a Radiotherapy Delineation Margin for Whole Breast 276
E M Pogson, L Bell, V Batumalai, E S Koh, G Delaney, P Metcalfe, L Holloway

012058 Automatic Treatment Planning with Convex Imputing 282
G A Sayre, D Ruan

012059 Modeling Clinical Radiation Responses in the IMRT Era 286
J L Schwartz, D Murray, R D Stewart, M H Phillips

012060 Evaluation of Atlas Based Auto-segmentation for Head and Neck Target Volume Delineation in Adaptive/Replan IMRT 291
R Speight, E Karakaya, R Prestwich, M Sen, R Lindsay, R Harding, J Sykes

012061 Two-dimensional Inverse Planning and Delivery for Precision Preclinical Radiobiological Investigations 295
J M P Stewart, P E Lindsay, D A Jaffray

012062 Planjury: Probabilistic Plan Evaluation Revisited 299
M Witte, J-J Sonke, M van Herk

012063 PyCMSXiO: An External Interface to Script Treatment Plans for the Elekta® CMS XIO Treatment Planning System 305
A Xing, S Arumugam, L Holloway, G Goozee

012064 Commissioning of SharePlan: The Liverpool Experience 310
A Xing, S Deshpande, S Arumugam, A George, L Holloway, G Goozee

012065 Inverse Planning in the Age of Digital Linacs: Station Parameter Optimized Radiation Therapy (SPORT) 316
L Xing, R Li

012066 Interactive Dose Shaping - Efficient Strategies for CPU-based Real-time Treatment Planning 322
P Ziegenhein, C P Kamerling, U Oelfke

VERIFICATION, RISK ASSESSMENT AND IGRT

012067 A Comprehensive EPID-based 3D Validation Technique for TrueBeam-delivered VMAT Plans 328
W Ansbacher, I M Gagne, C-L Swift

012068 A Comprehensive Tool to Analyse Dynamic Log Files from an Elekta-Synergy Accelerator 332
S Arumugam, A Xing, C Pagulayan, L Holloway

012069 A Decision Support Tool to Optimize IMRT QA Workflow in a Multi-vendor Equipment Environment 338
S Arumugam, A Xing, P Vial, L Holloway

012070 Automatic Image Registration Performance for Two Different CBCT Systems; Variation with Imaging Dose 344
J Barber, J R Sykes, L Holloway, D I Thwaites

012071 Evaluation of the Ability of a 2D Ionisation Chamber Array and an Epid to Detect Systematic Delivery Errors in Imrt Plans 348
O Bawazeer, A Gray, S Arumugam, P Vial, D Thwaites, J Descallar, L Holloway

012072 Impact of Image Noise on Gamma Index Calculation 353
M Chen, X Mo, D Parnell, G Olivera, D Galmarini, W Lu

012073 Assessing Local Outcomes in Heterogeneous Gliomas 357
N Dowson, P Thomas, Y Gal, M Fay, R L Jeffree, C Winter, A Coulthard, J Smith, P Bourgeat, O Salvado, S Crozier, S Rose

012074 Constructing a Clinical Decision-making Framework for Image-guided Radiotherapy Using a Bayesian Network 361
C Hargrave, M Moores, T Deegan, A Gibbs, M Poulsen, F Harden, K Mengersen

012075 An Automatic Dose Verification System for Adaptive Radiotherapy for Helical Tomotherapy 365
X Mo, M Chen, D Parnell, G Olivera, D Galmarini, W Lu

012076 Segmentation of Cone-beam CT Using a Hidden Markov Random Field with Informative Priors 369
M Moores, C Hargrave, F Harden, K Mengersen

012077 Deformable Image Registration for Geometrical Evaluation of DIBH Radiotherapy Treatment of Lung Cancer Patients 373
W Ottosson, J A L Andersen, S Borrisova, A Mellemgaard, C F Behrens

012078 Development of A Novel Image Guidance Alternative for Patient Localization using Topographic Images for TomoTherapy 379
X S Qi, B White, D A Low

012079 The Reconstruction Toolkit (RTK), An Open-Source Cone-Beam CT Reconstruction Toolkit Based on the Insight Toolkit (ITK) 383
S Rit, M V Oliva, S Brousmiche, R Labarbe, D Sarrut, G C Sharp

012080 Potential Interoperability Problems Facing Multi-Site Radiation Oncology Centers in the Netherlands 387
J Scheurleer, P Koken, R Wessel

012081 Target Point Correction Optimized Based on the Dose Distribution of Each Fraction in Daily IGRT 392
M Stoll, K Giske, E M Stoiber, M Schwarz, R Bendl

012082 Improving Accuracy of Markerless Tracking of Lung Tumours in Fluoroscopic Video by Incorporating Diaphragm Motion 396
M Schwarz, H Teske, M Stoll, R Bendl

012083 Towards Adaptive Radiotherapy for Head and Neck Patients: Validation of an In-house Deformable Registration Algorithm 400
C Veiga, J McClelland, S Moinuddin, K Ricketts, M Modat, S Ourselin, D D'Souza, G Royle

012084 Streamlining EPID-based IMRT Quality Assurance: Auto-Analysis and Auto-Report Generation 404
A Xing, S Arumugam, S Deshpande, A George, L Holloway, G Goozee, A Gray, P Vial

012085 Fast Regional Readout CMOS Image Sensor for Dynamic MLC Tracking 410
H Zin, E Harris, J Osmond, P Evans

TRIALS, REGISTRIES AND MORE

012086 The Use of a Virtual Reality Simulator to Explore and Understand the Impact of Linac Mis-calibrations 414
A W Beavis, J W Ward

012087 An MLE Method for Finding LKB NTCP Model Parameters Using Monte Carlo Uncertainty Estimates 420
M Carolan, B Oborn, K Foo, A Haworth, S Gulliford, M Ebert

012088 Semi-automated Contour Recognition Using DINCOMautomaton 426
H Clark, J Wu, V Moiseenko, R Lee, B Gill, C Duzenli, S Thomas

012089 Development of a Multi-Centre Clinical Trial Data Archiving and Analysis Platform for Functional Imaging 431
B Driscoll, D Jaffray, C Coolens

012090 Derivation and Representation of Dose-volume Response from Large Clinical Trial Data Sets: An Example from the RADAR Prostate Radiotherapy Trial 436
M A Ebert, K Foo, A Haworth, S L Gulliford, R Kearvall, A Kennedy, S Richardson, M Krawiec, N Stewart, D J Joseph, J W Denham

012091 Combining Dosimetry & Toxicity: Analysis of Two UK Phase III Clinical Trials 441
S L Gulliford

012092 Total Centralisation and Optimisation of an Oncology Management Suite Via Citrix® 447
C James, J Frantzis, L Ripps, P Fenton

012093 An Investigation of the Impact of Variations of DVH Calculation Algorithms on DVH Dependant Radiation Therapy Plan Evaluation Metrics 451
A M Kennedy, J Lane, M A Ebert

012094 Trust, But Verify – Accuracy of Clinical Commercial Radiation Treatment Planning Systems 456
J Lehmann, J Kenny, J Lye, L Dunn, I Williams

012095 Browser Based Platform in Maintaining Clinical Activities – Use of The iPads in Head and Neck Clinics 460
W Y Yang, J Moore, H Quon, K Evans, A Sharabi, J Herman, A Hacker-Prietz, T McNutt

012096 Organism-level Models: When Mechanisms and Statistics Fail Us 465
M H Phillips, J Meyer, W P Smith, J K Rockhill

012097 a Database Paradigm for the Management of DICOM-RT Structure Sets Using a Geographic Information System 471
W Shao, P A Kupelian, J Wang, D A Low, D Ruan

012098 Has the Use of Computers in Radiation Therapy Improved the Accuracy in Radiation Dose Delivery? 476
J V Dyk, J Battista

012099 Integration of a Clinical Trial Database with a PACS 482
M van Herk

012100 A DICOM Based Radiotherapy Plan Database for Research Collaboration and Reporting 487
J Westberg, S Krogh, C Brink, I R Vogelius

Author Index

Preface

Editorial

Dear colleagues,

From a professional perspective there is nothing quite as enjoyable as attending a good conference with colleagues that share not only interest for the same topic but also a similar vision for their field of work. Proceedings cannot replace the direct contact and lively conversations between participants who are at the same time at the same place sharing ideas and discussion. However, proceedings complement the actual conference by

1 giving the ideas, research findings and debates which characterise the conference lasting presence
2 giving participants an opportunity to refresh their memory about issues they would like to follow up on and
3 providing an opportunity for others who were not able to attend the meeting to share in the thoughts and issues discussed at the conference.

For the proceedings of the International Conferences on the Use of Computers in Radiation Therapy (ICCR) all this has been important in the past as the high citation rates for ICCR proceedings show. We hope that also the present proceedings will appeal to participants and others who are interested in cancer and ways to treat patients affected by it. It is exciting to note that what sounded like a small niche field ("Computers in Radiation Therapy") has become a very broad forum to discuss all aspects of cancer diagnosis and therapy. Of particular interest for the 17th ICCR have been data and how to collect, organise and use it effectively. This is relevant for most areas in medicine and we believe that radiation therapy with its focus on evidence based practice and measurable and standardised activities has an important and possibly leading role to play.

The 100 manuscripts combined in the present proceedings can hopefully contribute to this. The proceedings are organised into five different streams which reflect the foci of the conference:

- Dose Calculation
- Imaging for treatment planning and Motion Management
- Treatment planning and optimisation
- Verification, Risk assessment and IGRT
- Trials, registries and more

Each of the streams had its own theme editor and we would like to thank Peter Greer, Sarah Gulliford, Tomas Janssen, Joerg Lehmann and Peter Metcalfe and for their invaluable contributions. I believe we all shared the joy of reading the huge variety of manuscripts which reflect a dynamic field and a truly international authorship. All manuscripts have been peer reviewed by at least two independent referees. Still, proceedings are different from journal articles. The focus of the proceedings is on interest for others, novelty and creativity. While trying to ensure a common format to all contributions we have also aimed to maintain the style and approach of the individual authors. We would like to thank all authors and reviewers for their contribution to these ICCR proceedings.

On behalf of the editorial team we would like to invite you to explore the proceedings and hope that you enjoy reading them as much as we did editing them.

Yours sincerely

Annette Haworth and Tomas Kron
Editors, Proceedings of the 17th ICCR

ICCR 2013 in Images

Pictorial Essay of the

17th International Conference on the Use of Computers in Radiation Therapy

Melbourne 2013

by Joerg Lehmann

Tomas Kron, Convenor ICCR 2013

MELBOURNE
ICCR 2013, Melbourne Australia
GOD PARTICLE

ICCR 2013, Melbourne Australia

The BIG Machine
D.A. Jaffray
Princess Margaret Cancer Centre
Techna Institute/Ontario Cancer Institute
University Health Network
University of Toronto
Toronto, Ontario, CANADA
UNIVERSITY OF TORONTO
TECHNA
UHN

OUR NUMBERS ARE MANY,
OUR INTENT IS SINGULAR.
BY PURPOSE TO
FIGHT CANCER

AUSTRALIAN

Centre
Biofouling
105
ICCR 2013 - Concurrent
Breakout Session
Monday, May 06, 2013
ICCR 2013, Melbourne Australia
www.ICCR2013.org
6 - 9 May 2013

ICCR 2013, Melbourne Australia

ICCR 2013
Update/teaching Session 1: Image Guidance
Tuesday 7 May 2013 | 7:30 - 8:50
Plenary 3
D. Jaffray
P. Keall

NOVALIS
RADIOSURGERY
Novalis
ICCR 2013, Melbourne Australia

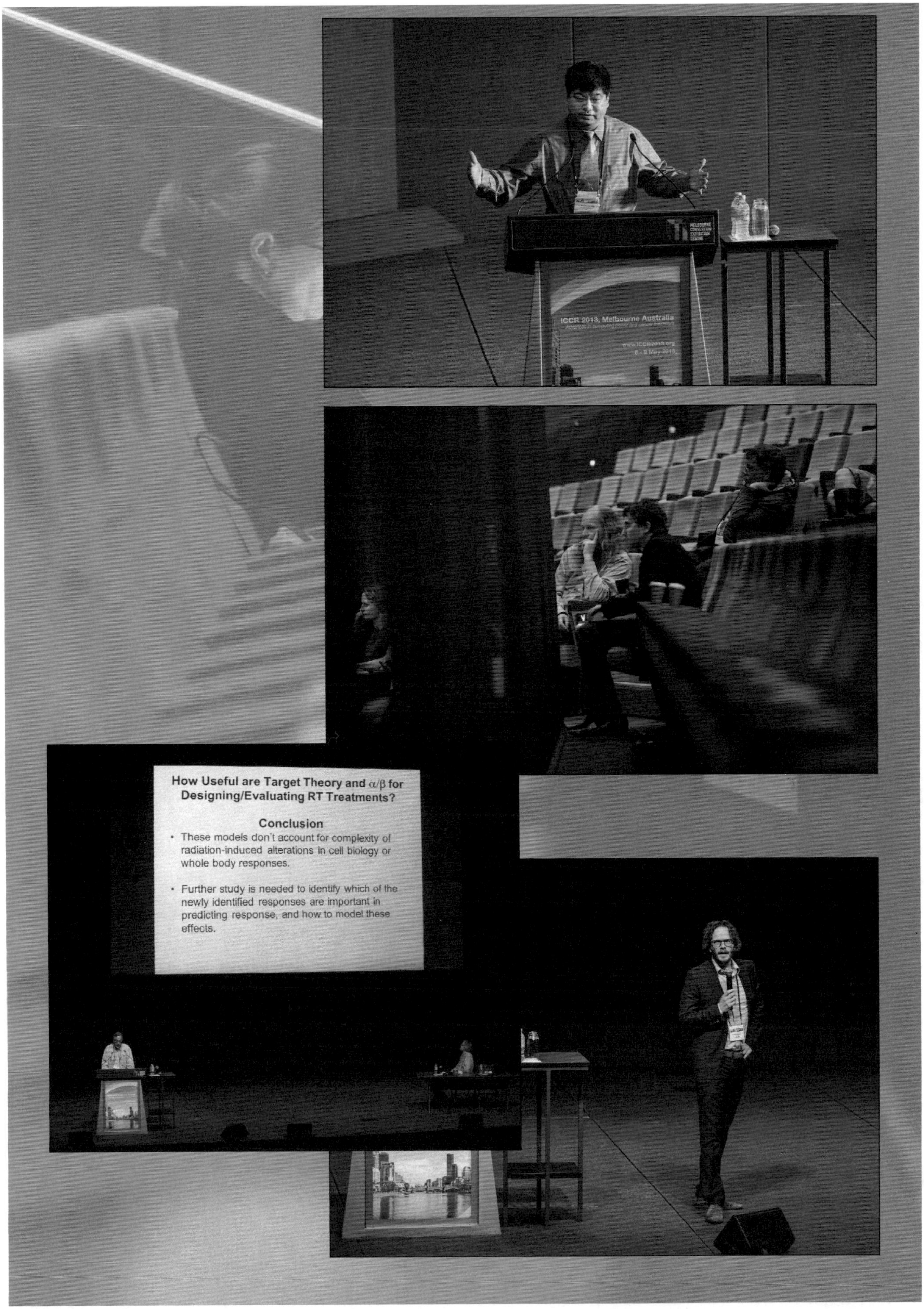
ICCR 2013, Melbourne Australia
How Useful are Target Theory and α/β for Designing/Evaluating RT Treatments?
Conclusion
• These models don't account for complexity of radiation-induced alterations in cell biology or whole body responses.
• Further study is needed to identify which of the newly identified responses are important in predicting response, and how to model these effects.

ORAL POSTER PRESENTATION 2
Objective
o develop an organ motion phantom
o evaluate the accuracy of radiation dose
lelivery during gated radiation therapy with the
Varian Real-time Position Management (RPM)
System using an in house moving phantom and
Radiochromic film dosimetry.

Poster # 106

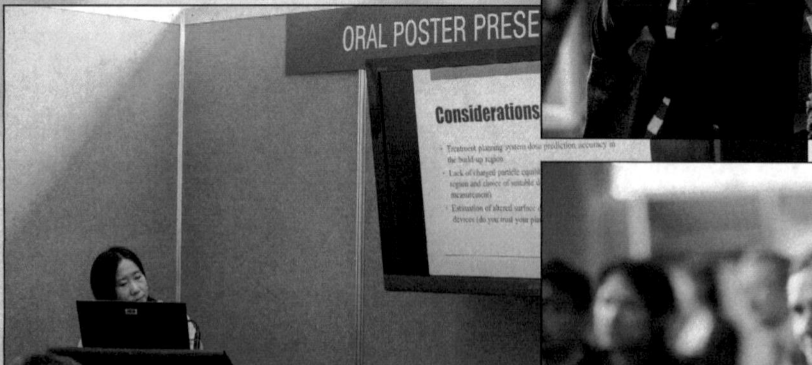
ORAL POSTER PRESE
Considerations

126
An EMR-based Radiation Oncology Registry
for Comparative Effectiveness Research
Purpose
Results
59
libdEdx – a Stopping Power Library
82
Trust, but verify – Accuracy of clinical commercial
Radiation Treatment Planning Systems
common platform of Monte Carlo dose calculation on
universal grid interface with Geant4 based particle
therapy simulation framework

MELBOURNE CONVENTION EXHIBITION CENTRE
ICCR 2013, Melbourne Australia
www.ICCR2013.org
6 - 9 May 2013

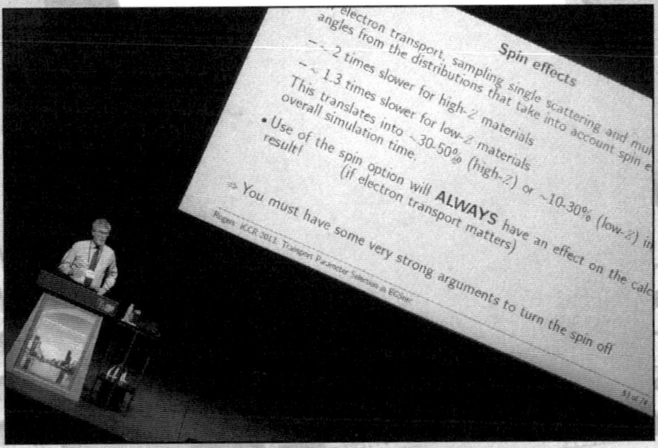
Spin effects
(if electron transport matters)
You must have some very strong arguments to turn the spin off

IOP PUBLISHING

114
ICR The Institute of Cancer Research
The ROYAL MARSDEN
Experimental investigation of VMAT dynamic MLC tracking delivery with the Elekta Agility MLC

Patient Specific Scatter Prediction in Cone Beam CT Imaging Based on Monte Carlo Simulations

VARIAN
medical systems
A partner for life

MELBOURNE
CONVENTION
EXHIBITION
CENTRE
ICCR 2013, Melbourne Australia

MELBOURNE
CONVENTION
EXHIBITION
CENTRE
ICCR 2013, Melbourne Australia

MELBOURNE
CONVENTION
EXHIBITION
CENTRE

VARIAN
medical systems

MELBOURNE
CLUB
Drug Testing
ENTRY WITHOUT
SECURITY ESCORT
EXT: 8881
Celebrating
175 Years

MELBOURNE CRICKET

3
pitality.com.au
TOLL
citi
Bank of Melbourne

Carleton
UNIVERSITY
Canada's Capital University

Proudly sponsored by
Australian Clinical Dosim
An Australian Governmen
www.iccr2013.org

RBM
adipose
YBM
Beaulieu et al (TG-186), Med Phy
Nucletron
AN ELEKTA COMPANY
ommitted to
Brachytherapy

ICCR 2013, Melbourne Australia
ICCR 2013, Melbourne Australia

On behalf of:
113
Institute of Cancer Research, University of London
Gray Laboratory, Oxford University
ICR The Institute of Cancer Research
GRAY INSTITUTE FOR RADIATION ONCOLOGY & BIOLOGY

ICCR 2016
Proposal from the West Coast USA
Stanford University
University California Los Angeles
University California San Diego
University of Washington
San Diego, California

EXIT

ICCR 2013, Melbourne Australia
Advances in computing power and cancer treatment

Advances in Computing Power and Cancer Treatment
ICCR 2013
International Conference on the Use of Computers in Radiation Therapy
6 - 9 May 2013, Melbourne, Australia
Confirmed Keynote Speakers
David A. Jaffray, Ph.D.
Princess Margaret Hospital, Canada
Vincent Khoo, M.D.
The Royal Marsden NHS Foundation Trust, UK
Dave Rogers, Ph.D.
Carleton University, Canada
HERALD SUN
Host Organisations
ACPSEM
Australasian College of Physical Scientists & Engineers in Medicine
The Royal Australian and New Zealand College of Radiologists
Convener Tomas Kron
Contact iccr2013@arinex.com.au
+61 3 9417 08888
www.iccr2013.org
arinex
follow us on twitter
www.twitter.com/iccr2013
Find us on Facebook
www.facebook.com/iccr2013
Platinum Sponsor VARIAN medical systems | A partner for life
Gold Sponsor BRAINLAB
Silver Sponsor ELEKTA
Bronze Sponsor Australian Clinical Dosimetry Service
An Australian Government Initiative
photography and design: www.JoergLehmann.com

Peer review statement

All papers published in this volume of *Journal of Physics: Conference Series* have been peer reviewed through processes administered by the proceedings Editors. Reviews were conducted by expert referees to the professional and scientific standards expected of a proceedings journal published by IOP Publishing.

Common platform of Monte Carlo dose calculation on universal grid interface with Geant4 based particle therapy simulation framework

G Iwai[1], W Takase[1], T Aso[1,2], Y Watase[1], T Sasaki[1], T Akagi[3], T Yamashita[3], Y Maeda[4] and T Nishio[5]

1 Computing Research Center, High Energy Accelerator Organization, 1-1 Oho, Tsukuba, Ibaraki 305-0801, Japan
2 Department of Electrical Engineering, Nagaoka University of Technology, 1603-1 Kamitomioka, Nagaoka, Niigata 940-2188, Japan
3 Hyogo Ion Beam Medical Center, 1-2-1 Koto, Shingu-cho, Tatsuno, Hyogo 679-5165, Japan
4 Fukui Prefectural Hospital Proton Therapy Center, 2-8-1 Yotsui, Fukui, Fukui 910-8526, Japan
5 National Cancer Center Hospital East, 6-5-1 Kashiwanoha, Kashiwa, Chiba 277-8577, Japan

E-mail: `aso@nagaokaut.ac.jp`

Abstract. While Monte Carlo (MC) simulation is believed to be the most reliable method of dose calculation in particle therapy, the simulation time is critical in attaining sufficient statistical accuracy for clinical applications. Therefore, parallelization of simulations is essential. This paper describes a common platform of MC dose calculation in grid-distributed computing environments. The platform is flexible and effective for dose calculation in both clinical and research applications for particle therapy. The platform consists of the universal grid interface (UGI) and the Geant4-based particle therapy simulation framework (PTSIM). The UGI, written in Python, provides a command-line interface for job submission, file manipulation, and monitoring in multiple-grid middleware environments. The PTSIM is a single software application for modeling a treatment port with patient data obtained from CT images. The common platform was constructed in grid computing environments using the computing resources in five institutions. The platform utilized these resources through the NAREGI grid middleware under UGI to provide stable computing resources and a common environment for MC dose calculation in particle therapy.

1. Introduction

Geant4 [1, 2] is a software toolkit for simulating the interaction of particles in matter. Monte Carlo (MC) simulation using Geant4 takes into account the physics of particle interactions by using theoretical models or experimental cross-section data. Therefore, MC simulation is believed to be the most reliable and useful application in the design of particle treatment facilities and the quality assurance of treatment plans. However, the simulation time is critical in performing dose calculation with sufficient statistical accuracy [3]. There are many techniques to reduce simulation time such as variance reduction techniques [4, 5] and parallelization using grid computing environments or computer graphics processing units (GPUs) [6, 7, 8]. The parallelization of simulations in grid computing environments is a straightforward way to gain computation speed with validated physics processes according to fundamental physics principles. Therefore, parallelization of simulations with high computing resources is essential.

Many reports about medical grid projects have been reported over the past few years [9, 10, 11]. These projects developed regulated grid environments for sharing bioinfomatics resources and analyzing standardized data such as CT and MRI images for patient diagnostics. The development of such environments was boosted because the data format is well standardized. In contrast to these grid environments, the development of a common platform for MC dose calculation has had the problems of high simulation costs and nonstandard data formats dependent on machine parameters. Furthermore, the confidential information of the machine parameters belongs to the vendor, and so the parameters are not open to the public. Therefore, a common platform is desirable to provide high-performance computing resources as well as a policy of data management.

In this paper, we describe our strategy for a common platform of MC dose calculation in multiple-grid environments. The platform should be flexible and effective for dose calculation in clinical applications, research, and education. We discuss the operational strategy in terms of simulation time and data resource sharing between members in order to facilitate the widespread use of MC dose calculation.

2. Materials and methods

2.1. Particle therapy simulation framework

The Geant4 based particle therapy simulation framework (PTSIM) [3, 12, 13] is a single software application program for modeling a treatment port with patient data obtained from CT images. Online commands are provided for modifying the configuration parameters of the treatment port, so that users do not need to develop a programming code.

In hospital information systems, CT images and treatment plans are exchanged in the Digital Imaging and Communications in Medicine (DICOM) format [14] and the extension of DICOM format for radiotherapy (DICOM-RT). The dataset of a treatment plan is called RTIonPlan and includes the configuration parameters of the treatment machine. The PTSIM has a DICOM interface for CT images and an RTIonPlan for coordinating with hospital information systems. Therefore, the PTSIM performs dose calculation by using the CT images and the RTIonPlan data to ensure the same conditions in the treatment planning system and the treatment machine. Although DICOM-RT is well established, the use of RTIonPlan is still limited because the parameters might not have the same meaning in different treatment facilities. The current PTSIM provides an online command for switching the RTIonPlan interface in accordance with the particle therapy facility. Therefore, a single PTSIM program is used among users in different treatment facilities.

Additionally the PTSIM supports an interface for reading and writing track information in the IAEA phase space data format [15]. The data include the particle type, kinetic energy, position, and direction of each track at a certain plane in the simulation volume. The data are prepared in advance by simulating particles from the beam extract window to the isocenter in the treatment port. The track information of photons, electrons, positrons, neutrons and protons are recorded. The simulation time is reduced by starting simulations from the phase space data recorded in front of the patient area. All particles in the data are generated for the simulation.

2.2. Universal grid interface

The universal grid interface (UGI) [13] provides a platform for sharing affordable CPU power through multiple-grid environments. It provides a command-line interface and Web interface for job submission, file manipulation, job monitoring, and a simple tool for result checking. The UGI is developed on top of Python binding of the Simple API for Grid Applications (SAGA)

[16, 17] framework, the Resource Namespace Service (RNS) [18], and the XML RPC [19]. By using plug-in adapters in the UGI, connections to multiple-grid environments are established by various grid middleware packages such as NAREGI [20] and gLite [21]. Figure 1 illustrates a conceptual diagram of the multiple-grid computing environment using UGI.

We assumed two usage scenarios of the UGI environment for the PTSIM. In the first scenario, the PTSIM is preinstalled in each grid environment by administrators, so that users run the PTSIM in the UGI environment. In the second scenario, the UGI handles the transfer of the PTSIM code and data from the UGI portal site, and then builds and runs the PTSIM. The first scenario is applied as a regular operation for PTSIM application users to reduce the maintenance costs, whereas the second scenario is for PTSIM developers who have to examine the modified PTSIM code before release. UGI environments have been examined for both scenarios.

2.3. Common platform and operational strategy

The common platform consists of the PTSIM and the UGI. It utilizes the multiple-grid computing environment to provide high computing resources and to share data resources with collaborating members.

For clinical applications, we mask private information with DICOM data to assure a patient's anonymity. Access to DICOM data is managed according to the permission flag of the file under a set of resource-sharing rules and conditions.

The PTSIM needs the machine parameters of the treatment port to ensure the same condition for the treatment machine. However, the machine parameters have to be kept confidential except for authorized members because the information belongs to the vendor of the treatment port. This restriction limits the widespread use of the PTSIM for research and educations.

We propose to construct a database of particle phase space data in the IAEA format for each configuration of validated beam delivery systems. The database allows hospital members, vendors, and research institutes to share the simulation of a beam delivery system for quality assurance, research, and education without exploring the machine parameters. Figure 2 shows the conceptual diagram of the common platform at the point of data management. For example, the phase space data recorded at the upstream of the multi-leaf collimator are used with the machine parameters for the verification of treatment plan in clinical applications by authorized members. The phase space data are also applicable without the machine parameters for researchers who are engaging to develop new beam devices in the treatment head. For the educational purpose, the phase space data are applied for simulating dosimetry experiments to optimize the experimental conditions.

Since all beam devices are referred to as beam modules in the PTSIM, only beam modules selected by users are placed in the beam line and configure a treatment port. By using the phase space data, the beam modules of the beam delivery system are removed from the beam line without modifying the PTSIM code. Therefore, the machine parameters are separated from the distribution of the PTSIM code. In addition, starting simulations by using the phase space data consolidates redundant simulations of the beam delivery system that had been performed by individual users. The validations and productions of the phase space data are performed by the authorized members such as medical physicists in treatment facilities and the PTSIM developers. Consequently, the use of phase space data reduces simulation time of database users.

3. Results and discussions

3.1. Simulation time

In our previous research [13], we examined the simulation time by simulating a 150 MeV proton therapeutic beam. The 9×10^8 primary protons were simulated from the beam extract window to the water phantom from CT images at the isocenter. The configuration parameters of the treatment port were taken from RTIonPlan data, which mimic the treatment plan for a

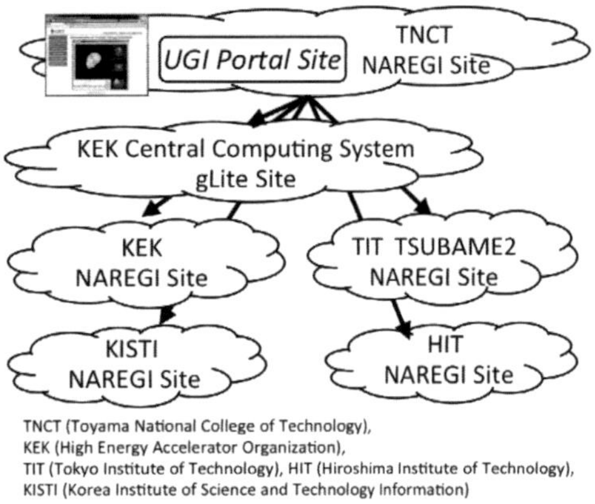

Figure 1. Schematic diagram of multiple-grid environments.

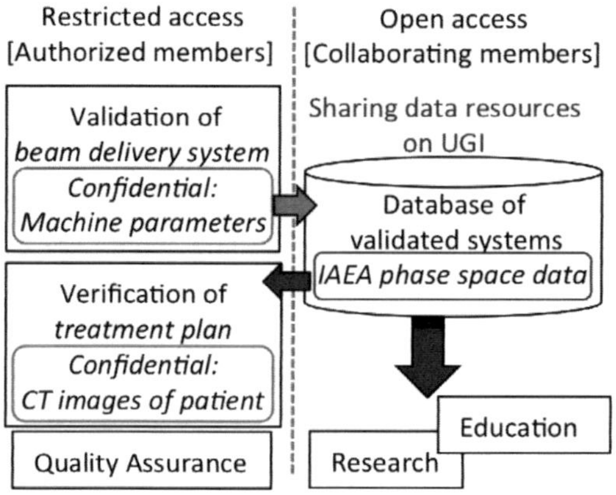

Figure 2. Data management in the common platform.

patient. The treatment head included a multi-leaf collimator, a bolus compensator, and a patient collimator.

In this research, we repeated simulations in the shared central computing system in KEK (KEKCC) with the same condition used in our previous research. We compared average simulation times with and without phase space data. The phase space data were recorded at the entrance of patient area, which corresponds to the upstream of the multi-leaf collimator. By separating the job into 500 tasks, the average simulation time of each task without phase space data was approximately 100 minutes, while it was 37 minutes with phase space data.

3.2. Data resource sharing

The phase space data need to be prepared for each set of typical treatment parameters, e.g., combination of beam energies, spread-out Bragg peak (SOBP), and range shifter. The data size was approximately 12 GB for 9×10^8 primary protons in the example treatment parameters. If we assume a treatment port in the Fukui prefectural hospital proton therapy center (FPHPTC) in Japan, the gantry treatment port has a variety of three beam energies, twelve SOBPs, and five range shifters. Currently, 146 combinations of those parameters are registered in the treatment planning system for clinical applications in the gantry treatment port. Therefore, 146 times more disk space and CPU hours are needed. This amount corresponds to a total of approximately 2 TB disk space and 120,000 CPU hours. These simulations insist large scale high computing resources for the validation of the beam delivery system with confidential machine parameters. Apparently, such preparation of phase space data by individual users is inefficient. Therefore, the common platform with the operational strategy is suitable to make and share IAEA phase space data among members for efficient usage of computing resources.

4. Conclusions

The common platform of MC dose calculation in multiple-grid environments provides stable computing resources and a common environment particle therapy in clinical applications, research, and education, while protecting the information of machine parameters and patient privacy.

Acknowledgments
The authors would like to thank the members of the National Institute of Informatics, and the Global Scientific Information and Computing Center of the Tokyo Institute of Technology. Development of the PTSIM was supported in part by the Japan Science and Technology Agency as part of the Core Research for Evolutional Research and Technology program. The present study was supported in part by grants from JSPS KAKENHI (No. 23540350 and No. 23390308).

References

[1] Agostinelli S et al. 2003 Geant4, a simulation toolkit, *Nucl. Instr. Meth.* A **506** 250
[2] Allison J et al. 2006, Geant4 developments and applications *IEEE Trans. Nucl. Sci.* **53 No. 1** 270
[3] Akagi T et al. 2011 The PTSim and TOPAS projects, bringing Geant4 to the particle therapy clinic *Prog. in Nucl. Sci. Technol.* **2** 912
[4] Kawrakow I et al. 2000 Investigation of variance reduction techniques for Monte Carlo photon dose calculation using XVMC *Phys. Med. Biol.* **45** 2163
[5] Ma C et al., 2002 A Monte Carlo dose calculation tool for radiotherapy treatment planning *Phys. Med. Biol.* **47** 1671
[6] Kohno R et al. 2011 Clinical implementation of a GPU-based simplified Monte Carlo method for a treatment planning system of proton beam therapy *Phys. Med. Biol.* **56** N287.
[7] Hissoiny S et al. 2011 GPUMCD: a new GPU-oriented Monte Carlo dose calculation platform *Med. Phys.* **38** 754
[8] Xun J et al. 2012 GPU-based fast Monte Carlo dose calculation for proton therapy *Phys. Med. Biol.* **57** 7783
[9] Blanchet C et al. 2006 Integrating bioinfomatics resources on the EGEE grid platform *Sixth IEEE Int. Sympo. on CCGrid* 48
[10] Eschenbach A C V et al. 2006 calBIG: Cancer Informatics Vision *Cancer Informatics* **2** 22
[11] Krefting D et al. 2009 MediGRID: Towards a user friendly secured grid infrastructure *Future Generation Computer Systems* **25** 326
[12] Aso T et al. 2007 Geant4 based simulation framework for particle therapy system *IEEE NSS/MIC Conf. Rec. NSS'07* 2564
[13] Aso T et al. 2012 Extension of the particle therapy simulation framework to hospital information systems and multi-grid environments *IEEE Int. Conf. on Comp. and Eng. (CSE) IEEE Comp. Soc.* 229
[14] DICOM Digital Imaging and Communications in Medicine *The National Electrical Manufacturers Association* [Online] http://medical.nema.org/
[15] IAEA INDC (International Nuclear Data Committee) phase-space database for external beam radiotherapy 2005 INDC(NDS) 0484 1.
[16] SAGA Simple API of grid application [online] http://saga.cct.lsu.edu/
[17] OGF Open Grid Forum OGF document:OGF-R-O.90 [Online] http://www.ggf.org/
[18] RNS Resource Namespace Specification OGF document: Recommendation [Online] http://www.ggf.org/
[19] XML-RPC [Online] http://xmlrpc.scripHng.com/
[20] NAREGI National Research Grid Initiative [online] http://www.naregi.org
[21] gLite Lightweight middleware for grid computing [online] http://glite.cern.ch/

Journal of Physics: Conference Series **489** (2014) 012002 doi:10.1088/1742-6596/489/1/012002

Analytical probabilistic proton dose calculation and range uncertainties

M Bangert[1], P Hennig[2], and U Oelfke[1]
[1] German Cancer Research Center - DKFZ
Im Neuenheimer Feld 280, 69120 Heidelberg, Germany
[2] Max Planck Institute for Intelligent Systems
Spemannstraße 38, 72076 Tübingen, Germany

E-mail: m.bangert@dkfz.de

Abstract. We introduce the concept of analytical probabilistic modeling (APM) to calculate the mean and the standard deviation of intensity-modulated proton dose distributions under the influence of range uncertainties in closed form. For APM, range uncertainties are modeled with a multivariate Normal distribution $p(\boldsymbol{z})$ over the radiological depths $\boldsymbol{z}$. A pencil beam algorithm that parameterizes the proton depth dose $d(\boldsymbol{z})$ with a weighted superposition of ten Gaussians is used. Hence, the integrals $\int dz\ p(\boldsymbol{z})\ d(\boldsymbol{z})$ and $\int dz\ p(\boldsymbol{z})\ d(\boldsymbol{z})^2$ required for the calculation of the expected value and standard deviation of the dose remain analytically tractable and can be efficiently evaluated. The means μ_k, widths δ_k, and weights ω_k of the Gaussian components parameterizing the depth dose curves are found with least squares fits for all available proton ranges. We observe less than 0.3% average deviation of the Gaussian parameterizations from the original proton depth dose curves. Consequently, APM yields high accuracy estimates for the expected value and standard deviation of intensity-modulated proton dose distributions for two dimensional test cases. APM can accommodate arbitrary correlation models and account for the different nature of random and systematic errors in fractionated radiation therapy. Beneficial applications of APM in robust planning are feasible.

1. Introduction

The quality of radiation therapy treatment simulations may be compromised by uncertainties originating from inter- and intra-fraction motion, setup errors, or inaccuracies during target volume definition, among others. Approaches to quantify the influence of these uncertainties on the resulting dose distribution and derivative treatment plan quality indicators include worst case simulations and variance calculations based on dose blurring or sampling [1, 2, 3, 4]. However, these approaches remain computationally challenging or suffer from inherent limitations [5].

In order to quantify the influence of range uncertainties on intensity-modulated proton dose distributions, we apply the concept of analytical probabilistic modeling (APM). Therefore, we describe the uncertainty in the radiological depths $\boldsymbol{z}$ with a multivariate Normal distribution $\mathcal{N}(\boldsymbol{z};\bar{\boldsymbol{z}},\Sigma)$ with mean vector $\bar{\boldsymbol{z}}$ and covariance matrix Σ and approximate the proton dose $d(\boldsymbol{z})$ with a functional form that can be integrated analytically against Gaussian probability distributions. Hence, it is possible to calculate the mth moment of the dose distribution d at depth z in closed form and thereby the mean $E[d(\boldsymbol{z})]$ and the variance $E[d(\boldsymbol{z})^2] - E[d(\boldsymbol{z})]^2$ of the dose distribution.

$$E[d(\boldsymbol{z})^m] = \int d\boldsymbol{z}\ \mathcal{N}(\boldsymbol{z}; \bar{\boldsymbol{z}}, \Sigma)\ d(\boldsymbol{z})^m \tag{1}$$

2. Material and methods

The proton depth dose $Z_j(z_{ij})$ of a pencil beam j at voxel i with the radiological depth z_{ij} is approximated with a weighted superposition of Gaussians. We found that ten components are sufficient to describe proton depth dose curves with ranges $R \leq 35\,\text{cm}$.

$$Z_j(z_{ij}) = \sum_{k=1}^{10} \frac{\omega_{jk}}{\sqrt{2\pi}\delta_{jk}}\ e^{-\frac{(z_{ij}-\mu_{jk})^2}{2\delta_{jk}^2}} \tag{2}$$

For this study, the weights ω_{jk}, widths δ_{jk}, and means μ_{jk} are determined with a least squares fit of analytically calculated proton depth dose curves [6] but it is also possible to use measured depth dose data. The least squares fit is performed with an open source MATLAB optimizer[1]. The gradient components used for optimization are

$$\frac{\partial Z_j}{\partial \omega_{jk}} = \frac{1}{\sqrt{2\pi}\delta_{jk}}\ e^{-\frac{(z-\mu_{jk})^2}{2\delta_{jk}^2}}\ ; \quad \frac{\partial Z_j}{\partial \mu_{jk}} = \frac{\omega_{jk}}{\sqrt{2\pi}\delta_{jk}}\ e^{-\frac{(z-\mu_{jk})^2}{2\delta_{jk}^2}}\ \frac{(z-\mu_{jk})}{\delta_{jk}^2}; \quad \frac{\partial Z_j}{\partial \delta_{jk}} = \frac{\omega_{jk}}{\sqrt{2\pi}\delta_{jk}}\ e^{-\frac{(z-\mu_{jk})^2}{2\delta_{jk}^2}} \left\{\frac{(z-\mu_{jk})^2}{\delta_{jk}^2} - \frac{1}{\delta_{jk}}\right\}$$

Figures 1 (a) and (b) show the approximation of proton depth dose curves for four different ranges. We observe less than 0.3% average deviation of the Gaussian parameterization from the analytically calculated proton depth dose curve for proton ranges $R \leq 35$ cm.

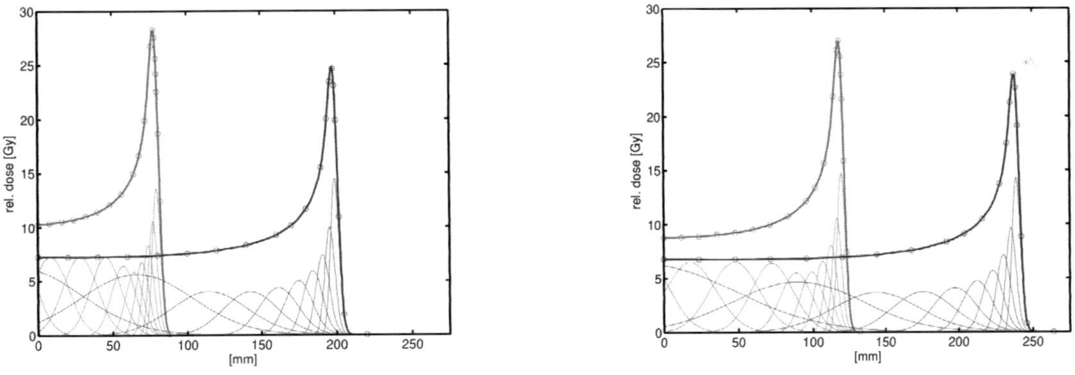

(a) (b)

Figure 1. Analytically calculated proton depth dose curves (circles) [6] and functional approximation (bold solid lines) with a weighted superposition of ten Gaussian components (thin solid lines) for (a) ranges $R = 8\,\text{cm}$ (red), $R = 20\,\text{cm}$ (blue) and (b) ranges $R = 12\,\text{cm}$ (red), $R = 24\,\text{cm}$ (blue).

We apply the Gaussian approximation for the proton depth dose in combination with a conventional pencil beam algorithm where the proton dose factorizes in lateral direction and in

[1] We found that *minimize* (available online at `http://learning.eng.cam.ac.uk/carl/code/minimize/`) is considerably more efficient than built-in solvers.

depth [7]. Hence, the three dimensional dose in an intensity modulated field can be described by

$$d(x_i, y_i, z_i) = \sum_j w_j L_{ij}(x_{ij}, y_{ij}, z_{ij}) \cdot \sum_k \frac{\omega_{jk}}{\sqrt{2\pi}\delta_{jk}} e^{-\frac{(z_{ij}-\mu_{jk})^2}{2\delta_{jk}^2}} \tag{3}$$

where w_j corresponds to the weight of pencil beam j and L_{ij} specifies the lateral proton dose profile of pencil beam j at voxel i with the lateral coordinates x_{ij} and y_{ij}.

The influence of range uncertainties can be modeled with an offset Δ_j^z of the radiological depth z_{ij} according to

$$d(z_i) = \sum_{jk} w_j L_{ij} \cdot \frac{\omega_{jk}}{\sqrt{2\pi}\delta_{jk}} e^{-\frac{(z_{ij}-\mu_{jk}+\Delta_j^z)^2}{2\delta_{jk}^2}}. \tag{4}$$

If we assume that the offsets Δ_j^z are distributed according to a multivariate Normal distribution

$$p(\boldsymbol{\Delta}) = p\begin{pmatrix} \Delta_1^z \\ \Delta_2^z \\ \vdots \end{pmatrix} = \mathcal{N}(\boldsymbol{\Delta}^z; \bar{\boldsymbol{\Delta}}^z, \Sigma^z) \tag{5}$$

the expected value of the dose is given by

$$\begin{aligned} E[d(z_i)] &= \int d\boldsymbol{\Delta}\, \mathcal{N}(\boldsymbol{\Delta}^z; \bar{\boldsymbol{\Delta}}^z; \Sigma^z) \sum_{jk} w_j L_{ij} \frac{\omega_{jk}}{\sqrt{2\pi}\delta_{jk}} e^{-\frac{(z_{ij}-\mu_{jk}+\Delta_j^z)^2}{2\delta_{jk}^2}} \\ &= \sum_{jk} w_j L_{ij} \int d\Delta_j^z\, \mathcal{N}(\Delta_j^z; \bar{\Delta}_j^z, \Sigma_{jj}^z) \frac{\omega_{jk}}{\sqrt{2\pi}\delta_{jk}} e^{-\frac{(z_{ij}-\mu_{jk}+\Delta_j^z)^2}{2\delta_{jk}^2}} \\ &= \sum_{jk} w_j L_{ij} \frac{\omega_{jk}}{\sqrt{2\pi(\delta_{jk}^2 + \Sigma_{jj}^z)}} e^{-\frac{(z_{ij}-\mu_{jk})^2}{2(\delta_{jk}^2+\Sigma_{jj}^z)}} \end{aligned} \tag{6}$$

where Σ_{jj}^z corresponds to the variance of Δ_j^z. The quadratic term $E[d(z_i)^2]$ required for the calculation of the dose variance $E[d(z_i)^2] - E[d(z_i)]^2$ is given by

$$\begin{aligned} E[d(z_i)d(z_l)] &= \int d\boldsymbol{\Delta}\, \mathcal{N}(\boldsymbol{\Delta}^z; \bar{\boldsymbol{\Delta}}^z, \Sigma^z)\, d(z_i)d(z_l) \\ &= \int d\boldsymbol{\Delta}\, \mathcal{N}(\boldsymbol{\Delta}^z; \bar{\boldsymbol{\Delta}}^z, \Sigma^z) \cdot \left\{ \sum_{jk} w_j L_{ij} \cdot \frac{\omega_{jk}}{\sqrt{2\pi}\delta_{jk}} e^{-\frac{(z_{ij}-\mu_{jk}+\Delta_j^z)^2}{2\delta_{jk}^2}} \right\} \\ &\quad \cdot \left\{ \sum_{mn} w_m L_{lm} \cdot \frac{\omega_{mn}}{\sqrt{2\pi}\delta_{mn}} e^{-\frac{(z_{lm}-\mu_{mn}+\Delta_m^z)^2}{2\delta_{mn}^2}} \right\}. \end{aligned} \tag{7}$$

To rewrite this equation as a quadratic form over two dimensional Gaussians we define

$$\boldsymbol{z}_{ijlm} = \begin{pmatrix} z_{ij} \\ z_{lm} \end{pmatrix},\ \boldsymbol{\mu}_{jkmn}^z = \begin{pmatrix} \mu_{jk} \\ \mu_{mn} \end{pmatrix},\ \boldsymbol{\Delta}_{jm}^z = \begin{pmatrix} \Delta_j^z \\ \Delta_m^z \end{pmatrix},\ \Theta^{jkmn} = \begin{pmatrix} \delta_{jk}^2 & 0 \\ 0 & \delta_{mn}^2 \end{pmatrix},\ \Sigma^{jm} = \begin{pmatrix} \Sigma_{jj}^z & \Sigma_{jm}^z \\ \Sigma_{mj}^z & \Sigma_{mm}^z \end{pmatrix} \tag{8}$$

to obtain

$$
\begin{aligned}
E[d(z_i)d(d_l)] &= \sum_{jknm} w_j w_m \; L_{ij} L_{lm} \int d\boldsymbol{\Delta}^z_{jm} \; \mathcal{N}(\boldsymbol{\Delta}^z_{jm}; \bar{\boldsymbol{\Delta}}^z_{jm}, \Sigma^{jm}) \\
&\quad \cdot \frac{\omega_{jk}\omega_{mn}}{2\pi\sqrt{|\Theta^{jkmn}|}} \; \mathrm{e}^{-\frac{1}{2}(\boldsymbol{z}_{ijlm}-\boldsymbol{\mu}^z_{jkmn}+\boldsymbol{\Delta}^z_{jm})^\top (\Theta^{jkmn})^{-1}(\boldsymbol{z}_{ijlm}-\boldsymbol{\mu}^z_{jkmn}+\boldsymbol{\Delta}^z_{jm})} \qquad (9) \\
&= \sum_{jknm} w_j w_m \; L_{ij} L_{lm} \; \frac{\omega_{jk}\omega_{mn}}{2\pi\sqrt{|\Theta^{jkmn}+\Sigma^{jm}|}} \; \mathrm{e}^{-\frac{1}{2}(\boldsymbol{z}_{ijlm}-\boldsymbol{\mu}^z_{jkmn})^\top (\Theta^{jkmn}+\Sigma^{jm})^{-1}(\boldsymbol{z}_{ijlm}-\boldsymbol{\mu}^z_{jkmn})}
\end{aligned}
$$

If we are to account for the different nature of random and systematic errors [2] within APM we have to include an additional summation over multiple fractions F as exercised in [5]. For the construction of Σ^{jm} we have to consider that systematic errors are also correlated over different fractions, while random errors of different fractions are uncorrelated.

3. Results

Figure 2 (a) compares the analytically calculated and sampled mean dose and standard deviation for a spread-out Bragg peak with range R = 20 cm and modulation M = 5 cm assuming a perfectly correlated range error of all underlying proton pencil beams of 3%. The analytical calculation is exact; analytically calculated and samples values are identical up to sampling error.

Figure 2 (b) visualizes the different influence of random and systematic errors in the context of fractionated radiation therapy. While the influence of random errors averages out when increasing the number of fractions from F = 1 to F = 35, the influence of systematic errors remains.

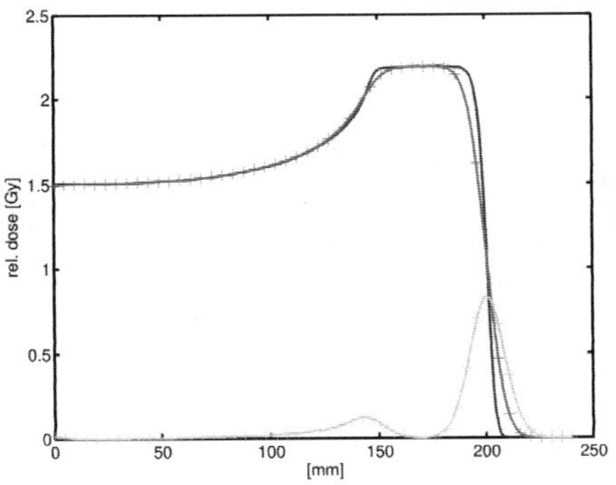

(a)

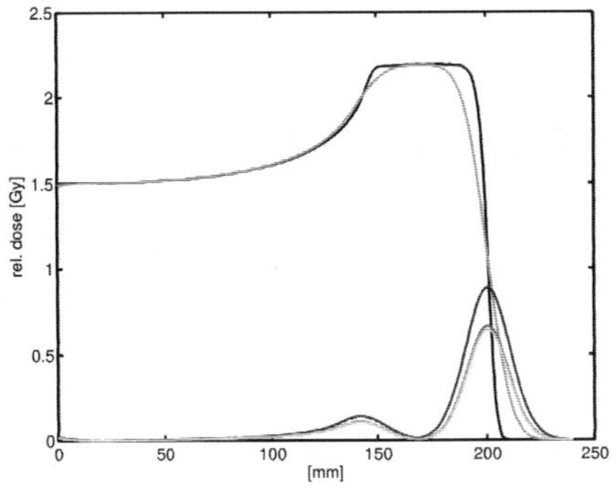

(b)

Figure 2. (a) Nominal proton dose of a spread-out Bragg peak (blue), expected value (red), and standard deviation (green) assuming a 3% range error. Analytical calculations are shown as solid lines, sampled values as crosses. (b) Nominal proton dose (black), expected value (gray), and standard deviation assuming a systematic range error of 3% and a random range error of 5 mm for 1 (blue), 5 (red), and 35 fractions (green).

The influence of different correlation assumptions is visualized in figure 3. We simulate a single field applying a uniform dose to a C-shaped target volume surrounding an organ at risk within a homogeneous water phantom[2]. For the calculation of the standard deviation in figure

[2] Note that APM is generally applicable to full-fledged multi-field IMPT treatments with appropriate weights of the constituting pencil beams $w_{j/m}$.

3 (a), we assumed that the range errors of all constituting pencil beams are perfectly correlated. In this case we observe dose standard deviations of up to 9 Gy. For figure 3 (b), we assumed that only the range errors of pencil beams impinging at the same lateral position and thereby penetrating the same tissue are perfectly correlated. This relaxed correlation assumption yields a reduction of the dose standard deviation.

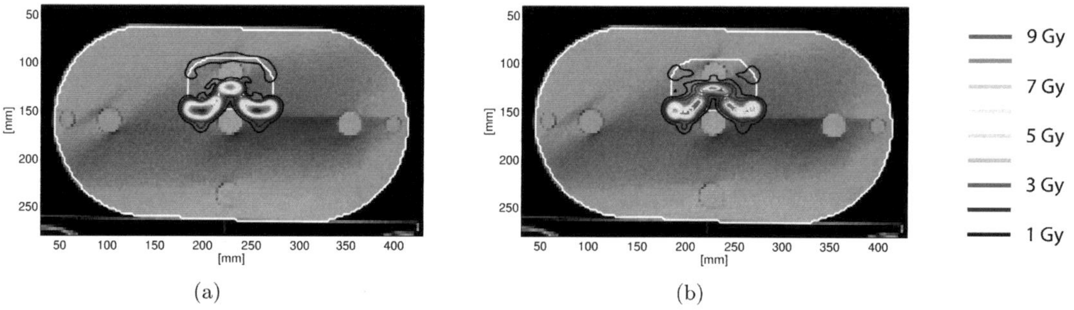

Figure 3. Dose standard deviation [Gy] for a single field uniform dose assuming that (a) the range errors of all pencil beams are perfectly correlated and (b) only range errors of pencil beams impinging at the same lateral position are correlated.

4. Discussion and conclusion

We introduced the concept of analytical probabilistic modeling (APM) for the quantification of range uncertainties in intensity-modulated proton therapy.

Based on a pencil beam algorithm that factorizes in lateral direction and in depth [7] we present closed form expressions for the calculation of the expected value and the standard deviation of the dose. For our derivations to remain analytically tractable, we parameterize the proton depth dose curve with a weighted superposition of ten Gaussians. This gives an approximation of very high numerical quality and allows analytical Gaussian error propagation through the nonlinearities in the proton dose calculations. Potential inaccuracies of APM in heterogeneous geometries due to the applied pencil beam model are part of ongoing research and might be resolved through sub pencil beam decomposition [8].

APM allows for the explicit quantification of the non-trivial interplay of random and systematic errors in the context of fractionated radiation therapy and for the incorporation of structured correlation assumptions about the range errors of the individual pencil beams. While these aspects are often neglected by existing approaches to quantify the influence uncertainties, our preliminary study shows that the correlation assumptions and the different nature of random and systematic errors have a critical impact on the resulting dose uncertainty, as depicted in figures 2 (b) and 3.

APM also applies, with minor variations, to lateral proton dose profiles, other ion species, and photons [5]. In combination with a quadratic objective function [9], such an analytical probabilistic model may lower the computational complexity of robust planning [5].

References

[1] Goitein M 1985 *Med. Phys.* **12** 608
[2] van Herk M, Remeijer P and Lebesque J 2002 *Int. J. Rad. Onc. Biol. Phys.* **52** 1407
[3] Baum C, Alber M, Birkner M and Nüsslin F 2004 *Phys. Med. Biol.* **49** 5475
[4] Maleike D, Unkelbach J and Oelfke U 2006 *Phys. Med. Biol.* **51** 2237

[5] Bangert M, Hennig P and Oelfke U 2013 *Phys. Med. Biol.* **58** 5401
[6] Bortfeld T 1997 *Med. Phys.* **24** 2024
[7] Hong L, Goitein M, Bucciolini M, Comiskey R, Gottschalk B, Rosenthal S, Serago C and Urie M 1999 *Phys. Med. Biol.* **41** 1305
[8] Soukup M, Fippel M and Alber M 2005 *Phys. Med. Biol.* **50** 5089
[9] Oelfke U and Bortfeld T 2001 *Med. Dosim.* **26** 113

Journal of Physics: Conference Series **489** (2014) 012003 doi:10.1088/1742-6596/489/1/012003

Improvements in the stopping power library libdEdx and release of the web GUI dedx.au.dk

J Toftegaard[1], A Lühr[1,2,3], N Sobolevsky[4,5], N Bassler[1,2]

[1] Department of Physics and Astronomy, Aarhus University, Aarhus, DENMARK
[2] Department of Experimental Clinical Oncology, Aarhus University Hospital, Aarhus, DENMARK
[3] German Cancer Consortium (DKTK), Dresden, GERMANY;
OncoRay – National Center for Radiation Research in Oncology, Medical Faculty and University Hospital Carl Gustav Carus, Technische Universität Dresden, GERMANY; and German Cancer Research Center (DKFZ), Heidelberg, GERMANY
[4] Institute for Nuclear Research of the Russian Academy of Sciences (INR), Moscow, RUSSIA
[5] Moscow Institute of Physics and Technology, State University, RUSSIA

E-mail: `bassler@phys.au.dk`

Abstract. Purpose: In ion beam therapy electronic stopping power data enter in different disciplines, e.g., dose planning, dosimetry, and radiobiology. However, relevant stopping power data are only known within an accuracy of 2%-10%. We started the software library project libdEdx to unify data from several well-known stopping power sources into one ready-to-use package being 1) freely available and 2) easy accessible via a web-based front end.

Methods: Currently, stopping power data from PSTAR, ASTAR, MSTAR and ICRU49+73 are implemented along with a version of the Bethe formula. The library is programmed in the language C to provide broad portability and high performance. A clean API provides full access to the underlying functions and thread safety in multi-threaded applications. The possibility to define arbitrary materials complements the list of predefined ICRU materials. Furthermore, we introduced a collection of tools, e.g., inverse stopping power look-up as well as CSDA range calculation and its inverse.

Results: On a standard desktop PC libdEdx calculates 22 million look-ups/sec. A web GUI (available at `http://dedx.au.dk`) provides easy access to libdEdx and download of stopping data and graphs. For compounds, we observe that stopping power data are robust for variations in the mean excitation potential of the constituents as long as the total mean excitation potential is fixated.

Conclusion: We released libdEdx (version number 1.2.1: `http://sf.net/projects/libdedx/`) with a web-based GUI. Future development will focus on implementing further stopping powers sources (e.g., for electrons and nuclear stopping) and relativistic effects.

1. Introduction

The energy loss of charged particles has extensively been researched for almost a century. Medical physics rely on electronic stopping powers in several ways. In ion therapy calculations, penetration ranges can be derived in the continuous slowing down approximation (CSDA), but uncertainties in the mean excitation potential lead to range uncertainties [1]. In clinical settings, dose is reported as dose-to-water while Monte Carlo particle transport codes calculate the dose deposited in the simulated medium, which might be different from water. The translation is done by calculating the stopping power ratio of the two media in the corresponding fluence spectrum. Such a situation arises, e.g., for ionization chamber dosimetry, in the evaluation of beam quality correction factors, where the dose in the ionization chamber wall material is translated to dose-to-water using stopping power ratios, as described in dosimetry protocols such as IAEA TRS-398 [2]. The impact of various stopping power tables in this context was investigated in [3–5]. Furthermore, radiobiology models that describe the response of *in vitro* cell cultures exposed to ion or neutron radiation rely on stopping powers [6, 7] in similar ways as numerous detector response models [8–11] do.

To date, several stopping power tables and algorithms are available, with various ranges of applicability with respect to ions and target combinations. The aim of the libdEdx project is to provide a unified interface to several stopping power algorithms and tables to enable users to test their specific application against the sensitivity of stopping powers and to easily migrate their code to new stopping power tables if these become available. Additionally, users should be able to conveniently look up stopping powers and particle ranges for various media/ion types using several algorithms and tables by means of a web front end `dedx.au.dk`.

2. Methods

The computer library libdEdx is coded in ANSI-C for platform independence, targeting, e.g., Windows, Linux/Unix and embedded systems such as mobile phones or dedicated hardware for real time data processing in scientific laboratory environments. libdEdx features two modes of operation, "fast" and "simple". The "simple" mode is intended for rapid application development and less experienced users. "Fast" mode is meant for CPU bound applications such as Monte Carlo codes. The installer is based on the CMAKE (`http://www.cmake.org`) build environment which provides a platform independent build system following de-facto standards on Unix/Linux and Windows systems. A single common programming interface (API) provides access to all implemented stopping power tables and algorithms, covering energy ranges from 10 keV/u up to 1 GeV/u. Functions for CSDA range calculation and inverse look-ups are provided as well. The library is extensible with new tables and algorithms providing stopping powers.

So far only electronic mass stopping powers are provided by libdEdx. PSTAR and ASTAR [12] tables handle proton and helium ion stopping powers, respectively, and are identical to those provided by the ICRU 49 report [13] by the International Commission on Radiation Units & Measurements (ICRU). However, libdEdx discriminates between these in case of future updates of PSTAR and ASTAR tables. The National Institute of Standards and Technology (NIST) allows the use of these data tables for free. MSTAR [14] is implemented for heavier ions (from Li to Ar), not covered by PSTAR and ASTAR. The implementation was made in agreement with Helmut Paul, Linz. ICRU 73 [15] tables are provided which also cover heavier ions from Li to Ar. Here, we also included the deprecated first version with the erroneous stopping power table for water to retain full compatibility. In addition, a version of the Bethe formula was implemented, which is sewed with the Lindhard-Scharff formula for low energies (below and around the Bragg peak).

The complete ESTAR [12] material composition list for compounds is included, following the ICRU material number and composition scheme. While this list covers 279 materials PSTAR, ASTAR, MSTAR, and the ICRU tables do only provide stopping powers for selected materials. libdEdx complements this by applying Braggs additivity rule when a compound is requested that is not in the list and all constituents are available. ICRU material definitions provide also mean excitation energies (I-values) which are applied by default for the extended Bethe algorithm. Additionally, I-values can be overridden if requested by the user. For compounds, ICRU 49 rules for gas phase or liquid/solid state apply.

The extended Bethe algorithm is identical to the implementation found in SHIELD-HIT [16–18]. The basis for the algorithm is the original Bethe from 1930 [19] together with a density correction and a charge scaling function by Hubert *et al.* [20]. The resulting modified Bethe formula is only applicable for energies above 1 MeV/nucleon. Energies below ∼ 1 MeV/nucleon are treated in accordance with the Lindhard-Scharff (LS) formula [21]. A result from the LS description is that the stopping power is proportional to the projectile velocity, $S = k\sqrt{E}$, where k is a constant depending on projectile and target charge and E is the energy of the projectile. Instead of calculating k analytically, k is found along the sewing point for the low-energy LS expression and the modified Bethe-equation. The sewing point is determined by the criteria that the combined stopping power function and its first derivative must be continuous, while the low-energy stopping power function must reach zero at $E = 0$. The sewing point is located between the point where the modified Bethe-equation turns negative (which is nonphysical) at low energies, and the global maximum found at slightly higher energies. A cost function for the sewing point is defined as the absolute difference between the first derivative of the low-energy function and the modified Bethe-equation, both evaluated in the sewing point interval. The actual sewing point is the minimum of the cost function, which is found using the golden section search method [22].

For compounds with more than one atomic element, Bragg's additivity rule,

$$\ln(I_c) = \sum_i w_i \ln(I_i)\,, \tag{1}$$

is applied, where w_i are the individual normalized compound fractions by mass for each constituent i. If required, the mean excitation potential of a compound, I_c, can be set by the user either by directly specifying I_c or by specifying all I_i for each of the constituents. If only I_c is specified by the user, equation 1 is under-determined. In this case, the default ICRU I-values are used for the constituents I_i and a common dimensionless scaling factor a is included so that

$$\ln(I_c) = \sum_i w_i \ln(aI_i)\,, \tag{2}$$

holds. Equation 2 is solved for a and the stopping power for each component is calculated using aI_i instead of I_i. Robustness of this method is tested by applying an asymmetric scaling factor a_i for the individual components, where $\max(a_i) = 1.2 \min(a_i)$.

Computation speed of libdEdx was tested on an Intel i7-2670QM CPU with a clock frequency of 2.20 GHz. libdEdx was compiled with GCC-4.7.2 on Debian 7.0 using -O2 optimization. Stopping power for helium ions in water was calculated with ASTAR and five different energy steps are chosen: 0.1, 1, 10, 100 and 200 MeV/nucleon.

Figure 1 illustrates how libdEdx is intended to be used in the "fast" mode. A workspace is allocated holding necessary memory for the data structures. Several material/ion/data set configurations can be set, and stopping powers can be retrieved from each or these configurations. After use, the workspace should be freed again.

3. Results and Discussion

An automated test script for libdEdx has been created that compares values determined by libdEdx with stopping power values taken directly from the corresponding original sources for more than 250 different ion, target and energy configurations. libdEdx reproduces the original sources with an accuracy better than 0.3 %. This test script is shipped with the libdEdx distribution and let the user verify the installation at any time.

The mean execution time for 100 million stopping power look-ups is 4.54 seconds, which corresponds to approximately 22 million look-ups per second making libdEdx attractive for CPU-intensive computations with many calls of stopping power routines.

In the case that an I-value for a compound, I_c, is specified by the user, the extended Bethe algorithm shows less than 0.1 % deviation between stopping power values obtained either with

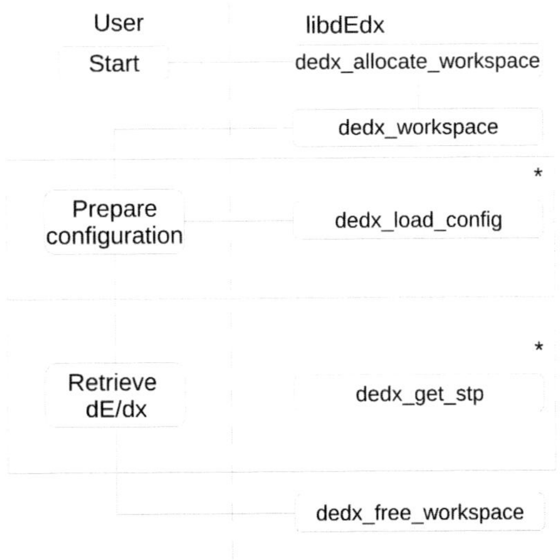

Figure 1. Diagram of how libdEdx can be used in the "fast" mode. Boxes with a star in the upper right corner can be invoked multiple times within the same workspace.

a common or with an asymmetric scaling factor a or a_i, respectively. This means that even if equation 2 is under-determined, there is only little sensitivity to relative changes of the I-values of the constituents, as long as the value of I_c for the compound is fixed.

Generally, the choice of a certain source of stopping powers may introduce a systematic error in a theoretical calculation. One can try to estimate the effect size by comparing to results obtained by applying alternative stopping power tables. The magnitude of deviations between stopping power sources depends on the considered combination of projectile, target, and energy. For instance, the stopping power algorithm implemented in SHIELD-HIT is for most purposes in good agreement with, e.g., ICRU 49, but deviates considerably at low energies for protons on aluminium, as pointed out in [23]. This encourages the use of different stopping power tables. A careful choice of stopping power tables is relevant when calculating, e.g., stopping power ratios which may differ in the order of some percent [5]. And also in the process of determining the quantities dose-to-media and dose-to-water [24, 25] it has been observed that the choice of stopping power tables might give +/- 2% systematic error to calculated dose.

4. Conclusion

libdEdx provides a programming library with a fast, clean and easy to use programming API that gives access to multiple stopping power tables and routines. This enables the user to perform comparative studies and, e.g., to determine uncertainties that arise from the choice of stopping power tables. A web front end to libdEdx is provided at `http://dedx.au.dk`.

For the future, we would like to include nuclear stopping power tables and implement further algorithms such as ATIMA [26], SRIM/TRIM [27], ESTAR, PASS [28], and possibly also radiative stopping power implementations.

Acknowledgments

We thank Helmut Paul from the Johannes Kepler University Linz, Austria, for allowing us to implement the MSTAR code in libdEdx, and for valuable discussions. This project was supported by CIRRO - The Lundbeck Foundation Center for Interventional Research in Radiation Oncology, The Danish Council for Strategic Research (`http://www.cirro.dk`), Novo Nordisk Fonden and by the Danish Cancer Society (`http://www.cancer.dk`) project ID# DP08023.

References

[1] Andreo P 2009 *Phys. Med. Biol.* **54** N205–N215
[2] IAEA TRS-398 2001 Absorbed dose determination in external beam radiotherapy an international code of practice for dosimetry based on standards of absorbed dose to water Tech. rep. International Atomic Energy Agency
[3] Henkner K, Bassler N, Sobolevsky N and Jäkel O 2009 *Med. Phys.* **36** 1230–5
[4] Geithner O, Andreo P, Sobolevky N, Hartmann G and Jäkel O 2006 *Phys. Med. Biol.* **51** 2279–2292
[5] Lühr A, Hansen D C, Jäkel O, Sobolevsky N and Bassler N 2011 *Phys. Med. Biol.* **56** 2515–2533
[6] Sørensen B S, Overgaard J and Bassler N 2011 *Acta Oncol.* **50** 757–762
[7] Krämer M and Scholz M 2000 *Phys. Med. Biol.* **45** 3319–3330
[8] Herrmann R, Carl J, Jäkel O, Bassler N and Petersen J B B 2010 *Acta Oncol.* **49** 1160–1164
[9] Herrmann R, Greilich S, Grzanka L and Bassler N 2011 *Radiat. Meas.* **46**(12) 1551–1553
[10] Edmund J M, Andersen C E and Greilich S 2007 *Nucl. Instrum. Methods Phys. Res. B* **262** 261–275
[11] Greilich S, Grzanka L, Bassler N, Andersen C E and Jäkel O 2010 *Radiat. Meas.* **45** 1406–1409
[12] Berger M J, Coursey J S, Zucker M A and Chang J 2005 ESTAR, PSTAR, and ASTAR: Computer Programs for Calculating Stopping-Power and Range Tables for Electrons, Protons, and Helium Ions (version 1.2.3) National Institute of Standards and Technology, Gaithersburg, MD.
[13] International Commision on Radiation Units and Measurements 1993 Stopping powers and ranges for protons and alpha particles Tech. Rep. 49 ICRU
[14] Paul H and Schinner A MSTAR URL `http://www.exphys.uni-linz.ac.at/stopping/`
[15] International Commision on Radiation Units and Measurements 2005 Stopping of ions heavier than helium Tech. Rep. 73 ICRU
[16] Gudowska I, Sobolevsky N, Andreo P, Belkić D and Brahme A 2004 *Phys. Med. Biol.* **49** 1933–1958
[17] Bassler N, Hansen D C, Lühr A and Sobolevsky N 2012 SHIELD-HIT12A [online] Available at: http://www.shieldhit.org [Accessed 28 October 2013]
[18] Hansen D C, Lühr A, Sobolevsky N and Bassler N 2012 *Phys. Med. Biol.* **57** 2393–2409
[19] Bethe H 1930 *Ann. Phys.* **5** 325–400
[20] Hubert F, Rimbot R and Gauvin H 1989 *Nucl. Instrum. Methods Phys. Res. B* **36** 357
[21] Lindhard J and Scharff M 1961 *Phys. Rev.* **124** 128–130
[22] Bazaraa M S and Shetty C M,1979 *Nonlinear Programming* (New York: John Wiley & Sons)
[23] Paul H and Sánchez-Parcerisa D 2013 *Nucl. Instrum. Methods Phys. Res. B* **312** 110–117
[24] Ableitinger A, Vatnitsky S, Herrmann R, Bassler N, Palmans H, Sharpe P, Ecker S, Chaudhri N, Jäkel O and Georg D 2013 *Radiother. Oncol.* **108** 99 – 106
[25] Lühr A, Hansen D C, Sobolevsky N, Palmans H, Rossomme S and Bassler N 2011 *Acta Oncol.* **50** 797–805
[26] Geissel H, Scheidenberger C, Malzacher P and Kunzendorf P ATIMA, calculating ATomic Interaction with MAtter URL `http://www-linux.gsi.de/~weick/atima/`
[27] Ziegler J F 2013 SRIM URL `http://www.srim.org`
[28] Sigmund P and Schinner A 2002 *Nucl. Instrum. Methods Phys. Res. B* **195** 64

XVII International Conference on the Use of Computers in Radiation Therapy (ICCR 2013) IOP Publishing
Journal of Physics: Conference Series **489** (2014) 012004 doi:10.1088/1742-6596/489/1/012004

SHIELD-HIT12A - a Monte Carlo particle transport program for ion therapy research

N Bassler[1,2], D C Hansen[2], A Lühr[1,2,3], B Thomsen[1], J B Petersen[4], N Sobolevsky[5,6]

[1] Department of Physics and Astronomy, Aarhus University, Aarhus, DENMARK
[2] Department of Experimental Clinical Oncology, Aarhus University Hospital, Aarhus, DENMARK
[3] German Cancer Consortium (DKTK), Dresden, GERMANY;
OncoRay – National Center for Radiation Research in Oncology, Medical Faculty and University Hospital Carl Gustav Carus, Technische Universität Dresden, GERMANY; and German Cancer Research Center (DKFZ), Heidelberg, GERMANY
[4] Department of Medical Physics, Aarhus University Hospital, Aarhus, DENMARK
[5] Institute for Nuclear Research of the Russian Academy of Sciences (INR), Moscow, RUSSIA
[6] Moscow Institute of Physics and Technology, State University, RUSSIA

E-mail: bassler@phys.au.dk

Abstract. Purpose: The Monte Carlo (MC) code SHIELD-HIT simulates the transport of ions through matter. Since SHIELD-HIT08 we added numerous features that improves speed, usability and underlying physics and thereby the user experience. The "-A" fork of SHIELD-HIT also aims to attach SHIELD-HIT to a heavy ion dose optimization algorithm to provide MC-optimized treatment plans that include radiobiology.

Methods: SHIELD-HIT12A is written in FORTRAN and carefully retains platform independence. A powerful scoring engine is implemented scoring relevant quantities such as dose and track-average LET. It supports native formats compatible with the heavy ion treatment planning system TRiP. Stopping power files follow ICRU standard and are generated using the libdEdx library, which allows the user to choose from a multitude of stopping power tables.

Results: SHIELD-HIT12A runs on Linux and Windows platforms. We experienced that new users quickly learn to use SHIELD-HIT12A and setup new geometries. Contrary to previous versions of SHIELD-HIT, the 12A distribution comes along with easy-to-use example files and an English manual. A new implementation of Vavilov straggling resulted in a massive reduction of computation time. Scheduled for later release are CT import and photon-electron transport.

Conclusions: SHIELD-HIT12A is an interesting alternative ion transport engine. Apart from being a flexible particle therapy research tool, it can also serve as a back end for a MC ion treatment planning system. More information about SHIELD-HIT12A and a demo version can be found on http://www.shieldhit.org.

1. Introduction

The Monte Carlo particle transport code SHIELD-HIT12A is designed to precisely simulate therapeutic beams of protons and ions in biological tissue relevant for ion beam cancer therapy. SHIELD-HIT (Heavy Ion Therapy) evolved from the common SHIELD code, which models interactions of hadrons and atomic nuclei in complex extended targets in the energy range up to TeV/nucleon. All SHIELD versions are based on the same models to simulate nuclear interactions, which are developed at JINR Dubna[1] and INR RAS (Moscow)[2]. More information about SHIELD can be found at `http://www.inr.ru/shield`.

The development of a medical adapted version of SHIELD started in 2002 [1]. The main SHIELD-HIT line is still developed, however, in 2010 the -A branch was forked from SHIELD-HIT08 with the purpose to improve physics models, computation speed and usability. Updates to the physics models in the A-fork are shared with the main development branch of SHIELD-HIT. Before SHIELD-HIT12A was conceived, the developments were done in SHIELD-HIT10A, which, however, was not released for official distribution. We foresee that SHIELD-HIT12A will be the first version of SHIELD-HIT that will be officially distributed. A demonstration version of SHIELD-HIT12A is already available, which is only limited in the total number of particles which can be simulated and a fixed random seed.

2. Methods

SHIELD-HIT12A is capable of transporting hadrons including nucleons, any ions and pions. Antiprotons and antineutrons are also transported, and the inelastic cross sections for antinucleons were recently revised. Inelastic cross sections are calculated by the formalism presented in [2–4] with modifications by [5–7] which include recently published data from Gesellschaft für Schwerionenforschung (GSI), Darmstadt, Germany, regarding interaction with carbon-12 projectiles. Every secondary particle is tracked and processed until it stops, decays, interacts destructively or leaves the simulation universe. Electrons and photons can be created through extra-nuclear cascades and π^0 decay, but are not transported, as no interaction physics is included for these particles.

The nuclear models are grouped together in the Multi Stage Dynamical Model (MSDM-generator) [8] which simulates nuclear reactions in the exclusive approach. The MSDM generator is composed of a fast, cascade stage of nuclear reaction, where — depending on the projectile energy — the Dubna cascade model (DCM) [9] or the quark-gluon string model (QGSM) [10,11] is selected. By the end of the cascade stage, nucleons which are close to each other in the momentum space can coalesce to form a complex particle according to the coalescence model [9]. Precompound emission of nuclei is handled by the Cascade-exciton model [12]. Subsequent equilibrium de-excitation is handled by Fermi break-up [13], evaporation/fission competition [13, 14] and multifragmentation [15] of highly excited nuclei.

Energy loss of primary particles is either calculated by the internal implementation of the Bethe equation — which is sewed with the Lindhard-Scharff low-energy description [16] — or can, alternatively, be imported from the external stopping power library libdEdx [17–19], which provides, e.g., ICRU recommended stopping powers. Secondary electrons are not generated.

3. Results and Discussion

Most updates to the physics engine are described in the references [5–7] and concern the development version SHIELD-HIT10A. Extensive tests of the physics models are covered in the same references. Specific to SHIELD-HIT12A, a new beam model was implemented that allows focused and defocused beams, following the formalism described in [20], motivated by the fact that several treatment facilities focus the beam at the treatment isocenter, as it is done, e.g., at the heavy ion treatment facility in Marburg, Germany.

We have hard-coded the complete ICRU material composition database with 279 default materials into SHIELD-HIT12A, so users conveniently can select materials by a single line in the configuration file. Yet, default mean excitation potentials can be overridden, and users also have the possibility to specify arbitrary chemical and isotope compositions. The 28-group ABBN

[1] `http://www.jinr.ru/`

[2] `http://www.inr.troitsk.ru/`

neutron cross section data [21] are still the same used in SHIELD-HIT08, except that detailed data for Argon were added, which may be relevant for simulations of ambient air.

A completely new scoring system has been developed for the "-A" fork of SHIELD-HIT to obsolete the old scoring system used in SHIELD-HIT. The old scoring system has some limitations, namely, it generates output in formats not intended for summing multiple runs. Also scoring parameters had to be hard-coded, such as the range and number of energy bins when scoring a particle spectrum. The new scoring system implemented in SHIELD-HIT12A resolves these issues. All parameters relevant for scoring can be altered by the user via configuration files, specific to each simulation. Within such a file, a "card" — containing scoring parameters — is defined for each scoring detector, along with options of scoring type and geometry. This approach is similar in style to what is used by several other Monte Carlo codes such as FLUKA [22, 23]. This gives the user maximum flexibility in scoring without the need to access the source code and allows to add new estimators and detectors in the future without breaking compatibility to previous versions.

Scoring can be made to specific geometric entities as well as to independent Cartesian and cylindrical grids which are overlaid over arbitrary regions of the simulation geometry. The grid scoring is performed via Sidon's algorithm [24] resulting in a minimal impact on the Monte Carlo performance even for large and fine-meshed scoring grids. A wide range of detectors are implemented for scoring basic quantities such as energy, dose, and fluence but also planar fluence as well as track- and dose-averaged linear energy transfer (LET). It is possible to override the target material in a scorer being useful for, e.g., calculating stopping power ratios, which are commonly needed for precisely simulating ionization chambers or for translating dose in material to dose in water. The alanine detector response model described in [25] is implemented for easy simulation of alanine dosimeter behavior in arbitrary ion fields. All estimators provide results in a binary format, which can be read by a supplied script for data merging when the computation is parallelized or when statistical inferences are extracted. The merged data are converted to human readable ASCII format and can then be plotted and further analyzed.

As a new feature, SHIELD-HIT12A is interfaced to the treatment planning system TRiP98 [26]. SHIELD-HIT12A generates spectrum files and depth-dose kernels in native TRiP format, so base data for beam modeling can be simulated for arbitrary beam lines and ion species. Vice versa, SHIELD-HIT12A can import the raster scan control files from TRiP optimization, and use these as beam sources to simulate spread out Bragg peaks.

Since the fork from SHIELD-HIT08 we realized a substantial speedup. First, SHIELD-HIT12A uses a faster random number generator [7]. In addition, the new scoring system enables convenient parallel computation. SHIELD-HIT12A provides scripts for submitting jobs on TORQUE or CONDOR based scheduling systems. Furthermore, the original SHIELD-HIT08 code contains a few functions concerning the Moliere scattering and Vavilov straggling which originate from Geant3.21, licensed under GPL. These functions were reimplemented, and in particular an entirely new algorithm for the Vavilov straggling was developed which performs 5-6 times faster than the original code by Rotondi and Montana [27,28]. Details on this improvement are in preparation and will be published elsewhere.

To date SHIELD-HIT12A has been used for assessing the impact of nuclear fragmentation models in clinical settings [29,30] and calculation of fluence correction factors [31,32]. The beam application and monitoring system (BAMS) at the heavy ion particle therapy facility at Marburg was simulated with SHIELD-HIT12A using a new generation of ripple filters that potentially may decrease treatment time by a factor of 2 [33,34]. SHIELD-HIT12A is used — in connection with TRiP98 — to investigate the concept of LET-painting [35] that may help to improve the treatment of hypoxic tumors [36]. We have experienced that users new to ion beam Monte Carlo simulations quickly get acquainted to the user interface, and quickly can setup their own geometries and simulations.

4. Conclusion and Outlook

SHIELD-HIT12A is a Monte Carlo particle transport program which is developed for precise simulation of ion beams in particle therapy. It is interfaced with the treatment planning tool TRiP, which makes it a flexible particle therapy research tool, and could serve as a back end for a MC ion treatment planning system in general. A demo version is available for download at `http://www.shieldhit.org`, and full versions are available upon request.

Future efforts on SHIELD-HIT12A will concentrate on implementing voxelized geometries (such as CT scans).

Acknowledgments

This project was supported by CIRRO - The Lundbeck Foundation Center for Interventional Research in Radiation Oncology, The Danish Council for Strategic Research (`http://www.cirro.dk`), Novo Nordisk Fonden and by the Danish Cancer Society (`http://www.cancer.dk`) project ID# DP08023. We thank Dr. Uli Weber from the Universitätsklinikum Gießen und Marburg, Germany for discussions on the implemented beam models and ripple filters.

References

[1] Gudowska I, Sobolevsky N, Andreo P, Belkić D and Brahme A 2004 *Phys. Med. Biol.* **49** 1933–1958
[2] Barashenkov V S 1993 *JINR, Dubna* 1–346 (In Russian)
[3] Barashenkov V S and Polanski A 1994 Electronic guide for nuclear cross sections Tech. Rep. E2 94 417 JINR Dubna
[4] Sychev B S 1999 Cross sections of interaction of high-energy hadrons with atomic nuclei Tech. rep. MRTI RAS Moscow (In Russian)
[5] Hansen D C 2011 *Improving the Nuclear Models in SHIELD-HIT for Heavy Ion Therapy* Master's thesis Dept. of Physics and Astronomy, Aarhus University, Aarhus, Denmark
[6] Hansen D C, Lühr A, Sobolevsky N and Bassler N 2012 *Phys. Med. Biol.* **57** 2393–2409
[7] Hansen D C, Lühr A, Herrmann R, Sobolevsky N and Bassler N 2012 *Int. J. Radiat. Biol.* **88** 195–199
[8] Botvina A S, Dementyev A V, Smirnova O N, Sobolevsky N M and Toneev V D 1997 *International Codes and Model Intercomparison for Intermediate Energy Activation Yields,(by R. Michel, P. Nagel), NSC/DOC (97)-1, NEA/P&T* **14**
[9] Toneev V D and Gudima K K 1983 *Nuclear Physics A* **400** 173–189
[10] Amelin N S, Gudima K K and Toneev V D 1990 *Sov. J. Nucl. Phys* **52** 172–178
[11] Amelin N S, Gudima K K, Sivoklokov S Y and Toneev V D 1990 *Yadernaya Fizika* **52** 272 (In Russian)
[12] Gudima K K, Mashnik S G and Toneev V D 1983 *Nuclear Physics A* **401** 329–361
[13] Botvina A S, Iljinov A S, Mishustin I N, Bondorf J P, Donangelo R and Sneppen K 1987 *Nuclear Physics A* **475** 663–686 ISSN 0375-9474
[14] Adeev G D, Botvina A S, Iljinov A S, Mebel M V, Piscasov N I and Serdyuk O I 1993 A method of calculation of mass and energy distributions of fission residuals in reactions induced by intermediate energy particles Tech. Rep. 816/93 INR RAS Moscow (In Russian) Preprint
[15] Botvina A S and Iljinov I N 1990 *Nuclear Physics A* **507** 649–674
[16] Lindhard J, Scharff M and Schiøtt H E 1963 *Mat. Fys. Medd. Dan. Vid. Selsk* **33** 1–42
[17] Toftegaard J, Lühr A and Bassler N 2010 *libdEdx* [Online] Available: http://libdedx.sf.net [28 October 2013]
[18] Lühr A, Toftegaard J, Kantemiris I, Hansen D C and Bassler N 2012 *Int. J. Radiat. Biol.* **88** 209–212
[19] Toftegaard J, Lühr A, Sobolevsky N and Bassler N 2013 *J. Phys.: Conf. Ser.* Submitted
[20] Gottschalk B 2012 Techniques of proton radiotherapy: Transport theory Tech. rep. Harward University Laboratory for Particle Physics and Cosmology http://arxiv.org/abs/1204.4470v2
[21] Abagyan L P, Bazazyants N O, Nikolaev M N and Tsibulya A M 1981 *Neutron group constants for reactors and shielding calculations* (Moscow: Energoizdat) (In Russian)
[22] Fassò A, Ferrari A, Roesler S, Sala P R, Battistoni G, Cerutti F, Gadioli E, Garzelli M V, Ballarini F, Ottolenghi A, Empl A and Ranft J 2003 *Computing in High Energy and Nuclear Physics 2003 Conference (CHEP2003)* (La Jolla, CA, USA) (paper MOMT005), eConf C0303241 (2003), arXiv:hep-ph/0306267
[23] Fassò A, Ferrari A, Ranft J and Sala P R 2005 FLUKA: a multi-particle transport code CERN-2005-10, INFN/TC_05/11, SLAC-R-773
[24] Siddon R L 1985 *Med. Phys.* **12** 252–255
[25] Bassler N, Hansen J W, Palmans H, Holzscheiter M H, Kovacevic S and the AD-4/ACE Collaboration 2008 *Nucl. Instrum. Methods Phys. Res. B* **266** 929–936

[26] Krämer M, Jäkel O, Haberer T, Kraft G, Schardt D and Weber U 2000 *Phys. Med. Biol.* **45** 3299–3317

[27] Rotondi A and Montagna P 1990 *Nucl. Instrum. Methods Phys. Res. B* **47** 215–223

[28] CERN Program Library Long Writeup W5013 1994 *GEANT, Detector Description and Simulation Tool* section PHYS325: Moliere scattering

[29] Lühr A, Hansen D C, Teiwes R, Sobolevsky N, Jäkel O and Bassler N 2012 *Phys. Med. Biol.* **57** 5169

[30] Lühr A, Priegnitz M, Fiedler F, Sobolevsky N and Bassler N 2013 *Appl. Radiat. Isotopes* In press

[31] Lühr A, Hansen D C, Sobolevsky N, Palmans H, Rossomme S and Bassler N 2011 *Acta Oncol.* **50** 797–805

[32] Palmans H, Al-Sulaiti L, Andreo P, Shipley D, Lühr A, Bassler N, Martinkovič J, Dobrovodský J, Rossomme S, Thomas R *et al.* 2013 *Phys. Med. Biol.* **58** 3481

[33] Ringbæk T P, Weber U, Petersen J B, Thomsen B and Bassler N 2013 *Acta Oncol.* In press

[34] Ringbaek T P 2013 *Monte Carlo Simulations of New 2D Ripple Filters for Particle Therapy Facilities* Master's thesis Dept. of Physics and Astronomy, Aarhus University, Aarhus, Denmark

[35] Bassler N, Jäkel O, Søndergaard C S and Petersen J B 2010 *Acta Oncol.* **49** 1170–1176

[36] Bassler N, Toftegaard J, Lühr A, Sørensen B S, Scifoni E, Krämer M, Jäkel O, Mortensen L S, Overgaard J and Petersen J B 2013 *Acta Oncol.* In press

Journal of Physics: Conference Series **489** (2014) 012005 doi:10.1088/1742-6596/489/1/012005

Monte Carlo simulation of the transit dosimetric response of an *a*-Si electronic portal imaging device

S J Blake[1,2], A L McNamara[1,2], P Vial[1–3], L Holloway[1–5], P B Greer[6,7] and Z Kuncic[1]

[1] Institute of Medical Physics, School of Physics, University of Sydney, Sydney, NSW 2006, Australia
[2] Ingham Institute for Applied Medical Research, Liverpool, NSW 2170, Australia
[3] Department of Medical Physics, Liverpool and Macarthur Cancer Therapy Centres, NSW 2170, Australia
[4] Centre for Medical Radiation Physics, University of Wollongong, Wollongong, NSW 2522, Australia
[5] South Western Sydney Clinical School, University of New South Wales, Sydney, NSW 2052, Australia
[6] School of Mathematical and Physical Sciences, University of Newcastle, Newcastle, NSW 2300, Australia
[7] Department of Radiation Oncology, Calvary Mater Newcastle Hospital, Newcastle, NSW 2298, Australia

E-mail: `s.blake@physics.usyd.edu.au`

Abstract.
Amorphous silicon (*a*-Si) electronic portal imaging devices (EPIDs) are x-ray detectors frequently used in radiotherapy imaging and dosimetry applications. EPIDs employ a copper plate and gadolinium oxysulfide phosphor screen with an array of *a*-Si photodiodes to indirectly detect incident radiation. In this study, a previously developed Monte Carlo (MC) model of an *a*-Si EPID has been extended for transit dosimetry. The GEANT4 MC toolkit was used to integrate an *a*-Si EPID model with two phantoms and a 6 MV x-ray source. A solid water phantom was used to simulate EPID transmission factors, field size output factors and relative dose profiles and results were compared to experimental measurements. An anthropomorphic head phantom was used to qualitatively compare simulated and measured portal images of humanoid anatomy. Calculated transmission factors and field size output factors agreed to within 2.0% and 1.9% of experimental measurements, respectively. A comparison of calculated and measured relative dose profiles yielded >98% of points passing a gamma analysis with 3%/3 mm criterion for all field sizes. The simulated anthropomorphic head phantom image shows macroscopic anatomical features and qualitatively agrees with the measured image. Results validate the suitability of the MC model for predicting EPID response in transit dosimetry.

1. Introduction

Amorphous silicon (*a*-Si) electronic portal imaging devices (EPIDs) serve a number of important clinical applications in modern radiotherapy. EPIDs are routinely used to image patient anatomy and verify patient setup prior to treatment. EPIDs are also suitable dosimeters since the pixel values of acquired images correlate to the absorbed dose in the detector. One method of performing dose verification using EPIDs is therefore by comparing portal dose images to dose distributions predicted using an EPID model. EPID dosimetric characteristics and their various clinical uses for dosimetry have been reviewed by van Elmpt *et al.*[1]

A number of arguments support the integration of EPID dosimetry into routine clinical practice. Linear accelerator (linac) vendors typically supply *a*-Si EPIDs with the necessary hardware mounted directly to the gantry, in line with the megavoltage (MV) treatment x-ray source. This configuration provides a readily available mechanism to detect the MV beam and enables direct monitoring of both patient position and dose delivery from the beam's-eye view. When compared to alternative 2D dosimeters such as arrays of diodes or ion chambers, *a*-Si EPIDs offer increased spatial resolution (typically 0.4×0.4 mm^2) and real-time data acquisition capabilities. Additionally, EPIDs are resilient to radiation-induced damage and respond both linearly with dose and independently of dose rate[2, 3]. One centre has reported on the routine use of EPIDs for pre-treatment and *in vivo* dosimetry, including the EPIDs ability to detect errors in treatment delivery[4].

The goal of this study is to extend the functionality of an EPID model that we previously developed for non-transit dosimetry[5] by integrating phantom geometries into the model. In doing so, we may investigate the EPID response in a transit dosimetry configuration that is more representative of clinical treatment situations. Furthermore, we aim to validate the transit dosimetric response of this model against experimental measurements.

2. Monte Carlo model and detector geometry

2.1. 6 MV photon source

The MC radiation transport code EGSnrc[6] (V4 2.3.1) with user code BEAMnrc[7] (V4 2.3.1) were used to create a 6 MV photon source model of an Elekta Synergy linac (Elekta, Crawley, UK). A description and validation of the source model has been previously reported[5]. The simulation of 10^9 primary histories was performed to generate phase space files for square fields ranging in size from 2×2 to 9×9 cm^2 (defined at the isocentre, 100 cm from the target).

2.2. EPID geometry and physics processes

The GEANT4 MC simulation toolkit[8] (version 9.4) was previously used to develop a model of an *a*-Si EPID and validate its dosimetric response in a non-transit configuration[5]. A complete description of the EPID geometry may be found in the previous study as only a brief overview is given here.

The EPID model (Figure 1) consists of a series of uniform slab layers with geometries and material compositions based on specifications provided by the manufacturer of a research EPID (XRD 1640 AN CS) used in the validation stage of this study (PerkinElmer, Santa Clara, CA). The EPID model has a cross-sectional area of 41×41 cm^2 and was positioned at a source to detector distance (SDD) of 160 cm for all simulations. It incorporates a 1 mm Cu buildup layer, a 133 mg cm^{-2} Gd_2O_2S:Tb phosphor screen (Lanex Fast Back, Carestream Health, Inc. Rochester, USA) and a 0.1 mm thick layer of *a*-Si supported by a 1 mm SiO_2 substrate.

The standard GEANT4 electromagnetic physics models were used to simulate radiation transport within the MC model. The transport of optical photons originating in the phosphor screen was not explicitly simulated. We previously found that optical transport does not significantly change calculated dosimetric quantities relative to those calculated using only standard electromagnetic physics[5]. Simulated processes included Compton scattering, pair production, photoelectric absorption, impact ionization, Bremsstrahlung radiation, electron/positron annihilation and multiple scattering.

2.3. Phantom definitions

This study incorporated two distinct phantom geometries into the MC model to investigate separate EPID dosimetric characteristics. The first phantom was a simple homogeneous box of solid water with a cross-sectional area of 40×40 cm^2 and a thickness along the central axis that varied depending on the quantity being simulated. The second phantom was an anthropomorphic head that was defined by integrating a set of CT images into the MC model using functions distributed with the GEANT4 source code. Both phantoms were centred about the isocentre and the head phantom was oriented with its anterior-posterior (AP) axis aligned with the beam central axis.

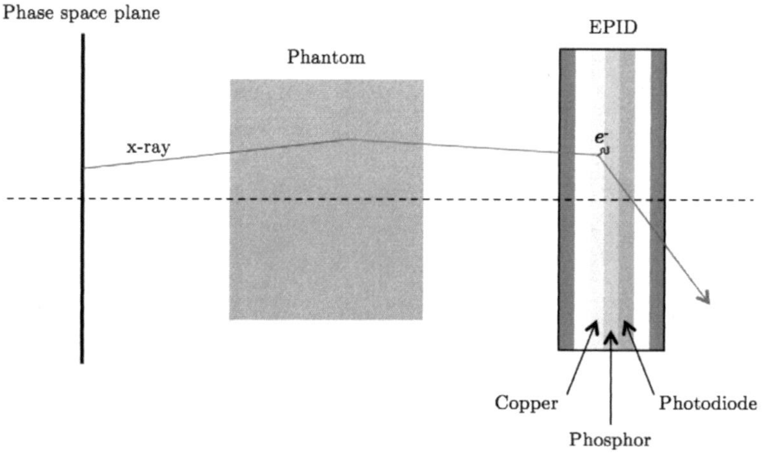

Figure 1. Schematic of the model components. An x-ray track incident from the left Compton scatters first in the phantom and then in the copper, creating an electron track that deposits energy in the phosphor.

3. Simulated dosimetric quantities

The EPID dose response characteristics investigated in this study include transmission factors, field size output factors and relative beam profiles. An image of an anthropomorphic head phantom was also simulated for qualitative evaluation. All quantities were calculated by tracking particles from the source phase space files and scoring the energy deposited in the phosphor layer of the EPID in a 2D histogram. Each histogram contained 1024×1024 bins (0.4×0.4 mm^2 pixels), equal in number and size to the pixels of the research EPID. All MC simulations were performed using a computer cluster of 252 2.67 GHz CPUs and the open source message passing interface OpenMPI (www.open-mpi.org) was used to facilitate parallel processing. ROOT[9] (version 5.28.00) was used for all post-processing analysis.

3.1. Transmission factors

Transmission factors were calculated by varying the solid water phantom thickness from 0 to 40 cm in 10 cm increments with a fixed beam field size of 9×9 cm^2. The mean response within the central 1×1 cm^2 region of each 2D histogram was calculated, normalized to the response for the phantom thickness of 0 cm. Uncertainties in all output factor calculations are quoted as the standard deviation of the response within the central region.

3.2. Field size output factors and relative dose profiles

Field size output factors and relative dose profiles were calculated by varying the beam field size from 2×2 to 9×9 cm^2 with a fixed solid water phantom thickness of 20 cm. Output factors were calculated as the mean response within the central 1×1 cm^2 region of each 2D histogram, normalized to the 9×9 cm^2 field response. Uncertainties in all output factor calculations are quoted as the standard deviation of the response within the central region.

Dose profiles were first normalized to a central axis response of 100%. 1D relative profiles were then obtained by extracting the response along a slice through the centre of the 2D histograms in the cross-plane direction. Agreement between simulated and measured profiles was evaluated by calculating the percentage of data points with a γ-index ≤ 1 based on 3%/3 mm criteria (with dose differences calculated globally relative to the dose at the central-axis and considering only those points above a minimum threshold relative dose of 10%)[10].

3.3. Projection phantom portal dose image

A static 9×9 cm^2 beam field size was used to generate an AP projection portal image of an anthropomorphic head phantom using the EPID model.

4. Experimental measurements and model validation

Experimental measurements to validate the MC model were made using the research *a*-Si EPID described in Section 2.2. An Elekta (Elekta, Crawley, UK) Synergy 6 MV linac with the MLCi multi-leaf collimator was used for all measurements. Images were acquired by averaging 50 frames when delivering a nominal dose rate of 500 MU/min. To minimize backscatter from the treatment couch, the EPID was positioned vertically (i.e. on its side) on the couch and centered on the collimator axis of rotation at a SSD of 160 cm, with the gantry rotated to 90 degrees. Phantoms (as described in Section 2.3) positioned on the couch were centred at the isocentre.

The XIS software package (PerkinElmer, Santa Clara, CA) was interfaced with the research EPID to acquire all images. A gain setting of 4 pF was used with a frame integration time of 499 ms. Images acquired for validation of the transmission factors and field size output factors were both dark-field and flood-field corrected. Flood-field corrections were not applied for validation of EPID relative dose profiles or the anthropomorphic head phantom image as this correction would remove the well-known off-axis detector response[11].

5. Results and Discussion

5.1. Transmission factors

Transmission factors calculated using the MC model and measured using the research EPID are shown in Figure 2. Calculated and measured transmission factors are in excellent agreement, with a maximum percent difference of only 2.0% (occurring for the 40 cm phantom thickness).

5.2. Field size response

Figure 3 shows the calculated and measured variation in EPID response with beam field size when a 20 cm thick solid water phantom is used. The calculated and measured field size responses are in close agreement with the greatest difference of 1.8% occurring for the 3×3 cm^2 field size.

5.3. Relative dose profiles

Relative dose profiles calculated using the MC model and measured using the research EPID are presented in Figure 4 for selected beam field sizes between 2×2 and 9×9 cm^2 when a 20 cm thick solid water phantom is used. The subplot shows the results of a γ comparison between the calculated and measured profiles for each field size using 3%/3 mm criterion. 98% and 99% of profile data points had $\gamma \leq 1$ for the 2×2 and 3×3 cm^2 field sizes respectively, whereas 100% of points had $\gamma \leq 1$ for the remaining field sizes. These results demonstrate excellent agreement between the calculated and measured EPID off-axis response.

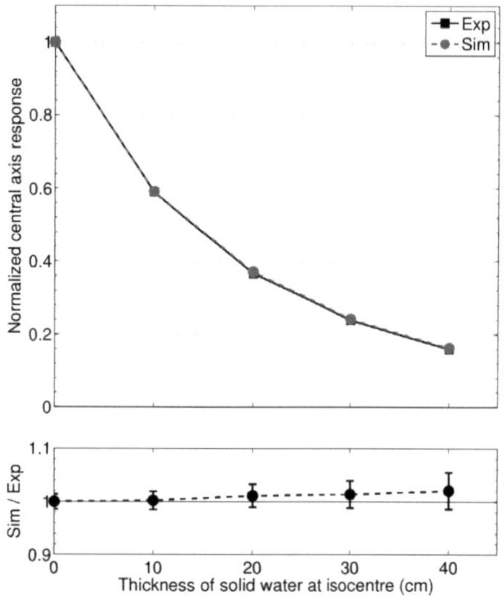

Figure 2. Measured (Exp) and calculated (Sim) EPID transmission factors.

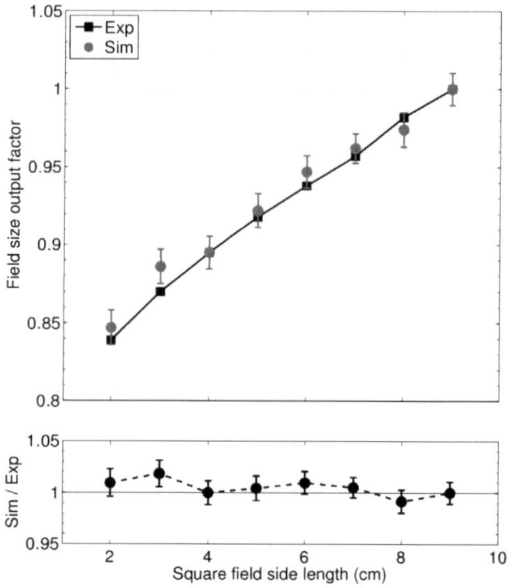

Figure 3. Measured (Exp) and calculated (Sim) EPID field size output factors.

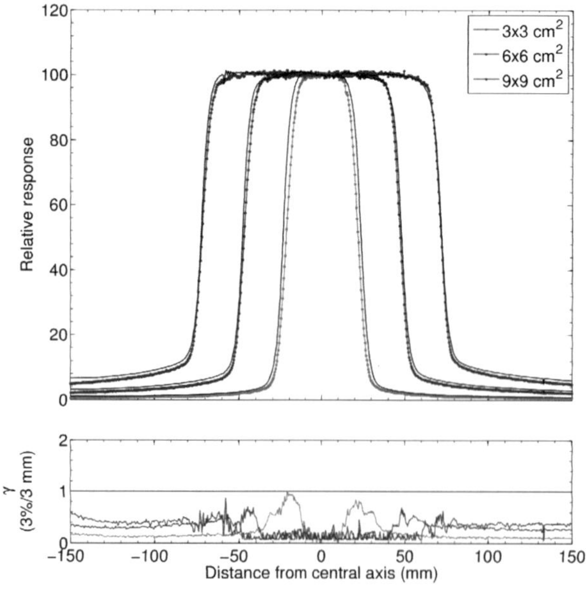

Figure 4. Calculated relative dose profiles from the standard EPID model using a 20 cm thick phantom of solid water (top panel) and corresponding γ-values for 3%/3 mm agreement with measured profiles (bottom panel).

5.4. Projection phantom portal dose image

Measured and calculated portal images of an anthropomorphic head phantom are presented in Figure 5. A qualitative comparison of these images demonstrates that the MC model is able to simulate spatial variations in detector response resulting from the use of an inhomogeneous phantom representative of human anatomy. The statistical noise present in the calculated image made it difficult to resolve the fine anatomical structures and slight changes in relative density within the phantom. However, macroscopic features such as the orbits and nasal cavity are discernable in the calculated image.

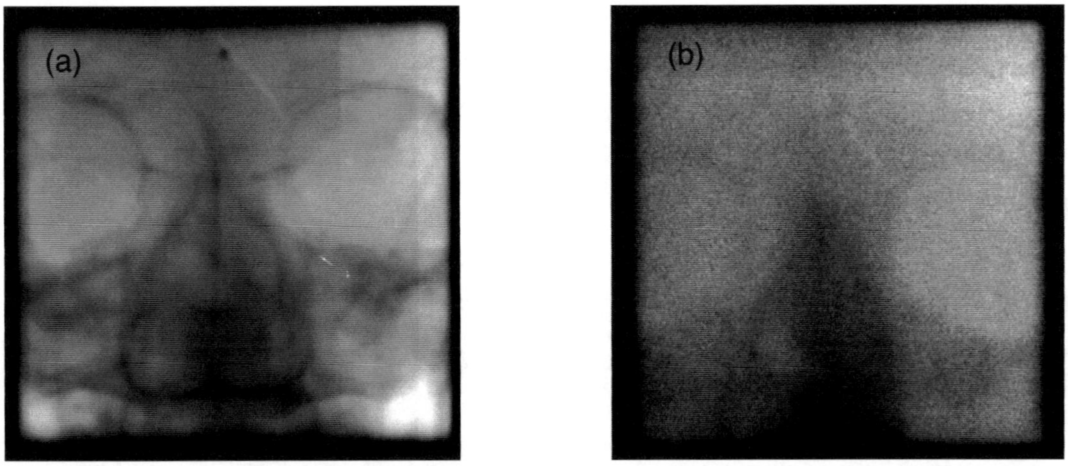

Figure 5. Measured (a) and simulated (b) portal images of an anthropomorphic head phantom.

6. Conclusions

A Monte Carlo model of a standard *a*-Si EPID that was previously developed for non-transit dosimetry has been extended to transit dosimetry applications. Transmission factors, field size output factors and relative dose profiles were calculated using the model and validated against experimental measurements with excellent agreement. The simulation of an anthropomorphic head phantom portal dose image provides a demonstration for applying this model to predicting EPID images of humanoid anatomy.

Acknowledgments

Financial support was provided by the Cancer Institute NSW (Research Equipment Grants 10/REG/1-20 and 10/REG/1-10), Cancer Council NSW (Grant ID RD 11-06) and the National Health and Medical Research Council of Australia (Project Grant ID569211). SJB acknowledges The University of Sydney and the Institute of Medical Physics for scholarship support, as well as the Liverpool and Macarthur Cancer Therapy Centres for additional financial support.

References

[1] van Elmpt W *et al.* 2008 *Radiother. Oncol.* **88** 289 – 309
[2] Boudry J M and Antonuk L E 1996 *Med. Phys.* **23** 743–754
[3] El-Mohri Y *et al.* 1999 *Med. Phys.* **26** 1530–1541
[4] Mans A *et al.* 2010 *Med. Phys.* **37** 2638–2644
[5] Blake S J *et al.* 2013 *Med. Phys.* **40** 041708
[6] Kawrakow I 2000 *Med. Phys.* **27** 485–498
[7] Rogers D W O *et al.* 1995 *Med. Phys.* **22** 503–524
[8] Agostinelli S *et al.* 2003 *Nucl. Instrum. Methods Phys. Res.* A **506** 250 – 303
[9] Brun R and Rademakers F 1997 *Nucl. Instrum. Methods Phys. Res.* A **389** 81 – 86
[10] Low D A *et al.* 1998 *Med. Phys.* **25** 656–661
[11] Greer P B 2005 *Med. Phys.* **32** 3558–3568

The influence of Monte Carlo source parameters on detector design and dose perturbation in small field dosimetry

P H Charles[1], S B Crowe[1], T Kairn[1,2] R Knight[2], B Hill[2], J Kenny[3,4], C M Langton[1] and J V Trapp[1]

[1] School of Chemistry, Physics and Mechanical Engineering, Queensland University of Technology, Brisbane, GPO Box 2434, Brisbane, Qld 4001, Australia.
[2] Premion, The Wesley Medical Centre, Suite 1, 40 Chasely St, Auchenflower, Qld 4066, Australia
[3] The Australian Clinical Dosimetry Service, Australian Radiation Protection and Nuclear Safety Agency, 619 Lower Plenty Road, Yallambie, Victoria 3085, Australia
[4] Radiation Oncology Queensland, St Andrew's Toowoomba Hospital, Toowoomba, Qld 4350, Australia

p.charles@qut.edu.au

Abstract. To obtain accurate Monte Carlo simulations of small radiation fields, it is important model the initial source parameters (electron energy and spot size) accurately. However recent studies have shown that small field dosimetry correction factors are insensitive to these parameters. The aim of this work is to extend this concept to test if these parameters affect dose perturbations in general, which is important for detector design and calculating perturbation correction factors.

The EGSnrc C++ user code cavity was used for all simulations. Varying amounts of air between 0 and 2 mm were deliberately introduced upstream to a diode and the dose perturbation caused by the air was quantified. These simulations were then repeated using a range of initial electron energies (5.5 to 7.0 MeV) and electron spot sizes (0.7 to 2.2 FWHM).

The resultant dose perturbations were large. For example 2 mm of air caused a dose reduction of up to 31% when simulated with a 6 mm field size. However these values did not vary by more than 2 % when simulated across the full range of source parameters tested.

If a detector is modified by the introduction of air, one can be confident that the response of the detector will be the same across all similar linear accelerators and the Monte Carlo modelling of each machine is not required.

1. Introduction and aim

Small air gaps upstream to a diode active volume heavily perturb small photon fields causing a large reduction in dose to the active volume[1,2]. Therefore air can be deliberately introduced into diode designs to negate the inherent over-response of diodes in small photon fields[2]. The amount of air required to produce a diode which responds uniformly at all field sizes depends on the diode design[2].

The aim of this work is to study the dependence of the dose reduction caused by controlled amounts of air on the Monte Carlo initial source parameters (incident electron spot size and energy) of a linear accelerator. Hence this study reveals if a new diode design involving any amount of air can be used with confidence across a population of similar linear accelerators.

2. Background

Dosimetry of small photon fields presents many challenges, one of the largest being the perturbation of the beam by the detector used[3-5]. Traditional radiation detectors such as ionization chambers and diodes become relatively large compared to the size of the field, and therefore the composition of these detectors becomes important when calculating the response of the detector in a small field.

Alfonso *et al*[4] proposed a formalism where the change in sensitivity of a detector as a function of small field size is accounted for by applying an additional factor $k^{f_{clin},f_{msr}}_{Q_{clin},Q_{msr}}$ to standard output factor measurements.

$k^{f_{clin},f_{msr}}_{Q_{clin},Q_{msr}}$ has been systematically calculated for diodes using Monte Carlo modelling[6-9] and its value depends strongly on the composition of the detector[10]. Diodes have been shown to have increased sensitivity in small fields due to the high density of the silicon active volume[11,12] as well as high density metals which may be upstream to the active volume[10]. It has been shown that $k^{f_{clin},f_{msr}}_{Q_{clin},Q_{msr}}$ values calculated via Monte Carlo simulations for diodes are independent of the initial source parameters used to model the linear accelerator (incident electron energy and focal spot size)[8-10]. However in each case the exact composition of the detector has been modelled, so the results are very specific to those detectors.

In a recent study Charles *et al*[2] used Monte Carlo simulations to study the effect of deliberately introducing air into diodes such that the dose reduction caused by the air eliminated the field size dependence of the diode. The authors successfully calculated the amount of air that would be required for a shielded diode, an unshielded diode and a theoretical unenclosed silicon chip to respond uniformly at all field sizes. In each case the air was simulated to be on the upstream end on the diode.

3. Methodology

3.1. Monte Carlo simulation overview

All linear accelerator Monte Carlo simulations were performed using the user code BEAMnrc[13] and a previously modeled Varian iX (Varian Medical Systems, Palo Alto, CA)[2,14,15]. The field size used throughout this study was a 6 mm × 6 mm square.

All diode simulations were performed using the EGSnrc C++ user code cavity[16]. The unmodified diode used in this study was a simple cylindrical silicon chip (see figure 1). All detector simulations were performed in water at a depth of 5 cm and source to detector distance of 100 cm. All dose calculations were performed on the central axis of the beam.

In the phantom material, the electron and photon cut off energies were chosen to be 521 and 10 keV respectively. Inside the volume of the simulated diode they were reduced to 512 and 1 keV respectively for increased accuracy within the detector volume and air gaps. Sufficient histories were simulated so that the statistical uncertainty of the Monte Carlo calculated correction factors was maintained at less than 0.5%.

3.2 Modification of the diode with air

Air was introduced immediately upstream to the silicon active volume (see figure 1). For each case the dose to active volume of the diode was calculated, as was the dose reduction caused by the air gap. Here dose reduction refers to the percent change in dose when compared to the dose to the detector with no air present. That is:

$$\text{Dose reduction }(x) = [(D_{Det,air(x)} - D_{Det,air(0)}) / D_{Det,air(0)}] \times 100$$

$D_{Det,air(x)}$ is the dose to the detector with x mm thickness of air placed upstream.

The simulations were initially performed with an incident electron energy onto the target of 5.8 MeV

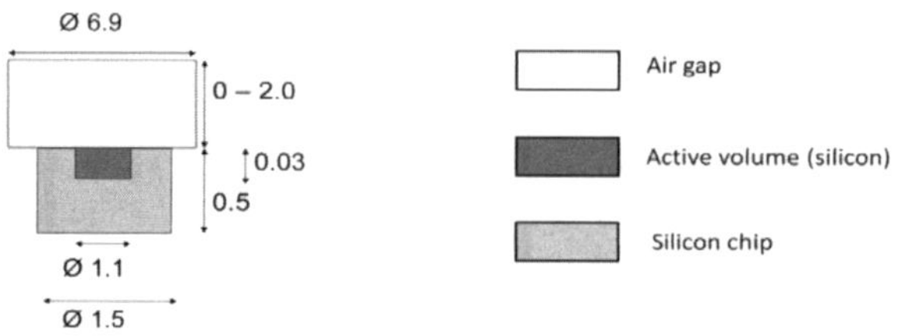

Figure 1. The simulation geometry of the detector used in this study (a basic silicon chip) and the upstream air gap. The dimensions are shown in mm. The diagram is not to scale.

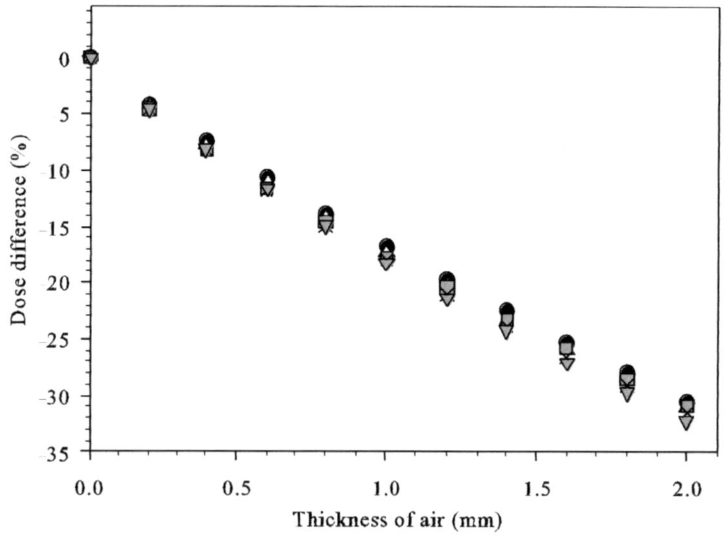

Figure 2. The dose reduction as a function of upstream air thickness. Dose reduction refers to the change in dose relative to a simulation with no air gap. Shown here are the results using an incident electron energy of 5.8 MeV and an incident electron spot size of 0.4 mm (black circles), 0.7 mm (white triangles), 1.2 mm (grey squares) and 2.2 mm (crosses). Also shown is the results for an incident electron energy of 7.0 MeV (grey upside-down triangles) (spot size = 1.2 mm).

and a focal spot size of 1.2 mm. The simulations were repeated for an incident electron energy of 7.0 MeV (the focal spot size remained 1.2 mm) and the dose differences caused by the air were compared. Keeping the energy constant at 5.8 MeV, the simulations were then repeated using the following focal spot sizes: 0.4, 0.7, and 2.2 mm. Once again the dose reduction caused by the air was compared across the different focal spot sizes.

4. Results and discussion

The dose difference as a function of air gap thickness is presented in figure 2 for the two different incident electron energies and for the various focal spot sizes. It can be seen that while the air strongly perturbs the dose to the detector, this dose difference does not vary greatly with the initial source parameters.

Air causes a large dose reduction to the active volume due to the filtering of electrons with a high angle of incidence[1]. However the dose reduction results were consistent across all input parameter variations. The consistency of these results suggests that the incident source parameters do not significantly affect the subsequent angular distribution of the dose depositing electrons in water. One can therefore be confident that introducing any amount of air into a detector design will result in a consistent effect across multiple linear accelerators.

5. Conclusions

This study showed the influence on initial source parameters on the dose perturbation caused by air placed immediately upstream to a silicon active volume at the central axis of the beam. If a detector is modified by the introduction of air, one can be confident that the response of the detector will be the same across all similar linear accelerators and the Monte Carlo modeling of each machine is not required.

6. Acknowledgements

This work was funded by the Australian Research Council in partnership with the Queensland University of Technology, the Wesley Research Institute and Premion (Linkage Grant No. LP110100401). Computational resources and services used in this work were provided by the High Performance Computing and Research Support Unit, Queensland University of Technology (QUT), Brisbane, Australia.

7. References

[1] Charles P H, Crowe S B, Kairn T, Kenny J, Lehmann J, Lye J, Dunn L, Hill B, Knight R T, Langton C M and Trapp J V. 2012 The effect of very small air gaps on small field dosimetry. *Phys Med Biol* **57** (21), 6947-6960.

[2] Charles P H, Crowe S, Kairn T, Knight R T, Hill B, Kenny J, Langton C M and Trapp J V. 2013 Monte Carlo-based diode design for correction-less small field dosimetry. *Phys Med Biol* **58** (Accepted for publication).

[3] Das I J, Ding G X and Ahnesjö A. 2008 Small fields: Nonequilibrium radiation dosimetry. *Medical Physics* **35** (1), 206.

[4] Alfonso R, Andreo P, Capote R, Huq M S, Kilby W, Kjäll P, Mackie T R, Palmans H, Rosser K, Seuntjens J, Ullrich W and Vatnitsky S. 2008 A new formalism for reference dosimetry of small and nonstandard fields. *Medical Physics* **35** (11), 5179.

[5] Taylor M L, Kron T and Franich R D. 2011 A contemporary review of stereotactic radiotherapy: inherent dosimetric complexities and the potential for detriment. *Acta Oncol* **50** (4), 483-508.

[6] Cranmer-Sargison G, Weston S, Evans J A, Sidhu N P and Thwaites D I. 2011 Implementing a newly proposed Monte Carlo based small field dosimetry formalism for a comprehensive set of diode detectors. *Medical Physics* **38** (12), 6592-6602.

[7] Francescon P, Cora S and Satariano N. 2011 Calculation of k(Q(clin),Q(msr)) (f(clin),f(msr)) for several small detectors and for two linear accelerators using Monte Carlo simulations. *Medical Physics* **38** (12), 6513-6527.

[8] Francescon P, Kilby W, Satariano N and Cora S. 2012 Monte Carlo simulated correction factors for machine specific reference field dose calibration and output factor measurement using fixed and iris collimators on the CyberKnife system. *Phys Med Biol* **57** (12), 3741-3758.

[9] Czarnecki D and Zink K. 2013 Monte Carlo calculated correction factors for diodes and ion chambers in small photon fields. *Phys Med Biol* **58** (8), 2431-2444.

[10] Cranmer-Sargison G, Weston S, Evans J A, Sidhu N P and Thwaites D I. 2012 Monte Carlo modelling of diode detectors for small field MV photon dosimetry: detector model simplification and the sensitivity of correction factors to source parameterization. *Phys Med Biol* **57** (16), 5141-5153.

[11] Scott A J, Kumar S, Nahum A E and Fenwick J D. 2012 Characterizing the influence of detector density on dosimeter response in non-equilibrium small photon fields. *Phys Med Biol* **57** (14), 4461-4476.

[12] Fenwick J D, Kumar S, Scott A J and Nahum A E. 2013 Using cavity theory to describe the dependence on detector density of dosimeter response in non-equilibrium small fields. *Phys Med Biol* **58** (9), 2901-2923.

[13] Rogers D W O, Faddegon B A, Ding G X, Ma C M and We J. 1995 BEAM: A Monte Carlo code to simulate radiotherapy treatment units. *Medical Physics* **22** (5), 503 - 524.

[14] Kairn T, Aland T, Franich R D, Johnston P N, Kakakhel M B, Kenny J, Knight R T, Langton C M, Schlect D, Taylor M L and Trapp J V. 2010 Adapting a generic BEAMnrc model of the BrainLAB m3 micro-multileaf collimator to simulate a local collimation device. *Phys Med Biol* **55** (17), N451-463.

[15] Kairn T, Kenny J, Crowe S B, Fielding A L, Franich R D, Johnston P N, Knight R T, Langton C M, Schlect D and Trapp J V. 2010 Technical Note: Modeling a complex micro-multileaf collimator using the standard BEAMnrc distribution. *Medical Physics* **37** (4), 1761.

[16] Kawrakow I, Mainegra-Hing E, Tessier F and Walters B R B, 2009. The EGSnrc C++ class library (Ottowa: National Research Council) *NRC Report* PIRS-898 (rev A).

Journal of Physics: Conference Series **489** (2014) 012007 doi:10.1088/1742-6596/489/1/012007

Recommendations for dose calculations of lung cancer treatment plans treated with stereotactic ablative body radiotherapy (SABR)

S Devpura, M S Siddiqui, D Chen, D Liu, H Li, S Kumar, J Gordon, M Ajlouni, B Movsas and I J Chetty[1]

Henry Ford Health System, Detroit MI 48202 USA

Email: ichetty1@hfhs.org

Abstract. The purpose of this study was to systematically evaluate dose distributions computed with 5 different dose algorithms for patients with lung cancers treated using stereotactic ablative body radiotherapy (SABR). Treatment plans for 133 lung cancer patients, initially computed with a 1D-pencil beam (equivalent-path-length, EPL-1D) algorithm, were recalculated with 4 other algorithms commissioned for treatment planning, including 3-D pencil-beam (EPL-3D), anisotropic analytical algorithm (AAA), collapsed cone convolution superposition (CCC), and Monte Carlo (MC). The plan prescription dose was 48 Gy in 4 fractions normalized to the 95% isodose line. Tumors were classified according to location: peripheral tumors surrounded by lung (lung-island, N=39), peripheral tumors attached to the rib-cage or chest wall (lung-wall, N=44), and centrally-located tumors (lung-central, N=50). Relative to the EPL-1D algorithm, PTV D95 and mean dose values computed with the other 4 algorithms were lowest for "lung-island" tumors with smallest field sizes (3-5 cm). On the other hand, the smallest differences were noted for lung-central tumors treated with largest field widths (7-10 cm). Amongst all locations, dose distribution differences were most strongly correlated with tumor size for lung-island tumors. For most cases, convolution/superposition and MC algorithms were in good agreement. Mean lung dose (MLD) values computed with the EPL-1D algorithm were highly correlated with that of the other algorithms (correlation coefficient =0.99). The MLD values were found to be ~10% lower for small lung-island tumors with the model-based (conv/superposition and MC) vs. the correction-based (pencil-beam) algorithms with the model-based algorithms predicting greater low dose spread within the lungs. This study suggests that pencil beam algorithms should be avoided for lung SABR planning. For the most challenging cases, small tumors surrounded entirely by lung tissue (lung-island type), a Monte-Carlo-based algorithm may be warranted.

1. Introduction

Stereotactic body radiation therapy (SBRT) or stereotactic ablative body radiotherapy (SABR) as coined by Loo and Timmerman *et al* [1] is a common therapeutic option available for patients with medically inoperable lung tumors. McGarry *et al* [2] and Timmerman *et al* [3,4] reported an excellent tumor control for early stage lung tumors when treated with SABR. Among the existing treatment planning systems, the Monte Carlo method is one of the most accurate algorithms for lung cancer dose calculations in the SABR setting [5]. In low density tissues, such as lung, the contribution of dose from electrons becomes significant because electrons which deposit dose within the tumor scatter laterally into the surrounding lung tissues carrying dose away from the tumor [6,7]. If the dose algorithm does not properly predict electron scattering, significant inaccuracies can result in the dose distributions, as in the case when pencil beam algorithms (employing equivalent path length, EPL corrections) are used.

In this retrospective study, five clinically commissioned dose calculation algorithms were used to compute lung SABR treatment plans for 133 non-small-cell lung cancer patients. The patients were planned and treated using a pencil beam algorithm (with a 1-D equivalent path length heterogeneity correction). We have investigated the dose accuracies in each algorithm by comparing the dose-volume histograms and dose parameters, and analyzed the impact of the field size and location comprehensively.

2. Method and Materials

The lung tumors were categorized by location into three groups as follows: island-type peripheral tumors surrounded by lung tissues (lung-island; N=39), tumors attached to the chest-wall (lung-wall; N=44), and tumors located in the central area (lung-central; N=50). Average equivalent square field sizes (FS, side of equivalent square) were used as a surrogate of tumor length in the beam's eye view (BEV) and were subdivided into three groups, 3≤FS<5, 5≤FS<7, and 7≤FS<10 cm. These lung patients were originally treated with 6 MV photons using IMRT. The dose regime was 48 Gy in 4 fractions normalized to 95% isodose line. Five clinically commissioned treatment planning systems (TPS) were used. They are BrainLAB iPlan 4.1.2 pencil beam (equivalent path length (EPL-1D)), Varian Eclipse 8.6 pencil beam algorithm (EPL-3D) and anisotropic analytical algorithm [8,9] (AAA), Phillips Pinnacle 8.1 collapsed cone convolution superposition [10,11] (CCC), and BrainLAB iPlan 4.1.2 Monte Carlo (MC) method [12,13,14]. Each of the above algorithms was commissioned using measurements in water phantoms as well as slab-phantoms with low-density, lung-equivalent media [15]. Beam models were within 2%/2 mm agreement vs. measurements in water phantoms and within 3% agreement in slab phantoms with lung-equivalent media [15] for all except the pencil-beam algorithms, were large discrepancies were observed.

For patient treatments IMRT plans were created using the iPlan EPL-1D algorithms. Using these same treatment plans (MUs and other parameters), dose was recomputed using the following algorithms: EPL-3D, AAA, CCC, and MC. To confirm that the beam models were in agreement with each other, uniform density (UD) plans (with densities assigned to 1.0) were first generated. Subsequently, heterogeneity corrected (HC) plans of all the treatment plans were calculated. The dose-volume histograms (DVHs) of the UD plans were among the five algorithms were found to be in agreement within a few percent for all plans. The dose to 95% of the planning target volume (PTV D95) was determined for all the dose calculation algorithms. The PTV D95 values were normalized to EPL-1D method (e.g. MC/EPL-1D). PTV mean doses were also reported. Mean lung dose (MLD) values were computed; the lung volume excluded the ITV from the sum of both lung volumes. The results presented here show the effects of tumor location and the field size in lung SABR-based treatment planning.

3. Results and Discussions

Tables 1 summarizes the PTV D95 (dose covering 95% of the PTV) and mean lung dose results with standard deviations for three field size categories. Note that N represents the number of patients in a particular category.

Table 1. PTV D95 and MLD for EPL-3D, AAA, CCC, and MC algorithms in three field size groups relative to the EPL-1D (100%) method.

FS (cm)	N	PTV D95 (%)				MLD (%)			
		EPL-3D	AAA	CCC	MC	EPL-3D	AAA	CCC	MC
3≤FS<5	50	95.3±1.9	82.4±5.1	82.6±6.2	82.3±6.0	94.4±7.4	90.7±5.6	91.8±6.1	90.9±5.7
5≤FS<7	62	95.8±2.1	85.3±5.2	85.7±6.1	85.6±5.8	100.5±2.5	95.3±2.7	96.1±2.4	96.1±1.8
7≤FS<10	21	95.9±1.7	90.4±3.7	90.6±3.8	90.8±3.8	102.0±2.8	96.1±2.1	95.0±2.4	97.3±3.2

The 3D pencil beam method (EPL-3D) showed approximately the same PTV D95 and PTV mean dose values regardless of the field size (within 4-5%). This illustrates that the EPL-3D algorithm is not accurate enough to estimate electron transport (lateral electron scattering and conditions of charged-particle disequilibrium) necessary for SABR-based dose calculation of lung tumors [16]. When the tumor and associated planning field size are small (< 5 cm) the impact of the dose reduction to the tumor is exacerbated due to the added effect of lateral electronic disequilibrium (resulting in more energy leaving versus being deposited in the tumor), which occurs when the field size is close to the range of the secondary electrons[5,7]. This requires more accurate accounting of electron transport in dose calculations for lung SABR [7]. Pencil beam algorithms lack the capability of account for loss of charge particle equilibrium and heterogeneity of the low density media. AAA and CCC dose values are in good agreement with MC and showed significantly lower doses (PTV D95~20%, PTV mean~12%, and MLD ~10%) for small field sizes. As the tumor (field) sizes increase, the PTV D95, mean dose, and MLD dose differences are reduced to ~10%, ~5%, and 3-4%, respectively.

Table 2. PTV D95 and MLD for EPL-3D, AAA, CCC, and MC algorithms for lung-island, lung-wall and lung-central tumors relative to the EPL-1D (100%) method.

Location	*N*	PTV D95 (%)				MLD (%)			
		EPL-3D	AAA	CCC	MC	EPL-3D	AAA	CCC	MC
Lung-island	39	95.2±2.0	81.6±4.4	81.4±5.8	81.4±5.8	97.2±6.1	92.1±5.1	92.9±5.6	92.9±5.3
Lung-wall	44	96.5±1.8	86.8±4.9	87.4±5.6	86.9±5.7	98.6±6.7	94.3±4.8	94.4±4.3	94.9±4.9
Lung-central	50	95.2±1.8	86.2±5.9	86.5±6.3	86.7±6.0	99.3±4.8	94.5±3.7	95.4±3.8	95.0±3.8

Table 2 summarizes the PTV D95 and normal MLD dose results for lung-island, lung-wall, and lung-central tumors. The EPL-3D method was not able to capture the dose effects due to heterogeneity of the lung tissue. Thus, the dose values for all the tumor locations were somewhat similar. There were dramatic dosimetric differences in AAA, CCC, and MC algorithms due to location of the lung tumors. AAA and CCC media algorithms can handle the tissue inhomogeneity and secondary electron transport within the lung better than pencil beam methods [9]. Thus, these convolution superposition type algorithms predict the dose in lung media with much better accuracy relative to pencil beam algorithms. The largest dose differences were observed for island type tumors computed with AAA, CCC, and MC algorithms relative to the EPL-based methods. The tumors located in the central area and tumors attached to the rib cage showed similar dose results for AAA, CCC, and MC algorithms.

Table 3. PTV D95 dose for each algorithm for different tumor locations and field sizes relative to the EPL-1D (100%) method.

Location	FS (cm)	*N*	EPL-3D (%)	AAA (%)	CCC (%)	MC (%)
	3≤FS<5	21	95.1±2.1	80.2±4.3	80.0±6.0	79.7±6.0
Lung-island	5≤FS<7	16	95.7±1.9	83.0±4.3	82.7±5.4	83.0±5.1
	7≤FS<10	2	92.8±0.3	84.5±0.8	85.3±0.7	85.7±1.4
	3≤FS<5	13	96.2±1.2	84.7±4.4	85.4±5.4	84.5±5.3
Lung-wall	5≤FS<7	22	96.4±2.3	86.2±4.8	86.7±5.6	86.3±5.6
	7≤FS<10	9	97.1±1.1	91.4±2.8	92.1±3.2	91.9±3.2
	3≤FS<5	16	94.8±1.8	83.6±5.5	83.7±6.1	84.0±5.7
Lung-central	5≤FS<7	24	95.3±2.1	86.1±5.8	86.8±6.5	86.8±6.1
	7≤FS<10	10	95.4±1.1	90.7±3.9	90.4±3.8	90.9±3.9

The combined impact of the location and the field size was reflected in table 3 for PTV D95 and Figure 1, in this case showing the comparison for the PTV mean dose. AAA, CCC, and MC algorithms predicted the lowest dose differences (PTV D95~20%, PTV mean dose~15%, and MLD~10%) for the lung-island tumors with small field sizes. The lung-wall and lung-central tumors with large field (tumor) sizes showed relatively lower dose differences (PTV D95~9%, PTV mean dose~5%, and MLD~4%). However, large lung-island tumors showed 15% PTV D95 dose differences and 8-10% PTV mean dose differences for AAA, CCC, and MC relative to EPL-1D. Note that the number of patients in 7≤FS<10 category of the island type tumors is much smaller (*N*=2).

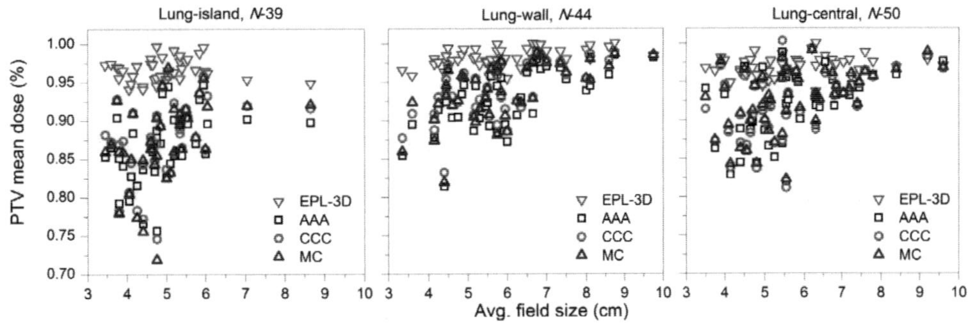

Figure 1. PTV mean dose vs. average field size for 133 lung cancer patients.

Jones *et al* also reported algorithms using radiological density scaling do not predict the dose near or within inhomogeneities accurately [6]. Large inaccuracies sometimes noted with radiological-path-length-based algorithms have the potential to impact clinical outcomes, as reported in documents, such as the AAPM Task Group Report No. 85 on tissue heterogeneity corrections [17]. This current study provides evidence of the magnitude of the PTV D95, mean dose and normal lung mean dose differences computed with five algorithms for lung tumors of different sizes and locations. Results are highly suggestive that, in addition to the EPL-1D algorithm being inappropriate for SABR-based lung cancer planning [18], the EPL-3D algorithm is also inaccurate and should not be used for SABR.

4. Conclusion

Amongst all locations, dosimetric differences were most strongly correlated with tumor size for lung-island tumors. Significant changes in tumor dose values for 133 lung cancer patients were observed. As the field size becomes small (< 5 cm), the predicted dose to the PTV with pencil beam algorithms showed unrealistic overestimations. In fact, the actual dose delivered is much less than expected based on convolution-type and MC-based calculations. For "island" type small tumors this under-dosage was in the range of 20-25% of the prescribed dose. For most cases, convolution/superposition and MC algorithms were in good agreement. For the most challenging cases, small tumors surrounded entirely by lung tissue, a Monte-Carlo-based algorithm may be warranted [18]. The inability to account for electron transport and lateral electronic dis-equilibrium in the low-density lung medium, result in inaccurate dose distributions using either EPL-1D or EPL-3D dose algorithms. Such algorithms should be avoided for SABR-based dose calculations of patients with lung cancers.

References

[1] Loo B W, Chang J Y, Dawson L A, Kavanagh B D *et al* 2011 *Practical Rad Onc.* **1** 38–9
[2] McGarry R C, Papiez L, Williams M *et al* 2005 *Int. J Radiat. Oncol. Biol. Phys.* **63**(4) 1010-15
[3] Timmerman R, McGarry R, Yiannoutsos C *et al* 2006 *J of Clin Oncology* **24** 4833-39
[4] Timmerman R, Paulus R, Galvin J *et al* 2010 *JAMA* **303**(11) 1070–76
[5] Chetty I J, Curran B, Cygler J E *et al* 2007 AAPM TG 105 *Med. Phys.* **34**(12) 4818-53
[6] Jones A O, Das I J and Jones F L Jr 2003 *Med. Phys.* **30**(3) 296-300
[7] Das I J, Ding G X and Ahnesjö A 2008 *Med. Phys.* **35**(1) 206-215
[8] Ulmer W and Harder D 1995 *Med. Phys.* **5** 25–30
[9] Sievinen J, Ulmers W and Kaissl W 2005 Varian Medical Systems, Palo Alto, CA
[10] Mackie T R, Bielajew A F, Rogers D W *et al* 1988 *Phys. Med. Biol.* **33**(1) 1-20
[11] Papanikolaou N, Mackie T R, Meger-Wells C *et al* 1993 *Med. Phys.* **20**(5) 1327-36
[12] Kawrakow I, Dippel M and Friedrich K 1996 *Med. Phys.* **23**(2) 445-457
[13] Fippel M 1999 *Med. Phys.* **26** 1466-75
[14] Fippel M, Laub W, Huber B *et al* 1999 *Phys. Med. Biol.* **44** 3039-54
[15] Fragoso M, Wen N, Kumar S *et al* 2010 *Phys. Med. Biol.* **55**(16) 4445-64
[16] Ahnesjö A and Aspradakis M M 1999 *Phys. Med. Biol.* **44** R99–R155
[17] Papanikolaou N, Battista J J, Boyer A L *et al* 2004 Medical Physics, Madison, WI 1-135
[18] Benedict S H, Yenice K M, Followill D *et al* 2010 AAPM TG 101 *Med. Phys.* **37** 4078-4101

In vivo and phantom measurements versus Eclipse TPS prediction of near surface dose for SBRT treatments

Gwi A Cho[1,2*], Anna Ralston[1], Mo Mo Tin[1], Darren Martin[1], Sheila Pickard[1], J-H Kim[2] and Regina Tse[1]

[1] Department of Radiation Oncology, Royal Prince Alfred Hospital, Camperdown, NSW, Australia
[2]Institute of Medical Physics, School of Physics, University of Sydney, NSW, Australia

*gwi.cho@sswahs.nsw.gov.au

Abstract. This study reports on Gafchromic EBT2 film skin dose measurements for lung stereotactic body radiotherapy. These measurements were compared to near surface skin doses predicted by Eclipse treatment planning system (TPS) using the Analytical Anisotropic Algorithm (AAA) for a 6 MV photon beam. The accuracy of the predicted near surface dose for 3×3, 5×5 and 10×10 cm^2 fields was assessed using an Attix chamber and EBT2 film in a Virtual Water phantom and compared to Monte Carlo calculation. The maximum near surface dose and its location were identified from the patient's treatment plan. For phantom measurements, the TPS dose (nominal 0 mm depth) was higher than the Attix chamber data by up to 24.0 % but in closer agreement with the EBT2 film measurement, which was up to 10.3 % higher than the Attix data. The MC calculated dose values were higher than the Attix data by up to 3.5% for the depths up to 2 mm. The maximum patient skin dose estimated from *in vivo* EBT2 film measurements was 7.5 – 19.5 Gy per course and depended on the number of overlapping fields, beam weight and/or contact with immobilisation devices. The TPS predicted dose for patient plans was mostly higher than the *in vivo* dose, by as much as 69.3%, but in two cases was lower, by as much as -23.1%.

1. Introduction

For patients undergoing stereotactic body radiotherapy (SBRT), also known as stereotactic ablative body radiotherapy (SABR), the skin is one of the critical organs at risk that requires dose monitoring due to the use of potentially overlapping multiple treatment fields in conjunction with highly hypo-fractionated dose prescription scheme [1, 2].

The estimation of the skin dose received from a megavoltage (MV) photon treatment requires specific dosimetric considerations due to the lack of charged particle equilibrium present in the build-up region. In this study, we refer to the clinically relevant nominal skin depth as 70 μm based on ICRP definitions [3]. The dose to patient's skin may be altered by build-up effects, dose from beams exiting the patient and/or scatter from the immobilization devices or beam modifiers required for patient set-up and treatment [1, 4, 5].

A patient's treatment plan from a TPS provides a calculated dose given to the target. However, the accuracy of the dose prediction by a TPS in the build-up region is reduced leading to broader tolerance limits [6, 7]. Several factors influence the TPS near surface dose prediction: dose calculation algorithm and its electron contamination model, initial commissioning of TPS build-up region dose calculation, clinical dose calculation parameters and treatment plan complexity [8].

Eclipse™ analytical anisotropic algorithm (AAA, V10.0.29, Varian Medical Systems Inc, Palo Alto, CA) uses triple source models comprised of primary photons, extra-focal photons and contaminating electrons to approximate the phase space parameters (particle fluence, energy spectrum) of the clinical beam. During the configuration step, the physical parameters that model each source are optimized using measured beam data. For volumetric dose calculation, the clinical beam is represented by 2-D fluence distributions comprised of photon and contaminating electrons that are divided into small beamlets. A 3-D matrix of calculation voxels, each with an associated mean electron density, and the voxel resolution determined by the dimension of calculation grid is formed from the patient body volume. The energy distribution is obtained from separate convolutions for two photon sources and electron contamination source. The dose at a calculation point is obtained by summing up the individual beamlet contributions by global superposition and finally converting absorbed energy to dose by dividing by the local (relative) electron density.

The fluence of contaminating electrons is obtained by convolving the primary photon fluence with a Gaussian (target plane) and the dose deposition by these contaminating electrons is calculated by the convolution with a second Gaussian, multiplied by the depth dependent energy deposition function. An empirical curve determined from the difference between the measured depth dose and the dose calculated by AAA without contribution from contaminating electrons for the largest field size (open beam), models the total energy deposition. More accurate modeling of electron contamination could contribute to improved accuracy of field size dependent build-up region dose prediction [9].

For typical SBRT treatment field arrangements, the limited accuracy in the TPS entrance dose prediction could add further uncertainty to the estimated skin dose where entry beam(s) may overlap with exit beam(s) and additional build-up effect and/or scatter arises from the presence of immobilization devices [10].

In this study, near surface doses were calculated by the AAA for a 6 MV photon beam and compared with measurements using an Attix plane parallel chamber (RMI 449, Middleton, WI) and Gafchromic EBT2 film (International Specialty Products, Wayne, NJ). These results were also compared to Monte Carlo (MC) calculations using BEAMnrc/EGSnrc. We also report on EBT2 film *in vivo* doses for nine lung SBRT treatments and compare with Eclipse™ dose predictions.

2. Materials and methods

2.1. *Eclipse™ phantom dose calculations*

For a 6 MV photon beam, the dose calculation was performed on an Eclipse™ water phantom at a 100 cm source to surface distance (SSD) for 3×3, 5×5 and 10×10 cm^2 field sizes at two gantry angles (0° and 30°). For dose calculations, a constant 100 monitor unit (MU) was prescribed and routine clinical AAA calculation parameters were used (calculation grid size 0.25×0.25×0.25 $cm^3$and heterogeneity correction on). Point doses were extracted at depths of 0, 2, 4 and 6 mm along the central axis.

2.2. *Attix chamber and EBT2 film phantom dose measurements*

Surface dose measurements were performed using an Attix parallel plate chamber and EBT2 films in a Virtual Water™ (Med-Cal, Verona, WI) phantom using a Novalis Tx linear accelerator (Varian Medical Systems Inc, Palo Alto, CA).

The reference point of an Attix chamber is located at a water equivalent depth of 48 μm being the inner surface of the chamber entrance window, which is less than the ICRP skin depth of 70 μm. For dose conversion, reference output measurements for the Attix chamber were performed in Virtual Water™ at 100 cm SSD for 10×10 cm^2 field size and at the depth of dose maximum (1.4 cm) and with 100 MU. The electrometer (PTW Unidos, Freiburg) readings were taken for two bias voltages with the chamber positioned at depths of 0, 2, 4 and 6 mm at the central axis while maintaining a fixed 100 cm SSD to the phantom surface. A total of six readings from both polarity settings (-300V and +300V) were averaged and converted to dose prior to normalisation at the depth of 1.4 cm. The chamber readings were not corrected by water-air stopping power ratios.

The cross section of EBT2 film is asymmetric, with the 30 μm thick active layer starting at a physical depth of 80 μm. For dose conversion, EBT2 films were calibrated at 1 Gy intervals up to maximum dose of 6 Gy with the 80 μm side facing the X-ray source. The same reference set-up in section 2.2.1 for the Attix chamber was used. EBT2 film depth dose measurement was performed at two gantry angles and at four depths as described in section 2.2.1, with each film piece placed on the central axis. The films were irradiated individually at 300 MUs.

2.3. Eclipse skin dose estimation and in vivo skin dose monitoring using EBT2 films

The clinical SBRT protocol utilizes a maximum dose limit of 30 Gy to the chest wall and skin based on the TROG 0902 clinical trial protocol recommendations [11]. A 5 mm thick inwardly expanding artificial skin structure was generated as part of contouring to estimate the maximum near surface skin dose via the dose volume histogram (DVH) statistics [2]. The reported DVH maximum dose was used as a conservative upper estimate of the near surface skin dose. Where the maximum DVH dose was found, the overlying near surface dose was evaluated for comparison with the *in vivo* film measurement. In addition, the overall dose distribution was evaluated using the Eclipse™ dose colour wash tool. Multiple overlap regions from entry and the exit beams and resulting high dose regions were also identified for film positioning and evaluation. The skin dose from Eclipse patient plan was recorded at a point closest to the skin and less than 2.5 mm distance inward from the surface, assuming a coverage by one calculation voxel. For patient skin dose estimation, calibration was performed for each individual sheet of *in vivo* EBT2 films up to 8 Gy prior to patient measurement. On the first day of treatment, *in vivo* dose measurements were performed using EBT2 films if the skin DVH maximum dose was greater than 20 Gy or for audit. *In vivo* EBT2 films were marked consistently to identify the film orientation for positioning and scanning.

2.4. EBT2 film scanning and dose evaluation

All EBT2 films were scanned approximately after 18-24 hours using an Epson 10000XL flatbed scanner. The films were scanned in a landscape orientation at the scanning resolution of 72dpi and in 48bits RGB format. The scanned film images were imported into RIT113 software (Radiological Imaging Technology Inc, Colorado Springs, CO) using the red channel. Calibration curves were generated and the pixel values of the films were converted to absorbed dose. The converted dose was sampled using a region of interest (ROI) dimension set to 7×7 mm^2 at the centre of each film pieces.

2.5. Monte Carlo simulation

The MC calculation was performed for 3×3, 5×5 and 10×10 cm^2 field sizes using the BEAMnrc/DOSXYZnrc Monte Carlo (MC) code. The first step involved generating the phase space file for the linear accelerator using BEAMnrc. In this simulation, a total of 8×10^7 incident electrons of 6.5 MeV energy specified as an elliptical beam with varying FWHM depending on the field size. The EGSnrc transport parameters ECUT/PCUT were set to 0.521 and 0.010 MeV and AE/AP were set to 0.512 and 0.001 MeV. Directional bremsstrahlung splitting was turned on with a splitting factor of 1000 and the XCOM photon cross section and NIST bremsstrahlung cross-section data were used. Boundary crossing algorithm was set to EXACT and for electron step algorithm, PRESTA-II was used. For surface dose calculation, a total of 8×10^9 particle histories were used incident upon a large water phantom. The dose in water was scored in voxels with 10 μm in thickness and 0.5 cm in radius for the first 0.1 mm in a water phantom that includes the relevant skin depth (70 μm) [12, 13].

3. Results and discussion

3.1. Comparison between phantom measurements and the Eclipse AAA predicted dose

Table 1 and 2 show the comparison between the Attix chamber, EBT2 film measurements and AAA predicted doses at two gantry angles 0° and 30°, respectively. The results show that AAA consistently overestimates the surface dose (nominal 0 mm depth) compared to the measured data (Attix and EBT2). In contrast, at 4 and 6 mm depths where maximum skin DVH dose is estimated, Eclipse AAA underestimates the dose. At the phantom surface and gantry angle 0°, the EBT2 film dose was consistently higher than that of the Attix chamber.

Table 1. Phantom measurements for gantry angle 0°. Attix chamber and EBT2 phantom measurement and Eclipse predicted near surface (skin) dose, all data normalized at nominal d_{max}.

Field Size (cm^2)	3x3			5x5			10×10	
Nominal Depth (mm)	EBT2 (%)	AAA (%)	Attix (%)	EBT2 (%)	AAA (%)	Attix (%)	EBT2 (%)	AAA (%)
0	17.8	28.8	10.1	20.4	31.2	16.0	23.9	40.0
2	56.7	47.5	55.9	60.6	49.1	59.9	60.3	55.1
4	76.6	68.5	77.9	82.0	69.4	80.6	80.7	72.9
6	86.2	83.2	87.9	85.6	83.6	89.8	90.4	85.7

Table 2. Phantom measurements for gantry angle 30°.Attix chamber and EBT2 phantom measurement and Eclipse predicted near surface (skin) dose, all data normalized at nominal d_{max}.

Field Size (cm^2)	3x3			5x5		10×10	
Nominal Depth (mm)	EBT2 (%)	AAA (%)	Attix (%)	EBT2 (%)	AAA (%)	EBT2 (%)	AAA (%)
0	20.6	22.7	12.4	20.9	25.6	18.3	35.0
2	62.5	43.6	61.2	60.5	45.9	64.3	51.8
4	82.0	68.1	82.7	85.6	68.6	84.0	71.8
6	88.2	84.8	91.9	91.6	84.6	92.3	86.5

Table 3 shows the reference depths of the Attix chamber and EBT2 film in water equivalent depths at nominal 0 mm depth. Table 4 shows the MC calculated doses. The MC calculated dose values were higher than the Attix data by up to 3.5%.

Table 3. Reference depths of Attix chamber and EBT2 film in water equivalent depths. The reference depth of EBT2 film is at the centre of the active layer and the material density for each layer is taken from Sutherland *et al.* [14].

Detector	Water Eq. Depths (mm)
Attix	0.0480
EBT2	
95 μm side	0.122
190 μm side	0.254

Table 4. Monte Carlo calculated build-up region dose for gantry angle 0°. (* Interpolated)

Field Size (cm^2)	3x3	5x5	10x10
Depth in water (mm)			
*0.048	10.7	13.2	18.7
0.070	12.3	14.6	20.2
*0.122	14.9	17.2	22.5
*0.254	19.6	21.8	26.8
1.000	42.0	43.4	47.3
2.000	58.3	59.4	62.6

3.2. Comparison between in vivo dose measurement and Eclipse AAA predicted near surface dose

Table 5 shows the comparison between EBT2 film measurements and Eclipse AAA predicted near surface doses. The maximum patient skin dose estimated from EBT2 film was between 7.5-19.5 Gy per course, which differed by -23.1% to +69.3% from the Eclipse prediction. Underestimation of the surface dose by Eclipse (for patient 8 and 9) occurred for posterior-oblique beams passing through the immobilization device.

Not all the *in vivo* measurements were carried out with the thinner polyester layer (75 μm) facing the beam. The dose difference due to asymmetric EBT2 film construction can be estimated from the MC calculation data, where up to +4.8% is found for two measurement depths representing different film orientations for the entrance beam (Table 4). The effect could be more significant for oblique beam entry. For the *in vivo* films placed between the skin and the immobilization vacuum bag, the dosimetric effect of film orientation would be insignificant.

The near surface dose estimated from Eclipse plan is subject to inherent uncertainty associated with the definition of mean electron density at the calculation voxel at the skin/air interface and spatial resolution and position of the dose matrix from which dose is interpolated at a particular point. The dosimetric resolution of AAA calculation (~2.5 mm) needs to be considered when comparing with the near surface dose values reported from the Attix, EBT2 and MC measurements/calculation (48 μm – 1 mm). For build-up region dosimetry, the estimated uncertainties of 2% and 2.7% have been reported for the Attix chamber and EBT2 film, respectively [8,12].

Table 5. Comparison of Eclipse AAA algorithm predicted near surface dose and *in vivo* skin dose estimated from the EBT2 film measurement for nine lung SBRT treatments (2011 – 2012)

Patient	Eclipse AAA (Gy)	*In vivo* EBT2 film	Difference (%)
1	15.0	10.4	+44.2
2	15.0	13.6	+10.3
3	17.0	17.0	0.0
4	15.7	15.0	+4.7
5	12.7	7.5	+69.3
6	10.0	9.4	+6.4
7	18.3	16.8	+8.9
8	15.0	19.5	-23.1
9	18.0	19.4	-7.2

4. Conclusion

Discrepancies have been noted between the measurements (Attix, EBT2 film) and the Eclipse AAA predicted doses. Further study is required for film dosimetry to be fully applicable for *in vivo* skin dose measurement for SBRT treatment, which will include EBT3. To better model the electron contamination, further investigation is necessary to evaluate the effect of more accurate build-up region depth dose data, together with optimization of electron contamination source models within AAA. The Eclipse dose prediction in the presence of immobilization devices and the modeling accuracy of build up and build-down effects needs further investigation and future work will also include the Acuros XB algorithm.

References

[1] Hoppe B *et al.* 2008 *Int. J. Radiat. Oncol. Biol. Phys.* **72** 1283
[2] Radiation Therapy Oncology Group RTOG0618 Phase II trial of stereotactic body radiation therapy (SBRT) in the treatment of patients with operable stage I/II non-small cell lung cancer.
[3] ICRP 1992 *The biological basis for dose limitation in the Skin. ICRP Publiction 59. Ann. ICRP* 22(2)
[4] Mellenberg D 1995 *Int. J. Radiat. Oncol. Biol. Phys.* **32** 1193
[5] Chiu-Tsao S and Chan M 2010 *Med. Phys.* **37** 3611-20
[6] Fraass B, Doppke K, Hunt M, Kutcher G, Stern R and Van Dyke J 1998 *Med. Phys.* **25** 1773
[7] IAEA Technical report series 430 2004 *Commissioning and quality assurance of computerized planning systems for radiation treatment of cancer.* (Vienna:IAEA)
[8] Hsu S, Moran J, Chen Y, Kulasekere R and Roberson P 2010 *Med. Phys.* **37** 2043
[9] Van Esch A, Tillikainen L, Pyykkonen J, Tenhunen M, Helminen H, Siljamaki S, Alakuijala J, Paiusco M, Loir M, Huyskens D 2006 *Med. Phys.* **33** 4130
[10] Devic S, Seuntjens J, Abdel-Rahman W, Evans M, Olivares M and Podorsak E 2006 *Med. Phys.* **33** 1116
[11] Trans tasman Radiation Oncology Group TROG0902 A randomised phase III trial of highly conformal hypofractionated image guided ("stereotactic") radiotherapy(HypoRT) versus

conventionally fractionated radiotherapy (ConRT) for inoperable early stage I non-small cell lung cancer (CHISEL)
[12] Kim J, Hill R and Kuncic Z 2012 *Australas. Phys. Eng. Sci. Med.* **35** 271
[13] Kim J, Hill R and Kuncic Z 2012 *Phys. Med. Biol.* **57**, N267
[14] Sutherland J and Rogers D 2010 *Med. Phys.* **37** 1110

Journal of Physics: Conference Series **489** (2014) 012009 doi:10.1088/1742-6596/489/1/012009

Automated segmentation and dose-volume analysis with DICOMautomaton

H Clark[1,2], S Thomas[3], V Moiseenko[2,4], R Lee[2], B Gill[2], C Duzenli[1,2] and J Wu[2]

[1] University of British Columbia, Vancouver, B.C., Canada.
[2] British Columbia Cancer Agency - Vancouver Centre, Vancouver, B.C., Canada.
[3] British Columbia Cancer Agency - Fraser Valley Centre, Surrey, B.C., Canada.
[4] University of California San Diego, San Diego, Ca., USA.

E-mail: `hdclark@phas.ubc.ca`

Abstract. Purpose: Exploration of historical data for regional organ dose sensitivity is limited by the effort needed to (sub-)segment large numbers of contours. A system has been developed which can rapidly perform autonomous contour sub-segmentation and generic dose-volume computations, substantially reducing the effort required for exploratory analyses.
Methods: A contour-centric approach is taken which enables lossless, reversible segmentation and dramatically reduces computation time compared with voxel-centric approaches. Segmentation can be specified on a per-contour, per-organ, or per-patient basis, and can be performed along either an embedded plane or in terms of the contour's bounds (e.g., split organ into fractional-volume/dose pieces along any 3D unit vector). More complex segmentation techniques are available. Anonymized data from 60 head-and-neck cancer patients were used to compare dose-volume computations with Varian's Eclipse™ (Varian Medical Systems, Inc.).
Results: Mean doses and Dose-volume-histograms computed agree strongly with Varian's Eclipse™. Contours which have been segmented can be injected back into patient data permanently and in a Digital Imaging and Communication in Medicine (DICOM)-conforming manner. Lossless segmentation persists across such injection, and remains fully reversible.
Conclusions: `DICOMautomaton` allows researchers to rapidly, accurately, and autonomously segment large amounts of data into intricate structures suitable for analyses of regional organ dose sensitivity.

1. Purpose

Dose-volume response for normal tissue complications has been extensively studied. Typically, organs at risk are demarcated on CT images as whole organs and dose-volume analysis is performed based on dose-volume histograms (DVHs) which preclude positional dose dependence. The assumption underlying this treatment is that function is homogeneous throughout the organ volume. Evidence to the contrary has been demonstrated. Combining mean lung dose with positional dependence leads to improved predictive power for incidence of radiation pneumonitis [1]. Incidence of subjective xerostomia in head-and-neck cancer patients correlates more strongly with dose to the lateral cranial regions of the parotid compared with mean dose to the whole parotid [2]. The search for regional effects requires sub-segmenting normal tissue contours based on anatomical (e.g., liver lobes) or geometrical features. This search cannot conceivably be performed manually. Herein we present a computational system, `DICOMautomaton`, which can rapidly perform autonomous contour sub-segmentation and generic dose-volume computations, substantially reducing the human effort required for exploratory dose-volume analyses.

2. Methods

`DICOMautomaton` offers a variety of segmentation routines which can be applied identically from single contour loops to entire patient organ sets. The most intuitive are those which split data along an embedded $\mathbb{R}^2$ plane. For example, a researcher requiring equal volume lateral halves of a given organ need only specify that the division should occur along a sagittal plane. Not all segmentation is limited to organs, as per-contour or per-patient segmentation is available.

Alternatively, a user might not want a flush, planar boundary between segments, instead preferring to sub-segment in terms of distance from an organ's surface. `DICOMautomaton` can perform this *ray-casting* segmentation along a specified direction. Planar and casting segmentation are sufficient for producing segments of equal volume, contour (planar) area, or number in a variety of orientations and configurations. All techniques are composable; they can be chained together sequentially or mixed selectively on a per-segment basis.

More complex segmentation is possible including fractal-like branching, which is useful for imitating branching vessels, core-and-peel, for separating inner and outer portions, and constrained segmentation, which can partition voxels by means of a user-provided heuristic. A canonical use for the latter is enclosing all voxels within an organ receiving dose above a certain threshold with an exclusive contour. The heuristic function is a stateful closure which can be passed in and altered at run-time, admitting highly complex selection heuristics.

It is often desirable to test a variety of segmentations for some given suitability. This feedback allows for automatic searching of regional organ dose sensitivity. `DICOMautomaton` provides rudimentary facilities for the automation of such searching: given a set of segmentation rules (i.e., a "schedule"), the system will attempt to optimize a user-provided cost function by traversing a sub-segmentation graph. The schedule can be provided at run-time.

To encourage interoperability, contours which have been segmented can be injected back into anonymized patient data permanently in a DICOM-conforming manner. Lossless segmentation persists across such injection, and remains fully reversible.

To provide a useful platform for automation of dosimetric analyses, `DICOMautomaton` offers the ability to compute standard dosimetric quantities on any set of contour (sub-)segments. DVH, minimum/mean/median/maximum voxel dose, and statistical moment routines exist, while the implementation of additional routines is straightforward. All dosimetric routines avoid resampling and interpolation where possible by working directly with DICOM contour and voxel dose data.

3. Results

To demonstrate the segmentation capabilities, we have produced a variety of images using patient contours. In figure 1 we demonstrate simple planar segmentation of a whole left parotid. Figure 2 shows mixed ray-casting and planar segmentations proceeding in sequence. Finally, more complex segmentations are shown in figure 3, including random-orientation sequential planar, core-and-peel, and dose-threshold heuristic segmentations of various contours.

System accuracy was measured by comparing minimum, mean, and maximum doses from 60 head-and-neck cancer dosimetric data sets to Varian's Eclipse™. The mean dose to the left parotid varied from 2 to $65\,Gy$. The average absolute percent discrepancy was 0.8% ($\sigma = 1.2\%$) with the largest occurring at low mean doses ($< 10\,Gy$). Total computation time, including file input and output, was 80 seconds on an Intel® Pentium® E2200. DVHs produced by `DICOMautomaton` were found to strongly agree with those produced by Varian's Eclipse™. An example comparison is shown in figure 4.

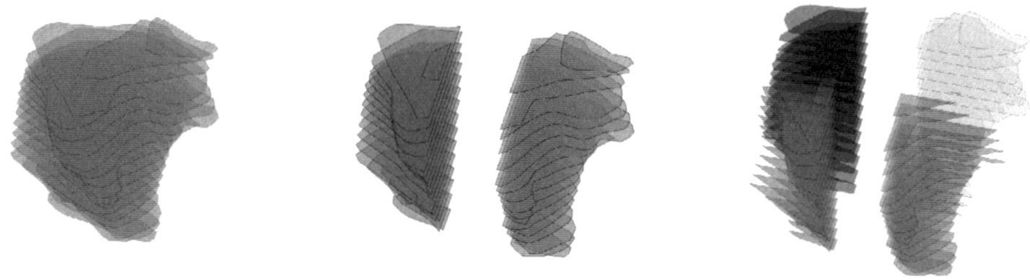

Figure 1. Sequentially performed planar segmentation on a whole left parotid. Per-organ (per-patient) planar boundaries are identical for all individual contours within an organ (patient) whereas per-contour planar boundaries vary from contour to contour.

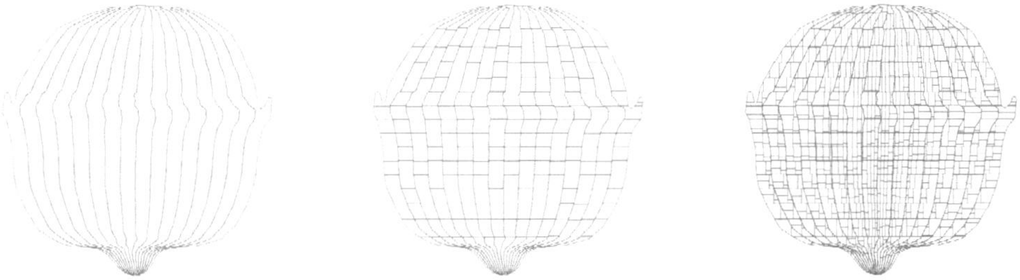

Figure 2. Demonstration of mixed contour segmentations of a body contour at the facial level in sequence. From left to right: simple lateral ray-casting, planar (coronal) per-contour, and several mixed planar per-organ and per-contour segmentations. This sequence demonstrates the composability of segmentation operations, which can easily produce highly complex subdivisions.

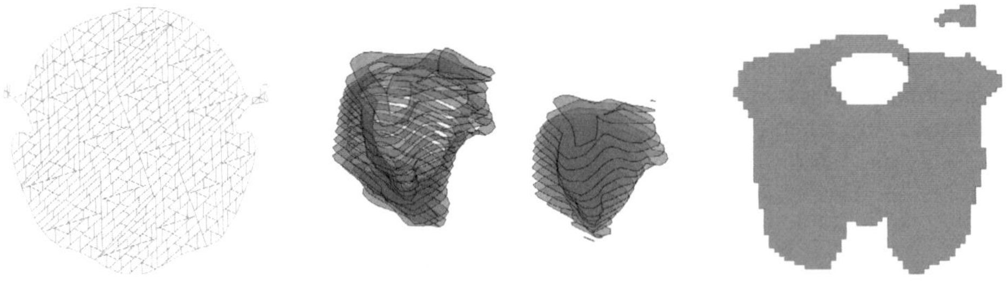

Figure 3. Demonstration of complex segmentations. From left to right: random-orientation sequential planar segmentation on a body contour at the upper neck level, core-and-peel segmentation on a whole left parotid, and dose-threshold heuristic segmentation of a body contour at the shoulder level. In the latter, voxels with high dose are exclusively encircled with a contour; spinal cord sparing is evidenced by the omitted (low-dose) circular region. Sub-segments can be further sub-segmented well below the typical voxel size, if desired.

A typical segmentation workflow using popular analysis software takes a voxel approach whereby contours are resampled onto legitimately voxel (e.g., image or dosimetric) data for the convenience of having a single, unified structure or to simply fit into a voxel paradigm. This approach requires irreversible resampling and interpolation to project the contours onto a voxel grid and is wasteful in the sense that algorithms working on contour lines embedded in planes decrease considerably in computational efficiency when adapted to run on voxels in $\mathbb{R}^3$. Efficient

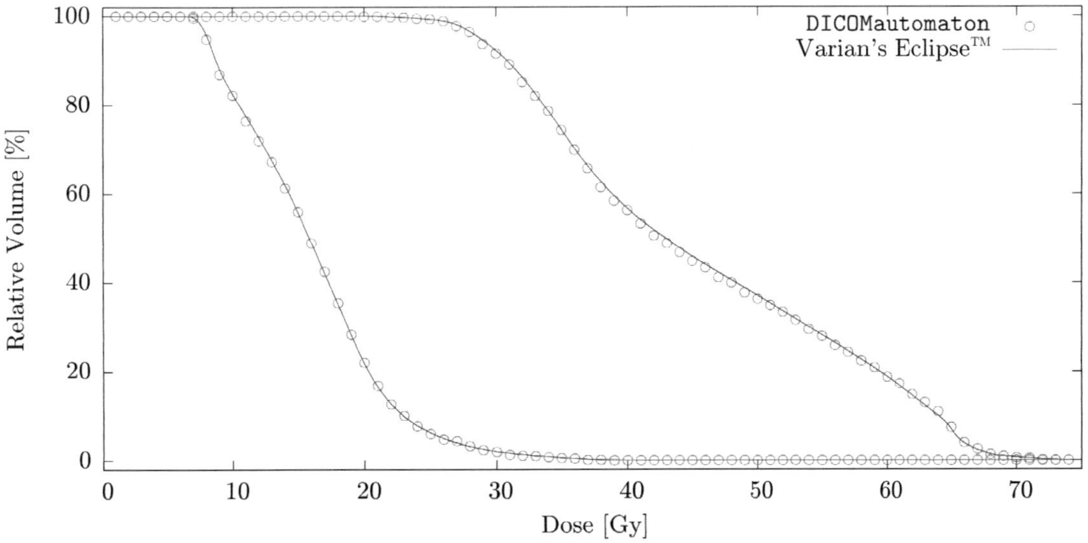

Figure 4. Example DVHs as produced by `DICOMautomaton` and Varian's Eclipse™ for a typical left (lower) and right (upper) parotid. Strong agreement is observed.

voxel image segmentation algorithms run in $O(n \log n)$ time where n is the number of voxels [3], which, due to the dimensionality, makes n unfortunately large at even modest resolutions (e.g., 256^3 voxels). The popular marching cubes algorithm [4], which is more directly adaptable to voxel-based contour segmentation, not only suffers from the aforementioned dimensionality issue but also notoriously exhibits topological inconsistency and produces fragmented volume boundaries [5]. Attempts to address these limitations are numerous [6]. To avoid such problems `DICOMautomaton` forgoes the voxel paradigm, instead working directly with DICOM-style contours. This enables lossless, reversible segmentation, reduces resampling and truncation errors to near-machine precision, and dramatically reduces computation time.

4. Conclusions

`DICOMautomaton` allows users to rapidly, accurately, and autonomously (sub-)segment large amounts of contour data into intricate structures suitable for analyses of regional organ dose sensitivity. Dosimetric routines are robust; quantities computed with `DICOMautomaton` strongly agree with those produced by Varian's Eclipse™.

Acknowledgments

The authors acknowledge financial support from The Ride to Conquer Cancer and the BCCA.

References

[1] Marks L B, Bentzen S M, Deasy J O, Kong F M S, Bradley J D, Vogelius I S, El Naqa I, Hubbs J L, Lebesque J V, Timmerman R D *et al.* 2010 *IJROBP* **76** S70–S76

[2] Buettner F, Miah A, Gulliford S, Hall E, Harrington K, Webb S, Partridge M and Nutting C 2012 *Radiotherapy and Oncology* **103** 82

[3] Felzenszwalb P F and Huttenlocher D P 2004 *IJCV* **59** 167–181

[4] Lorensen W E and Cline H E 1987 Marching cubes: A high resolution 3D surface construction algorithm *ACM Siggraph Computer Graphics* vol 21 (ACM) pp 163–169

[5] Montani C, Scateni R and Scopigno R 1994 Discretized marching cubes *Proceedings of the conference on Visualization '94* (IEEE Computer Society Press) pp 281–287

[6] Newman T S and Yi H 2006 *Computers & Graphics* **30** 854–879

XVII International Conference on the Use of Computers in Radiation Therapy (ICCR 2013) IOP Publishing
Journal of Physics: Conference Series **489** (2014) 012010 doi:10.1088/1742-6596/489/1/012010

GMC ['gɪmɪk]: a one-variable Monte Carlo dose algorithm for proton therapy.

N. Depauw[1,2], B. Clasie[1], T. Madden[1], A. Rosenfeld[2], and H. Kooy[1]
[1]Francis H. Burr Proton Therapy Center, Department of Radiation Oncology, Massachusetts General Hospital (MGH), Boston MA 02114, USA
[2]Centre for Medical Radiation Physics, University of Wollongong

E-mail: ndepauw@partners.org

Abstract. This work presents the CPU implementation of GMC ['gɪmɪk]: a fast yet accurate one-variable Monte Carlo dose algorithm for proton therapy to be incorporated into our in-house treatment planning system, Astroid. GMC is based on a simple mathematical model using the formulated proton scattering power and tabulated data of empirical depth-dose distributions. These Bragg peaks determine the energy deposited along the particle's track. The polar scattering angle is based on the particle's local energy and the voxel's density, while the azimuthal component of that scattering angle is the single variable in GMC, uniformly distributed from 0 to 2π. The halo effect of the beam, currently not implemented, will consider large scattering angles and secondary protons for a small percentage of the incident histories. GMC shows strong agreement with both the empirical data and GEANT4-based simulations. Its current CPU implementation runs at $\sim$300 m.s^{-1}, approximately ten times faster than GEANT4. Significant speed improvement is expected with the upcoming implementation of multi-threading and the portage to the GPU architecture. In conclusion, a one-variable Monte Carlo dose algorithm was produced for proton therapy dose computations. Its simplicity allows for fast dose computation while conserving accuracy against heterogeneities, hence drastically improving the current algorithms used in treatment planning systems.

1. Introduction

Proton therapy treatment planning systems (TPS) are all currently relying on a Pencil Beam Algorithm (PBA) for dose computation in patient. PBA, while relatively fast, fails to model heterogeneities' effects which could be critical in Pencil Beam Scanning delivery (PBS)[1]. For accurate results it is necessary to turn towards full Monte Carlo (MC) systems such as GEANT4. The latter, however, present a large overhead and can be significantly time consuming.

We therefore propose an extremely simplified MC algorithm for proton dose calculation. The latter, known as GMC ['gɪmɪk] (*A unique or quirky special feature that makes something "stand out" from its contemporaries.*), intends to be extremely fast yet accurate against tissue inhomogeneities. Moreover, GMC will be implemented at the lowest level as part of our in-house PBS TPS, Astroid, simply replacing the current PBA.

2. Materials and methods

Figure 1 shows a representation of the particle transport in GMC at the voxel level.

A particle enters a voxel (density ρ, volume V) with certain input parameters: current energy E_i, relative position in the voxel (x_i, y_i, z_i), direction (u_i, v_i, w_i), and initial energy E_0 (energy at the entrance of the phantom/patient). A cord length L is computed from a non-scattered transport of the particle (straight line) through the voxel. The mean radial scattering angle is subsequently computed from the scattering power described by Gottschalk [2], as shown in equation 1:

$$\langle\theta^2\rangle = \int_0^L T_{dM}\mathrm{d}x \approx T_{dM_i} \times L \qquad (\mathrm{rad}^2), \tag{1}$$

with

$$T_{dM} = f_{dM}(E_i, E_0) \times \left(\frac{E_s}{E_i}\right)^2 \times \frac{1}{X_s}, \tag{2}$$

where

$$\begin{aligned} f_{dM} = \quad 0.5244 \quad + \quad 0.1975 \times \log_{10}(1 - \left(\frac{E_i}{E_0}\right)^2) \quad + \\ 0.2320 \times \log_{10}(\frac{E_i}{MeV}) \quad - \quad 0.0098 \times \log_{10}(\frac{E_i}{MeV}) \times \log_{10}(1 - \left(\frac{E_i}{E_0}\right)^2) \qquad (\mathrm{MeV}) \end{aligned} \tag{3}$$

where θ is the average radial scattering angle, x the distance already traveled along the track in cm, $Es = 15.0$ MeV, and $X_s = 46.88$ cm.

The azimuthal scattering angle is the only random variable in GMC, between 0 and 2π.

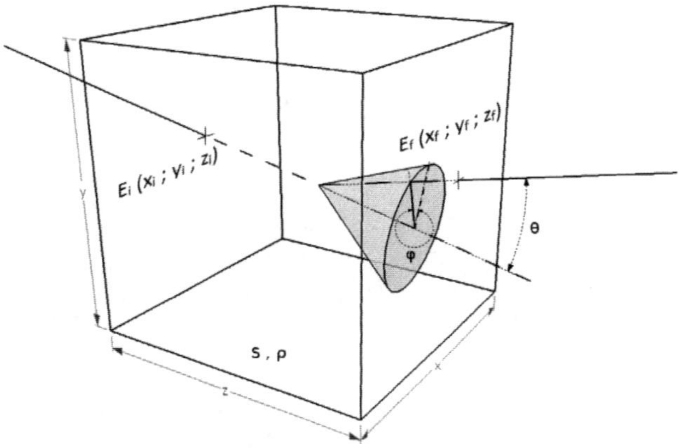

Figure˜1. GMC model at the voxel level.

The mean energy loss ΔE is based on the initial proton energy E_0 and the trapezoidal approximation of the integral under the curve of its corresponding Bragg Peak, TPS beam data as described in Clasie [3]. The local deposited dose ΔD is then computed from equation 4 assuming a mass stopping power ratio equal to unity. This integral is shown in figure 2.

$$\Delta D = \frac{\Delta E}{\rho V} \qquad (\mathrm{Gy}), \tag{4}$$

where,

$$\Delta E = S_{medium}^{water} \times L \times \left(\frac{\mathrm{d}E}{\mathrm{d}x}\right)_{water} = \int_{X_1}^{X_2} f(X)\mathrm{d}X \qquad (\mathrm{MeV}), \tag{5}$$

and,

$$\int_{X_1}^{X_2} f(X)\mathrm{d}X \approx \rho L \frac{f(X_2) + f(X_1)}{2} \tag{6}$$

where S_{medium}^{water} is the relative stopping power ratio, and $\int_{X_1}^{X_2} f(X)\mathrm{d}X$ the integral below the Bragg peak curve in which X_1 and X_2 are based on the current particle energy E_i; $f(X)$ is given in units of $Gy.cm^2$, and X_1 and X_2 in g/cm^2).

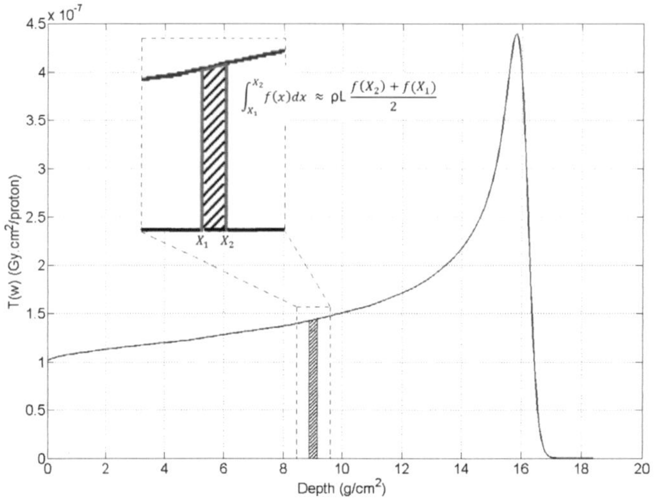

Figure˜2. Dose deposition based on TPS beam data [3].

Not currently taken into account, the Halo effect [3], characterized by larger scattering angles and secondary protons, will later be implemented based on further experimental results.

3. Results

The presented implementation of GMC ran at ~300 m.s^{-1}, approximately ten times faster than GEANT4.9.0 (a full MC system) for a similar setup.
Figure 3 shows two simulations in a water phantom that highlights the expected depth-dependence of an incident beam with *a*) zero lateral spread and *b*) uniform square distribution.

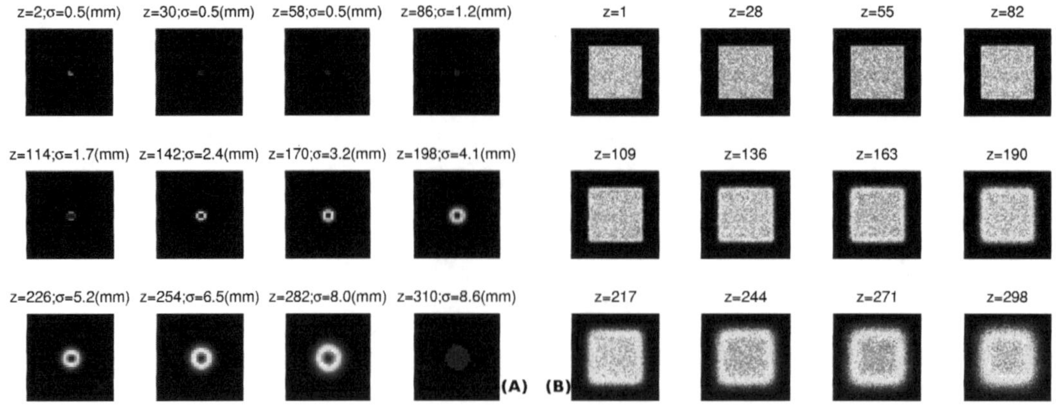

Figure˜3. GMC simulations: a) no initial spread nor entrance angle, b) a uniform field.

Fig·re 4 stresses that the longitudinal dose algorithm was implemented correctly within

GMC[2]. Simulations were performed in water and compared to the original TPS (Astroid) Bragg peaks that were used to tune the model. Figure 4 is an example of such a Bragg peak comparison, zoomed in on the end of range.

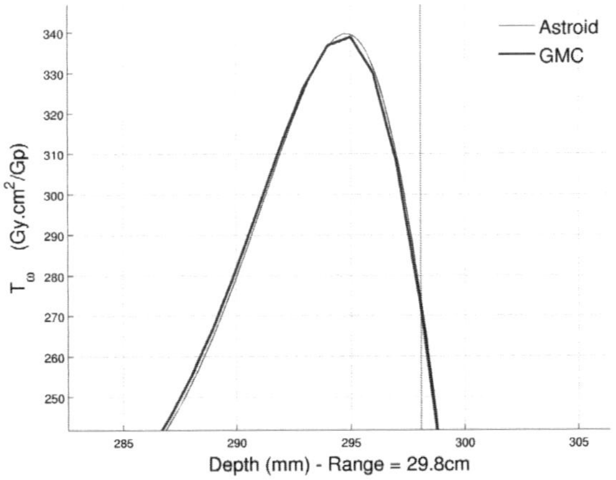

Figure˜4. Comparison of depth dose deposition between the original TPS Bragg peak data (Astroid) and a GMC simulation.

Finally, and most importantly, Figure 5 highlights GMC's behavior against heterogeneities. In that simulation, a water phantom ($\rho = 1$) is considered with blocks of lung material ($\rho = 0.4$, white overlay) and bone material ($\rho = 1.71$, black overlay); two pencil beams go through the inhomogeneities in a different order. PBA would result in similar scattering and dose deposition for each pencil beam. GMC, however, properly models the expected physics and dissimilarities in shape and dose are observed after the heterogeneities.

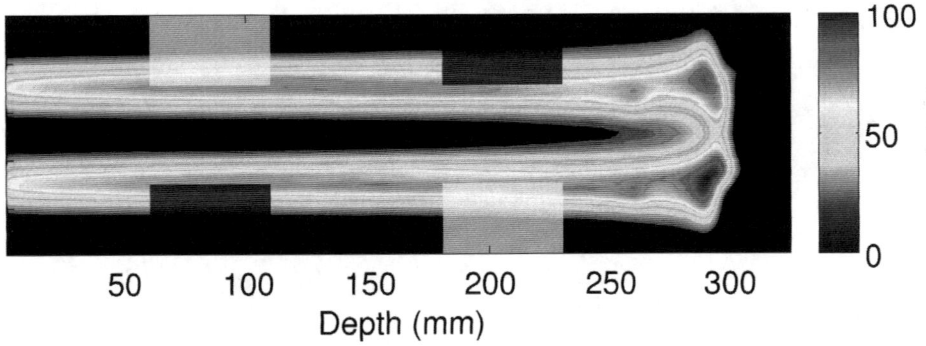

Figure˜5. GMC's behavior against heterogeneities.

4. Conclusion

GMC, a fast yet accurate proton dose algorithm, has been successfully implemented. Expected dose distributions were obtained, and the algorithm exhibited appropriate behavior when dealing with heterogeneities. Future work will involve systematic comparison/analysis of beam scattering and dose deposition against expected results, as well as significant speed improvement by implementing multi-threading and portage to a GPU architecture.

5. References

[1] Hong L, Goitein M, Bucciolini M, Comiskey R, Gottschalk B, Rosenthal S, Serago C, Urie M 1996,"A pencil beam algorithm for proton dose calculations" *Phys. Med. Biol.*, **41**, 1305.

[2] Gottschalk G 2009, "On the scattering power of radiotherapy protons", *Med. Phys.*, **37**, 352-367.
[3] Clasie B, Depauw N, Fransen M, Gom C, Panahandeh HR, Seco J, Flanz JB, Kooy HM 2012, "Golden beam data for proton pencil-beam scanning", *Phys. Med. Biol*, **57**, 1147.

In silico radiobiology: Have we reached the limit of Monte Carlo simulations?

Y Gholami[1], M Toghyani[1], C Champion[2], Z Kuncic[1]
[1]Institute of Medical Physics, School of Physics, The University of Sydney, Sydney, NSW, Australia
[2]Universite Bordeaux 1, CNRS/IN2P3, Centre dEtudes Nucleaires de Bordeaux Gradignan, France

E-mail: y.gholami@physics.usyd.edu.au

Abstract. Monte Carlo radiation transport models are increasingly being used to simulate biological damage. However, such radiation biophysics simulations require realistic molecular models for water, whereas existing Monte Carlo models are limited by their use of atomic cross-sections, which become inadequate for accurately modelling interactions of the very low-energy electrons that are responsible for biological damage. In this study, we borrow theoretical methods commonly employed in molecular dynamics simulations to model the molecular wavefunction of the water molecule as the first step towards deriving new molecular cross-sections. We calculate electron charge distributions for molecular water and find non-negligible differences between the vapor and liquid phases that can be attributed to intermolecular bonding in the condensed phase. We propose that a hybrid Monte Carlo – Molecular Dynamics (MC-MD) approach to modelling radiation biophysics will provide new insights into radiation damage and new opportunities to develop targeted molecular therapy strategies.

1. Introduction

Classical radiobiology has provided a remarkably successful framework for understanding cell killing by ionising radiation and linear-quadratic models, in particular, have proven to be invaluable for quantifying and predicting survival of tumor cells exposed to typical therapeutic doses of radiation [1]. While empirical radiobiological models are now gradually being used to biologically optimise radiotherapy treatment, many Monte Carlo (MC) models continue to be developed to gain further insight into the biophysics of DNA damage inflicted by radiation. Most *in silico* models focus on direct damage to nuclear DNA [2, 3], rather than the more prevalent indirect damage caused by the free radicals produced by radiolysis. New MC models are also emerging that address non-targeted damage to healthy cells caused by radiation at low-doses [4, 5], where radiobiological models predicting normal tissue complications have had limited success in describing clinically observed normal tissue reactions [6].

A major challenge confronting radiation biophysics simulations is that existing MC models use atomic interaction cross-sections. While this is adequate for high incident energies and tissue-scale dose volumes, the atomic target approximation is inadequate for determining energy and momentum transfer on sub-cellular scales [7]. At the very low energies ($\ll 100\,\mathrm{eV}$) of the electrons that are ultimately responsible for biological damage, the molecular nature of the target cannot be ignored. Thus, it is necessary to consider the nanoscale dynamics of electrons

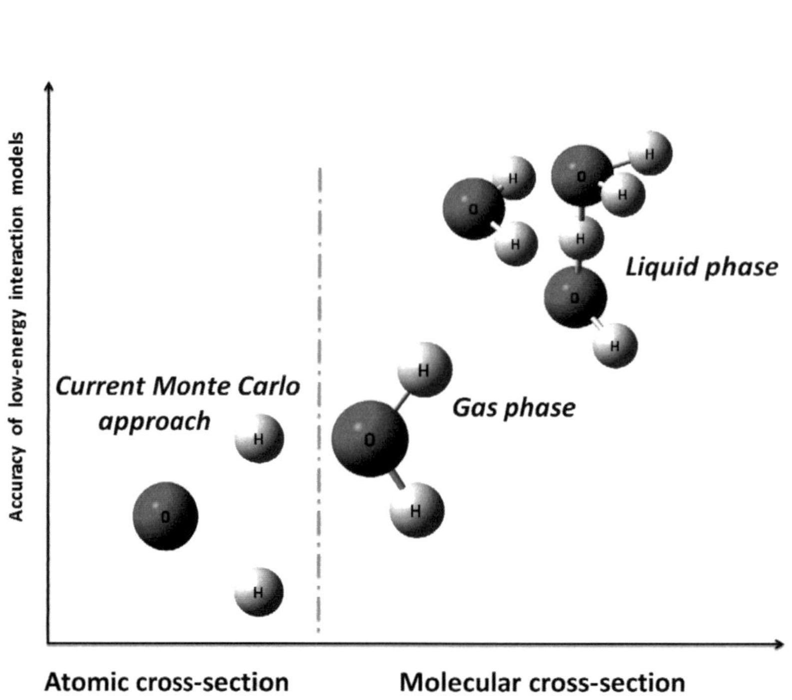

Figure 1. Schematic illustration depicting the difference between an atomic water target, as currently used in Monte Carlo radiation transport simulations, and molecular targets for water in vapor and liquid phases.

in soft, condensed matter. This requires new models to be developed for molecular interaction cross-sections (see Figure 1).

Several previous studies investigating low-energy electron inelastic scatter in water have used a semi-classical approach, deriving a differential cross-section in terms of the dielectric response function for water [8, 9]. An alternate approach is to derive the differential cross-section using a quantum mechanical approach, which requires an accurate description of the molecular wavefunction for water [10]. Here, we briefly report on our study using this first principles approach and we present preliminary results on the radial distribution of electron charge in molecular water for both the vapor and liquid phases, a key test of the molecular wavefunction.

2. Approach and Methodology

In this study, we used an *ab initio* method based on the Hartree-Fock level of theory for multi-electron systems. We constructed atomic orbitals using the same basis set and contracted gaussian functions as that used by Champion [10]. To calculate the molecular orbitals and construct the water molecule wavefunction, however, we used the linear combination of atomic orbitals approximation, whereas Champion used a single-center partial wave expansion method. All quantum chemistry calculations were performed using the Gaussian 03 program [11]. Calculations for liquid water used the polarizable dielectric continuum model available within Gaussian 03. The radial distribution of electron charge density and the partial atomic charges were calculated to evaluate the derived molecular wavefunction for water in the two distinct thermodynamic phases.

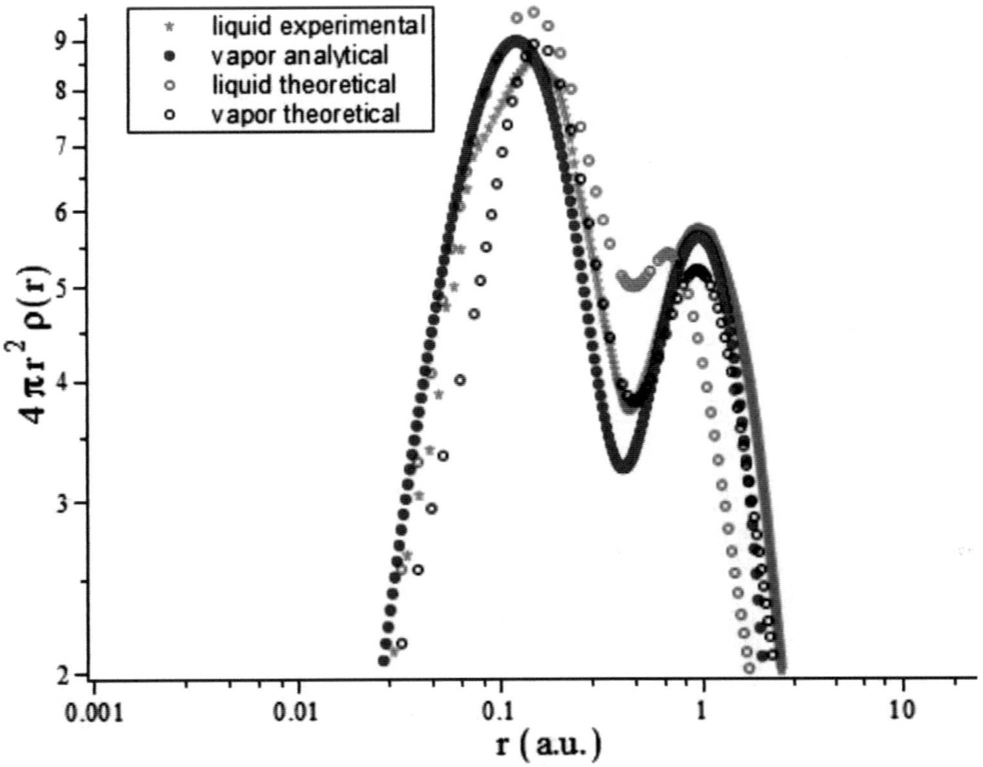

Figure 2. Spherical electron charge density radial distributions for molecular water for the liquid and vapor phases (with origin centred on the oxygen atom). Theoretical results from the present work (open circles) are shown against the analytical prediction of Katase *et al.* [12] for water vapor (blue circles) and experimental data (stars) from Neuefeind et al. [13].

3. Results and Discussion

3.1. Electron charge density distribution for molecular water

Figure 2 shows the spherical electron charge density distribution $4\pi r^2\rho(r)$ calculated for molecular water in the vapor and liquid phases. The main peak in the distribution (at $r \simeq 0.1$ a.u.) is governed by the electronic distribution around the oxygen atom. The second peak corresponds to the influence of the hydrogen atoms. Our theoretical results, calculated using contracted gaussian functions for the atomic orbitals, are plotted against the analytical prediction of Katase *et al.* [12] for water vapor and experimental data for liquid water [13]. Noticeable differences are evident in Fig. 2, particularly before the first peak, between our results for water vapor (black circles) and those of Katase *et al.* [12]. This can be attributed to the different descriptions of atomic orbitals, since the analytical prediction is based on Slater-type orbitals. Our theoretical results for the liquid phase (red circles) also differ from the corresponding experimental data (stars), especially between the two peaks. This may be attributed to some of the simplifications made in the theoretical model, such as the number of gaussian functions used to approximate the atomic orbitals. On the other hand, the differences in $4\pi r^2\rho(r)$ between the vapor and liquid phases show a similar trend for our results to that exhibited in the experimental and analytical data. In particular, the liquid phase exhibits a more pronounced peak in charge density near 0.1 a.u. (oxygen atom) compared to the vapor phase,

Journal of Physics: Conference Series **489** (2014) 012011 doi:10.1088/1742-6596/489/1/012011

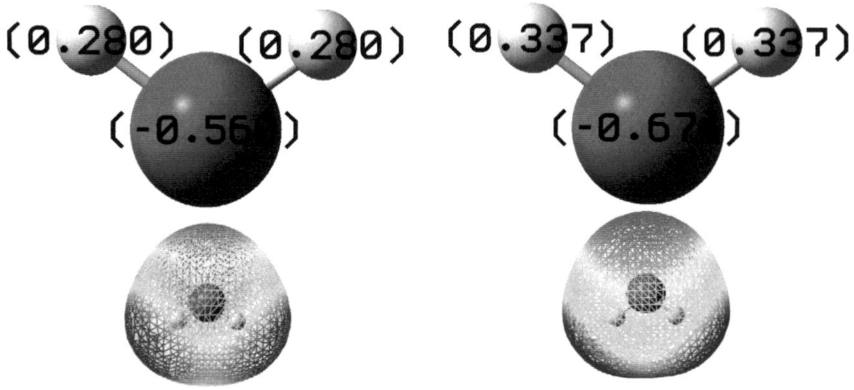

Figure 3. Partial atomic charges and corresponding total electron surface charge distributions for a water molecule in vapor (left) and liquid (right) phases.

which reflects the influence of the stronger dipole moment on the valence electron distribution. The enhanced dipolar nature of molecular water in the liquid phase can be attributed to intermolecular hydrogen bonding, which is also responsible for the difference in $4\pi r^2 \rho(r)$ near $r \simeq 0.3 - 0.4$ a.u. between the vapor and liquid phases.

3.2. Surface charge analysis

Figure 3 shows the partial charges of each atom as well as the total surface charge distribution in a water molecule for the vapor and liquid phases. The partial charges were calculated within Gaussian 03 by dividing the total electron distribution according to overlaps of the atomic orbitals. As is evident in Fig. 3, oxygen has a more negative partial charge in a liquid water molecule compared to a vapor molecule, while the hydrogen atoms are more positive. This indicates a stronger polarity of the water molecule in its liquid phase compared to its vapor phase, which can be attributed to the hydrogen bonding with surrounding molecules. This is also reflected in the surface charge density distributions shown in Fig. 3, where red indicates maximum negative charge, while blue indicates minimum negative charge. This supports the interpretation of the differences in the radial charge distribution $\rho(r)$ being due to the intermolecular hydrogen bonds present in liquid water.

4. Conclusion

Our results demonstrate that there are non-negligible differences in the molecular orbitals of water in the vapor and liquid phases that can be attributed to the increase in the dipole moments due to hydrogen bonding. This will in turn affect the cross-section for low-energy electron inelastic scatter in biological tissue. Work is currently in progress to calculate new cross-sections and determine the effect on existing models for low energy electron transport, including track structure. We propose that a hybrid Monte Carlo – Molecular Dynamics (MC-MD) approach will extend the current capability of radiation transport simulations, enabling more realistic simulation of biological damage and hence, the development of new targeted molecular therapy strategies into the future.

References

[1] Chapman JD, Gillespie CJ 2012 *The power of radiation biophysics – let's use it* Int. J. Rad. Onc. Biol. Phys. **84**, 309

[2] Bernal MA, et al. 2011 *The invariance of the total direct DNA strand break yield* Med. Phys. **38**, 4147

[3] McNamara AL, Guatelli S, Prokopovich DA, Reinhard MI, Rozenfeld AB 2012 *A comparison of x-ray and proton beam low energy secondary electron track structures using the low energy models of Geant4* Intl. J. Rad. Biol. **88**, 164

[4] Kuncic Z, Byrne HL, McNamara AL, Guatelli S, Domanova W, Incerti S 2012 *In silico nanodosimetry: New insights into nontargeted biological responses to radiation* Comp. Math. Meth. Med. 147252

[5] Byrne HL, McNamara AL, Domanova W, Guatelli S, Kuncic Z 2013 *Radiation damage on sub-cellular scales: beyond DNA* Phys. Med. Biol. **58**, 1251

[6] Denham JW et al. 1996 *Mucosal regeneration during radiotherapy - Trans Tasman Radiation Oncology Group (TROG)* Radiother. Oncol. **41**, 109

[7] Incerti S *et al.* 2010 *Comparison of GEANT4 very low energy cross section models with experimental data in water* Med. Phys. **37**, 4692

[8] Emfietzoglou D, Nikjoo H 2007 *Accurate electron inelastic cross sections and stopping powers for liquid water over the 0.1-10 keV range based on an improved dielectric description of the Bethe surface* Rad. Res. **167**, 110

[9] Dingfelder M, Ritchie RH, Turner JE, Friedland W, Paretzke HG, Hamm RN 2008 *Comparisons of calculations with PARTRAC and NOREC: Transport of electrons in liquid water* Rad. Res. **169**, 584

[10] Champion C 2010 *Electron impact ionization of liquid and gaseous water: a single-center partial-wave approach* Phys. Med. Biol. **55**, 11

[11] Gaussian 03, Revision C.01, 2004 Frisch MJ *et al*

[12] Katase A *et al.* 1986 *Elastic scattering of electrons by water molecules over the range 100-1000 eV* J. Phys. B: At. Mol. Phys. **19**, 2715

[13] Neuefeind J, Benmore CJ, Tomberli B, Egelstaff PA 2002 *Experimental determination of the electron density of liquid* H_2O *and* D_2O J. Phys.: Condens. Matt. **14**, L429

Journal of Physics: Conference Series **489** (2014) 012012 doi:10.1088/1742-6596/489/1/012012

An improved Monte-Carlo model of the Varian EPID separating support arm and rear-housing backscatter

M E Monville[1], Z Kuncic[2] and P B Greer [1,3]

[1] University of Newcastle, Newcastle, NSW, Australia.
[2] University of Sydney, NSW, Australia
[3] Calvary Mater Newcastle, Newcastle, NSW, Australia

E-mail: peter.greer@newcastle.edu.au

Abstract. Previous investigators of EPID dosimetric properties have ascribed the backscatter, that contaminates dosimetric EPID images, to its supporting arm. Accordingly, Monte-Carlo (MC) EPID models have approximated the backscatter signal from the layers under the detector and the robotic support arm using either uniform or non-uniform solid water slabs, or through convolutions with back-scatter kernels. The aim of this work is to improve the existent MC models by measuring and modelling the separate backscatter contributions of the robotic arm and the rear plastic housing of the EPID. The EPID plastic housing is non-uniform with a 11.9 cm wide indented section that runs across the cross-plane direction in the superior half of the EPID which is 1.75 cm closer to the EPID sensitive layer than the rest of the housing. The thickness of the plastic housing is 0.5 cm. Experiments were performed with and without the housing present by removing all components of the EPID from the housing. The robotic support arm was not present for these measurements. A MC model of the linear accelerator and the EPID was modified to include the rear-housing indentation and results compared to the measurement. The rear housing was found to contribute a maximum of 3% additional signal. The rear housing contribution to the image is non-uniform in the in-plane direction with 2% asymmetry across the central 20 cm of an image irradiating the entire detector. The MC model was able to reproduce this non-uniform contribution. The EPID rear housing contributes a non-uniform backscatter component to the EPID image, which has not been previously characterized. This has been incorporated into an improved MC model of the EPID.

1. Introduction

Dosimetry with electronic portal imaging devices (EPIDs) can provide a highly efficient means to perform pre-treatment quality assurance for complex delivery techniques, as well as transit-type dosimetry for patient dose verification [1, 2]. For the Varian EPID system the backscatter from the EPID supporting arm, cables, and layers underneath the sensitive Gd_2O_2S layer contributes additional non-uniform signal to the dose image. This has been quantified as up to 7% additional signal for a flood-field irradiation of the EPID [3, 4]. It therefore affects the reliability of the EPID as a dosimetry device [1, 2]. Currently there are several different methods for quantifying, modelling, and removing the backscatter from the EPID supporting arm when performing EPID dosimetry. Some authors have regarded the backscatter material as uniform in composition. Rowshanfarzad et al. [5] developed an analytical model of back-scatter with a backscatter kernel derived from the results of pencil-beam backscatter measurements acquired with the EPID on and off the support arm. Siebers et al. [6] developed a comprehensive MC model of the EPID and established that a uniform water slab, added underneath the EPID original layers was necessary to account for the field-size dependence of EPID signal at central axis. Wang et al. [7] extended this method by determining specific water-equivalent thicknesses to account for different field size response of imagers. Other authors have regarded the backscatter material as non-uniform. Cuffin et al. [8] added water-equivalent slabs of non-uniform thickness at the bottom of the original EPID layers for their MC model of extended fields. To improve MC modeling of EPIDs we have applied a novel approach of optimizing the MC model using EPID data recorded without the support arm present. This decouples the arm backscatter artifact from the EPID model and enables model parameters and model components for EPID dose deposition and arm backscatter to be independently optimized. As with previous studies we initially modeled the rear plastic housing of the EPID as a uniform layer of plastic. However comparison of the measured and the predicted doses resulted in some disagreement, particularly in the low dose region. This suggested that some contributions of backscatter were still present in the measurements. An indentation in the EPID rear housing layer was therefore investigated. In this paper we experimentally measure the effect of the rear housing with this indentation on EPID signal and we incorporate the indentation into our MC model for comparison to measurements. The results shed new light on the sources and effects of EPID backscatter and how to effectively model EPIDs using Monte-Carlo for dosimetric applications.

2. Materials and Methods

The EPID used in this study was the Varian IDU20 aS500 model (Varian Medical Systems, Palo Alto, CA, USA). The EPID was no longer used due to damage to some rows at the top and bottom of the EPID that are not able to contribute signal. This restricted the signal that was obtained in the in-plane direction from 10.04 cm below the central axis to 10.03 cm above the central axis on the EPID. The bottom housing is non-uniform with an 11.9 cm width and 1.75 cm depth indented section running in the cross-plane direction in the upper half of the EPID (Figure 1). The indented section is 1.75 cm closer to the EPID sensitive layer than the rest of the housing. The thickness of the plastic housing is approximately 0.5 cm in all directions (determined from CT scan of the EPID) including the indentation walls.

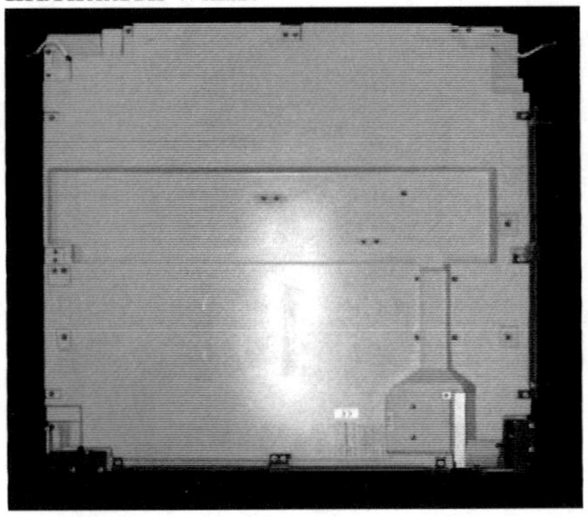

Figure 1. EPID viewed from underneath showing the indentation in the rear housing

In order to quantify the contribution of the rear housing to the measured dose, the EPID was positioned on a low density styrofoam plate on the linac couch. The EPID was connected to a separate digitization unit and acquisition computer used for research. Images for jaw-defined open fields of sizes 5×5, 10×10, 20×20 cm^2 and an image covering the entire detector were acquired with 6 MV energy on a Varian Trilogy linac at 105 cm to the EPID sensitive layer. The EPID layers and electronics were then removed from the EPID housing and repositioned on the styrofoam and the measurements repeated. Images were acquired without any corrections applied to the images to examine the raw EPID signals. To compare to the MC model a separate measurement was performed with the same model of EPID but in clinical use. This EPID was removed from the support arm and supported at the sides without a treatment couch underneath. Measurements with a 20×20 cm^2 field were obtained again with raw EPID signal. The BEAMnrc MC code [9] was used to simulate photon and electron propagation from the target through the linac head for 6 MeV photons down to the phase space plane which is defined at the linac isocenter. The energy of the incident electron beam was 6.02 MeV with a FWHM of 0.12 cm. The linear accelerator The DOSXYZnrc MC code [10] was used to simulate propagation of particles from the phase space down through the air gap and the α-Si EPID modeled as 40×30 cm^2 layers with 0.1 cm pixel resolution. The center of the EPID sensitive layer was positioned at 105 cm from the target for consistency with the experimental set-up. The surrounding medium was defined to be uniform air. The MC EPID model comprised 22 layers, as described in the manufacturer's specification data sheets and in Siebers et al. They included the Cu buildup plate, the Gd_2O_2S phosphor screen, the detector front cover and the rear housing. The standard EPID model, which comprises uniform homogeneous slabs, was modified to accommodate the rear-housing indentation. The first 19 layers were modeled as a sequence of uniform material slabs. The last 3 layers were partially filled with the back-housing cover and partially with air therefore reproducing the indentation at the bottom. The model results for a 20×20 cm^2 field were compared to raw EPID data measures with the EPID removed from the support arm as described above.

3. Results

An in-plane profile through the central axis of the ratio of the entire detector image recorded with and without the rear housing present is shown in Figure 2. The maximum additional backscatter signal contribution of the housing is approximately 3% and is 2.5% at central axis. Furthermore, the contribution is non-uniform in the in-plane direction with about 2% asymmetry across the central 20 cm of the field irradiating the entire detector. The backscatter contribution from the support arm for a field irradiating the entire detector has been previously quantified from measurements on and off the arm. This contribution does not include rear housing backscatter as this is present in both on and off arm measurements and therefore cancels. The separate backscatter from the rear housing has been determined in this work.

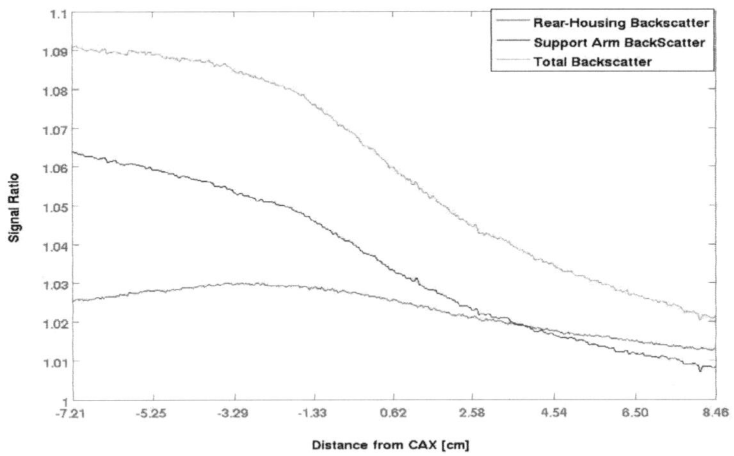

Figure 2. EPID rear housing backscatter. A profile in the in-plane direction of the ratio of EPID signal with and without housing present is plotted. Also plotted for comparison is the support arm backscatter along with the total backscatter from these sources.

In Figure 2**Error! Reference source not found.** the two separate backscatter components (ratio of with and without backscatter) are shown using in-plane profiles along with the total backscatter contribution (housing + support am) to the EPID signal. This is a maximum of 9% with a gradient of over 6% across the in-plane distance. Profiles for a 20×20 cm^2 field are shown in Figure 3. The effect of the rear housing is insignificant for smaller fields, showing that the area of irradiation of the housing is important. The MC code is able to reproduce the non-uniform contribution of the indented rear-housing to the measured dose in the in-plane direction as shown in Figure 4 where the raw measured dose is compared against the predicted. In spite of modeling the effect of the indentation on the measured signal, some disagreement is still noticeable over the tails of the profiles in the in-plane direction. The EPID housing case study was finalized by analyzing also the relative response or field size factor at central axis for the open fields normalized to the 10×10 cm^2 field. The MC model agreed with the measured factors to within 0.5%.

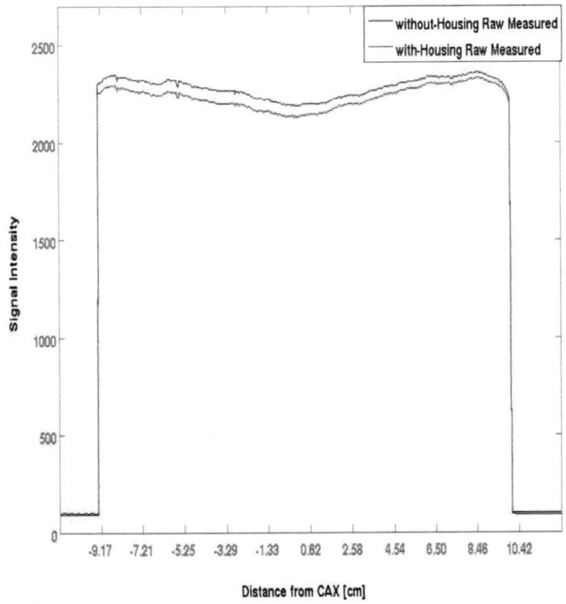

Figure 3. Open-field EPID profiles with and without the rear-housing present for a 20×20 cm^2 field. The raw EPID in-plane direction through central axis is plotted in each case

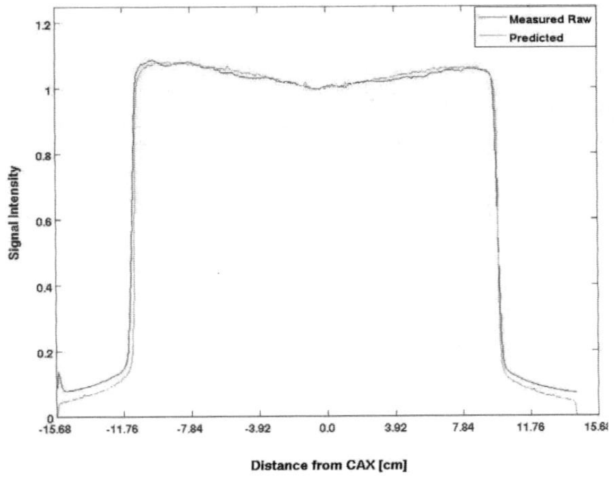

Figure 4. Comparison of MC model with the indented rear-housing and measured raw image. The normalized profile in the in-plane direction through central axis for a 20x20 cm^2 field is plotted.

4. Discussion

Some details of the EPID are not present in the model including screws, nuts, and metallic parts that are also sources of backscatter. These could contribute to some of discrepancies in the low dose tails observed in Figure 4. The source of these discrepancies is not known but are believed to be back-scatter related. Indirect evidence of this was obtained by comparing MC model to measurement for a backscatter shielded EPID model where a layer of lead shields the sensitive layer from backscatter radiation contamination [11]. This comparison did not produce a measured tail region higher than the predicted. The results here are in broad agreement with the findings of Schach von Winnenau et al. [12] who concluded that radiation scattered from non-detector components of the detector itself, such as the front and back covers, are significant sources of image blurring. When an image acquired with the entire EPID irradiated is flood-field or pixel sensitivity matrix corrected, the backscatter from the rear housing (and support arm) is not apparent as the same backscatter is present in both the acquired and the correction image and hence cancels. However when the field size changes, there will be a mismatch in the scatter in both images and backscatter artifacts become apparent. It is clear from this work that previous methods to account for this backscatter artifact are in fact simultaneously modeling or correcting both support arm and rear housing non-uniform backscatter artifacts. The next stage of this work will be to model the support arm backscatter contribution to the EPID.

5. Conclusion

The EPID rear plastic housing contributes a non-uniform backscatter component to the measured dose which has been characterized here using measurements with the EPID layers removed from the housing. This additional source of dosimetric image contamination has been incorporated into an improved Monte Carlo model of the Varian EPID.

6. Acknowledments

This work was funded by the Cancer Council New South Wales grant RG10-03. We are grateful to Dr. Siebers for providing the EPID details, Varian Medical Systems for providing the linear accelerator specifications. CPU hours used in this work were provided by Intersect Australia Ltd. We wish to thank Dr. Joachim Mai (Intersect Organization) for granting access to their HPC center, and all NCI staff for their assistance with our specific computational needs and issues.

References

[1] van Elmpt W et al. 2008 A literature review of electronic portal imaging for radiotherapy dosimetry *Radiotherapy and Oncology* **88** 289-309

[2] Greer P B Vial P Rosenfeld A B et al. 2011 Epid Dosimetry *Concepts and Trends in Medical Radiation Dosimetry* 129-144

[3] Ko L Kim J O and Siebers J V 2004 Investigation of the optimal backscatter for an α-Si electronic portal imaging device *Phys. Med. Biol.* **49** 1723-1738

[4] Greer P B et al. 2009 An energy-fluence convolution model for amorphous silicon EPID dose prediction *Med. Phys.* **36** 547-555

[5] Rowshanfarzad P et al. 2010 Measurement and modelling of the effect of support arm backscatter on dosimetry with a Varian EPID *Med. Phys.* **37** 2269-2278

[6] Siebers J V et al. 2004 Monte Carlo computation of dosimetric amorphous silicon electronic portal images *Med. Phys.* **31** 2135-2146

[7] Wang S et al. 2009 Monte Carlo-based adaptive EPID dose kernel accounting for different field size responses of imagers *Med. Phys.* **36** 3582-3595

[8] Cufflin R S et al. 2010 An investigation of the accuracy of Monte Carlo portal dosimetry for verification of IMRT with extended fields *Physics in Medicine and Biology* **55** 4589-4600

[9] Rogers D W O et al. 1995 BEAM: A Monte Carlo code to simulate radiotherapy treatment units *Med. Phys.* **22** 503-524

[10] Walters B R B Kawrakow I and Rogers D W O 1996 DOSXYZ Users' Manual *NRCC Report PIRS-794 Rev. B (*Ottawa: *NRC).*

[11] King B W Morf D and Greer P B 2012 Development and testing of an improved dosimetry system using a backscatter shielded electronic portal imaging device *Med. Phys.* **39** 2839-2847

[12] Schach von Wittenau A E et al. 2002 *Blurring artifacts in megavoltage radiography with a flat-panel imaging system:* Comparison of Monte Carlo simulations with measurement *Med. Phys.* **29** 2559-2570

Journal of Physics: Conference Series **489** (2014) 012013 doi:10.1088/1742-6596/489/1/012013

A Monte Carlo tool for raster-scanning particle therapy dose computation

U Jelen[1], M Radon[1,2], A Santiago[1], A Wittig[1] and F Ammazzalorso[1]

[1] Department of Radiotherapy and Radiation Oncology - Particle Therapy Center, University of Marburg, Marburg, Germany

E-mail: `Urszula.Jelen@med.uni-marburg.de`

Abstract. Purpose of this work was to implement Monte Carlo (MC) dose computation in realistic patient geometries with raster-scanning, the most advanced ion beam delivery technique, combining magnetic beam deflection with energy variation. FLUKA, a Monte Carlo package well-established in particle therapy applications, was extended to simulate raster-scanning delivery with clinical data, unavailable as built-in feature. A new complex beam source, compatible with FLUKA public programming interface, was implemented in Fortran to model the specific properties of raster-scanning, i.e. delivery by means of multiple spot sources with variable spatial distributions, energies and numbers of particles. The source was plugged into the MC engine through the user hook system provided by FLUKA. Additionally, routines were provided to populate the beam source with treatment plan data, stored as DICOM RTPlan or TRiP98's RST format, enabling MC recomputation of clinical plans. Finally, facilities were integrated to read computerised tomography (CT) data into FLUKA. The tool was used to recompute two representative carbon ion treatment plans, a skull base and a prostate case, prepared with analytical dose calculation (TRiP98). Selected, clinically relevant issues influencing the dose distributions were investigated: (1) presence of positioning errors, (2) influence of fiducial markers and (3) variations in pencil beam width. Notable differences in modelling of these challenging situations were observed between the analytical and Monte Carlo results. In conclusion, a tool was developed, to support particle therapy research and treatment, when high precision MC calculations are required, e.g. in presence of severe density heterogeneities or in quality assurance procedures.

1. Introduction

Scanned beam particle therapy, the most advanced ion beam delivery technique, combining magnetic beam deflection with active energy variation, offers the highest degree of target dose conformation attainable in external beam radiotherapy. To fully and safely exploit such delivery precision, accurate dose computation is required.

However, analytical algorithms employed in the treatment planning systems (TPS), typically involve some simplifications, e.g. in modelling of density effects or handling of lateral scattering [1], resulting in dose prediction uncertainties which may affect clinically relevant situations [2]. Furthermore, analytical algorithms offer limited facilities to model delivery-time uncertainties, like beam profile distortions [3].

[2] Present address: Cyclotron Centre Bronowice, Institute of Nuclear Physics Polish Academy of Sciences, Krakow, Poland

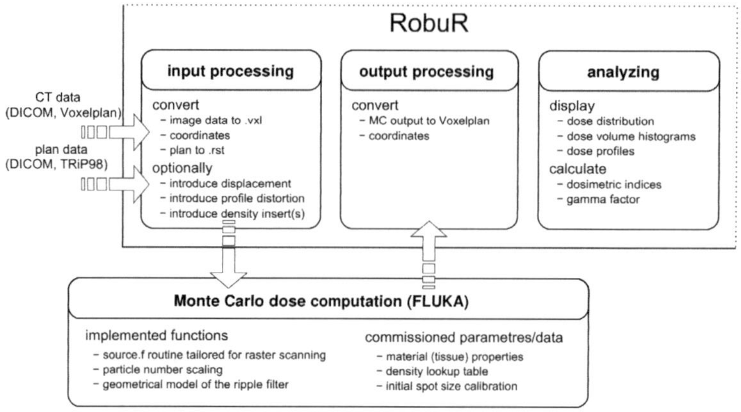

Figure 1. The workflow steps: data input from external sources, conversion and preprocessing, Monte Carlo dose computation with FLUKA, conversion and processing for display and analysis.

Monte Carlo (MC) methods have the potential to validate and complement analytical dose computations in complex situations, when high precision is required, supporting particle therapy scientific and clinical applications, as well as quality assurance procedures. Therefore, purpose of this work was to enable MC particle dose computation in patient geometries with raster-scanning and combine it, in a single platform, with flexible data input and preprocessing (to support a range of diverse computations) and comprehensive results display and analysis (e.g. to uniformly compare MC with analytical calculations in terms of DVHs or dosimetric indices).

2. Material and Methods

2.1. Monte Carlo tool

A general purpose Monte Carlo engine FLUKA [4], well-established in particle therapy applications [5, 6], was employed as core of our dose computation tool. For all calculations the default settings optimized for ion beam therapy applications (HADROTHErapy [7]) were used, which ensure the detailed transport of primary and secondary particles (e.g. 2% kinetic energy loss step, delta-ray production threshold 100 keV, particle transport threshold at 100 keV, except for neutrons at 1e-5 eV, multiple scattering threshold at minimum allowed energy).

A complex beam definition was implemented, using the Fortran programming language, to model raster-scanning, i.e. delivery by means of multiple sources with variable spatial distributions, energies and numbers of particles. The beam source was plugged into the MC engine through the user hook system provided by FLUKA (stub routine `source.f`). Additionally, for carbon ion dose calculations, a geometrical model of the ripple filter [8], an element used to broaden the pristine Bragg peaks in depth, was defined.

Finally, the beam properties were adjusted to match the base data (definitions and values) used in comparative analytical computations, e.g. initial spot sizes were associated with the corresponding values at isocentre in air, used as references by the TRiP98 TPS [9]. Also, a Hounsfield unit lookup table, compatible with the one used for TRiP98, was defined, adopting material composition according to Schneider *et al* [10] expanded to account for metallic implants similarly to Parodi *et al* [5].

2.2. Data processing

The dose computation tool was interfaced (figure 1) with our radiotherapy toolkit RobuR [11], which handles input/output and processing of data in various formats, including converting computerized tomography (CT) data into FLUKA's VXL format and populating the beam source with treatment plan data (DICOM RTPlan or TRiP98's RST format). Additional input

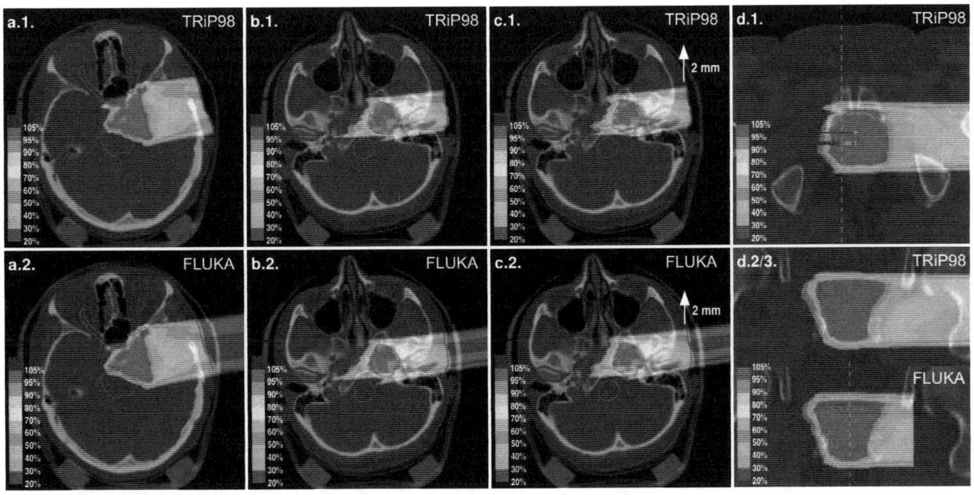

Figure 2. Dose distributions: a.1-2. and b.1-2. for skull base case (see also: figure 4), c.1-2. for a skull base case recomputed with a 2 mm set-up error (see also: figure 4), d.1. for prostate case with with two high density inserts introduced (see also: figure 5, a) and d.2/3. for prostate case with pencil beam distortions introduced (see also: figure 5, b).

data modification functionalities were provided, allowing to e.g. simulate (translational and rotational) positioning errors, introduce density inserts (by Hounsfield unit override in a specified contour or for selected pixels) and model beam profile distortions (e.g. width variations or profile asymmetry) through a user-controlled multiple-Gaussian parametrization, enabling accurate investigation of the resulting dosimetric effects. A direct comparison between MC and TRiP98 (or other systems adopting the DICOM format) dose distributions is possible through RobuR display and analysis capabilities, which were extended to directly support FLUKA data.

2.3. Comparison with analytical computations
The Monte Carlo tool was demonstrated on the recomputation of two representative carbon ion treatment plans (absorbed dose), a skull base and a prostate case, prepared with the TRiP98 TPS, employing analytical dose calculation models. Selected, clinically relevant issues influencing the dose distributions were chosen: (1) presence of positioning errors in skull base treatments (simulated by rigid CT displacement), (2) influence of high-density fiducial markers implanted in the prostate and (3) distortions of pencil beam profiles at extraction time.

3. Results
Figure 2 (a-c) shows dose distributions obtained for a skull base plan and demonstrates very good qualitative agreement between analytical and MC calculations in the homogeneous tissue regions (a), similarly to what was observed in water phantom calculations (figure 3), and slight discrepancies in the regions of heterogeneous tissue (b-c). The corresponding DVHs are presented in figure 4 showing differences in the order of 2%.

In figure 2 (d.1) the dose distribution is shown obtained through recalculation with the analytical TPS of the original prostate case plan, with two 1x1x3 mm^3 high-density inserts (obtained by Hounsfield unit override for selected pixels), simulating gold seeds. Due to limited modelling of the lateral scatter, nearly no dose is registered (distally) behind the inserts. The same case, recomputed with the MC tool, shows instead reduced, but substantial dose deposition

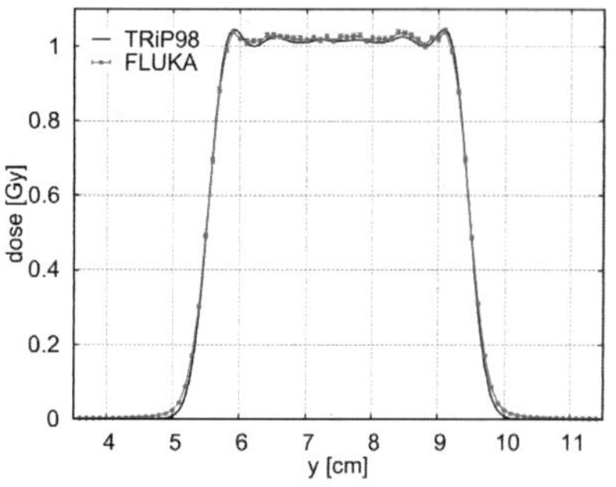

Figure 3. Dose profile of a 3.5x3.5x3.5 cm^3 target in a 7x7x7 cm^3 water phantom.

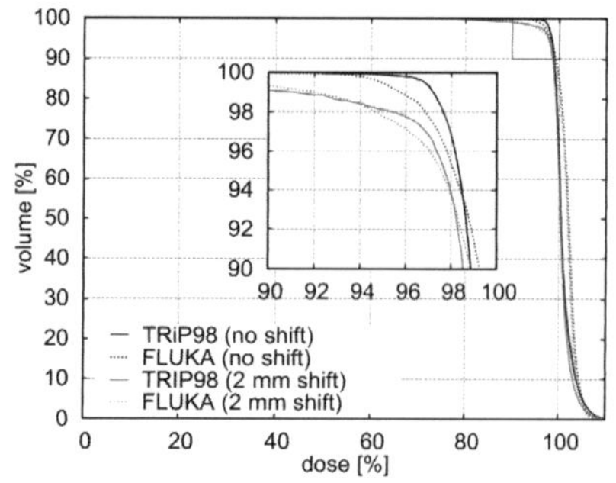

Figure 4. Dose volume histograms for the clinical target volume (CTV) for the skull base case.

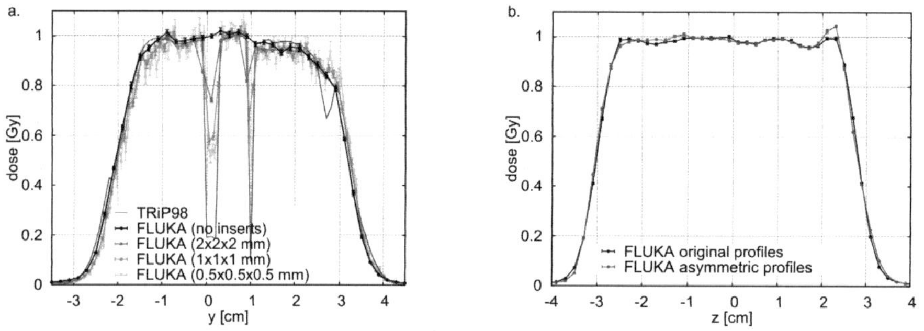

Figure 5. Dose profiles for the prostate case (indicated with dashed lines in figure 2, d): a. original plan and recomputation with high density inserts (for 3 scoring grids), b. recomputation with/without pencil beam distortions.

past the inserts, as visible in the dose profiles (figure 5, a). The MC computation results were scored at various dose grid resolutions (2x2x2, 1x1x1 and 0.5x0.5x0.5 mm^3) to illustrate partial volume effects.

In figure 2 (d.2/3.) the coronal slice of the original prostate case plan (obtained with the analytical TPS) is compared with a recomputation, performed with the MC tool, with horizontal pencil beam profile distortions introduced (impossible to model within the TPS). The resulting dose deterioration, occurring predominately at field edge, is visualized in the cross-profiles in figure 5 (b).

4. Discussion and Conclusions

A tool was developed to enable dose recomputation of clinical scanned beam particle plans in patient and phantom geometries, with the FLUKA Monte Carlo code, and uniform comparison of the results within a single platform, also in terms of clinical parameters (DVH, dosimetric indices). Our work stems from the practical need, for our upcoming ion therapy centre, of including MC in QA procedures and from the research interest in the limitations of analytical dose computation algorithms. Therefore, the current implementation supports only computation of absorbed dose. However coupling of FLUKA with a radiobiological model has been demonstrated [12] and represents one of the next priorities in the extension of our framework.

While the major features of the current implementation were demonstrated on relevant tests cases, where differences between analytical and MC results were observed, a quantitative comparison between dose calculations approaches is beyond the purpose of this study and is instead one of the planned future applications of the presented tool.

Other groups have worked on enabling MC recomputations of clinical treatment plans for QA and research purposes, including applications of FLUKA to carbon ion therapy. Notable is the work by Parodi *et al* [6], who investigated the suitability of FLUKA as clinical tool and applied it to relevant issues in proton and carbon ion therapy. Recently, integration of FLUKA within a planning software tool has been reported by Mairani *et al* [13]. The tight integration represents a common trait with the software platform we present in this manuscript, while the two projects appear to currently differ in focus. Where Mairani *et al* [13] privileged plan optimization and integration with a radiobiological model, creating a full-fledged clinical planning system, we have favoured flexibility in input data manipulation and output data analysis, to directly enable a broad range of simulations.

Finally, implicit in the design of our MC framework is the possibility to easily distribute it to other institutions, which could directly benefit from it at the sole cost of introducing their own facility-specific base data. It is in fact entirely written in standard Fortran (FLUKA interfaces) and C++, targeting GNU/Linux systems. Currently, its sole dependency is our radiotherapy software toolkit RobuR, which is scheduled for release under a permissive software licence [11].

Acknowledgments

This work has been supported by a Research Grant of the University Medical Center Giessen and Marburg (UKGM) 35/2010 MR and by the Anneliese Pohl Stiftung. Conference attendance was enabled by Verein zur Förderung der Strahlentherapie und Radioonkologie im Universitätsklinikum Marburg e.V.
The authors are grateful to Dr M. Krämer for enabling the TRiP98 computations and to Dr G. Iancu and Dr U. Weber for providing technical information on beam spot variations.

References

[1] Paganetti H 2012 *Phys. Med. Biol.* **57** R99–117
[2] Ammazzalorso F, Jelen U, Strassman G and Zink K 2010 *Strahlenther. Onkol.* **186** 50
[3] Chanrion M-A, Ammazzalorso F, Wittig A, Engenhart-Cabillic R and Jelen U 2013 *Phys. Med. Biol.* **58** 3979–93
[4] Battistoni G, Muraro S, Sala P R, Cerutti F, Ferrari A, Roesler S, Fasso A and Ranft J 2007 *AIP Conf. Proc.* **896** 31–49
[5] Parodi K, Paganetti H, Cascio E, Flanz J B, Bonab A A, Alpert N M, Lohmann K and Bortfeld T 2007 *Med. Phys.* **34** 419–35
[6] Parodi K, Mairani A, Brons S, Hasch B G, Sommerer F, Naumann J, Jäkel O, Haberer T and Debus J 2012 *Phys. Med. Biol.* **57** 3759–84
[7] FLUKA online manual, www.fluka.org (last accessed: 7-09-2013)
[8] Weber U and Kraft G 1999 *Phys. Med. Biol.* **44** 2765–75
[9] Krämer M, Jäkel O, Haberer T, Kraft G, Schardt D and Weber U 2000 *Phys. Med. Biol.* **45** 3299–317
[10] Schneider W, Bortfeld T and Schlegel W 2000 *Phys. Med. Biol.* **45** 459–78
[11] Ammazzalorso F, Chanrion M-A, Graef S and Jelen U 2013 *52nd Annual Conf. of the Particle Therapy Co-Operative Group* (Essen, Germany) P302
[12] Mairani A, Brons S, Cerutti F, Fasso A, Ferrari A, Krämer M, Parodi K, Scholz M and Sommerer F 2010 *Phys. Med. Biol.* **55** 4273–89
[13] Mairani A, Böhlen TT, Schiavi A, Tessonnier T, Molinelli S, Brons S, Battistoni G, Parodi K and Patera V 2013 *Phys. Med. Biol.* **58** 2471–90

Using narrow beam profiles to quantify focal spot size, for accurate Monte Carlo simulations of SRS/SRT systems

T Kairn[1, 2], S B Crowe[2], P H Charles[2] and J V Trapp[2]
[1] Premion Cancer Care, Brisbane, Australia
[2] Science and Engineering Faculty, Queensland University of Technology, Brisbane, Australia

E-mail: t.kairn@gmail.com

Abstract. This study investigates the variation of photon field penumbra shape with initial electron beam diameter, for very narrow beams. A Varian Millenium MLC (Varian Medical Systems, Palo Alto, USA) and a Brainlab m3 microMLC (Brainlab AB. Feldkirchen, Germany) were used, with one Varian iX linear accelerator, to produce fields that were (nominally) 0.20 cm across. Dose profiles for these fields were measured using radiochromic film and compared with the results of simulations completed using BEAMnrc and DOSXYZnrc, where the initial electron beam was set to FWHM = 0.02, 0.10, 0.12, 0.15, 0.20 and 0.50 cm. Increasing the electron-beam FWHM produced increasing occlusion of the photon source by the closely spaced collimator leaves and resulted in blurring of the simulated profile widths from 0.24 to 0.58 cm, for the MLC, from 0.11 to 0.40 cm, for the microMLC. Comparison with measurement data suggested that the electron spot size in the clinical linear accelerator was between FWHM = 0.10 and 0.15 cm, encompassing the result of our previous output-factor based work, which identified a FWHM of 0.12 cm. Investigation of narrow-beam penumbra variation has been found to be a useful procedure, with results varying noticeably with linear accelerator spot size and allowing FWHM estimates obtained using other methods to be verified.

1. Introduction

The sensitivity of output factors for small photon fields to the diameter of the linear accelerator's initial electron beam has received increasing attention over recent years, with several authors recognising that the characteristics of small fields can be used for fine-scale optimisation of electron spot parameters used in Monte Carlo simulations of photon beams [1, 2, 3].

Despite the high spatial resolution, near water-equivalence and relative energy independence demonstrated by radiochromic films [4, 5, 6], the use of radiochromic film measurements to aid in the optimisation of the initial electron beam parameters used in Monte Carlo simulations has been minimal [7]. Reported problems with radiochromic film uniformity [8, 9] have led to observed localised inaccuracies in dose measurements [2, 3]. However, it has been shown that the use of repeated measurements, in combination with the use of an appropriate handling, irradiation, scanning and analysis procedure, can produce reliable and informative results [10, 11].

This study aims to demonstrate that the accuracy of the electron focal spot diameter used in a Monte Carlo model of a linear accelerator operating with or without an external microMLC can be

[1] To whom any correspondence should be addressed.

broadly verified with reference to narrow-beam dose profiles obtained from radiochromic film measurements. This work is therefore expected to assist in the accurate Monte Carlo modelling of both stereotactic radiosurgery and radiotherapy (SRS/SRT) delivery systems and more conventional linear accelerator designs.

2. Method

A Varian Millenium MLC (Varian Medical Systems, Palo Alto, USA) and a Brainlab m3 microMLC (Brainlab AB. Feldkirchen, Germany) were used, with one Varian iX linear accelerator, to produce 6 MV photon fields that were (nominally) 0.20 cm across. Dose profiles for these fields were measured at the isocentre using 5 x 5 cm^2 squares of Gafchromic EBT2 radiochromic film (International Specialty Products, Wayne, USA) that were placed at 3 cm depth in Virtual Water (Standard Imaging, Middleton, USA). To obtain adequate film darkening to provide a measurable result, for the narrow beams used in this study, 500 MU irradiations were used. Each film measurement was repeated three times and the results were averaged. The handing, scanning and calibration of the EBT2 film was completed as previously described [9, 10, 11], with scans obtained before and after irradiation and results analysed using the red channel only. The only deviation from the published procedure was that the scanning resolution was increased from 72 DPI to 150 DPI (resulting in 0.015 cm pixel widths), in order to obtain suitably detailed results.

A series of Monte Carlo simulations were completed, using BEAMnrc and DOSXYZnrc [12, 13], with established, geometrically and dosimetrically accurate models of the Varian iX linear accelerator, Millennium MLC and Brainlab m3 microMLC [14, 15, 16]. The film and water-equivalent phantom were modelled as a simple volume of water with voxels 0.1 cm thick in the beam-axis direction and 0.01 x 0.01 cm^2 across in the lateral directions. For these narrow fields, BEAMnrc runs using 10^9 histories (with a directional bremsstrahlung splitting number of 1000) produced 10^6 particles at the exit plane of the linear accelerator. DOSYZnrc runs of 10^9 histories (with a photon splitting number of 10 and with phase-space particle recycling permitted as necessary) were required to achieve dose calculation results with a statistical precision within 0.5%.

To investigate the effects of source occlusion on the resulting narrow beam profiles, the width of the initial electron beam in the simulations was varied. The BEAMnrc simulations, for the MLC-collimated and microMLC-collinated systems, were completed using electron beam full widths at half maxima (FWHM) of 0.02, 0.10, 0.12, 0.15, 0.20 and 0.50 cm.

3. Results

Figures 1(a) to (d) and table 1 illustrate the results of varying the FWHM of the initial electron beam used in the simulations of the MLC-collimated beam (figures 1(a) and (b)) and the microMLC-collimated beam (figures 1(c) and (d)). Results normalised to the central axis dose (figures 1(a) and (c) show that increasing the FWHM from 0.02 to 0.50 cm resulted increasing photon source occlusion, leading to blurring of the simulated profile widths from 0.24 to 0.58 cm, for the MLC, from 0.11 to 0.40 cm, for the microMLC (also see table 1).

In figures 1(a) to (d), the profiles obtained using electron spot diameters from FWHM = 0.10 to FWHM = 0.15 all closely overlie each other and are all in close agreement with the profiles obtained from the radiochromic film measurements. Data listed in Table 1 confirm that the required electron spot size to be used in simulating these beams clearly falls into the range $0.10 \leq \text{FWHM} \leq 0.15$ cm. This result is consistent across collimation systems and applies in both the radial direction (pictured in figure 1) and the transverse direction (not pictured). This range encompasses the result of our previous diode-based examination of the beam produced by this linear accelerator, which identified an optimal FWHM of 0.12 cm [17].

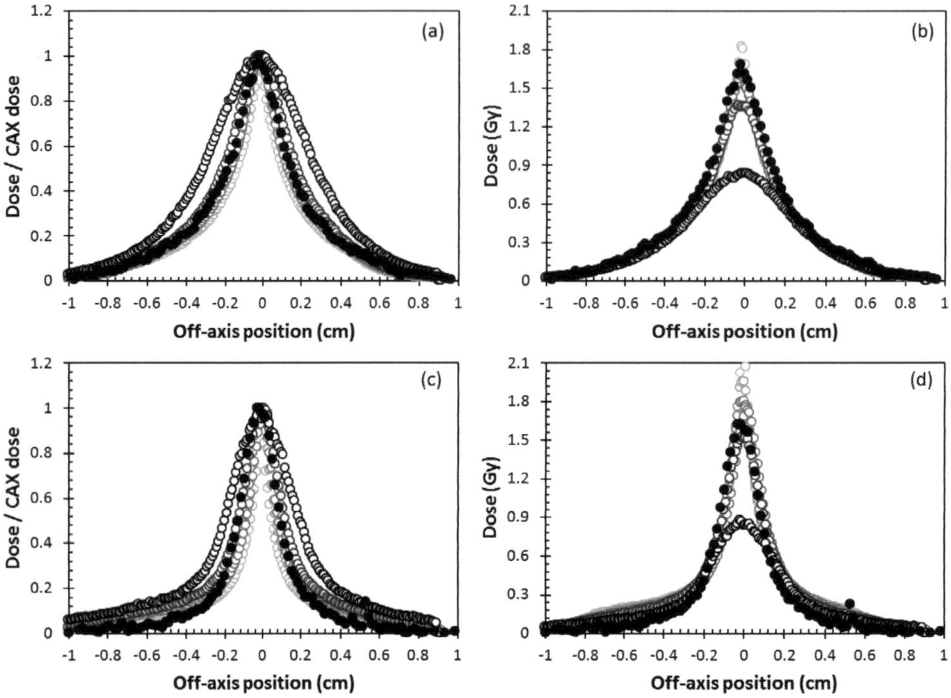

Figure 1. Dose profiles for the MLC-collimated fields (a) normalised to the central axis (CAX) dose and (b) measured and calculated in Gy, and for the microMLC-collimated fields (c) normalised to the CAX dose and (d) measured and calculated in Gy. Filled data points represent dose from film measurement. Open data points represent dose from Monte Carlo simulations using the different FWHM values listed in section 2. The shading of the data points representing the Monte Carlo results increases in saturation with FWHM, from light grey for FWHM = 0.02 cm to black for FWHM = 0.50 cm.

Table 1. Field widths, measured at the 50% isodose (see figure 1(a) and (c)), and central axis doses (see figure 1 (b) and (d)) for narrow fields collimated using MLC and microMLC, simulated using the listed FWHM values. Data shown in italicised bold-face type are from the EBT2 film measurements.

	MLC-collimated fields		microMLC-collimated fields	
FWHM (cm)	Field width (cm)	CAX dose (Gy)	Field width (cm)	CAX dose (Gy)
0.50	0.58 ± 0.01	0.84 ± 1.36	0.397 ± 0.004	0.88 ± 0.01
0.20	0.387 ± 0.004	1.36 ± 0.01	0.253 ± 0.003	1.52 ± 0.02
0.15	0.300 ± 0.003	1.61 ± 0.02	0.199 ± 0.002	1.75 ± 0.02
0.12	0.319 ± 0.003	1.56 ± 0.02	0.185 ± 0.002	1.81 ± 0.02
0.10	0.305 ± 0.003	1.60 ± 0.02	0.164 ± 0.002	1.96 ± 0.02
0.02	0.243 ± 0.002	1.82 ± 0.02	0.113 ± 0.001	2.44 ± 0.02
Film result:	***0.33 ± 0.02***	***1.65 ±0.05***	***0.23 ± 0.02***	***1.63 ± 0.05***

4. Conclusion

The use of radiochromic film to investigate narrow-beam profile variation with electron focal spot size has been found to be a useful procedure, with results varying sufficiently to allow identification of an optimal electron FWHM to within 0.05 cm. This result was consistent, whether a standard MLC or a microMLC was used to collimate the field, and indicates that this procedure could be used to provide independent verification of FWHM estimates obtained using diodes or other dosimeters.

Acknowledgements

This study was supported by the Australian Research Council, the Wesley Research Institute, Premion and the Queensland University of Technology (QUT), through linkage grant number LP110100401. Computational resources and services used in this work were provided by the HPC and Research Support Unit, QUT.

References

[1] Scott A D J, Nahum A E and Fenwick J D 2009 *Med. Phys.* **36**(7) 3132-3144
[2] Francescon P, Cora S and Cavedon C 2008 *Med. Phys.* **35**(2) 504-513
[3] Cranmer-Sargison G, Weston S, Evans J A, Sidhu N P and Thwaites D I (2011) *Med. Phys.* **38**(12) 6592-6602
[4] Butson M J, Yu P K N, Cheung T and Alnawaf H 2010 *Radiat. Meas.* **45** 836-839
[5] Devic S 2011 *Physica Medica* **27**(3) 122-134
[6] Kim J H, Hill R and Kuncic Z 2012 *Phys. Med. Biol.* **57**(14) N267-N278
[7] Scott A J D, Nahum A E and Fenwick J D 2008 *Med. Phys.* **35**(10) 4671-4684
[8] Richley L, John A C, Coomber H and Fletcher S 2010 *Phys. Med. Biol.* **55**(9) 2601-2617
[9] Aland T, Kairn T and Kenny J 2011 *Australas. Phys. Eng. Sci. Med.* **34**(2): 251-260
[10] Kairn T, Aland T and Kenny J 2010 *Phys. Med. Biol.* **55**(15) L37-L42
[11] Kairn T, Hardcastle N, Kenny J, Meldrum R, Tome W and Aland T 2011 *Australas. Phys. Eng. Sci. Med.* **34**(3) 333-343
[12] Rogers D W O, Faddegon B A, Ding G X, Ma C M, We J and Mackie, T R 1995 *Med. Phys.* **22** 503-522
[13] Kawrakow I and Walters B R B 2006 *Med. Phys.* **33** 3046-3056
[14] Kairn T, Kenny J, Crowe S B, Fielding A L, Franich R D, Johnston P N, Knight R, Langton C M, Schlect D and Trapp J V 2010 *Med. Phys.* **37**(4) 1761-1767
[15] Kairn T, Aland T, Franich R D, Johnston P N, Kakakhel M B, Kenny J, Knight R, Langton C M, Schlect D, Taylor M L and Trapp J V 2010 *Phys. Med. Biol.* **55**(17) N451-N463
[16] Charles P H, Crowe S B, Kairn T, Kenny J, Lehmann J, Lye J, Dunn L, Hill B, Knight R T, Langton C M and Trapp J V 2012 *Phys. Med. Biol.* **57** 6947-6960
[17] Charles P H, Crowe S B, Kairn T, Knight R T, Hill B, Kenny J, Langton C M and Trapp J V 2013 *Phys. Med. Biol.* **58** 4501- 4512

XVII International Conference on the Use of Computers in Radiation Therapy (ICCR 2013) IOP Publishing
Journal of Physics: Conference Series **489** (2014) 012015 doi:10.1088/1742-6596/489/1/012015

In vivo verification of radiation dose delivered to healthy tissue during radiotherapy for breast cancer

P Lonski*[1,2], ML Taylor[2], W Hackworth[1], A Phipps[3], RD Franich[2] and T Kron[1,2]

1. Physical Sciences, Peter MacCallum Cancer Centre, Melbourne, Australia
2. School of Applied Sciences, RMIT University, Melbourne, Australia
3. Radiation Therapy, Peter MacCallum Cancer Centre, East Melbourne, Australia

*corresponding author: email Peta.Lonski@petermac.org

Abstract. Different treatment planning system (TPS) algorithms calculate radiation dose in different ways. This work compares measurements made *in vivo* to the dose calculated at out-of-field locations using three different commercially available algorithms in the Eclipse treatment planning system. LiF: Mg, Cu, P thermoluminescent dosimeter (TLD) chips were placed with 1 cm build-up at six locations on the contralateral side of 5 patients undergoing radiotherapy for breast cancer. TLD readings were compared to calculations of Pencil Beam Convolution (PBC), Anisotropic Analytical Algorithm (AAA) and Acuros XB (XB). AAA predicted zero dose at points beyond 16 cm from the field edge. In the same region PBC returned an unrealistically constant result independent of distance and XB showed good agreement to measured data although consistently underestimated by ~0.1 % of the prescription dose. At points closer to the field edge XB was the superior algorithm, exhibiting agreement with TLD results to within 15 % of measured dose. Both AAA and PBC showed mixed agreement, with overall discrepancies considerably greater than XB. While XB is certainly the preferable algorithm, it should be noted that TPS algorithms in general are not designed to calculate dose at peripheral locations and calculation results in such regions should be treated with caution.

1. Introduction

The accuracy of treatment planning system (TPS) dose calculations is crucial to the overall integrity of radiotherapy treatment delivery. An established appreciation of the importance of dose calculation accuracy[1] has motivated the continual development of calculation techniques which, when used in conjunction with improved computing power, has led to increasingly sophisticated dose calculation methods. Various algorithms are commercially available for dose calculation in radiotherapy. Each takes a different approach to the radiation transport problem and as such will return a different result to other algorithms under identical calculation conditions. The accuracy of these algorithms is usually verified at regions within and near the treatment beam however is less well documented at peripheral regions.

At present, convolution superposition and collapsed cone algorithms are commonly used in clinical practice for complex scenarios since their accuracy within heterogeneous media is generally regarded as superior to pencil beam convolution[2,3,4,5,6].These algorithms have begun to replace pencil beam convolution in many centres. More recently a deterministic grid-based Boltzmann solver, Acuros XB, has been released for the Varian Eclipse treatment planning system (Varian Medical Systems, Palo Alto, CA) which has shown promising results in complicated calculation scenarios. Agreement between Acuros and Monte Carlo simulations within +/- 3% has been demonstrated within regions of rapidly changing media by various groups[7,8,9], where superposition algorithms showed errors in the order of 10%. TPS algorithms are not designed nor commissioned to calculate dose at peripheral regions well. With modern algorithms continually improving dose calculation accuracy in-field the accuracy out-of-field warrants investigation.

Dose calculations from the TPS are often the only means of estimating the radiation dose reaching out-of-field locations in routine radiotherapy. However, very little data is available on the performance of these algorithms in such regions. Furthermore, TPS commissioning usually only requires data up to a few centimetres beyond the treatment field, so dose calculations at more distant regions are not supported by measured data. This work compares dose calculations from different TPS algorithms at the contralateral breast of 5 patients who underwent photon beam radiotherapy for breast cancer. Point doses from PBC, AAA and Acuros XB are compared at identical measurement points. *In vivo* TLD data is used as a benchmark for assessing the accuracy of each algorithm.

2. Methods

High-sensitivity thermoluminescent dosimeter (TLD) chips (Harshaw 100H, LiF:Mg,Cu,P material) were placed on the skin surface of 5 patients who underwent tangential beam radiotherapy for breast cancer at the Peter MacCallum Cancer Centre. TLDs were annealed in a TLD oven prior to use. Individual TLD sensitivity factors were used to correct the TL signal from each chip following a measurement. Sensitivity factors were assigned based on the relative signal of each individual chip to the batch average following 3 calibration cycles. Several 'standard' TLD chips from the same batch were irradiated to a known dose in a 6 MV 10 x 10 cm^2 field at depth of maximum dose for each *in vivo* measurement. The TL signal from *in vivo* TLDs was compared with TLD standards allowing conversion to dose. Three TLD chips were used at each measurement point and placed in custom-made acrylic buildup domes (Figure 1). Six measurement points were used on each of the 5 patients: one in each cardinal direction to the contralateral breast, one at the mid axilla and one at the suprasternal notch.

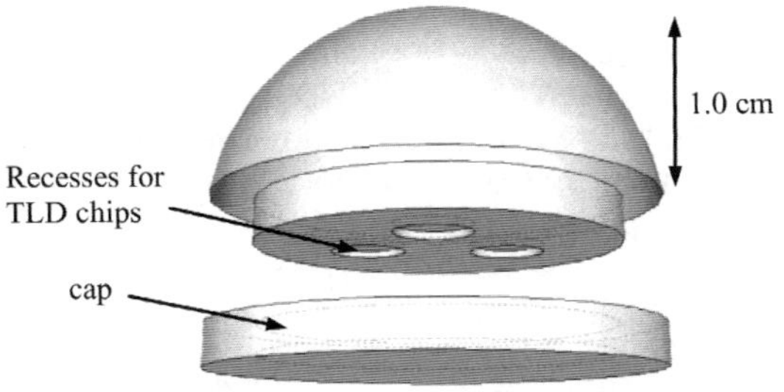

Figure 1 – Schematic depicting the buildup domes used for *in vivo* TLD measurements. Three TLD chips were housed in a single holder which was then taped to the patient's skin. The Perspex (ρ=1.18 g·cm^{-3}) domes offer up to 1.0 cm thickness of buildup material.

All patients were prescribed 50 Gy in 25 fractions; TLD measurements were conducted for a single 2 Gy fraction and results extrapolated to a complete 50 Gy treatment. The point dose as predicted by the Eclipse treatment planning system was compared to *in vivo* dose measurements at identical locations for each patient. Dose at each point was re-calculated in Eclipse using Anisotropic Analytical Algorithm (AAA), Pencil Beam Convolution (PBC) and Acuros XB (XB).

3. Results

The results of *in vivo* dose measurements are shown in Figure 2a along with the corresponding TPS dose calculation at each point, expressed as a function of distance from the field edge (cm). TLD error bars are indicative of a 10 % uncertainty associated with buildup dome positioning and patient setup errors (the uncertainties associated with relative dosimetry using TLDs are less, +/- 2 % on the 95 % confidence interval with careful calibration and handling[10]. The ratio of calculated (TPS) / measured (TLD) dose at the same locations for each algorithm is shown in Figure 2b.

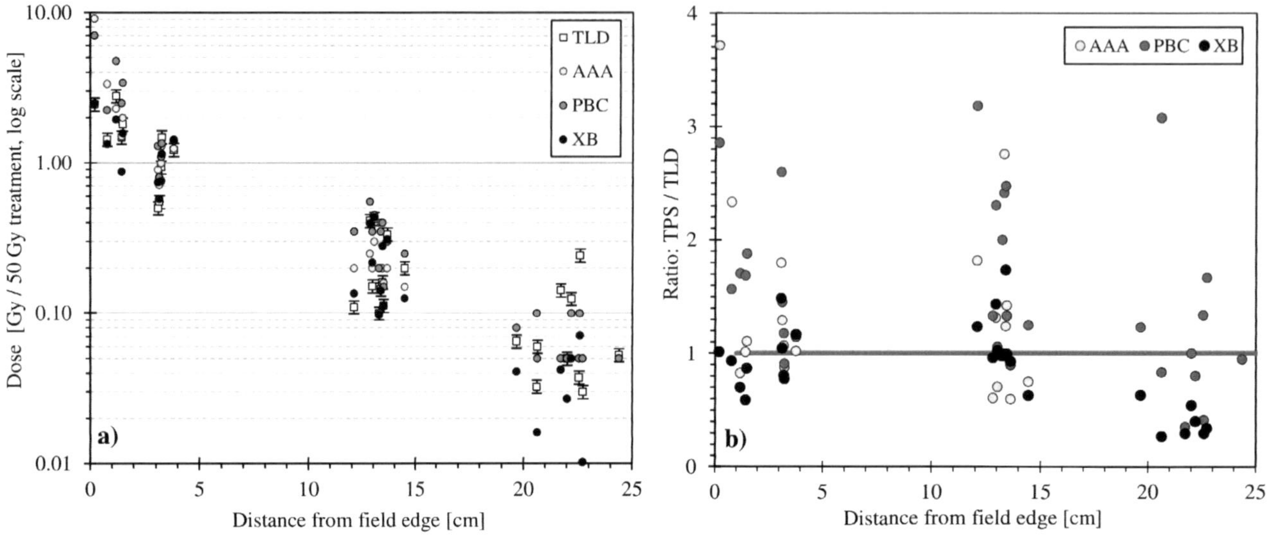

Figure 2 – a) Results of *in vivo* TLD point dose measurements comparing TPS calculations at the same locations for a 50 Gy tangential beam breast treatment, expressed as a function of distance from the field edge, **b)** ratios of TPS calculations to measured data at each measurement point as a function of distance from the field edge.

From Figure 2, measurements were conducted at several distances that can be grouped into 3 distinct regions having different levels of agreement with TPS calculations: 'near' (<5 cm from field edge), 'medium' (11 to 15 cm) and 'far' (>19 cm). The performance of each algorithm is assessed using the average ratio of calculated to measured dose along with its relative standard deviation, expressed as a percentage (%RSD). Each algorithm can be compared within each region. In the near region the average ratios (TPS/TLD) for each algorithm were 1.36 (%RSD = 58), 1.56 (%RSD = 48) and 0.85 (%RSD = 44) for AAA, PBC and XB, respectively. In the medium region average ratios were 1.5 (%RSD = 60), 1.7 (%RSD = 37) and 0.94 (%RSD = 27). In the far region AAA returned no dose result (for any point beyond ~ 15 cm). The average ratio in this region for PBC was 1.17 (%RSD = 67) and 0.4 (%RSD = 35) for XB. The PBC dose calculations within this region were a constant value irrespective of distance beyond 20 cm (either 0.1 % or 0.05% of the prescription dose).

4. Discussion

An exponential dose falloff with increasing distance from the radiation field is shown in Figure 2a. It can be seen that TPS calculations underestimate dose at more distant regions (beyond ~ 20 cm from the field edge), where in the same region TLD measurements hint at a dose 'bath' of ~0.1% of the prescription dose, although with some degree of variability.

Figure 2b shows the ratios of calculated (TPS) to measured (TLD) dose as a function of distance from the treatment field edge. These results demonstrate considerable variation between algorithms. Convolution superposition algorithms (i.e., AAA and PBC) locally scale pre-defined (Monte Carlo-derived) dose kernels to calculate radiation dose distributions. In the case of PBC, this expansion is conducted along 2 dimensions while AAA utilises additional degrees of freedom within heterogeneous regions. Acuros XB employs a fundamentally dissimilar approach in that it explicitly models particle interactions within different media to yield a numerical solution to the linear Boltzmann transport equation.

It was noted that although XB underestimated dose at larger distances from the field edge (> 20 cm) it was able to correctly predict the *behaviour* of the dose distribution, unlike AAA or PBC. The magnitude of this shortfall was on average 0.1 % of the prescription dose, presumably attributable to leakage radiation from the linac. Retrospectively correcting XB data for a 0.1 % leakage dose contribution at peripheral regions improves the XB to TLD ratio to 1.16 (%RSD = 23). This suggests that the use of Acuros XB in conjunction with a leakage dose correction may lend the ability to calculate peripheral dose to within +/- 15 % up to 24 cm beyond the field edge. For assessment of stochastic risks, such an uncertainty may be acceptable. More work is required to validate this correction-based approach.

5. Conclusions

Treatment planning system dose calculation algorithms are not designed to accurately predict peripheral dose, nor are data within these regions verified upon commissioning. This work has demonstrated that dose calculations should be used with caution as large errors have been shown in peripheral regions. Using the most sophisticated algorithms available in conjunction with an applied correction for leakage dose from the linac may improve dose calculations to achieve accuracy within +/- 15 %. Further validation work is required to accurately quantify leakage dose from different linac models.

6. References

[1] Papanikolaou N, Battista J J, Boyer A L, Kappas C, Klein E, Mackie T R, Sharpe M and Van Dyk J 2004 'Report of the Task Group No. 65 of the Radiation Therapy Committee of the American Association of Physicists in Medicine: Tissue inhomogeneity corrections for megavoltage photon beams' Medical Physics Publishing, Madison, WI

[2] Van Esch A, Tillikainen L, Pyykkonen J, Tenhunen M, Helminen H, Siljamäki S, J. Alakuijala J, Paiusco M, Iori M and Huyskens D 2006 'Testing of the analytical anisotropic algorithm for photon dose calculation, *Med Phys* **33** 11

[3] Fogliata A, Nicolini G, Vanetti E, Clivio A and Cozzi L 2006 'Dosimetric validation of the anisotropic analytical algorithm for photon dose calculation: fundamental characterization in water,' *Phys Med Biol.* **51** 1421-1438

[4] Ahnesjö A 1989 'Collapsed cone convolution of radiant energy for photon dose calculation in heterogeneous media,' *Med Phys* **16** 4130 577-592

[5] Ahnesjö A and Aspradakis M 1999 'Dose calculations for external photon beams in radiotherapy,' *Phys Med Biol* **44**

[6] Knöös T, Wieslander E, Cozzi L, Brink C, Fogliata A, Alberts D, Nyström H and Lassen S 2006 'Comparison of dose calculation algorithms for treatment planning in external photon beam therapy for clinical situations,' *Phys Med Biol.* **51** 5785-5807

[7] Han T, Mikell J K, Salehpour M and Mourtada F 2011 'Dosimetric comparison of Acuros XB deterministic radiation transport method with Monte Carlo and model-based convolution methods in heterogeneous media,' *Med Phys* **38**:5 2651-2663

[8] Vassilev O N, Wareing T A, McGhee J, Failla G, Salehpour M R and Mourtada F 2010 'Validation of a new grid-based Boltzmann equation solver for dose calculation in

radiotherapy with photon beams,' *Phys Med Biol.* **55** 581-598 doi:10.1088/0031-9155/55/3/002.

[9] Bush K, Gagne I M, Zavgorodni S, Ansbacher W and Beckham W 2011 'Dosimetric validation of Acuros XB with Monte Carlo methods for photon dose calculations,' *Med Phys* **38**:4 2208-2220

[10] Kron T 1994 'Thermoluminescence dosimetry and its applications in medicine – Part 1: physics, materials and equipment', *Australasian Physical and Engineering Sciences in Medicine* **17** 4 175 – 199

Fast in vivo volume dose reconstruction via Reference Dose Perturbation

Weiguo Lu*, Mingli Chen, Xiaohu Mo, Donald Parnell, Gustavo Olivera, and Daniel Galmarini

21st Century Oncology, USA

*wlu@rtsx.com

Abstract. Purpose: Accurate on-line reconstruction of in-vivo volume dose that accounts for both machine and patient discrepancy is not clinically available. We present a simple reference-dose-perturbation algorithm that reconstructs in-vivo volume dose fast and accurately. Methods: We modelled the volume dose as a function of the fluence map and density image. Machine (output variation, jaw/leaf position errors, etc.) and patient (setup error, weight loss, etc.) discrepancies between the plan and delivery were modelled as perturbation of the fluence map and density image, respectively. Delivered dose is modelled as perturbation of the reference dose due to change of the fluence map and density image. We used both simulated and clinical data to validate the algorithm. The planned dose was used as the reference. The reconstruction was perturbed from the reference and accounted for output-variations and the registered daily image. The reconstruction was compared with the ground truth via isodose lines and the Gamma Index. Results: For various plans and geometries, the volume doses were reconstructed in few seconds. The reconstruction generally matched well with the ground truth. For the 3%/3mm criteria, the Gamma pass rates were 98% for simulations and 95% for clinical data. The differences mainly appeared on the surface of the phantom/patient. Conclusions: A novel reference-dose-perturbation dose reconstruction model is presented. The model accounts for machine and patient discrepancy from planning. The algorithm is simple, fast, yet accurate, which makes online in-vivo 3D dose reconstruction clinically feasible.

1. Introduction

The goal of treatment planning is to find an optimal set of machine control parameters (CP), including the beam, gantry, couch, jaw, multi-leaf modulators (MLC), etc., so that the calculated dose distribution satisfies the prescription. Treatment planning requires accurate modeling of treatment components, such as the machine, beam and patient. The machine model includes physical and geometric properties of various LINAC components. The beam model includes curves extracted from the commissioning measurements. The patient model includes both patient setup information and the patient electron density map, which is typically converted from a CT image.

However, even with an optimal plan that involves much hard work, patients may not be treated as planned. There could be patient setup error or uncertainty caused by operators. Patient could experience inter-fraction anatomical changes, such as weight loss or tumor response, or exhibit intra-fraction motion. Additionally, machine behavior, such as beam output, gantry, couch, and/or MLC motion may vary fraction by fraction and different from what was planned.

The modern LINAC is equipped with many hardware/software components that can record in-vivo information during each treatment. For example, machine log files that record information from the sensors, treatment records that contain setup information, online or real-time volume imaging with megavoltage CT (MVCT) or cone beam CT (CBCT), and transit dosimetry using Electronic Portal Imaging Devices (EPID) or linear detectors. Those in-vivo signals provide invaluable information for treatment verification. However, they do not directly translate to dosimetric impact on the patient anatomy, which clinicians are most concerned.

The goal of in vivo dose reconstruction is to translate the delivery errors back to the patient anatomy with a clinically relevant accuracy. This level of accuracy was reported in [1] when using a well calibrated Image Value to Density Table (IVDT) that converts MVCT to a density map to calculate patient dose. The implementation needs to be fast and fully automatic so that it can be used routinely and meet requirements of time-sensitive applications, such as online verification or real-time adaptation. Therefore, we cannot afford to start from scratch as in the treatment planning. On the other hand, the rich information about the machine, beam and patient representations is already contained in the plan dose that is the result of hard work in treatment planning, and it would be a natural candidate as our starting point. Indeed, we are able to incorporate the plan dose as a reference base (reference dose) in a perturbation framework. In this work, we demonstrate a fast in vivo dose reconstruction by perturbing the reference dose with the in-vivo information of each treatment. Note that we assume the availability of machine data and anatomical change for the purpose of this paper; however, that assumption is yet advanced and it requires much effort to realize. Moreover, the dose reconstruction relies on the patient density information retrieved from on-line CT image, such as TomoTherapy MVCT or cone-beam CT (CBCT), which requires comprehensive Image Value to Density Table (IVDT) calibration. As reported from others [1] and verified by ourselves, dose calculation using TomoTherapy MVCT with well calibrated IVDT can fall within 1% accuracy.

2. Methods

The idea of reference dose perturbation (RDP) went back to the adaptive full dose correction as used in plan optimization iterations in the Non-Voxel-based Broad-Beam (NVBB) framework [2]. The reference dose is the base, and the perturbation is some variation, which accounts for fluence and density changes. In fact, the perturbation term is computed as the difference of the approximate dose for two different conditions: the treatment condition and the planned condition.

Let $D(f_0, R_0, \boldsymbol{x})$ denote the reference dose (plan dose) at location $\boldsymbol{x}$ for the planned fluence f_0 and the planning image R_0. When the fluence and density change to f and R, respectively, the dose incurs a change $\Delta = D(f, R, \boldsymbol{x}) - D(f_0, R_0, \boldsymbol{x})$. Rather than calculating the true difference Δ, which involves long computation of D, we calculate an approximation to the difference $\tilde{\Delta} = \tilde{D}(f, R, \boldsymbol{x}) - \tilde{D}(f_0, R_0, \boldsymbol{x})$ instead, where $\tilde{D}$ is an approximation to D and ideally, fast to calculate. It was proved that the perturbed dose $D(f_0, R_0, \boldsymbol{x}) + \tilde{\Delta}$ approximates the true dose $D(f, R, \boldsymbol{x})$ with improved accuracy than the approximation $\tilde{D}(f, R, \boldsymbol{x})$ does alone by an order of magnitude [2]. The work flow of the RDP algorithm is illustrated in Figure 1.

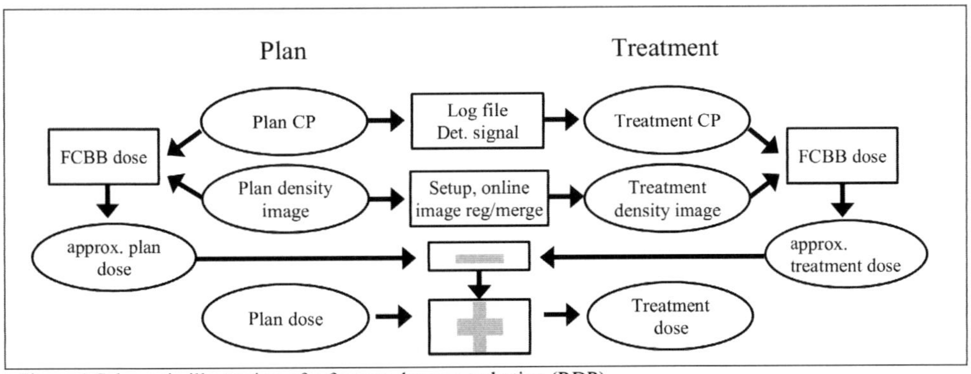

Figure 1 Schematic illustration of reference dose perturbation (RDP).

In principle, the reference plan dose is usually calculated by treatment planning system (TPS) with (time-consuming) model based algorithm, such as the Monte Carlo method or the Collapsed Cone Convolution/Superposition (CCCS) algorithm. Because of the small magnitude in fluence and density perturbation, the accuracy demand on the dose calculation algorithm for the perturbation term is lower than that on TPS dose calculation algorithms, but the speed demand is much higher. As an example of fast approximation, we describe below the Fluence-Convolution Broad Beam (FCBB) dose calculation. As the name suggests, the calculation has a filtering component and a ray-tracing component [3]:

$$\tilde{D}(f,R,\boldsymbol{x}) = \underbrace{(f \otimes k)}_{\text{filtered fluence map}} \cdot \underbrace{C(R(\boldsymbol{x}))}_{\text{attenuation}} \cdot \underbrace{(s_0/s)^2}_{\text{inverse square}} \tag{1}$$

Here $\tilde{D}(f,R,\boldsymbol{x})$ denotes the FCBB dose at a point $\boldsymbol{x}$ for the fluence f and radiological image R. s_0 and s are the source-to-axis distance and distance of $\boldsymbol{x}$ from the source, respectively. The fluence is filtered by a lateral kernel k, which is derived from an infinitesimal beam [3], and the term $C(R(\boldsymbol{x}))$ for attenuation is calculated based on the radiological distance via direct ray-tracing.

3. Results

In Figure 2 and Figure 3, we show the progressive outcome from plan dose, FCBB dose, to RDP, against treatment dose due to change of the fluence map. The plan fluence was a circular field of diameter 10 cm, but the treatment fluence had additional annular modulation (Figure 2) to simulate fluence perturbation. The radiation field was applied on slabs of different densities to simulate inhomogeneity. Then we calculated the treatment dose using FCBB and RDP and compared the results in isodose to the treatment dose [4] calculated by the CCCS method, which is regarded as the gold standard. Figure 3 (a) compares the plan dose with the treatment dose. Without correction, the discrepancy can be as large as 10%. Figure 3 (b) compares the treatment dose calculated by FCBB and by CCCS. The FCBB dose has apparent higher dose in the lung region, as the fluence convolution kernel does not have an accurate model of scattering for inhomogeneity. However, the RDP (Figure 3 (c)) approaches the gold standard within 1% when the change of delivery condition is small from the reference condition.

We retrospectively studied various TomoTherapy treatment cases. The CT, sinogram, plan dose are retrieved from the TomoTherapy archive. The ground truth fraction doses were re-calculated using the CCCS dose calculator as in TomoTherapy TPS, and took 0.5-2 minutes per fraction. The RDP doses were calculated using the presented algorithm, which took about 1-5 seconds per fraction for a 3D volume dose reconstruction.

Figure 4 shows a head and neck case with weight loss and machine output variation. Figure 4 (a) shows the registered online MVCT in green and plan kVCT in red. The red portion in the neck and shoulder region indicates the mismatch between MVCT and kVCT as a result of patient weight loss. Figure 4 (b) shows that there is a +/- 2% difference in plan sinogram and delivery sinogram due to output variation. Both the anatomical change and machine parameter change caused delivery discrepancy from the plan. Figure 4 (c) shows the difference of the treatment dose and plan dose with ~ 3% dose difference in most region and up to ~5% in the boundary region (skin and air). Using RDP to calculate the treatment dose (Figure 4 (d)), the difference from the ground truth generally falls within 1%, with exceptions in the air region, which is of little clinical significance. Figure 4 (e-f) shows the Gamma Index between the ground truth treatment dose and PDP treatment dose. Figure 4 (e) uses 1%/1mm criteria and Figure 4 (f) uses 3% / 3mm criteria. It is shown that for the most body region, the PDP treatment dose is well within 1% /1mm difference compared with the ground truth. When counting voxels with dose above 10% of the prescribed dose, the Gamma pass rates are above 90% for 1%/1mm criteria and above 95% for 3%/3mm criteria. The voxels with Gamma fail appear mainly outside the body region.

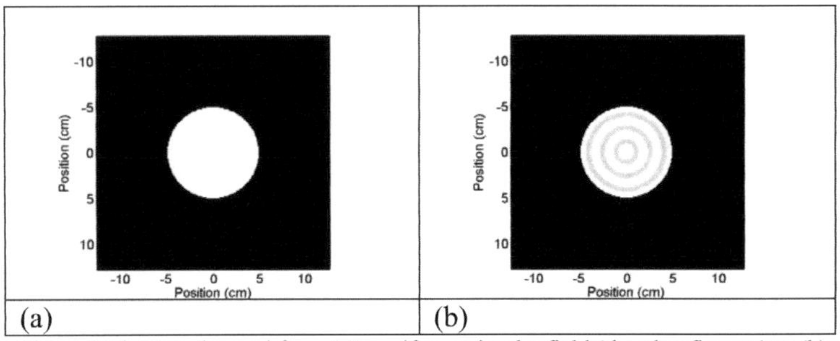

Figure 2 Fluence changed from (a) a uniform, circular field (the plan fluence) to (b) an annularly modulated field (the treatment fluence).

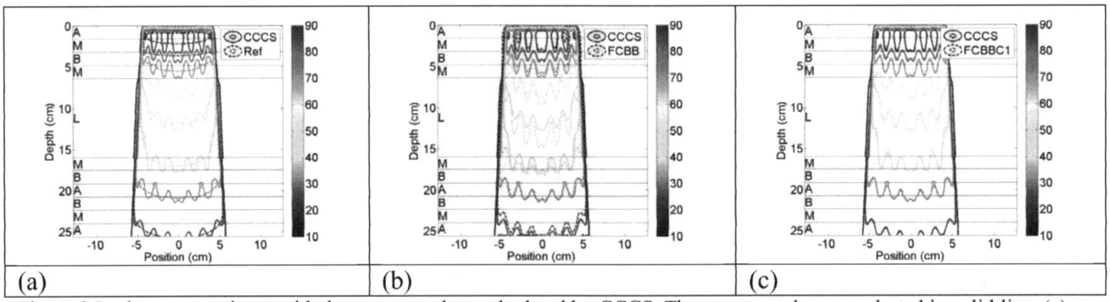

Figure 3 Isodose comparisons with the treatment dose calculated by CCCS. The treatment dose are plotted in solid line. (a) The plan dose, (b) FCBB, and (c) RDP are plotted in dotted line. The "A", "M", "B", and "L" indicate adipose, muscle, bone, and lung, respectively, for the simulated density of different slabs.

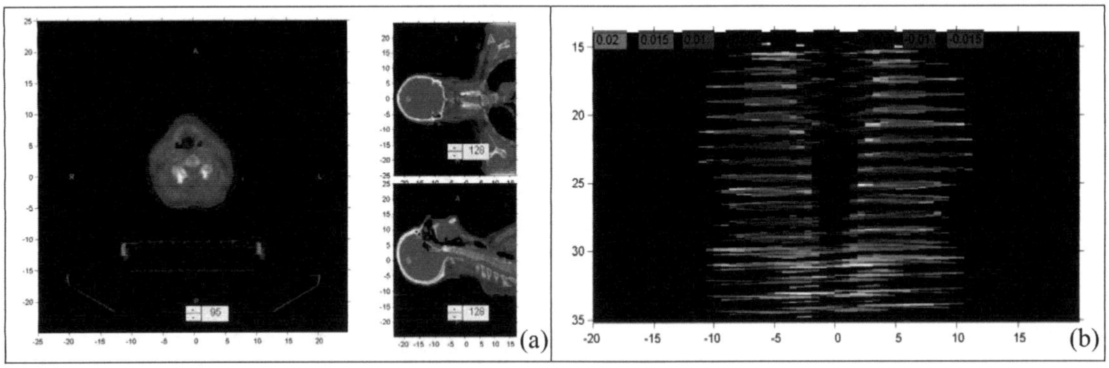

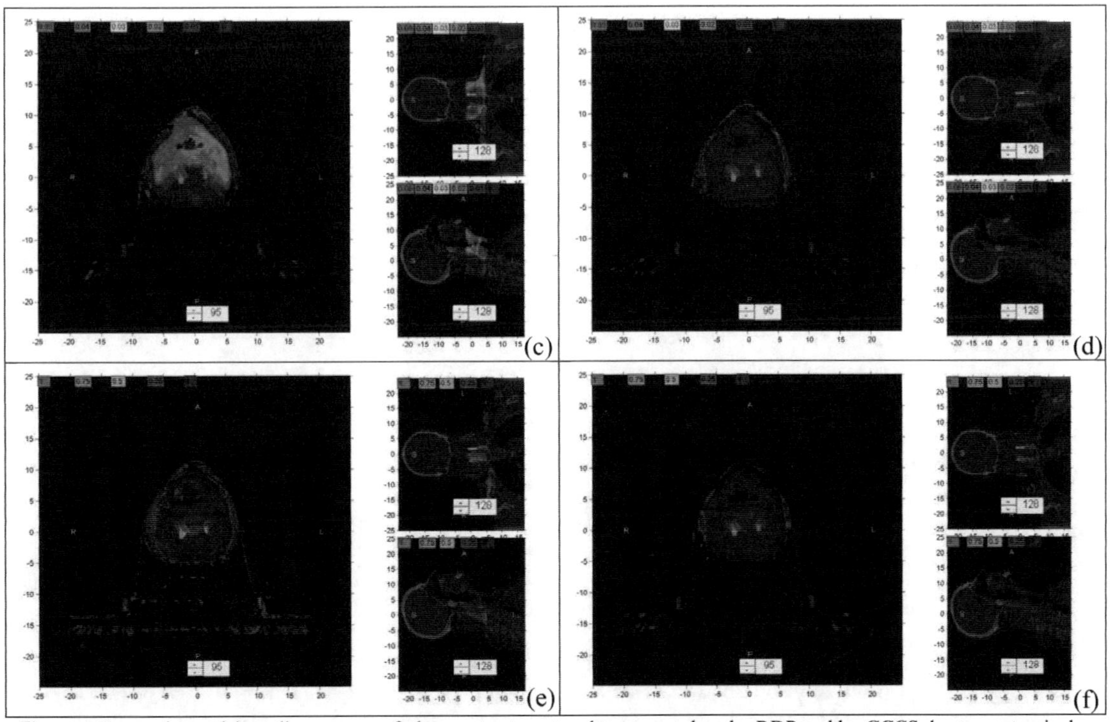

Figure 4 Comparison of dose discrepancy of plan vs. treatment and treatment dose by RDP and by CCCS due to anatomical and machine parameter changes. (a) Registered MVCT (in green) and KVCT (in red) for a head and neck patient indicate patient weight loss. (b) Difference of the delivery sinogram and plan sinogram overlaid on the plan sinogram. (c) Difference of the treatment dose and plan dose overlaid on the merged density image. (d) Difference of the treatment dose by RDP and by CCCS, overlapped on the merged density image. (e) Gamma index (with criteria 1% and 1mm) between ground truth and RPD treatment dose, overlapped on the merged density image. (f) Gamma index (with 3% and 3mm criteria) between the ground truth and PDP treatment dose, overlapped on the merged density image. In (a), (c), and (d), the left plots are the transverse view, the upper right plots are the coronal view, and the lower right plots are the sagittal view. The color boxes on the top of (b-f) indicate the level of deviation. In (b), the boxes vary from -1.5% to 2% at an increment of 0.5%. In (c) and (d), the boxes vary from 0% to 5% at an increment of 1% (absolute difference). In (e) and (f), the boxes vary from 0.25 to 1 at an increment of 0.25 (Gamma index).

4. Conclusions

We presented a fast algorithm for in vivo volume dose reconstruction via reference dose perturbation. It effectively models machine data and anatomical changes into in vivo dosimetry. The automatic and efficient dose reconstruction workflow, combined with an automatic flagging system, enables time-sensitive reviews and clinical decision making.

References

[1] Langen, K. M., Meeks, S. L., Poole, D. O., Wagner, T. H., Willoughby, T. R., Kupelian, P. A., Ruchala, K. J., Haimerl, J. and Olivera, G. H. The use of megavoltage ct (mvct) images for dose recomputations 2005 *Phys Med Biol* **50** 18 4259-76

[2] Lu, W. A non-voxel-based broad-beam (nvbb) framework for imrt treatment planning 2010 *Phys Med Biol* **55** 23 7175-210

[3] Lu, W. and Chen, M. Fluence-convolution broad-beam (fcbb) dose calculation 2010 *Phys Med Biol* **55** 23 7211-29

[4] Lu, W., Olivera, G. H., Chen, Q., Ruchala, K. J., Haimerl, J., Meeks, S. L., Langen, K. M. and Kupelian, P. A. Deformable registration of the planning image (kvct) and the daily images (mvct) for adaptive radiation therapy 2006 *Phys Med Biol* **51** 17 4357-74

Journal of Physics: Conference Series **489** (2014) 012017 doi:10.1088/1742-6596/489/1/012017

Open-Source Medical Devices (OSMD) Design of a Small Animal Radiotherapy System

S Prajapati[1,2], T R Mackie[1,2], and R Jeraj[2,3]
[1]Morgridge Institute for Research, Madison, WI, USA
[2]University of Wisconsin-Madison, WI, USA
[3]Jozef Stefan Institute, Ljubljana, Slovenia

E-mail: prajapati@wisc.edu

Abstract. Open-Source Medical Devices (OSMD) was initiated with the goal of facilitating medical research by developing medical technologies including both hardware and software on an open-source platform. Our first project was to develop an integrated imaging and radiotherapy device for small animals that includes computed tomography (CT), positron emission tomography (PET) and radiation therapy (RT) modalities for which technical specifications were defined in the first OSMD conference held in Madison, Wisconsin, USA in December 2011. This paper specifically focuses on the development of a small animal RT (micro-RT) system by designing a binary micro multileaf collimator (bmMLC) and a small animal treatment planning system (SATPS) to enable intensity modulated RT (IMRT). Both hardware and software projects are currently under development and their current progresses are described. After the development, both bmMLC and TPS will be validated and commissioned for a micro-RT system. Both hardware design and software development will be open-sourced after completion.

1. Introduction

Open-source has emerged as a popular concept in the world of technology, especially in software development. However, interest is also growing in hardware open-source initiatives. There is a need to design and build medical instruments that are inexpensive and affordable to research groups. The open-source model is also a playground for innovation. This leading principle directed to the initiation of the Open-Source Medical Devices (OSMD) concept at the Morgridge Institute for Research in cooperation with the University of Wisconsin. The main goal of OSMD is to collaborate globally to promote medical research by making medical instruments and technology available as an open-source resource to research groups around the world. Furthermore, the concept of OSMD is ideally suited to facilitate accelerated introduction of medical technology to developing countries and places with limited research resources. OSMD initiative aims to create a forum for free sharing and development of ideas and existing resources. Importance of the open-source model in the medical field has been discussed in Technology Quarterly of the Economist in June of 2012, where our work with the OSMD initiative was also featured along with other open-source medical device software projects [1].

It should also be noted that the open-source initiatives do not preclude protecting intellectual property with patents or copyrights as they can coexist. Valuable and novel ideas generated from the use of the open-source platform do not have to be themselves put into open-source but can be formally protected with a patent or copyright. Furthermore, the open-source model can be a way of ensuring that good ideas are preserved at the expiration of a patent. The open-source platform also becomes a test-bed for proving feasibility of intellectual property and making them ready for demonstration and licensing. Some novel ideas will not pass the hurdle of cost and effort of applying for a patent but may be still excellent and useful ideas worth memorializing in an open-source code or hardware design. Such a symbiotic relationship is demonstrated in figure 1.

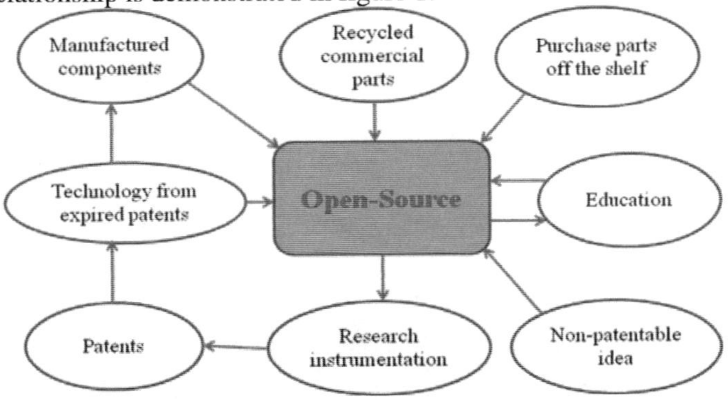

Figure 1. Significance of open source platform and its relation with patents and education.

While initiating OSMD, we have realized the importance of preclinical research in the field of medical research. For this reason, we chose to initiate OSMD with the development of a small animal imaging and therapy system (SAITS) that includes micro-CT, micro-PET and micro-RT modalities. This is a first of a kind integrated system that facilitates RT with both anatomical (CT) and functional (PET) imaging. The SAITS should be readily scalable to enable transition from preclinical/research medical device to clinical devices.

The design specifications of the SAITS were determined during the first OSMD Conference in December, 2011, which was held in Madison, Wisconsin, USA [2]. As the project is big and multifaceted, we started focussing on the micro-RT part of the SAITS. The first part of this project is to design a binary micro multileaf collimator (bmMLC) with the potential for use in IMRT applications. The second part of this project is to develop a small animal treatment planning system (SATPS) using kV x-rays to supplement the existing in-house MV TPS called WiscPlan [3,4]. Hence, the WiscPlan TPS will extend from the MV to kV energy range and will be available as open-source software.

2. Methods

The OSMD design should be simple and easily reproducible. Modular designs are encouraged. For each OSMD project, the technical specifications and preliminary design are discussed with interested collaborative groups before finalizing the design and building it. Several iterations of design might be necessary to address any potential problems. The final design needs to be validated before it can be available as open-source technology. Progress of these projects can be monitored on the OMSD website [2].

For this specific project, the small animal imaging and RT field was surveyed and customer requirements were discussed with potential system users. This helped to create an engineering requirements document with technical specifications for the system. The specifications were updated after the brainstorming discussion at the first OSMD Conference [2]. Solidworks™ software was used to design the 3D model of the system. To develop the SATPS, WiscPlan was used as a starting point. The convolution/superposition based TPS, WiscPlan, was developed throughout the last 10 years [3,4]. For the kV extension, kV energy deposition kernels (EDK) and kV x-ray fluence spectra need to be added to the current MV WiscPlan. The EDKs at kV energies will be created using the EGSnrc user-code, EDKnrc [5].

Journal of Physics: Conference Series **489** (2014) 012017 doi:10.1088/1742-6596/489/1/012017

3. Results

The schematic of SAITS system was designed in Solidworks™ drawing software that is shown in figure 2. The body of the system was designed as a slip ring gantry system for continuous rotation, facilitating micro-CT and micro-RT. With a slip ring gantry, both CT imaging and RT treatment have more freedom as well as better efficiency. The custom built gantry is 100 cm in diameter with 12 cm diameter bore which facilitates imaging and treating small animals up to the size of rats. A 250 kVp x-ray tube with a dual focal spot has been considered so that both RT treatment and CT imaging is possible with the same x-ray tube. The source to isocenter distance will be 30-40 cm, with the maximum imaging and treatment field size at the isocenter being 10 cm x 10 cm. For beam collimation, the bmMLC is designed to provide IMRT treatments with the treatment resolution of 1 mm x 1 mm at isocenter. Two separate designs of the bmMLC have been pursued.

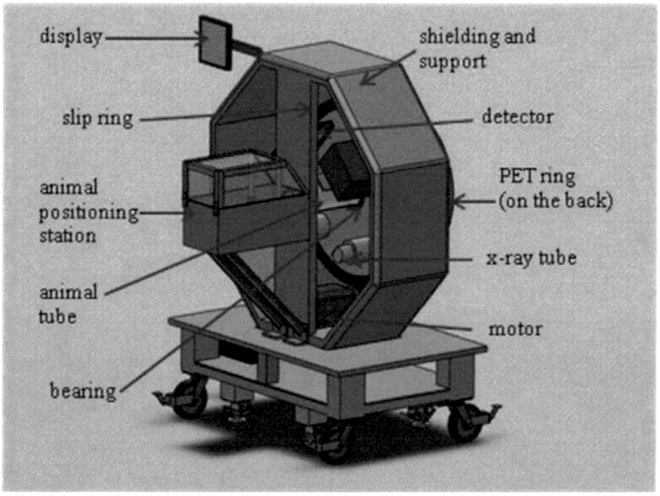

Figure 2. Schematic of the integrated micro-CT, micro-PET and micro-RT system.

3.1 The bmMLC Design 1

The bmMLC system was designed using Flexinol™ actuator wire that features muscle-like contraction when electrically driven. The design consisted of 0.5 mm thick brass collimator leaves, a customized collimator support structure, Flexinol™ and an open-source Arduino microcontroller. When electricity was passed through the Flexinol™, it contracted, pulling the assembly against the spring on the supporting tube to open the leaves. The bmMLC leaves come to a default closed position when power is turned off. The opposing collimator leaves were stacked in an interleaved position. Figure 3 illustrates the physical set up of the 20 (10+10) leaf binary collimation systems. The actuation system in the prototype worked, however, we moved on with a different design due to slow response of Flexinol™ wire and lack of sufficient force to overcome interleaf friction between the collimator leaves for full actuation.

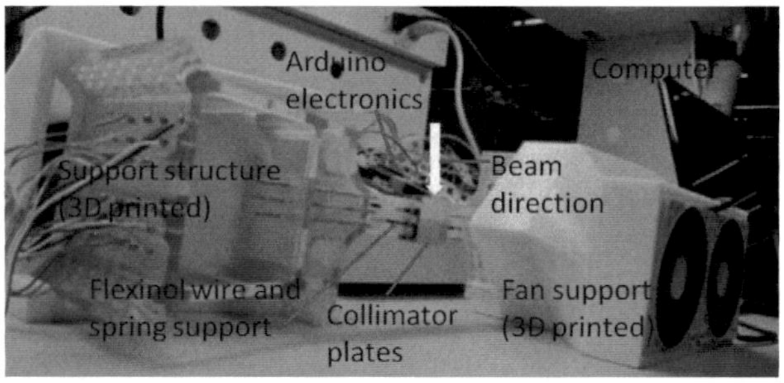

Figure 3. bmMLC prototype using Flexinol™ wire.

3.2 The bmMLC Design 2

This design implements servomotors for actuation. The rotation of a servomotor horizontally pulls and rotates the collimator plates via a pivot point that leads to opening and closing of the beam path. This design is more robust and has better reproducibility than the previous design. A similar actuation mechanism was proposed by Cadman et al. in 2006 for TomoTherapy™ beam delivery [6]. The 0.5 mm thick, custom-designed brass collimator-plate is connected to the motor via wire cable, which spools around the servomotor to open the collimator (physically 0.75 cm wide). The tension spring brings the plate to a closed position as the motor releases. The spring will also act as a fail-safe mechanism closing the beam when not powered. The collimator-plates are interleaved so that consecutive plates move in opposite directions. The schematic of the design was completed in Solidworks™ and it is shown in figure 4(a).

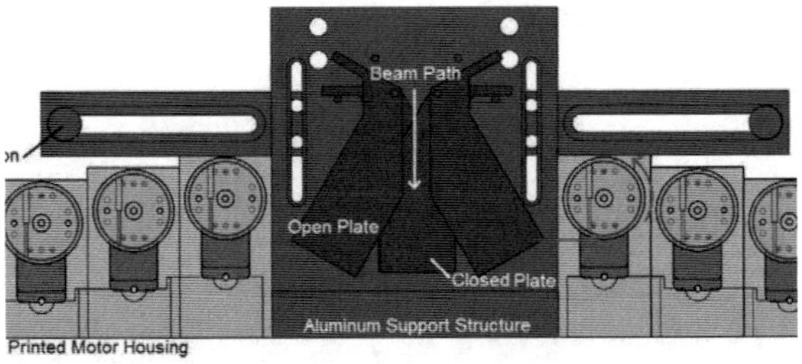

Figure 4(a). Solidworks™ model of the bmMLC design. The red arrows demonstrate the movement of the collimator plate and the motor. The yellow arrow points in the beam direction.

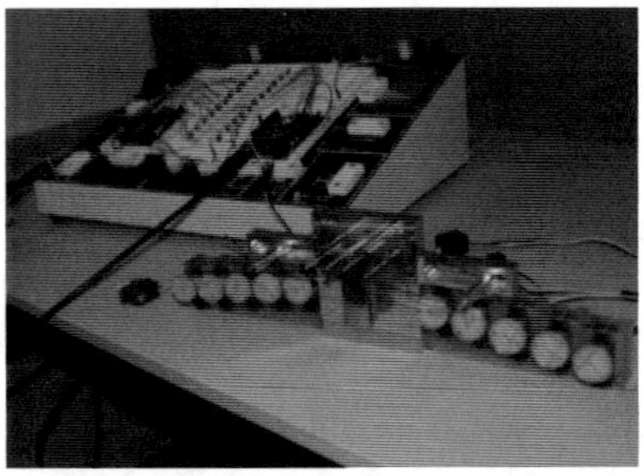

Figure 4(b). Setup for mechanical testing, actuation controlled by Arduino microcontroller.

As shown in figure 4(b) the new bmMLC prototype has been built and tested using ten interleaved collimator-plates which create a maximum field size of 1.5 x 2cm^2 at isocenter (with magnification = 2). Each micro servomotor has 1.5 kg·cm torque and rotates 170° in less than 0.5 seconds. The motor force was enough to overcome the leaf-to-leaf friction. The materials are easily available and cost less than \$300 (USD). Both bmMLC designs were created and assembled at the Morgridge Institute for Research utilizing the machine shop and 3D printing technologies for custom parts.

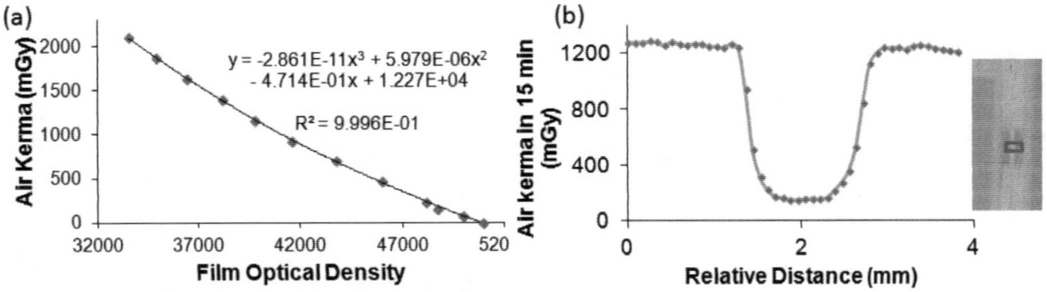

Figure 5. (a) Gafchromic™ EBT2 film calibration for 250 kVp x-ray (b) Horizontal air kerma profile (through blue box in the film picture) in the bmMLC via the film measurement for one leaf closed.

The dosimetric characteristics of the bmMLC (Design 2) were evaluated using a fully-characterized 250 kVp x-ray beam (HVL: 18.5 mm Al or 3.15 mm Cu) from UW-Accredited Dosimetry Calibration Lab (ADCL) with reference to the AAPM Report No. 72 [7]. The air kerma rate for the 250 kVp beam (with added filtration: 0.99 mm Al + 3.22 mm Cu) was 1.30 mGy/sec at 1 m. The bmMLC was set up directly in front of a Gafchromic™ EBT2 film for measurement. The film was placed 100 cm from the x-ray source where the field size was 10 x 10 cm^2. Primary collimators were not used in this set up. The Gafchromic™ EBT2 film calibration curve associates the optical density and air kerma to the film, which is shown in figure 5a. The films were scanned after 48 hours post-irradiation with the same orientation and face, and saved as 'TIFF' images at 300 dpi resolution (0.072 mm/pixel) using a professional grade flatbed scanner (Epson Expression 10000 XL). A third degree polynomial was a good fit for the calibration curve.

The calibration curve was used to measure the dosimetric parameters as follows: transmission: < 0.1%; leakage: 6.5 ± 0.1%; horizontal (80%-20%) penumbra: 2.6 ± 0.2 mm; vertical penumbra: 0.4 ± 0.1 mm. Figure 5b shows the horizontal profile through a single, closed collimator leaf, which depicts the blue box on the image of the EBT2 film. It shows that the dose at the closed collimator leaf is reduced to about 7% of the non-collimated area. The minimal amount of measured dose in the closed leaf is due to scatter through other leaves and the support structures. The dosimetric performance can be improved by further optimizing the design and set-up as shown in figure 6.

The bmMLC design has been updated to make the system more robust and reproducible. The motion mechanism is the same, but new custom-leaves (similar shape but bigger size to improve rotation) have been fabricated and placed in interleaved position with 1 cm physical beam opening. And, 20 interleaved collimator plates will provide 2 cm x 2 cm beam opening at isocenter. Custom-designed and 3D printed comb (to guide the leaves precisely) and brass-spacers (to space the leaves exactly 0.5 mm to avoid swaying away from the leaf path) have also been printed and machined. The revised design is illustrated in figure 6. When the design and assembly is complete, full mechanical and dosimetric tests will be performed on this updated collimator version.

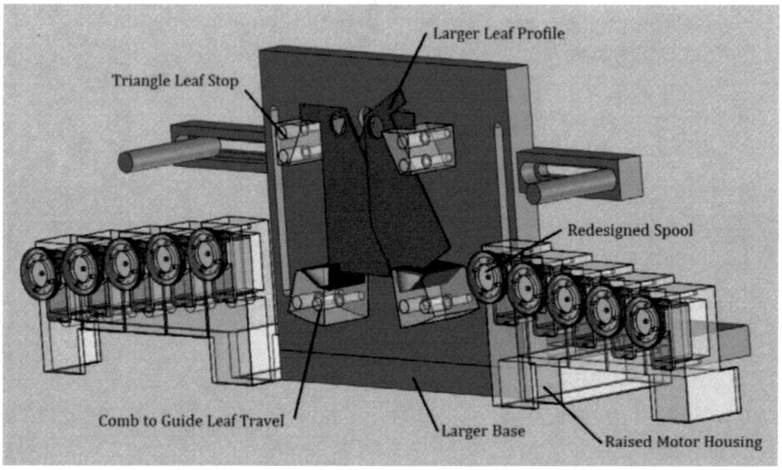

Figure 6. Solidworks™ design of revised mechanical design of bmMLC

3.3. SATPS software development

To date, the in-house TPS, WiscPlan can be used only for megavoltage external photon beams. The WiscPlan TPS workflow consists of four main steps. After running the WiscPlan TPS code, a MATLAB user interface opens up. The first step is to load the DICOM-RT geometry with the CT images and contours, and then select the target region of interest (ROI) and patient directory. Contouring on the CT image can be done in Pinnacle, Eclipse or other contouring software packages that are compatible with DICOM-RT images, where the structures are saved. The loaded geometry is updated and displayed in the MATLAB interface. The second

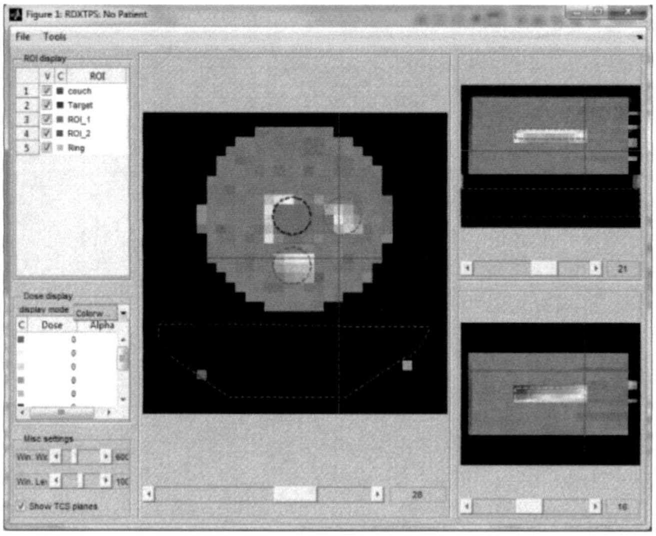

Figure 7. MATLAB interface in WiscPlan showing calculated dose distribution in a phantom

step is to perform beamlet dose calculation using an in-house MATLAB code that prompts a command window to run the dose calculation. After the beamlets are calculated, the WiscPlan optimization code is run in a command window that prompts the user to select the target and all ROIs. After the optimization window opens, dose parameters are adjusted if necessary. The optimization cycle can be paused and resumed until the user is satisfied with the optimization. The last step is to save the optimization results. In the MATLAB interface, the dose distribution can be displayed for the selected regions. This workflow is currently implemented for a 6 MV Tomotherapy^{TM} beam. A screenshot of WiscPlan's MATLAB interface showing the dose calculation distributions for a cylindrical test phantom using a 6 MV beam is shown in figure 7.

4. Discussion

The WiscPlan TPS needs to be updated so that it generates the planning files that will communicate with the bmMLC controller and the x-ray source. In addition, software to connect WiscPlan with the bmMLC and x-ray tube also needs to be developed. In order to expand WiscPlan for low energies, EDKs for 300 keV and lower energy are needed. Some work has been done to create EDKs at low energies in the past [5,8,9]. Those projects used EGS4 Monte Carlo Codes to obtain the EDKs where incident photons of a given energy are forced to interact at the centre of a 60 cm radius water sphere and the secondary particles are followed within that geometry. The sphere is divided into 24 radii shells and 48 cones [10]. We have received the EGS4 user-code EDKnrc from NRC-Canada [5]. We intend to follow their code to create additional low energy kernels from 10 keV to 300 keV. Once the EDKs are obtained, the kernel file in WiscPlan will be updated along with other necessary changes like fluence spectra. We received fluence spectra for the well-characterized x-ray tube at the UW-ADCL for 250 kVp energy. This 250 kVp beam will be utilized to initially check and validate the WiscPlan TPS at low energies.

In addition to dose calculation and planning for small animals, the WiscPlan extension at low energies can also be applicable to calculate dose for imaging applications in the clinic, such as radiography, fluoroscopy, mammography and CT imaging.

5. Conclusion

The OSMD concept has been initiated with the design of a small animal imaging and therapy system. Currently, the micro-RT system is designed on a slip-ring gantry. Work on both hardware and software part of this project has started. A bmMLC has been designed and tested. In addition, a MV to kV treatment planning system is also being developed within the WiscPlan TPS. After completion, both hardware and software aspects of the project will be validated, and will be available as open-source resources at the OSMD website.

References

[1] Harris M 2012 Open-source medical devices: *The Economist Technology Quarterly Q2 2012, June 2, 2012*
[2] OSMD Wisconsin Institutes for Discovery *http://discovery.wisc.edu/osmd*
[3] Flynn R 2007 *PhD Thesis* University of Wisconsin - Madison
[4] Westerly, David C. 2009 *PhD Thesis* University of Wisconsin - Madison
[5] Mainegra-Hing E, Rogers D W O and Kawrakow I 2005 *Med. phys.* **32** 685–99
[6] Cadman P, Mackie T and Reckwerdt P 2006 *Med. Phys.***33** 2124
[7] Boyer A, Biggs P, Galvin J, Klein E, LoSasso T, Low D, Mah K and Yu C 2001 *AAPM Report No. 72* Med Phys Publishing, Madison.
[8] Modrick J M 2000 *PhD Thesis* University of Wisconsin - Madison
[9] Alaei P 2001 *Med. Phys.***28** 403
[10] Mackie T R, Bielajew A F, Rogers D W O and Battista J J 1988 *Phys. Med. Biol.***33** 1–20

Journal of Physics: Conference Series **489** (2014) 012018 doi:10.1088/1742-6596/489/1/012018

Revealing the underlying mechanism of microbeam radiation therapy with low energy Monte Carlo simulations

A L McNamara[1], U Oelfke[2] and Z Kuncic[1]

[1] Institute of Medical Physics, School of Physics, University of Sydney, NSW, Australia
[2] Department of Medical Physics Radiation Oncology, DKFZ Heidelberg, University of Heidelberg, Germany

E-mail: aimee.mcnamara@sydney.edu.au

Abstract. Microbeam radiation therapy (MRT) is a new experimental oncological modality, intended for the treatment of inoperable brain tumours, particularly in difficult cases where conventional radiation therapy can cause irreversible damage. MRT consists of an array of highly collimated, quasi-parallel x-ray microbeams aimed at the tumour tissue, delivering high dose within the beam path and low doses in regions between the beams. For reasons still not fully understood, healthy tissue exposed to the microbeam array is able to regenerate while tumour volumes are significantly reduced. Low energy Monte Carlo radiative transport simulations provide new insight into understanding the underlying mechanisms of MRT. In particular, predicting the ionisation cluster distribution, which is a significant cause of lethal damage to cells, would provide insight into the biological responses. Geant4-DNA was used to model an x-ray microbeam of width $20\,\mu$m in liquid water. Secondary electrons, predominately responsible for ionisation clustering, were tracked to predict damage to cells within and adjacent to the beams. We find that higher energy beams (100 keV) produce less secondary electrons in the regions outside the beam than low energy beams (30-50 keV).

1. Introduction

A new innovative radiation modality called Microbeam Radiation Therapy (MRT) could potentially revolutionise brain malignancy radiotherapy in paediatric cases. In MRT nominally supralethal doses ($>$ 100 Gy/s) are delivered to the patient in micron-sized beams, based on the observation that healthy tissue is exceptionally resistant to high dose in small volumes, while cancerous tissue is not [1, 2, 3]. Despite promising results in tumour-bearing animals, MRT is yet to be applied to human oncological treatment since the underlying biological mechanism for preferential tumour damage is not known.

MRT consists of an array of intense, parallel synchrotron x-ray beams, each with a typical width of $20 - 50\,\mu$m and with a $200 - 400\,\mu$m centre-to-centre distance [1, 2]. The microbeam array is aimed at the tumour volume, delivering high radiation dose in the beam paths (peaks) and low dose in the regions in between the beams (valleys). The success of the treatment in animal studies has been shown to be highly dependent on the microbeam width and beam-to-beam spacing as well as the peak-to-valley dose ratio [4, 5, 6, 7].

Computational modelling of ionising particles down to very low energies ($\sim$ few eV), where biological damage predominantly occurs, has advanced our understanding of the interaction of ionising radiation with living cells [8, 9]. To date, both *in silico* and *in vivo* studies have mostly focused on the effect of radiation on the DNA molecule because of its central role to continuing cell viability [10, 11]. Although DNA strand breaks as a direct result of ionising radiation can be used to measure the efficacy of damage, it is well known that *indirect* damage caused by free radicals produced in the immediate surroundings of DNA is considerably more prevalent than direct radiation in causing damage to the DNA molecule itself [12]. Furthermore, there is mounting evidence to suggest that free radical production and damage in other regions of the cell (e.g. cytoplasm, mitochondria) are also important for understanding inexplicit biological responses e.g. bystander non-targeted effects [13, 14, 15]. Indeed, non-DNA focused models [16, 17] could be particularly important for understanding the success of MRT.

In this investigation, the distribution of secondary electrons is investigated in the peak and valley regions for MRT. The low-energy secondary electrons, produced through the interaction of ionising radiation with matter, play a key role in the production of biological damaging ionising events, particularly electrons with energies less than $\sim$ 1 keV. Predicting the number of ionisation and excitation events occurring within molecular sized volumes can infer information about the biological effectiveness of the primary ionising particle. Monte Carlo simulations have shown that ionisation clustering can be correlated with the biological effectiveness of radiation on the nanometre scale [8] and may be a more pertinent quantity than the absorbed dose, currently used to determine radiation damage.

2. Monte Carlo Simulation

An x-ray microbeam of width $20 \times 20\,\mu\text{m}^2$ incident on a liquid water cube of dimension $40\times40\times40\,\mu\text{m}^3$ was simulated using the Monte Carlo simulation toolkit Geant4 (version 9.6.p01). X-rays were modelled using the Low Energy Electromagnetic Physics models, including the following processes: Compton scatter, Rayleigh scatter, pair production and the photoelectric effect. Secondary electrons were modelled with the Geant4-DNA physics models [18, 19] including the following processes: elastic scattering, electronic excitation, ionisation, vibrational excitation and attachment. Monoenergetic microbeams with 30, 50 and 100 keV energies were investigated and in each case 10^6 incident photons were simulated.

3. Results and Discussion

The left panel in figures 1, 2 and 3 show the 3D electron track structure for 30, 50 and 100 keV x-ray microbeams of width $20\,\mu$m centred at 0 on the yz plane, respectively. The right panels show the total amount of elastic scattering (process number 11 on x-axis), excitation (12), ionisation (13) and excitation vibrational (15) processes occurring in the entire phantom volume. In each case, the number of excitation and ionisation process counts are significantly high, however the distribution of the secondary electrons differ for each energy case. Ionisation and excitation processes are primarily caused by low energy secondary electrons and are predominately responsible for inducing direct and indirect (free radical) biological damage. In the 30 and 50 keV cases (fig. 1 and 2), a large amount of electrons are found both in the peak and valley regions, indicating damage may occur in both regions. In the 100 keV case however, the electrons are mostly confined to the peak region. Animal studies have indicated that the success of MRT is somewhat dependent on confining cell death to the peak regions [1], if this is the case, our results indicate that high energy beams would be most suitable for successful MRT.

These preliminary results are limited by the size of the phantom volume. The high energy case is more indicative of an entrance dose, while the lower energy beams would represent the beam at a greater depth within a realistic phantom, most likely closer to a deep-seated tumour volume. In a future study, a phantom of realistic dimension, corresponding to a human head, will be modelled.

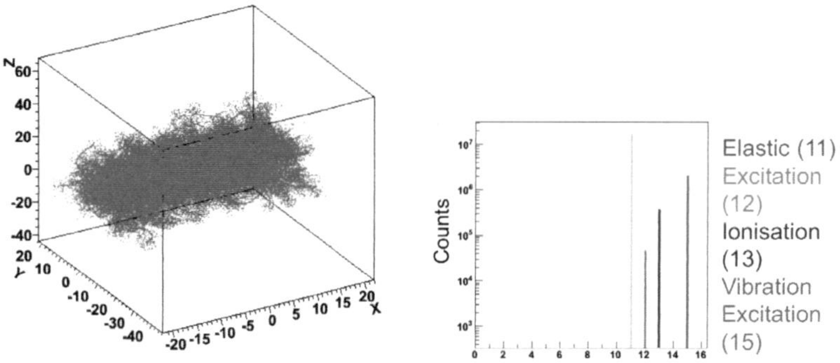

Figure 1. The left panel shows the 3D electron track structure for a 30 keV monoenergetic X-ray microbeam of width $20\,\mu$m incident at the centre of the yz plane of the $40 \times 40 \times 40\,\mu$m liquid water phantom, centred at the origin. The right panel shows the total number of each physical process (elastic scatter - green (11), excitation - orange (12), ionisation - blue (13) and vibration excitation - red (15)) occurring within the entire phantom volume. 10^6 incident photons was simulated.

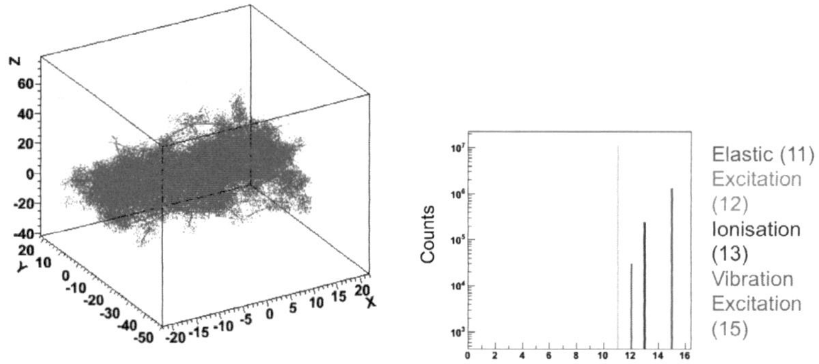

Figure 2. The left panel shows the 3D electron track structure for a 50 keV monoenergetic X-ray microbeam of width $20\,\mu$m incident at the centre of the yz plane of the $40 \times 40 \times 40\,\mu$m liquid water phantom, centred at the origin. The right panel shows the total number of each physical process (elastic scatter - green (11), excitation - orange (12), ionisation - blue (13) and vibration excitation - red (15)) occurring within the entire phantom volume. 10^6 incident photons was simulated.

4. Conclusion

The success of MRT in eliminating cancerous tissue, while preserving surrounding healthy tissue has been experimentally shown to be dependent on the beam width, spacing between the beams and the peak-to-valley dose ratio. The underlying biological mechanism for the success of the treatment is however still unknown. Investigating the effect of ionising radiation on a nanometer/molecular level could provide more insight into MRT.

Three different energy x-ray microbeams, incident on a $40 \times 40 \times 40\,\mu$m phantom, were investigated: 30, 50 and 100 keV. Large numbers of ionisation and excitation processes, mostly responsible for biological damage, occurred in each case. However the distribution of the secondary electrons responsible for these processes, differed with incident photon microbeam energy. Electrons in the 100 keV case were mostly confined to the beam (peak) region while

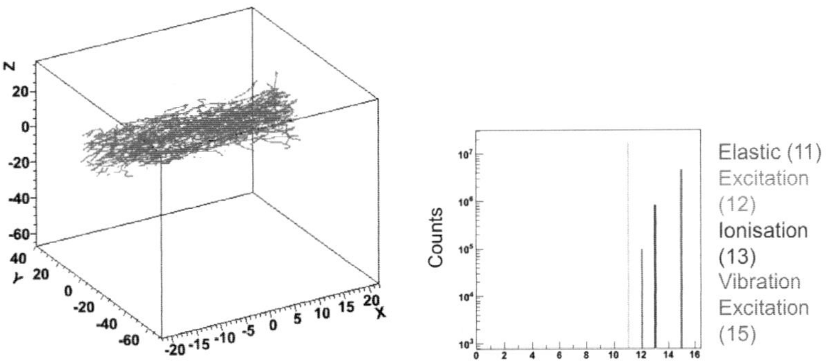

Figure 3. The left panel shows the 3D electron track structure for a 100 keV monoenergetic X-ray microbeam of width 20 μm incident at the centre of the yz plane of the $40 \times 40 \times 40\,\mu$m liquid water phantom, centred at the origin. The right panel shows the total number of each physical process (elastic scatter - green (11), excitation - orange (12), ionisation - blue (13) and vibration excitation - red (15)) occurring within the entire phantom volume. 10^6 incident photons was simulated.

in the 30 and 50 keV cases, large numbers of electrons were found in both the peak and valley regions. Results in animals studies have indicated that the therapy is most successful in cases where cells are killed only in the peak regions. From these preliminary results, higher energy beams will result in cell death in the peak regions and possibly less biological damage in the valley regions. In a future investigation, a realistic phantom will be modelled to further investigate the spread of electrons in a larger volume.

4.1. Acknowledgments
The authors acknowledge the Cancer Institute of NSW for funding the Advanced Computing Facility for Cancer Research which was used in this study.

5. References
[1] Laissue J A *et al* 1999 *SPIE Proc. Medical Applications of Penetrating Radiation* **3770** 38
[2] Slatkin DN, Spanne P, Dilmanian FA and Sandborg M 1992 *Med. Phys.* **19** 1395
[3] Crosbie J C *et al* 2010 *IJROBP* **77** 886
[4] Dilmanian F A *et al* 2002 *Neuro. Oncol.* **4** 26
[5] Laissue H A *et al* 2001 *Proc SPIE Penetrating Radiation Systems and applications II* **4508** 65
[6] Orion I *et al* 2000 *Phys. Med Biol.* **45** 2497
[7] Martínez-Rovira I *et al* 2010 *Phys. Med. Biol.* 55
[8] Friedland W, Jacob P, Paretzke H G and Stork T 1998 *Rad. Res.* **150** 170
[9] Nikjoo H and Goodhead DT 1991 *Phys. Med. Biol.* **36** 229
[10] Brenner D J and Ward J F 1992 *Int. J. Rad. Biol.* **61** 737
[11] Nikjoo H, O'Neill P, Terrissol M and Goodhead DT 1999 *Radiat Environ Biophys* **38** 31
[12] Zeman EM 2007 *Biological basis of radiation oncology, Clinical Radiation Oncology 2nd ed* eds LL Gunderson, JE Tepper (Philadelphia, PA: Churchill Livingstron) chapter 1
[13] Prise K M, Schettino G, Folkard M and Held K D, 2005 *Lanc. Oncol.* **6** 520
[14] Morgen W F 2003 *Rad. Res.* **159** 567
[15] Kam W W Y *et al* 2013 *Mitochondrion* in press
[16] Byrne HL *et al Phys. Med. Biol.* **58** 1251
[17] Kuncic Z *et al* 2012 *Comp. Math. Meth. Med.* 147252
[18] Incerti S *et al* 2010 *Med. Phys.* 37 4692
[19] Karamitros M, Incerti S and Champion C 2012 *Rad. Onc.* **102** S191

Journal of Physics: Conference Series **489** (2014) 012019 doi:10.1088/1742-6596/489/1/012019

An Improved Method of Heterogeneity Compensation for the Convolution / Superposition Algorithm

Robert Jacques[1], Todd McNutt[2]

1) Elekta / IMPAC Medical Systems, Inc., Maryland Heights, Missouri, USA

2) Department of Radiation Oncology and Molecular Radiation Sciences, Johns Hopkins University, Baltimore, Maryland, USA

Electronic mail: Robert.Jacques@elekta.com

Abstract.

Purpose: To improve the accuracy of convolution/superposition (C/S) in heterogeneous material by developing a new algorithm: heterogeneity compensated superposition (HCS).

Methods: C/S has proven to be a good estimator of the dose deposited in a homogeneous volume. However, near heterogeneities electron disequilibrium occurs, leading to the faster fall-off and re-buildup of dose. We propose to filter the actual patient density in a position and direction sensitive manner, allowing the dose deposited near interfaces to be increased or decreased relative to C/S. We implemented the effective density function as a multivariate first-order recursive filter and incorporated it into GPU-accelerated, multi-energetic C/S implementation. We compared HCS against C/S using the ICCR 2000 Monte-Carlo accuracy benchmark, 23 similar accuracy benchmarks and 5 patient cases.

Results: Multi-energetic HCS increased the dosimetric accuracy for the vast majority of voxels; in many cases near Monte-Carlo results were achieved. We defined the per-voxel error, %|mm, as the minimum of the distance to agreement in mm and the dosimetric percentage error relative to the maximum MC dose. HCS improved the average mean error by 0.79 %|mm for the patient volumes; reducing the average mean error from 1.93 %|mm to 1.14 %|mm. Very low densities (i.e. <0.1 g / cm3) remained problematic, but may be solvable with a better filter function.

Conclusions: HCS improved upon C/S's density scaled heterogeneity correction with a position and direction sensitive density filter. This method significantly improved the accuracy of the GPU based algorithm reaching the accuracy levels of Monte Carlo based methods with performance in a few tenths of seconds per beam.

Acknowledgement: Funding for this research was provided by the NSF Cooperative Agreement EEC9731748, Elekta / IMPAC Medical Systems, Inc. and the Johns Hopkins University. James Satterthwaite provided the Monte Carlo benchmark simulations.

1. Overview

It is challenging for the convolution/superposition (C/S) algorithm to maintain accuracy in regions of electronic disequilibrium. Electronic disequilibrium occurs in areas where there are gradients in both TERMA and density. This leads to significant errors near material interfaces for small or highly modulated fields. Empirically, these errors appear as an overestimation of the dose after a high-density to low-density transition and an underestimation of the dose after a low-density to high-density transition (Figure 1).

We propose to improve C/S by filtering the density (ρ) used by the superposition component of the calculation in a position and direction sensitive manner. The use of an effective density (ρ_{eff}) value allows the dose deposited to be empirically increased or decreased, relative to traditional C/S. This change in the deposited dose occurs due to the change in the integration width of the superposition kernel across the voxel; the superposition kernels themselves are not modified. Therefore, the total integral of the superposition kernel never changes and both heterogeneity compensated superposition (HCS) and C/S compute similar dose distributions in regions of electronic equilibrium.

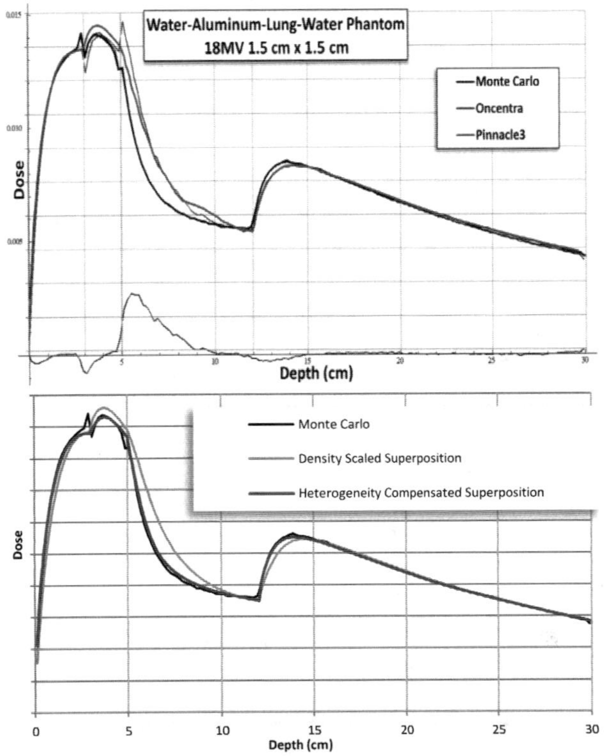

Figure 1. (top)Depth dose for 2 commercial C/S methods vs Monte Carlo for the ICCR benchmark showing the error in dose near heterogeneities: (bottom) GPU accelerated superposition with normal density scaling and HCS showing accuracy improvements.

2. Materials and Methods

HCS was implemented as an extension to our previous work in GPU accelerated convolution/ superposition (C/S).[1,2] Previously, we leveraged the hardware functionality of the GPU texture unit to allow our C/S implementation to efficiently perform multiple superpositions simultaneously. By computing the superposition for each single energy bin independently (multi-energetic) and summing the resulting doses, the kernel is effectively hardened with depth as the lower energy bins will have less contribution to the TERMA at depth. For efficiency, we experimented with simply dividing the spectrum into 2 (dual-energetic) and 4 (quad-energetic) energy sub-spectra and performing the superposition independently for each to more efficiently handle the kernel hardening. These divisions were applied after the TERMA was computed using the exact, full-spectrum attenuation. Use of multi-energetic transport substantially increased the accuracy of C/S in homogeneous phantoms when the source model used the true Monte-Carlo spectrum as opposed to a manually commissioned spectrum.

In order to further improve the accuracy near density changes, we formulated the ρ_{eff} function at position t as a first-order exponential linear recursive filter with a varying time constant as it can be implemented using a small, fixed number of GPU registers and can handle the irregular signal-sampling rate produced by our incremental ray-tracing algorithm. The time constant, τ_t, is dependent on the voxel width, Δ_t, an adaptive α term and a decaying β term.

Equation 1 shows the traditional form of the convolution function using density scaling of the water kernels. Equation 2 shows how it has been adapted to accommodate the ρ_{eff} function defined in Equations 3-8. The coefficients were optimized to best fit the low and high dose gradient regions of the 18MV Water-Lung-Bone-Water and the 24MV Water-Bone-Lung-Water benchmarks. Table 1 indicates the optimized values of the coefficients used in the ρ_{eff} function.

$$D_{C/S}(r) \cong \iiint_V \int \mathbf{T}(r+tv)\mathbf{K}_{dv}\left(\int_r^{r+tv} \rho(r')dr'\right)dt \tag{1}$$

$$D_{HCS}(r) \cong \iiint_V \int \mathbf{T}(r+tv)\mathbf{K}_{dv}\left(\int_r^{r+tv} \rho_{eff}(r,v,r')dr'\right)dt \tag{2}$$

$$\rho_{eff,t+\Delta_t}(r') = (1-\tau_t)\rho_t + \tau_t\rho_{eff,t} \tag{3}$$

$$\tau_t = \max\left(0, 1-\Delta_t\left(\alpha_t(1+\beta_t)\frac{\rho_{eff,t}}{\rho_{eff,t}+\rho_t}\right)\right) \tag{4}$$

$$\alpha(\rho) = \left(C_{\alpha,0} + C_{\alpha,1}\left|\rho - C_{\rho,\alpha}\right| + C_{\alpha,2}\left|\rho - C_{\rho,\alpha}\right|^2\right) \tag{5}$$

$$\beta(\rho) = \left(C_{\beta,0} + C_{\beta,1}\left|\rho - C_{\rho,\beta}\right| + C_{\beta,2}\left|\rho - C_{\rho,\beta}\right|^2\right) \tag{6}$$

$$\beta_{t+\Delta_t} = \max\left(0, \beta_t - \Delta_t\left(C_{\tau_\beta,0} + C_{\tau_\beta,1}\beta(\rho_{eff,t}) + C_{\tau_\beta,2}\frac{\rho_{eff,t}}{\rho_t+\rho_{eff,t}}\right)\right) \tag{7}$$

$$\tau_{\alpha_t} = \max\left(0, 1-\Delta_t\left(C_{\tau_\alpha,0} + C_{\tau_\alpha,1}\alpha_t + C_{\tau_\alpha,2}\frac{\rho_{eff,t}}{\rho_t+\rho_{eff,t}}\right)\right) \tag{8}$$

$C_{\rho,\alpha}$	$C_{\alpha,0}$	$C_{\alpha,1}$	$C_{\alpha,2}$	$C_{\tau\alpha,0}$	$C_{\tau\alpha,1}$	$C_{\tau\alpha,2}$	$C_{\rho,\beta}$	$C_{\beta,0}$	$C_{\beta,1}$	$C_{\beta,2}$	$C_{\tau\beta,0}$	$C_{\tau\beta,1}$	$C_{\tau\beta,2}$
0.18	0.28	3.26	0.89	1.00	0.65	0.18	1.58	1.76	0.75	0.75	0.32	0.00	1.22

Table 1: Optimized coefficients used to calculate effective density.

Figure 2 shows the effects of the filter on ρ_{eff} and the resulting effective distance used in the kernel lookup during the superposition ray-tracing. When entering a high density region from a lower density, the filter causes the ρ_{eff} to gradually increase, and similarly when going from a high to low density, it will smoothly decrease. This alters the kernel lookup from the more traditional straight density scaling.

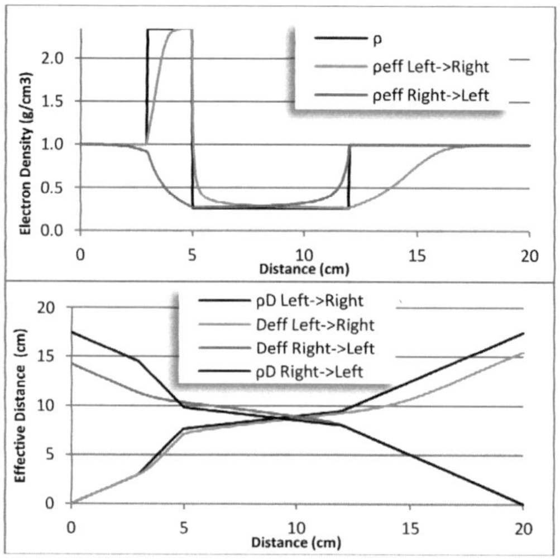

Figure 2. Plots of the effective density (ρ_{eff}) and distance when traversing change in actual densities (ρ) from the left or right side of the plot using the filter function.

3. Results

Accuracy of the method was tested using both the ICCR benchmark Monte Carlo phantoms [3] and human CT scans. Comparisons were made for the identical spectra and source model for the HCS and EGS3 Monte Carlo dose calculation algorithms.

3.1. Slab Phantom Benchmark Results

Figure 3 shows the results from selected slab phantoms for the dual and quad energetic HCS methods. The improvement in the model by breaking it up into the multiple energy groups is evident in the upper left plot where the traditional single spectrum poly-energetic HCS is displayed. Agreement with Monte Carlo is within 1% for the majority of the profiles in the various field sizes and phantoms displayed. Some error remained in the drop off from bone to lung in the 4MV results using the 1.5 cm field size.

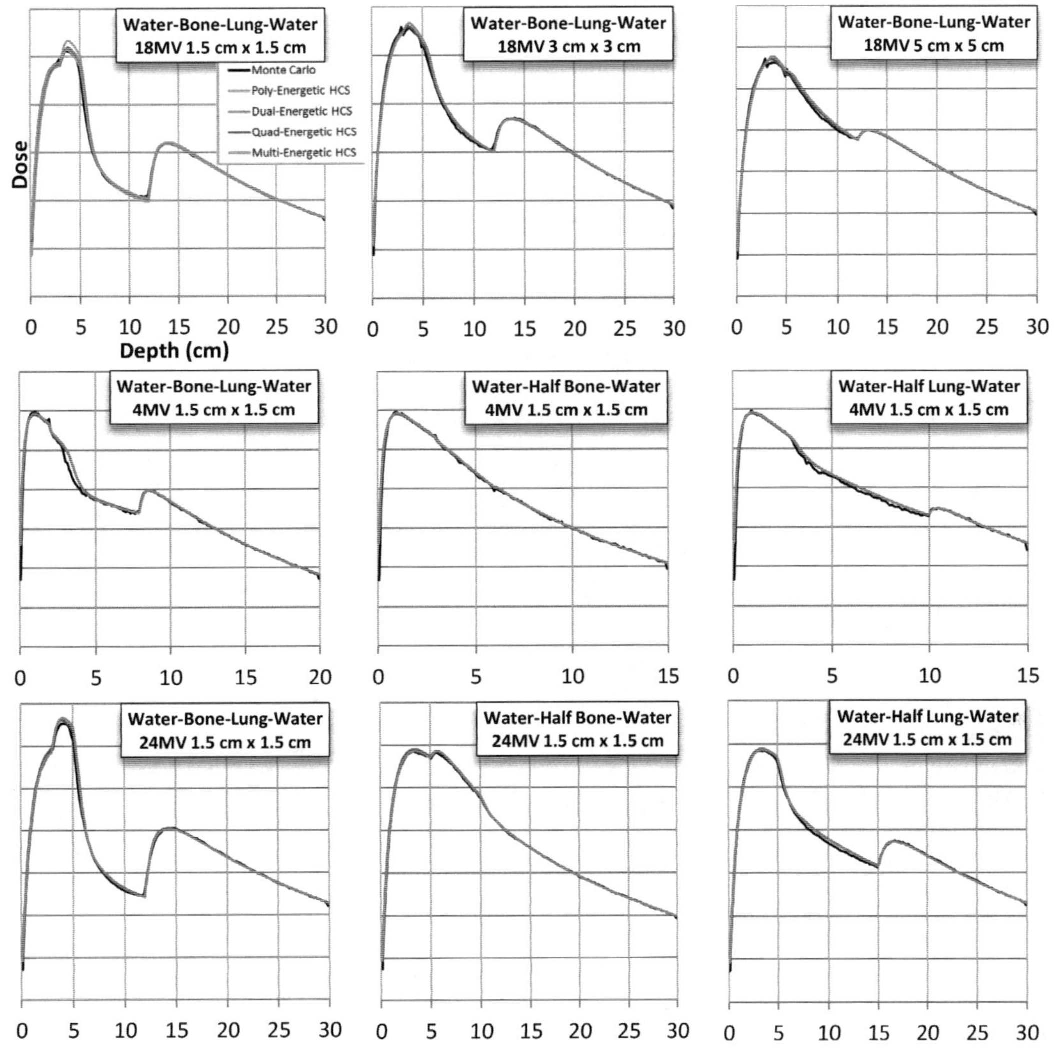

Figure 3. Phantom results comparing HCS to Monte Carlo for 18, 4 and 24MV for both Dual (green) and Quad (red) Energetic modes. Upper left: the single poly-energetic (orange) shows the impact of not breaking up the spectrum. Top row shows field sizes of 1.5, 3 and 5 cm for an 18 MV beam on the water-bone-lung-water phantom. Bottom 2 rows shows 4 and 24 MV on the water-bone-lung-water, water-half_bone-water, and water-half_lung-water phantoms for a 1.5 cm field size.

3.2. Patient Benchmark Results

Figure 4 shows patient accuracy benchmarks comparing C/S, HCS indicating the improvement in accuracy between the C/S and HCS methods in the regions of electronic disequilibrium due to heterogeneities. All fields were 18MV and 5 cm x 5 cm at the volume's surface. HCS improved the average mean error by 0.79 %|mm for the patient volumes; reducing the average mean error from 1.93 %|mm to 1.14 %|mm.

3.3. Discussion

HCS was compared to traditional C/S using 29 slab phantom benchmarks, 5 patient cases and 1 water phantom (not all presented here). In many cases the accuracy of HCS was comparable to Monte-Carlo methods. However, discrepancies were observed between HCS and Monte Carlo in regions of low density (ρ <= 0.1 g/cm3). This was due in part to the geometric scale of electronic disequilibrium being the largest at low densities and in part due to our formula for the HCS parameters not robustly extrapolating to very low densities. The build-up regions of the patient datasets were also systematically overestimated. This was due to a layer of low density voxels surrounding the patient that caused the build-up region to be treated as a re-build-up region. We intend to introduce a depth dependence to the HCS filter function to correct for this issue.

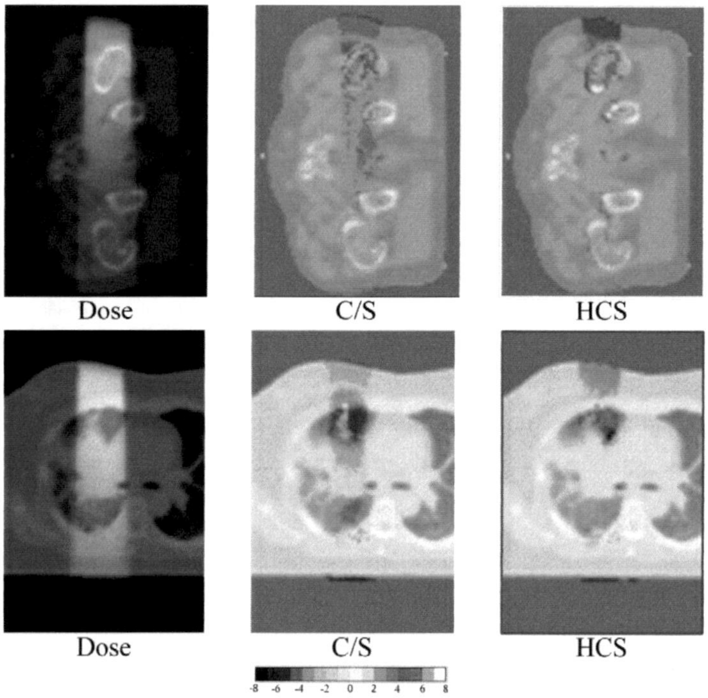

Figure 4. Comparison of C/S and HCS methods to Monte Carlo on patient CT scans showing increased accuracy of HCS in the regions near heterogeneities for an 18MV beam and a 5x5 cm field.

4. Conclusions

We have developed a novel improvement to the C/S that better accounts for the electronic disequilibrium caused by patient heterogeneity. This work shows that the convolution/superposition algorithm can rival Monte Carlo in accuracy when the traditional approximations are removed from the C/S algorithm and improved models for handling kernel hardening and kernel scaling with density have been applied.

5. References

[1] Jacques R, Wong J, Taylor R, and McNutt T 2010 Towards Real-Time Radiation Therapy: GPU Accelerated Superposition/Convolution. *Computer Methods and Programs in Biomedicine Volume 98, Issue 3, June 2010*, pg. 285-292 HP-MICCAI 2008, doi:10.1016/j.cmpb.2009.07.004

[2] Jacques R, Wong J, Taylor R, and McNutt T 2011 Real-time dose computation: GPU-accelerated source modeling and superposition/convolution *Medical. Physics vol 38*, pp 294-305 (2011).

[3] Rogers DWO, Mohan R 2000 Questions for comparison of clinical Monte Carlo codes *XIIIth International Conference on the Use of Computers in Radiation Therapy* Schlegel, Bortfeld (Editors) Heidelberg, Germany May 22-25, 2000 pp 120

Journal of Physics: Conference Series **489** (2014) 012020 doi:10.1088/1742-6596/489/1/012020

IMRT treatment Monitor Unit verification using absolute calibrated BEAMnrc and Geant4 Monte Carlo simulations

B. M. Oborn[1,2], **M. Williams**[1,2], **M. Bailey** [1,2], **M. G. Carolan**[1,2]

[1]Illawarra Cancer Care Centre, Wollongong Hospital, Wollongong NSW 2500, Australia
[2]Centre for Medical Radiation Physics, University of Wollongong, Wollongong NSW 2522, Australia

E-mail: brad.oborn@gmail.com

Abstract.
Intensity Modulated Radiation Therapy (IMRT) treatments are some of the most complex being delivered by modern megavoltage radiotherapy accelerators. Therefore verification of the dose, or the presecribed Monitor Units (MU), predicted by the planning system is a key element to ensuring that patients should receive an accurate radiation dose plan during IMRT. One inherently accurate method is by comparison with absolute calibrated Monte Carlo simulations of the IMRT delivery by the linac head and corresponding delivery of the plan to a patient based phantom. In this work this approach has been taken using BEAMnrc for simulation of the treatment head, and both DOSXYZnrc and Geant4 for the phantom dose calculation. The two Monte Carlo codes agreed to within 1% of each other, and these matched very well to our planning system for IMRT plans to the brain, nasopharynx, and head and neck.

1. Introduction

IMRT is performed commonly in modern radiotherapy centres around the world on a daily basis. It involves delivering many multileaf collimator (MLC) modulated beams to build up a complex 3D sculpted dose distribution to a patient's tumor while sparing as much healthy tissue as possible. It is therefore comprised of many regions of high dose gradient change, and so quality assurance of IMRT plans is a detailed and constantly evolving practice. The first step in this quality assurance chain is the prediction of the dose by the planning system. One such common method to perform this task is with Monte Carlo simulations, where the same plan is delivered to a CT-based Monte Carlo patient phantom and the doses compared. This however is not an absolute dose method as there is no fundamental connection made between the number of histories in the simulations and dose delivered to a phantom.

In this work we have performed absolute calibrated Monte Carlo simulations, where we simulate the linac head and calibrate our model to replicate exactly what the linac delivers when it is told to deliver 1 MU. The first step taken is to calibrate the linac head simulation by determining the number of electrons required to produce 1 MU (1 cGy) for the reference setup of a 10x10 cm^2 field, as well as to record the amount of charge collected in the linac head monitor chamber. Once known, all future simulations get corrected by a factor which is directly

Figure 1. Left: Image of the Varian monitor chamber. Right: Implementation of the monitor chamber in BEAMnrc using two CHAMBER component modules.

the ratio of monitor chamber dose to that of the reference field. This replicates what the linac is delivering, i.e. it knows what charge collected in the monitor chamber is producing 1 cGy.

2. Methods

2.1. BEAMnrc Monte Carlo simulations

The linac head (Varian 2100C) was modeled using the EGSnrc user code BEAMnrc[2]. This simulation was used to generate IAEA format phase space files below the MLC's (558 mm from the target), as well as to model the dose scored (including the backscatter) in the treatment head monitor or ion chamber. All the typical component modules were used to model the 2100C linac head; x-ray target (SLABS), primary collimator (CONS3R), exit window (SLABS), flattening filter (FLATFILT), ion chamber (CHAMBER, CHAMBER), mirror (MIRROR), Y and X-jaws (JAWS, JAWS), Millenium 120 leaf MLC (DYNVMLC), and the reticle (SLABS). Particular attention however was paid to the monitor chamber to get the best possible match to the real chamber geometry. This was achieved by coupling two CHAMBER modules back-to-back. This allowed a very close match by including all the sensitive layers and surrounding frame material, see figure 1. For all BEAMnrc simulations the global ECUT= 0.521 MeV, and global PCUT= 0.01 MeV. The low ECUT value is important for tracking secondary electrons in the monitor chamber accurately[3].

2.2. Absolute Calibration Simulations: DOSXYZnrc and Geant4

At our centre 1 MU (6MV beam) corresponds to 1 cGy at D_{max} where SSD = 100 cm. In practice the measurement is done at 10 cm depth and related back to D_{max} by: $D_{10} = 0.661 \times D_{max}$ as per the PDD characteristics of our centres' 6MV beam. This setup was replicated in both Geant4 (version 9.6.p01)[4] and DOSXYZnrc[3] (version V4-r2-4-0) and both codes used the same phase space file generated by the BEAMnrc simulation as input. Simulation were run to determine how many electrons would produce 1 cGy at D_{max}.

2.3. Patient Phantom Simlations: DOSXYZnrc and Geant4

CT based phantoms were setup in both DOSXYZnrc and Geant4. These simulations read the corresponding IAEA phase space files from the BEAMnrc treatment head simulations. Similar to the BEAMnrc simulations, no variance reduction techniques where used in either code. For the DOSXYZnrc simulations the global ECUT = 0.7 MeV, while the global PCUT= 0.01 MeV. The remaining transport settings followed the default approach outline in the DOSXYZnrc

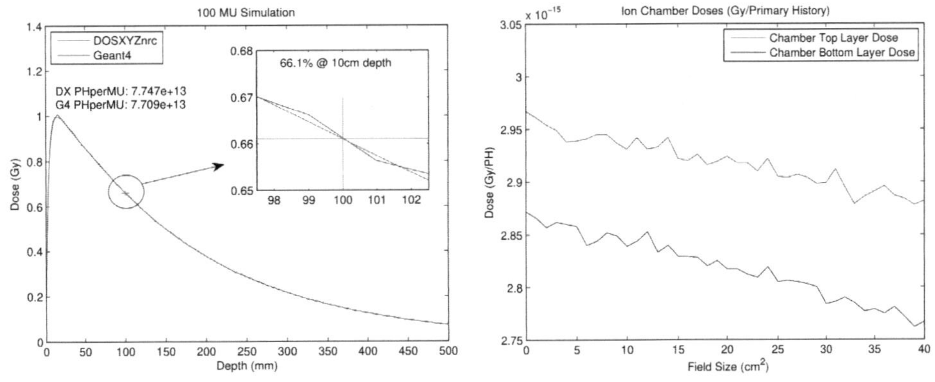

Figure 2. Left: Absolute calibration simulation results. An excellent match is seen between Geant4 and DOSXYZnrc. Right: Monitor chamber doses for symmetric beams from 0x0 cm^2 to 40x40 cm^2.

manual. For Geant4, the Livermore Low-Energy physics models were uses with 1 mm range cuts throughout the phantom and surrounding.

In terms of implementing a Patient CT based phantom "ctcreate" was used for the DOSXYZnrc simulations while for Geant4 in-house code was used to replicate the ctcreate approach. In both codes only air and water were employed to somewhat match what Pinnacle performs. The conversion from CT number to water density was done in 0.05 g/cc bins ranging from 0.3 to 5.5 g/cc. Below 0.3 g/cc the material was set to air. In both codes the voxel resolution was set to 2 CT pixel x 1 CT slice resolution, or 1.95x1.95x2 mm in x, z, y directions accordingly.

3. Results

3.1. Benchmark Linac Model, Absolute Calibration, and Chamber Dose

Linac commissiong data was used to benchmark our Monte Carlo model. This included ion chamber and diode measurements for 5x5, 10x10, 20x20, 30x30 and 40x40 cm^2 field sizes at depths of 15 mm, 50 mm, 100 mm, 200 mm, and 300 mm. A 2%/2 mm gamma match between the Monte Carlo simulations and measured data was achieved for the entire range of field sizes and depths for the following settings of the electron beam striking the x-ray target: Spot size = 0.9x1.2 mm, FWHM = 1%, Peak = 6.00 MeV.

The absolute calibration simulation results (see fig 2) concluded that the number of electrons required to produce 1 MU at our reference conditions was: $7.709 \times 10^{+13} \pm 1\%$ (Geant4) and $7.747 \times 10^{+13} \pm 1\%$ (DOSXYZnrc). The 0.5% match is within simulation error and so these values could be considered identical.

Figure 2 also shows the monitor chamber doses for symmetric beams from 0x0 cm^2 to 40x40 cm^2. Dose was scored in both sensitive layers and the average taken. It is clear from this plot that there is around 3% change from 0x0 cm^2 to 40x40 cm^2, enforcing the need to correct for the backscatter to the monitor chamber inside the Monte Carlo simulation.

To the best of our knowledge the only comparison (for a 6MV Varian 2100C beam) we can draw with from the literature is from the 2005 work by Popescu et al [1]: $8.124 \times 10^{+13} \pm 1.0\%$. In this setup the SSD was 98.5 cm rather than 100 cm as in our case. There has also been a change to the 6MV flattening filter since 2006 and so we estimate the differences to be a combination of both different setup and different geometry. Regardless of this value, it is the monitor chamber dose ratio to the reference field that is the most important for this work.

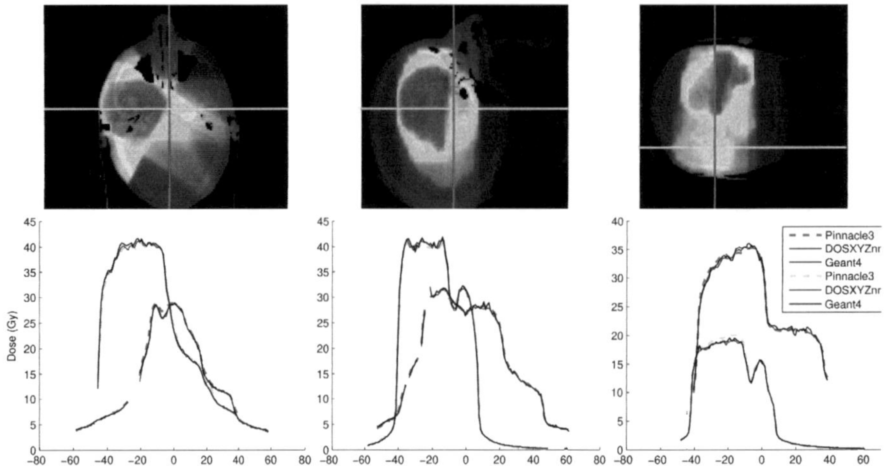

Figure 3. IMRT plan comparison for a 4 field (54 segment) brain IMRT plan.

3.2. IMRT Plan comparisons: Brain, Nasopharyx, and Head and Neck

Figures 3-5 show comparisons between Pinnacle and DOSXYZnrc/Geant4 for a brain, nasopharynx, and a head and neck cancer IMRT case. Both the brain and nasopharynx plans are 4 fields with 56 segments in total while the head and neck is 9 fields with 129 segments. The monitor chamber backscatter correction factors for each plan were brain: 1.005, 0.996, 0.999, 1.007, nasopharynx: 0.994, 1.003, 0.999, 0.997, and head and neck: 1.005, 1.009, 1.000, 1.009, 1.009, 1.001, 1.006, 1.000, 0.999. At most there is only about a 1% correction as these beams all were of similar shape to to the reference field of 10x10 cm^2. In reality the largest corrections take place for either the largest open fields, or blocked fields where a Y-jaw (closest to monitor chamber) is mostly closed to allow off-axis treatments.

In each of figures 3-5 the top row shows the Pinnacle dose overlayed on the CT slice and two corresponding profiles of interest through each orientation. The bottom plots show the dose profiles as described by the lines drawn in the top plots. In all plots the dose to air has been removed to improve clarity. In the 4 field brain and nasopharynx cases there is excellent agreement between the two Monte Carlo codes (within error estimates of $< \pm 2\%$) across the profiles shown. These match Pinnacle very well also, with only a minor deviation in the brain case for one profile far from the tumor site (right bottom plot). For the more advanced 9-field head and neck case there is again excellent agreement between the Monte Carlo codes. We can see a slight overprediction of the dose by Pinnacle as compared to Monte Carlo across some sections of the green (horizontal) profile shown on the right. As can be seen this lies along a sharp dose gradient and so we would naturally attribute this to endleaf transmission effects not being modeled as accurately as the Monte Carlo approach.

4. Conclusions

Absolute calibrated Monte Carlo simulations have been successfully set up and tested for comparing our planning system predictions on the MU's for IMRT treatment plans. Excellent agreement is seen between Geant4 and DOSXYZnrc for dose calculation in water only phantoms, and generally excellent agreement between Monte Carlo and Pinnacle where expected. Future work will include benchmarking the MLC Monte Carlo model against experimental transmission measurements.

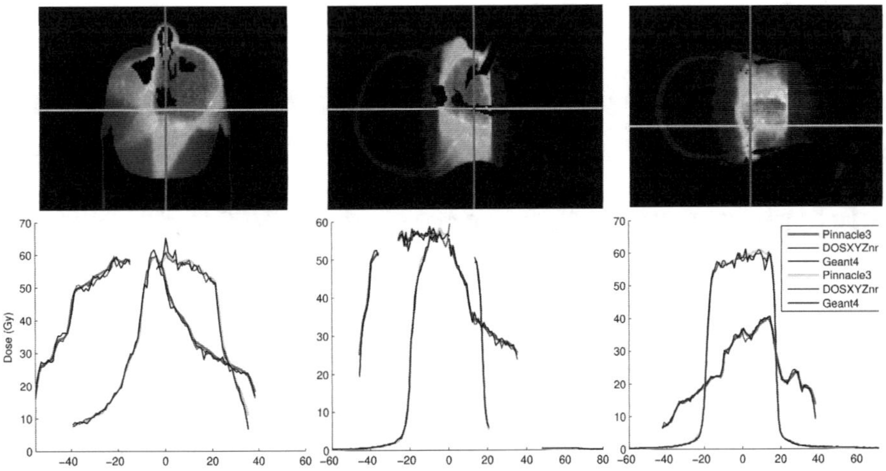

Figure 4. IMRT plan comparison for a 4 field (54 segment) nasopharynx IMRT plan.

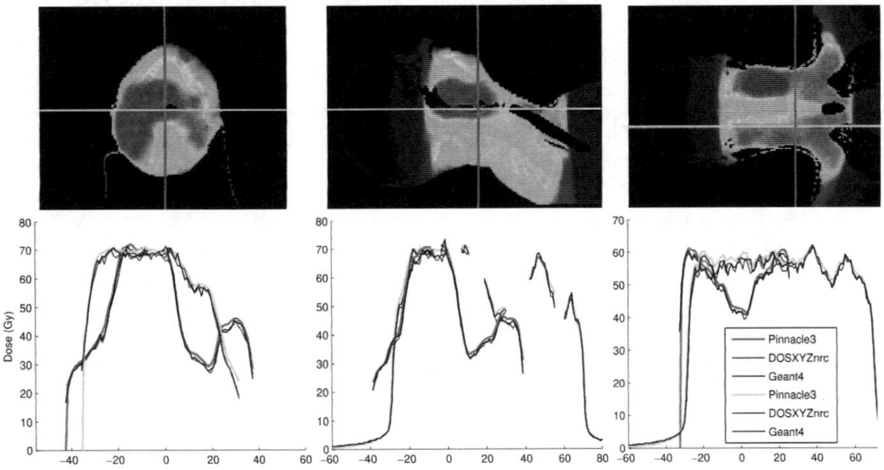

Figure 5. IMRT plan comparison for a 9 field (129 segment) head and neck IMRT plan.

References

[1] I. A. Popescu et al., "Absolute dose calculations for Monte Carlo simulations of radiotherapy beam", Phys. Med. Biol, 50, 3375-3392 (2005).

[2] D. W. O. Rogers et al, "BEAM: A Monte Carlo code to simulate radiotherapy treatment units", Med. Phys. 22, p503-524 (1995).

[3] B. R. B. Walters et al, "DOSXYZnrc users manual", National Research Council of Canada, PIRS-794 (2003).

[4] S. Agostinelli et al, "Geant4-a simulation toolkit", Nucl. Instrum. Methods Phys. Res. A, 506, 250-303 (2003).

Journal of Physics: Conference Series **489** (2014) 012021 doi:10.1088/1742-6596/489/1/012021

Formal quality control for a proton Monte Carlo system in radiation therapy

J Perl[1]
SLAC National Accelerator Laboratory, Menlo Park, CA, USA

E-mail:perl@slac.stanford.edu

Abstract: TOPAS (TOol for PArticle Simulation) is a Monte Carlo particle transport tool being released to a wide variety of proton therapy users worldwide. Because TOPAS provides unprecedented ease in 4D placement of geometry components, beam sources and scoring, including options to place geometry components, beam sources or scorers within each other, Quality Control (QC) for TOPAS is both critical and challenging. All simulation details (geometry, particle sources, scoring, physics settings, time-dependent motions, gating, etc.) are specified in the TOPAS Parameter Control System (which catches many user errors). QC includes Unit and End-to-End Testing. Each code unit is tested (each geometry component, particle source option, scoring option, etc.) and these unit testing procedures are shared with end users so they can reproduce tests. End-to-End testing of several full clinical setups is routinely performed. End-to-End testing presents a challenge since one cannot anticipate all the ways users will combine TOPAS flexible units for their specific project. Automated checking catches geometry overlaps and some other problematic setups, but one can never rule out the potential for problems when users combine units in new setups. QC is ultimately a partnership between the tool developer and the user. Key is that the developer be clear to the end user about what has been tested and what has not.

1. Introduction

Monte Carlo particle transport simulation (MC) codes have become an important tool for radiation therapy research and practice. As the variety and complexity of radiation therapy modes has increased, and as therapies have moved from simple to highly conformal to four dimensional (4D), 3D plus time, both the value of and the requirements on MC have increased. We present some thoughts on quality control (QC) for highly configurable and 4D MC codes. This work has grown out of five years of effort to develop the TOPAS Tool for Particle Simulation [1], a project to make Monte Carlo particle transport simulation faster and easier to use for proton therapy. The conclusions we draw here are not specific to TOPAS. Rather we use the TOPAS experience to raise QC issues that apply to any simulation code, especially those that are highly configurable and 4D.

TOPAS is an NIH funded project of the Massachusetts General Hospital, the University of California at San Francisco and the SLAC National Accelerator Laboratory. TOPAS lets you model a treatment head, import a patient, simulate all kinds of beams, has advanced scoring and graphics, and, because proton therapy involves so many moving parts, TOPAS is, from the ground up, 4D. Because TOPAS provides unprecedented ease in 4D setup, QC for TOPAS is both critical and challenging.

[1] To whom any correspondence should be addressed.

2. 4D Behaviors

TOPAS lets you place geometry components, particle sources and scoring regions in 3D, that is, each component can have its own translation and rotation specified however the user wishes relative to its mother component. TOPAS allows one to simulate an infinite variety of setups with just one pre-compiled software application, but different control files. All simulation details (geometry, particle sources, scoring, physics settings, time-dependent motions, gating, etc.) are specified within a TOPAS Parameter Control System, which was designed using principles of safety engineering to help prevent user errors. Automated checking provided by the underlying Geant4 toolkit [2] catches some geometry overlaps and some other problematic setups, but TOPAS adds many layers of additional safety. The system not only provides a unified repository for all parameters, but provides a unified manner in which parameters can be set to change over time, that is, a unified approach to 4D behavior [3].

Supported 4D behaviors can go beyond simple motion to include multiple moving parts, motions within motions, time-dependent shape changes, patient motion, time-dependent electric and magnetic fields, beam current modulation, sources (even multiple sources) that turn on and off, and gated scoring (figure 1). All of this can be put it together into some very complex systems, like the full IBA double-scattering nozzle at the Francis H. Burr Proton Therapy Center (figure 2) which combines moving range modulator wheels (RMW), beam current modulation, adjustable second scatterers and jaws and patient-specific apertures and compensators built on the fly, different for every field and every patient.

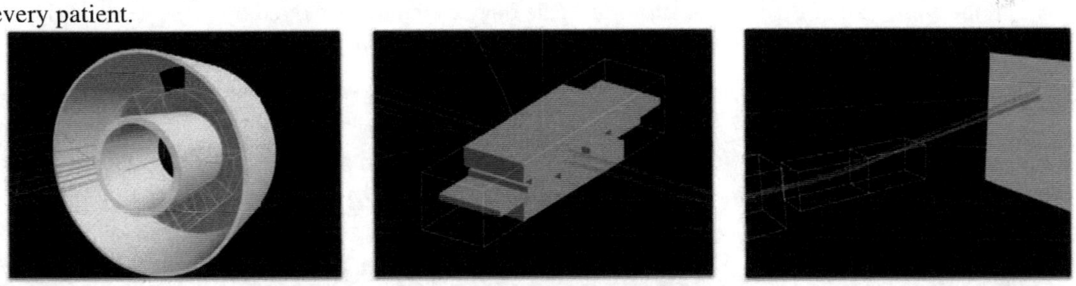

Figure 1. Components with 4D behaviors. a) Rotating Range Modulator Wheel. b) Dynamic Multileaf Collimator (simplified here to only three leaves). c) Time-varying dipole for magnetic beam steering.

For a fully configurable 4D simulation system, the user should be able to vary almost any parameter over time. The goal is to allow extreme flexibility of 4D behaviors so users can test entirely novel hypotheses. So, for example, TOPAS allows a user to change the angular extent of a cylinder over time even though we have yet to find a reason that one might wish to do this.

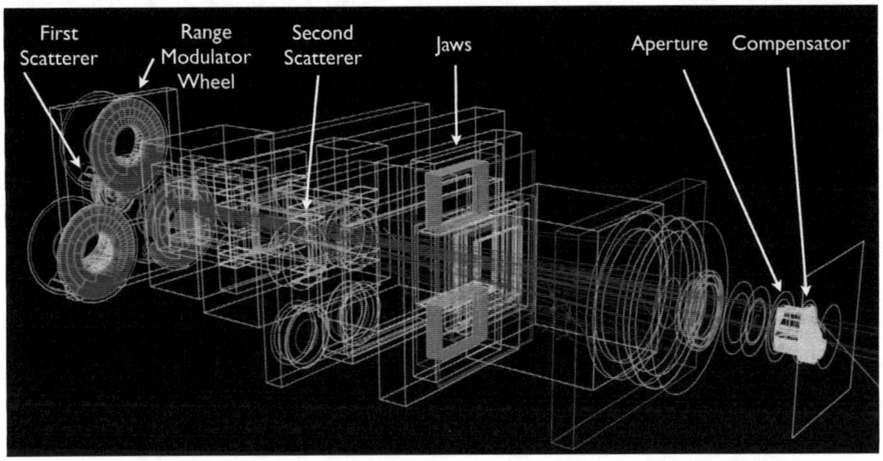

Figure 2. Proton gantry at the Francis H Burr Proton Therapy Center in double-scattering mode. Protons move left to right (blue lines). Secondary particles include electrons (red) and gammas and neutrons (green).

3. Limitations of 2 and 3D Visualization

4D behaviors are more difficult to test than 3D. There are ways in which an MC may appear to work correctly in 4D while in reality it does not. An example is in the use of visualization tools. TOPAS' underlying general-purpose MC toolkit, Geant4, contains an extensive visualization system [4] and includes capabilities for 4D OpenGL viewing. While Geant4's 4D graphics can be compelling, visualization of moving components do not necessarily prove the correctness of 4D scored results. Key is that visualization code is not the same as MC particle transport code. Visualization has the luxury of working slowly, at the sluggish pace of the human visual system. To produce visualizations, Geant4 follows the most straightforward approach, recalculating the whole image after every motion. But Monte Carlo transport needs to work much faster. One rarely wants to simulate just a few histories, but rather millions or billions. So MCs incorporate speedup strategies. We will discuss one that is particular to Geant4, the "Smart Voxel" system, but every code fast enough for therapy calculations has speedup techniques, and all such techniques must be re-considered in 4D.

Any general purpose transport code spends a lot of time trying to figure out where a particle will go when it leaves a volume - which of many other volumes will be the next one hit. There can be thousands of such volumes, and their shapes can make the "Am I going to hit you next?" question mathematically complex. Figure 3 shows an example of such a collection of volumes (the figure does not represent any actual therapy system but is just a random set of somewhat complicated objects - a few of the many solids supported by Geant4). The geometrical situation is complicated by factors beyond the complexity of the individual solids - the trajectory may be curved by a magnetic or electric field, there may be coulomb scattering, and all of this has to be solved in 3D.

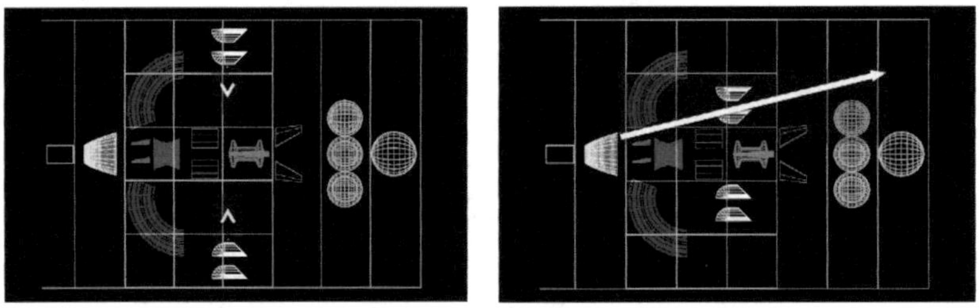

Figure 3. Geant4 smart voxel system.
Left: Arrows indicate volume moving across boundaries.
Right: Particle shown as yellow track will not "see" white volumes.

Codes use tricks to limit how many volumes must be interrogated. For the Geant4 case, before the first particle is tracked, Geant4 slices the world into what it calls "smart voxels" (the vertical lines slicing figure 3). If more than a few volumes are in any given "smart voxel", Geant4 slices the world further along different axes (the horizontal lines slicing figure 3). A lookup table is generated recording which volumes are in which of these voxels. At any given time, Geant4's tracking need only interrogate the few volumes in the current voxel. Or if the trajectory seems to exit the current voxel, it just interrogates the volumes in the next voxel. Generating the lookup table takes some time, but it only needs to be done once at the start of the simulation, so the time burden is not significant.

The Geant4 smart voxel system works very well. Not only does it provide efficient tracking in complex arrangements of varied 3D solids, but it also works as well in regular 3D grid geometries such as a set of DICOM pixels [5]. But if a volume moves into a voxel that it wasn't in when the lookup table was generated, the trajectory will pass straight through the volume without interacting. As far as the smart voxel system is concerned, the volume is still where it was at the start of the simulation. This means one could, for example, have an MLC leaf move into the beam path yet have the beam continue to propagate as if the leaf were not there. Of course this can be fixed. The solution is to re-calculate the relevant part of the lookup table when a motion has occurred. In Geant4, it is left to the user code to explicitly trigger this re-calculation at appropriate times. The underlying Geant4 toolkit does not automatically recognize that such a change might be needed (as it would be time-consuming for Geant4 to continually check for such changes).

4. Implications for Caching

The Geant4 smart voxel system described above is just one example of "caching," a speedup technique critical to all fast MC codes. Code developers note which operations are costly and repetitive and cache results to avoid re-computation. An example comes when a code needs to check whether a particular scoring volume has been intersected. Performing a string comparison to check the name of a volume is a slow operation compared to checking the pointer to this volume in memory (an integer comparison). Accordingly, codes may look up pointers to volumes of interest only once, then cache these pointers for later rapid comparison. A 4D code may at times need to delete and re-create volumes to account for the widest possible set of 4D variations. After such changes, while the new volume may still have the same name, the memory pointer may be changed.

Another example of caching could be for storing the result of a volume calculation. Dose calculations require knowing the volume of the scored section. Such a calculation may involve time-consuming trigonometry if the volume is, for example, an arbitrary wedge of a cylinder. Because the same scoring area may be used many times, code may be written to calculate this volume only once, then cache this result. But if a 4D behavior alters this volume, the cached result may become incorrect. 4D code must therefore take care with caching. Caching itself is a 4D behavior.

Particle splitting (such as Bremsstrahlung Splitting), directional biasing and geometrical biasing [6-9] are additional examples of speedup techniques where extra care must be taken in the 4D case. Such techniques are likely to involve special actions to be taken in certain regions of a moving geometry or may involve calculations with direction vectors that may be varying over time.

5. Importance of 4D Tests

In the smart voxel case described above, the potential errors from mishandling 4D were obvious - a particle either encountered a volume or missed the volume completely. But in real world applications, effects are more subtle. A classic example from proton therapy concerns ribs motion during 4D lung treatment. A beam targeted on a lung tumor may encounter only soft tissue on its way to the lung in some breathing phases, but rib on some other breathing phases. Not only is the presence or absence of rib dependent on breathing phase, but even the angle of the rib relative to the treatment beam may vary. As the proton range is highly dependent on atomic structure of materials encountered, this subtle motion can have profound effects on the end of proton range. A 4D visualization tool that simply confirms that anatomy is moving is not a sufficient QC aid. There must be tools that confirm that the code correctly accounts for all 4D changes. Figure 4 shows how TOPAS 4D scoring can effectively confirm 4D behavior, in this case from a rotating modulation propeller in an eye treatment beamline.

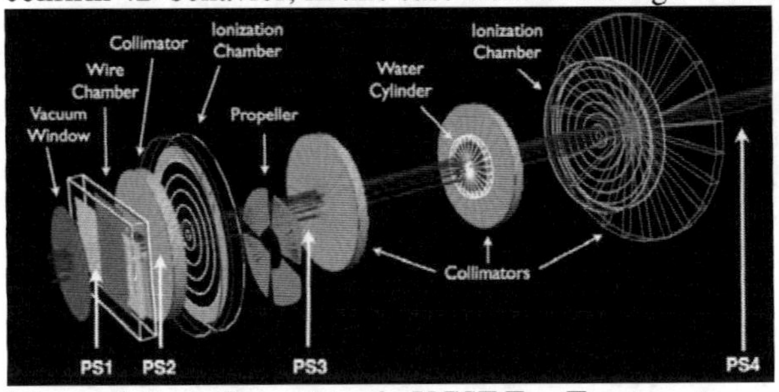

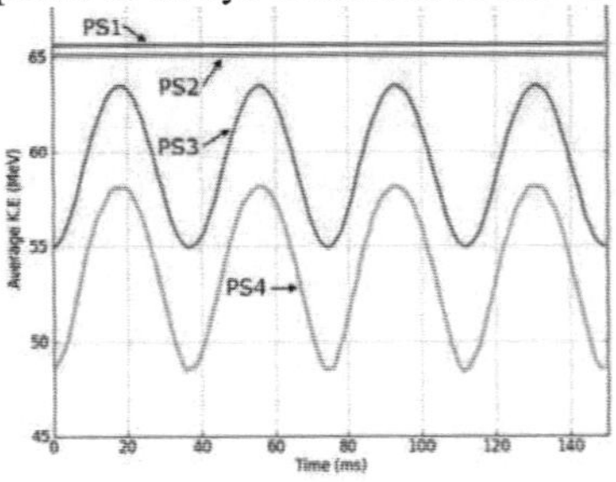

Figure 4. Propeller rotates in UCSF Eye Treatment beamline. Protons (blue lines) travel left to right.

TOPAS scores energy as function of time at four phase space output planes PS1 to PS4.

Energy delivered to a water phantom located at right (beyond PS4) for four different angular intervals of propeller rotation.

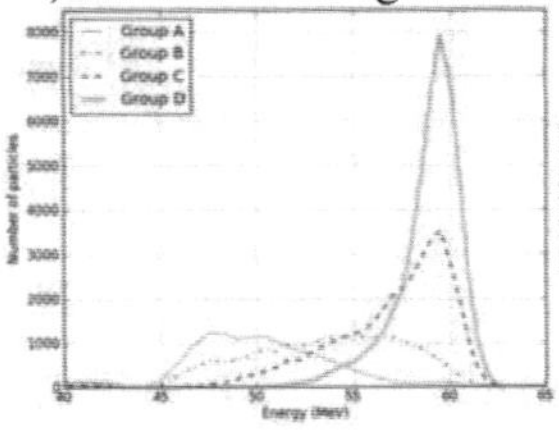

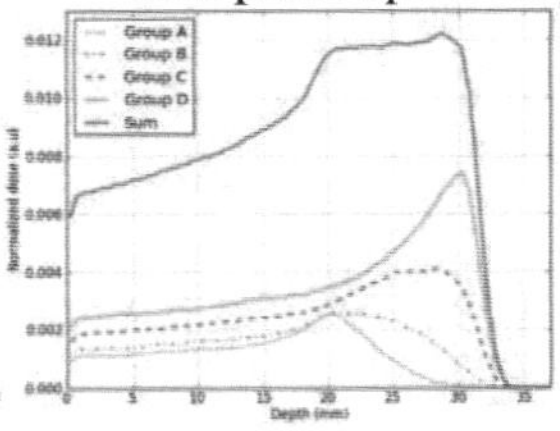

SOBP for the same four angular intervals of propeller rotation (lower four lines) and the overall SOBP (upper line).

To make visualization more trustworthy in 4D, we gave TOPAS a feature whereby scored volumes visibly darken as dose is accumulated (figure 5).

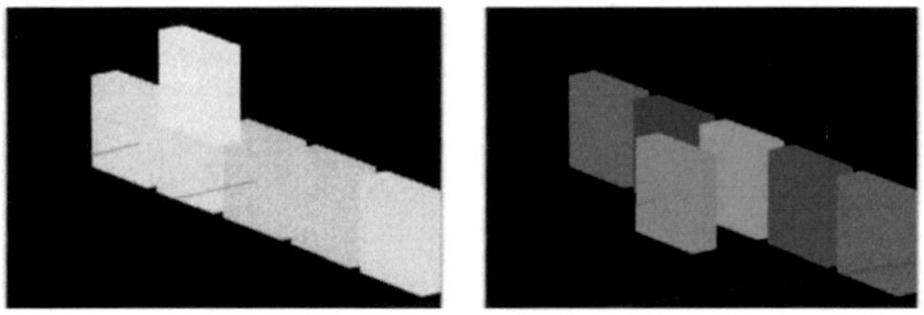

Figure 5. A beam of protons hits an array of sensitive films. Top film starts above beam path, moves downwards during irradiation. Final position, at right, shows moving film has darkened and has shielded film behind it.

6. Unit Testing versus End-to-End Testing

The more complex and flexible a simulation tool, the more formal its testing must be. Formal testing must include both Unit Testing and End-to-End testing. "Unit" refers to the smallest piece of the code, one particle source, one treatment head component, one scorer. To make the distinction, we step into metaphor. We show a single unit from the world of railroads: one brake shoe from one railroad car (figure 6). For a unit test, we use the minimum combination of parts to be able to test some fundamental behavior. Put a brake and the wheel together to test the stopping power.

Figure 6. Unit test. One wheel. One brake. Stopping power can be measured.

(Klotzbremse MaK 450 C.jpg, ©MdE Wikimedia Commons, CC-BY-SA 3.0 German)

Figure 7. Unit testing is not sufficient in complex systems, and end-to-end testing can never anticipate all situations.
The train wreck at Montparnasse, 1895 (Studio Lévy & fils).

When we combine many units into a larger system, that is an end-to-end test. End to end tests can get very complicated and include a huge number of independently moving parts or even human interaction. And though we get the stopping power right in our unit test, we may still get an unexpected result from an end-to-end test. End-to-End testing of a very complex system can never be perfect (figure 7) as there are too many possible overall states for each to be tested and human factors must be considered (this topic is addressed in detail in the growing field of "system safety" [10,11]).

Unit tests should be included in simulation codes. An example of such a test from TOPAS is as follows: since TOPAS scoring allows the user to segment a cylinder into rho, phi and z sections, we need to confirm that dose computed in one of those sections uses the correct volume of that section. The calculation involves trigonometry, correct counting of the number and extent of segments and involves caching values to avoid repeated calls to trigonometric functions during the MC's frequently-visited stepping loop. As these complications provide areas where a programming mistake could creep in, we provide a unit test - we shine a broad flat beam on that section and confirm that the scored dose equals what we expect from first principles of energy, density and volume. TOPAS has many such tests, from checking fluences to checking that the direction of our plus phi angle is as intended. The tests are distributed to our users as part of our included examples so they can replicate the results.

As End-to-End testing, we measure dose output from several entire beamlines in realistic setups. We don't rely on these tests to tell us that the units work; we rely on them to tell us that our process of putting modules together works. In an end-to-end test, two wrongs can make a right. For example, if we have a beam energy too high, but it passes through a block that is too thick, the dose beyond the block may be correct. But if you move one of those components, the errors then manifest.

7. Conclusion - QC is a Partnership

Developers of highly configurable, 4D MC simulation tools should unit test as many parts of their system as possible. They should End-to-End test some complicated setups, but must make it clear to end users that the tool developer cannot end-to-end test everything. QC is ultimately a partnership between the tool developer and the user. Key is that the developer be clear to the end user about what has been tested and what has not. As TOPAS developers, we have been fortunate that our users are medical physicists - by the nature of the profession they are the most skeptical of all scientists. As codes move to increasing flexibility, increasing 4D, medical physicists must stay skeptical, must insist on formal unit testing and must understand the severe limits of any one particular end-to-end test.

8. References

[1] Perl J, Shin J, Schümann J, Faddegon B and Paganetti H 2012 TOPAS: An innovative proton Monte Carlo platform for research and clinical applications. *Med. Phys.* **39,** 6818–6837

[2] Agostinelli S et al. 2003 Geant4—a simulation toolkit. *Nucl. Instrum. Methods A* **506,** 250–303

[3] Shin J, Perl J, Schümann J, Paganetti H and Faddegon B A 2012 A modular method to handle multiple time-dependent quantities in Monte Carlo simulations. *Phys. Med. Biol.* **57,** 3295–3308

[4] Allison J, Asai M, Barrand G, Donszelmann M, Minamimoto K, Perl J, Tanaka S, Tcherniaev E and Tinslay J 2008 The Geant4 visualisation system. *Comp. Phys. Comm.* **178**, 331-365

[5] Schümann J, Paganetti H, Shin J, Faddegon B A and Perl J 2012 Efficient voxel navigation for proton therapy dose calculation in TOPAS and Geant4. *Phys. Med. Biol.* **57,** 3281–3293

[6] Kawrakow I and Fippel M 2000 Investigation of variance reduction techniques for Monte Carlo photon dose calculation using XVMC. *Phys. Med. Biol.* **45**, 2163-2183

[7] Kawrakow I, Rogers D W O and Walters B R B 2004 Large efficiency improvements in BEAMnrc using directional splitting. *Med. Phys.* **31**, 2883– 2898

[8] Brualla L, Salvat F and Polanco-Zamora R 2009 Efficient Monte Carlo simulation of multileaf collimators using geometry-related variance-reduction techniques. *Phys. Med. Biol.* **54**, 4345–4360

[9] Ramos-Méndez J, Perl J, Faddegon B, Schümann J and Paganetti H 2013 Geometrical splitting technique to improve the computational efficiency in Monte Carlo calculations for proton therapy. *Med. Phys.* **40,** 041718

[10] Chou D 2012 Health IT and Patient Safety: Building Safer Systems for Better Care. *JAMA: The Journal of the American Medical Association, 308*(21), 2282-2282

[11] Leveson N G 2012 *Engineering a safer world: Systems thinking applied to safety*. MIT Press

Acknowledgments

This work was supported by NIH/NCI under R01 CA 140735-01.

Journal of Physics: Conference Series **489** (2014) 012022 doi:10.1088/1742-6596/489/1/012022

Accounting for the effects of carbon fibre treatment couches in the XiO treatment planning system.

M Roche, S M Mahou-Lago and T Crabtree

The Townsville Hospital, Qld, Australia,
IMB 29, The Townsville Hospital, 100 Angus Smith Drive, 4814, Qld, Australia

E-mail: michael_roche@health.qld.gov.au

Abstract. An investigation was conducted to determine the impact of two carbon fibre couch tops on treatment planning and to establish a method by which these couch tops may be accounted for in the XiO treatment planning system. In this study, the effects of the Civco Universal and the Medical Intelligence iBeam evo carbon fibre couch tops were investigated. Customised templates were developed in Focal to mimic the effects of these patient support devices and each template was then included in calculations performed by the XiO treatment planning system. The accuracy of modelling the couch tops in this manner was investigated and it was shown that, while using the customised couch templates, XiO modelled the increase in the surface dose due to treating through the couch tops at 180° to 1.0% ± 1.7%. The attenuation due to the presence of the couches was modelled to within 0.5% ± 0.4% for angles that pass through the flat central region of the couch tops and to within 1.4% ± 1.2% for angles that passed through their rapidly varying edges. When the couch templates were not included in XiO, the calculated dose at depth was recorded to be, in some cases, up to 5.3% ± 1.0% more than corresponding measured values. It was concluded that using the method described in this study it is possible to accurately model the effect of carbon fibre couch tops in the XiO treatment planning system.

1. Introduction

In the past carbon fibre therapy couch tops were assumed to be radio transparent because of their low density foam and carbon fibre composition. However, a number of studies have indicated that dose modification due to these devices can be significant [1-4]. The effects expected to be observed when treating through a carbon fibre couch with a mega-voltage photon beam include a reduction in the skin sparing effect, a shift in the point of maximum dose (d_{max}) and attenuation of the beam at depth.

Modern treatment techniques including stereotactic body radiation therapy (SBRT) [5], volumetric modulated arc therapy (VMAT) and intensity modulated radiation therapy (IMRT) have highlighted the negative effect carbon fibre couch tops can have on patient planning when the couch top isn't accounted for in the treatment planning system (TPS) [6-8]. This work quantifies the effects due to treating through these devices and examines the ability of a commercial treatment planning system (XiO v.4.70.02) in modeling these effects.

2. Materials and Methods

2.1. Carbon fibre treatment couches
Two independently manufactured therapy couch tops, the iBeam evo carbon fibre couch top and the Civco Universal couch top were investigated. The iBeam evo couch top manufactured by Medical Intelligence (Schwabmünchen, Germany) consists of two thin carbon fibre plates, 2 mm thick, sandwiching 46 mm of foam. The carbon fibre thickness increases to 4.5 mm towards the edges of the couch (iBEAM evo Couch top User Manual, Version 11 of 2007-11-15). The Civco Universal table top is of similar construction with two thin carbon fibre plates approximately 1.8 mm thick, sandwiching approximately 46.5 mm of foam. To determine the effect due to the two couch tops, percentage depth dose (PDD) and attenuation measurements were made on both Elekta Axesse and Elekta Synergy linear accelerators.

2.2. Surface dose measurements
All PDD measurements were made in 30x30 cm^2 slabs of Virtual Water™ using a PTW Advanced Markus type 34035 parallel-plate ionisation chamber. The central axis of the beam was normal to both the couch and phantom surface and PDD measurements were made at a source to surface distance (SSD) of 100 cm, at both 6 and 10 MV, and a field size of 10x10 cm^2. The ionisation current was measured with a PTW Webline electrometer. Results are displayed as the average ionisation at positive and negative polarity normalised to the point of maximum dose.

2.3. Attenuation measurements
Attenuation measurements were made in Virtual Water™ (30x30 cm^2) using an IBA Farmer type ionisation chamber (FC65-G) and PTW Webline electrometer. The chamber was placed at the centre of the Virtual Water™ phantom and positioned so that the centre of the chamber was at the linear accelerator's isocentre. A source to axis distance (SAD) of 100 cm was used for both 6 and 10 MV and field sizes used included 5x5 cm^2 and 10x10 cm^2. Measurements were taken at 10° intervals from 0°-350° and the attenuation due to the couch was calculated as the percentage difference of the charge measured with the couch in the path of the beam, to the charge measured with the couch out of the beam's path.

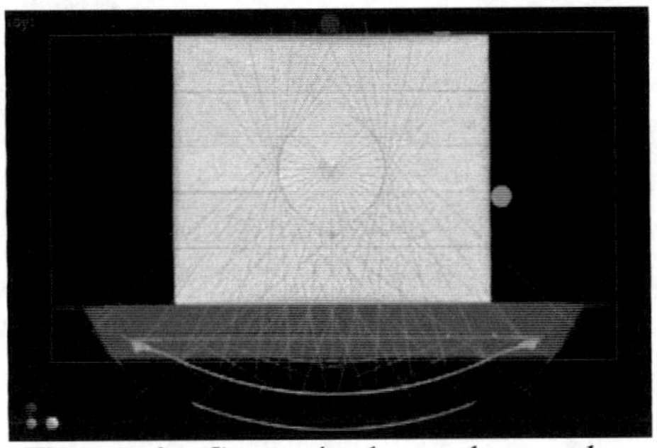

Figure 1: Customised couch template developed in Focal to mimic the MI iBeam evo Universal couch top.

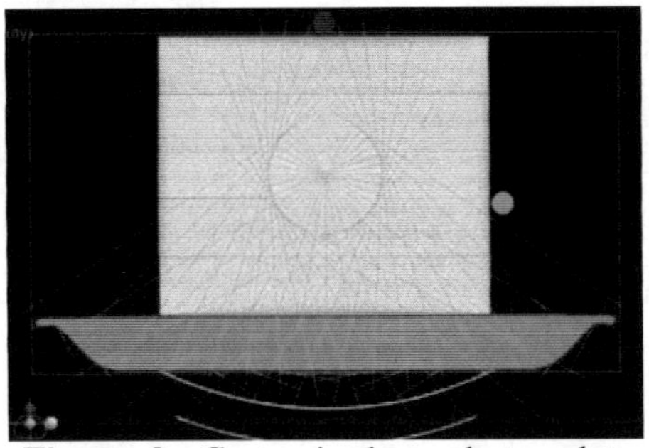

Figure 2: Customised couch template developed in Focal to mimic the Civco Universal couch top.

2.4. Treatment Planning System
To mimic the effects of the couch tops in the XiO treatment planning system, customized couch templates were developed (Figures 1 and 2). To construct the customised couch templates Elekta's distributed planning system Focal v.4.70.00 was used. A CT dataset of the iBeam evo and Civco

couch tops were obtained on a Siemens Sensation Open scanner and two contours were created in Focal, one for the foam core and one for the carbon fibre. The average relative electron density of the carbon fibre (0.55) and foam (0.10) sections of the couch tops were obtained from the CT data set and each contour was forced to the relevant value. To ensure that the couch structure is included in XiO's calculations, the customised couch templates were included in the external patient contour before exporting a plan from Focal to XiO.

2.5. Template Validation

To validate the customised templates, PDD and attenuation values were calculated using the superposition algorithm in the XiO treatment planning system and compared to those measured. Three prostate IMRT plans were then calculated in XiO before and after including the customised couch templates in the patient plan. Dose maps calculated were compared with measurements taken using a 2-D diode array (MapCHECK™).

3. Results and discussion

3.1. Percentage Depth Dose

For 6 MV an increase in surface dose (Ds) of 74.7% ± 1.8% was measured for the iBeam evo couch, while for 10 MV the increase in Ds was measured to be 67.6% ± 1.7%. Using the customised iBeam evo couch template XiO was able to calculate the increase in Ds due to passing through the couch at 180° to within 1.0% ± 1.7%. Similar results were obtained for the Civco Universal couch template. When treating through the couch tops, a shift towards the surface in d_{max} was also observed. This shift is illustrated in figure 3(a) and (b) and was measured to be 8.0 mm ± 1.4 mm for 6 MV and 7.0 mm ± 1.4 mm for 10 MV. Again this shift was accounted for when the couch templates were included in XiO's calculation.

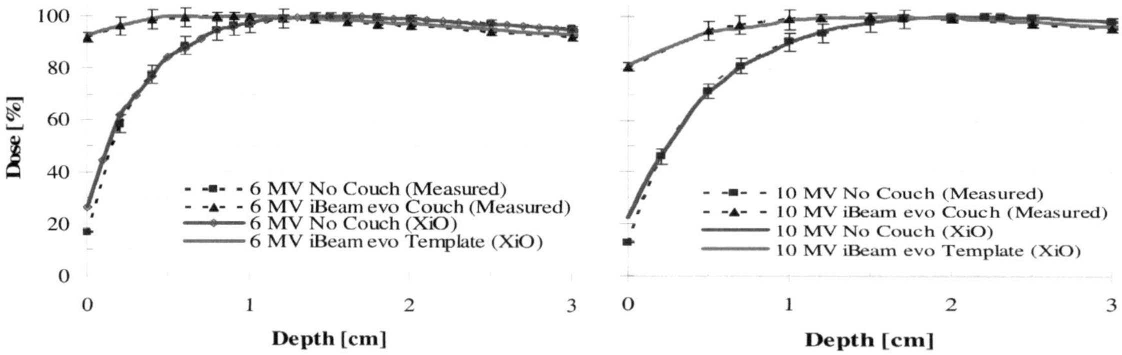

Figure 3(a), (b): 6 MV and 10 MV PDD measured (Gantry angle: 0°, SSD: 100 cm, Field Size: 10 x 10 cm^2) and calculated in XiO before and after transmission through the iBeam evo couch and iBeam evo couch template.

3.2. Attenuation

When the iBeam evo couch was in the path of the beam the maximum attenuation observed at 12 cm depth occurred at a gantry angle of 130° for a 5 x 5 cm^2 field size. This attenuation was 5.3% ± 1.0% for 6 MV and 4.2% ± 0.9% for 10 MV. At the same angle and depth but for a 10 x 10 cm^2 field size the attenuation recorded was 4.9% ± 0.9% for 6 MV and 3.8% ± 0.8% for 10 MV. Similar results were measured for the Civco Universal couch. From these results it was possible to conclude that attenuation due to the couch tops increases as field size is decreased, as energy is decreased and at posterior oblique angles.

When modelling the attenuation using the customised couch templates, calculated values through the flat central regions of the couch templates were within 0.5% ± 0.4% of measured values, while calculated values through the larger varying edges of the templates were within 1.4% ± 1.2% of measured values. This was attributed to the rapid change in the structure at the edge of the couches.

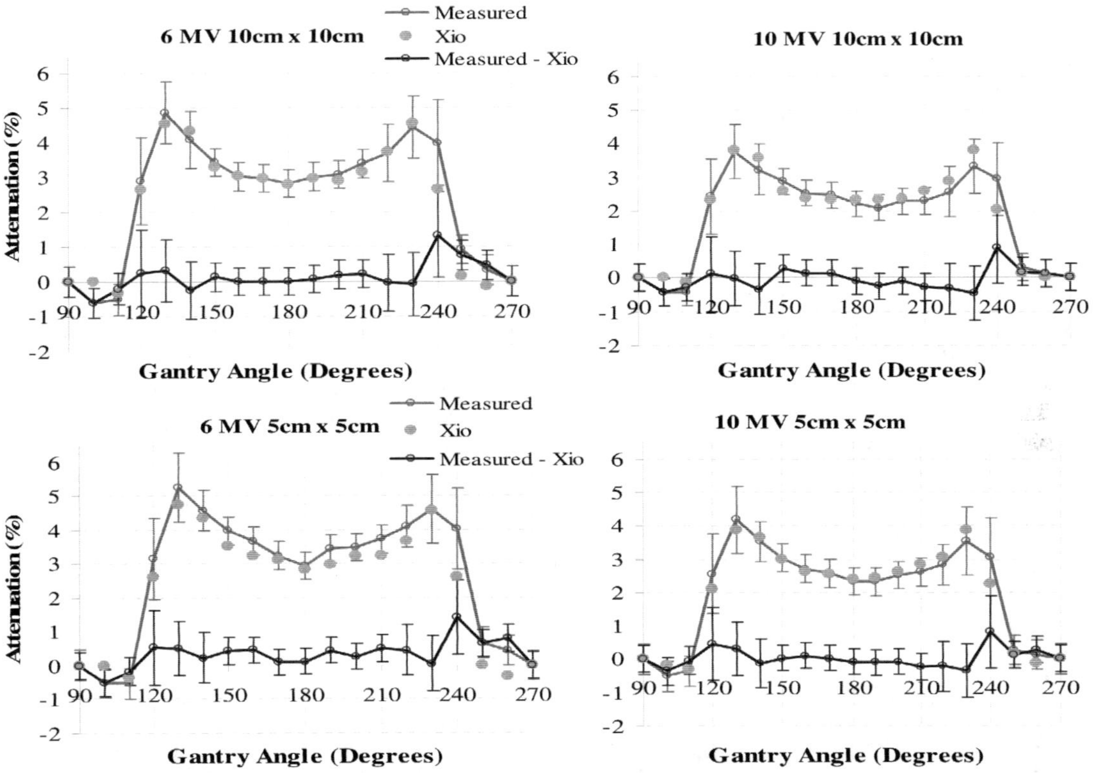

Figure 4: 6 MV and 10 MV attenuation measured in a phantom and calculated in XiO. Also displayed is the difference between measured and calculated attenuation. (Gantry angles: 0° to 360°, SAD: 100 cm, Measurement Depth: 12 cm, Field Sizes: 10x10 cm^2 and 5x5 cm^2).

3.3. Validation: IMRT Analysis
Results showed that calculated dose maps differed from those measured when the iBeam evo couch was not accounted for in XiO. Both planned and measured dose maps were compared using the gamma criteria of 3% dose difference and 2 mm distance to agreement (DTA), with a gamma pass rate of 90.0%. When the couch was accounted for in the measurements but not in XiO, the pass rate was 85.5%. However, when the couch was taken into account in XiO, the pass rate was 90.1%. Similar results were obtained for the Civco couch top.

4. Conclusions
This investigation has highlighted the clinically significant effect the iBeam evo and Civco Universal couch tops can have on patient planning. The dosimetric properties of both couches were quantified and a method of accounting for the couches in the XiO treatment planning system was discussed. If the customised couch templates are included in the patient plan, it has been shown that XiO v4.70.02 accurately models the attenuation due to the presence of the couch to within 1.4% ± 1.2%. It has also been shown that using this process XiO modelled the reduction in the skin sparing effect, observed by treating through the couch at 180°, to within 1.0% ± 1.7%. Furthermore, dose map analyses of IMRT

plans indicate a further clinical benefit of including the customised couch templates during treatment planning.

References

[1] Smith D, Christophides D, Dean C, Naisbit M, Manson J and Morgan A 2010 Dosimetric characterization of the iBeam evo carbon fiber couch for radiotherapy *Med. Phys.* **37** 3595-3606

[2] Mihaylov I, Corry P, Yan Y, Ratanatharathorn V and Moros E 2008 Modeling of carbon fiber couch attenuation properties with a commercial treatment planning system *Med. Phys.* **35** 4982-88

[3] Simpson J and Godwin G 2011 The effect of the iBeam evo carbon fiber tabletop on skin sparing *Med. Dos.* **36** 330-3

[4] Gerig L, Niedbala M and Nyiri B 2009 Dose perturbations by two carbon fiber treatment couches and the ability of a commercial treatment planning system to predict these effects *Med. Phys.* **37** 322-8

[5] Hope B, Laser B, Kowalski A, Fontenla S, Pena-Greenberg E, Yorke E, Lovelock M, Hunt M and Rosenzweig K 2008 Acute skin toxicity following stereotactic body radiation therapy for stage 1 non-small-cell lung cancer: Who's at risk? *Int. J. Rad. Onc. Biol. Phys.* **72** 1283-6

[6] Li H, Lee A, Johnson J, Zhu R and Kudchadker R 2011 Characterization of dose impact on IMRT and VMAT from couch attenuation for two Varian couches *J. Appl. Clin. Med. Phys.* **12** 23-31

[7] Mihaylov I, Bzdusek K and Kaus M 2011 Carbon fiber couch effects on skin dose for volumetric modulated arcs *Med. Phys.* **38** 2419-23

[8] Pulliam K, Howell R, Followill D, Luo D, White R and Kry S 2011 The clinical impact of the couch top and rails on IMRT and arc therapy *Phys. Med. Bio.* **56** 7435-47

Acknowledgments

This work was kindly supported by the Pump Priming Research Fund for Health Practitioners, The Townsville Hospital, Townsville, Australia. The authors wish to acknowledge all assistance provided by the staff of the Department of Medical Physics and the Department of Radiation Therapy, The Townsville Hospital, Townsville, Australia.

Calculation of the characteristics of clinical high-energy photon beams with EGS5-MPI

M. Shimizu, Y. Morishita, M. Kato, T. Kurosawa, T. Tanaka, N. Takata, and N. Saito
Quantum Radiation Division, National Metrology Institute of Japan (NMIJ), AIST, Tsukuba, Ibaraki 305-8568, Japan
morihito-shimizu@aist.go.jp

Abstract
A graphite calorimeter has been developed as a Japanese primary standard of absorbed dose to water in the high-energy photon beams from a clinical linac. To obtain conversion factors for the graphite calorimeter, the beam characteristics of the high-energy photon beams from the clinical linac at National Metrology Institute of Japan were calculated with the EGS5 Monte Carlo simulation code. To run the EGS5 code on High Performance Computing machines that have more than 1000 CPU cores, we developed the EGS5 parallelisation package "EGS5-MPI" by implementing a message-passing interface. We calculated the photon energy spectra, which are in good agreement with those previously calculated by D. Sheikh-Bagheri and D. W. O. Rogers (*Med. Phys.* **29** 3). We also estimated the percentage-depth-dose distributions of photon beams from the linac using the calculated photon energy spectra. These calculated percentage-depth-dose distributions were compared with our measured distributions and were found they are in good agreement as well. We will calculate conversion factors for the graphite calorimeter using our results.

1. Introduction

The Japanese primary absorbed dose standard in ^{60}Co-γ ray field was established in 2011 by using a graphite calorimeter. As the next step in the absorbed dose standard development, the National Metrology Institute of Japan (NMIJ) has been developing the absorbed dose standard in high-energy photon beams from a clinical linac using a graphite calorimeter. The calorimetric method using a graphite calorimeter requires a conversion factor from the dose to graphite to the dose to water. Further, evaluation of the characteristics of photon beams plays an important role in the determination of the conversion factor [1].

The characteristics of the high-energy photon and electron beams from a clinical linac are important for estimating the various correction factors of clinical radiation dosimetry. The beam characteristics have been estimated experimentally and numerically [2–4]. Kawrakow *et al.*, Rogers *et al.*, and Hirayama *et al.* developed Monte Carlo (MC) calculation codes [5–7]. These codes can be used for calculating the characteristics or the dose distribution of the high-energy photon and electron beams from a clinical linac. In particular, the BEAM code developed by Rogers *et al.* is a very suitable MC code for calculating photon beams from a clinical linac [6]. BEAM codes are widely used for the dosimetry on the radiation therapy using a clinical linac. However, these calculation codes do not support the message-passing interface (MPI) or any other parallel calculation techniques. Further, it is very difficult to run BEAM codes on a cluster machine, which has many nodes and CPU cores, because users have to manually manage the random seed number and other parameters of each calculation process.

In this work, we have developed the "EGS5-MPI," which is the extended package of the MC calculation code Electron Gamma Shower version 5 (EGS5), for parallel calculation using MPI. To obtain the conversion factor of a graphite calorimeter, we calculated the characteristics of the high-energy photon beams from the clinical linac at NMIJ. We then compared the percentage-depth-dose (PDD) distributions estimated experimentally and numerically.

2. Method

To obtain the conversion factors for the graphite calorimeter, we calculated the beam characteristics of the high-energy photon beams from the clinical linac at NMIJ. To calculate as many cases as possible to reduce statistical uncertainty, we developed the extended EGS5-MPI package for using MPI on the EGS5 MC code.

2.1. EGS5-MPI

The EGS5-MPI provides the subroutines and common variables for managing the random seed number and the MPI environment. In the EGS5-MPI calculation code, users can call the MPI subroutines. Therefore, we used the MPI subroutines for final calculation results or for changing the calculation parameters during each process. The most important function of EGS5-MPI is the random seed number management. The random seed number (*inseed*) of EGS5 for each MPI process is generated automatically by the root process (which has an MPI process rank number of 0) and provided through the MPI communication protocol. The root process has the main random seed number, which is only input manually by user. The EGS5-MPI calculation code can be executed on UNIX-like operating systems (Linux, Cygwin, Free-BSD, Mac OSX, etc.). In many cases, the EGS5-MPI code can be executed on a large cluster machine that has more than 1000 CPU cores.

2.2. Calculation of the beam characteristic

We calculated the beam characteristics of 6, 10, and 15 MV high-energy photon beams from the clinical linac at the NMIJ by using EGS5-MPI. We used the geometry and material data of the clinical linac from the information provided by Elekta K. K. The energy distribution and beam width of the incident electron beam at its X-ray target is also input along with the information provided by Elekta K. K. The electron cut-off energy (*ECUT*) is 0.7 MeV, and the range rejection (*ESAVE*) of the X-ray target and other linac parts are 0.7 MeV and 2.0 MeV, respectively. Other calculation parameters are listed in Table 1. The energy spectra of the photon beams were calculated on our home-made cluster machine (AMD Phenom II, 3.2 GHz, six cores, four nodes). We simulated the photon energy spectra with about 10^{10} incident electrons, and all information (charge, energy, position, direction) about the photons and the electrons at the detector plane (Source to Surface Distance (*SSD*) = 80 or 100 cm) was saved as binary data.

We also calculated the PDD with 10^{10} photons by using the HPC machines at the Univ. of Tokyo and Kyushu Univ., which have more than 1000 CPU cores. In the depth-dose calculation, *ECUT* and *ESAVE* were 0.521 MeV and 0.6 MeV, respectively. Other parameters were essentially the same as those of the beam-characteristics calculation. To reduce the calculation time, we assumed that the photon beam had a uniform distribution in a plane perpendicular to the photon beam direction, and the secondary electrons and scattered photons were not included in the photon beam. The calculation time of the PDD was about 5 h. The parallelisation efficiency of the PDD calculation was 99.999%. The EGS5-MPI is considered very suitable for calculation on the HPC machines. To compare the calculated and experimental results, we measured the PDD distribution of the NMIJ linac by using a 3D water phantom (Blue Phantom, IBA Inc.) and a pinpoint ionisation chamber (CC-13, Scanditronix Wellhofer).

Table 1. Calculation parameters used in EGS5 for the beam characteristics calculation

ECUT = 0.70	Energy cut-off for electrons (MeV)
PCUT = 0.01	Energy cut-off for photons_ (MeV)
ESAVE (Target) = 0.70	Range rejection on x-ray target (MeV)
ESAVE (Other parts) = 2.0	Range rejection (MeV)
IBRDST = 1	Koch and Motz formula for Bremsstrahlung photons
IPRDST = 1	Motz, Olsen, and Koch formula for pair production
IPTHER = 1	Switches for PE-angle sampling
IEDGFL = 1	K & L-edge fluorescence
IAUGER = 1	K & L-Auger
IRAYLR = 1	Rayleigh scattering
LPOLAR = 1	Linearly polarised photon scattering
INCOHR = 1	S/Z rejection
IPROFR = 1	Doppler broadening
IMPACR = 1	Electron impact ionisation

3. Results and Discussion

The calculated energy spectrum of a 6 MV photon beam from the clinical linac at NMIJ is shown in Fig. 1. The x-axis represents the photon energy, and the y-axis represents the photon counts per 100 incident electrons. For comparison, the energy spectrum for an Elekta linac calculated by Sheikh-Bagheri and Rogers using the BEAM code [4] is also shown in Fig. 1. As can be seen clearly, the overall shape of the present calculated energy spectrum is in good agreement with that by Sheikh-Bagheri. This agreement suggests that our parallel MC calculation code using EGS5-MPI can provide the same results as that obtained using the BEAM code.

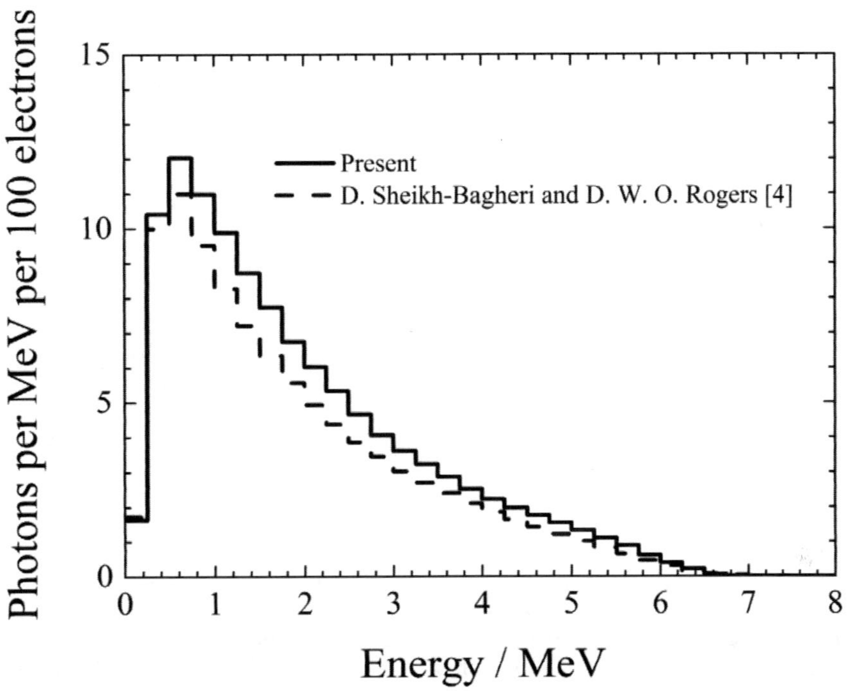

Figure 1. Energy spectra of 6 MV photon beams from an Elekta clinical linac. The solid line shows the present calculation result. The dashed line shows Sheikh Bagheri's calculation result (*Med. Phys.* **29** 3) [4]. These data show the energy spectra for the region $0 < r < 2.25$ cm at the detector plane ($SSD = 100$ cm).

We have calculated the PDD distributions of photon beams. The calculated and experimental PDD distributions of the 15 MV photon beam are shown in Fig. 2. The x-axis represents the water depth, and the y-axis represents the normalised dose. The differences at each depth point are also shown in Fig. 2. As can be seen, the experimental PDD distribution agrees well with the calculated one. This result strongly supports the validity of our calculation method. However, there are small systematic differences near the depth of the maximum dose. These differences arise from the electrons emitted from the linac head [4]. The contribution of the secondary electrons on the surface dose distribution is only a few percent of the photon beam contribution, and the range of the secondary electrons in water is 3–5 cm.

To take the secondary electrons into account, we have calculated PDD distributions more precisely with the binary data of photon beams, which include the data of all particles (secondary electrons and scattered photons, etc.) from the linac head. The precise calculations are also shown as the open symbols in Fig. 2. As can be seen, the systematic differences between the calculated and the experimental PDD distributions disappeared. However, the statistical uncertainty increased because the number of calculation events in the binary data is smaller than 10^9 cases due to the calculation time.

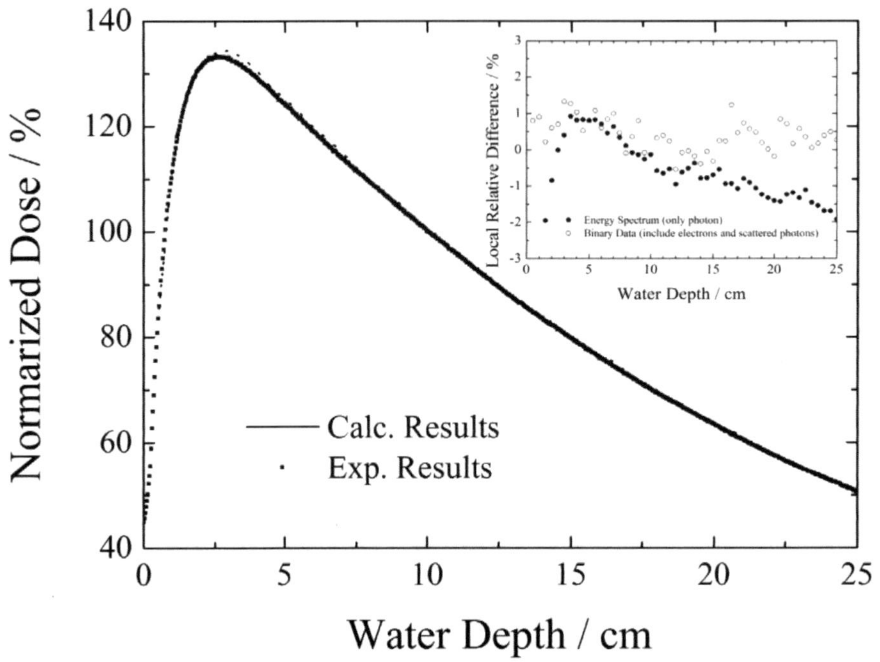

Figure 2. Percentage depth dose (PDD) distribution of a 15 MV photon beam (*SSD* =90 cm). The dotted line shows the experimental distribution. The solid line shows the calculated result. The differences between the experimental and calculated distributions are shown in top-right inset. The filled circles show the differences for the PDD distribution calculated using the energy spectrum of the photon beam. The open circles show the differences for the PDD distribution using the binary data of the photon beam, which include the secondary electrons and scattered photons.

4. Conclusion

The beam characteristics of high-energy photon beams from the clinical linac at NMIJ have been calculated with EGS5-MPI. The calculated photon energy spectra are in good agreement with the previous ones by Sheikh-Bagheri and Rogers [4]. For comparison with the experimental results, we also calculated the percentage depth dose distribution using the energy spectra of photon beams. These calculated PDD distributions also agree with the experimental ones. EGS5-MPI is a very powerful tool for EGS5 parallelization on the HPC machine. EGS5 users can develop the parallel MC calculation on their desktop PCs and execute it on the HPC machines. In future works, we will calculate the conversion factors for our graphite calorimeter using our present energy spectra results. The new primary absorbed dose standard for high-energy photons from a clinical linac is going to be established in 2013.

This work is partially supported by HPCI Systems Research Projects (Project ID hpci120020) and the support program for young and women researchers of University Tokyo.

References

[1] Y. Morishita, M. Kato, N. Takata, T. Kurosawa, T. Tanaka and N .Saito, *Rad. Prot. Dosimetry* **154** (2012) 331–339.
[2] "*Dosimetry of High-Energy Photon Beams based on Standards of Absorbed Dose to Water*", *ICRU Report* **64** (2001).
[3] R. Mohan, C. Chui, and L. Lidofsky, *Med. Phys.* **12** (1985) 592–597.
[4] D. Sheikh-Bagheri and D. W. O. Rogers, *Med. Phys.* **29** (2002) 3.
[5] I. Kawrakow, E. Mainegra-Hing, D. W. O. Rogers, F. Tessier and B. R. B. Walters, *NRCC Report PIRS-***601** (2003)
[6] D. W. O. Rogers, B. A. Faddegon, G. X. Ding, C-M. Ma, J. We and T. R, Mackie, *Med. Phys.* **22** (1995) 503–524.
[7] H. Hirayama, Y. Namito, A. F. Bielajew, S. J. Wilderman and W. R. Nelson, *SLAC-R-***730** and **KEK Report 2005-8**.2005.

XVII International Conference on the Use of Computers in Radiation Therapy (ICCR 2013) IOP Publishing
Journal of Physics: Conference Series **489** (2014) 012024 doi:10.1088/1742-6596/489/1/012024

Motion artifact detection in four-dimensional computed tomography images

G Bouilhol[1,2], M Ayadi[2], R Pinho[2], S Rit[1,2] and D Sarrut[1,2]

[1] University of Lyon, CREATIS; CNRS UMR 5220; Inserm U1044; INSA-Lyon; University Lyon 1, Villeurbanne, France

[2] Department of Radiation Oncology, Leon Berard Cancer Center, Lyon, France

E-mail: gauthier.bouilhol@gmail.com

Abstract. Motion artifacts appear in four-dimensional computed tomography (4DCT) images because of suboptimal acquisition parameters or patient breathing irregularities. Frequency of motion artifacts is high and they may introduce errors in radiation therapy treatment planning. Motion artifact detection can be useful for image quality assessment and 4D reconstruction improvement but manual detection in many images is a tedious process. We propose a novel method to evaluate the quality of 4DCT images by automatic detection of motion artifacts. The method was used to evaluate the impact of the optimization of acquisition parameters on image quality at our institute. 4DCT images of 114 lung cancer patients were analyzed. Acquisitions were performed with a rotation period of 0.5 seconds and a pitch of 0.1 (74 patients) or 0.081 (40 patients). A sensitivity of 0.70 and a specificity of 0.97 were observed. End-exhale phases were less prone to motion artifacts. In phases where motion speed is high, the number of detected artifacts was systematically reduced with a pitch of 0.081 instead of 0.1 and the mean reduction was 0.79. The increase of the number of patients with no artifact detected was statistically significant for the 10%, 70% and 80% respiratory phases, indicating a substantial image quality improvement.

1. Introduction

Motion artifacts appear in four-dimensional computed tomography (4DCT) images because of suboptimal acquisition parameters or patient breathing irregularities. Occurrence frequency of motion artifacts is high and they may introduce errors in radiation therapy treatment planning. Motion artifacts generally appear as discontinuities in voxel intensities perpendicularly to the scanning direction. They appear at the interface between groups of slices called stacks and composed of slices acquired during the same breathing cycle and attributed to the same respiratory phase.

Motion artifact detection can be useful for image quality assessment [1, 2, 3] and 4D reconstruction improvement [4, 5] but manual detection in many 4DCT images [6] is a tedious process and it is observer-dependent. Various algorithms have been developed to automatically detect motion artifacts. Discontinuities induced by motion artifacts can be detected by measuring the mean squared gray value difference between adjacent slices [4]. They can also be detected by measuring the difference between observed and predicted slices at the interface between stacks [5]. However, these methods have a limited efficiency because of their sensitivity to anatomic variations in the scanning direction. It is also possible to predict the occurrence of artifacts directly from the respiratory signal but according to Han *et al.* [2] poor efficiency is associated with this method. For a given interface between two stacks, the method described by Han *et al.* consists of finding among the other reconstructed phases, a bridge stack containing common slices with each of the two considered stacks. A rigid spatial registration is then performed between each stack and the bridge stack. The registration result (contiguous, overlapping or separated stacks) then gives an indication of the presence or absence of an artifact at the interface of the two stacks. The advantage of this method relies on the ability to identify the artifact type (duplication or overlapping), but it has several drawbacks and limitations. First, access to the entire 4DCT dataset is required. In addition, as noted by the authors, the correct identification of bridge stacks is essential but it may not be possible in case of large irregular respiratory motions (leading to high magnitude artifacts). This method is not sensitive to anatomic changes in the scanning direction but it can be influenced by the deformation of structures between the respiratory phases. Finally, it can only be used for helical acquisitions since in cine mode the stacks contain the same slices whatever the phase so that it is impossible to find a stack with common sections with each of the two considered stacks.

Given the limitations of existing algorithms, a novel method for automatically detecting motion artifacts in 4DCT images has been proposed. The optimization of acquisition parameters at our institute led to a change in the pitch value used for treatment planning. The initial value of 0.1 has been changed to a value of 0.081. Our method was used to assess the image quality and thus quantify the impact of this optimization on the presence of motion artifacts in a large patient database.

2. Materials and methods

2.1. Artifact detection

The proposed method is based on Ehrhardt *et al.* [4]. It consists of computing the p-th power of the n-th order absolute difference of the mean gray value between adjacent slices:

$$D(z) = \left|\Delta^n \overline{I}(z)\right|^p \tag{1}$$

where $\overline{I}(z)$ is the mean gray value in the slice z. In a preliminary study, n = 3 and p = 7 were found to be suitable for thoracic acquisitions with the parameters used at our institute (see *Patient database* section). This brings out the strongest gray value discontinuities and lower the smallest ones such as those due to anatomical changes.

To improve the sensitivity to artifacts and further limit the sensitivity to anatomical changes, images were first automatically cropped in the left-right and antero-posterior directions to fit the lung boundaries and the detection was performed separately for each side (right and left) of the cropped region. This prevents the false detection of artifacts at the top of the diaphragmatic dome or at the apex of the lungs when these structures are located at the same height on both sides. Separation of right and left also improves the sensitivity to artifacts clearly appearing in only one of the two sides. In addition, only the voxels within the limits of the patient are considered to avoid the influence of objects outside of the patient (e.g. immobilization systems). This improvement step relies on the segmentation of the lungs and the patient boundaries. It could be automatically performed using previously described methods [7, 8].

Furthermore, the detection was performed using two grayscale windows excluding bone structures. Voxels of intensity between -2000 and 220 HU (lung window) were considered between the superior-inferior limits of the lungs, and voxels of intensity between -180 and 220 HU (soft tissue window) below the lungs. This way, sudden changes in voxel intensity due to the bone structures or gas in the abdomen are ignored.

Artifacts are then detected and located by applying two thresholds, s_1 (lung window) and s_2

(soft tissue window), above which the values of D correspond to discontinuities considered as artifacts.

Finally, a minimum distance of 3 pixels (in the scanning direction) between two artifacts is introduced in order to limit the detection of more than one artifact within an area smaller than a stack. The principle of this artifact detection method is shown in Figure 1.

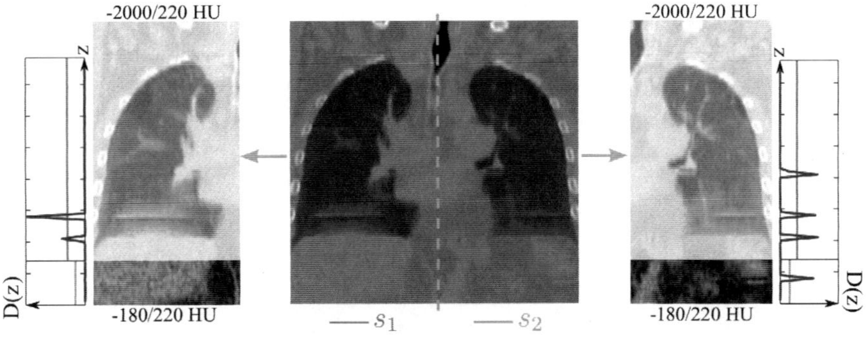

Figure 1. Principle of the artifact detection method. The red area represents the result of the automatic segmentation of the lungs used to crop the image. Blue spikes indicate where artifacts occur.

2.2. Threshold adjustment

The adjustment of the detection thresholds, s_1 and s_2, is automatic and specific to each image. Based on a heuristic, the adjustment method takes into account that the number of detected artifacts not only depends on the thresholds and the number of artifacts in the image, but also on the patient's morphology and the scanned length. Discontinuities appear at the interface between structures of different densities. Consequently, the variation of the average intensity $I(z)$ between two slices is not the same for thin or obese patients (contrast variation, especially in the abdomen), or for patients with many or few dense structures in the lungs. It is therefore possible to adjust the detection thresholds based on the number of detected artifacts. The first step is to determine the numbers of artifacts detected with the two grayscale windows, $h_1(s)$ and $h_2(s)$ respectively, for thresholds ranging from 1 to 10^{2p} on a logarithmic scale. The choice of s_1 and s_2 is then carried out in different ways for the two windows. Low and high threshold limits are introduced to improve the detection for patients of unusual morphology.

For the lung window, the threshold s_1^{max} corresponding to the largest number of detected artifacts is determined. If multiple values match, the smallest one is selected. The decreasing cumulative number of artifacts, $H_1(s)$, is then calculated for s ranging from 10^{2p} to s_1^{max}. The line passing through $H_1(s_1^{max})$ and the previous value, $H_1(s_1^{max+})$, is then determined and its intersection with the x-axis was found to be a good estimate of s_1. If s_1 is smaller than a minimum value s_1^{inf} or if no artefact is detected above an upper limit s_1^{sup}, then no artifact is detected. Figure 2 illustrates the s_1 adjustment method.

For the soft tissue window, the threshold s_2^{max} corresponding to the largest number of detected artifacts is also determined. If multiple values match, the largest one is selected and $s_2 = s_2^{max}$. If s_2 is smaller than a minimum value s_2^{inf} then $s_2 = s_2^{inf}$, and if no artifact is detected above an upper limit s_2^{sup}, an intermediate value s_2^{int} is chosen as s_2. Figure 3 illustrates the s_2 adjustment method.

We can see that when the threshold value decreases, the number of detected artifacts increases to a maximum and then decreases (see Figures 3 and 4). This is due to the minimum distance

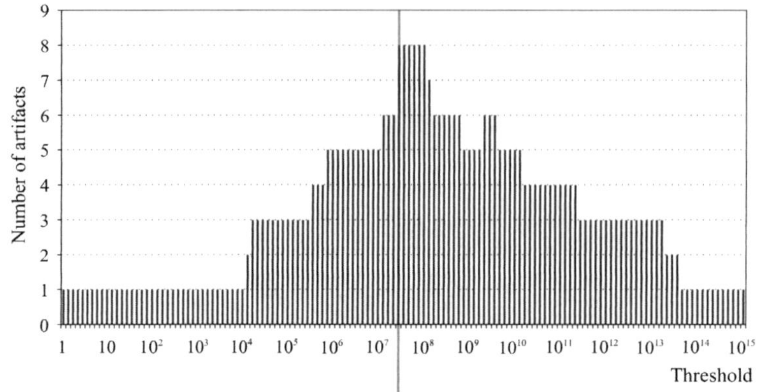

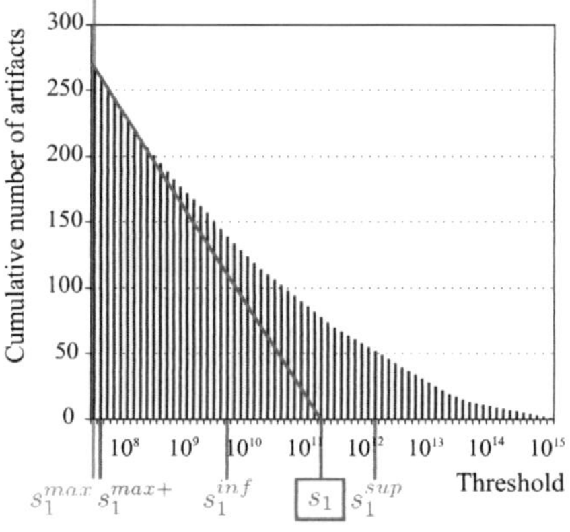

Figure 2. Adjustment method of s_1. The line passing through $H_1(s_1^{max})$ and $H_1(s_1^{max+})$ is drawn in red.

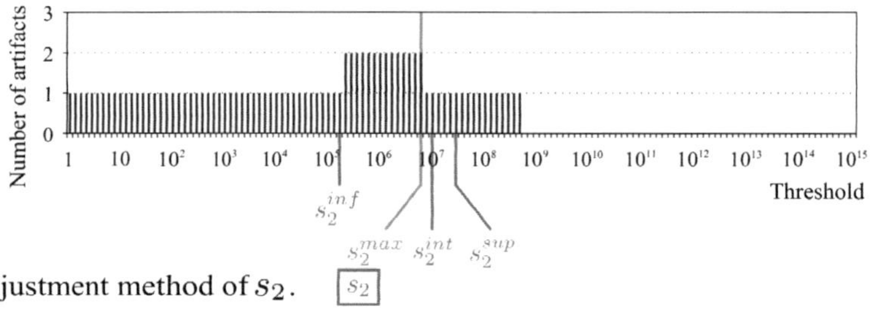

Figure 3. Adjustment method of s_2.

of 3 pixels imposed between two consecutive detected artifacts. Indeed, for low threshold values, the number of sets of three consecutive slices for which the value of D is smaller than the threshold is low. The determination of optimal values of s_1^{inf}, s_1^{sup}, s_2^{inf}, s_2^{int} and s_2^{sup} was performed using two leave-one-out cross-validations, one for each window. To do so, artifacts were visually identified in 27 respiratory phases randomly selected from the 4DCT scans of 25 patients in the database. For each image, the analysis consisted of assessing the sensitivity Se and specificity Sp for the 26 other images and for a large number of values s_1^{inf}, s_1^{sup}, s_2^{inf}, s_2^{int} and s_2^{sup}. To measure the specificity, the number of true negatives was determined assuming a stack size equal to the collimation width. It corresponds to the maximum stack size [2] and thus a low estimate of the specificity. The product $Se \times Sp$ was calculated to determine the optimal values for each of the 27 combinations.

2.3. Patient database

4DCT images of 114 lung cancer patients were analyzed with the proposed method. Each patient dataset consisted of 10 respiratory phases reconstructed with a 3 mm slice thickness. Acquisitions were performed with a Brillance Big Bore CT scanner (Philips) in helical mode using 400 or 800 mAs/slice, a collimation width of 16 $\times$ 1.5 mm and a rotation period of 0.5 seconds. For 74 patients, a pitch of 0.1 was used and for 40 patients, the pitch was 0.081. Given a maximum acquisition time of 120 seconds, the maximum scanned length is about 45 cm with a pitch of 0.081, which is suitable for the acquisition of the entire lung.

3. Results

3.1. Threshold adjustment

Optimal values $s_1^{inf} = 7 \times 10^9$ and $s_1^{sup} = 3 \times 10^{12}$ were obtained for each of the 27 combinations and $s_2^{inf} = 3 \times 10^5$, $s_2^{int} = 1 \times 10^7$ and $s_2^{sup} = 5 \times 10^7$ were obtained for 22 of the 27 combinations. With these values, a sensitivity of 0.73 and a specificity of 0.97 were observed. Hence the corresponding product $Se \times Sp$ was 0.70.

3.2. Patient database

Automatic detection of motion artifacts in the 4DCT images of 114 patients showed that 92% of patients had at least one artifact in one of the phases with a pitch of 0.1, and 85% with a pitch of 0.081. In addition, the average number of respiratory phases affected by artifacts was 6.2 with a pitch of 0.1, and 5.2 with a pitch of 0.081. The 40%, 50% and 60% respiratory phases were the least affected by artifacts. These phases are indeed the most stable since they are located near end-exhale. In phases where motion speed is high (10%, 20%, 30%, 70%, 80% and 90%), the number of detected artifacts was systematically reduced with a pitch of 0.081 instead of 0.1 and the mean reduction was 0.79 (SD 0.23). In the other phases where motion speed is low, the mean reduction was 0.09 (SD 0.17). The reduction of the number of detected artifacts induced an increase of the number of patients with no artifact detected. This increase was statistically significant for the 10% (p = 0,041), 70% (p = 0,004) and 80% (p = 0,025) respiratory phases (see Figure 4).

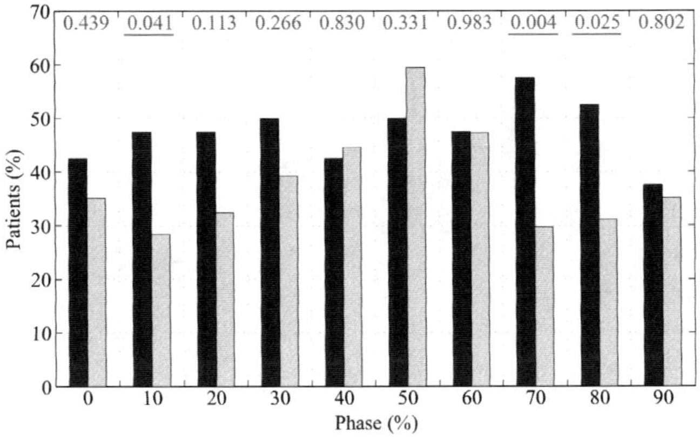

Figure 4. Percentage of patients with no artifact detected according to the respiratory phase for a pitch of (light blue) 0.1 and (dark blue) 0.081. In green are presented the p-values and those reflecting a significant difference are underlined.

4. Discussion

A motion artifact detection method has been proposed. The basic principle is similar to the method described by Ehrhardt *et al.* [4] but several improvements are proposed by our method. The $Se \times Sp$ score of 0.70 obtained in this study is comparable to the results obtained by Han *et al.* [2] who reported a mean $Se \times Sp$ value of 0.72 (5 patients) but our method does not require access to the entire 4DCT dataset and it is not limited to helical acquisitions. Despite a slightly lower sensitivity (0.73 vs. 0.87), the advantage of the method proposed in this paper is its high specificity (0.97 vs. 0.82). Moreover, false negative results corresponded to low magnitude artifacts. This method requires automatic segmentation of the lungs but only basic segmentation is needed since it is used to determine the lung bounding box. Motion artifacts in 4DCT images could be automatically detected in a large database. Results compare favorably to visual detection reported in literature (90% of patients have at least one artifact [6]).

5. Conclusion

4DCT image quality was improved with a pitch of 0.081 instead of 0.1. In respiratory phases with high motion speed, the increase of the number of patients with no artifact detected was statistically significant. This increase can be considered as clinically relevant since motion artifacts may impact the treatment planning quality. Breathing irregularities remain the main cause of motion artifacts in 4DCT images. Reducing the pitch allows to improve the robustness with respect to low frequency respiration cycles [2]. Methods such as patient training, data sorting optimization or data interpolation with deformable registration can also help to reduce frequency and magnitude of motion artifacts.

References

[1] Langner U W and Keall P J 2010 *Int J Radiat Oncol Biol Phys* **76** 1242–1250
[2] Han D, Bayouth J, Bhatia S, Sonka M and Wu X 2011 *Med Phys* **38** 2074–2087
[3] Smith D W, Dean C and Lilley J 2012 *Radiother Oncol* **102** 393–398
[4] Ehrhardt J, Werner R, Sring D, Frenzel T, Lu W, Low D and Handels H 2007 *Med Phys* **34** 711–721
[5] Georg M, Souvenir R, Hope A and Pless R 2008 *IEEE Computer Society Workshop on Mathematical Methods in Biomedical Image Analysis MMBIA* pp 1–8
[6] Yamamoto T, Langner U, Loo B W, Shen J and Keall P J 2008 *Int J Radiat Oncol Biol Phys* **72** 1250–1258
[7] Pinho R, Delmon V, Vandemeulebroucke J, Rit S and Sarrut D 2011 *Medical Imaging and Computer-Assisted Intervention MICCAI, Fourth International Workshop on Pulmonary Image Analysis* pp 225–232
[8] Vandemeulebroucke J, Bernard O, Rit S, Kybic J, Clarysse P and Sarrut D 2012 *Med Phys* **39** 1006–1015

XVII International Conference on the Use of Computers in Radiation Therapy (ICCR 2013) IOP Publishing
Journal of Physics: Conference Series **489** (2014) 012025 doi:10.1088/1742-6596/489/1/012025

A comparison of rigid registration methods for prostate localization on CBCT and the dependence on rectum distension *

C Boydev[1,2], D Pasquier[3,4], F Derraz[1,5], L Peyrodie[6], A Taleb-Ahmed[1] and JP Thiran[2,7]

[1]Laboratoire d'Automatique, de Mécanique et d'Informatique Industrielles et Humaines (LAMIH), Université de Valenciennes et du Hainaut-Cambrésis, France
[2]Signal Processing Laboratory (LTS5), École Polytechnique Fédérale de Lausanne, Switzerland
[3]Centre de Radiothérapie et d'Oncologie Galilée de Lille, France
[4]Département Universitaire de Radiothérapie, Centre Oscar Lambret, Lille, France
[5]Unité de Traitements de Signaux Biomédicaux (UTSB), Faculté Libre de Médecine, Lille, France
[6]Unité de Traitements de Signaux Biomédicaux (UTSB), Hautes Études d'Ingénieur, Lille, France
[7]Department of Radiology, University Hospital Center (CHUV) and University of Lausanne (UNIL), Switzerland
{christine.boydev, jp.thiran}@epfl.ch, d-pasquier@o-lambret.fr,
{foued.derraz,taleb}@univ-valenciennes.fr, Laurent.PEYRODIE@hei.fr

E-mail: `christine.boydev@epfl.ch`

Abstract. We evaluated automatic three-dimensional intensity-based rigid registration (RR) methods for prostate localization on CBCT scans and studied the impact of rectum distension on registration quality. 106 CBCT scans of 9 prostate patients were used. Each one was registered to the planning computed tomography (CT) scan using different methods: (a) global registration, (b) pelvis bony structure registration, (c) bony registration refined by a local prostate registration using the CT clinical target volume (CTV) expanded with 1, 3, 5, 8, 10, 12, 15 or 20-mm margin. Automatic CBCT contours were generated after propagation of the manual CT contours. To evaluate results, a radiation oncologist was asked to manually delineate the CTV on the CBCT scans (gold standard). The Dice similarity coefficients between propagated and manual CBCT contours were calculated.

1. Introduction

Daily image guidance, such as cone-beam computed tomography (CBCT) image-guided radiotherapy (IGRT) systems [1], has become a widely-used tool for patient repositioning in the treatment of prostate cancer. The prostate gland is known to be a moving and deformable gland under influence of rectal and bladder filling changes [2, 3, 4], which limits the effectiveness of skin marks in patient setup. Prostate localization on CBCT scans is challenging due to the relatively poor image quality [5, 6].

*This work was financially supported by ELEKTA SAS, Boulogne Billancourt, France.

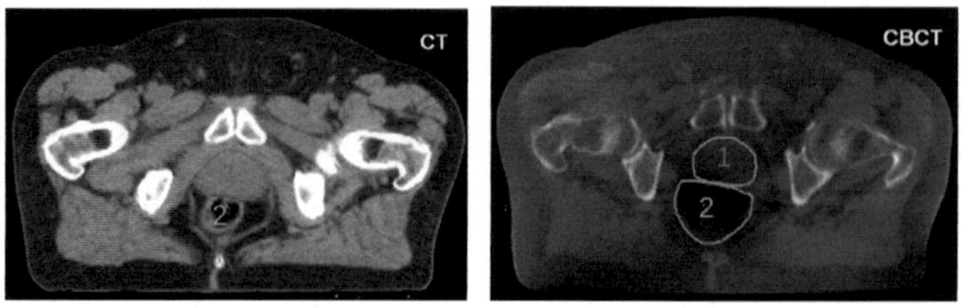

Fig. 1: *Comparison of the image quality between (left) a CT scan and (right) a CBCT scan. The manual delineated contours of the prostate (number 1) and the rectum (number 2) are displayed. The display software used is VV (*`http://vv.creatis.insa-lyon.fr/`*).*

This paper aims at evaluating different rigid registration (RR) methods for the purpose of prostate position verification. CBCT scans were rigidly registered to the planning CT scan and the manual delineated prostate contours were propagated from the CT scan to each CBCT scan to match the treatment anatomy. Deurloo *et al.* reported that the deformation of prostate during the course of radiotherapy is small compared to organ motion, and therefore in IGRT of prostate cancer, in first order, only setup error and organ motion need to be corrected for, whereas prostate deformation can be considered as a second-order effect [7]. RR accounts for first-order inter-fraction prostate motion. In our work, we tested different types of CT/CBCT RRs: global, bony, and local soft-tissue RRs. We also evaluated the impact of rectal distension on registration quality. Finally we drew up a couple of recommendations for clinical practice for the use of automatic RR for prostate localization on CBCT scans.

In this paper, we will use the terms clinical target volume (CTV) and planning target volume (PTV) as defined by the ICRU [8].

2. Material and methods

To automatically localize the prostate on the daily treatment CBCT scan, a CT/CBCT RR was performed and the resulting displacement was then applied to the contours manually drawn on the planning CT scan to generate the automatic CBCT contours. Three types of RR methods were tested:

(a) global RR,

(b) RR of the pelvis bony structures of CT and CBCT scans,

(c) bony RR followed by local soft-tissue RR. The latter was conducted using a region of interest defined by the CT CTV expanded with a margin among 1, 3, 5, 8, 10, 12, 15 and 20 mm. The CTV represents the whole prostate gland and was manually delineated by the physician in the planning process, prior to treatment.

In the following, for the sake of simplicity, the combination of bony RR with local RR (method (c)) will be referred to as local RR.

2.1. Data collection

In total, 106 images of 9 prostate cancer patients were analyzed. All patients were instructed to follow a dietary protocol in order to have a full bladder and an empty air-free rectum at the time of the planning CT acquisition and during treatment. The planning CT data was acquired using a General Electrics Light Speed scanner. The treatment system was an ELEKTA Synergy linear accelerator (LINAC) equipped with CBCT imaging. During CT (CBCT) acquisition, the peak-voltage, X-ray tube current and exposure time were 120 kVp (120 kVp), 300 mA (40 mA or 64 mA) and 1000 ms (40 ms), respectively.

For clinical requirements, the prostate CTV was manually delineated on each planning CT scan by a radiation oncologist (this step is always required in clinical practice for treatment planning). For the purposes of this study, for validation, the same radiation oncologist manually delineated the CTV on each CBCT scan. The CT contours were used in the definition of the local registration mask. The CBCT contours were only used as ground truth in quantitative RR validation.

2.2. Registration Algorithm

Although the context is not strictly monomodal image registration, the relationship between the intensities on the CT image and those on the CBCT image is given by a linear equation. Hence the normalized-cross correlation metric was chosen as a suitable advanced cost function. The similarity between images was intensity-based, allowing registration to be fully automatic. Optimization was performed with the regular step gradient method. Transformations were rigid (six degrees of freedom, allowing translations and rotations). Linear interpolation was used in all our experiments. Three resolution levels were used.

For this study, all the data processing and visualization were performed on a Linux computer with distribution openSUSE 11.4 x86_64, with an Intel Dual Core i5-560M 2.66 GHz processor, 3MB L2 cache, 4 threads, and 8GB RAM.

For our implementation, the following open-source software, based on C++, was used:

- the Insight Toolkit ITK [9], freely available at `www.itk.org`,
- the ITK-based Command Line Image Toolkit clitk, freely available at `http://www.creatis.insa-lyon.fr/rio/clitk`.

The software versions used were ITK 3.20.1, CMake 2.8.3 and gcc 4.5.1.

2.3. Validation

To evaluate the RR results, we calculated the Dice similarity coefficient between the propaged CBCT contours and the manual CBCT contours (referred to as ground truth) [10]. Ideally, when the two volumes perfectly overlap, the Dice coefficient equals 1. A null Dice coefficient corresponds to two disjoint volumes.

2.4. Statistical analysis

Differences in the Dice results across the multiple RR methods were tested for significance using the inferential non-parametric Friedman statistical test (with α set to 0.05). The Wilcoxon-Nemenyi-McDonald-Thompson post-hoc test was conducted to decide which methods are significantly different from each other [11, page 295]. Software R was used [12].

2.5. Impact of rectal distension on local RR quality

The performance of RR is deteriorated when the size or the shape of an organ changes. When performing local RR on the prostate region of interest (ROI), the mask we use necessarily includes a portion of the rectum. However the rectum is highly prone to changes in size and shape due to its ever-changing filling (gaseous and solid contents). Our hypothesis we wish to validate is that unsuccessful local RRs are caused by the difference in rectal distension between the CT and the CBCT scans. In this study, we correlate unsuccessful local RRs to the difference in rectum filling between the images. For this purpose, on the CT scan, we calculated the average intensity, $\overline{I}_{CT,r}$, in the rectum portion included in the registration mask (or ROI). We used the manual segmentations to generate this region, $R_{partial\_rectum}$, corresponding to the intersection of the ROI and the rectum volume on the CT scan. The rest of the rectum will not influence the registration process as only voxels that are inside the mask will be considered for the calculation of the metric. We also calculated the average intensity, $\overline{I}_{CBCT,r}$, inside region $R_{partial\_rectum}$ on the CBCT scan after rigidly aligning the bony structures of the CBCT and the CT scans. Because the overall range of intensities on a reconstructed CBCT scan can be shifted, we calculated $\overline{I}_{CBCT,p}$ that represents the CBCT average intensity inside the region corresponding to the CT manual prostate segmentation after rigidly aligning the bony structures of the CBCT and the CT scans, and we substracted from the $\overline{I}_{CBCT,r}$. We used the following variable to quantify rectum filling variation:

$$F = |(\overline{I}_{CBCT,r} - \overline{I}_{CBCT,p}) - (\overline{I}_{CT,r} - \overline{I}_{CT,p})| \quad (1)$$

The F number given by Equation 1 consistently reflects the rectum distension. On a scan, an air-free empty rectum and a prostate, being both soft tissue, have the same range of pixel intensities. The rectum volume increases if its filling increases, i.e. when gas and solid contents appear. Gas and solid contents in the rectum have lower intensities than those in an air-free empty rectum.

We plotted the cumulative number of failed registrations, arranging the 106 CBCT scans in order of increasing F number. A registration was assessed as not successful if the Dice coefficient after registration was found to be lower than 95% of the Dice coefficient without registration.

3. Results

3.1. Statistical analysis

The statistical analysis showed that there was a highly significant difference between the following RR methods: (c)5mm vs (a) ($p = 2.0\,10^{-5}$), (c)5mm vs (b) ($p = 6.6\,10^{-8}$), (c)5mm vs (c)1mm ($p = 1.1\,10^{-6}$), (c)8mm vs (a) ($p = 2.0\,10^{-4}$), (c)8mm vs (b) ($p = 4.0\,10^{-7}$), (c)8mm vs (c)1mm ($p = 2.8\,10^{-5}$), and (c)10mm vs (b) ($p = 6.1\,10^{-3}$). All RR methods were found to yield Dice results significantly different from those obtained without registration. Statistically, RR gave the best agreement between the manual and the propagated contours when performed locally on soft tissue, with 5-mm or 8-mm CTV expansion. Table 1 shows the Dice medians, standard deviations (SD) and the number of failed registrations for each RR method. The Dice median without registration was found to be 0.742. The highest Dice medians were obtained with local RR with 5-mm and 8-mm margins. Conversely, bony RR appeared to be more robust than local RR methods as it counted the lowest number of failed registrations (5 cases over 106). When local RR with small margins failed, it could be caused by the lack of contrast and/or the frequently observed presence of (moving or not) gas pockets situated in the rectum and contiguous with the prostate membrane. We examined the few cases where bony RR failed, and it turned out that the Dice coefficients were all between 90% and 95% of the Dice coefficients without registration. We chose to focus, in the following, on the local RR with 8-mm margin as it produced the second best median after local RR with 5-mm margin and the second lowest number of failed registrations after bony RR.

RR method	Global	Bony	Local							
			1 mm	3 mm	5 mm	8 mm	10 mm	12 mm	15 mm	20 mm
Dice median	0.786	0.785	0.787	0.806	0.823	0.819	0.806	0.796	0.799	0.796
Dice SD	0.071	0.066	0.097	0.140	0.088	0.075	0.084	0.064	0.076	0.068
Failed registrations	5	5	22	20	11	7	9	11	12	7

Table 1: *Registration results. The median and the standard deviation (SD) of the Dice coefficients without registration were 0.742 and 0.109, respectively.*

3.2. Impact of rectal distension on local RR quality

We investigated in which cases local RR with 8-mm margin failed. Figure 2 illustrates the impact of the variation of rectal filling between the images to be registered on the quality of local RR with 8-mm margin. We observed with our database that if the F factor as defined in Eq. 1 is lower than or equal to 112.6, registrations were all successful. All failed registrations appeared for values of F higher than 147.6. The range of F values obtained was from 0.2 to 406.6.

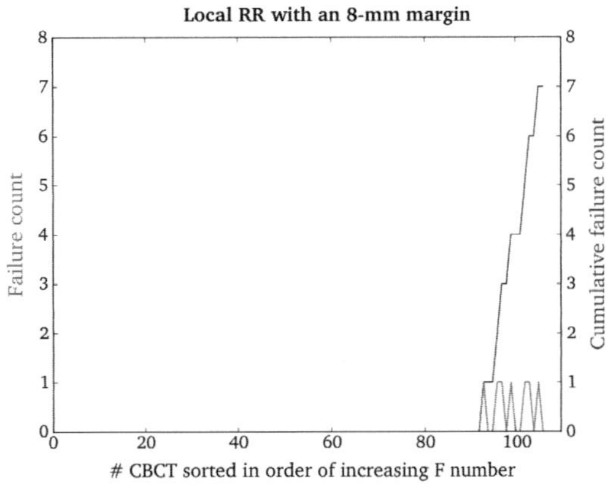

Fig. 2: *Plot of (blue) the cumulative number of failed registrations and (red) its derivative, i.e. the failure frequency, obtained with local RR with 8-mm margin, against the CBCT scans arranged in order of increasing F number.*

4. Discussion

Local RR with 8-mm margin was able to improve upon the Dice results obtained with bony RR as long as the rectal distension, that is the difference in the rectum anatomy between the planning CT scan and the treatment CBCT scan, is limited. For low values of F, results exceeded global RR or bony RR results, the latter being the current clinical standard for prostate repositioning. Smitsmans *et al.* found that 5-mm was the optimum margin for prostate CT/CT local RR [13]. In our study, we showed that indeed this method yielded the highest Dice median but it counted 11 failed registrations over 106, against only 7 failed registrations for local RR with 8-mm margin. Local RR with 8-mm has also a smaller SD than and a Dice median very close to those obtained with the 5-mm margin. That is why we preconize to use an 8-mm margin. In a later study, Smitsmans *et al.* also reported that local CT/CBCT RRs with 5-mm margin mainly failed because of streaks in the CBCT scans caused by moving gas pockets in the rectum during CBCT acquisition [14]. They implemented some improvements to their local RR algorithm, including the removal of the gray values of the pubic bone from the ROI. In our study, we could not find any correlation between the F factor defined in Eq. 1 and registration failures for local RR with margins lower than or including 5 mm. This is consistent with the fact that the lower the margin, the smaller the rectum portion included in the mask, and so the smaller the influence of the rectum portion on the registration process. Our assumption is that for margins lower than or including 5 mm, registration is not influenced by rectum distension.

Geometrical deviations occuring during treatment lead to target underdosage. To account for these geometrical deviations such as target volume delineation, organ motion and setup errors, a margin between the CTV and the PTV must be added. Van Herk *et al.* derived a margin recipe for prostate cancer radiotherapy, separating geometrical deviations into treatment execution (random) and treatment preparation (systematic) variations [15]. An accurate daily verification and correction protocol of the patient setup before each treatment session should allow reducing treatment margins, after evaluating on clinical data the SD of random and systematic variations due to organ motion and residual setup errors after setup correction. Currently, CBCT-based daily verification and correction are widely performed using a bony RR in order to reduce the setup error. Furthermore a bony RR refined by a local RR with an 8-mm margin around the CTV as described in this paper should also reduce inter-fraction organ motion without the use of implanted fiducials markers.

5. Conclusions

With this study, we aim to provide guidance for good practice in the use of CBCT scans for prostate position verification and correction using RR. We recommend to start from the current clinical standard for prostate repositioning, that is, a RR on the pelvis bony structure. The next step is to determine whether local RR with 8-mm margin can be performed on top of the bony RR to improve upon registration quality. To do so, the user should evaluate the difference in rectum anatomy between the planning CT scan and the treatment CBCT scan. We propose, in this paper, a way to conduct such an evaluation, easily applicable in clinical practice, which uses the manual CT contours only and requires calculations of mean intensities in the prostate and in the portion of the rectum that is included in the registration mask. If the difference in rectum anatomy is limited, local RR performed on the CTV extended by an 8-mm margin will improve registration quality and prostate targeting. If not, the local RR may deteriorate registration quality and hence should not be applied. We highly recommend that the user should always visually assess the final registration quality, particularly when local RR is applied.

6. Conflict of interest statement

This work was financially supported by Elekta SAS, Boulogne-Billancourt, France.

References

[1] Jaffray D A, Siewerdsen J H, Wong J W and Martinez A A 2002 *International Journal of Radiation Oncology*Biology*Physics* **53** 1337 – 1349 ISSN 0360-3016 1
[2] Schild S, Casale H, Bellefontaine L *et al.* 1993 *Medical dosimetry: official journal of the American Association of Medical Dosimetrists* **18** 13 1
[3] Roeske J C, Forman J D, Mesina C F, He T, Pelizzari C A, Fontenla E, Vijayakumar S and Chen G T 1995 *Int J Radiat Oncol Biol Phys* **33** 1321–1329 1
[4] de Crevoisier R, Tucker S L, Dong L, Mohan R, Cheung R, Cox J D and Kuban D A 2005 *International Journal of Radiation Oncology*Biology*Physics* **62** 965 – 973 ISSN 0360-3016 1
[5] Siewerdsen J H and Jaffray D A 2001 *Med Phys* **28** 220–231 1
[6] Smitsmans M H P, Pos F J, de Bois J, Heemsbergen W D, Sonke J J, Lebesque J V and van Herk M 2008 *Int J Radiat Oncol Biol Phys* **71** 1279–1286 1
[7] Deurloo K E, Steenbakkers R J, Zijp L J, de Bois J A, Nowak P J, Rasch C R and van Herk M 2005 *International Journal of Radiation Oncology*Biology*Physics* **61** 228 – 238 ISSN 0360-3016 1
[8] Reviewer D J 1994 *Medical Physics* **21** 833–834 1
[9] Ibanez L, Schroeder W, Ng L and Cates J 2005 *The ITK Software Guide* Kitware, Inc. ISBN 1-930934-15-7 http://www.itk.org/ItkSoftwareGuide.pdf 2nd ed 2.2
[10] Dice L R 1945 *Ecology* **26** 297–302 2.3
[11] Hollander M and Wolfe D A 1999 *Nonparametric statistical methods* (Wiley-Interscience) 2.4
[12] R Core Team 2012 *R: A Language and Environment for Statistical Computing* R Foundation for Statistical Computing Vienna, Austria ISBN 3-900051-07-0 2.4
[13] Smitsmans M H, Wolthaus J W, Artignan X, de Bois J, Jaffray D A, Lebesque J V and van Herk M 2004 *International Journal of Radiation Oncology* Biology* Physics* **60** 623–635 4
[14] Smitsmans M H P, de Bois J, Sonke J J, Betgen A, Zijp L J, Jaffray D A, Lebesque J V and van Herk M 2005 *Int J Radiat Oncol Biol Phys* **63** 975–984 4
[15] van Herk M, Remeijer P, Rasch C and Lebesque J V 2000 *International Journal of Radiation Oncology*Biology*Physics* **47** 1121 – 1135 ISSN 0360-3016 4

Journal of Physics: Conference Series **489** (2014) 012026 doi:10.1088/1742-6596/489/1/012026

In-room breathing motion estimation from limited projection views using a sliding deformation model

V Delmon[1], J Vandemeulebroucke[2,3], R Pinho[1], M Vila Oliva[1,4], D Sarrut[1] and S Rit [1]

[1] Université de Lyon, CREATIS ; CNRS UMR5220 ; Inserm U1044 ; INSA-Lyon ; Université Lyon 1 ; Léon Bérard Cancer Center ; F-69373 Lyon, France
[2] Vrije Universiteit Brussel (VUB), Dept. ETRO, Pleinlaan 2, B-1050 Brussels, Belgium
[3] iMinds, Dept. FMI, Gaston Crommenlaan 8 (box 102), B-9050 Ghent, Belgium
[4] iMagX project, ICTEAM Inst., Université catholique de Louvain, Louvain-la-Neuve, Belgium

E-mail: `vivien.delmon@creatis.insa-lyon.fr`

Abstract.
Purpose: To estimate in-room breathing motion from a limited number of 2D cone-beam (CB) projection images by registering them to a phase of the 4D planning CT.
Methods: Breathing motion was modelled using a piecewise continuous B-spline representation [1], allowing to preserve the sliding along the thoracic wall while limiting the degrees of freedom. The deformed target 3D image was subsequently used to generate Digitally Reconstructed Radiographs (DRR). The Normalized Correlation Coefficient (NCC) between the measured projection images and the DRR was computed in the 2D projection space. However, the partial derivatives of the NCC relative to the transform parameters were backprojected into the 3D space, avoiding the projection of the transform Jacobian matrix which is computationally intractable [2].
Results: The method was quantitatively evaluated on 16 lung cancer patients. 40 CB projection images were simulated using the end-exhale phase of the 4D planning CT and the geometric parameters of a clinical CB protocol. The end-inhale phase was deformed to match these simulated projections. The Target Registration Error (TRE) decreased from 8.8 mm to 2.0 mm while the TRE obtained from the 3D/3D registration of the reconstructed CBCT was significantly worse (2.6 mm), due to view aliasing artefacts. We also provide the motion compensated image reconstructed from a real CB acquisition showing the quality improvement brought by the in-room deformation model compared to the planning motion model.
Conclusions: We have developed a 2D/3D deformable registration algorithm that enables in-room breathing motion estimation from cone-beam projection images.

1. Introduction

In-room imaging systems are widely used in radiotherapy and provide a better knowledge of the patient motion during and between treatment fractions. Nowadays, the cone-beam (CB) system is mostly used to check the patient positioning before each fraction by reconstructing a 3D CBCT and registering it to the 3D planning CT. The set of projections contains information about the breathing motion and could be used to adapt the treatment. Breathing motion can be estimated from a respiratory correlated 4D CBCT [3], but 4D CBCT images are impacted by the reduced number of projections that lead to streak artefacts. Instead, we propose a method to estimate the breathing motion directly from the set of CB projections.

2. Method

Problem description From the planning of the radiotherapy treatment, we dispose of a 4D CT image of the patient from which we retain the 3D image corresponding to the end-exhale $f(\boldsymbol{x})$. From the in-room CB system, we dispose of N 2D projections $g_{\boldsymbol{t}}(\boldsymbol{u})$ with $\boldsymbol{t} \in [0..N-1]$. The acquisition geometry is known, providing the CB projection operator

$$\mathcal{P}_{\boldsymbol{t}} f(\boldsymbol{u}) = \int f(\boldsymbol{s} + l \cdot \zeta(\boldsymbol{u}))dl \tag{1}$$

with ζ the direction of the line going from $\boldsymbol{s}$ the source to $\boldsymbol{u}$ a pixel of the detector.

The goal is to estimate the 4D non-rigid transformation $\boldsymbol{T} : \mathbb{R}^3 \times [0..N-1] \mapsto \mathbb{R}^3$ that best describes the motion of the patient during the CB acquisition. To simplify the problem, we assume a cyclic motion consisting of ten phases as the 4D CT. The projection subset corresponding to each phase $\boldsymbol{P_p} \subset [0..N-1]$ is obtained by extracting the breathing signal from the complete projection set [4] and selecting one projection for each phase in each breathing cycle. To obtain a 4D deformation field, each subset is registered to the end-exhale CT image.

Mapping function The motion between the reference phase of the planning CT and a subset of projections is represented by a multi B-spline transform modeling sliding [1] $\mathcal{B}_{\boldsymbol{p}} : \mathbb{R}^3 \mapsto \mathbb{R}^3$ and parameterized by $\delta_{\boldsymbol{p}}$. This mapping function is a linear combination of B-spline transforms allowing sliding to occur at the motion mask interface [5] while preserving a smooth deformation elsewhere. The motion mask encompasses organs with the largest displacements during breathing, comprising the lung, the mediastinum and the abdomen.

Similarity metric To measure the similarity between the deformed 3D CT image and the subset of projections corresponding to a single phase, we use the known projection geometry to generate a new subset of projections from the deformed 3D CT and compute the NCC with the CB projections.

$$\begin{aligned} \tilde{f}(\boldsymbol{x}) &= f(\mathcal{B}_{\boldsymbol{p}}(\boldsymbol{x})) \\ NCC_{\boldsymbol{p}} &= \frac{1}{|\boldsymbol{P_p}|} \sum_{t \in \boldsymbol{P_p}} NCC(g_t(\boldsymbol{u}), \mathcal{P}_t \tilde{f}(\boldsymbol{u})) \end{aligned} \tag{2}$$

Optimization We used a gradient based approach to find the set of parameters $\delta^*_{\boldsymbol{p}}$ that minimizes the NCC. This approach requires the partial derivatives according to the parameters of the transformation $\frac{\partial NCC_{\boldsymbol{p}}}{\partial \delta_{\boldsymbol{p}}}$. By using the chain rule these partial derivatives are split into two terms,

$$\frac{\partial NCC_{\boldsymbol{p}}}{\partial \delta_{\boldsymbol{p}}} = \frac{1}{|\boldsymbol{P_p}|} \sum_{t \in \boldsymbol{P_p}} \frac{\partial NCC(g_t, \mathcal{P}_t \tilde{f})}{\partial \mathcal{P}_t \tilde{f}} \frac{\partial \mathcal{P}_t \tilde{f}}{\partial \delta_{\boldsymbol{p}}}. \tag{3}$$

The first term is the partial derivative of the NCC in each pixel of the projection, while the second term is the projection of a vectorial image with a large number of components equal to $|\delta_{\boldsymbol{p}}|$

$$\frac{\partial \mathcal{P}_t \tilde{f}(\boldsymbol{u})}{\partial \delta_{\boldsymbol{p}}} = \mathcal{P}_t \frac{\partial \tilde{f}(\boldsymbol{x})}{\partial \boldsymbol{x}} \frac{\partial \mathcal{B}_{\boldsymbol{p}}(\boldsymbol{x})}{\partial \delta_{\boldsymbol{p}}}. \tag{4}$$

Depending on the smoothness of the sought deformation the number of parameters can vary from hundreds to thousands of parameters, which makes the projection of this vectorial image intractable. The solution proposed by Vandemeulebroucke and Zeng [2, 6] is to integrate over the 3D space instead the 2D space of each projection. By rearranging the sum they obtain

$$\frac{\partial NCC_{\boldsymbol{p}}}{\partial \delta_{\boldsymbol{p}}} = \frac{1}{|\boldsymbol{P_p}|} \sum_{t \in \boldsymbol{P_p}} \mathcal{P}_t^* \Big(\frac{\partial NCC}{\partial \mathcal{P}_t \tilde{f}} \Big) \frac{\partial \tilde{f}}{\partial \boldsymbol{x}} \frac{\partial \mathcal{B}_{\boldsymbol{p}}}{\partial \delta_{\boldsymbol{p}}} \tag{5}$$

with $\mathcal{P}_t^*$ the backprojection operator. The projection of the 3D vectorial image of $|\delta_{\boldsymbol{p}}|$ components is replaced by the backprojection of a 2D scalar image.

3. Experiment

In order to dispose of a ground truth, we evaluated the proposed method on 16 4D CT images of the thorax using the end-inhale phase to simulate a CB acquisition and registering the end-exhale phase to it. The evaluation was done using the TRE computed on manually defined anatomical landmarks.

Image data The first 6 4D CT image were acquired at our hospital. The resolution was approximately $1 \times 1 \times 2$ mm and $512 \times 512 \times 150$ voxels [7]. The next 10 patients were obtained from the DIR-labs (DL) database `www.dir-lab.com` [8, 9]. Their spatial resolution was between $0.97 \times 0.97 \times 2.5$ mm and $1.16 \times 1.16 \times 2.5$ mm.

To simulate each CB acquisition, we extracted the respiratory signal of a real acquisition made at our hospital, containing 632 projection images and 40 breathing cycles. We extracted the geometric parameters associated to the end-inhale projections. Then we generated each simulated CB acquisition by projecting the 3D CT phase corresponding to the end-inhale using the RTK projection tool [10] and the extracted geometric parameters.

Implementation The proposed method was implemented as a new component of elastix (`http://elastix.isi.uu.nl/`), which is a toolbox for intensity-based medical image registration [11]. This component also uses RTK [10] to handle projection, backprojection and CB geometry.

We compared the end-inhale to end-exhale registration using the proposed method (2D/3D) with one that first reconstructs the 3D image from the simulated CB projections (3D/3D FDK). To obtain a sufficiently smooth result, we used a multi resolution approach, reducing the B-spline control point spacing and image resolution in subsequent levels. For the finest resolution, the spacing of the B-spline control points was 32 mm in every direction, which is large enough to impose spatially smooth deformations without additional regularization.

Quantification To measure the quality of the registration, we used the TRE which is the mean Euclidean distance between corresponding landmarks defined in the reference image and in the target image. These landmarks correspond to recognizable structures like bronchial tree bifurcations, manually selected by experts, on both the reference and the target image. For the first 6 patients, 100 landmarks were obtained with a semiautomatic method [7]. For DL patients, 300 landmarks were also extracted with a semiautomatic method [8, 9].

4. Results

Target registration error Before the registration, the mean Target Registration Error (TRE) was 8.8 ± 5.1 mm. After the registration using 3 resolutions, the 3D/3D FDK obtains a mean TRE of 2.6 ± 2.3 mm and the 2D/3D a mean TRE of 1.8 ± 1.3 mm. The 3D/3D FDK registration is affected by the streak artefacts present in the reconstructed CB due to the limited number of projections. The TRE obtained with the proposed method is always better than the one obtained with the 3D/3D FDK method.

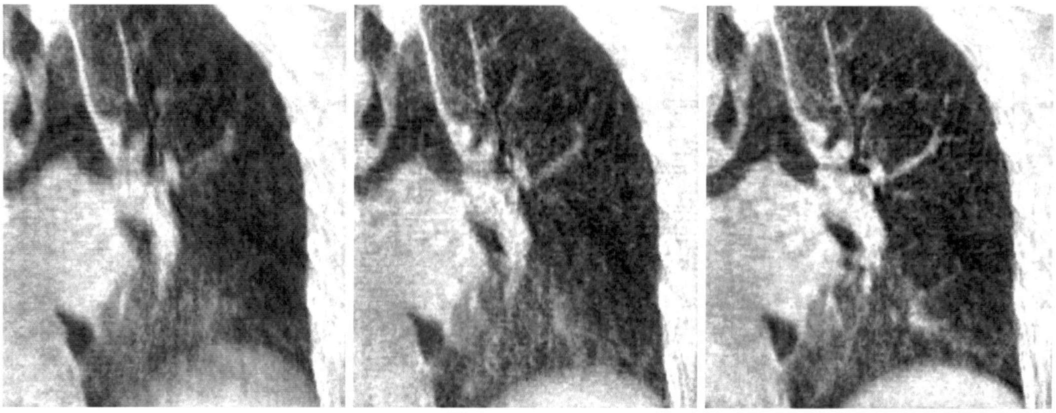

Figure 1. Cone-beam motion compensated reconstructions with the FDK algorithm, assuming no motion (left), with an *a priori* motion (middle) and with the in-room motion (right)

Motion compensated reconstruction To assess the quality of the deformation fields estimated with the proposed method on real data, we registered the end-exhale phase of a 4D planning CT with the 10 phases of a real CB acquisition. The average time per iteration of our unoptimized implementation was 35 min and we used 50 iterations per phase. The amplitude of the found breathing motion was 32 mm around the diaphragm. Figure 1 shows the motion compensated reconstructions of this CBCT without motion compensation, an *a priori* motion extracted from the 4D planning CT [12], and the motion of the day found with the proposed method. On this selected example, the reconstruction compensated with an *a priori* model is visually better than the reconstruction assuming no motion but we can still see blurred structures. The proposed method corrects most of these blurred structures and reveal some bronchi that where not distinguishable with the previous method.

5. Discussion

These results show the potential benefits of using the motion information contained in the CB acquisition. In this study, it was applied to motion compensated CB reconstruction, but it can also serve to adapt treatment, to derive population-based treatment margins, etc. By computing an independent deformation for each phase, the proposed method does not enforce the temporal smoothness on the sought deformation field. It would be interesting to evaluate the improvement that can be brought by a temporal regularization. The signal extraction is also a concern, the quality of the motion compensated reconstructions that have been made using the proposed method were highly correlated to the quality of this signal.

6. Conclusion

We have proposed a 2D/3D registration method that extracts motion information from limited projection views. The method was quantitatively evaluated on simulated data and qualitatively evaluated on real cone-beam acquisitions. Motion compensated reconstruction using the motion estimated from cone-beam projections showed an improved image quality over non-compensated reconstruction and motion compensated reconstruciton using an *a priori* motion model.

Acknowledgments

This work was supported in part by the Labex PRIMES (ANR), the Lyric grant INCa-4664, the Association Nationale de la Recherche Technique (ANRT) and Elekta Oncology Systems.

Journal of Physics: Conference Series **489** (2014) 012026 doi:10.1088/1742-6596/489/1/012026

References

[1] Delmon V, Rit S, Pinho R and Sarrut D 2013 *Physics in Medicine and Biology* **58** 1303
[2] Vandemeulebroucke J 2010 *Lung Motion Modelling and Estimation for Image Guided Radiation Therapy* phd INSA de Lyon
[3] Sonke J J, Zijp L, Remeijer P and van Herk M 2005 *Med Phys* **32** 1176–1186
[4] Zijp L, Sonke J J and van Herk M 2004 *International Conference on the Use of Computers in Radiation Therapy (ICCR)* (Seoul, Republic of Korea: Jeong Publishing) pp 507–509
[5] Vandemeulebroucke J, Bernard O, Rit S, Kybic J, Clarysse P and Sarrut D 2012 *Med Phys* **39** 1006
[6] Zeng R, Fessler J and Balter J 2005 *Med Phys* **32** 984–991
[7] Vandemeulebroucke J, Rit S, Kybic J, Clarysse P and Sarrut D 2011 *Medical Physics* **38** 166–178
[8] Castillo R, Castillo E, Guerra R, Johnson V E, McPhail T, Garg A K and Guerrero T 2009 *Physics in Medicine and Biology* **54** 1849
[9] Castillo E, Castillo R, Martinez J, Shenoy M and Guerrero T 2010 *Physics in Medicine and Biology* **55** 305
[10] Rit S, Oliva M V, Brousmiche S, Labarbe R, Sarrut D and Sharp G C 2013 *International Conference on the Use of Computers in Radiation Therapy (ICCR)* (Melbourne, Australia)
[11] Klein S, Staring M, Murphy K, Viergever M and Pluim J 2010 *IEEE Trans Med Imaging* **29** 196–205
[12] Rit S, Wolthaus J, van Herk M and Sonke J J 2009 *Med Phys* **36** 2283–2296

Journal of Physics: Conference Series **489** (2014) 012027 doi:10.1088/1742-6596/489/1/012027

Fast cine-magnetic resonance imaging point tracking for prostate cancer radiation therapy planning

J Dowling[1], K Dang[2], Chris D Fox[2], S. Chandra[1], Suki Gill[2], T. Kron[2], D. Pham[2], F. Foroudi[2]
[1]CSIRO Australian e-Health Research Centre, Royal Brisbane and Women's Hospital, Australia
[2]Peter MacCallum Cancer Centre, Melbourne, Australia

E-mail: jason.dowling@csiro.au

Abstract. The analysis of intra-fraction organ motion is important for improving the precision of radiation therapy treatment delivery. One method to quantify this motion is for one or more observers to manually identify anatomic points of interest (POIs) on each slice of a cine-MRI sequence. However this is labour intensive and inter- and intra- observer variation can introduce uncertainty. In this paper a fast method for non-rigid registration based point tracking in cine-MRI sagittal and coronal series is described which identifies POIs in 0.98 seconds per sagittal slice and 1.35 seconds per coronal slice. The manual and automatic points were highly correlated (r>0.99, p<0.001) for all organs and the difference generally less than 1mm. For prostate planning peristalsis and rectal gas can result in unpredictable out of plane motion, suggesting the results may require manual verification.

1. Introduction

Organ motion is an important factor in radiation therapy treatment, which aims to precisely deliver radiation dose to a target while limiting the amount of radiation to organs at risk. In prostate treatment planning the patient's clinical target volume (CTV) is expanded with an internal margin covering potential intra-fraction organ movement to generate the Internal Planning Volume (IPV). An addition margin is then added for patient setup error to generate the planning target volume (PTV). The internal margin is necessary as during treatment the prostate and seminal vesicles can still move independently. Understanding the internal motion during each treatment delivery fraction is important to ensure that the intended and actual dose is the same.

Inclusion of the seminal vesicles in the CTV significantly increase the risk of acute and late rectal toxicity (1). Although a number of papers evaluate prostate motion during treatment less research has focused on seminal vesicle movement. This has motivated a clinical trial at the Peter MacCallum Cancer Centre, Melbourne, Australia, which aims to measure appropriate CTV to PTV treatment margins for the prostate and seminal vesicles based on cine-MRI point displacement data over a typical image guided radiation therapy treatment duration (15 minutes).

[1] To whom any correspondence should be addressed.

The traditional method for quantifying motion from these scans is for one or more expert observer to manually identify points of interest (POIs) on each slice (2,3). However this manual process is labour intensive and inter- and intra-observer variation can introduce uncertainty. This paper evaluates the use of fast deformable registration to map an initial set of points over the set of entire set of cine-MRI images.

2. Method

2.1.1. Image data

For this study cine-MRI scans were acquired prior to treatment for five patients prescribed radical radiation therapy to a dose of 78 Gy in 39 fractions using 3DCRT or IMRT techniques. Patients were advised to apply the same bowel and bladder preparation as for treatment (patients were asked to empty both bladder and bowels 30 minutes before treatment and then drink four small cups of water. Scanning commenced 30 minutes after the last cup of water). Patients were scanned in the treatment position on a flat table with the same stabilization devices as during treatment.

The Cine-MRI datasets were acquired with a true-FISP T2w sequence on a Siemens 3T Tim Trio. 2D images were stored at four second intervals along sagittal (n=235, size=416x512 pixels, total time=940 seconds) and coronal (n=200, size=512x512 pixels, total time=800 seconds) planes. The total scan time each session was 30 minutes (15 minutes for the sagittal followed by 12 minutes for the coronal plane).

2.1.2. Manual point collection

Manual points were carefully marked on each cine-MRI series by an experienced radiation therapist using custom in-house software which recorded the name and co-ordinates of each point. The points of interest marked are shown in Figure 1 and were used for to calculate the displacement of either the organ centroid (C) or the absolute position (A) of a landmark. For the sagittal images the points comprised the bladder (C), prostate (C), seminal vesicles (C), rectum (C), pubis symphysis (C), and external surface marker (A) and the anterior edge of the coccyx (A). For the coronal images the points consisted of the bladder (C), prostate (C), right and left seminal vesicles (C) and right and left femoral head (A).

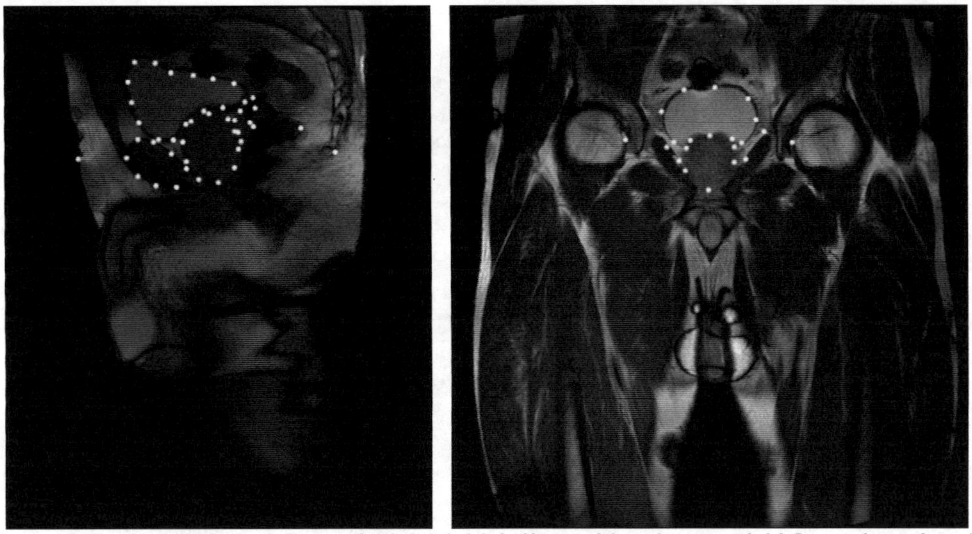

Figure 1. Sagittal (left) and Coronal (right) initial slices with points overlaid for study patient #2 (points have been enlarged for visibility in this paper)

2.1.3. Automatic point tracking
The aim of the automatic point tracking was to use the initial manual points from the first slice and map these to all remaining slices in that series. The automatic point tracking system was written in C++ using the ITK 4.2 library (www.itk.org) (4) and requires two inputs: the list of initial points with physical co-ordinates (a CSV file output from the custom in-house software) and the corresponding initial DICOM image. Three main steps are performed:

1. The DICOM images for the initial image plane are identified and sorted by sequence number (tag 0020:0013). Centroids for organs were also calculated (if required).
2. The initial image is registered to each other image in the series. The method used was the 2D diffeomorphic demons (5) non rigid registration (ITK 4 implementation) with the following parameters: 4 level multi resolution pyramid (40,40,32,22 iterations); 256 histogram levels, 21 match points, threshold at mean intensity on, smooth deformation field turned on, standard deviations = 2.0, and smooth update field turned on.
3. The deformation field from step 2 was then used to map the list of initial points (and centroids) across to the corresponding target images. The mapped co-ordinates were also stored in a CSV file for analysis. The initial points were mapped to all other images (rather than the points from each image mapped to the next) to reduce error propagation throughout the series.

3. Results

3.1.1. Speed
Manual identification of points for all sagittal images (~235 images) for a single patient required between 120 to 240 minutes. The coronal images (~200 images) required an additional 120 to 180 minutes for an expert to contour. In comparison the automatic point propagation mapped all POIs from the initial image in ~230 seconds for all of the sagittal images (0.98s per slice) and ~269 seconds for all of the coronal images (1.35s per coronal slice).

3.1.2. Validation of automatic points
Validation results comparing automatic points versus manual points are summarised in Tables 1 and 2. Although validation was possible as manual points had been recorded for each image for each patient, the number of manual points varied on each image and points were recorded in an inconsistent order, therefore in the sagittal images centroids were compared for the bladder, prostate, seminal vesicles, rectum, and public symphysis; and the absolute position for the coccyx and skin marker. For the coronal images centroids were used for the bladder, prostate and seminal vesicles (absolute position was used for the femoral heads). The largest difference between manual and the automatic method occurred in the sagittal images, particularly for the rectum (AP), coccyx (AP) and bladder (SI). These differences generally appear to be registration errors from out of plane motion. Despite these challenges the difference between manual and automatic centroids was generally less than 1mm and were significantly correlated (r>0.99, p<0.001) for all organs in both the sagittal and coronal cine-MRI sequences.

Table 1. Mean difference ± standard deviation (mm) between manual and automatic sagittal centroid or landmark displacements (landmark points are marked with an asterisk). SV = Seminal Vesicles; Pubic S = Pubic Symphysis. Skin = skin marker.

	AP	SI
Bladder	0.68±2.41	-1.64±2.57
Prostate	0.77±1.86	0.91±3.13
SV	0.06±2.41	-1.40±4.52
Rectum	-2.40±4.46	-0.87±4.11
Pubic S.	-0.02±1.43	0.32±1.34
Coccyx*	-1.76±7.57	0.19±4.05
Skin*	-0.55±1.31	-1.66±3.51

Table 2. Mean difference ± standard deviation (mm) between manual and automatic coronal centroid or landmark displacements (landmark points are marked with an asterisk). SV = Seminal Vesicles; Pubic S = Pubic Symphysis.

	LR	SI
Bladder	1.42±3.22	0.27±2.28
Prostate	-1.22±1.34	1.85±4.13
RT SV	-0.66±1.42	-0.12±1.70
LT SV	-0.04±1.69	1.32±2.90
RT Femur*	-0.15±1.39	-0.83±1.73
LT Femur*	-0.10±2.88	-0.87±1.66

3.1.3. Automatic results
Automatic tracking results for all organ points were generated (based on the set of initial points) and are summarised in Tables 3 - 6. Patient 1 was found to have noticeably higher amounts of internal organ motion (particularly at the edges of the prostate and seminal vesicles).

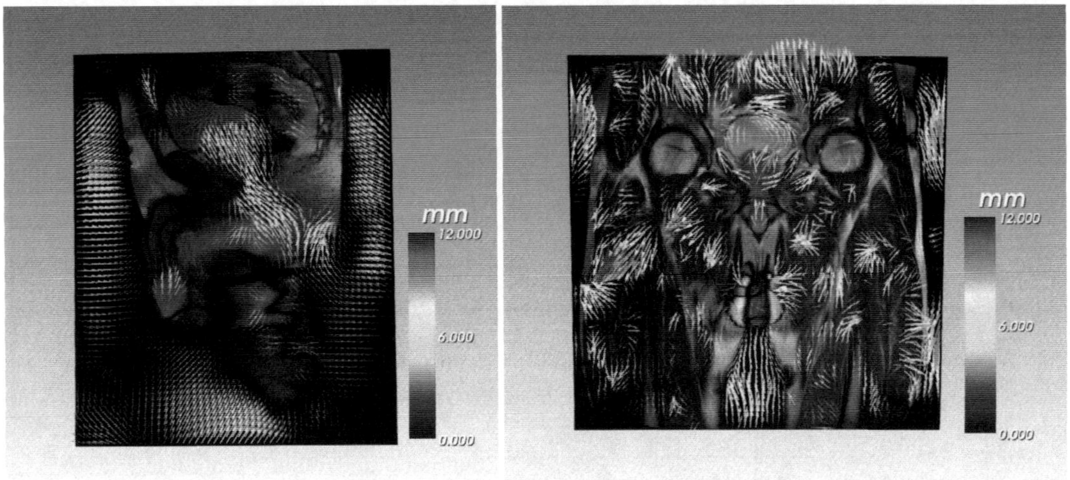

Figure 2. Sagittal (left) and Coronal (right) accumulated deformation fields for study patient #2. The prostate, seminal vesicles and anterior part of the bladder show the most motion on the sagittal slice overlay (see also Table 3 and 4).

Journal of Physics: Conference Series **489** (2014) 012027 doi:10.1088/1742-6596/489/1/012027

Table 3. Mean ± standard deviation (mm) point tracking results (Sagittal AP direction) for each of the five study patients over all slices. SV = Seminal Vesicles; Pubic S = Pubic Symphysis; Skin = skin marker. The results are either organ centroid or landmark displacements (landmark points are marked with an asterisk).

Patient ID:	1	2	3	4	5
Bladder	3.88±2.67	-0.15±1.53	-0.71±1.66	1.74±2.01	1.42±2.14
Prostate	7.17±2.41	-1.49±1.01	-1.35±1.10	2.11±1.96	-1.19±1.11
SV	5.60±2.01	-2.02±1.08	-1.66±1.23	3.14±1.97	-0.80±1.40
Rectum	3.48±2.46	-1.41±0.94	-1.01±1.82	0.86±2.37	-1.48±0.67
Pubic S.	4.81±2.13	-0.51±0.52	-0.71±1.19	-0.12±0.57	0.02±0.55
Coccyx*	-2.45±2.43	-0.59±0.41	-0.83±0.62	-0.12±0.13	-0.90±0.39
Skin*	2.63±1.05	0.28±0.30	-1.66±0.77	-0.18±0.98	0.11±0.84

Table 4. Mean ± standard deviation (mm) point tracking results (Sagittal SI direction) for each of the five study patients over all slices. SV = Seminal Vesicles; Pubic S = Pubic Symphysis; Skin = skin marker. The results are either organ centroid or landmark displacements (landmark points are marked with an asterisk).

Patient ID:	1	2	3	4	5
Bladder	5.44±2.78	-0.98±2.02	-0.03±1.29	2.31±2.36	0.65±2.05
Prostate	6.76±2.83	-3.08±2.07	-0.36±1.22	-0.02±1.37	-0.70±1.14
SV	6.31±3.42	-3.67±1.71	0.80±1.60	2.32±1.95	0.04±1.14
Rectum	5.58±3.14	-2.22±2.58	1.47±1.32	0.78±1.27	0.32±0.55
Pubic S.	3.82±1.83	-0.56±0.35	-1.02±1.12	-0.02±0.55	-0.21±0.55
Coccyx*	5.56±3.43	-0.21±0.13	1.59±0.52	-0.13±0.40	-0.08±0.26
Skin*	4.24±2.00	-0.61±0.66	-2.31±1.40	-0.36±0.90	-0.50±1.40

Table 5. Mean ± standard deviation (mm) point tracking results (Coronal LR direction) for each of the five study patients over all slices. SV = Seminal Vesicles. The results are either organ centroid or landmark displacements (landmark points are marked with an asterisk).

Patient ID:	1	2	3	4	5
Bladder	-0.86±2.39	1.06±3.37	-1.73±2.93	0.27±2.20	0.48±2.07
Prostate	-1.01±2.30	-1.83±2.12	0.29±1.52	0.78±1.36	0.24±1.64
RT SV	-1.63±2.06	-1.07±1.28	0.92±1.70	1.14±1.73	0.26±0.88
LT SV	0.61±1.65	0.27±0.93	-0.72±1.10	-0.25±1.14	-0.79±1.01
RT Femur*	-1.13±1.06	0.36±0.15	0.67±1.11	0.08±0.07	-0.29±0.13
LT Femur*	-0.93±0.56	-0.32±0.20	0.31±0.78	-0.39±0.10	-0.51±0.23

Table 6. Mean ± standard deviation (mm) point tracking results (Coronal SI direction) for each of the five study patients over all slices. SV = Seminal Vesicles. The results are either organ centroid or landmark displacements (landmark points are marked with an asterisk).

Patient ID:	1	2	3	4	5
Bladder	1.08±3.23	-1.73±2.22	1.04±1.92	2.89±2.84	0.36±1.79
Prostate	3.27±3.36	-1.16±1.34	-0.34±1.13	2.55±1.54	0.09±1.73
RT SV	2.71±2.93	-2.19±1.45	0.66±1.80	1.38±1.45	0.73±1.10
LT SV	2.68±2.49	-1.78±0.92	0.49±1.25	0.50±1.02	-0.66±1.24
RT Femur*	-2.08±1.10	-1.59±0.97	0.50±0.96	0.46±0.15	0.15±0.22
LT Femur*	-0.43±0.59	-0.52±0.53	0.23±1.45	0.64±0.38	-1.70±0.47

Journal of Physics: Conference Series **489** (2014) 012027 doi:10.1088/1742-6596/489/1/012027

4. Discussion

The proposed system was in general an order of magnitude faster than manual point identification (reducing the time required from 4-7 hours per patient to 8 minutes). The automatic method enables fast quantification and visualization (Figure 2) of deformation during the cine-MRI acquisition time period. Another advantage of the method is that the motion of all voxels in the volume can be analysed, rather than just the user selected points. This could enable the discovery and quantification of areas of significant motion. One limitation of the current system is the requirement for manual points to be placed on the initial slice. Combining the tracking with an anatomical atlas (6) could be useful to initialise the set of points and to automatically analysis motion within all contoured structures within the atlas. As Tables 3-6 show, patient specific differences can be large (for example, patients 3 and 5 had much lower organ motion than patient 1). The method is easily adaptable to 3D cine-MRI by using 3D registration (7).

The system was validated successfully against manual points (generally <1mm difference) and was useful in detecting some input errors from the custom manual point recording software. A limitation of the proposed method when using 2D cine-MRI is unpredictable out of plane motion due to peristalsis, rectal gas and bladder related internal organ motion which led to some registration errors (this error occurs as image features being tracked can disappear from the target image). This difficulty could be overcome by verification from an expert observer or with 3D cine-MRI acquisition and registration.

Acknowledgments
Jason Dowling acknowledges funding support from the Cancer Council NSW (RG 11-05), Prostate Cancer Foundation of Australia (YI2011), Movember and Cure Cancer Australia. Cine MRI scans were funded by the CASS foundation.

References

1. Bayman NA, Wylie JP. When should the seminal vesicles be included in the target volume in prostate radiotherapy? Clinical oncology. 2007 Jun;19(5):302–7.

2. Chan P, Dinniwell R, Haider MA, Cho Y-B, Jaffray D, Lockwood G, et al. Inter- and intrafractional tumor and organ movement in patients with cervical cancer undergoing radiotherapy: a cinematic-MRI point-of-interest study. International journal of radiation oncology, biology, physics. 2008 Apr 1;70(5):1507–15.

3. Ghilezan MJ, Jaffray DA, Siewerdsen JH, Van Herk M, Shetty A, Sharpe MB, et al. Prostate gland motion assessed with cine-magnetic resonance imaging (cine-MRI). International journal of radiation oncology, biology, physics. 2005 Jun 1;62(2):406–17.

4. Ibanez L, Schroeder W. The ITK Software Guide 2.4. Kitware, Inc.; 2005.

5. Vercauteren T, Pennec X, Perchant A, Ayache N. Diffeomorphic demons: efficient non-parametric image registration. NeuroImage. 2009 Mar;45(1 Suppl):S61–72.

6. Dowling JA, Fripp J, Chandra S, Pluim JPW, Lambert J, Parker J, et al. Fast automatic multi-atlas segmentation of the prostate from 3D MR images. Prostate Cancer Imaging. Image Analysis and Image-Guided Interventions. Springer; 2011. p. 10–21.

7. Chandra S, Dowling J, Shen K, Raniga P, Pluim J, Greer P, et al. Patient Specific Prostate Segmentation in 3D Magnetic Resonance Images. IEEE Transactions on Medical Imaging. 2012 Aug 2;31.

Journal of Physics: Conference Series **489** (2014) 012028 doi:10.1088/1742-6596/489/1/012028

Contribution of FDOPA PET to radiotherapy planning for advanced glioma

Nicholas Dowson[1], Michael Fay[2,3,4], Paul Thomas[2,3], Rosalind Jeffree[2,3], Robert McDowall[2], Craig Winter[2], Alan Coulthard[2,3], Jye Smith[2,3], Yaniv Gal[3], Pierrick Bourgeat[1], Olivier Salvado[1,3], Stuart Crozier[3], Stephen Rose[1,3]
[1]Australian e-Health Research Centre, CSIRO. [2]Departments of Nuclear Medicine, Radiology, Radiation Oncology, and Neurosurgery, Royal Brisbane and Women's Hospital. [3]University of Queensland, St Lucia, Australia. [4]Eberhard Karls Universität Tübingen

nicholas.dowson@csiro.au

Abstract. Despite radical treatment with surgery, radiotherapy and chemotherapy, advanced gliomas recur within months. Geographic misses in radiotherapy planning may play a role in this seemingly ineluctable recurrence. Planning is typically performed on post-contrast MRIs, which are known to underreport tumour volume relative to FDOPA PET scans. FDOPA PET fused with contrast enhanced MRI has demonstrated greater sensitivity and specificity than MRI alone. One sign of potential misses would be differences between gross target volumes (GTVs) defined using MRI alone and when fused with PET. This work examined whether such a discrepancy may occur. **Materials and Methods**: For six patients, a 75 minute PET scan using 3,4-dihydroxy-6-18F-fluoro-L-phynel-alanine (18F-FDOPA) was taken within 2 days of gadolinium enhanced MRI scans. In addition to standard radiotherapy planning by an experienced radiotherapy oncologist, a second gross target volume (GTV) was defined by an experienced nuclear medicine specialist for fused PET and MRI, while blinded to the radiotherapy plans. The volumes from standard radiotherapy planning were compared to the PET defined GTV. **Results**: The comparison indicated radiotherapy planning would change in several cases if FDOPA PET data was available. PET-defined contours were external to 95% prescribed dose for several patients. However, due to the radiotherapy margins, the discrepancies were relatively small in size and all received a dose of 50 Gray or more. **Conclusions**: Given the limited size of the discrepancies it is uncertain that geographic misses played a major role in patient outcome. Even so, the existence of discrepancies indicates that FDOPA PET could assist in better defining margins when planning radiotherapy for advanced glioma, which could be important for highly conformal radiotherapy plans.

1. Introduction

Despite radical treatment with surgery, radiotherapy and chemotherapy, advanced gliomas recur within months (2). Radiotherapy planning is performed on post-surgical contrast enhanced MRI (1) or CT. The contrast agent used in MRI is a marker for blood brain barrier breakdown. Active regions of tumours are enhanced because tumours damage or destroy the blood brain barrier. However, tissue disruption and reactive inflammation are not sufficient to damage the local blood vessels in all regions of active tumour. One example of this is on the "advancing fronts" at the tumour periphery where tumour cells, infiltrating along white matter tracts, are not yet of sufficient bulk to disrupt the blood brain barrier. For this reason MRI is well known to under-report tumour volumes (3)(4). Even so, the use of PET for routine radiotherapy planning has only been explored relatively recently (5), and then mainly using FluoroDeoxyglucose (6). The planning of radiotherapy on a modality known to under-report tumour volumes raises the question of whether recurrence occurs because regions of active tumour cells may be missed in the planning of radiotherapy and hence do not receive the radiation dose necessary to kill them.

The aim of this study is to address whether radiotherapy planning would change given the availability of PET images. Such discrepancies could be indicative of potential geographic misses due to the lack of metabolic information, although this would need to be verified histologically. PET imaging was performed with 3,4-dihydroxy-6-18F-fluoro-L-phynel-alanine (18F-DOPA) PET four weeks after a post-surgery course of chemoradiotherapy. FDOPA rather than FDG was chosen because superior tumour-to-background ratios are obtained by FDOPA in the brain (3). FDOPA is avidly taken up by active tumour tissue along the amino acid transporters (3) and correlates well with the Ki-67 index (7). FDOPA has proven to be sensitive for detecting tumour (8).

2. Materials and methods

The six patients (5 males, age range 52 to 71 years) used in this study had a single primary brain tumour without metastases. The study was approved by the Ethics Committee for Royal Brisbane and Women's Hospital and all patients gave written informed consent to participate in the imaging study. In all patients the tumours were newly diagnosed. MRI analysis was used to diagnose glioblastoma-multiforme, which was confirmed by subsequent histopathological analysis.

Radiotherapy planning was performed using a MRI scan taken 24-48 hours after surgery, by an experienced radiation oncologist (Michael Fay). Gross, clinical and planned target volumes (respectively GTV, CTV, & PTV) were generated as was a dose plan. The PTV was generated by adding 5mm to the CTV but radiosensitive tissue was spared. The CTV was based on visible oedema combined with a 5mm margin where safe. Following a six-week course of chemoradiotherapy (external beam radiotherapy was given at 60Gy/30 fractions or 40Gy/15 fractions, with concurrent temozolomide) and a four week recovery period, patients received follow-up PET and MRI scans, taken within two days of each-other. MRI data was acquired on a 3T Siemens Trio employing a Magnetization-Prepared Rapid Acquisition Gradient-echo (MPRAGE) sequence with the following parameters (FOV 24x25.6x17.6 cm, TR/TE/TI 2300/2.26/900 ms, flip angle of 90, 1 mm isotropic resolution). Images were acquired before and after administration of contrast agent Gadovist, (Bayer HealthCare Pharmaceuticals).

18F-FDOPA was synthesised according to a previously reported procedure (9). PET imaging was performed using a Philips Gemini GXL scanner. A transmission CT scan was acquired first. Subsequently, an FDOPA bolus with a mean activity of 151MBq (range 138 to 164MBq) was administered intravenously and a 75-minute acquisition initiated. The images were reconstructed using ordered subset expectation maximisation with corrections for attenuation and scatter. The final volume has a matrix size of 128x128, consisting of 90 planes of 2x2x2mm3 voxels. The PET intensities were normalised to the cerebellum to enable reporting of Standardized Uptake Value Ratio (SUVR) values. MRI and PET images were fused using multi-scale mutual information rigid registration. The MRI taken immediately post-surgery was also fused to the post-therapy MRI, as was the CT used to plan the radiotherapy dose. The registration was propagated from the planning CT to the dose map.

In addition, an experienced nuclear medicine physician (Paul Thomas) manually contoured each tumour on the FDOPA PET image fused with post-contrast MR, while blinded to the radiotherapy plan. The overlap between the PET-defined volume and available volumes defined during conventional radiotherapy planning was computed. The overlap with the 95% of prescribed dose isocontour was also calculated. Note that the PET images used were taken 3 months after the post-surgery MRIs used for conventional radiotherapy planning.

3. Results

A summary of the results is shown in Table 1. The regions enclosed by the 95% of prescribed dose contours, referred to as the *therapeutic dose volumes* (TDV), range from 197-400ml in size. In two cases, the TDV completely encloses the CTV, and in the remaining four cases at most 2.4ml of the CTV is found outside the TDV. For the PET defined volumes the volumes ranged from 8 to 96ml, with regions of PET-defined volumes outside of their corresponding TDVs ranging from 0 to 2.2ml. The PET defined volumes outside of the conventionally defined volumes ranged from 2.4 to 84ml for GTVs and from 0 to 27ml for CTVs.

Table 1: Therapeutic, i.e. 95% of prescribed dose, therapeutic dose volume (TDV) compared to gross, clinical, & planned target volumes (GTV, CTV, PTV) and PET-derived volume. All units are in ml. Dashes indicate unavailable volumes.

	Radiotherapy Volumes			95% prescribed dose (TDV)				PET-defined volume					Comment
				Total	GTV	CTV	PTV	Total	Outside				
Pat.	GTV	CTV	PTV	Vol	outside of TDV			Vol	**GTV**	CTV	PTV	TDV	
1	-	-	194	278	-	-	3.7	54	-	-	3.4	0.4	G/CTV unavailable
2	14	142	231	313	0	0	1.8	96	**84**	27	9.5	2.2	
3	25	141	253	337	0.3	1.4	17	28	**22**	11	4.6	0.8	Near organ at risk
4	14	129	217	324	0	0	0.7	71	**57**	3.5	0.2	0	
5	-	196	317	400	-	0.3	14	53	-	1.9	0.2	0	GTV unavailable
6	27	107	186	197	0	2.4	30	8	**2.4**	0	0	0	

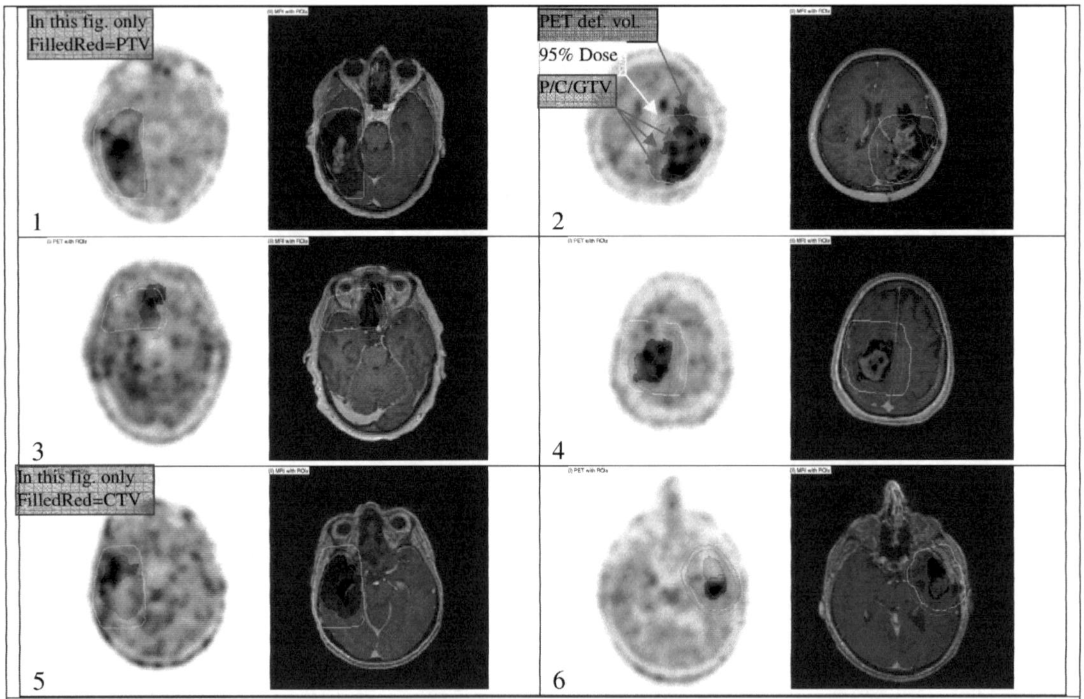

Fig. 1. Slices from each scan showing therapeutic dose volume (yellow), and PET defined volume (blue) for each patient. The clinical target volume is also shown (red). Patient numbers are on left.

Examples of one slice from each patient are shown in Fig. 1, where the therapeutic dose volume is shown in yellow, the GTV, CTV and PTV are shown in red, and the PET defined volume is shown in blue. For patient 1 only a PTV was available. For patient 5 the GTV was not available. The smallest available radiotherapy volume is filled. The contours of each volume are displayed on the PET image and the follow-up MRI (not the post-surgery MRI, as this shows same information as is contained in the contours).

4. Discussion

Arguably, the PET-derived contours probably are most equivalent to the definition for GTV, i.e. visible tumour. However PET images are highly sensitive and have some, limited, capability to detect infiltration that would not be visible on contrast enhanced MRI. So it may be fairer to make a comparison to the CTV. Although PET is sensitive to tumour compared to other modalities, microscopic infestation below a certain threshold is not visible to this modality either. Hence comparisons against both GTV and CTV are made. This implies that PET defined contours would also need margins to account for microscopic infiltration and patient motion. How large such margins would be is an open question that is beyond the scope of this work and no margin was applied to the PET volume in this work.

The volume of the PET-derived contours outside of the smallest available radiotherapy volume is 3.5ml or less for three patients (GTV for patient 6, CTV for patient 5, PTV for patient 1), indicating concordance between the MRI and PET images. However for three patients, the discrepancies ranged from 22 to 84 ml (all GTVs) indicating that the MRI scans substantially under-reported the tumour volume. These large discrepancies imply that the treatment plans would probably have been different if PET data had been available during radiotherapy planning. These values may seem large, but it is important to re-emphasise that the PET images were taken ~3 months after the post-surgery MRIs used for radiotherapy planning. And although chemo-radiotherapy was given in the intervening period, advanced glioma is known to be resistant to treatment and may have even progressed during treatment. Hence the one-to-one comparison results in Table 1 should be treated as indicative only.

On the other hand, Fig. 1 shows that although the GTVs from the follow-up and post-surgical MRIs would differ in some cases (Patients 1, 2 & 4) regions of PET uptake are seen some distance beyond the boundary of MRI enhancing tumour. This implies that the availability of PET would result in some GTVs increasing in size, and possibly commensurate increases in the CTV and PTV margins.

Despite the high PET activity beyond the bounds of both the post-surgery defined GTVs and the enhancement in follow-up MRI, all of the PET-defined volumes received relatively high doses of radiation. This is demonstrated by a comparison between the PET-defined volume and the 95% of prescribed dose contour (therapeutic dose volume). In three cases, the PET defined contours were entirely encompassed by the 95% of max isocontours. For the remaining patients, excluded PET-defined volumes ranged from 0.4 to 2.2ml. In comparison, the GTVs outside of the 95% max isocontour were zero in three cases and 0.3ml in one case, due to limitations imposed by avoiding organs at risk. Hence, the results indicate that radiotherapy plans could change in some cases, if PET were available.

Note however that all regions within the PET defined contours received a 50Gy radiation dose or more. Also, recurrence occurred in every case except patients 3 & 6. Hence it cannot be concluded from this data that a lack of metabolic information is a key reason for subsequent recurrence. Even if GTVs were enlarged, and given the palliative objective of treatment, it is unclear that every part of the GTV should receive a full dose of radiation when near organs at risk. An additional consideration is that treatment failure also stems from radio-resistance arising from local hypoxia and boosting the standard radiation dose to such regions to induce complete cell death to occur (10) (11), may not be practical in many regions near to radiosensitive tissue.

5. Conclusion

Large discrepancies between GTVs defined on MRI alone and contours defined using FDOPA PET in some cases, imply that the availability of FDOPA images could result in larger gross tumour volumes being generated for advanced glioma. However the large time-gap between the MRI used for planning and the PET scans mean these results are indicative only. Even with the opportunity for tumour growth granted by the gap between scans, all PET-defined regions received a 50Gy dose or more. It cannot be concluded that the lack of metabolic information during planning is a primary reason for later recurrence. Even so, the discrepancies between the margins of contrast enhancing MRI and high activity PET motivate for the use of PET during radiotherapy planning.

Acknowledgements

Substantial assistance was provided by the staff in various departments at the Royal Brisbane and Women's Hospital. Thank you. This research was supported by a Queensland Government Smart State NIRAP grant (MedTeQ) and the Australian National Health and Medical Research Council (NHMRC) grant number 631567.

References

[1] Newton HB, Jolesz FA, editors. Handbook of Neuro-Oncology Neuroimaging. Academic Press; 2007.
[2] Greene FL, editor. AJCC Cancer Staging Manual. Springer; 2002. p. 387–390.
[3] Chen W, Silverman DHS, Delaloye S, Czernin J, Kamdar N, Pope W, et al. 18F-FDOPA PET imaging of brain tumors: comparison study with 18F-FDG PET and evaluation of diagnostic accuracy. J. Nucl. Med. 2006 Jun;47(6):904-911.
[4] Grosu A-L, Weber WA. PET for radiation treatment planning of brain tumours. Radiother Oncol. 2010 Sep;96(3):325-327.
[5] Thorwarth D, Geets X, Paiusco M. Physical radiotherapy treatment planning based on functional PET/CT data. Radiother Oncol. 2010 Sep;96(3):317-324.
[6] Ford EC, Herman J, Yorke E, Wahl RL. 18F-FDG PET/CT for Image-Guided and Intensity-Modulated Radiotherapy. J Nucl Med. 2009 Oct;50(10):1655-1665.
[7] Fueger BJ, Czernin J, Cloughesy T, Silverman DH, Geist CL, Walter MA, et al. Correlation of 6-18F-fluoro-L-dopa PET uptake with proliferation and tumor grade in newly diagnosed and recurrent gliomas. J. Nucl. Med. 2010 Oct;51(10):1532-1538.
[8] Ledezma CJ, Chen W, Sai V, Freitas B, Cloughesy T, Czernin J, et al. 18F-FDOPA PET/MRI fusion in patients with primary/recurrent gliomas: initial experience. Eur J Radiol. 2009 Aug;71(2):242-248.
[9] F. Füchtner, J. Zessin, P. Mäding, and F. Wüst, "Aspects of 6-[18F]fluoro-L-DOPA preparation. Deuterochloroform as a substitute solvent for Freon 11," Nukl. Nucl. Med., vol. 47, no. 1, pp. 62–64, 2008.
[10] Swanson KR, Chakraborty G, Wang CH, Rockne R, Harpold HLP, Muzi M, et al. Complementary but distinct roles for MRI and 18F-fluoromisonidazole PET in the assessment of human glioblastomas. J. Nucl. Med. 2009 Jan;50(1):36-44.
[11] Hall EJ. Radiobiology for the Radiologist. 4th ed. Lippincott Williams & Wilkins; 1993.

Journal of Physics: Conference Series **489** (2014) 012029 doi:10.1088/1742-6596/489/1/012029

Fully automated shape model positioning for bone segmentation in whole-body CT scans

A Fränzle[1], M Sumkauskaite[2], J Hillengass[2,3], T Bäuerle[4], R Bendl[1,5]

[1] Department of Medical Physics in Radiation Oncology, German Cancer Research Center (DKFZ), Heidelberg, Germany
[2] Department of Radiology, German Cancer Research Center (DKFZ), Heidelberg, Germany
[3] Department of Internal Medicine V, University of Heidelberg, Heidelberg, Germany
[4] Department of Medical Physics in Radiology, German Cancer Research Center (DKFZ), Heidelberg, Germany
[5] Department of Medical Informatics, Heilbronn University, Heilbronn, Germany

E-mail: a.fraenzle@dkfz-heidelberg.de

Abstract. Analysing osteolytic and osteoblastic bone lesions in systematically affected skeletons, e.g. in multiple myeloma or bone metastasis, is a complex task. Quantification of the degree of bone destruction needs segmentation of all lesions but cannot be managed manually. Automatic bone lesion detection is necessary. Our future objective is comparing modified bones with healthy shape models. For applying model based strategies successfully, identification and position information of single bones is necessary. A solution to these requirements based on bone medullary cavities is presented in this paper.

Medullary cavities are useful for shape model positioning since they have similar position and orientation as the bone itself but can be separated more easily. Skeleton segmentation is done by simple thresholding. Inside the skeleton medullary cavities are segmented by a flood filling algorithm. The filled regions are considered as medullary cavity objects. To provide automatic shape model selection, medullary cavity objects are assigned to bone structures with pattern recognition. To get a good starting position for shape models, principal component analysis of medullary cavities is performed. Bone identification was tested on 14 whole-body low-dose CT scans of multiple myeloma patients.

Random forest classification assigns medullary cavities of long bones to the corresponding bone (overall accuracy 90%). Centroid and first principal component of medullary cavity are sufficiently similar to those of bone (mean centroid difference 21.7 mm, mean difference angle 1.54° for all long bones of one example patient) and therefore suitable for shape model initialization.

This method enables locating long bone structures in whole-body CT scans and provides useful information for a reasonable shape model initialization.

1. Introduction

Bone tumor diseases like bone metastasis or multiple myeloma have bone destructive effects on bone and cause osteolytic lesions. To assess overall disease progression or treatment response exact segmentations of all lesions are needed. Since the whole skeleton can be affected, segmentation of all lesions cannot be managed manually since it is too time-consuming. Therefore an automatic computerized approach is needed. To find modified regions in bones and to be able to find missing bone mass also in case of breached cortical bone, a reconstruction of the outer bone contour is necessary first. Recent publications about bone segmentation present interactive methods [1] or are atlas based and defined for particular body regions [2], [3], [4]. Because of large image data in case of whole-body CT scans in multiple myeloma, we want to avoid any user interaction and provide global lesion detection on whole-body scans. For the reconstruction of the outer bone contour of single bones, statistical shape models [5] are to be used in our case. These shape models are able to deal with poor edge contrast in the joints and to reconstruct the former bone contour, since these models do not tend to leak into osteolytic modified bone contours. For successful segmentation, these models need a reasonable initial positioning. To segment the correct structure, the shape model has to be positioned near the corresponding bone. To prevent wrong alignment of the model during the adjustment process, the model has to be roughly aligned. In this paper, we focus on the automatic initialization of shape models as a necessary step before segmentation and present a method to get the information required for automatic positioning of shape models for segmentation of long bones. For this purpose an automatic segmentation of the medullary cavities is used. Medullary cavities can be segmented more easily compared to the bone structures themselves, since poor contrast and complex structures in the joints impede delineation of single bones with simple image processing methods, like thresholding. Medullary cavities have similar position and orientation as the bones themselves and therefore provide information necessary for model initialization. With automatic segmentation of medullary cavities and assignment to corresponding bone structures with pattern recognition we are able to locate bone structures in whole body CT scans without any user interaction. All information needed to select the appropriate shape models and to find a good starting position is provided with the presented method.

2. Material and Methods

Data, segmentation of medullary cavities and identification of segmented medullary cavity objects with a random forest classifier are described in detail in the following sections.

2.1. Data

Medullary cavity segmentation and bone identification were tested on 14 whole-body CT scans of multiple myeloma patients (voxel size 0.98 mm $\times$ 0.98 mm $\times$ 3.0 mm). The comparison of position and orientation of bones and their medullary cavities was performed exemplarily on one patient.

2.2. Automatic segmentation of medullary cavities

The segmentation of medullary cavities mainly consists of the following steps, which are shown in figure 1 for the femur bones (fig. 1 a):

- binarization of the bones with a global threshold (fig. 1 b)
- filling the segmented skeleton with a flood filling (fig. 1 c)
- determining the filled regions as difference of filled skeleton and skeleton segmentation.

The filled regions (fig. 1 d), can be considered as segmentation of the medullary cavities.

2.3. Bone identification with pattern recognition

To be able to select the appropriate shape model for a specific bone for further developments the medullary cavity objects are assigned to the corresponding bone structure using random forests [6]. A classification is performed based on object volume and position parameters (distance of centroid to median plane, distance of centroid to mid-coronal plane, relative position of centroid in cranial-caudal direction). The classification performance was assessed by training 14 random forests with 100 trees, leaving one patient out as test case in each run.

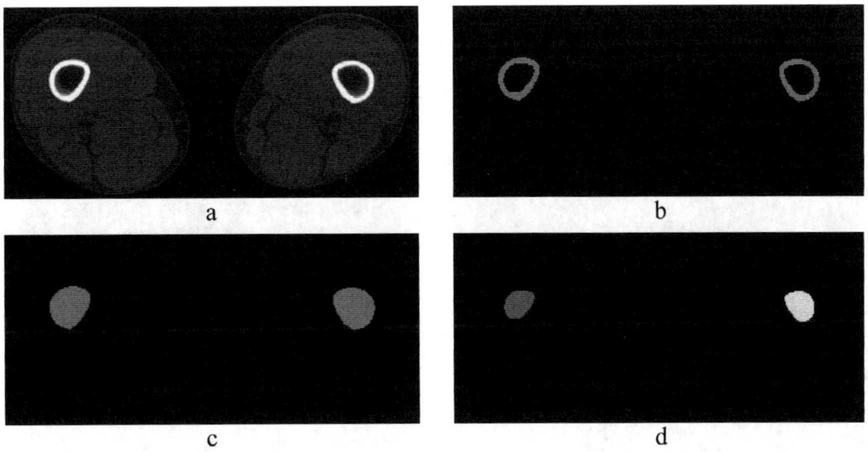

Figure 1. Process of segmentation of medullary cavities. The skeleton (left and right femur shown here in a) is binarized (b) and filled (c). d shows the filled and labelled regions.

3. Results
Figure 2 shows the automatically segmented medullary cavities.

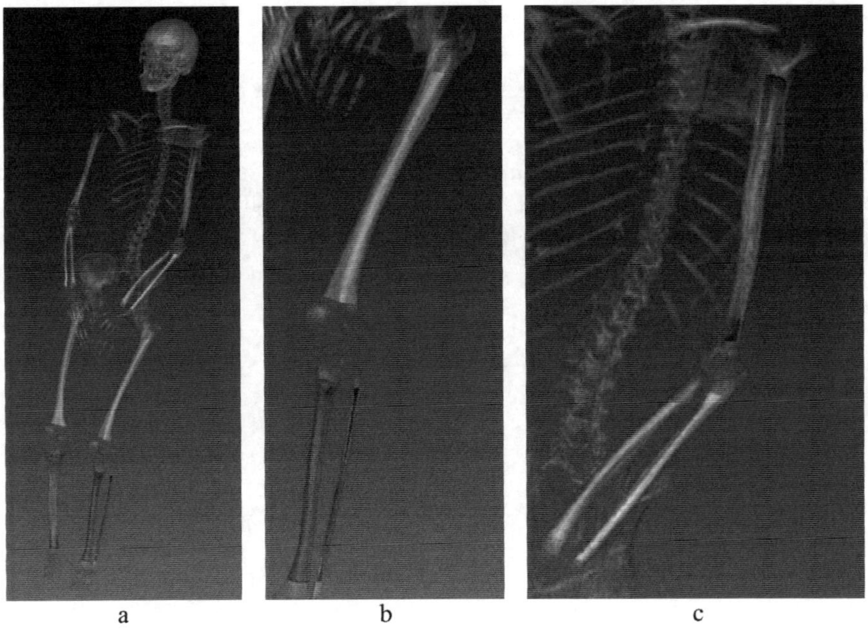

Figure 2. Automatically segmented medullary cavities in the skeleton (a) and detailed in leg bones (b) and arm bones (c).

Figure 3 shows centroids of bone (blue) and medullary cavities (red). It can be seen, that bone centroids and medullary cavity centroids are located close to each other. To quantify deviations in location and orientation the mean centroid difference and mean angle between first principal components of bone and medullary cavity are calculated exemplarily for all long bones of one patient. The mean centroid difference is 21.7 mm (SD 15.37 mm) and the mean angle is 1.54° (SD 0.94°), which is accurate enough for model initialization.

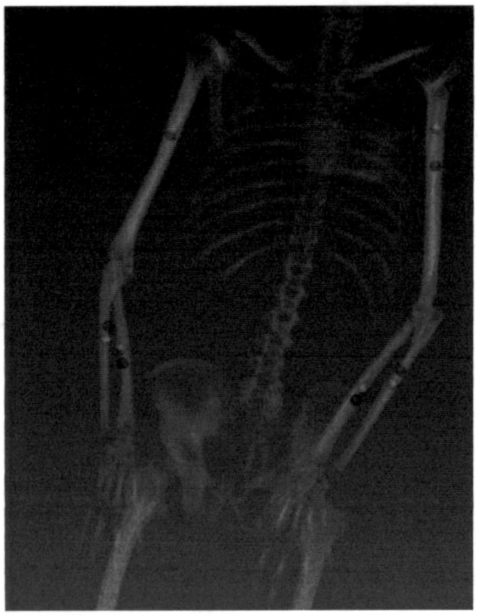

a

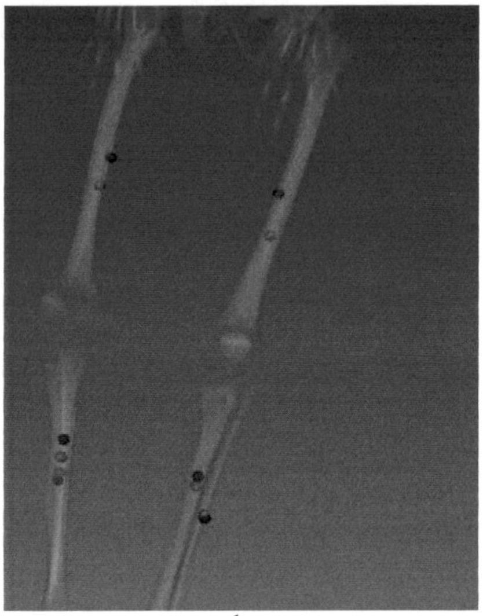

b

Figure 3. Centroids of bone (blue) and medullary cavities (red) in arm bones (a) and leg bones (b).

To assess classifier performance the classification accuracy for the different long bones is determined: femur 90%, tibia 79%, fibula 79%, humerus 93%, radius 69%, ulna 46%, other 99%. The class "other" helps to exclude objects segmented in the medullary cavity segmentation process but not belonging to a long bone from further processing. The classification accuracy over all objects including those defined as "other" is 90%.

4. Conclusions

Classification of medullary cavities enables identification of long bone structures in whole-body CT scans. The accuracy of identifying the ulna is low, but this bone could be identified after the identification of humerus and radius. The centroid of a medullary cavity can be used for initial positioning of a shape model and the first principal component can be used for initial orientation. In conclusion, the presented method provides all information needed for the automatic selection and initialization of a statistical shape model for long bone segmentation.

Acknowledgments

This project is funded by the German Research Foundation (DFG), Project BE 2523/3-1.

References

[1] Liu L, Raber D, Nopachai D, Commean P, Sinacore D, Prior F, Pless R and Ju T 2008 *Med Image Comput Comput Assist Interv* **11** 296–304
[2] Ehrhardt J, Handels H, Malina T, Strathmann B, Plötz W and Pöppl S J 2001 *Int J Med Inform* **64** 439–447
[3] Haas B, Coradi T, Scholz M, Kunz P, Huber M, Oppitz U, André L, Lengkeek V, Huyskens D, van Esch A and Reddick R 2008 *Phys Med Biol* **53** 1751–1771
[4] Wu C, Murtha P E and Jaramaz B 2010 *Int J Med Robot* **6** 1–17
[5] Heimann T and Meinzer H P 2009 *Med Image Anal* **13** 543–563
[6] Breiman L 2001 *Machine Learning* **45** 5–32

Journal of Physics: Conference Series **489** (2014) 012030 doi:10.1088/1742-6596/489/1/012030

Investigation of the accuracy of breast tissue segmentation methods for the purpose of developing breast deformation models for use in adaptive radiotherapy

P Juneja[1], EJ Harris[1], and PM Evans[2]
[1] Joint Department of Physics, Institute of Cancer Research, Sutton SM2 5PT, UK
[2] CVSSP, University of Surrey, Guildford GU2 7XH, UK

E-mail: Prabhjot.Juneja@icr.ac.uk

Abstract. Realistic modelling of breast deformation requires the breast tissue to be segmented into fibroglandular and fatty tissue and assigned suitable material properties. There are a number of breast tissue segmentation methods proposed and used in the literature. The purpose of this study was to validate and compare the accuracy of various segmentation methods and to investigate the effect of the tissue distribution on the segmentation accuracy. Computed tomography (CT) data for 24 patients, both in supine and prone positions were segmented into fibroglandular and fatty tissue. The segmentation methods explored were: physical density thresholding; interactive thresholding; fuzzy c-means clustering (FCM) with three classes (FCM3) and four classes (FCM4); and k-means clustering. Validation was done in two-stages: firstly, a new approach, supine-prone validation based on the assumption that the breast composition should appear the same in the supine and prone scans was used. Secondly, outlines from three experts were used for validation. This study found that FCM3 gave the most accurate segmentation of breast tissue from CT data and that the segmentation accuracy is adversely affected by the sparseness of the fibroglandular tissue distribution.

1. Introduction

A common treatment for breast cancer involves surgical removal of the tumour, followed by radiotherapy to the remaining breast tissue. This reduces the risk of local recurrence. The region of the breast close to where the tumour is removed is known as the 'tumour bed', and is the most likely place for disease relapse. It is therefore desirable to preferentially irradiate this region. However the tumour bed undergoes significant shape and volume changes over the time-frame of radiotherapy planning and delivery [1,2]. The localization of the tumour bed is difficult because it comprises soft tissue and hence has little x-ray imaging contrast. These factors may result in under-dosage of the clinical target volume or over-dosage of healthy adjacent tissues. These issues become more concerning in the context of partial breast irradiation (PBI). Modelling of tissue mechanics could be used to accurately quantify changes in the breast during the course of radiotherapy and could therefore be developed for use in adaptive radiotherapy (ART).

For realistic modelling the breast has to be segmented into its various components, including fibroglandular and fatty tissue, and assigned suitable material properties. Several authors have proposed algorithms for segmentation of breast tissues [3-6]. A study that validates and compares these methods is required. However, in the absence of true segmentation, validation becomes a challenging task. We have proposed a two-stage approach for evaluating the performance of tissue segmentation methods. The first stage is to evaluate the performance of segmentation algorithms based on the knowledge that the measured breast composition should remain the same when measured from image data acquired in different positions. The second stage was comparison with segmentation using the tissue outlines from the experts. Figure 1 illustrates two different distributions (sparse and non-sparse) of fibroglandular tissue in the breast. We tested the hypothesis that the sparseness of the fibroglandular tissue distribution affects the accuracy of breast tissue segmentation.

The aims of this study were, (1) to validate and compare various segmentation methods for computed tomography (CT) data and to determine the best method; (2) to determine the effects of the tissue distribution on the segmentation accuracy.

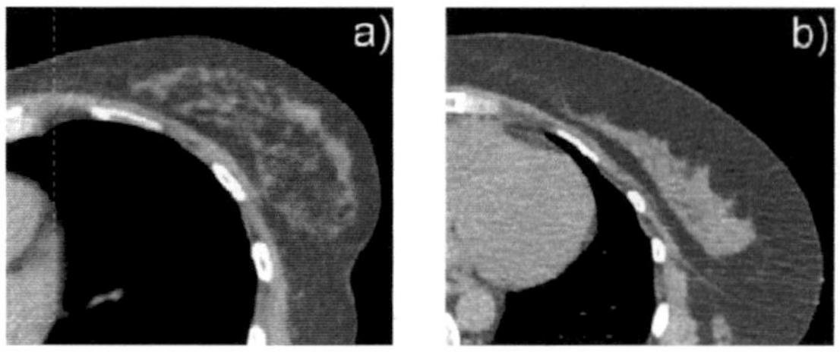

Figure 1. Example middle-breast CT images: a. Breast with sparse distribution of fibroglandular tissue; b. Breast with non-spare distribution of fibroglandular tissue.

2. Methods

A set of algorithms was used to segment datasets into fibroglandular and fatty tissues, and in some cases also into 'other' (breast tissue that cannot be classified either as fat or fibroglandular tissue) and background (a voxel that does not lie within the whole breast). The whole breast was first segmented from the CT dataset using clinician outlining and then tissue segmentation methods were applied.

2.1. Patient data

A clinical pilot study, approved by the local ethics committee, to compare prone and supine positioning for breast radiotherapy was carried at the Royal Marsden Hospital [7]. Patients underwent CT imaging in both supine and prone positions on the same day. Datasets of 24 patients from that study were used in this work. CT data consisted of axial slices. On each slice whole breast were delineated by a single clinician [7]. An observer (EH), visually assessed the breast datasets and ranked the sparseness of fibroglandular tissue distribution on a scale of 1 to 5, (where 5 was the most sparse, i.e. very thin strands of fibroglandular tissue). The patient data were divided into two groups based on the ranks: non-sparse (rank 1-3) and sparse group (rank 4-5) [8].

2.2. Segmentation Methods

2.2.1. Physical density thresholding. The physical density values corresponding to the CT number were assigned to each voxel in a CT dataset by linear interpolation of the CT to density conversion table of the scanner. Physical density ranges for different breast tissues were obtained from the literature [9]. Voxels with physical density values that fall within the ranges were classified as fat or fibroglandular tissue. This will be called the hard range method. An expanded tissue range (the soft range method) was also investigated.

2.2.2. Interactive Thresholding. In this method the user interactively sets intensity CT number threshold and all the voxels with CT number higher than this threshold are labelled as fibroglandular and other voxels as fat, until the user judges the best segmentation is achieved.

2.2.3. K-means clustering. This method results in clustering of the dataset. Segmentation was performed using 3 classes: background, fibroglandular and fat [6].

2.2.4. Fuzzy C-Means (FCM) clustering. This method results in fuzzy clustering, the voxels are assigned probabilities (membership values) of belonging to the class (fat, fibroglandular) [4]. After fuzzy clustering, the fibroglandular class was thresholded at various probabilities to yield the segmentation. Segmentation was performed with three classes (FCM3) i.e. background, fibroglandular and fat and with four classes (FCM4) by adding another class labelled as 'other'.

2.3. Validation of Segmentation

Analysis of variance (ANOVA)) followed by post-hoc comparisons (Bonferroni, $p<0.05$) were used to test for statistical differences between methods. The Wilcoxon rank sum test was used to compare the accuracy of the segmentation methods for non-sparse and sparse groups.

2.3.1. Prone-supine evaluation. The segmentation methods were used to find the volume of fibroglandular tissues and volumetric breast density (VBD, the percentage of breast that is composed of fibroglandular tissue) was calculated for all patient datasets. For each patient, relative difference in VBD between prone and supine positions was also calculated.

2.3.2. Expert validation. Mid-breast CT slices from 12 of the patients were selected randomly and outlined by the three experts. These 36 outlines were used individually for pair-wise comparison with algorithmic segmentation using the Dice similarity coefficient (DSC). The DSC is defined as the ratio of the volume of intersection of two volumes to the mean of the two volumes. For identically segmented slices DSC=1.

3. RESULTS

3.1. Prone-supine validation

Mean (averaged over all patients) VBD measured using FCM3 and FCM4 decreases with an increase in threshold level as shown in figure 2a. VBD decreased with increasing threshold because the fibroglandular class assignment was made more stringent. VBD was smaller for FCM4 than FCM3 at a given threshold because in FCM4 the membership values for fibroglandular class will always be less than or equal to that for FCM3. Mean relative difference between supine and prone VBD for various thresholding criteria was 8%-9% and 16%-17% with 3 and 4 classes, respectively. The measured mean VBD using the various methods are shown in figure 2b. The various methods give different VBD values and the mean relative difference between supine and prone values from the respective methods was between 8% and 21%. Mean relative difference was lowest for FCM3. ANOVA failed to indicate significant differences (p=0.08) between the various methods. Physical density methods performed poorly and were not considered for expert validation.

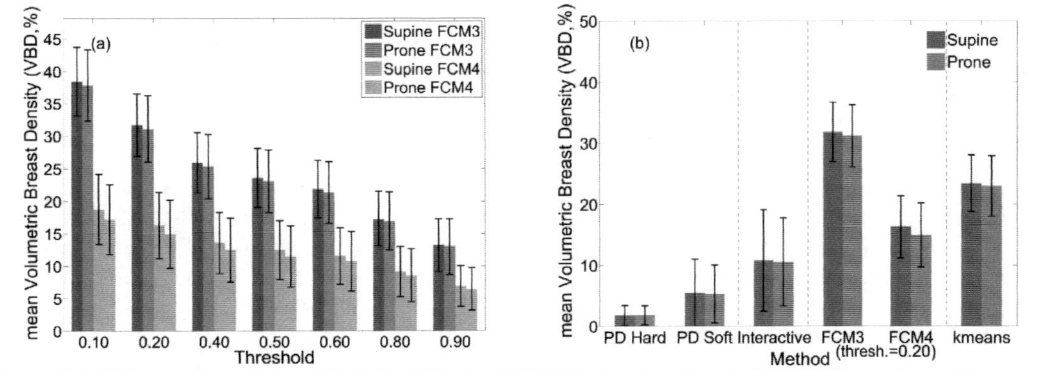

Figure 2. (a) Comparison of mean VBD for FCM3 and FCM4 at various thresholds. (b) Comparison of mean VBD for various methods. thresh. is thresholding value; Error bars shows ±1 standard deviation.

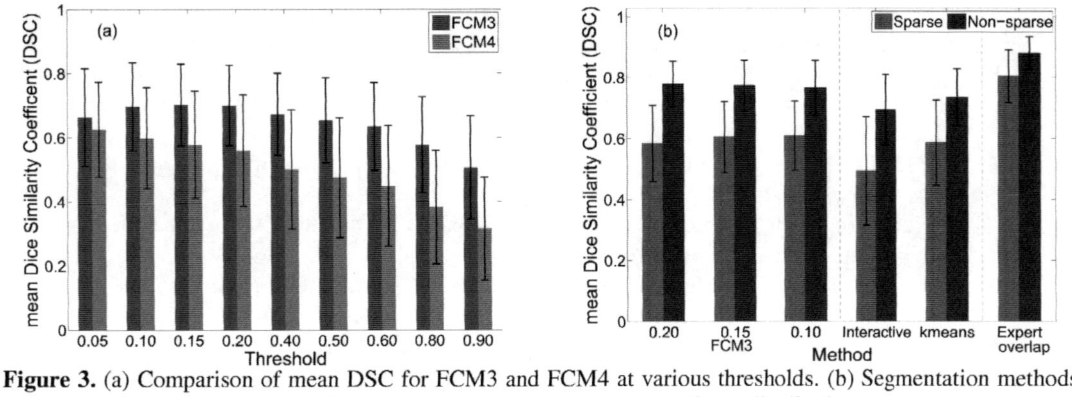

Figure 3. (a) Comparison of mean DSC for FCM3 and FCM4 at various thresholds. (b) Segmentation methods accuracy and experts overlap for the two groups: non-sparse and sparse tissue distribution.

3.2. Expert validation

The Dice similarity coefficient was calculated for FCM3, FCM4, interactive thresholding, and kmeans clustering methods. There was good agreement between expert outlines with mean DSC of 0.85 (standard deviation was 0.08) for the pairs of expert outlines. The Dice similarity coefficient (DSC) for validation of FCM segmentation with three classes (FCM3) and with four classes (FCM4) as a function of threshold is presented in figure 3a. It was found that for any threshold, the mean DSC for segmentation with FCM3 was higher than for FCM4. ANOVA on the DSC of FCM3 and FCM4 methods with different thresholding criteria indicated that FCM3 performed significantly better ($p<0.001$) than FCM4. The FCM3 thresholded at 0.10; 0.15; and 0.20 had the highest mean Dice similarity coefficient (DSC) of 0.70. ANOVA on the DSC of FCM3 at different threshold values indicated that there were significant differences between different thresholding levels ($p<0.001$). However, post-hoc analysis (Bonferroni) did not give any significant pair wise differences.

4. Discussion

In the present study, the results demonstrated that fuzzy c-means methods with three classes (FCM3) thresholded at 0.10, 0.15, and 0.20 generated segmentation results closest to expert segmentation. This is compatible with the work of Ertas et al [3] who noted for breast MRI that the correlation between breast densities estimated based on interactive thresholding and FCM3 is highest when a threshold of 0.20 is used.

The prone-supine evaluation of segmentation algorithms did not show significant differences between the various methods. In 7 out of 24 cases, large relative differences (>10%) between prone and supine VBD were found. The likely reason for these large relative differences is the poor accuracy of breast tissue segmentation algorithms for breasts with sparse distribution of fibroglandular tissue. Otherwise our findings were similar with those of Nie et al [10] who observed in their study, using breast MRI, a 3-6% variation in measurements of VBD with body positioning.

This study demonstrated: (1) fuzzy c-means clustering with three classes gives the most accurate segmentation of breast tissue; (2) the distribution of tissues within the breast significantly affects the segmentation accuracy. The results of this study is expected to aid development of adaptive breast radiotherapy based on modelling of tissue mechanics [8].

Acknowledgements

This work was supported by Cancer Research-UK under programme grant reference no C46/A10588 and we would like to acknowledge the NIHR for funding of the biomedical research imaging centre.

References

[1] Harris EJ, Donovan EM, Yarnold JR, Coles CE, Evans PM. 2009 Characterization of target volume changes during breast radiotherapy using implanted fiducial markers and portal imaging. *Int J Radiat Oncol Biol Phys* **73** 958-966.

[2] Kim LH, Vicini F, Yan D. 2008 What do recent studies on lumpectomy cavity volume change imply for breast clinical target volumes? *Int J Radiat Oncol Biol Phys* **72** 1-3.

[3] Ertas G, Reed S, Leach MO. 2009 Comparison of Breast Density Assessments Based on Interactive Thresholding and Automated Fast Fuzzy c-means Clustering in Three-Dimensional MR Imaging. *IFMBE Proc* **25** 1893-1896.

[4] Bezdek JC, Ehrlich R. 1984 FCM: The fuzzy c-means clustering algorithm. *Computers & Geosciences* **10** 191-203.

[5] Glide-Hurst CK, Duric N, Littrup P. 2007 A new method for quantitative analysis of mammographic density. *Med Phys* **34** 4491-4498.

[6] MacQueen JB. 1967 Some Methods for Classification and Analysis of Multivariate Observations. *Proc Fifth Berkeley Symp on Math Statist and Prob* **1** 281-297.

[7] Kirby AM, Evans PM, Donovan EM, Convery HM, Haviland JS, Yarnold JR. 2010 Prone versus supine positioning for whole and partial-breast radiotherapy: A comparison of non-target tissue dosimetry. *Radiotherapy and Oncology* **96** 178-184.

[8] Juneja P, Harris EJ, Kirby AM, Evans PM. 2012 Adaptive Breast Radiation Therapy Using Modeling of Tissue Mechanics: A Breast Tissue Segmentation Study. *Int J Radiat Oncol Biol Phys* **84** e419-e425.

[9] Johns PC, Yaffe MJ. 1987 X-ray characterisation of normal and neoplastic breast tissues. *Phys Med Biol* **32** 675-695.

[10] Nie K, Chen JH, Chan S, Chau MKI, Yu HJ, Bahri S, Tseng T, Nalcioglu O, Su MY. 2008 Development of a quantitative method for analysis of breast density based on three-dimensional breast MRI. *Med Phys* **35** 5253-5262.

XVII International Conference on the Use of Computers in Radiation Therapy (ICCR 2013) IOP Publishing
Journal of Physics: Conference Series **489** (2014) 012031 doi:10.1088/1742-6596/489/1/012031

Graded-threshold parametric response maps: towards a strategy for adaptive dose painting

A Lausch[1], N Jensen[2], J Chen[3], TY Lee[1], M Lock[3], E Wong[1,2]

[1] Department of Medical Biophysics, The University of Western Ontario, London, ON, N6A 3K7, Canada
[2] Department of Physics and Astronomy, The University of Western Ontario, London, ON, N6A 3K7, Canada
[3] London Regional Cancer Program, London, ON, N6A 4L6, Canada

Email: alausch2@uwo.ca

Purpose: To modify the single-threshold parametric response map (ST-PRM) method for predicting treatment outcomes in order to facilitate its use for guidance of adaptive dose painting in intensity-modulated radiotherapy.
Methods: Multiple graded thresholds were used to extend the ST-PRM method (Nat. Med. 2009;15(5):572-576) such that the full functional change distribution within tumours could be represented with respect to multiple confidence interval estimates for functional changes in similar healthy tissue. The ST-PRM and graded-threshold PRM (GT-PRM) methods were applied to functional imaging scans of 5 patients treated for hepatocellular carcinoma. Pre and post-radiotherapy arterial blood flow maps (ABF) were generated from CT-perfusion scans of each patient. ABF maps were rigidly registered based on aligning tumour centres of mass. ST-PRM and GT-PRM analyses were then performed on overlapping tumour regions within the registered ABF maps.
Main findings: The ST-PRMs contained many disconnected clusters of voxels classified as having a significant change in function. While this may be useful to predict treatment response, it may pose challenges for identifying boost volumes or for informing dose-painting by numbers strategies. The GT-PRMs included all of the same information as ST-PRMs but also visualized the full tumour functional change distribution. Heterogeneous clusters in the ST-PRMs often became more connected in the GT-PRMs by voxels with similar functional changes.
Conclusions: GT-PRMs provided additional information which helped to visualize relationships between significant functional changes identified by ST-PRMs. This may enhance ST-PRM utility for guiding adaptive dose painting.

1. Introduction

Traditionally, the aim of radiation therapy (RT) is to deliver a large homogeneous dose of radiation to the tumour while limiting the dose to surrounding critical structures. With intensity modulated radiation therapy (IMRT) and image-guidance dose can be delivered with more conformality and precision than ever before. This has led to significant interest in the idea of using non-homogeneous dose distributions to take advantage of heterogeneity in tumour radio-sensitivity. This concept has been termed “dose-painting” and involves the use of functional imaging as a surrogate for indicating tumour radio-sensitivity or risk of relapse after treatment [1]. Approaches have typically involved delivering boost doses to predicted radio-resistant tumour sub volumes or heterogeneous doses to the entire tumour which scale with local radio-resistance ("dose-painting-by-numbers") [1].

Recently, Galban et al [2] have demonstrated a promising new method for predicting treatment outcomes which may be well-suited towards informing dose painting techniques. Parametric response maps (PRMs) are generated which indicate whether each tumour voxel has undergone a significant increase or decrease in function as indicated by analysis of functional imaging acquired before and during treatment. This information was found to be predictive of overall survival among a cohort of patients treated for glioma and imaged with perfusion MRI [1]. Therefore tumour response and progression could potentially have been visualized during treatment on a voxel-by-voxel basis using the PRMs. Consequently the PRM method may be ideal for informing adaptive dose painting techniques within contexts in which its predictive utility has been verified.

To date, the PRM method has been investigated within a few different contexts [2-5]. Galban et al. are currently investigating the use of a PRM biomarker to identify non-responding patients treated for glioma who might benefit from salvage therapy with bevacizumab and dose painting IMRT [6]. However, these results are not yet available and to our knowledge no other studies discussing the PRM method as applied to dose painting have appeared in the literature.

The PRM method employs a single threshold value to distinguish if tumour functional changes are significant with respect to treatment outcomes. In this study we present a natural extension of this method which uses multiple graded thresholds in order to improve PRM utility within adaptive dose painting contexts. Hereafter the original and proposed PRMs are referred to as single-threshold PRMs (ST-PRM) and graded-threshold PRMs (GT-PRM) respectively. We motivate and discuss the underlying logic of GT-PRMs by comparing them to traditional ST-PRMs through application of both methods to CT-perfusion scans of patients treated with RT for hepatocellular carcinoma.

2. Materials and methods

2.1. Single-threshold parametric response maps

To generate ST-PRMs, baseline functional imaging acquired before treatment is compared to followup imaging acquired during treatment. Since the comparison is voxel-based, the scans must be spatially aligned using image registration. A tumour voxel is classified as significantly increasing or decreasing in function if its functional change (ΔF) is such that $\Delta F > +T$ or $\Delta F < -T$ respectively, where T is a single threshold value. Voxels for which $|\Delta F| \leq T$ are classified as "insignificant change". ST-PRMs correspond to images of the tumour that indicate which voxels significantly increased, decreased, or did not significantly change in function after treatment.

T is determined by the functional changes observed within similar healthy tissue [2]. Here, the ST-PRM thresholds were determined by investigating the functional changes in the remaining healthy portions of the liver. The function of each healthy liver voxel in the followup scan is plotted against the function of each corresponding voxel in the baseline scan. $[-T, T]$ is then defined to be the 95% confidence interval for the residuals that result from fitting a line to this plot. Consequently, changes in tumour function are considered to be significant if they exceed the magnitude of most of the changes observed in similar healthy tissue that is untreated.

2.2. Graded-threshold parametric response maps

GT-PRMs are generated in much the same way as ST-PRMs except multiple thresholds are used in order to classify voxels by the magnitude of their functional changes. Let $\{T_i\}$ for i = 1...10 represent a set of thresholds. Each $[-T_i, T_i]$ corresponds to a different confidence interval for the residuals from the same plot used to determine the threshold for the original PRM method. In this study, the T_i for i = 1...10 corresponded to the 5%, 15%, 25%, 35%, 45%, 55%, 65%, 75%, 85%, and 95% confidence intervals respectively. Accordingly, T_{10} is the same threshold used to generate the ST-PRMs

To create the GT-PRM, each tumour voxel is classified by the maximum threshold that its associated functional change exceeds. This classification is represented visually by assigning a value of $\pm i$ to each voxel in the map where $+i$ indicates $\Delta F > +T_i$ and $-i$ indicates $\Delta F < -T_i$. For example, a value of 5 indicates that $T_5 \leq \Delta F < T_6$ whereas a value of -5 indicates $-T_6 < \Delta F \leq -T_5$.

2.3. *Example cases*

The abdominal CT-perfusion scans of 5 patients with hepatocellular carcinoma were used for this study. Scans were acquired pre and post-RT using a 64 slice scanner (Discovery VCT or CT750HD, GE Healthcare, Waukesha, WI) with a spatial resolution of 0.7 mm by 0.7 mm by 5 mm and a temporal resolution of 2.8 seconds. Each image volume contained 512x512x16 voxels. Arterial blood flow (ABF) functional image maps were generated from each 4D acquisition.

2.4. *Image registration*

To estimate ST-PRM and GT-PRM thresholds, the healthy liver must be accurately aligned within the pre and post-RT ABF maps. The average pre and post-RT 4DCT scans were non-rigidly registered to one another using an elastic, spline-based, multi-resolution registration algorithm [7]. The resulting transformations were then used to align the pre and post-RT ABF functional maps.

While the above method is suitable for registering the healthy liver, its accuracy cannot be assumed within the tumour region due to potential volume changes that are unrelated to motion. Consequently ST-PRM and GT-PRM analysis was performed on rigidly registered ABF maps instead of the non-rigidly registered maps used for threshold estimation. The tumours were delineated in the pre and post-RT scans and then rigidly registered by aligning their centers of mass. Subsequent analysis was performed only on the overlapping regions of the tumour within the rigidly registered maps. This approach is similar to Galban et al [2].

3. Results and discussion

ST-PRM and GT-PRM maps were computed and then superimposed on the associated average pre-RT 4DCT for each patient (Figure 1). For further comparison, maps indicating the relative difference in tumour ABF (RD-ABF) were also generated. RD-ABF was defined as the difference between the pre and post-RT voxels divided by the magnitude of the maximum ABF difference within the tumour.

The ST-PRMs contained many disconnected clusters of voxels that were classified as significantly increasing or decreasing in function. Since a single threshold is used to determine functional change significance, ST-PRMs do not contain information regarding the lesser magnitude functional changes within the voxels surrounding these disconnected clusters. While surrounding voxels may not be predictive of outcomes, their values help to visualize the spatial functional change relationships between the disconnected clusters of significantly changing voxels. Since GT-PRMs are generated using multiple thresholds they can convey this information along with all of the ST-PRM information.

For example, the ST-PRM for patient 4 contains a large circular sub region with mostly insignificant functional changes and some significant increases (red) and decreases (blue) speckled throughout. However, the GT-PRM indicates that these clusters are connected by voxels that are mostly increasing in function with many of the changes exceeding the T_5 (55% CI) threshold. This demonstrates a systemic regional functional increase which may be useful for defining a radiation boost volume for adaptive radiotherapy. Similarly for the other ST-PRMs, regional functional change relationships become clearer within the corresponding GT-PRMs.

The RD-ABF maps also provided insight into regional functional change relationships. However the GT-PRMs illustrate this information while putting it into context with respect to the changes that might be expected due to normal physiological variation. The observer can immediately identify and stratify the potential treatment related significance of tumour functional changes. Therefore, GT-PRMs may be amenable to informing adaptive dose-painting-by-numbers whereby local tumour dose is scaled by the significance of local functional changes.

In this study, pre and post-RT ABF maps were used to demonstrate the GT-PRM method. As a result of treatment or progression related tumour volume changes, registration may have been less accurate within the tumours. In practice, followup imaging would be acquired during treatment

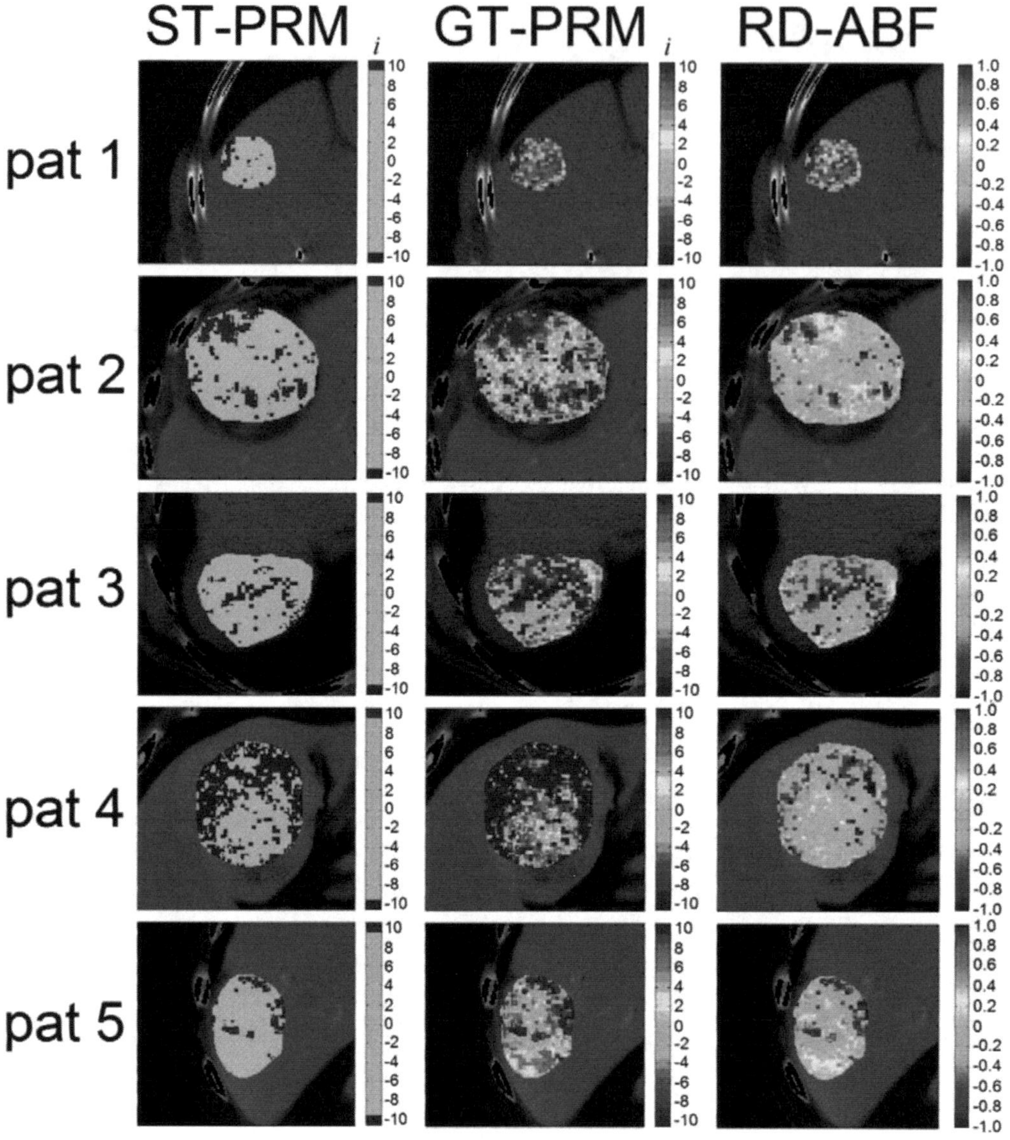

Figure 1. Representative slice from patient 3 simulations. Maps were cropped down to a 10x10 cm field of view and values outside the tumour were suppressed for improved visualization. A) S_0P_3 and S_iP_3 (no RE), B) $S_iP_3M_8$ (~4.3 mm average RE) and C) $S_iP_3M_{20}$ (~7.5 mm average RE)

reducing volume changes and associated registration uncertainty. Registration errors could lead to tumour voxel misclassification for both the ST-PRM and GT-PRM methods. However, the severity of misclassification within GT-PRMs should be reduced since a spectrum of graded thresholds are employed. For example, a significantly increasing voxel misclassified as unchanging by a ST-PRM, could be misclassified as a somewhat less significantly increasing voxel in a GT-PRM rather than completely unchanging. Consequently, GT-PRMs may be more robust to registration error with respect to visualization of regional functional change relationships.

Ultimately, the predictive utility of the PRM method needs to be verified for each combination of imaging modality and pathology. This work has not yet been done for liver tumours. Liver tumour examples were selected within the present study for several reasons. First, due to RT delivery constraints, dose-painting by numbers or delivering boosts to tumour subvolumes may only be feasible in larger tumours. Liver tumours referred for RT tend to be large and unresectable, and therefore potentially amenable to this treatment approach. Second, prognosis for patients with liver tumours remains poor [8] and so there is a push to develop new treatment techniques which offer improved outcomes. In principle, the ST-PRM and GT-PRM methods can be applied to any pathology with repeat imaging. In the absence of verified predictive utility, the GT-PRM still provides an effective means of visualizing functional changes with respect to estimated normal physiological variations.

4. Conclusion

The ST-PRMs contained many disconnected clusters of voxels classified as having a significant change in function. While this may be useful to predict global treatment response, it may pose challenges for identifying boost volumes or for informing dose-painting by numbers strategies. The GT-PRMs included all of ST-PRM information but also visualized the full tumour functional change distribution. GT-PRMs provided additional information which helped to visualize relationships between significant functional changes identified by ST-PRMs. This may enhance PRM utility for guiding adaptive dose painting.

References

[1] Bentzen S M, Gregoire V. 2011 *Semin. Radiat. Oncol.* **21** 101-10
[2] Galban C J et al. 2009 *Nature Medicine* **15**(5) 572-76
[3] Galban C J et al. 2009 *Transl. Oncol.* **2**(3) 184-90
[4] Hoff B A et al. 2012 *Bone* **51**(1) 78-84
[5] Harris R J, Cloughesy T F, Pope W B, Nghiemphu PL, Lai A, Zaw T, Czernin J, Phelps ME, Chen W, Ellingson BM. 2012 *Neuro-Oncology* **14**(8) 1079-89
[6] Galban C J et al. 2011 *Clin. Cancer. Res.* **17**(14) 4751-60
[7] Modersitzki J 2009 *FAIR: Flexible Algorithms for Image Registration.* Philadelphia, USA: Society for Industrial and Applied Mathematics.
[8] Ellison LF, Wilkins K. 2010 *Health Reports* **21**(3) 55-60

Journal of Physics: Conference Series **489** (2014) 012032 doi:10.1088/1742-6596/489/1/012032

Effects of registration error on parametric response map analysis: a simulation study using liver CT-perfusion images

A Lausch[1], NKG Jensen[2], J Chen[3,4], TY Lee[1], M Lock[3,4], E Wong[1,2]

[1] Department of Medical Biophysics, The University of Western Ontario, London, ON, N6A 3K7, Canada
[2] Department of Physics and Astronomy, The University of Western Ontario, London, ON, N6A 3K7, Canada
[3] London Regional Cancer Program, London, ON, N6A 4L6, Canada

Email: alausch2@uwo.ca

Purpose: To investigate the effects of registration error (RE) on parametric response map (PRM) analysis of pre and post-radiotherapy (RT) functional images.
Methods: Arterial blood flow maps (ABF) were generated from the CT-perfusion scans of 5 patients with hepatocellular carcinoma. ABF values within each patient map were modified to produce seven new ABF maps simulating 7 distinct post-RT functional change scenarios. Ground truth PRMs were generated for each patient by comparing the simulated and original ABF maps. Each simulated ABF map was then deformed by different magnitudes of realistic respiratory motion in order to simulate RE. PRMs were generated for each of the deformed maps and then compared to the ground truth PRMs to produce estimates of RE-induced misclassification.
Main findings: The percentage of voxels misclassified as decreasing, no change, and increasing, increased with RE For all patients, increasing RE was observed to increase the number of high post-RT ABF voxels associated with low pre-RT ABF voxels and vice versa. 3 mm of average tumour RE resulted in 18-45% tumour voxel misclassification rates.
Conclusions: RE induced misclassification posed challenges for PRM analysis in the liver where registration accuracy tends to be lower. Quantitative understanding of the sensitivity of the PRM method to registration error is required if PRMs are to be used to guide radiation therapy dose painting techniques.

1. Introduction

The ability to predict tumour response with high spatial precision during or after radiotherapy (RT) could provide for opportunities to adapt treatment to a patient's unique tumour biology or predict their risk of recurrence. Galban et al [1] have developed a promising method for predicting tumour response by analyzing tumour functional changes on a voxel-by-voxel basis. Parametric response maps (PRMs) are generated which classify each voxel as having a significant decrease, no change, or significant increase in function. The method's predictive utility has been demonstrated for a number of different sites and pathologies [1-4].

Ultimately, we would like to use the PRM method to develop adaptive dose painting techniques informed by repeat CT-perfusion imaging. However, PRM analysis is influenced by the ability to identify corresponding voxels in repeat scans and therefore accuracy may be reduced within imaging contexts where registration accuracy tends to be lower (e.g. abdomen). In this study we investigate the effects of registration error (RE) on PRM analysis by simulating known functional changes and RE on repeat CT-perfusion scans of liver tumours. PRMs are produced for different combinations of functional changes and RE and then compared to ground truth to produce estimates of RE-induced misclassification.

2. Materials and methods

2.1. Image data

The abdominal CT-perfusion scans of 5 patients with hepatocellular carcinoma were used for this study. Scans were acquired pre and post-RT using a 64 slice scanner (Discovery VCT or CT750HD, GE Healthcare, Waukesha, WI) with a spatial resolution of 0.7 mm by 0.7 mm by 5 mm and a temporal resolution of 2.8 seconds. Each image volume contained 512x512x16 voxels. Arterial blood flow (ABF) maps were generated from each 4D acquisition. A consistent image acquisition and contrast agent injection protocol was used for all patients. The imaging protocol, perfusion analysis methods, and registration technique applied to the 4D-CT scans are further described in [5].

2.2. Simulating functional changes

Seven distinct post-RT functional change scenarios (S_i , i = 1...7) were simulated per patient by modifying the pre-treatment ABF maps. Each functional change scenario is described by $S_iP_j(\mathbf{x}) = S_0P_j(\mathbf{x}) + N(\mu, \sigma)$ where $S_iP_j(\mathbf{x})$ is the simulated ABF map at position **x** for the *i*-th functional change scenario for patient *j*. $S_0P_j(\mathbf{x})$ is the pre-treatment ABF map at position **x** for patient *j* and $N(\mu, \sigma)$ is a value sampled from a normal distribution with mean $\mu = c_iT_j$ and standard deviation $\sigma = T_j/3$. T_j is the significant change threshold for patient j and the multiplicative constants c_i vary for each scenario. The T_j were determined using the technique described in [1]. The constants $\{c_i\}$ = {0, -1.2, -1.5, -1.8, 1.2, 1.5, 1.8} were selected to produce a spectrum of different PRM classification outcomes which predict no change (c_i = 0) or small, medium and large tumour response ($c_i < 0$) and progression (c_i > 0). For simplicity, tumour size was left unchanged in the simulated post-RT ABF maps. Each S_iP_j was blurred using a 3x3x1 averaging filter to reduce pixelation. An example of the ABF functional change simulations is shown in Figure 1A.

Functional changes were simulated under the premise that high-ABF tumour regions are more likely to decrease in function after successful RT and low ABF-regions are more likely to increase in function after unsuccessful RT. Consequently for response and progression scenarios, tumour voxel values within the top 70% and bottom 70% of all tumour ABF values were modified using the non-zero c_i while the remaining 30% of voxels were modified using $c_i = 0$. At the conditional boundary, c_i was scaled linearly between zero and the non-zero values to eliminate ABF value discontinuity.

2.3. Simulating image registration error

In this study, simulated deformable motion was used to emulate the residual image misalignment caused by RE. Each simulated ABF map S_iP_j was deformed to produce twenty-one copies ($S_iP_jM_k$, k = 0...20) associated with different magnitudes of inter-scan RE. Abdominal structures move in the anterior and inferior directions when a patient lies supine [6]. To simulate this motion, the diaphragm was modeled as a half sinusoid with varying amplitude similar to [7]. $S_iP_jM_k$ was generated by displacing S_iP_j voxels according to,

$$\Delta SI(x,k) = -\Delta SI_{max}|\sin(\pi k/k_{max})|\sin(\pi x/x_{max})$$
$$\Delta AP(x,k) = \Delta AP_{max}|\sin(\pi k/k_{max})|\sin(\pi x/x_{max})$$

where ΔSI_{max} = 10 mm and ΔAP_{max} = 5 mm were the maximum SI and AP displacements respectively. x_{max} is the LR extent of the image volume, and $k_{max} = 20$. Finally, a small rotation

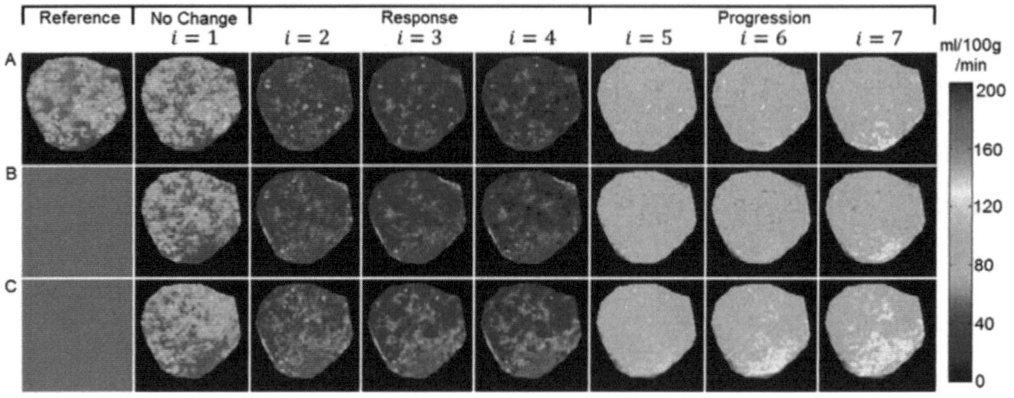

Figure 1. Representative slice from patient 3 ABF map simulations. ABF maps were cropped down to a 10x10 cm field of view and values outside the tumour were suppressed for improved visualization. A) S_0P_3 and S_iP_3 (no RE), B) $S_iP_3M_8$ (~4.3 mm average RE) and C) $S_iP_3M_{20}$ (~7.5 mm average RE)

about the SI axis was added and was specified by $\theta(k) = \theta_{max}|\sin(\pi k/k_{max})|$ where $\theta_{max} = 1.5°$. Figure 1B and 1C illustrates an example of the influence of simulated RE on the ABF maps.

2.4. Parametric response maps

PRMs are generated by comparing two functional imaging scans to one another. A voxel of tissue is classified as significantly increasing or decreasing if its change in function (ΔF) is such that $\Delta F > +T$ or $\Delta F < -T$ respectively, where T is a threshold determined by investigating the change in function within similar healthy tissue [1]. Here, the threshold for each patient ($\pm T_j$) was determined by comparing their pre and post-RT ABF maps within the healthy liver. The maps were first non-rigidly registered to one another by applying a transformation derived from registering the average of the 4DCT scans used to create each map. Elastic, multi-resolution registration was performed using the Flexible Algorithms for Image Registration Toolkit [8]. These thresholds were used to inform functional change simulations since functional changes derived from a direct comparison of the pre and post- treatment ABF maps would be confounded by RE. The ground truth and RE influenced PRMS were generated by comparing $S_iP_jM_0$ to S_0P_j and $S_iP_jM_k$ to S_0P_j ($k \neq 0$) respectively.

3. Results and discussion

Similar to [1], scatter plots of $S_iP_jM_k$ values versus corresponding S_0P_j values were created to visualize changes in tumour ABF values with respect to $\pm T_j$. Figure 2 illustrates representative plots for patient 3, where k = 0 (no RE) and k = 20 (~7.5mm average tumour RE). Average tumour RE refers to the average Euclidean distance between corresponding tumour voxels within compared scans. The plots for k = 0 represent the ground truth functional changes.

The percentage of voxels classified as increasing/decreasing deviated away from the ground truth as RE increased. For all patients, increasing RE was observed to increase the number of high post-RT ABF voxels associated with low pre-RT ABF voxels (and vice versa) within compared maps. This phenomenon produced a characteristic artifact within the scatter plots which appeared as a spreading out of the points contained by a triangular shaped envelope. In practice, this phenomenon may be difficult to distinguish from scenarios where parts of the tumour with low functional values actually increase and parts with high functional values actually decrease.

The true and RE influenced PRMs were compared to compute the percentage of tumour voxels that were misclassified due to varying magnitudes of RE. The average percent misclassification among the 5 patients was computed for all combinations of functional change scenarios and simulated RE and was then plotted against the average tumour voxel RE (figure 3).

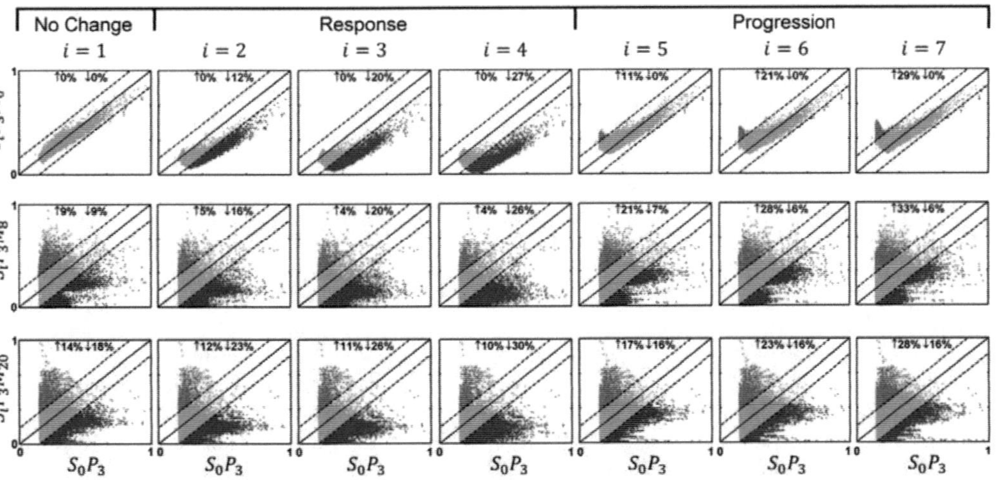

Figure 2. PRM scatter plots, $S_iP_3M_k$ versus S_0P_3 tumour ABF values (normalized). k = 0, 8, and 20 correspond to 0 mm, 4.3 mm, and 7.5 mm of average tumour RE respectively. The hatched lines indicate the significant change thresholds. The percent of tumour voxels classified as significantly increasing/decreasing is also indicated.

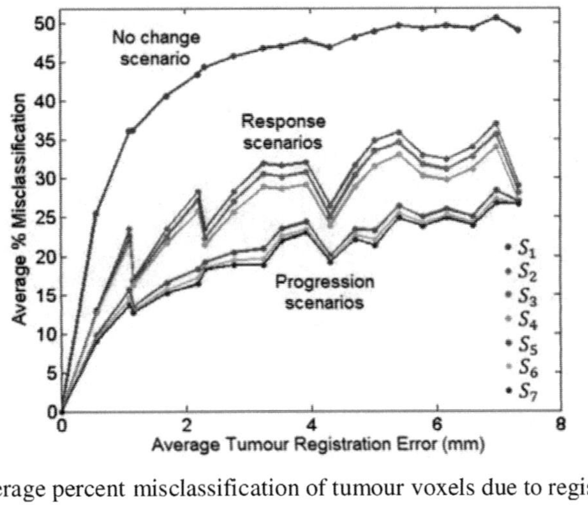

Figure 3. Average percent misclassification of tumour voxels due to registration error.

For all scenarios, misclassification steeply increased and then gradually plateaued. We hypothesize that the severity of the increase is influenced by the interplay between the magnitude of RE and the size and heterogeneity of the features within the compared ABF maps. For example, if features in the compared ABF maps are very small and heterogeneous (particularly in the direction of maximum RE), then small amounts of RE can lead to complete misalignment and result in very high rates of misclassification. As the RE exceeds the feature size, the misclassification rate begins to plateau since feature misalignment has been maximized. Consequently the exact relationship between misclassification and RE will differ for every tumour depending on feature size and heterogeneity within the interrogated functional maps. This helps to explain the lower misclassification rates observed for the response and progression scenarios since these scenarios were defined in a manner which increased tumour ABF homogeneity. Increased homogeneity decreases classification sensitivity to RE since features are effectively increased in size.

4. Conclusion

In this study we have simulated known functional changes and RE to investigate the effects of RE on parametric response map analysis using liver tumour CT-perfusion images. 3 mm of average tumour RE was observed to result in 18-45% voxel misclassification rates. This presents a major challenge towards the development of PRM-guided adaptive dose painting techniques since RE-induced misclassification could lead to incorrect radiation targeting. However, this challenge is not unique to the PRM response prediction method. All approaches involving a voxel-to-voxel comparison of registered images could be highly sensitive to RE depending on the interplay between RE magnitude/direction and the size, complexity, and orientation of features within the analyzed images. Since this interplay is unique to every image set and registration, methods need to be developed which can characterize and potentially mitigate RE-induced uncertainty in the analysis of each image voxel. The analysis uncertainty can then be visualized and taken into consideration during treatment planning. Without this information, voxel-based analyses like the PRM should not be used to guide locally adaptive dose painting unless it has been carefully verified that the maximum registration error within the images is much smaller than the image features.

References

[1] A Galban C J et al. 2009 *Nat. Med.* **15**(5) 572-76
[2] A Galban C J et al. 2009 *Transl. Oncol.* **2**(3) 184-90
[3] M Hoff et al. 2012 *Bone* **51**(1) 78-84
[4] Harris R J, Cloughesy T F, Pope W B, Nghiemphu PL, Lai A, Zaw T, Czernin J, Phelps ME, Chen W, Ellingson BM. 2012 *Neuro-Oncology* **14**(8) 1079-89
[5] Jensen N K G, Lock M, Fisher B, Kozak R, Chen X, Chen J, Wong E and Lee TY. 2013 *Acad. Radiol.* **20**(4) 414-22
[6] Brandner E D, Wu A, Chen H, Heron D, Kalnicki S, Komanduri K, Gerszten K, Burton S, Ahmed I, Shou Z. 2006 *Int. J. Radiat. Oncol. Biol. Phys.* **65**(2) 554-60
[7] Lausch A, Ebrahimi M, Martel A 2011 Image registration for abdominal dynamic contrast-enhanced magnetic resonance images. *Proc. 8th IEEE Int. Symposium on Biomedical Imaging: From Nano to Macro (Chicago, USA, March 30-April 2 2011).* pp 561-65
[8] Modersitzki J 2009 *FAIR: Flexible Algorithms for Image Registration.* Philadelphia, USA: Society for Industrial and Applied Mathematics.

Journal of Physics: Conference Series **489** (2014) 012033 doi:10.1088/1742-6596/489/1/012033

Audiovisual biofeedback improves image quality and reduces scan time for respiratory-gated 3D MRI

D. Lee[1*], P.B. Greer[2,3], J. Arm[4], P. Keall[1] and T. Kim[1,5]

[1]Radiation Physics Laboratory, Sydney Medical School, The University of Sydney, Camperdown, NSW, Australia
[2]The University of Newcastle, Newcastle, NSW, Australia
[3]Calvary Mater Newcastle, Newcastle, NSW, Australia
[4]Hunter New England Health Service, Newcastle, NSW Australia
[5]Radiological Physics, University of Virginia Health System, Charlottesville, US

*Email: danny.lee@sydney.edu.au

Abstract. The purpose of this study was to test the hypothesis that audiovisual (AV) biofeedback can improve image quality and reduce scan time for respiratory-gated 3D thoracic MRI. For five healthy human subjects respiratory motion guidance in MR scans was provided using an AV biofeedback system, utilizing real-time respiratory motion signals. To investigate the improvement of respiratory-gated 3D MR images between free breathing (FB) and AV biofeedback (AV), each subject underwent two imaging sessions. Respiratory-related motion artifacts and imaging time were qualitatively evaluated in addition to the reproducibility of external (abdominal) motion. In the results, 3D MR images in AV biofeedback showed more anatomic information such as a clear distinction of diaphragm, lung lobes and sharper organ boundaries. The scan time was reduced from 401±215 s in FB to 334±94 s in AV (*p-value* 0.36). The root mean square variation of the displacement and period of the abdominal motion was reduced from 0.4±0.22 cm and 2.8±2.5 s in FB to 0.1±0.15 cm and 0.9±1.3 s in AV (*p-value* of displacement <0.01 and *p-value* of period 0.12). This study demonstrated that audiovisual biofeedback improves image quality and reduces scan time for respiratory-gated 3D MRI. These results suggest that AV biofeedback has the potential to be a useful motion management tool in medical imaging and radiation therapy procedures.

1. Introduction

Respiratory-related motion blurring and ghost artifacts[1] can be reduced using respiratory-gating techniques with RF navigator[2], respiratory bellows belt[3], or real-time position management system (RPM).[4] However, variations in cycle-to-cycle breathing can cause inadequate respiratory-gating in image acquisition, resulting in image artifacts and increased scan time.[5]

Audiovisual (AV) biofeedback was proposed to monitor real-time respiratory motion using a marker on the abdomen with feedback to the human subject for respiratory motion management.[6-9] A number of respiratory cycles are acquired at the beginning of AV biofeedback to prepare a guiding waveform for each subject and the guiding waveform is displayed. The regular respiration can be reproduced in that the subject matches the red ball corresponding to the present respiratory position of the subject to the guiding waveform. AV biofeedback has been previously demonstrated to improve breathing regularity in 2D MRI[8] but not in 3D MRI. The aim of this study is thus to investigate whether AV biofeedback improves image quality and reduces scan time for respiratory-gated 3D MRI.

2. Methods

2.1 AV biofeedback system setup in MRI

An AV biofeedback system has been employed to provide respiratory guidance during MR scans. Figure 1 shows the experimental setup of the AV biofeedback system for MRI. The respiratory motion signals were obtained using the real-time position management (RPM) system (Varian, Palo Alto, USA) consisting of an infrared camera and a marker block on the abdomen.

To guide real-time patient breathing in a 3T system (Skyra, Siemens Healthcare Erlangen, Germany), the visual display of the guiding waveform and red ball was displayed on a screen during MR imaging. An 18-channel body matrix coil for thoracic imaging and a head mounted mirror for the patient's view of the visual display were used.

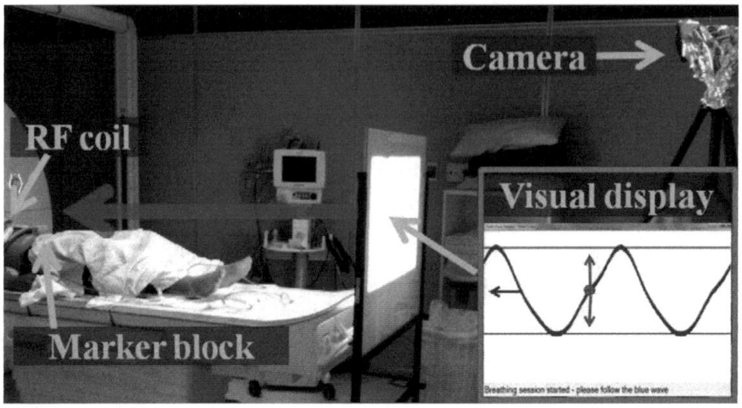

Figure 1. AV biofeedback system for respiratory-gated 3D MR imaging in a 3T Skyra Siemens MRI.

2.1 Respiratory-gated 3D MRI

The improvement in respiratory-gated 3D MR images using the AV biofeedback system combined with thoracic MRI was investigated with five healthy male subjects (aged 33 ± 6). For thoracic imaging, T2-weighted 3D SPACE (Sampling Perfection with Application optimized Contrast using different angle Evolutions) MR pulse sequence with an RF navigator placed on liver dome was employed. Typical parameters were TR/TE = 2200/89 ms, flip angle = 170^0, FOV = 380 × 380 mm^2, voxel size = 1.19 × 1.19 × 4 mm^3 and image matrix = 320 × 320 × 52.

In this study, each subject underwent two sessions to assess the image quality and gating efficiency with (AV: AV biofeedback) and without (FB: Free breathing). In order to reduce the effect of AV training on the FB results, a 15-20 minute AV training session was performed after the FB session and before the AV session. Respiratory-gated 3D MR images were acquired at three gating target positions with designated acceptance windows in the MR pulse sequence. The gating target positions at 10% (near maximum exhalation), 50% (middle) and 90% (near maximum inhalation) were set with a ±2 mm or ±4 mm acceptance window range.

The improvement of respiratory-gated 3D MR images using the AV biofeedback has been evaluated in terms of respiratory-related artifacts. The gating efficiency and the 3D MR data acquisition time have been compared. The abdominal motion was evaluated using the root mean square error (RMSE) in displacement and period obtained from the RPM system.

3. Results and Discussion

3.1 Image quality and gating efficiency improvement

Using the AV biofeedback system, respiratory-related blurring artifacts have been noticeably improved and scan time was considerably reduced as shown in Figure. 2.

3D MR images with FB (subject 1 and subject 2) were significantly blurred due to the variation of the baseline shift and amplitude in respiration. In contrast, there was noticeable reduction of blurring artifacts due to the more regular respiratory motion in the same subjects with AV (subject 1 and subject 2). 3D MR images in AV included more anatomic information such as a clear distinction of diaphragm and lung lobes. In addition, the edge and intersection of organs on 3D images were sharpened.

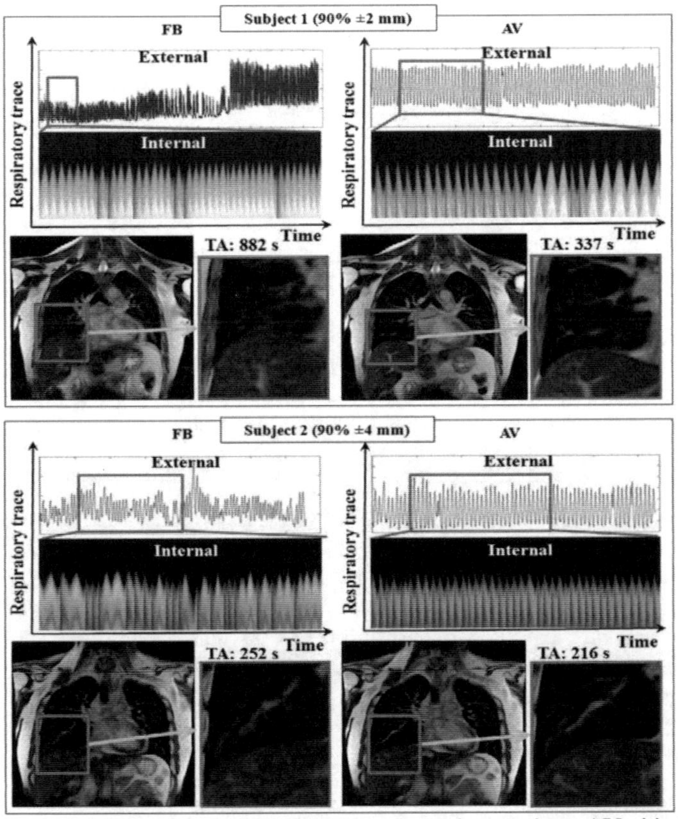

Figure 2. An example of improved image quality and total acquisition time (TA) when AV biofeedback was employed during gated 3D MR imaging for subject 1 (top) and subject 2 (bottom). Free-breathing results (left) and AV results (right) are shown. The external respiratory signal, internal diaphragm signal and images are shown. Gating was triggered at 90% of the breathing cycle (near maximum inhalation) with ±2 mm (subject 1) and ±4 mm (subject 2) gating thresholds. The regular external (abdomen) respiration correlated with the regular internal (diaphragm) respiration, leading to better image quality and shorter scan time.

In addition, regular respiratory motion using AV biofeedback reduced the incidence of unsuccessful gating, leading to a reduction of MR scan time. The scan time reduced up to 545 s and 36 s during respiratory-gated 3D MRI at 90% acceptance window in subject 1 and 2, respectively. A ±2 mm range required relatively longer MR scan time, compared to ±4 mm range, to acquire the same number of images due to the small range of acceptance window. In other volunteer cases, the considerable gating efficiency in AV compared to FB was found at 10% and 90% acceptance windows but it was quite similar at 50% acceptance window between the two breathing conditions.

The impact of AV biofeedback for gating target positions corresponding to the three acceptance positions is shown in Figure 3. A red dashed line at the top of the diaphragm for the 10% gated scan shows only small differences in the diaphragm positions for FB, however there is clearer evidence of inspiration for the MR scan where the inferior motion of the diaphragm with inspiration is apparent.

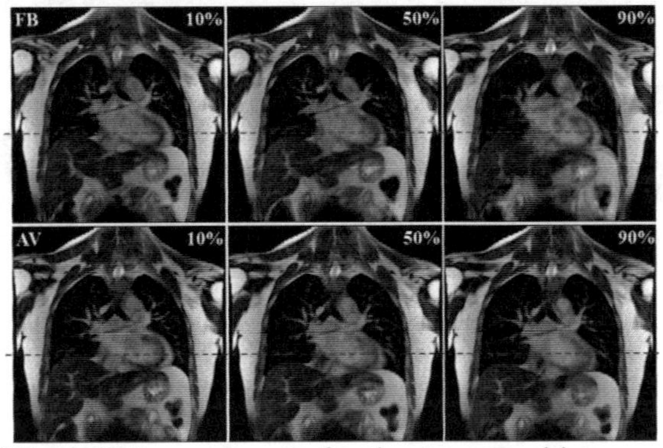

Figure 3. Improvement of gating target positions corresponding to the three acceptance windows in AV compared to FB (subject 1).

3.2 Scan time and breathing variation

Scan time was reduced (or similar) at all gating target positions due to the improvement of breathing regularity: the reduction of root mean square error (RMSE) in displacement and period in Table 1.

Table 1. Scan time and breathing variation of respiratory-gated 3D MRI.

Session	Average scan time (s)	RMSE in Average variation in displacement (cm)	RMSE in Average variation in Period (s)
FB	401±215	0.4±0.22	2.8±2.5
AV	334±94	0.1±0.15	0.9±1.3
% reduction with AV	17%	75%	68%
p-values	0.36	<0.01	0.12

An average reduction in scan time was from 401 s in FB to 334 s in AV, coming from the reduction of breathing variation in displacement (-75%) and period (-68%).

This study demonstrated the improvement of respiratory-gated 3D MR images with AV biofeedback due to improved respiratory motion reproducibility, leading to regular internal organ displacement. In addition, scan time was simultaneously reduced. The images in this study spanned the thorax and abdomen, indicating that AV biofeedback can be broadly applicable to imaging sites affected by respiration, provided that both a respiratory signal and a patient display system are available.

A limitation of the current study was that healthy human subject volunteers, and not patients, were used. Therefore, without further testing, the application of this study can only be extended to patients with similar age and lung function characteristics to the volunteers, for example some lymphoma, breast, pancreas and kidney cancer patients. In order to investigate the potential benefits of AV biofeedback for a quite different population, lung cancer patients, a clinical study has been initiated.

A potential problem for real-time audiovisual biofeedback is the time delay between the patient breathing and the projected signal. Fortunately, using the 30 Hz RPM signal the time delay between the patient breathing and the projected signal is undetectable to the user. However, with a different respiratory monitor with either a lower frame rate and/or using more processing time the projected real-time respiratory signal could be delayed or discontinuous. For such systems it is likely that further processing algorithms, such as signal prediction or smoothing algorithms would be needed.

4. Conclusion

This study demonstrated, for the first time, that audiovisual biofeedback improves image quality and reduces scan time for respiratory-gated 3D MRI. These results suggest that AV biofeedback has the potential to be a useful motion management tool in medical imaging and radiation therapy procedures.

Acknowledgements

This project was supported by an NHMRC Australia Fellowship, NHMRC Program Grant APP1036078 and ARC Discovery Grant DP120100821.

References

[1] Wood, M. L. & Henkelman, R. M. MR image artifacts from periodic motion. *Medical physics* **12**, 143 (1985).
[2] Taylor, A. M. *et al.* MR navigator-echo monitoring of temporal changes in diaphragm position: Implications for MR coronary angiography. *Journal of Magnetic Resonance Imaging* **7**, 629 (1997).
[3] Santelli, C. *et al.* Respiratory bellows revisited for motion compensation: preliminary experience for cardiovascular MR. *Magnetic Resonance in Medicine* **65**, 1097 (2011).
[4] Kini, V. R. *et al.* Patient training in respiratory-gated radiotherapy. *Medical Dosimetry* **28**, 7-11 (2003).
[5] Hu, Y., Caruthers, S. D., Low, D. A., Parikh, P. J. & Mutic, S. Respiratory Amplitude Guided 4-Dimensional Magnetic Resonance Imaging. *International Journal of Radiation Oncology* Biology* Physics* (2013).
[6] George, R. *et al.* Audio-visual biofeedback for respiratory-gated radiotherapy: impact of audio instruction and audio-visual biofeedback on respiratory-gated radiotherapy. *International Journal of Radiation Oncology* Biology* Physics* **65**, 924 (2006).
[7] Venkat, R. B. *et al.* Development and preliminary evaluation of a prototype audiovisual biofeedback device incorporating a patient-specific guiding waveform. *Physics in medicine and biology* **53**, N197 (2008).
[8] Kim, T. *et al.* Audiovisual biofeedback improves diaphragm motion reproducibility in MRI. *Medical physics* **39**, 6921 (2012).
[9] Yang, J. *et al.* The impact of audio-visual biofeedback on 4D PET images: Results of a phantom study. *Medical physics* **39**, 1046 (2012).

Fully-deformable patient motion models from cone-beam CT for radiotherapy applications

J Martin[1], J McClelland[1], C Yip[2], C Thomas[3], C Hartill[2], S Ahmad[2], I Meir[4], D Landau[2] and D Hawkes[1]

[1] Centre for Medical Image Computing, UCL, London, WC1E 6BT
[2] Department of Oncology, Guy's and St. Thomas' NHS Foundation Trust, London
[3] Medical Physics, Guy's and St. Thomas' NHS Foundation Trust, London
[4] Vision RT Ltd., Dove House, Arcacia Avenue, London, N3 2JU

E-mail: james.martin.09@ucl.ac.uk

Abstract. We propose a method to build a fully deformable motion model directly from cone-beam CT (CBCT) projections. This allows inter-fraction variations in the respiratory motion to be accounted for. It is envisaged that the model be used to track the tumour, and monitor organs at risk (OAR), during gated or tracked radiotherapy (RT) treatment of lung cancer. The method is tested on CBCT projections from a simulated phantom in two cases. The simulations are generated from a patient respiratory trace and associated CBCT scanner geometry. Without and with motion correction, l^2 norm maximum errors were reduced from 24.5 to 0.698 mm in case 1, and 20.0 to 0.101 mm in case 2, respectively.

1. Introduction

Accurate identification of targets prior to and during the course of radiotherapy are essential for state-of-the-art RT treatments [1]. Advances have enabled dose escalation, conformal sparing of OAR and non-uniform dose distributions. In treatments such as stereotactic ablative body RT, the dose is delivered in fewer fractions (3-5) and with a higher dose per fraction. Tumour respiratory motion is an important consideration during these treatments [2]. It is common to use 4DCT planning scans to identify the extent of respiratory motion and add a margin onto the delineated tumour region. Gated and tracked [3, 4] RT treatments allow this margin to be reduced, but rely on an accurate means of determining tumour position throughout the fraction.

We have previously published a method of generating a motion model from cine-CT [5] and to fit a model of tumour motion directly to a CBCT scan [6, 7]. The motion was related to a surrogate signal which can be monitored throughout the fraction and hence used to estimate tumour position. In this work the method is extended to non-rigid motion, allowing the respiratory motion of a number of anatomical structures to be modelled. This approach would allow other regions beyond the tumour, such as the organs at risk (OAR), to be tracked during treatment.

Modelling non-rigid deformations is achieved by the use of weighting arrays, which provide local variation of the deformation. It is envisaged that the weighting arrays be built from the 4DCT planning scans. Utilising planning scans may also allow the clinical planning delineations to be adapted to the day of treatment. During a fraction of RT treatment the global parameters are fitted. These scale the local variation in each spatial dimension, hence accounting for inter-fraction changes in a patient-specific manner. Parameterising in this way (i.e. fitting the global parameters only during a fraction of treatment) reduces the complexity of the cost function, allowing the parameters to be found within seconds. In the method the iterative-based approach is described, including the form of the cost function to optimise. Once the motion model is fitted, only a new surrogate signal is needed to animate the OAR and tumour regions. This could be used, for example, to animate the 4DCT-derived clinical delineations, facilitating real-time monitoring during tracked RT treatments. Results are provided for two simulated cases, followed by conclusions and future work.

2. Modelling patient respiratory motion

During the CBCT scan, assume that the patient motion can be described by a series of deformations, F_n, of a motion-free image of the patient, $V_{ref}(x) : x \in \Omega$, where Ω is the region imaged during the CBCT scan. $n(= 0, 1, \ldots, N-1)$ is an index corresponding to the time of each CBCT projection. Let the deformed patient volume V_n be:

$$V_n(x) = V_{ref}\left(x + F_n(x)\right). \quad (1)$$

As with a previous publication [7], a surrogate signal is derived from an optical stereo-camera system (Vision RT, London, UK). A skin surface image is determined at the time of each projection, and the average height within a bounding box is used for the surrogate signal. The surrogate traces are normalised (mean subtracted; divided by standard deviation). The motion model used in this work is dependent on the surrogate signal, $s(n)$, and its temporal rate of change, $\dot{s}(n)$ [8]. As well as modelling hysteresis and changes in breath cycle length and depth during breathing, non-rigid deformations can be accounted for.

$$F_n(x) = s(n)D_1(x) \circ \phi_1 + \dot{s}(n)D_2(x) \circ \phi_2. \quad (2)$$

ϕ_1 and ϕ_2 are the motion model parameters (3D vectors). $\circ$ is an element-wise multiplication (Hadamard product). D_1 and D_2 are weighting arrays for ϕ_1 and ϕ_2, respectively. The weighting arrays are deformation fields, with a 3D vector at each point in the patient volume, x. These arrays provide spatially varying deformations, hence allowing non-rigid breathing motion to be modelled. The weighting arrays are static during parameter fitting. As only the global parameters are fitted, this facilitates a clinically feasible optimisation time of the cost function.

It is envisaged that the weighting arrays be determined prior to the fraction of treatment, for example using the 4DCT planning scan. D_1 could be constructed from registration of the end exhale and end inhale phases, and D_2 from mid-exhale and mid-inhale phases. A further registration of the mean 4DCT image to the CBCT reconstruction (on the day of treatment) would allow the weighting arrays to be aligned accordingly. If the final 4DCT to CBCT registration step can account for complex changes from planning, such as baseline shifts or tumour shrinkage, this method could also account for these changes. The weighting arrays determine relative motion of different parts of the anatomy whereas the motion model parameters control global changes to the magnitude of the overall LR, AP and SI motion. Hence some degree of inter-fraction variation can be accounted for by fitting the motion model parameters on the day of treatment.

3. Iterative approach to calculate motion compensated reconstruction (MCR) and motion model parameters

First presented in [6], an iterative approach is used to jointly determine the MCR and motion model parameters from the CBCT. An open-source, FDK-based software package (www.openrtk.org) was modified to perform the MCRs. MCRs are performed by back-projecting each projection through a deformed volume. The deformation used was the inverse of the forward transformation estimated by the motion model for that particular projection.

Beginning with zero motion ($\phi_1 = \phi_2 = \mathbf{0}$), an MCR is calculated. This is then used as the reference volume and updates ($\delta\phi_1$ and $\delta\phi_2$) to the motion model parameters are determined. The updates were calculated via minimisation of a cost function, which alters the motion of the surrogate-animated MCR to produce projections more similar to the actual projections of the CBCT. The cost function takes the following form:

$$\delta\phi_1, \delta\phi_2 = \underset{\delta\phi_1,\delta\phi_2}{argmin} \left[\sum_{n} \sum_{pixels} \left((p_n - P_n(V_n)) - P_n \left(\nabla V_n \cdot \delta F_n \right) \right)^2 \right]. \quad (3)$$

where P_n is the forward projection operator, p_n is the nth projection of the CBCT and the update to the deformation

$$\delta F_n = s(n) D_1(x) \circ \delta\phi_1 + \dot{s}(n) D_2(x) \circ \delta\phi_2. \quad (4)$$

An algebraically expanded version of (3) was used for this work. This allowed the motion model parameters to be factored out of the summations, reducing optimisation time. The more compact version of the cost function is included here for brevity. Please see [7] for more information on deriving this expanded form.

The new motion model parameters can then be calculated via the update equation:

$$\boldsymbol{\phi}_{new} = \boldsymbol{\phi} + \delta\boldsymbol{\phi}. \quad (5)$$

Recalculating the MCR will give the updated MCR. Repeating the update steps will result in more accurate estimates of the motion model parameters and MCR.

4. Results

The method was initially tested on simulated data. A modified 3D-Shepp logan phantom was generated in MatLab within a 100 mm sided cube, constituting voxels of side length 1 mm. The phantom was animated and using a motion model with manually defined weighting arrays and a real patient surrogate trace, so as to produce realistic motion. A CBCT acquisition of the phantom was simulated using real scanner geometry. Two simulations are included. Both had relatively large SI and AP motion, with the maximum deformation being up to 25 and 20 mm for simulations 1 and 2, respectively. With the first case, the deformations were centred on the middle of the phantom. The second had deformations centred inside the middle of two centrally located ovals within the phantom.

The weighting arrays were manually defined to produce plausible respiratory-like deformations. For these preliminary experiments, the same weighting arrays were used for both generating the motion and fitting the motion model parameters. This was to determine if the parameters could be recovered if the weighting arrays were known. In both cases, the algorithm was terminated if the update caused the mean l^2 norm of the estimated motion to change by less than 0.4 mm, within the phantom region.

For the first simulation, the maximum l^2 norm of the motion estimation errors (relative to the mean position) was reduced from 24.5 to 0.698 mm, within the phantom region. The mean l^2 norm error was reduced from 3.01 to 0.157 mm. For the second simulation, the maximum l^2 norm error was reduced from 20.0 to 0.101 mm, with the mean l^2 norm error reduced from 0.969 to 0.0086 mm. Figure 1 shows the improvement in the MCR before and after fitting the parameters for both simulations. The best possible MCR is also given for reference. It can be seen that the results from the fitted model parameters are very similar to those from the known model parameters, and in both cases the motion artefacts have been removed from the reconstructions. Simulations 1 and 2 terminated after 9 and 3 iterations respectively.

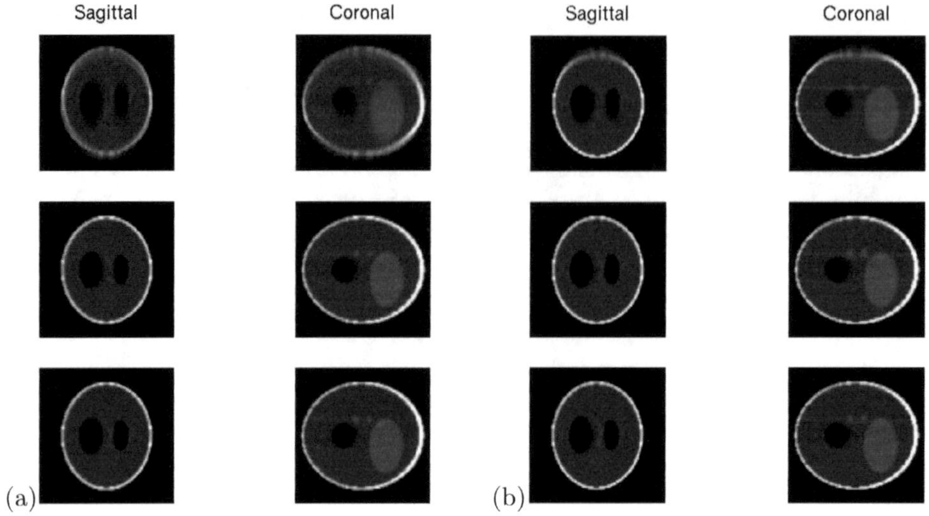

Figure 1. Select slices of MCR before (top) and after (middle) motion correction with fitted parameters, and with the known parameters (bottom) for (a) simulation 1 and (b) 2.

5. Conclusions and future work

A method is presented to fit a non-rigid motion model directly to CBCT projections. Promising results are given for a simulated case undergoing two different types of motion. The authors intend to further develop the method, including obtaining results for actual patient data. The robustness of the method to artefacts present in actual CBCT data will be evaluated. An important part of this is determining the best method of generating the weighting arrays from the 4DCT planning scans. The most suitable method of their calculation and alignment to the CBCT reconstruction still needs to be determined and is currently being investigated by the authors.

Acknowledgments

James Martin gratefully acknowledges the support of the EPSRC funded UCL VEIV EngD programme, Vision RT and the 1851 Royal Commission Industrial Fellowship.

References

[1] Verellen D, De Riddeer M, Linthout N, Tournel K, Soete G and Storme G 2007 *Nature Review*
[2] Benedict S H, Yenice K M, Followill D, Galvin J M, Hinson W, Kavanagh B, Keall P, Lovelock M, Meeks S, Papiez L, Purdie T, Sadagopan R, Schell M C, Salter B, Schlesinger D J, Shiu A S, Solberg T, Song D Y, Stieber V, Timmerman R, Tomé W A, Verellen D, Wang L and Yin F F 2010 *Medical Physics* **37** 4078–4101
[3] Kubo H D and Hill B C 1996 *Physics in Medicine and Biology* **41** 83
[4] Schweikard A, Shiomi H and Adler J 2004 *Medical Physics* **31** 2738–2741
[5] McClelland J R, Blackall J M, Tarte S, Chandler A C, Hughes S, Ahmad S, Landau D B and Hawkes D J 2006 *Medical Physics* **33** 3348–3358
[6] Martin J, McClelland J, Thomas C, Wildermuth K, Landau D, Ourselin S and Hawkes D 2012 *Mathematical Methods in Biomedical Image Analysis (MMBIA), 2012 IEEE Workshop on* pp 281 –286
[7] Martin J, McClelland J, Yip C, Thomas C, Hartill C, Ahmad S, OBrien R, Meir I, Landau D and Hawkes D 2013 *Physics in Medicine and Biology* **58** 1809
[8] McClelland J, Hawkes D, Schaeffter T and King A 2013 *Medical Image Analysis* **17** 19 – 42 ISSN 1361-8415

XVII International Conference on the Use of Computers in Radiation Therapy (ICCR 2013) IOP Publishing
Journal of Physics: Conference Series **489** (2014) 012035 doi:10.1088/1742-6596/489/1/012035

Optimized dynamic contrast-enhanced cone-beam CT for target visualization during liver SBRT

Bernard L Jones,[1] **Cem Altunbas, Brian Kavanagh, Tracey Schefter and Moyed Miften**
Department of Radiation Oncology, University of Colorado School of Medicine, Aurora CO, USA

E-mail: bernard.jones@ucdenver.edu

Abstract. The pharmacokinetic behavior of iodine contrast agents makes it difficult to achieve significant enhancement during contrast-enhanced cone-beam CT (CE-CBCT). This study modeled this dynamic behavior to optimize CE-CBCT and improve the localization of liver lesions for SBRT. We developed a model that allows for controlled study of changing iodine concentrations using static phantoms. A projection database consisting of multiple phantom images of differing iodine/scan conditions was built. To reconstruct images of dynamic hepatic concentrations, hepatic contrast enhancement data from conventional CT scans were used to re-assemble the projections to match the expected amount of contrast. In this way the effect of various parameters on image quality was isolated, and using our dynamic model we found parameters for iodine injection, CBCT scanning, and injection/scanning timing which optimize contrast enhancement. Increasing the iodine dose, iodine injection rate, and imaging dose led to significant increases in signal-to-noise ratio (SNR). Reducing the CBCT imaging time also increased SNR, as the image can be completed before the iodine exits the liver. Proper timing of image acquisition played a significant role, as a 30 second error in start time resulted in a 40% SNR decrease. The effect of IV contrast is severely degraded in CBCT, but there is promise that, with optimization of the injection and scan parameters to account for iodine pharmacokinetics, CE-CBCT which models venous-phase blood flow kinetics will be feasible for accurate localization of liver lesions.

Introduction

The efficacy of Stereotactic Body Radiation Therapy (SBRT) of the liver is hindered by the inability to accurately localize the lesion at the time of treatment. Liver metastases and the liver parenchyma are iso-attenuating, so these tumors cannot be seen directly with cone-beam CT (CBCT). The image quality of CBCT is also degraded by the large respiratory-induced motion of the abdomen, and it is difficult to visualize the surface of the liver. The lack of direct anatomical landmarks for localization necessitates larger safety margins, which in turn increases the mean dose to the liver and potentially increases the incidence of radiation-induced liver disease [1].

Despite these difficulties, the unique dual-supply architecture of the liver makes it possible to improve the quality of imaging with IV agents. This concept is exploited in Contrast-Enhanced CT (CECT), and can be used in diagnosis and/or treatment planning to visualize liver metastases [2]. If an image is acquired during venous perfusion of the liver, the parenchyma appears bright while the tumor is unchanged, and thus the lesion can be differentiated from the surrounding tissue.

[1] To whom any correspondence should be addressed.

In order to improve the accuracy of liver SBRT, we investigated the feasibility of Contrast-enhanced Cone-Beam CT (CE-CBCT) with the goal of using iodinated IV contrast to visualize liver metastases. CE-CBCT has been attempted at a handful of institutions, but has met with little success. Iodine washes in and out of the liver on a roughly one-minute timescale, while CBCT can take up to two minutes to acquire. Additionally, one of the difficulties in studying the effect of contrast-enhanced imaging is the pharmacokinetics of the contrast agent. IV agents exhibit dynamic behavior within a biological system, and the image quality can vary greatly between different injection/acquisition protocols. Moreover, it is difficult to manufacture a phantom that can duplicate this complex behavior [3]. In order to optimize the effect of iodine imaging, we have developed a model that allows for controlled study of CE-CBCT image quality using static phantoms. Using this model, we have identified several key properties of the iodine injection, the physics of the CBCT image, and the relative timing of the injection and image acquisition which optimize the image quality of CE-CBCT.

Methods and Materials

The most straightforward way to study image quality in CE-CBCT would be to mimic the conditions of patient imaging using a phantom. However, since the iodine concentration *in vivo* changes during the course of imaging, studying this scenario would require a phantom with variable iodine concentration. Instead of constructing such a phantom, we simulated CE-CBCT imaging by acquiring many CBCT images with different concentrations of iodine, and then re-assembled these projections in such a way as to mimic published patient data. The first step was to assemble a database of CBCT projections which encompass the full range of possible contrast parameters (such as iodine concentration) or CBCT conditions (such as kV/mA and geometry). This database was assembled by acquiring multiple static CBCT images of the phantom containing an iodine-loaded insert whose concentration varied between images. Next, Hepatic Enhancement vs. Time (HEVT) curves were retrieved from literature in order to model realistic dynamic behavior in CE-CBCT [4]. These HEVT curves are given for different injection parameters, such as iodine dose, injection rate, and timing. Finally, to generate the predicted CE-CBCT image, CBCT projections are chosen from the database such that the concentration of iodine in the phantom matches the expected HU value from the HEVT curve. These projections are assembled into a pseudo-scan and re-reconstructed to yield an image, the content of which is equivalent to the hypothetical expected image (were the phantom iodine concentration to vary as given by the patient-derived HEVT curve). In this way, our model translates a given HEVT curve into a CE-CBCT image that accounts for the dynamic behavior of IV contrast.

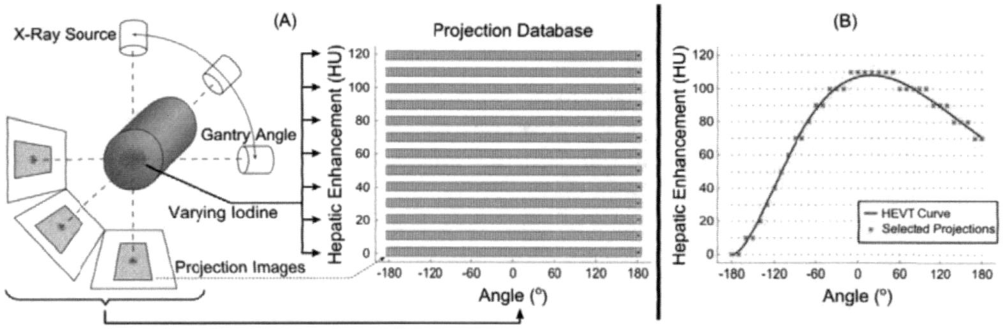

Fig 1: Dynamic Phantom Model: Selecting from the projection database. A) A projection database is built by acquiring numerous images, each containing various concentrations of iodine. The database consists of thousands of projections, with each one containing differing concentrations of iodine and acquired at different gantry angles. B) To reconstruct a dynamic image, a HEVT curve is used to select projections from the database. In this example, projections are chosen from the database such that at each moment in time the iodine concentration matches that expected from the HEVT curve.

A Gammex 467 Tissue Characterization Phantom was used to acquire static images with various concentrations of iodine. A series of 50 mL conical tubes were filled with a mixture of Isovue-370 (Iopamidol, 37% Iodine by weight) and water. The proportion of these mixtures was chosen such that

Journal of Physics: Conference Series **489** (2014) 012035 doi:10.1088/1742-6596/489/1/012035

the tubes corresponded to 10-120 Hounsfield units (HU) when imaged, and these values were verified using a Phillips Big Bore CT Scanner. Next, a projection dataset was assembled on a TrueBeam Accelerator by acquiring numerous CBCT scans, each acquired with various concentrations of iodine and under different CBCT conditions. In order to study the effect of a certain injection parameter (for instance, iodine dose or injection rate), the HEVT curve for that factor was retrieved from the published literature [4]. Likewise, to analyze the effect of imaging physics (e.g. x-ray scatter or kV/mA), a separate set of static CBCTs were acquired under these parameters, and our model was applied. To study the effects of the relative injection/CBCT imaging timing, appropriate adjustments were made in our model when converting from time (on the HEVT curve) to gantry angle. Signal-to-noise ratio (SNR) was chosen as the primary metric of image quality, which is suitable in the current early stage of development as a crude indicator of conspicuity [5]. HEVT curves were normalized to represent a hypothetical adult male (30 years old; weight, 70 kg) under various injection conditions. Also note that, while the HU of CT and CBCT are generally different, HEVT curves were derived entirely from CT studies. The HU in CBCT images were used only to calculate SNR, which requires only relative (and not absolute) accuracy.

Results

Fig. 2 shows the effects of various parameters of the iodine injection, CBCT acquisition, and relative injection/imaging timing on the SNR of the resulting image. As expected, the SNR increases as the average amount of iodine present during CBCT acquisition increases. Fig. 2A demonstrates the relationship between SNR and injection rate. As the injection rate increases, the peak hepatic enhancement increases, but the width of the peak is shorter [3]. For the full-fan (180°) acquisition, which takes 30 s to acquire, there is not much enhancement in SNR as the injection rate is increased above 3 mL/s. In the half-fan (360°) acquisition, there was no different in SNR with respect to injection rate due to the length of the CBCT acquisition (60s) compared to the width of the enhancement peak (30-60s). Fig. 2B shows that SNR increases as the iodine dose is increased. There is a larger relative increase in the full-fan acquisition, which is again a result of the shorter scan time.

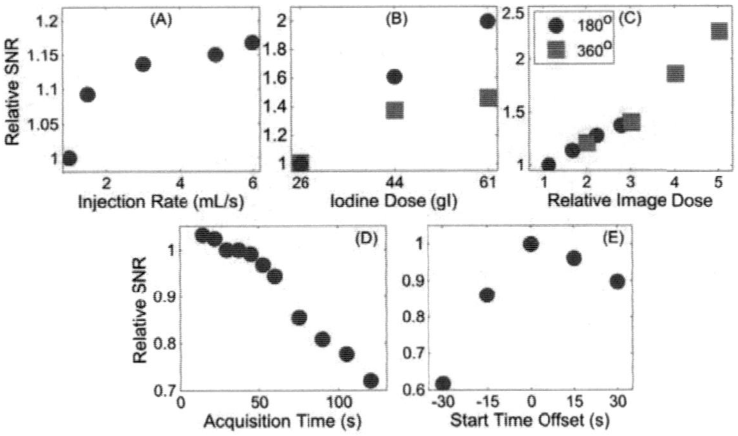

Fig. 2: Optimization of Signal-to-Noise Ratio (SNR) in CE-CBCT. A) Varying injection rate of a 125 mL injection of contrast medium (350 mgI/mL). B) Varying dose of contrast medium (350 mgI/mL) injected at 2 mL/s. C) Varying the imaging dose used in the CBCT acquisition. Results are normalized such that the dose per unit mAs is equal. D) Varying the time over which the CBCT is acquired. E) Varying the delay between the initiation of the iodine injection and initiation of the CBCT image acquisition. Time is shown relative to the theoretical best time (t=0).

Fig. 2C shows the relationship between SNR and CBCT imaging dose. As expected, SNR increases with higher dose, due to the reduction in noise in the reconstructed images. Results are shown for both the half- and full-fan acquisitions; one may note that since the half-fan acquisition is roughly twice as long, the number of projections acquired (and thus the image dose) is increased. This results in a reduction in noise and a higher SNR for half-fan images. No significant different in SNR was seen when the tube voltage changed from 125 kV to 100 kV. Since the attenuation coefficient for iodine is greater at lower photon energies, there was a larger signal observed in the iodine-loaded samples. However, this increase is offset by a reduction in imaging dose due to the lower bremsstrahlung production efficiency within the photon source of 100 keV electrons (compared to 125 keV). This increased the noise in the images, and resulted in no net SNR change.

Fig. 2D demonstrates the effect of total CBCT acquisition time for a 125 mL injection at 3 mL/s. As the scan time increases, a greater portion of the image is taken at a time where either the iodine has not arrived at the liver, or the iodine has washed out of the liver. As a result, the mean iodine concentration in the image decreases and the SNR is reduced. Fig. 2E demonstrates the magnitude of the effects of sub-optimal timing between the initiation of the iodine injection and initiation of the CBCT acquisition. SNR is degraded by as much as 40% if the CBCT acquisition is started at the incorrect time. The degradation is lessened for delays greater than the optimum time; this is due to the fact that iodine washes out of the liver slower than the initial bolus enters, or in other words, the slope of the HEVT curve is lower on the receding side.

Discussion and Conclusions

CE-CBCT takes advantage of the unique architecture of the liver to increase the visibility of hepatic malignancies. This, in turn, facilitates better treatment accuracy in intense therapies such as SBRT that rely on very large dose gradients to spare normal tissue. In order to improve the image quality of CE-CBCT, we have investigated factors which affect the amount of iodine present during imaging, and have developed a model to optimize these parameters. In general, SNR increases as the mean iodine concentration during imaging increases. Increasing the dose of iodine injected has a large effect (up to 100% SNR increase, Fig. 2B); increasing the rate of injection also plays a role.

Several challenges remain to be addressed before CE-CBCT can be successful clinically. One can optimize the CE-CBCT acquisition for a given HEVT curve, but these curves are known to vary based on patient-specific factors such as weight and cardiac output. Our results indicate that the resulting image quality can be very sensitive to errors in timing, and so it is crucial to know the time to peak enhancement for a given patient. Possible solutions are to use bolus-tracking during the simulation CECT, and record the relative timing during that acquisition, or to acquire liver perfusion images prior to treatment. A more comprehensive way may be to investigate an injection/acquisition protocol which is less sensitive to differences in the time to peak enhancement.

The ultimate goal of this work is to improve the accuracy of liver SBRT. In this preliminary study, we demonstrate techniques to increase the signal-to-noise ratio of CE-CBCT imaging; however, conspicuity of liver lesions can depend on many factors (tumor size, shape of the lesion, SNR, background characteristics, and other factors). The results of this study will be used to guide future clinical implementations of contrast-enhanced CBCT. A pilot study is underway in our department to assess the clinical impact of IV contrast on the visibility of liver lesions and the localization accuracy benefit of this imaging.

Acknowledgments

The authors would like to thank Varian Medical Systems for providing the iTools Reconstruction Software in support of this project.

References

[1] Dawson L A, Normolle D, Balter J M, McGinn C J, Lawrence T S and Ten Haken R K 2002 *Int J Rad Oncol* Biol* Phys* **53** 810
[2] Baron R L 1994 *American Journal of Roentgenology* **163** 323-31
[3] Driscoll B, Keller H and Coolens C 2011 *Medical Physics* **38** 4866
[4] Bae K T 2010 *Radiology* **256** 32-61
[5] Rose A 1948 *J. Opt. Soc. Am.* **38** 196-208

Construction of a patient observation system using KINECT™

Kazunori Miyaura [1,3], Yu Kumazaki [1], Chika Fukushima [2], Shingo Kato [1], Hidetoshi Saitoh [3]

[1] Saitama Medical University International Medical Center, Hidaka-shi Saitama 3501298 Japan
[2] Technology of Radiotherapy Corporation, Bunkyou-ku Tokyo 1120002 Japan
[3] Tokyo Metropolitan University Graduate School of Health Sciences, Arakawa-ku Tokyo 1160012 Japan

E-mail: kmiyaura@saitama-med.ac,jp

Abstract. Improvement in the positional accuracy of irradiation is expected by capturing patient motion (intra-fractional error) during irradiation. The present study reports the construction of a patient observation system using Microsoft® KINECT™. By tracking movement, we made it possible to add a depth component to the acquired position coordinates and to display three-axis (*X*, *Y*, and *Z)* movement. Moreover, the developed system can be displayed in a graph which is constructed from the coordinate position at each time interval. Using the developed system, an observer can easily visualize patient movement. When the body phantom was moved a known distance in the *X*, *Y*, and *Z* directions, good coincidence was shown with each axis. We built a patient observation system which captures a patient's motion using KINECT™.

1. Introduction

Reduction of the planning target volume (PTV) margin and the setup of exact internal target volume (ITV) are important in advanced radiotherapy [1]. Intra-fraction motion can be caused by the respiratory, musculoskeletal, cardiac, and gastrointestinal systems [2]. Improvement in irradiation position accuracy is expected by capturing the motion (intra-fractional error) during patient irradiation. A common patient observation system is expensive and more complicated program. The purpose of this study is to construct a patient observation system using Microsoft® KINECT™.

2. Method

KINECT™ is a tool that uses speech recognition, a motion sensor, frame pursuit, etc. [3]. A frame pursuit has the skeletal tracking method. This method can locate the joints of the tracked targets in space and track their movements over time. We constructed our KINECT™system in two stages.

Figure 1 shows the sensor array of the KINECT™ system. This size is 65 × 280 × 70 mm (length × width × height). It has a color camera and two sensors. The system runs on the Microsoft® Windows® Software Development Kit (SDK). KINECT™ has the following features [3].

- An RGB camera that stores 3-channel data at 1280×960 resolution.
- An IR emitter and an IR depth sensor.
- A multi-array microphone, which contains 4 microphones for capturing sound.
- Skeletal Tracking allows KINECT™ to recognize people and follow their actions.

Journal of Physics: Conference Series **489** (2014) 012036 doi:10.1088/1742-6596/489/1/012036

2.1. Development of the Interface

First we developed an interface to display body motion using Microsoft® Visual Studio 2010 and Visual C#. The input to the system is a 640 × 480 15 Hz YUV color image of the body phantom and a corresponding depth image provided by KINECT™.

2.2. Accuracy of the measurement

Second, we verified the accuracy of the measured value of the physical shift received by the KINECT™ mounted on the CyberKnife® gantry head. The body phantom on the couch was displaced in the *X*, *Y*, and *Z* directions, and the displacement was measured by KINECT™ using a skeletal tracking method. The method does not need a maker. The phantom was moved in 1-cm steps from −5 cm to 5 cm for all directions. Distance from KINECT™ to the target was based on 100 cm.

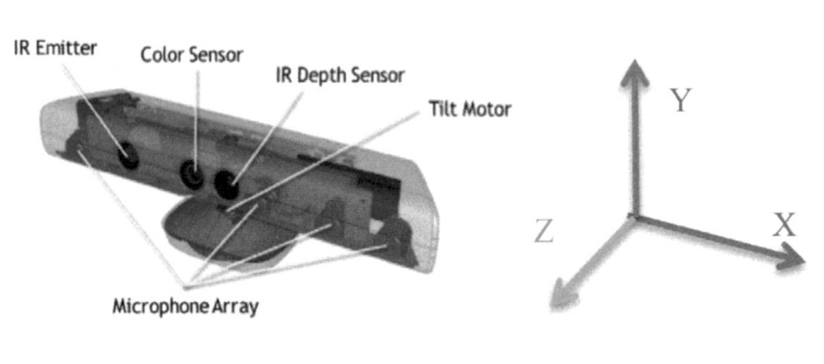

Figure 1. Sensor arrays of the KINECT™ system and directions of measurement.

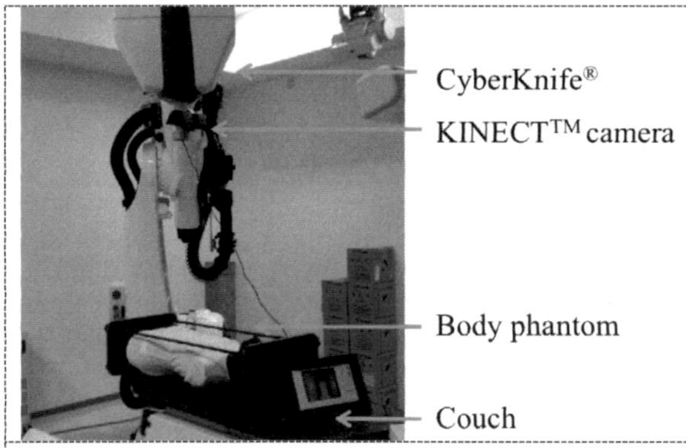

Figure 2. Setup for measurement.

This system runs on a computer equipped with an Intel® i5 quad-core 3550 CPU with 4 GB of RAM.

3. Results/Discussions

Figure 3 shows indicative results displayed on the interface. We developed the interface to display 3 directions of motion (*X*, *Y*, and *Z*) continuously as a graph.

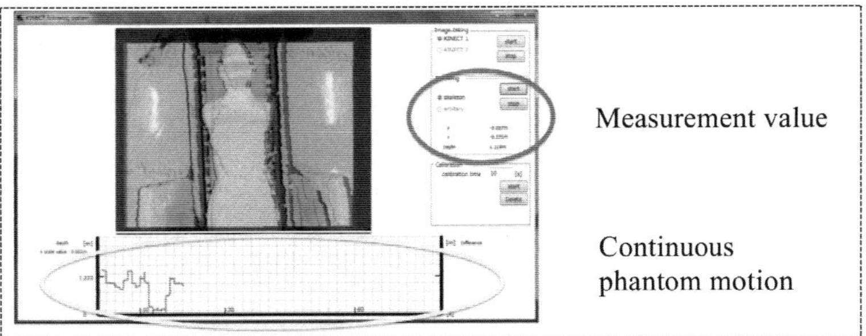

Figure 3. Indicative results displayed on the interface.

Measurements were performed 10 times for each point, and the average value is shown in the graphs. An error bar shows the standard deviation of each measurement.

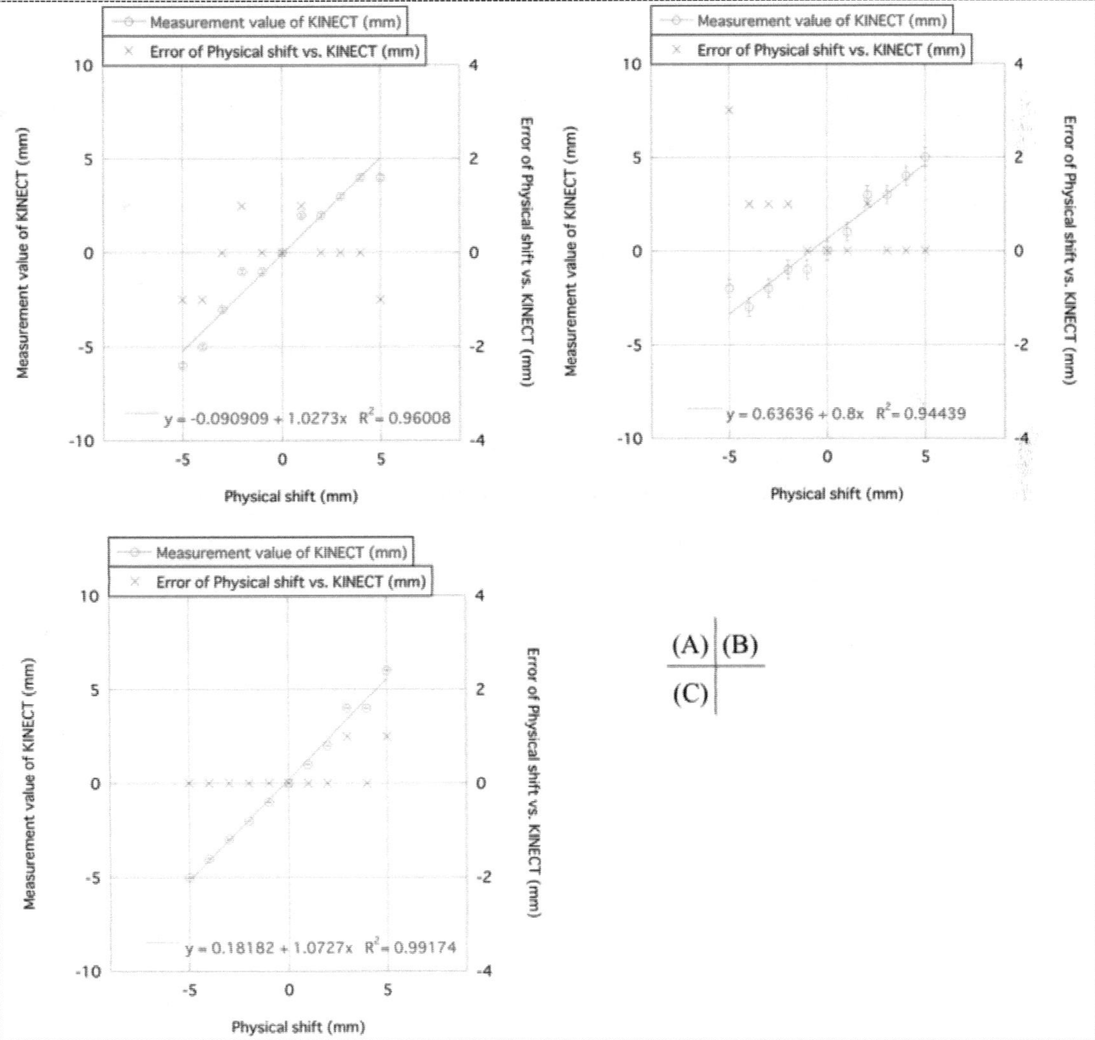

Figure 4. Quantitative evaluation of the performance of this system with respect to (A):X = *Lateral* (B):Y = *Longitudinal* (C):Z = *Vertical*

Figure 4 shows a quantitative evaluation of the performance of this system with respect to 3 directions. The upper left graph (A) shows the result in the *X* (lateral) direction, the upper right graph (B) showed the result of *Y* (longitudinal) direction, and, lower graph (C) shows the results of *Z* (vertical) direction. The horizontal axis displays the physical shift by the couch in the graph. The vertical axis displays the measurement value of KINECT™ and the error of physical shift vs. the measurement by KINECT™.

In the present study, the correlation coefficient was high in all directions. Errors of physical shift vs. KINECT™ were less than 1 mm in all directions. Standard deviations in the *Y*-direction at each point were larger than those in the *X* and *Z* directions because the effective field of view in the *Y* direction was narrower than that in the *X* direction.

Accuracy of position recognition in the *Z* direction was higher than in other directions, because this direction was measured with the IR camera. Further, its spatial resolution is higher than that of the color camera. It was considered that the detection sensitivity in the *Z* direction was higher than that in the other directions. The skeletal tracking method might depend on the direction of motion detection. In future work, we need to perform accuracy verification as shown in the report of AAPM Task Group 147 [4].

4. Conclusion

In the present study, we constructed a patient observation system using the Microsoft® KINECT™ system . Our system is simple and inexpensive and made it possible to track body motion.

5. References

[1] John L. Meyer 2011 IMRT, IGRT, SBRT-Advances in the Treatment Planning and Delivery of Radiationtherapy.
[2] Keall PJ, Mageras GS, Balter JM, et al. 2006 The management of respiratory motion in radiation oncology report of AAPM Task Group 76. *Med. Phys.* **33** 3874.
[3] http://www.microsoft.com/en-us/kinectforwindows/
[4] Willoughby T, Lehmann J, Bencomo JA, et al. 2012 Quality assurance for nonradiographic radiotherapy localization and positioning systems: Report of AAPM Task Group 147. *Med. Phys.* **39** 1728

Acknowledgments

We appreciate R-Tech, Inc., for their cooperation in this study.

2D-3D registration for brain radiation therapy using a 3D CBCT and a single limited field-of-view 2D kV radiograph

R Munbodh[1], **D J Moseley**[2]

[1] Department of Radiation Oncology, University of Pennsylvania, Philadelphia, PA, USA
[2] Radiation Medicine Program, Princess Margaret Hospital, Toronto, Canada

E-mail: Reshma.Munbodh@uphs.upenn.edu

Abstract. We report results of an intensity-based 2D-3D rigid registration framework for patient positioning and monitoring during brain radiotherapy. We evaluated two intensity-based similarity measures, the Pearson Correlation Coefficient (ICC) and Maximum Likelihood with Gaussian noise (MLG) derived from the statistics of transmission images. A useful image frequency band was identified from the bone-to-no-bone ratio. Validation was performed on gold-standard data consisting of 3D kV CBCT scans and 2D kV radiographs of an anthropomorphic head phantom acquired at 23 different poses with parameter variations along six degrees of freedom. At each pose, a single limited field of view kV radiograph was registered to the reference CBCT. The ground truth was determined from markers affixed to the phantom and visible in the CBCT images. The mean (and standard deviation) of the absolute errors in recovering each of the six transformation parameters along the x, y and z axes for ICC were ϕ_x: 0.08(0.04)°, ϕ_y: 0.10(0.09) °, ϕ_z: 0.03(0.03)°, t_x: 0.13(0.11) mm, t_y: 0.08(0.06) mm and t_z: 0.44(0.23) mm. For MLG, the corresponding results were ϕ_x: 0.10(0.04)°, ϕ_y: 0.10(0.09)°, ϕ_z: 0.05(0.07)°, t_x: 0.11(0.13) mm, t_y: 0.05(0.05) mm and t_z: 0.44(0.31) mm. It is feasible to accurately estimate all six transformation parameters from a 3D CBCT of the head and a single 2D kV radiograph within an intensity-based registration framework that incorporates the physics of transmission images.

1. Introduction

In the treatment of brain tumours with photon or proton radiation, accurate radiation delivery is critical for local tumour control and the sparing of healthy tissue. We have previously developed an automated 2D-3D rigid registration framework for patient positioning during prostate radiotherapy [1–5]. In this study, we extend our previous methods and develop an automated, accurate image-guided solution specific to radiation therapy of intracranial and skull base tumours. Patient setup errors are determined by registering 2D kV radiographs acquired during a particular treatment fraction to a 3D CBCT, which provides the desired position of the patient with respect to the treatment beams. By providing a framework for improved patient setup and, as a result, reduced geometrical (targeting) uncertainties in the delivery of the radiation with minimal additional imaging dose, we expect reduced radiation-induced toxicity while maintaining similar treatment outcomes. We will validate two similarity measures, the Pearson correlation coefficient and the maximum likelihood measure with Gaussian noise. These similarity measures make different assumptions about the nature of signal and noise in the CBCT images and 2D radiographs.

2. Methods

2.1. Pose determination experiments

Validation was performed on gold-standard data consisting of 3D kV CBCT scans and 2D kV radiographs of an anthropomorphic head phantom acquired on a robotic couch at each of 23 known poses with parameter variations along all six degrees of freedom (translations of up to 10 mm and rotations of up to 3.5°). The ground truth at each pose was determined by registering radio-opaque ball bearings affixed to the phantom and visible in the CBCT images to ball bearings in the reference CBCT. We used single, limited field-of-view (FOV) 2D kV radiographs ($\sim 10\times10$ cm^2 of the 17×20 cm^2 phantom head size, resolution: 1 mm^2) in the anterior-posterior view comprising the orbits and part of the nasal structures for registration to the reference 3D CBCT (resolution: 1 mm^3). Limited FOV kV radiographs permit reduced imaging dose and also ensure exclusion of any patient immobilisation devices from the registration. The x-axis was directed across the subject from right to left, the y-axis along the subject from toe to head and the z-axis through the subject towards the x-ray source. In-plane parameters were t_x, t_y and ϕ_z and out-of-plane parameters were ϕ_x, ϕ_y and t_z.

2.2. Similarity measures

Given a 3D kV CBCT and a 2D kV radiograph, our task, is to estimate the global rigid tranformation, ψ, which aligns the CBCT to the 2D radiograph. The transformation ψ is a vector consisting of three translations, t_x, t_y and t_z, and three rotations, ϕ_x, ϕ_y and ϕ_z, around the x, y and z axes. The optimal transformation, ψ, was found by iteratively shifting and rotating the CBCT until the similarity between the DRR, $\mathbf{v}_\psi$, which is a perspective projection of the CBCT at transformation ψ and the 2D radiograph, $\mathbf{u}$, was maximised. In this study, we evaluated two intensity-based similarity measures, the Pearson correlation coefficient (ICC) and the maximum likelihood measure with Gaussian noise (MLG) [5]. ICC measures the strength of the linear relationship between two normal random variables. MLG assumes that Poisson noise is a limiting source of noise in the 2D kV radiographs and for a large number of photons, the Poisson noise can be approximated by non-stationary Gaussian noise. It assumes that the pixel intensities in the 2D kV radiograph are obtained from a non-stationary Gaussian process with mean and variance given by the intensity values of the corresponding pixels in the 2D DRR. Previously, we showed that high-pass filtering the DRRs and radiographs reduced the bias and variance in the registration results [3]. Here, the cutoff frequency for the filter was determined from the bone-to-no-bone ratio projection images of a CT of the head phantom.

2.3. Quantification of registration error

In addition to examining the errors in the individual transformation parameters at every pose, we computed the final target registration error (TRE) as the mean distance, $\mathcal{E} = \frac{1}{n}\sum_{i=1}^{n}\|(T_{\hat{\psi}} - T_\psi)x_i\|$, moved by n uniformly distributed points on a virtual sphere of radius 3 cm centred at the simulated isocentre, and representing the tumour, from their intended position. $T_{\hat{\psi}}$ and T_ψ are the transformation matrices at the estimated and actual transformation parameters, respectively, and x are the 3D coordinates of the points on the surface of the sphere. We examined the final TRE as a function of initial TRE defined as $\mathcal{E}_o = \frac{1}{n}\sum_{i=1}^{n}\|T_\psi x_i\|$.

3. Results

Fig. 1(a) shows the final TRE, $\mathcal{E}$, as a function of the initial TRE, $\mathcal{E}_o$ for each of the 14 poses with translational shifts only for the two similarity measures. A single 2D kV radiograph in

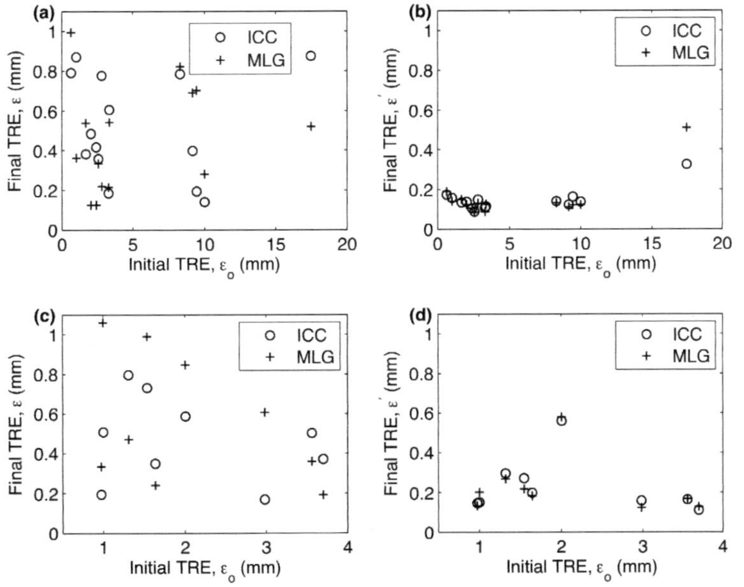

Figure 1. Final TRE as a function of initial TRE. The final TRE, $\mathcal{E}$, due to errors in estimating the six transformation parameters is shown in (a) and the final TRE, $\mathcal{E}'$, excluding the error in estimating t_z, is shown in (b) for poses with translational displacements only. (b) and (c) show $\mathcal{E}$ and $\mathcal{E}'$, respectively, for poses with rotations or combinations of rotations and translations.

the AP view was used for the registration. The initial TRE ranged from 0.63 mm to 17.5 mm. The final TRE was less than 1 mm for all 14 poses for both ICC and MLG. Even though the out-of-plane translation, t_z, was solved to better than 1 mm in all instances, errors in estimating t_z were the largest contributor to the overall registration error. Fig. 1 (b) shows the final TRE excluding errors in estimating t_z. The final TRE, $\mathcal{E}'$, in this case was smaller than 0.2 mm for 13 of the 14 poses. Figs. 1 (c) and (d) show similar plots for the nine poses that had either rotations only or combinations of rotations and translations of 3 mm along each axis. Once again, errors in estimating t_z were the largest as illustrated in the reduction in the final TRE in Fig. 1 (d). The presence of in-plane and out-of-plane rotations did not seem to adversely affect the performance of the registration algorithm. The magnitude of the final TRE for both the translational and rotational poses was not correlated to the magnitude of the initial TRE.

A box plot of the distribution of absolute errors in estimating each of the six transformation parameters and the final target registration errors, $\mathcal{E}$ and $\mathcal{E}'$,is shown in Fig. 2. The lines in the boxes denote the lower quartile, median and upper quartile values. The mean (and standard deviation) of the absolute errors in recovering each of the six transformation parameters along the x, y and z axes for ICC were ϕ_x: 0.08(0.04)°, ϕ_y: 0.10(0.09)°, ϕ_z: 0.03(0.03)°, t_x: 0.13(0.11) mm, t_y: 0.08(0.06) mm and t_z: 0.44(0.23) mm. For MLG, the corresponding results were ϕ_x: 0.10(0.04)°, ϕ_y: 0.10(0.09)°, ϕ_z: 0.05(0.07)°, t_x: 0.11(0.13) mm, t_y: 0.05(0.05) mm and t_z: 0.44(0.31) mm.

4. Discussion

The performance of ICC and MLG was found to be comparable. In-plane and out-of-plane parameters were estimated to better than 1 mm and 1° accuracy with both similarity measures, as quantified by the mean of the absolute registration error. Estimates of the out-of-plane translation, t_z, had the largest uncertainty. This is attributable to the imaging geometry. The results suggest that an accuracy of better than 0.5 mm can be obtained by using a second radiograph in the lateral view for the registration and where t_z would be an in-plane translation.

Most registration methods for setup verification are 2D and semi-automated. Semi-automated 2D-3D methods for the skull include [6]. Automated 2D-3D registration methods for

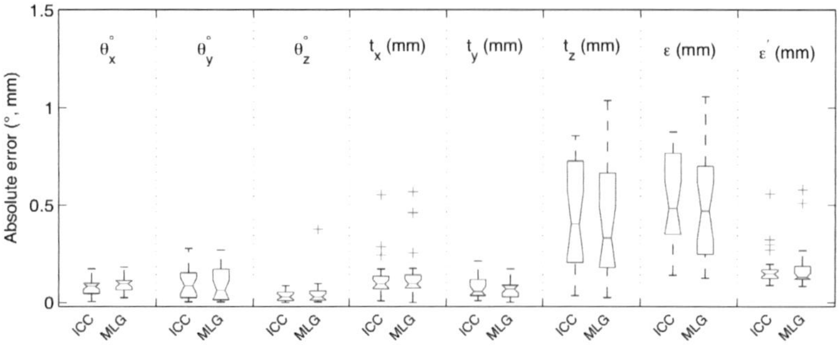

Figure 2. Box plot showing the distribution of absolute errors in estimating the six transformation parameters and the final target registration errors, $\mathcal{E}$ and $\mathcal{E}'$.

radiotherapy are few and employ two orthogonal 2D radiographs. They include intensity-based methods using pattern intensity [7], stochastic rank correlation [8], correlation-based measures and information theoretic measures [9–12]. Feature-based methods include [13–15], and hybrid methods include [16]. Automated 2D-3D methods applied to brain RT include an intensity-based method [17] and gradient-based approaches [18, 19]. The testing performed and results achieved in this study with a single 2D radiograph compare favourably with the above studies.

Our results suggest that at current imaging doses and levels of Poisson noise in 2D radiographs of the head, ICC and MLG are equivalent. The MLG method provides insight as to why ICC would work for these images. For larger levels of noise and lower images doses, we might expect MLG to provide better registration accuracy. Results from our study support the importance of understanding the image formation process and of modelling the relationship between intensity values in the two modalities to guide the choice of similarity measure in intensity-based registration of CT to x-ray images.

5. Conclusions

We have presented an automated 2D-3D registration framework for patient setup during brain radiation therapy. Sub-millimeter registration accuracy was obtained with two similarity measures and a single, limited FOV 2D kV radiograph.

References

[1] Munbodh R 2004 *Achieving accurate automated image registration for prostate radiotherapy* Ph.D. thesis Yale University
[2] Munbodh R, Jaffray D A, Moseley D J *et al.* 2006 *Med. Phys.* **33** 1398–1411
[3] Munbodh R, Chen Z, Jaffray D A *et al.* 2007 *Med. Phys.* **34** 3005–3017
[4] Munbodh R, Chen Z, Jaffray D *et al.* 2008 *Med. Phys.* **35** 4352–4361
[5] Munbodh R, Tagare H, Chen Z *et al.* 2009 *Med. Phys.* **36** 4555–4568
[6] Lujan A E, Balter J M and Ten Haken R K 1998 *Med. Phys.* **25** 703–8
[7] Penney G P, Batchelor P G, Hill D L *et al.* 2001 *Med. Phys.* **28** 1024–1032
[8] Birkfellner W, Stock M, Figl M *et al.* 2009 *Med. Phys.* **36** 3420–3428
[9] Penney G P, Weese J, Little J A *et al.* 1998 *IEEE Trans. on Medical Imaging* **17** 586–595
[10] Kim J, Fessler J A, Lam K L *et al.* 2001 *Med. Phys.* **28** 2507–2517
[11] Clippe S, Sarrut D, Malet C *et al.* 2003 *Int. J. Radiat. Oncol. Biol. Phys.* **56** 259–265
[12] Kim J and Fessler J A 2004 *IEEE Trans. Medical Imaging* **23** 1430–1444
[13] Tomaževič D, Likar B, T S and Pernuš F 2003 *IEEE Trans. Medical Imaging* **22** 1407–1416
[14] Livyatan H, Yaniv Z and Joskwicz L 2003 *IEEE Trans. Medical Imaging* **22** 1395–1406
[15] Gilhuijs K G, van de Ven P and van Herk M 1996 *Med. Phys.* **23** 389–399 PMID: 8815382
[16] Bansal R, Staib L H, Chen Z *et al.* 2003 *IEEE Trans. Medical Imaging* **22** 29–49
[17] Fu D, Kuduvalli G, Mitrovic V *et al.* 2005 *SPIE* **5744** 1605–7422
[18] Lemieux L, Jagoe R, Fish D R *et al.* 1994 *Med. Phys.* **21** 1749–1760
[19] Murphy M J 1997 *Med. Phys.* **24** 857–866

XVII International Conference on the Use of Computers in Radiation Therapy (ICCR 2013) IOP Publishing
Journal of Physics: Conference Series **489** (2014) 012038 doi:10.1088/1742-6596/489/1/012038

A novel marker enhancement filter (MEF) for fluoroscopic images

Olesya Peshko[1], Timothy N Davidson[1], Jan Modersitzki[2], Tamás Terlaky[3] and Douglas J Moseley[4]

[1] McMaster University, Hamilton, ON L8S 4K1, Canada
[2] University of Lübeck, Lübeck, 23562, Germany
[3] Lehigh University, Bethlehem, PA 18015, USA
[4] Princess Margaret Hospital, Toronto, ON M5G 2M9, Canada

E-mail: olesya.peshko@gmail.com, davidson@mcmaster.ca

Abstract. To enhance the measurements of radio-opaque cylindrical fiducial markers in low contrast x-ray and fluoroscopic images, a novel nonlinear marker enhancement filter (MEF) has been designed. It was primarily developed to assist in automatic initialization of a tracking procedure for intra-fraction organ motion analysis in fluoroscopic sequences. Conventional procedures were not able to provide sufficient improvement due to the complications of noise, small marker size, cylindrical shape and multiple orientations, intensity variations of the background, and the presence of overlaying anatomical measurements in this application. The proposed MEF design is based on the principles of linear scale space. It includes measures that assess the probability of each pixel to belong to a marker measurement, morphological operations, and a novel contrast enhancement function for standardization of the filter output. The MEF was tested on fluoroscopic images of two phantoms and three prostate patients, and was shown to perform better or comparable to the existing filters in terms of marker enhancement and background suppression, while performing significantly better in marker shape preservation.

1. Introduction

Image guidance with kilovoltage imaging units mounted on a linear accelerator has become a conventional clinical practice for radiation therapy [1]. One particular example is the monitoring of intra-fraction organ motion using fluoroscopic image sequences. Fluoroscopic tracking has the advantage of providing unassisted real-time organ motion information using existing equipment, but the desire to keep the imaging radiation dose as low as possible results in images that require significant postprocessing in order to extract the relevant content.

Often, radio-opaque fiducial markers are inserted prior to planning to improve the reliability of localizing organs with limited x-ray contrast, such as the prostate [2]. In order to measure the intra-fraction organ motion of prostate patients, we are designing a method for tracking the fiducial markers in lateral fluoroscopic image sequences [3][4]. Automatic tracking requires reliable initialization, i.e., finding markers in the first frame of a sequence. That is challenging due to the low contrast, noise, and the small marker size. In our case, the markers are gold cylinders 3 mm long and 0.8 mm in diameter that appear as about 12×3 pixel measurements. Automated tracking is initialized by preprocessing the fluoroscopic image to enhance the markers, and then registering it [5] to the 3D cone-beam computed tomography (CBCT) image.

Several approaches for marker enhancement and detection have been reported in the literature, such as morphological opening [6], template matching [7][8], and the marker extraction kernel (MEK) [9]. For cylindrical markers, these methods require the use of multiple templates, preliminary knowledge of marker orientations, and additional processing, such as noise reduction filters [6][9], edge enhancement [7], and intensity thresholding [7], for which the selection of parameters that will work well for different images can be challenging. In addition, these methods tend to distort the marker shape, e.g., by amplifying one side of the marker more strongly than the other. This distortion can degrade localization accuracy and compromise estimation of marker orientation.

In this paper, we propose a novel marker enhancement filter (MEF) that is tailored to the characteristics of the markers. The MEF amplifies the markers, suppresses the background and anatomical features, and preserves marker shapes. The proposed filter is designed within the linear scale-space framework [10] that has found wide utilization in the enhancement of vessels and other line-like and blob structures [11][12], and in feature detection and image matching applications for computer vision [13], due to its ability to achieve global contrast for the sought-for features. The MEF includes an adaptation of the vessel enhancement approach proposed by Frangi et al. [11], morphological operations [14], and a novel contrast enhancement function that is able to operate successfully on a variety of images with different intensity characteristics. While Frangi's method was targeted at enhancement of the tubular structures, the markers we seek to amplify are quite different, which leads to a different filter design. The fluoroscopic images are subject to temporal filtering with a Dolph-Chebyshev filter [15] prior to the MEF application, which provided the best contrast-to-noise ratio among other methods that we tested on a phantom with moving markers [4].

2. Marker enhancement filter design

2.1. Scale-space analysis

In computer vision, there is an accepted perspective that features in the images only meaningfully exist over limited ranges of scales [10]. To perform an automatic image interpretation, a multi-scale representation was proposed [10] (and references therein), in which the original image $\mathcal{A} : \mathbb{R}^2 \to \mathbb{R}$ is embedded into a one-parameter family of derived images $\mathcal{L} : \mathbb{R}^2 \times \mathbb{R}_+ \to \mathbb{R}$, where the additional non-negative scalar parameter is the scale. The scale-space representation is defined as the convolution of the image with the Gaussian functions of the appropriate scales:

$$\mathcal{L}(x,\sigma) = \mathcal{A}(x) * \mathcal{G}(x,\sigma) \text{ for } \sigma > 0, \text{ and } \mathcal{L}(x,0) = \mathcal{A}(x), \tag{1}$$

where the Gaussian is defined as $\mathcal{G}(x,\sigma) = \exp\left(-\frac{\|x\|^2}{2\sigma^2}\right)/(2\pi\sigma^2)$, and $*$ denotes two-dimensional convolution. Due to the fact that the fiducial markers in our application are all of equal or very similar size to each other, we only compute $\mathcal{L}$ for one fixed scale, $\sigma = \sigma_0$, that is selected to amplify the features close in size to the markers.

A common approach to analyze the local behaviour of an image intensity function is to consider its second order Taylor expansion in the neighborhood of a point x_0 [11][12]. Some methods directly utilize second order derivatives of $\mathcal{L}$. For example, the Laplacian of Gaussian (LoG) [10] method was used in several studies for marker enhancement in portal images [16][17]. However, in addition to enhancing the markers, the LoG is prone to enhancing anatomical features that have a dimension that is of a similar scale to the markers. In the following sections we describe the proposed filter and show how it is able to overcome this issue.

2.2. MEF principles: Magnitude-and-ratio versus magnitude-only image

One way to characterize the contrast and shape of features in the image at a particular scale is to analyze the sign and magnitude of the eigenvalues of the Hessian matrix that is constructed from

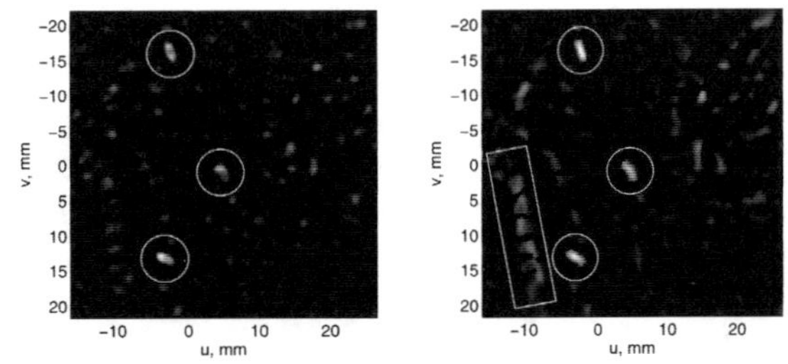

Figure 1. The magnitude-and-ratio $\mathcal{Y}_{\mathrm{mr}}$ (*left*) and magnitude-only $\mathcal{Y}_{\mathrm{m}}$ (*right*) images. While both attain high response in marker pixels (*encircled*), $\mathcal{Y}_{\mathrm{m}}$ is characterized with better marker shape preservation, but also undesirably amplifies long anatomical features (e.g., *in rectangle*).

the second order derivatives of $\mathcal{L}$. For the bright markers on a dark background, the intensity function should: (i) be concave, which requires negative eigenvalues; (ii) possess a reasonable contrast that is reflected in the magnitude of the eigenvalues; and (iii) describe a feature whose dimensions are not very different, which can be assessed in terms of the ratio of the eigenvalues. More formally, if we let λ_1 and λ_2 denote the eigenvalues of the Hessian of $\mathcal{L}(x, \sigma_0)$ at a given point $x = x_0$, the magnitude measure at that point is defined as $\mathcal{X}_{\mathrm{m}} = \sqrt{\lambda_1^2 + \lambda_2^2}$, and the ratio as $\mathcal{X}_{\mathrm{r}} = \lambda_1/\lambda_2$, where $|\lambda_1| \leq |\lambda_2|$ [11]. An effective marker enhancement filter should provide high response in the image pixels with $\lambda_1, \lambda_2 < 0$, and large $\mathcal{X}_{\mathrm{m}}$ and $\mathcal{X}_{\mathrm{r}}$.

Based on the insight from the work of Frangi et al. [11], we incorporate $\mathcal{X}_{\mathrm{m}}$ and $\mathcal{X}_{\mathrm{r}}$ into probability-like measures, based on which we define a *magnitude-and-ratio image*:

$$\mathcal{Y}_{\mathrm{mr}} = \begin{cases} \left(1 - \exp\left(-\frac{\mathcal{X}_{\mathrm{m}}^2}{2\gamma^2}\right)\right)\left(1 - \exp\left(-\frac{\mathcal{X}_{\mathrm{r}}^2}{2\beta^2}\right)\right), & \text{if } \lambda_1 < 0 \text{ and } \lambda_2 < 0, \\ 0, & \text{otherwise,} \end{cases} \tag{2}$$

where γ and β are used to normalize the values of $\mathcal{X}_{\mathrm{m}}$ and $\mathcal{X}_{\mathrm{r}}$. As suggested in [11], γ is chosen to be half of the maximal $\mathcal{X}_{\mathrm{m}}$ value. For circularly shaped blobs, $\beta = 0.5$ would be used [11]. We reduce β to 0.25 based on visual observations of the filtered images in order to produce high response for a range of shapes from circular to elliptical blobs. Although the marker dimensions are known, it is not practical to tune the filter to elliptical blobs exclusively, as the measurements are often corrupted by the noise and overlaying structures, and rarely possess exactly the same ratio as the markers. Observing that the markers may often have reasonably high contrast in fluoroscopic images, we have also considered a *magnitude-only image*:

$$\mathcal{Y}_{\mathrm{m}} = \begin{cases} 1 - \exp\left(-\frac{\mathcal{X}_{\mathrm{m}}^2}{2\gamma^2}\right), & \text{if } \lambda_1 < 0 \text{ and } \lambda_2 < 0, \\ 0, & \text{otherwise.} \end{cases} \tag{3}$$

Comparing $\mathcal{Y}_{\mathrm{mr}}$ and $\mathcal{Y}_{\mathrm{m}}$ in application to our fluoroscopic images (see Figure 1), we notice that while the markers possess high intensities in both of them, $\mathcal{Y}_{\mathrm{m}}$ offers better marker shape preservation, but also undesirably amplifies long anatomical features similar in width to the marker size. To gain from the desirable properties of both images, we combine them by using $\mathcal{Y}_{\mathrm{mr}}$ as a mask for $\mathcal{Y}_{\mathrm{m}}$, i.e., the intensities in the filtered image are derived from $\mathcal{Y}_{\mathrm{m}}$, but only from the high-intensity regions of $\mathcal{Y}_{\mathrm{mr}}$.

2.3. Contrast enhancement function

Intensity thresholding is often used as a part of multi-stage procedures to improve the filtering outcome by further separating the sought-for markers from the background. It is typically performed by setting all intensities in the image that fall below a preset value ϑ to the minimal background intensity, e.g., [7][17]. However, we prefer to use a smooth threshold function as it increases overall reliability of image registration that will be performed on the output of the MEF filter in the proposed application. In particular, given an image $\mathcal{Y}$ with values in $[0, 1]$ we will construct the contrast enhanced image $\mathcal{C}(\mathcal{Y}) = \mathcal{Y}^{\mathcal{Z}(\mathcal{Y})}$, where $\mathcal{Z}(\mathcal{Y}) = \frac{\vartheta^s \log_\vartheta \tau}{\vartheta^s - 1}\left(1 - \frac{1}{\mathcal{Y}^s}\right)$. In this expression ϑ is the threshold, τ is the minimum intensity of all marker points after contrast enhancement, and s is the sharpness of distinction between the marker and non-marker points. In our experiments we chose $\tau = 1/2$ and $s = 3$.

As in all thresholding methods, the choice of ϑ is critical. We have derived a marker-based model for choosing the threshold [4]: We seek to amplify as many bright pixels in the image, as there are expected to be occupied by the markers, and suppress everything else. For the cylindrical markers that are close to the isocentre and appear full-length in the lateral images, the number of pixels can be approximated by $m = n\lceil 2rh/\delta^2 \rceil$, where n is the number of markers, r and h are their radius and length, respectively, and δ is the size of the pixel side. The intensity values in the image are sorted in a descending order, and the threshold ϑ for this image is chosen to be the m-th element in this sorted sequence.

2.4. MEF algorithm
The MEF consists of the following steps, which are depicted in Figure 2:

- The scale-space representation is computed for a (temporally filtered) fluoroscopic image. We found $\sigma_0 = 0.7$ mm to be the most appropriate for our marker size.
- Images $\mathcal{Y}_{\mathrm{m}}$ and $\mathcal{Y}_{\mathrm{mr}}$ are computed as described in Section 2.2, and $\mathcal{Y}_{\mathrm{mr}}$ is contrast enhanced by the function $\mathcal{C}$ defined in Section 2.3.
- Computation of the mask $\mathcal{M}$. As markers in $\mathcal{Y}_{\mathrm{mr}}$ often appear smaller than in $\mathcal{Y}_{\mathrm{m}}$ (see Figure 1), the high-intensity regions in $\mathcal{Y}_{\mathrm{mr}}$ are enlarged by morphological dilation [14] with the circle of radius $h/2$ as the structuring element. To have the same intensity for all points within one contiguous region, we group them and assign the largest intensity value to all.
- The grayscale mask $\mathcal{M}$ is multiplied with $\mathcal{Y}_{\mathrm{m}}$, and the final MEF-image is produced after another contrast enhancement step with the function $\mathcal{C}$.

Our experimental implementation of this algorithm in Matlab produced the MEF-image in Figure 2 in about 3 seconds on a modestly powered desktop (Intel Quad CPU Q6700 @ 2.66Hz, 4 GB RAM). This enables us to foresee a real-time implementation for clinical use.

3. Filter validation
In addition to visually verifying, on multiple images, that the MEF produces desirable results in terms of marker amplification, background suppression, and marker shape preservation, we have conducted quantitative validation experiments. The fluoroscopic images were acquired at 5.5 Hz at a technique of 120 kVp, 32 mA, 40 ms per frame for phantoms, and 120 kVp, 64 mA, 20 ms for prostate patients. MV irradiation was off.

3.1. MEF success rates
The true positive rate (TPR) is defined as the percentage of the points enhanced by the MEF (those with intensity greater than τ) that belong to the markers. In the controlled environments we consider, the marker positions can be determined to high accuracy [4]. Table 1 shows median TPR for moving Quasar respiratory phantom (Modus Medical Devices Inc., London, ON; 23 frames, linear motion at 2.3 mm/sec), stationary pelvic RSVP phantom (The Phantom Laboratory, Salem, NY; 342 frames), and 3 prostate patients (200 frames each), for images obtained by combining ℓ frames with the Dolph-Chebyshev filter (i.e, temporal filtering). In

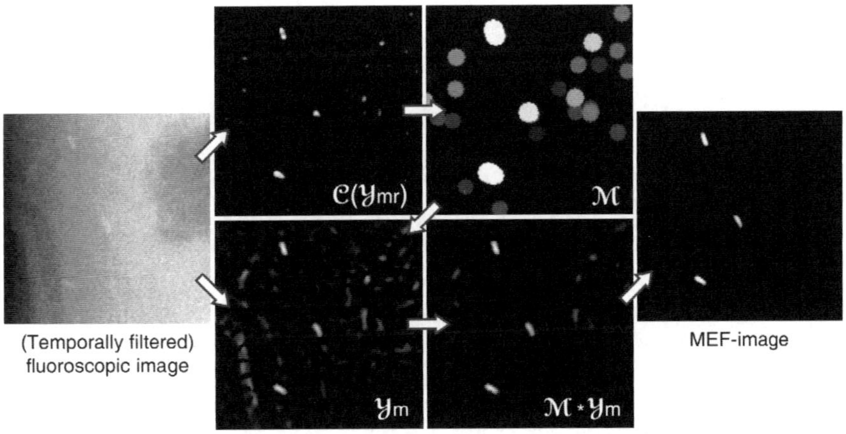

Figure 2. Steps of the MEF algorithm.

Table 1. TPR depending on the kernel size, ℓ, of the Dolph-Chebyshev temporal filter.

ℓ	Quasar	RSVP	Patient 1	Patient 2	Patient 3
1	100	78.17	76.85	92.59	73.15
3	100	95.63	97.22	100	88.89
7	100	99.6	100	100	90.74
11	100	100	100	100	93.52

general, the TPR increases for larger ℓ, and decreases as the region of interest (ROI) grows. Table 1 contains the results for ROIs that measure approximately 45×100 mm for the phantoms (include 7 markers) and 50×70 mm for the patients (include 3 markers). Note that substantial TPR improvement is observed from $\ell = 1$ (no temporal filtering) to $\ell = 3$, and that for the phantoms and the first two patients the TPR reaches 100% for $\ell = 11$.

3.2. Comparison to other filters

We have also compared the MEF to other filters reported in the literature, such as the MEK [9], LoG^2 [10], Sato's blob filter [12], and Frangi's line filter [11]. The data points in Figure 3 are obtained by computing the median true (TP) and false positives (FP) as a percentage of positives after applying a series of intensity thresholds to 342 images of the RSVP. As the results suggest, the MEF shows the least FP for high TPs.

4. Discussion

Although it was designed with a particular application in mind, the proposed filter can be used in a variety of marker localization and tracking applications. The main advantages of the MEF over other filters are its marker enhancement and shape preservation properties, and its successful background suppression. No preliminary information about the marker orientation is required. We also observe that the use of temporal filtering prior to the MEF application improves the filtering outcome and reduces the computational cost of constructing the mask $\mathcal{M}$.

The main limitation of the MEF comes from the assumption that among all features observed

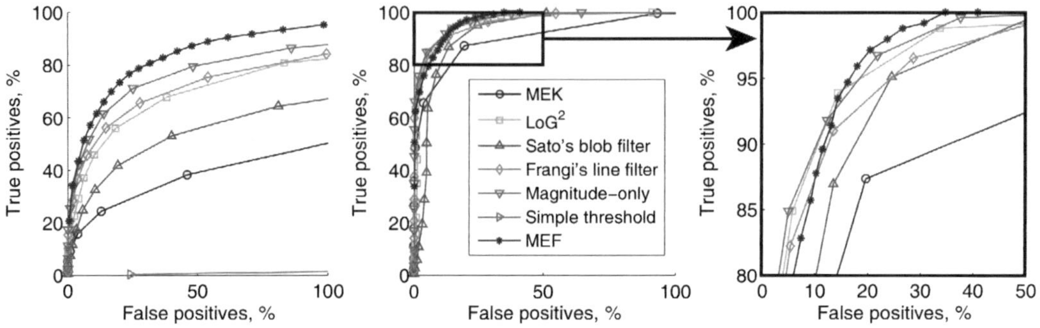

Figure 3. TP and FP percentages for different filters for $\ell = 1$ (*left*) and $\ell = 7$ (*centre*). The figure on the *right* shows a zoom-in into the high TP and low FP region for $\ell = 7$.

at scale σ_0, the markers possess the highest contrast. While a couple of features similar to the markers in size and contrast typically do not jeopardize marker localization, the presence of a higher-contrast feature with one of the dimensions corresponding to σ_0 and the other much larger is not desirable. In scenarios in which one of the markers has extremely low contrast, the MEF tends to allocate more high-intensity points to the remaining markers rather than amplifying anatomical features. In our application [4], the missing markers are recovered by subsequent registration to the daily 3D CBCT image.

Acknowledgments

We would like to thank Dr. Cynthia Menard and Dr. Tim Craig from Princess Margaret Hospital, Toronto, for the patient images provided to test the MEF. The work on this project was partially supported by Elekta.

References

[1] Jaffray D A, Siewerdsen J H, Wong J W and Martinez A A 2002 *Int. J. Radiat. Oncol. Biol. Phys.* **53** 1337–49
[2] Moseley D *et al.* 2007 *Int. J. Radiat. Oncol. Biol. Phys.* **67** 942–53
[3] Peshko O, Moseley D J, Craig T, Terlaky T and Menard C 2007 *Med. Phys.* vol 34 p 2392
[4] Peshko O 2013 *Automatic Detection and Tracking of Fiducial Markers in Medical Images (in preparation)* Ph.D. thesis McMaster University, Hamilton, Canada
[5] Modersitzki J 2004 *Numerical Methods for Image Registration* (New York: Oxford University Press)
[6] Keall P J, Todor A D, Vedam S S, Bartee C L, Siebers J V, Kini V R and Mohan R 2004 *Med. Phys.* **31** 3492–9
[7] Park J C, Park S H, Kim J H, Yoon S M, Kim S S, Kim J S, Liu Z, Watkins T and Song W Y 2011 *Med. Phys.* **38** 1028–36
[8] Mao W, Wiersma R D and Xing L 2008 *Med. Phys.* **35** 1942–49
[9] Nederveen A, Lagendijk J and Hofman P 2000 *Int. J. Radiat. Oncol. Biol. Phys.* **47** 1435–42
[10] Lindeberg T 1994 *Scale-Space Theory in Computer Vision* (Dordrecht: Kluwer Academic Publishers)
[11] Frangi A F, Niessen W J, Vincken K L and Viergever M A 1998 *Medical Image Computing and Computer-Assisted Interventation – MICCAI'98* ed Wells W M *et al.* (Berlin Heidelberg: Springer) pp 130–7
[12] Sato Y, Westin C F, Bhalerao A, Nakajima S, Shiraga N, Tamura S and Kikinis R 2000 *IEEE Trans. Visual. Comput. Graphics* **6** 160–80
[13] Tuytelaars T and Mikolajczyk K 2007 *Foundations and Trends in Computer Graphics and Vision* **3** 177–280
[14] Gonzalez R C, Woods R E and Eddins S L 2004 *Digital Image Processing using Matlab* (Upper Saddle River: Pearson Prentice Hall) pp 334–377
[15] Harris F J 1978 *Proc. IEEE* **66** 51–83
[16] Park S J, Ionascu D, Hacker F, Mamon H and Berbeco R 2009 *Med. Phys.* **36** 4536–46
[17] Buck D, Alber M and Nüsslin F 2003 *Phys. Med. Biol.* **48** 763–74

Inverse-consistent rigid registration of CT and MR for MR-based planning and adaptive prostate radiation therapy

David Rivest-Hénault[1], Nicholas Dowson[1], Peter Greer[2,3], and Jason Dowling[1]

[1] CSIRO, Royal Brisbane and Women's Hospital, Herston, Queensland, Australia
[2] Calvary Mater Newcastle Hospital, Newcastle, New South Wales, Australia
[3] University of Newcastle, Newcastle, New South Wales, Australia

E-mail: david.rivest-henault@csiro.au

Abstract. MRI-alone treatment planning and adaptive MRI-based prostate radiation therapy are two promising techniques that could significantly increase the accuracy of the curative dose delivery processes while reducing the total radiation dose. State-of-the-art methods rely on the registration of a patient MRI with a MR-CT atlas for the estimation of pseudo-CT [5]. This atlas itself is generally created by registering many CT and MRI pairs. Most registration methods are not symmetric, but the order of the images influences the result [8]. The computed transformation is therefore biased, introducing unwanted variability. This work examines how much a symmetric algorithm improves the registration.
Methods: A robust symmetric registration algorithm is proposed that simultaneously optimises a half space transform and its inverse. During the registration process, the two input volumetric images are transformed to a common position in space, therefore minimising any computational bias. An asymmetrical implementation of the same algorithm was used for comparison purposes.
Results: Whole pelvis MRI and CT scans from 15 prostate patients were registered, as in the creation of MR-CT atlases. In each case, two registrations were performed, with different input image orders, and the transformation error quantified. Mean residuals of 0.63 ± 0.26 mm (translation) and $(8.7\pm7.3)\times10^{-3}$ rad (rotation) were found for the asymmetrical implementation with corresponding values of 0.038 ± 0.039 mm and $(1.6\pm1.3)\times10^{-3}$ rad for the proposed symmetric algorithm, a substantial improvement.
Conclusions: The increased registration precision will enhance the generation of pseudo-CT from MRI for atlas based MR planning methods.

1. Introduction

Medical image registration is a computerised alignment process now well established in many prospective or curative clinical workflows. In particular, in image-guided radiation therapy (IGRT) the planning CT volume is aligned with the portal images of the linear accelerator (linac) in order to ensure the accurate delivery of the radiation dose to the tumour region. One weakness of the classical IGRT approach for the treatment of prostate cancer is that the planning CT might not be an accurate representation of the patient anatomy throughout the whole duration of the treatment, which might consist of around 35 fractions delivered over a span of 7 weeks. As it is not generally considered acceptable in clinical practice to perform multiple CT acquisitions during the course of the treatment, adaptive MRI-based prostate radiation therapy has been proposed as a way to adapt the treatment dose planning to a changing patient's anatomy at regular intervals. This approach promises to significantly increase the accuracy of the curative dose delivery processes while reducing overall patient exposure to radiation. With MRI-alone treatment planning, the planning CT itself is replaced by a MRI acquisition that serves as a reference for the calculation of the radiation plan. The main challenge in MRI-alone treatment planning and adaptive MRI-based prostate radiation therapy is to estimate the electron density map from MRI acquisitions.

As it is generally acknowledged, there is no simple mapping between any clinically relevant MRI intensity and the underlying electron density [3]. State-of-the-art methods address this challenge by registering the MR image with a CT volume representing the same anatomy, and then use the spatial mapping between the MRI and the CT to estimate the probable CT intensity, and thus the electron density, at each MRI voxel location. By repeating this process with multiple patients, a CT-MR atlas is built. In turn, a new MR scan, which does not have a corresponding CT, can be registered with this atlas in order to estimate the electron density [5]. With this approach, it is required to accurately register CT and MR scans of the pelvis, which is a hard multi-modal problem. In addition, most registration methods are not symmetric, but instead the order of the images influences the result [8]. The computed transformations are therefore biased, introducing unwanted variability in the process. This phenomenon is usually referred to as the *inverse consistency* property of an algorithm, and the inverse consistency problem refers to the definition of algorithms that are independent of the input image order.

While the inverse consistency problem was the subject of a certain number of articles [11, 10, 2, 8] in recent years, they were mostly concerned with overall flexible transformations [11, 10, 2]. The inverse consistency issue was largely ignored in the seemingly easier rigid transformation. One notable exception is the work of Reuteur *et al* [8] who implemented and published an inverse-consistent method as part of the FreeSurfer [1] package. Unfortunately, in our experience, this method is not robust enough to register CT and MR scans of the pelvis. This paper thus presents a new inverse consistent rigid registration method, and examines how much an inverse consistent algorithm improves the registration.

2. Methods

A robust symmetric registration algorithm is proposed that simultaneously optimises a half space transform and its inverse. During the registration process, the two input volumetric images are transformed to a common position in space, therefore minimising any computational bias. An asymmetric implementation of the same algorithm was used for comparison purposes.

2.1. Symmetric formulation of the registration problem

Let I_1 and I_2 be the two volumetric images to be registered. Classic registration methods usually involve finding the transformation T that minimise an energy function of the form

$$E_{\mathrm{C}} = \sum_{\forall \mathbf{x} \in \Omega} M(I_1(\mathbf{x}), I_2(T(\mathbf{x}))), \quad (1)$$

where M is a certain metric characterising the difference between two image intensities, and where $\mathbf{x} \in \Omega$ represent a spatial coordinate inside the domain of interest. Such energy function is minimal when the rigid transformation T reduces the overall distance between I_1 and I_2, as computed by the metric M, as much as possible. In practice, however, the transformation of a spatial coordinate $T(\mathbf{x})$ will generally not fall on the image grid, so some interpolation method must be used when computing image intensities $I_2(T(\mathbf{x}))$. In contrast, this is not the case with image intensities $I_1(\mathbf{x})$, which bias the result toward I_1. Hence, registering I_1 with I_2 is not

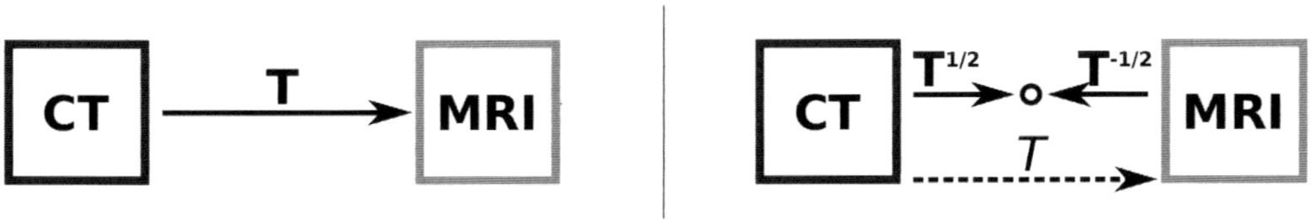

Figure 1. *Left:* classic registration scheme corresponding to (1), *Right:* inverse-consistent registration scheme corresponding to (2).

equivalent to registering I_2 with I_1, and significant differences can be observed in practice, as is demonstrated in section 3.

The inverse consistency of the method can be improved by adopting a fully symmetrical formulation, as follows:

$$E_{\text{Sym}} = \sum_{\forall \mathbf{x} \in \Omega} M(I_1(T^{-1/2}(\mathbf{x})), I_2(T^{1/2}(\mathbf{x}))). \tag{2}$$

In this case, the transformation T is split in two parts using a matrix square root transform such as $T = T^{1/2} \times T^{1/2}$, where $T^{-1/2}$ denotes the matrix inverse of $T^{1/2}$. With this formulation, a spatial coordinate $\mathbf{x}$ is transformed before sampling either I_1 or I_2. Therefore, the two images are given an equivalent treatment, and the formulation is perfectly symmetric and mostly unaffected by a switch in the input image order. The classic and the inverse-consistent registration schemes are represented in Fig. 1.

2.2. Robust image registration via block matching

The algorithm presented in this paper is based on a robust block matching approach that has been demonstrated as being effective at aligning difficult datasets [7]. The algorithm performs a local minimization of the image comparison metric, and iteratively updates a global rigid transformation. It is composed of three main steps:

(i) *Image resampling.* In the first step, the two input images are resampled in a way such that they share a common image grid (see discussion in section 2.3). Tri-linear interpolation is used for both images.

(ii) *Block matching.* After the images have been ressampled, the image domain is divided into an $N_1 \times N_2 \times N_3$ grid of N cubic blocks of size B^3. Then, a block from the grid covering I_1 and centred at $\mathbf{x}$ is matched against the blocks from I_2 that are centred at $\mathbf{x} + \epsilon$, where $\epsilon = [i, j, k]$, $i, j, k \in \{-L, \ldots, -l, 0, l, \ldots, L\} \subset \mathbb{Z}$. The position $\mathbf{x}^\star = \mathbf{x} + \epsilon^\star$ maximizing the block matching metric is kept if and only if: a) the matching score is higher than a small numerical epsilon, and b) the best matching score is significantly higher than the second best. In that case the pair $\{\mathbf{x}, \mathbf{x}^\star\}$ is stored in the set C_1. This process is repeated independently for all blocks of the grid covering I_1, resulting in a set C_1 of at most N matches. The whole block matching process is also repeated for all the blocks from the grid covering I_2, storing the correspondences in the set C_2. At the end of the block matching process, we thus have two sets of correspondences, C_1 and C_2. In this study, we used $B = 9$, $L = 28$, $l = 3$, and the block matching metric was normalized cross-correlation [7].

(iii) *Updating the rigid transformation.* In the last step, the rigid transformation T is updated by using a quaternion representation and a robust process: a) create the set of all correspondences $C = C_1 \cup C_2^-$, where C_2^- correspond to set C_2 but with the $\mathbf{x}$ and $\mathbf{x}^\star$ swapped; b) compute the rigid transformation that best aligns the set of all $\mathbf{x} \in C$ with

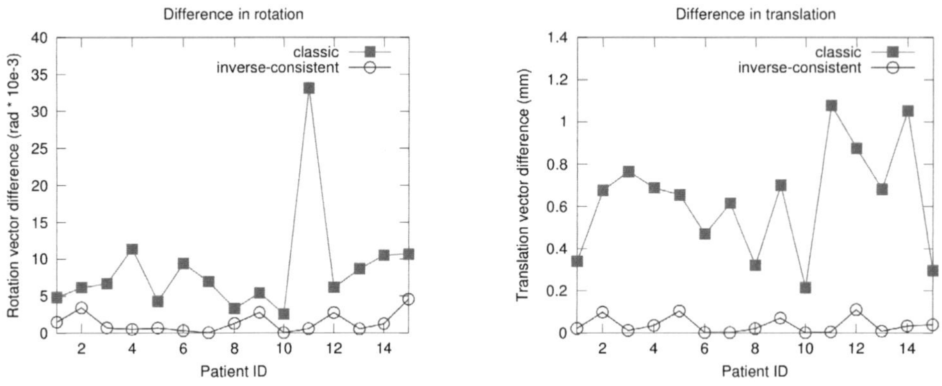

Figure 2. Evaluation of the consistency of the classic and inverse-consistent implementation of the presented registration algorithm. The graphs show the residual between the registration transformations computed using the forward ($MRI \rightarrow CT$) and backward ($CT \rightarrow MRI$) scheme.

the set of all $\mathbf{x}^\star \in C$ [6]; c) create $C^\star$, a subset of C, by computing the distance between the points of all pairs $\{\mathbf{x}, T \cdot \mathbf{x}^\star\} \in C$ and keeping the $\mathtt{size}(C)/2$ points with the smallest distance; and d) iterate steps b) and c) until convergence, but using $C^\star$ in place of C.

The registration algorithm iterates through steps (i), (ii), and (iii) until the computed transformation changes little or until a certain number of iterations have been performed.

2.3. Implementation details

The algorithm described above was implemented within a multiresolution approach. Given the two input images, the software automatically computes three levels of resolution: the first level is defined as the minimum size along each dimension of I_1 and I_2, and subsequent levels are defined by halving the image dimensions of the precedent level that have the highest value (if two dimensions have the same size, both are halved). The algorithm starts from the lowest resolution level (level 3), and then progressively refines the registration transformation at the higher resolution levels. The algorithm was implemented in C++, and the block matching step was parallelized on the CPU. Registering a CT with an MR scan took around 90 sec on a 6 core Intel® Xeon® W3670 at 3.20GHz desktop computer.

3. Experiments and results

Ethics approval for the study protocol was obtained from the local area health ethic committee, and informed consent was obtained from all patients. CT scans were acquired as part of the normal patient radiation therapy process with either a GE LightSpeed radiotherapy large bore scanner with 2.0 mm slices or a Toshiba Acquilion scanner with 2.5 mm slices. MRI scans were T2 scans of type FSE-x1 with a field of view encompassing the entire pelvis. TE was 15ms, TR was 675ms, and the slice thickness was 3.00 mm [5]. For the purpose of this evaluation, 15 CT-MRI scans pairs were randomly selected in the collected dataset.

The presented inverse-consistent registration algorithm (implementing (2)) was evaluated and compared to a classic formulation of the same algorithm (implementing (1)). The two versions of the software were rigorously identical in what concern the multiresolution approach, the image resampling techniques, and the block matching method used, and differed only in the management of the rigid registration transformation, as described in section 2.

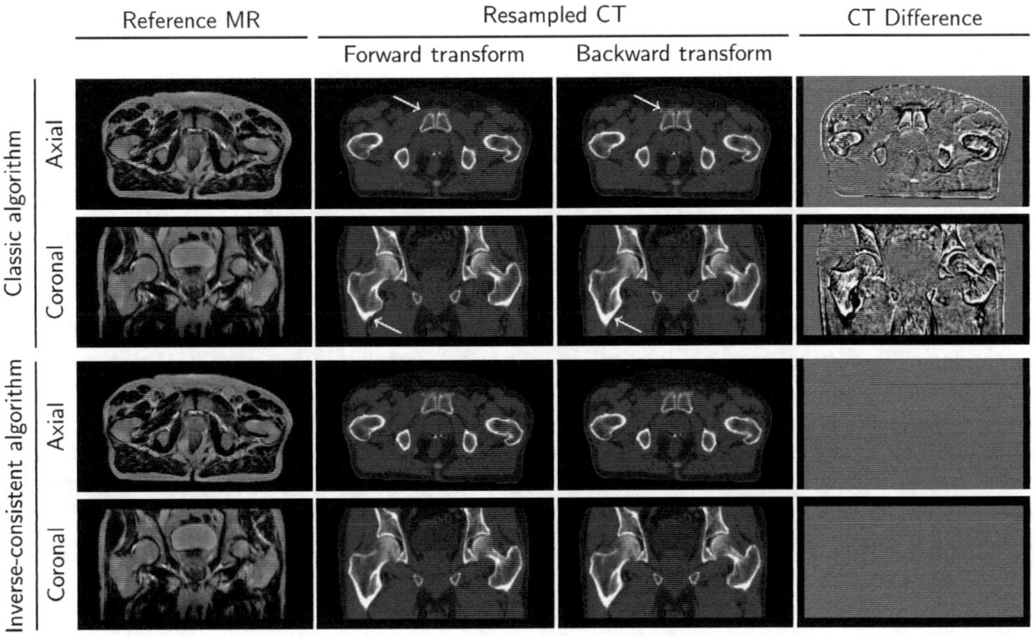

Figure 3. Example of CT–MR multimodal registration for patient 11. *Top:* results obtained with the classic implementation. *Bottom:* results obtained with the inverse-consistent implementation. Annotations indicate some areas of interest. The MR images are displayed for visual reference. The CT images have been resampled into the MR space using either the forward transform or the inverse of the backward transform. The difference images present the result of the subtract of the two CT images; black indicate a value ≤ -50, white a value ≥ 50, and grayscale are linearly interpolated between those two extreme. In the case of the inverse-consistent implementation, the all gray images indicate very small differences.

We evaluated the inverse consistency of the classic and inverse-consistent implementation of the algorithm by conducting two registrations for each dataset, for each implementation: 1) the registration of each MRI scan with the corresponding CT scan from the same patient (*forward registration*), and 2) the registration of each CT scan with the corresponding MRI scan (*backward registration* — the order is inverted). The result of the registration processes is a 4×4 rotation matrix in homogeneous coordinates [4], so the quantification of the difference between the forward and backward registration results need to be split in two parts. The translation component of the two rigid transformations can be compared directly, so that we have

$$\Delta_t = \|t_b - t_f\|,$$

where t_f and t_b are the translation component of the forward and backward rigid registration transformation, respectively. The rotation component of the rigid registration matrices are not linear and thus cannot be compared directly [4]. We thus converted the rotational part of the rigid transformation matrices into the forward and backward rotation vectors r_f and r_b [9]. The difference in rotation is then quantified as

$$\Delta_r = \|r_b - r_f\|.$$

Residuals (Mean±StdDev) of $\Delta_t = 0.63 \pm 0.26$ mm and $\Delta_r = (8.7 \pm 7.3) \times 10^{-3}$ rad were found for the classic implementation with corresponding values of $\Delta_t = 0.038 \pm 0.039$ mm and $(1.6 \pm 1.3) \times 10^{-3}$ rad for the proposed inverse-consistent algorithm, a substantial improvement. Detailed results on a per-patient basis can be seen on Fig. 2, sample qualitative results are presented in Fig. 3.

4. Conclusion

The proposed inverse-consistent algorithm improved the consistency of the rigid registration algorithm by an order of magnitude. However, the results indicate that even the inverse consistent formulation does not lead to totally consistent results at high accuracy. After further investigation in the software, it appears that this limitation is the result of the accumulation of very small numerical errors due to the limited precision of the computer platform used. Nonetheless, we believe that the precision currently achieved is suitable for most clinical applications, especially in those concerning difficult multimodal registration problems, such as in the creation of CT-MR atlases for MRI-alone treatment planning and adaptive MRI-based prostate radiation therapy. The increased registration precision will enhance the generation of pseudo-CT from MRI for atlas based MR planning methods.

Acknowledgments

The authors acknowledge financial support from the Cancer Council NSW (RG 11-05), Prostate Cancer Foundation of Australia (YI2011), Movember and Cure Cancer Australia.

References

[1] Freesurfer, www.freesurfer.net.

[2] B. B. Avants, C. L. Epstein, M. Grossman, and J. C. Gee. Symmetric diffeomorphic image registration with cross-correlation: evaluating automated labeling of elderly and neurodegenerative brain. *Medical Image Analysis*, 12(1):26–41, February 2008.

[3] Y. Berker, J. Franke, A. Salomon, M. Palmowski, H. C. W. Donker, Y. Temur, F. M. Mottaghy, C. Kuhl, D. Izquierdo-Garcia, Z. Fayad, F. Kiessling, and V. Schulz. MRI-based attenuation correction for hybrid PET/MRI systems: a 4-class tissue segmentation technique using a combined ultrashort-echo-time/Dixon MRI sequence. *Journal of Nuclear Medicine*, 53(5):796–804, May 2012.

[4] J. Boisvert, F. Cheriet, X. Pennec, H. Labelle, and N. Ayache. Geometric variability of the scoliotic spine using statistics on articulated shape models. *IEEE Transactions on Medical Imaging*, 27(4):557–68, April 2008.

[5] J. Dowling, J. Lambert, J. Parker, O. Salvado, J. Fripp, A. Capp, C. Wratten, J. W Denham, and P. B. Greer. An Atlas-based Electron Density Mapping Method for Magnetic Resonance Imaging MRI-alone Treatment Planning and Adaptive MRI-based Prostate Radiation Therapy. *International Journal of Radiation Oncology*Biology*Physics*, 83(1):e5–e11, February 2012.

[6] B. K. P. Horn. Closed-form solution of absolute orientation using unit quaternions. *J. Opt. Soc. Am. A*, 4(April):629–642, 1987.

[7] S. Ourselin, A. Roche, G. Subsol, X. Pennec, and N. Ayache. Reconstructing a 3D structure from serial histological sections. *Image and vision computing*, 19(1-2):25–31, 2001.

[8] M. Reuter, H. D. Rosas, and B. Fischl. Highly accurate inverse consistent registration: a robust approach. *NeuroImage*, 53(4):1181–96, December 2010.

[9] D. Rivest-Hénault, H. Sundar, and M. Cheriet. Nonrigid 2D/3D registration of coronary artery models with live fluoroscopy for guidance of cardiac interventions. *IEEE Transactions on Medical Imaging*, 31(8):1557–72, August 2012.

[10] O. Skrinjar, A. Bistoquet, and H. Tagare. Symmetric and transitive registration of image sequences. *International journal of biomedical imaging*, 2008:686875, January 2008.

[11] T. Vercauteren, X. Pennec, A. Perchant, and N. Ayache. Symmetric log-domain diffeomorphic Registration: a demons-based approach. *Medical image computing and computer-assisted intervention : MICCAI 2008*, 11(Pt 1):754–61, 2008.

Tissue motion tracking at the edges of a radiation treatment field using local optical flow analysis

P T Teo[1, 2] and S Pistorius[1,2,3]
[1]CancerCare Manitoba, 675 McDermot Avenue, Winnipeg, Manitoba, R3E 0V9, Canada.
[2]Department of Physics & Astronomy, University of Manitoba, Winnipeg, Manitoba, R3T 2N2, Canada.
[3]Department of Radiology, University of Manitoba, Winnipeg, Manitoba, R3A 1R9, Canada.

Email: troy.teo@cancercare.mb.ca

Abstract. This paper investigates the feasibility and accuracy of tracking the motion of an intruding organ-at-risk (OAR) at the edges of a treatment field using a local optical flow analysis of electronic portal images. An intruding OAR was simulated by modifying the portal images obtained by irradiating a programmable phantom's lung tumour. A rectangular treatment aperture was assumed and the edges of the beam's eye view (BEV) were partitioned into clusters/grids according to the width of the multi-leaf collimators (MLC). The optical flow velocities were calculated and the motion accuracy in these clusters was analysed. A velocity error of 0.4 ± 1.4 mm/s with a linearity of 1.04 for tracking an object intruding at 10mm/s (max) was obtained.

Keyword: adaptive image-guided radiotherapy, organ intrusion detection, optical flow

1. Introduction

Multi-leaf collimators with leaf positions which are dynamically modified to follow the motion of the target have been proposed [1, 2]. To eliminate errors between the correlation of internal and external surrogates, tracking using electronic portal images (EPI) during beam delivery has been studied [2, 3]. Various marker-less techniques have been proposed for real-time tracking of targets with EPI. Most of these methods require either prior knowledge of the target object or segmentation of the target, such that the delineated contours on the initial image can be used for tracking on subsequent images. To ensure sufficient radiation is delivered to the tumor, the emphasis of these methods is on tracking the tumor. However, studies have shown that discrepancies in the entire dose distribution can arise if the delivery of intensity modulated radiation therapy (IMRT) disregards the motion of the target and normal tissues [4, 5]. For example, during the irradiation of the mammary tissues of breast cancer patients, the ipsilateral lung and heart are often exposed to non-negligible dose. Although the breathing adapted radiotherapy and deep inspiration breath hold technique have been shown to reduce cardiac exposure, some patients were unable to comply with the requirements [6], resulting in unplanned intrusion of OARs into the treatment field. With setup errors and daily variations in organ motion, it is difficult to consistently maintain the OAR outside of the planned radiation field.

Detection and tracking of motion close to the leaf edges is challenging. This is partially due to the difficulty in identifying all OARs that could intrude into the treatment beam. As a result, current tracking methods based on apriori information such as template matching might not find an exact match of an intruding OAR. It has been shown that a dynamically-weighted optical flow algorithm can track the global motion of an un-contoured moving target on EPI with an accuracy of 0.4 ± 0.2 mm[7].

This paper aims to investigate the feasibility and accuracy of using a local variant of the optical flow analysis to track motions at the edges of a treatment field without the need to delineate OARs.

2. Methods and Materials

2.1. Image acquisition

A phantom's tumor target placed at a distance of 100 cm from the source was irradiated using a Varian Linac 2100iX 6 MV beam at a dose rate of 600 MU min^{-1}. The target was programmed to move in a periodic pattern with a maximum velocity of 10 mm s^{-1} as described in [7]. A total of 237 EPI, each with an image resolution of 0.391mm/pixel, were collected with a frame rate of 7.5 frames s^{-1}. A duplicate of the image sequence was flipped and concatenated to the original to simulate two objects traveling in the opposite direction and with the same velocity. To simulate the intrusion of objects into the treatment field, a region of 235 pixels by 178 pixels was cropped from the concatenated image matrix, leaving a portion of the flipped target out of the cropped image view (Figure 1).

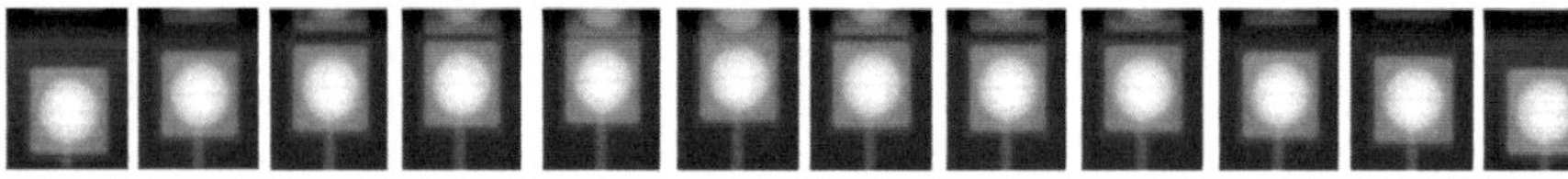

Figure 1. Images with simulated intrusion of object.

2.2. Local optical flow analysis

In this proof-of-concept study, a rectangular treatment aperture is assumed to be bounded by MLC leaf pairs at the top and bottom of the cropped image frame (figure 2). ***Step 1(figure 2):*** The optical flow algorithm was applied to consecutive image pairs to detect and track motions within the treatment aperture. To analyse the local motions near the edges of the treatment aperture, the top and bottom regions were partitioned with horizontal and vertical grids, creating clusters of optical flow vectors. ***Step 2:*** Since each MLC leaf has a width of 3.55 mm at the EPID which is 142.1 cm from source, a nine pixel wide column with a size of 3.52 mm is used for each vertical cluster. With an image size of 178 pixels horizontally, the edge of the image is partitioned into 20 vertical columns. ***Step 3.*** For this work each cluster was 9 × 9 pixels in size. Changing the number of pixels within a horizontal cluster would produce a different cluster size that yields different motion estimation. ***Step 4.*** The mean and standard deviation of the computed motion were compared as a function of position and velocity. While the tumor is always moving within the treatment aperture, the intruding object, i.e. OAR, only appears in the treatment aperture for a short period of time. The accuracy of tracking the intruding OAR at the edges was obtained by comparing the calculated results with the potentiometer readings when the OAR is present in the image.

3. Results

The estimated velocities for the 20 vertical clusters were observed to be symmetrical about the central column. By way of example, figure 3(a) depicts the velocities of the intruding OAR for the cluster located one row from the top edge of the treatment field and centred horizontally in the 11th vertical column, as shown in Figure 2. Figure 3(b) plots the differences (errors) between the optical flow estimation and potentiometer measurement. The errors between the potentiometer and the computed velocity vary around a mean of zero. A mean error of 0.3± 1.3mm s^{-1} was obtained for the entire image sequence. Figure 3(c) depicts the velocities of the intruding OAR determined for the cluster that is located on the 11th vertical column and two rows away from the edge of the treatment field. A mean error of 0.4± 1.3mm s^{-1} was obtained (figure 3(d)). Figure 4 plots the linear regression between the measured and computed velocities. A regression coefficient of 0.8 and 1.0 were obtained for the clusters located in row 1 and row 2 respectively.

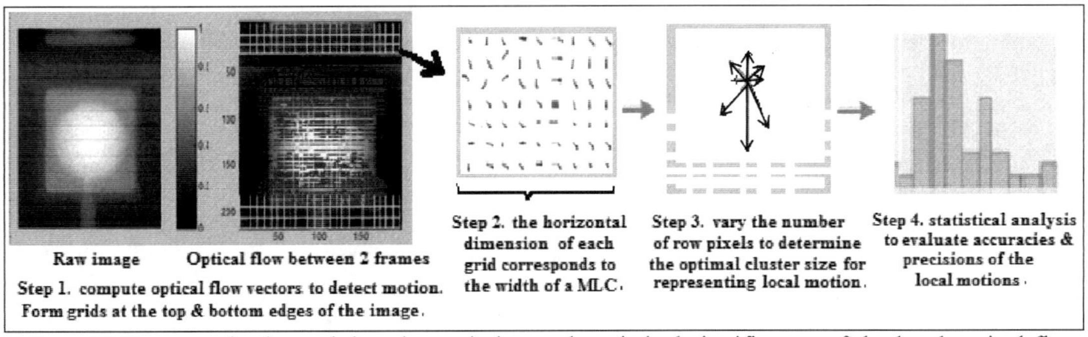

Figure 2. Key steps in determining the variation and statistical significance of the local optical flow vectors at the edges of a radiation treatment field. For this work, each cluster was 9 × 9 pixels in size. Analysis was performed for each of the clusters as a function of position and velocity.

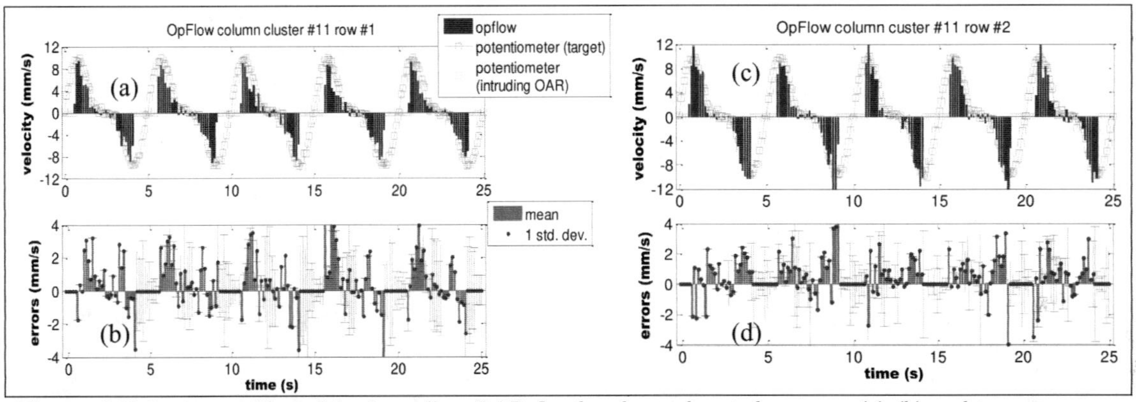

Figure 3. Velocity profiles of the intruding OAR for the cluster located one row (a)-(b) and two rows (c)-(d) from the top edge of the treatment field and centred horizontally in the 11th vertical column. (a) and (c) plots the velocity trajectories; (b) and (d) plots the differences (errors) between the optical flow (opflow) estimation and potentiometer measurement.

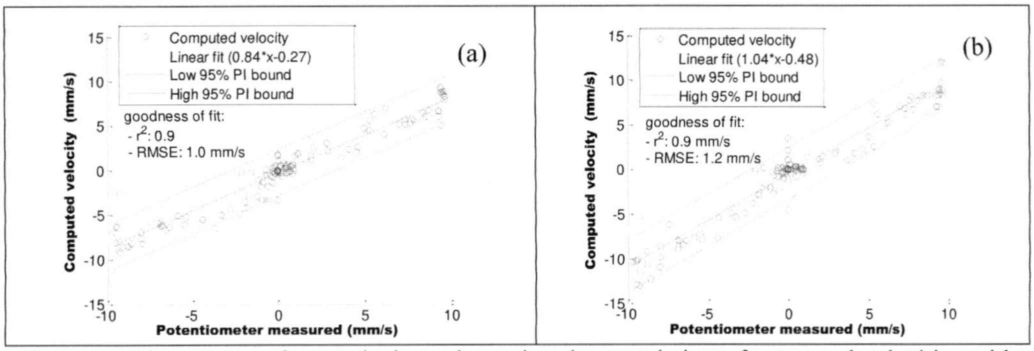

Figure 4. Linear regression analysis to determine the correlation of measured velocities with that detected by the optical flow algorithm for the11^{th} vertical cluster of the 1st row (a) and 2nd row away from the top edges of the treatment aperture respectively.

4. Discussions

Figure 3(b) and 3(d) illustrate that the optical flow tends to overestimate the velocity of the intruding OAR. The linearity between the computed and measured velocity improves for clusters that are further away from the edges (Figure 5). The detection of motion in the first row of clusters can be used to trigger the motion of the MLC leaves to adapt to an incoming OAR. Analysis of the 2nd cluster of pixels can further increase the accuracy with which the direction and magnitude of intrusion is determined.

Further work is required to define the optimal cluster size that would provide an efficient tradeoff between the aperture problem (high motion uncertainties associated with small detection window), noise and the systematic errors due to averaging of larger areas. The combination of local and global tracking [7] to account for main target motions and unplanned intrusions at the edges of a treatment field will be further studied.

For clinical implementation, several challenges can be identified. Firstly, the contrast of the patient anatomy will degrade with respect to the contrast of our phantom. However, the increase in errors, between using a weighted average optical flow to track high contrast objects and objects with a CNR = 5 (which is typical for a clinical portal image) at a velocity of 12 mm/s, is 0.25 mm/s [7]. This is within the uncertainties of this work, and thus the increase in errors for lower contrast clinical images is expected to be minimal. However the patient anatomy will be relatively inhomogeneous in comparison to our phantom object and this structural noise may improve the tracking accuracy. In clinical systems, latencies may be encountered. To realize an adaptive system based on real-time tracking at the treatment edges, prediction algorithms such as the neural network currently being studied by our group, must be implemented.

5. Conclusions

Local optical flow analysis yields a velocity error of 0.4 ± 1.4 mm s^{-1} with a linearity of close to 1 for tracking an object intruding at about 10mm s^{-1} (max). This compares to 0.0 ± 0.5 mm s^{-1} (yielding a mean positional error of 0.4 ± 0.2 mm) for global average tracking. The tradeoff between precision and accuracy can still be improved by optimizing the cluster size. This work demonstrates the feasibility of automatic detection and tracking of intrusion without any pre-defined OAR template or learning data.

References

[1] Keall P J, Kini V R, Vedam S S and Mohan R 2001 Motion adaptive x-ray therapy: a feasibility study *Phys. Med. Biol.* **46** 1–10

[2] Keall P J, Cattell H, Pokhrel D, Dieterich S, Wong K H, Murphy M J, Vedam S S, Wijesooriya K and Mohan R 2006 Geometric accuracy of a real-time target tracking system with dynamic multileaf collimator tracking system *Int. J. Radiat. Oncol. Biol. Phys.* **65** 1579–84

[3] Lin T, Cervino L I, Tang X, Vasconcelos N and Jiang S B 2009 Fluoroscopic tumor tracking for image-guided lung cancer radiotherapy *Phys. Med. Biol.* **54** 981-92

[4] Webb S 2006 Motion effects in (intensity modulated) radiation therapy: a review, *Phys. Med. Biol.* **51** R403-25

[5] McMahon R, Papiez L and Sandison G 2007 Relative motion of tumors and normal tissue during dynamic MIC tracking delivery *Australas. Phys Eng. Sci. Med.* **30 (4)** 331-6

[6] Nissen H D and Appelt A L 2013 Improved heart, lung and target dose with deep inspiration breath hold in a large clinical series of breast cancer patients *Radiother Oncol.* **106(1)** 28-32

[7] Teo P T, Crow R, Van Nest S, Sasaki D and Pistorius S 2013 Tracking lung tumour motion using a dynamically weighted optical flow algorithm and electronic portal imaging device *Meas. Sci. Technol* **24** 074012 (15pp)

Journal of Physics: Conference Series **489** (2014) 012041 doi:10.1088/1742-6596/489/1/012041

Establishing tumour tracking accuracy in free-breathing respiratory gated SBRT of lung cancer

Chuan-Dong Wen[1,2], C Wong[1], T Ackerly[1], J Ruben[1] and J Millar[1]

[1] William Buckland Radiotherapy Centre, The Alfred Hospital, Melbourne, VIC 3004 Australia

Abstract. Free-breathing respiratory gated SBRT of surgically inoperable lung cancer has been clinically commissioned. This study was to establish the tumour tracking accuracy under clinical conditions based on an implanted fiducial marker. A Visicoil™ marker embedded in tissue-equivalent material mounted in a phantom (ET Gating Phantom™ Brainlab) driven by a patient's breathing data was treated with the ExacTrac™ system. This one-dimensional moving marker represented a tumour motion in superior-inferior (S-I) direction measured through 4DCT study of the same patient. Both Gafchromic™ films and the stereoscopic kV images were used for tracking the position of the marker. For tumour motion at magnitudes of 10, 20 and 29 mm and treated with corresponding gate widths of 50%, 33% and 20% of free breathing amplitude, the implanted marker was able to be tracked with a deviation $\leq$1.53 mm to its planned position.

1. Introduction:

The goal of respiratory gated radiation therapy is to reduce the volume of healthy tissue irradiated while maintaining full coverage of a moving planned target volume (PTV). Hence, it is important to accurately measure and track tumours at sites in the thorax and abdomen affected by respiratory motion during radiotherapy planning and treatment with external beam [1]. Tenn *et al.* investigated the targeting accuracy of an image-guided gating technique for stereotactic body radiotherapy (SBRT) using phantoms [2]. However, that study simplified the clinical situation somewhat because it used a sinusoidal pattern to represent the patient's breathing trace. Clinically an actual breathing trace is not sinusoidal but complex, irregular and varied in both amplitude and period. This study examines the tracking accuracy based on a fiducial marker implanted in the gross tumour volume for respiratory

[2] Corresponding author at email: C.Wen@wbrc.org.au

gated delivery to a lung tumour in a commercial breathing phantom driven by the superior-inferior component of an actual patient's breathing trace.

2. Method:

A Visicoil™ 0.75 mm diameter gold coil marker was embedded in a Perspex plate that was mounted on ET Gating Phantom™ (Brainlab AG Germany) and moved along the superior-inferior (S-I) direction of the patient to simulate the predominant respiratory motion of a lung tumour (figure 1). A Gafchromic™ RT QA film was placed underneath the marker to record its position when a beam was delivered. To enhance the marker projection on the film, a ball bearing was added between the marker and the film.

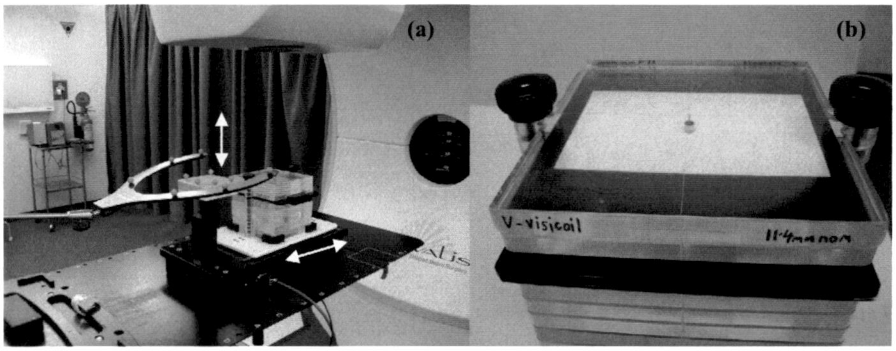

Figure 1. The setup of phantom and fiducial marker: (**a**) The block was moving in the S-I direction while the skin markers were moving up-and-down and the reference remained still; (**b**) A Visicoil™ embedded in a Perspex plate above a ball bearing and film placed in a cassette (black).

The horizontal motion of the phantom was obtained from the corresponding component of the lung tumour motion while the vertical motion was measured from the same patient's external chest. The magnitudes of both motion ranges were scaled to reflect the possible clinical measurements. The phantom was irradiated on a Novalis™ Classic linac equipped with ExacTrac™ (Brainlab AG Germany) robotic couch and in-room stereoscopic kV imager. During the setup and delivery

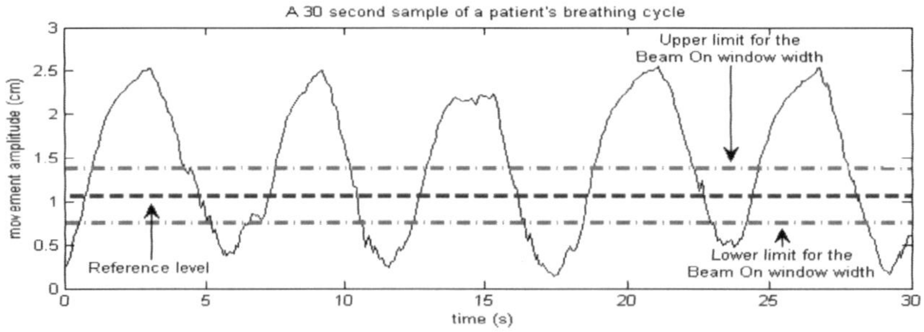

Figure 2. A 30-second sample of a patient's representative breathing trace. The beam-on gate was used in the region between the dash-dot lines. Beam-on gate widths of 50%, 33% and 20% of the average full breath window were utilised, centred on a reference level (dashed line).

processes, the gold marker was moving within the vicinity of the beam isocentre and its position was verified via X-ray images acquired by the ExacTrac™ kV imager. To enable the verification images to be acquired at the beginning and end of the beam-on gate, the gate was placed at the end of exhale but did not include the trough of this breathing trace. The displacement of the marker from the isocentre could be displayed and was used to define the MV beam-on gate width as illustrated in figure 2. In this study, the marker was set at isocentre at reference level of 33% of magnitude above the trough. The MV beam-on gate widths of 50%, 33% and 20% of full amplitude were used to satisfy the clinical requirements when treating a tumour with a relatively small, medium or large motion range respectively.

A 6MV MLC shaped beam from a clinical plan was delivered at gantry angle zero to the Visicoil™ marker positioned at isocentre in the phantom. Both non-gated motionless treatment and gated treatments at three motion ranges of 10, 20 and 29 mm were delivered sequentially. A dose of 200 MU at 800 MU/min. was delivered in each treatment. The spatial positions of the marker were recorded on films located immediately beneath. Figure 3 shows the motionless and gated films and the vertical dose profiles over the marker at the field centre.

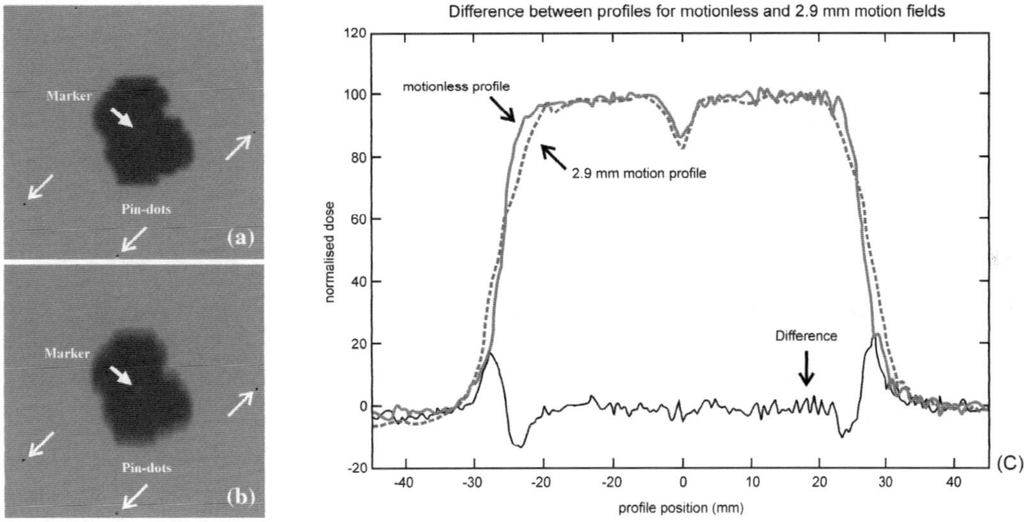

Figure 3. Visicoil™ marker projections in: (**a**) Motionless film; (**b**) Film of motion range of 29 mm and gate width of 33% of full amplitude; (**c**) Plot of vertical profiles showing displacement between motionless marker/isocentre (thick) and averaged position (dotted) in the gated field.

By using the technique described by Tenn *et al.*[2], films from both the motionless and the motion-gated deliveries were scanned and co-registered with pin-dots (shown in figure 3(**a**) and 3(**b**)) before a vertical profile of relative dose along the film motion direction was generated over the marker (ball bearing) projection. In the centre of these profiles, the separation in millimetres between local minimums was the displacement between marker positions on the gated and the motionless films, as displayed in figure 3(**c**).

3. Results:

The measured displacements of fiducial marker projections on films between gated fields and motionless field where the marker was at machine isocentre for three tumour motion ranges of 10 mm, 20 mm and 29 mm, at three gate widths of 20%, 33% and 50% of the full respiration amplitude are displayed in Table 1. The beam was only delivered in exhalation phase. Since the film was placed in a cassette below the marker and the ball bearing, a correction factor to scale the films to the isocentre level has been applied to the result.

Table 1. Measured displacements of Visicoil^{TM} to isocentre in S-I direction only for gated deliveries using 1D component of patient breathing trace.

Gate width in percentage of full respiration amplitude	Marker/tumour motion magnitude (mm)		
	10 mm	20 mm	29 mm
20%	1.53	1.53	0.0
33%	1.02	1.53	0.51
50%	0.0	0.51	1.02

4. Discussions:

The results in Table 1 above are mainly due to random distribution in multiples of 0.51 mm as the bin size in vertical profiles is 0.51 mm. The maximum deviation from the ideal tumour tracking is 1.53 mm, which is larger than the uncertainty of 0.51 mm arising from binning. Due to the limitation in this study that the phantom can move in one dimension, the fiducial marker (tumour) motion was examined along the superior/inferior direction only. With the same system a 3D accuracy of 1.7 mm was reported by Willoughby *et al.* from a phantom study [3]. Seppenwoolde *et al.* reported that thoracic tumour motion was greatest along the superior-inferior direction [4], hence the obtained result should reflect the maximal tracking uncertainty. However, these results indicate only a threshold of accuracies measured from a single clinical case and the breathing data will vary for each patient. Further study is therefore needed using the tumour motion data of more patients and more sites as well as taking into account the tumour hysteresis effect. A sub-millimetre systematic accuracy for stereotactic radiosurgery of cranial tumour using the same equipment has been reported by Ackerly *et al.* [5]. By following this method, the localisation accuracy in SBRT of tumours affected by respiration can be established. This accuracy is in line with that previously reported and is useful when considering margins for clinical treatment.

5. Conclusions:

With the method used in this clinical simulation study, the implanted fiducial marker in phantom can be tracked in the S-I direction with a deviation of equal to or less than 1.53 mm from its planned position during free-breathing respiratory gated deliveries of 20%, 33% or 50% of full expiration amplitude generated from skin surface surrogate of a real clinical case, for tumour moving in the range up to 29 mm.

References:

[1] Keall P J, Mageras G S, Balter J M, 2006 The management of respiratory motion in radiation

oncology report of AAPM Task Group 76 *Med. Phys.* **33(10)** 3874–900.

[2] Tenn S E, Solberg T D and Medin P M 2005 Targeting accuracy of an image guided gating system for stereotactic body radiotherapy *Phys. Med. Biol.* **50** 5443–62

[3] Willoughby T R, Forbes A R, Buchholz D, Langen K M, Wagner T H, Zeidan O A, Kupelian P A and Meeks S L 2006 Evaluation of an infrared camera and X-ray system using implanted fiducials in patients with lung tumors for gated radiation therapy *Int. J. Radiat. Oncol. Biol. Phys.* **66** 568-75

[4] Seppenwoolde Y, Shirato H, Kitamura K, Shimizu S, van Herk M, Lebesque J V and Miyasaka K 2002 Precise and real-time measurement of 3D tumor motion in lung due to breathing and heartbeat, measured during radiotherapy *Int. J. Radiation Oncology Biol. Phys.* **53(4)** 822–34.

[5] Ackerly T, Lancaster C M, Geso M and Roxby K J 2011 Clinical accuracy of ExacTrac intracranial frameless stereotactic system: paradigm shift to fractionation *Med. Phys.* **38 (9)** 5040-8

Computational Model to Simulate the Interplay Effect in dynamic IMRT delivery

SA Yoganathan[1], KJ Maria Das[2] and Shaleen Kumar[2]
[1] Department of Radiation Oncology, Dr. Ram Manohar Institute of Medical Sciences, Lucknow, Uttar Pradesh, India.
E-mail: sayoganathan@yahoo.co.in
[2] Department of Radiotherapy, Sanjay Gandhi Postgraduate Institute of Medical Sciences, Lucknow, Uttar Pradesh, India.

Abstract. The purpose of this study was to develop and experimentally verify a patient specific model for simulating the interplay effect in a DMLC based IMRT delivery. A computational model was developed using MATLAB program to incorporate the interplay effect in a 2D beams eye view fluence of dynamic IMRT fields. To simulate interplay effect, the model requires two inputs: IMRT field (DMLC file with dose rate and MU) and the patient specific respiratory motion. The interplay between the DMLC leaf motion and target was simulated for three lung patients. The target trajectory data was acquired using RPM system during the treatment simulation. The model was verified experimentally for the same patients using Imatrix 2D array device placed over QUASAR motion platform in CL2100 linac. The simulated fluences and measured fluences were compared with the TPS generated static fluence (no motion) using an in-house developed gamma evaluation program (2%/2mm). The simulated results were well within agreement with the measured. Comparison of the simulated and measured fluences with the TPS static fluence resulted 55.3% & 58.5% pixels passed the gamma criteria. A patient specific model was developed and validated for simulating the interplay effect in the dynamic IMRT delivery. This model can be clinically used to quantify the dosimetric uncertainty due to the interplay effect prior to the treatment delivery.

1. Introduction

Intensity Modulated Radiation Therapy (IMRT) is a state of art treatment modality in many complex sites. IMRT technique is an effective tool to achieve dose escalation with high conformal dose distribution around the target. Dose escalation is an essential process for lung cancers to increase the survival rate. But, the delivery of IMRT in lung is not considered in most of the clinics due to the respiratory motion of target which produces dosimetric errors in the delivery, especially when using Dynamic Multi Leaf Collimator (DMLC). These dosimetric errors are caused by blurring of dose and interplay between tumour motion and DMLC motion. The interplay motion can be divided into two orthogonal motions; parallel interplay motion where the target motion is parallel to DMLC motion; perpendicular interplay motion where the target motion is perpendicular to DMLC motion.

Various studies investigated the dosimetric uncertainties due to the intra-fractional respiratory motion of tumour. Most of the studies concluded that the respiratory motion induced dosimetric uncertainties were larger for single filed or fractions and averaged out over the entire fractions [(1-5)]; whereas one may expect significant errors in small fraction treatments such as stereotactic body radiation therapy (SBRT). However, this effect may be different for different patient respiratory parameters such as breathing amplitude, phase and period. Advanced techniques such as gating and tracking [(6-13)] can be employed to reduce these dosimetric uncertainties. Use of these techniques is encouraged only if considerable dosimetric uncertainties are expected over the entire treatment fractions. In order to have an idea about the dosimetric uncertainties due to interplay effect, a patient specific model is necessary.

Therefore the purpose of this work was to develop a patient specific computational model to simulate the parallel interplay effect in the delivery of dynamic IMRT.

2. Materials & Methods

2.1 Computational Model

A computational model was developed to simulate the dosimetric uncertainty introduced by parallel interplay effect in a two dimensional (2D) beams eye view (BEV) fluence of a dynamic IMRT field. This model was implemented in MATLAB (ver 7.0). The model requires two inputs; dynamic IMRT field (along with MU and dose rate information) and patient specific respiratory motion of target at BEV. Model extracts MLC leaf positions and fractional MU of each control point from the inputted DMLC file and by using this information original velocity (V_{MLC}) of each DMLC pair was calculated. The velocity of moving target (V_{tar}) was calculated using the patient specific respiratory motion. The interplay motion of target was incorporated into DMLC leaf motion by modifying the velocity of DMLC leaf pair (V_{Res}) as follows

$$V_{Res} = \begin{cases} V_{MLC} - |V_{tar}| & \text{for } V_{tar} > 0 \\ V_{MLC} + |V_{tar}| & \text{for } V_{tar} < 0 \end{cases}$$

The model subtracts the velocity of the moving target (V_{tar}) with V_{MLC} during the forward (positive) direction of target motion (i.e. both the target and DMLC are moving in the same direction). Conversely, the program adds the V_{tar} with the V_{MLC} during backward (negative) direction of target motion (i.e. both the target and DMLC are moving in opposite direction). During the forward target motion, the resultant direction of leaf motion is: towards backward when V_{tar} is greater than V_{MLC} (V_{Res} = negative) and towards forward when V_{tar} is less than V_{MLC} (V_{Res} = positive). The model finally derives the 2D BEV fluence for this modified DMLC file at isocenter.

2.2 Experimental Verification:

To demonstrate the functionality of this model, the interplay effect was simulated for dynamic IMRT plans of three lung patients. These plans had seven fixed fields and generated for a total dose of 60Gy in 30 fractions with 6MV X-ray photon beam at a dose rate of 400 MU/min using a commercial treatment planning system (Eclipse™, Varian Medical Systems, Palo Alto, CA). The DMLC leaf motions were calculated using leaf motion calculator (LMC- ver 8.6). Patient respiratory motion was acquired using Real-time position management (RPM) system (Varian Medical Systems, Palo Alto, CA). The RPM system uses a marker block and infrared camera to capture the chest movement of patient. It was assumed that the external surrogate - marker block motion simulated the internal tumour motion. Figure 1 shows the respiratory motion of a patient acquired using RPM system. Using the RPM generated file, the velocity of target along the DMLC motion direction was derived. LMC calculated DMLC leaf motions along with MU and dose rate information were imported into the model and the velocity of the each DMLC leaf pair was modified. The 2D BEV fluence at isocenter was derived using the modified DMLC file. This process was repeated for each patient.

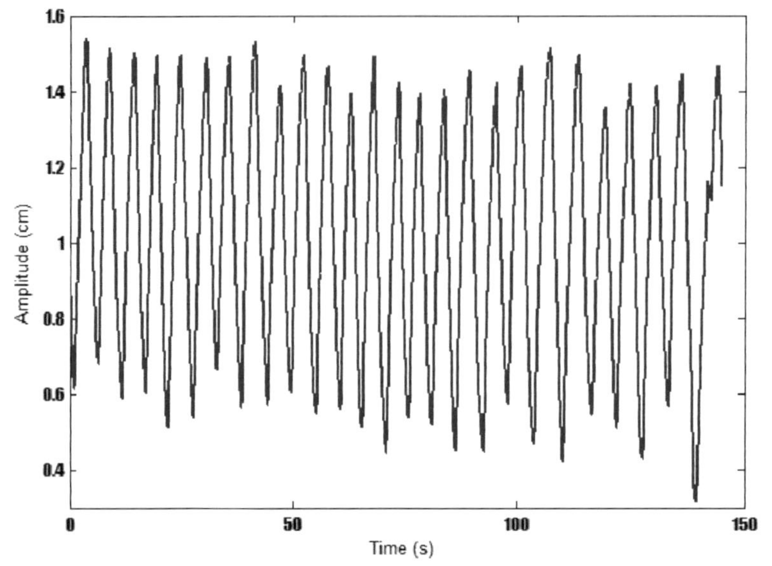

Figure 1. A patient respiratory motion acquired using RPM system.

To test the accuracy of this model, the experimental verification measurements were carried out in a medical linear accelerator (CL2100CD, Varian Medical Systems, Palo Alto, CA) equipped with millennium 120 DMLC system. The experimental setup of dosimetric verification is shown in Figure 2. The original dynamic IMRT plan was delivered on a moving Imatrixx 2D ion chamber array device (Scanditronix Wellhöfer, Freiberg, Germany) which was sandwiched between perspex slab phantoms. QUASAR motion platform (Modus Medical Devices Inc., Canada) was used to move the Imatrixx device. The RPM generated patient respiratory motion file was imported into the QUASAR motion platform software to simulate the patient specific motion. In order to simulate the parallel interplay effect, the Imatrixx device motion direction was kept along the DMLC leaf motion.

Figure 2. The experimental setup of dosimetric verification

The model simulated fluence and experimentally measured fluences were compared with the TPS static fluence using an in-house developed gamma evaluation program with 2% dose difference and 2mm distance to agreement criteria.

Journal of Physics: Conference Series **489** (2014) 012042 doi:10.1088/1742-6596/489/1/012042

3. Results & Discussion

Figure 3 show the 2D BEV fluence of a static IMRT field exported from TPS. Figure 4 show the computational model simulated 2D BEV fluence (interplay effect introduced). Figure 5 shows the gamma comparison of static fluence with model simulated delivered fluence.

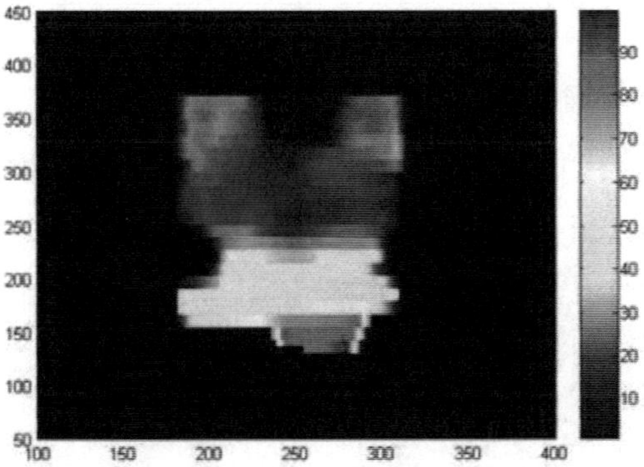

Figure 3. 2D BEV fluence of a static IMRT field

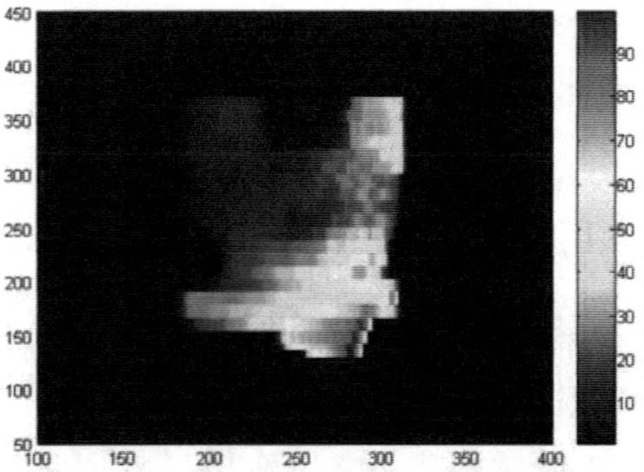

Figure 4. Delivered fluence simulated by the model

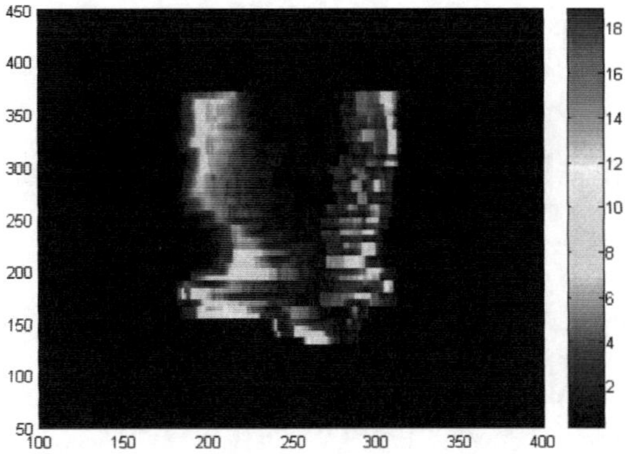

Figure 5. Gamma comparison of static fluence with the model simulated delivered fluence.

The simulated results were well agreement with the measured (table 1). Average percentage of pixels passed the gamma criteria were 59.3% & 57.6% for simulated and measured fluences.

Table 1. Percentage of pixels passed the gamma criteria 2% & 2mm

Patient Number	Measured	Simulated
1	55.3 %	58.5 %
2	53.3 %	59.2 %
3	64.2 %	60.2 %

Studies were already attempted to develop automatic computational models to simulate the interplay effect in dynamic treatment delivery. Mohn etal [14] developed a model for evaluating the dosimetric effects of interplay effect in dynamic IMRT dose delivery. Their model was able to calculate the three dimensional (3D) dose distributions both with and without the interplay effect. Jensen etal [15] developed a 4D computational tomography (CT) Monte Carlo simulation method for accounting the respiratory motion induced interplay effect in step-shoot IMRT using the MLC controller delivery log files. Rao etal [16] studied the dosimetric impact of respiratory motion in SBRT; they used deformable image registration to calculate the 4D dose distributions. In this present study, we also developed a computational model for simulating the dosimetric effect of interplay in dynamic IMRT delivery. Instead of using the 3D/4D CT data set, we directly modelled the interplay effect in the 2D BEV fluence. This method provides a reasonable estimation of dosimetric uncertainties due to interplay effect with trivial calculation time. The proposed model could simulate only parallel interplay effect, but for complete uncertainty evaluation 3D model is required.

4. Conclusion

A patient specific model was developed and validated for simulating the interplay effect in the dynamic IMRT delivery. This model can be clinically used to quantify the patient specific dosimetric uncertainty due to the interplay effect prior to the treatment delivery.

5. Reference

[1] Pemler P, Besserer J, Lombriser N, Pescia R, Schneider U. 2001 Med Phys. 28(11); 2234–40.
[2] Hugo G, Tenn S, Agazaryan N. Med Phys. 2003; 30:1470.
[3] Chen GTY, Kung JH, Beaudette KP. 2004; 14(1):19–26.
[4] Mechalakos J, Yorke E, Mageras GS, et al Radiat. Oncol. 2004; 71(2):191–200.
[5] Berbeco RI, Pope CJ, Jiang SB. J. Appl. Clin. Med Phys. 2006; 7(4):33–42.
[6] Ohara K, Okumura T, Akisada A, et al. 1989; 17(4):853–57.
[7] Kubo HD and Hill BC. Phys Med Biol. 1996; 41(1):83–91.
[8] Keall PJ, Mageras GS, Balter JM, et al. Med Phys. 2006; 33(10):3874–900.
[9] D'Souza WD, Naqvi SA, Yu CX. Phys Med Biol. 2005; 50(17):4021–33.
[10] Schweikard A, Glosser G, Bodduluri M, Murphy MJ, Adler JR. Computer Aided Surg. 2000; 5(4):263–77.
[11] Keal PJ, Kini VR, Vedam SS, Mohan R. Phys Med Biol. 2001; 46(1):1–10.
[12] SA Yoganathan, Maria Das KJ, Agarwal A, Senthil Kumar SK, Velmurugan J, Kumar S. In: Long M, editor. IFBME Proceedings; 2013; 39:1923–25.
[13] SA Yoganathan, Maria Das KJ, Agarwal A, Kumar S. J. Appl. Clin. Med Phys. 2013; 14(3):39-51.
[14] Mohn S, Wasbo E: Simulation of respiratory motion during IMRT dose delivery. Acta Oncol 50:935-43, 2011
[15] Jensen MD, Abdellatif A, Chen J, et al: Study of the IMRT interplay effect using a 4DCT Monte Carlo dose calculation. Phys Med Biol 57:N89-99, 2012.

[16] Rao M, Wu J, Cao D, et al: Dosimetric impact of breathing motion in lung stereotactic body radiotherapy treatment using image-modulated radiotherapy and volumetric modulated arc therapy. Int J Radiat Oncol Biol Phys 83:e251-6, 2012

Acknowledgments
The authors would like to acknowledge the assistance received from the Department of Science & Technology, Grant No: IR/SO/LS 02/2003, Government of India.

XVII International Conference on the Use of Computers in Radiation Therapy (ICCR 2013) IOP Publishing
Journal of Physics: Conference Series **489** (2014) 012043 doi:10.1088/1742-6596/489/1/012043

GPU-accelerated automatic identification of robust beam setups for proton and carbon-ion radiotherapy

F Ammazzalorso[1,2], T Bednarz[3] and U Jelen[1]

[1] Department of Radiotherapy and Radiation Oncology - Particle Therapy Center, University of Marburg, Marburg, Germany
[2] Department of Medical Physics in Radiation Oncology, German Cancer Research Center (DKFZ), Heidelberg, Germany
[3] CSIRO Mathematics Informatics & Statistics, Sydney, Australia

E-mail: filippo.ammazzalorso@uni-marburg.de

Abstract. We demonstrate acceleration on graphic processing units (GPU) of automatic identification of robust particle therapy beam setups, minimizing negative dosimetric effects of Bragg peak displacement caused by treatment-time patient positioning errors. Our particle therapy research toolkit, RobuR, was extended with OpenCL support and used to implement calculation on GPU of the Port Homogeneity Index, a metric scoring irradiation port robustness through analysis of tissue density patterns prior to dose optimization and computation. Results were benchmarked against an independent native CPU implementation. Numerical results were in agreement between the GPU implementation and native CPU implementation. For 10 skull base cases, the GPU-accelerated implementation was employed to select beam setups for proton and carbon ion treatment plans, which proved to be dosimetrically robust, when recomputed in presence of various simulated positioning errors. From the point of view of performance, average running time on the GPU decreased by at least one order of magnitude compared to the CPU, rendering the GPU-accelerated analysis a feasible step in a clinical treatment planning interactive session. In conclusion, selection of robust particle therapy beam setups can be effectively accelerated on a GPU and become an unintrusive part of the particle therapy treatment planning workflow. Additionally, the speed gain opens new usage scenarios, like interactive analysis manipulation (e.g. constraining of some setup) and re-execution. Finally, through OpenCL portable parallelism, the new implementation is suitable also for CPU-only use, taking advantage of multiple cores, and can potentially exploit types of accelerators other than GPUs.

1. Introduction

The finite range of protons and carbon ions (Bragg peak) contributes to their dose deposition precision, unmatched among external beam radiotherapy modalities [1], but also sets high requirements on patient setup reproducibility. Displacement of density interfaces (e.g. air-bone, common in the cranial region) can introduce changes in particle penetration depth, with consequent adverse dosimetric effects [2,3,4]. These may have extent larger than the positioning errors that caused them and may not be fully counteracted by the use of safety margins, a situation unknown to the more conventional treatment with high energy photons.

A possible solution, which well complements accurate patient immobilization and image-guided position verification, is the application of robust treatment planning approaches, for instance the choice of irradiation directions that minimize the potential effects of patient mispositioning, by avoiding severe inhomogeneities across entrance channels [2,3].

In real patient geometries the choice of robust beam setups is typically not straightforward and its verification through extensive dose computation and evaluation, with simulated uncertainties, computationally prohibitive, in the time-constrained clinical routine. An alternative is the use of computationally light surrogates, like the Port Homogeneity Index (PHI) [3], which can score the dosimetric stability of irradiation setups against mispositioning, by analyzing density patterns, prior to treatment plan optimization and computation.

2. Materials and methods

In the calculation of the PHI, running time is strongly dominated by the so-called *pencil homogeneity scoring*. During this phase, for each potential beam direction, non-divergent pencil beams from a high resolution raster are traced through the planning CT to the target volume and each is compared to the neighbouring ones using a distance metric that scores their relative density homogeneity [3]. To exploit the data-parallel nature of this and other radiotherapy computations, RobuR [5], a radiation therapy software toolkit developed at the University of Marburg, was recently extended with OpenCL [6] support and used to accelerate the PHI analysis on graphic processing units (GPU) (figure 1).

For 10 skull base cases, the new GPU-accelerated PHI analysis was employed to select particle irradiation setups that are robust against mispositioning. Numerical results and speed were compared against a previously available highly-optimized single-threaded native CPU implementation [5]. The test CPU was a 4-core Intel Xeon W3530 2.8 GHz on a workstation (host) with 12 GB RAM, while the test GPU was an Nvidia Tesla C2050 with 3 GB RAM. The software, written in C++, was compiled with the GCC version 4.8.0 and run under Linux using up-to-date stable OpenCL drivers from the respective vendors.

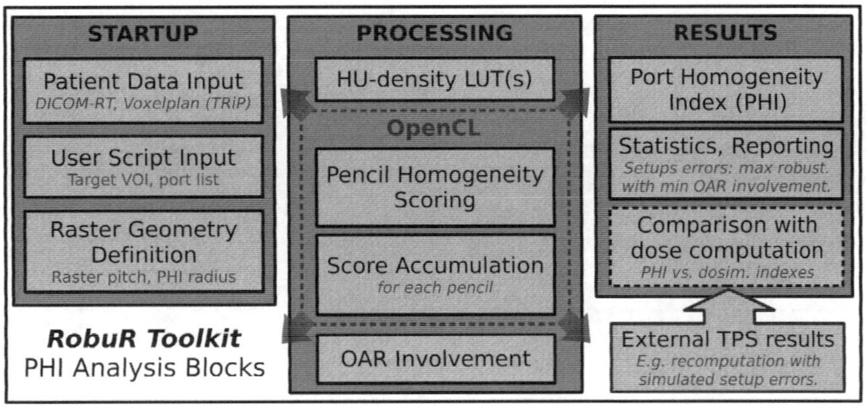

Figure 1. PHI analysis blocks within RobuR. The dashed box in the centre encompasses modules re-implemented with OpenCL and presented in this manuscript. The corner arrows indicate that OpenCL parallelization will be extended to other RobuR modules in the future.

3. Results

Numerical results were in agreement between OpenCL implementation and native CPU implementation. As previously validated for the CPU implementation, treatment plans adopting the selected beam setups were dosimetrically robust and maintained high mean target coverage and dose homogeneity values, when fully recomputed with the TRiP98 treatment planning system [7] in presence of simulated positioning errors (figure 2).

Performance-wise, acceleration on the GPU was on average more than 90-fold. Table 1 reports mean computation times and speedups, with respect to the native CPU implementation, for a single beam port (42 were tested for each patient) in the individual and pooled cases. Each port analysis was performed 10 times to reduce statistical errors. Since OpenCL enables parallelization on different devices (including multicore CPUs) the accelerated PHI analysis was also run without GPU support, with execution times and speedups included in table 1.

Results in table 1 refer to raster resolution and in-depth pencil step both of 0.5 mm (resulting in 4k-40k pencils of 120-350 samples each) and a PHI neighbourhood radius (i.e. max expected setup error) of 2 mm and indicate pure computational time with single-precision floating point operations [6], neglecting overhead for data allocation and data transfer between host and GPU.

For comparison, previous dosimetric verification of each beam angle value by extensive recomputation with TRiP98 (22 selected combinations of 1-2 mm setup errors in 3D) required from several hours to one day, depending on tumour size, on a single CPU core.

Table 1. Results for individual (rows) and pooled (last row) cases: mean(standard deviation) for a single port (in 420 computations). Second column: running times of the native CPU PHI analysis. Other columns: time and speedup (relative to the native CPU implementation) of the OpenCL (OCL) implementation, running respectively on GPU and on CPU (multi-threaded).

Patient number	Native CPU time [s]	OCL GPU time [s]	OCL GPU speedup	OCL CPU time [s]	OCL CPU speedup
P1	21.299 (6.628)	0.219 (0.109)	110.9 (32.6)	6.498 (1.996)	3.3 (0.0)
P2	9.891 (2.837)	0.122 (0.051)	91.4 (31.4)	2.552 (0.615)	3.8 (0.4)
P3	6.258 (1.485)	0.088 (0.038)	81.1 (27.2)	1.898 (0.438)	3.3 (0.0)
P4	15.768 (4.959)	0.148 (0.056)	114.2 (32.6)	4.810 (1.495)	3.3 (0.0)
P5	4.342 (1.136)	0.061 (0.024)	77.3 (22.7)	1.324 (0.331)	3.3 (0.0)
P6	10.888 (3.179)	0.160 (0.065)	75.2 (22.6)	3.280 (0.944)	3.3 (0.0)
P7	10.117 (2.849)	0.151 (0.064)	77.0 (27.8)	3.051 (0.841)	3.3 (0.0)
P8	5.464 (1.470)	0.052 (0.022)	116.0 (39.2)	1.396 (0.310)	3.9 (0.4)
P9	8.155 (2.195)	0.109 (0.046)	83.8 (28.3)	2.481 (0.649)	3.3 (0.0)
P10	11.388 (3.936)	0.126 (0.061)	99.5 (25.8)	2.935 (0.818)	3.8 (0.4)
Pool	10.357 (5.941)	0.124 (0.075)	92.6 (33.2)	3.022 (1.798)	3.5 (0.3)

4. Conclusions

Selection of robust particle therapy beam setups through the Port Homogeneity Index can be effectively accelerated on a GPU and become an unintrusive part of the particle therapy treatment planning workflow, instead of a pre-processing offline step, as previously envisioned [3]. This opens new usage scenarios, like performing brute-force fine-grained robustness analysis of an entire treatment site (e.g. $\geq$ 300 beam angles for the whole head [3]) or re-running the entire analysis many times, interactively manipulating the results [5], e.g. including constraints.

From the implementation point of view, OpenCL enables portability to other kinds of accelerators [6] and brings benefit to (multicore) CPU-only use, too, making the new implementation the only one needed, independently of the hardware.

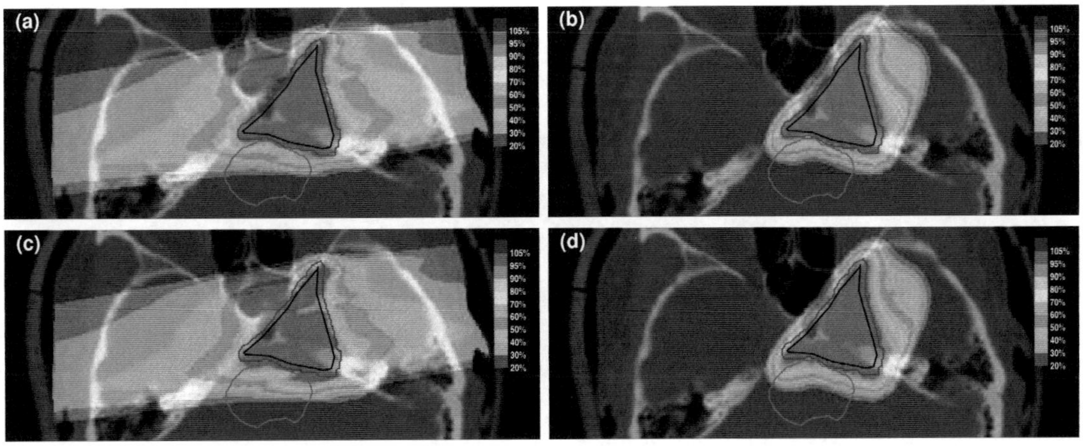

Figure 2. Exemplary dose distributions with a typical two lateral opposed beam setup (left) and a robust beam setup chosen through the PHI analysis (right). Effects on the planned dose (a, b) of a 2 mm patient positioning error in anterior direction are shown (c, d), demonstrating the greater dosimetric robustness of the PHI-selected plan.

The presented results do not include the overhead for data transfer between host and GPU, which can be non-negligible when compared to the pure computational time and affect the overall application performance, as observed for other radiotherapy GPU applications [8]. However, when processing a large number of irradiation setups, the impact of data transfer latency may be minimized by scheduling transfer operations concurrently with computations whose input data have already reached the GPU. This requires accurate transfer/execution synchronization implemented explicitly at application level, which is possible through OpenCL [6] and will be topic of further investigation. The project outlook includes the acceleration on GPU of additional high-level functionalities of RobuR (figure 1) that are complementary to the port robustness analysis, e.g. the assessment of OAR involvement [5].

Acknowledgments
Work supported by a research grant of the University Medical Center of Giessen and Marburg (no. 35/2010MR). ICCR2013 attendance enabled by Verein zur Förderung der Strahlentherapie und Radioonkologie im Universitätsklinikum Marburg e.V.

References

[1] Amaldi U and Kraft G 2005 *Rep. Prog. Phys.* **68** 1861–1882
[2] Lomax A J 2008 *Phys. Med. Biol.* **53** 1043–1056
[3] Ammazzalorso F, Jelen U, Krämer M, Straßmann G and Engenhart-Cabillic R 2009 *Radiother. Oncol.* **92** S109
[4] Hopfgartner J, Stock M, Knäusl B and Georg D 2013 *Acta Oncol.* **52** 570–579
[5] Ammazzalorso F, Jelen U, Kraemer M and Zink K 2010 *49th Annual Conf. of the Particle Therapy Co-Operative Group* (Gunma, Japan) P6-16
[6] Bednarz T 2010 Introduction to OpenCL *OzViz 2010 OpenCL Workshop* (Brisbane, Australia) `https://sites.google.com/site/ozvizworkshop/ozviz-2010/ozviz-2010-opencl-workshop-program`
[7] Krämer M, Jäkel O, Haberer T, Kraft G, Schardt D and Weber U 2000 *Phys. Med. Biol.* **45** 3299–3317
[8] Gu X, Jelen U, Li J, Jia X and Jiang S B 2011 *Phys. Med. Biol.* **56** 3337–3350

Journal of Physics: Conference Series **489** (2014) 012044 doi:10.1088/1742-6596/489/1/012044

Ultra-fast fluence optimization for beam angle selection algorithms

M Bangert, P Ziegenhein, and U Oelfke
German Cancer Research Center - DKFZ
Im Neuenheimer Feld 280, 69120 Heidelberg, Germany
E-mail: `m.bangert@dkfz.de`

Abstract. Beam angle selection (BAS) including fluence optimization (FO) is among the most extensive computational tasks in radiotherapy. Precomputed dose influence data (DID) of all considered beam orientations (up to 100 GB for complex cases) has to be handled in the main memory and repeated FOs are required for different beam ensembles. In this paper, the authors describe concepts accelerating FO for BAS algorithms using off-the-shelf multiprocessor workstations. The FO runtime is *not* dominated by the arithmetic load of the CPUs but by the transportation of DID from the RAM to the CPUs. On multiprocessor workstations, however, the speed of data transportation from the main memory to the CPUs is non-uniform across the RAM; every CPU has a dedicated memory location (node) with minimum access time. We apply a thread node binding strategy to ensure that CPUs only access DID from their preferred node. Ideal load balancing for arbitrary beam ensembles is guaranteed by distributing the DID of every candidate beam equally to all nodes. Furthermore we use a custom sorting scheme of the DID to minimize the overall data transportation. The framework is implemented on an AMD Opteron workstation. One FO iteration comprising dose, objective function, and gradient calculation takes between 0.010 s (9 beams, skull, 0.23 GB DID) and 0.070 s (9 beams, abdomen, 1.50 GB DID). Our overall FO time is < 1 s for small cases , larger cases take ~ 4 s. BAS runs including FOs for 1000 different beam ensembles take $\sim 15-70$ min, depending on the treatment site. This enables an efficient clinical evaluation of different BAS algorithms.

1. Introduction

Automated beam angle selection (BAS) incorporating fluence optimization (FO) requires repeated FOs in order to compare different beam ensembles. Relying on commercially available treatment planning software, which may require a couple of minutes per FO, however, would result in prohibitive computation times. In order to accelerate FO, BAS approaches including FO often compromise regarding the resolution in the dose deposition space [1] or regarding the number of evaluated beam ensembles [2]. Furthermore, the number of considered candidate beam orientations is often restricted [3] because precomputed dose influence data (DID) of all candidate beams has to be handled in higher memory.

We have recently investigated the structure of the combinatorial BAS problem [4] and the performance of different BAS algorithms for intracranial IMRT [5]. Therefore, we have developed an ultra-fast BAS framework operating on the same resolution like clinical planning systems, considering up to 1 400 candidate beams, and evaluating up to 10 000 beam ensembles per patient. Here, we present a detailed description of the underlying FO algorithm and the hard- and software implementation of our BAS framework.

2. Material and methods

BAS strategies including FO usually comprise two independent modules, as depicted in figure 1. A BAS module controls which beam ensembles B_i will be evaluated by the FO module in iteration i. In subsequent iterations $i+1$, the BAS module suggests new beam ensembles based on the preceding objective function values $F(B_i)$. Choices for BAS modules are manifold and include simulated annealing [3], genetic algorithms [6], and iterative search strategies [7], among others. Usually, the runtime of the BAS module is orders of magnitude shorter than the runtime of the FO module. Hence, accelerating the entire BAS algorithm is only possible by accelerating the FO.

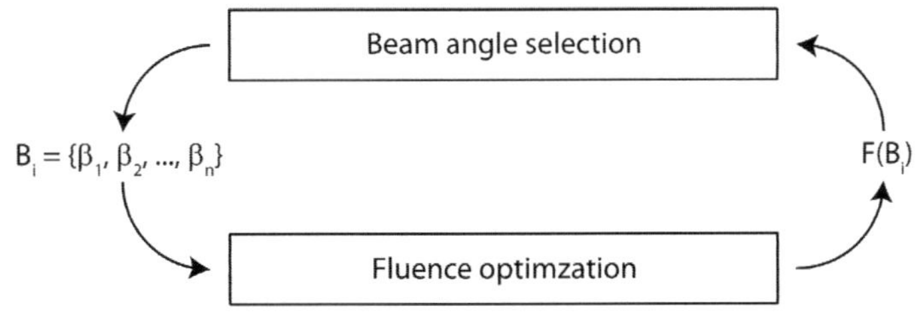

Figure 1. Schematic software design of beam angle selection algorithms.

2.1. Algorithmic implementation

We use a quadratic objective function F [8] for FO

$$F = \sum_{i \in Target} p_i (D_i - D_i^{pres})^2 + \sum_{i \in OAR/NT} p_i \lfloor D_i - D_i^{max} \rfloor_+^2 \quad (1)$$

where p_i, D_i, and $D_i^{pres/max}$ denote the penalty, the dose, and the prescribed/maximum dose of voxel i. The positivity operator ensures that only violated constraints contribute to the objective function, i.e. $\lfloor x \rfloor_+ = x$ for $x > 0$ and $\lfloor x \rfloor_+ = 0$ otherwise. The dose in voxel i is given by $d_i = \sum_j D_{ij} w_j$ with a DID matrix D_{ij} specifying the dose contribution from bixel j to voxel i. Minimizing equation 1 is a constrained convex optimization problem [9, Chapter 2.1.2.] as we require $w_j \geq 0$ for all beamlets j. Usually, such optimization problems are solved by sequential quadratic programming (SQP). However, we use a simple projected Quasi-Newton optimizer [10] as the improved algorithmic convergence of SQP is compromised by the additional arithmetic operations required during the solution of the quadratic subproblem. We apply an L-BFGS two-loop recursion algorithm to approximate the product of the inverse Hessian and the gradient [10]. An Armijo backtracking line search controls the step length in every iteration [10]. If a beam orientation β_k is used for the first time during BAS, its beamlet weights are uniformly initialized $w_{jk} = 1$. If β_k is used again during BAS as part of a different beam ensemble, its beamlet weights are initialized with the result of the last optimization. This accelerates the FO roughly by a factor of two. Even though the initial beamlet weights of β_k have been obtained in combination with other beams, this does not affect the result of the FO due to the convexity of the FO problem.

2.2. Hard- and software implementation

The objective function and gradient calculation of equation 1 correspond to simple matrix vector products yet on large DID sets D_{ij}. Hence, FO is a memory bound problem [11]. The arithmetic load of the CPU does *not* limit the runtime of the algorithm but the transportation of the dose influence data D_{ij} from the main memory to the CPU, which typically amounts to 100 MB - 1 GB for a single beam ensemble. On state-of-the-art multiprocessor workstations the speed of data transportation from the main memory to the CPUs is non-uniform across the RAM; every CPU has a dedicated memory location (node) with minimum access time. We apply a thread node binding strategy to ensure that CPUs only access DID from their preferred node. The DID of every candidate beam is distributed equally to all nodes. Hence, we guarantee an ideal load balancing for every potential beam ensemble that might be evaluated during BAS. Furthermore we sort the DID according to the access pattern during the objective function and gradient calculation in order to minimize the overall data transportation of base data between RAM and CPU [12]. Efficient cache memory filling is ensured by always processing chunks of 64 bytes, which corresponds to the standard cache line size, at the same time. We use the OpenMP standard for parallelization in C++. The BAS framework is implemented on an AMD Opteron workstation with 4 CPUs with 12 cores each and 128 GB main memory. Besides a small overhead required for the optimization, the entire main memory is available to store precomputed DID.

2.3. Beam angle selection study

We measure the runtimes of BAS runs comprising 10 000 FOs for three prostate, three pancreas, and three intracranial cases. We compare treatment plans applying nine non coplanar beam orientations that are found with a genetic algorithm [5] and treatment plans applying nine equi-spaced coplanar beams; a more detailed analysis of the clinical implications of our BAS framework for intracranial IMRT can be found in [5].

3. Results

Figures 2 (a)-(c) summarize the results of the performance measurements of BAS runs for nine different patient cases with 10 000 FOs each. As depicted in figure 2(a), we require 10-70 ms per FO iteration depending on the problem size, i.e., the amount of DID for a given beam ensemble. One iteration comprises one gradient calculation and on average 1.09 line searches. As depicted in figure 2(b), the number of FO iterations is usually around 50 if the FO is terminated after three successive iterations with an objective function decrease of less than 0.1%. The number of iterations does not depend on the amount of DID processed during the FO. The runtime of a single FO is 0.5-4.0 s depending on the problem size, as depicted in figure 2(c).

Figure 2(a) shows that the minimum runtime linearly depends on the amount of DID. The fluctuations can be explained by the individual locality of the DID in the RAM, by the size of the DID, and in figure 2(c) also by the number of iterations.

Figures 2 (d)-(f) show representative DVHs for one prostate, pancreas, and intracranial case. We observe modest reductions of the dose in organs at risk (OARs) for the prostate and pancreas cases at similar target coverage. For the intracranial cases BAS may enable substantial dose savings in OARs.

4. Discussion and conclusion

We have presented an ultra-fast FO implementation for BAS that requires about 15-70 min to compare 1 000 treatment plans. This enables an efficient evaluation of different BAS strategies for clinical application [5, 4].

The fast BAS runtimes originate from a CPU-based FO engine which needs less than 1 s for small cases. Men et al. report runtimes of 2.79 s on a GPU for FO problems of comparable size [13].

The developed ultra-fast FO engine may be useful for manifold applications beyond BAS as it is generally suited to accelerate inverse planning. Naturally, a speedup is especially desirable for time consuming treatment planning tasks involving large DID sets > 1 GB such as modulated arc therapy and particle therapy. For small data sets ~ 0.1 GB it is possible to enable a more

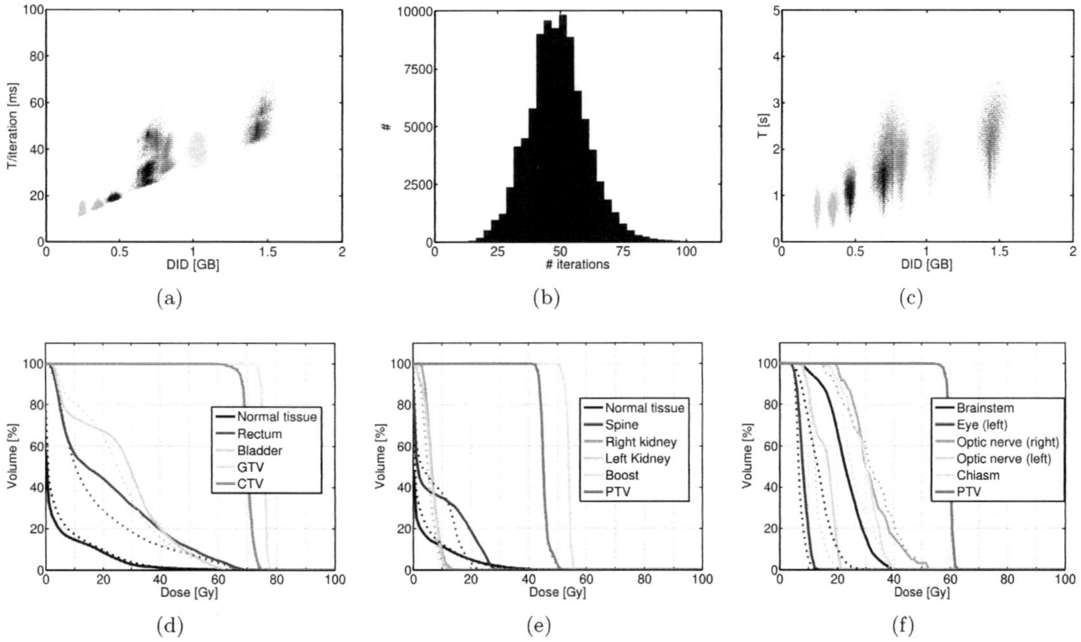

Figure 2. (a) Runtime of one FO iteration versus the amount of processed DID. (b) Number of required FO iterations. (c) Overall FO runtime versus the amount of processed DID. The nine different colors in figures (a) and (c) correspond to different patient cases. The point clouds comprise 10 000 dots each; a single dot corresponds to a single FO. DVHs for (d) a prostate case, (e) a pancreas case, and (f) an intracranial case. Solid lines correspond to a treatment plan applying nine equi-spaced coplanar beams and dotted lines correspond to nine non coplanar optimized beams.

interactive treatment planning experience - and maybe ultimately realtime FO. For adaptive radiation therapy, an acceleration of inverse planning is an indispensable prerequisite before clinical application because a treatment plan of the day has to be generated based on up-to-date imaging data while the patient is on the couch.

Even though our implementation relies on the standard quadratic objective function [8], the acceleration concepts are generally applicable to treatment planning approaches that use a matrix vector product for the dose calculation. This includes alternative objective functions, multicriteria optimization, and direct aperture optimization, among others.

In this study, we have used an AMD workstation with 128 GB main memory in order to cope with the DID of all candidate beams. However, our implementation can also be used to accelerate FO on conventional personal computers [11].

An efficient application of GPUs in the context of BAS is currently not possible. Due to the limited memory available on GPUs, the DID of all candidate beams cannot be loaded with a single GPU. An efficient application of GPU clusters is infeasible due to synchronization and load balancing issues. Futhermore GPU clusters are very expensive; the AMD workstation we applied for this study is available at US$ 4500.

References

[1] Engel K and Tabbert E 2005 *Optimization and Engineering* **6** 393–419

[2] Rowbottom C G, Khoo V S and Webb S 2001 *Medical Physics* **28** 1696
[3] Stein J, Mohan R, Wang X H, Bortfeld T, Wu Q, Preiser K, Ling C C and Schlegel W 1997 *Medical Physics* **24** 149
[4] Bangert M, Ziegenhein P and Oelfke U 2012 *Physics in Medicine and Biology* **57** 6707
[5] Bangert M, Ziegenhein P and Oelfke U 2013 *Medical physics* **40** 011716
[6] Hou Q, Wang J, Chen Y and Galvin J M 2003 *Medical Physics* **30**
[7] Woudstra E, Heijmen B J and Storchi P R 2005 *Radiotherapy and Oncology* **77**
[8] Oelfke U and Bortfeld T 2001 *Medical Dosimetry* **26** 113–124
[9] Bangert M 2011 *New concepts for beam angle selection in IMRT treatment planning : From heuristics to combinatorial optimization* Ph.D. thesis Universitätsbibliothek Heidelberg
[10] Kelley C T 1987 *Iterative methods for optimization* vol 18 (Society for Industrial and Applied Mathematics)
[11] Ziegenhein P, Kamerling C, Bangert M, Kunkel J and Oelfke U 2013 *Physics in medicine and biology* **58** 3705–15
[12] Ziegenhein P, Wilkens J, Nill S, Ludwig T and Oelfke U 2008 *Physics in medicine and biology* **53**
[13] Men C, Gu X, Choi D, Majumdar A, Zheng Z, Mueller K and Jiang S B 2009 *Physics in medicine and biology* **54** 6565

PyTRiP - a toolbox and GUI for the proton/ion therapy planning system TRiP

J Toftegaard[1], **J B Petersen**[2], **N Bassler**[1,3]

[1] Department of Physics and Astronomy, Aarhus University, Aarhus, DENMARK
[2] Department of Medical Physics, Aarhus University Hospital, Aarhus, DENMARK
[3] Department of Experimental Clinical Oncology, Aarhus University Hospital, Aarhus, DENMARK

E-mail: `bassler@phys.au.dk`

Abstract.
Purpose: Only very few treatment planning systems (TPS) are capable of handling heavy ions. Commercial heavy ion TPS are costly and normally restrict the possibility to implement new functionalities. PyTRiP provides Python bindings and a platform-independent graphical user interface (GUI) for the heavy ion treatment program TRiP, and adds seamless support of DICOM files. We aim to provide a front-end for TRiP which does not require any special computer skills.

Methods: PyTRiP is written in Python combined with C for fast computing. Routines for DICOM file import/export to TRiPs native file format are implemented. The GUI comes as an executable with all its dependencies including PyTRiP making it easy to install on Windows, Mac and Linux.

Results: PyTRiP is a comprehensive toolbox for handling TRiP. Treatment plans are handled using an object oriented structure. Bindings to TRiP (which only runs on Linux, either locally or on a remote server) are performed through a single function call. GUI users can intuitively create treatment plans without much knowledge about the TRiP user interface. Advanced users still have full access to all TRiP functionality. The user interface comes with a comprehensive plotting tool, which can visualize 2D contours, volume histograms, as well as dose- and linear energy transfer (LET) distributions.

Conclusion: We developed a powerful toolbox for ion therapy research using TRiP as backend. The corresponding GUI allows to easily and intuitively create, calculate and visualize treatment plans. TRiP is thereby more accessible and simpler to use.

1. Introduction

TRiP98 [1, 2] is a well known heavy ion treatment planning system for heavy ions, which was developed at GSI Helmoltzzentrum für Schwerionenforschung, Darmstadt, Germany. PyTRiP [3] started as a package for handling VOXELPLAN [4] formatted files, which are used by TRiP. Since then it has been extended with a series of functions for pre and post processing of TRiP generated-data [5]. In its current version it is possible to execute TRiP from simple python scripts. A graphical user interface (GUI) is developed as well, from which most features of TRiP can be accessed.

2. Methods

2.1. DICOM support

As TRiP98 only is capable of reading files in VOXELPLAN format, PyTRiP features a full converter of DICOM files to TRiP format. DICOM 2D CT images are recombined to a single 3D cube containing the Hounsfield units in floating point notation. A header file describes the format. Dose cubes are prepare in similar way, however here the dose is represented as integer ranging from 0 to 1000, specifying the relative dose in steps of tenths of a percent. All voxels can be shifted with the DICOM offset tag, which is retained in the VOXELPLAN format. The conversion of structures (or volumes of interest (VOIs)) is less trivial, though. A structure can consist of one or more separated contours, where the contours themselves are handled equally in DICOM and VOXELPLAN format. In both cases, the structure is divided into slices which match the placement of the respective CT slice. The problem however arises from a limitation in TRiP that in a given slice, there can only be a single contour per structure. In practice, this is handled by manually connecting contours with an area with the width of 0. The two contours can then in topological sense be regarded as a single contour. To automate this process and to avoid user interaction for this tedious and time consuming step, a function is written which fulfills the following criteria:

- If one contour is located inside another contour, then the inner contour is regarded as a hole in the outer contour.
- If two contours have no overlap, these must be connected with an infinitely thin area, i.e. two overlapping connecting lines, which combines the two contours to a single contour with a single surface.
- A connecting line between two contours may not pass a third contour.

To fulfill these requirements, an algorithm is conceived where first all contours are organized in a tree. The contours are organized within a top-level contour (K0), and all subsequent contours (Ki) are various generations of child nodes of the parent contour. This is illustrated for a case with 4 contours (K1-K4) in figure 1 and 2. Here, K3 and K4 are holes in K1, and K2 is an independent contour.

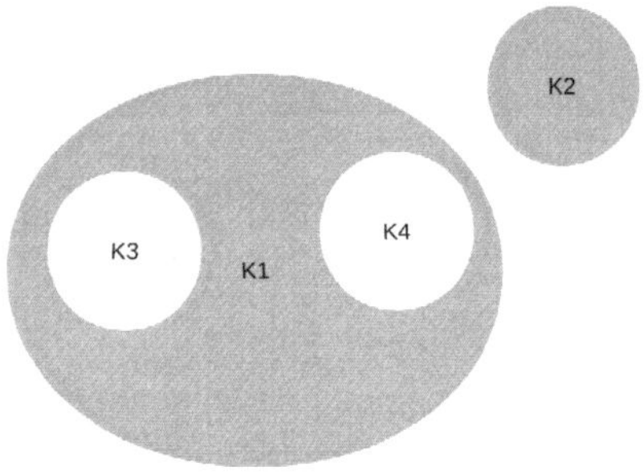

Figure 1. Example of a structure in a CT slice containing several contours.

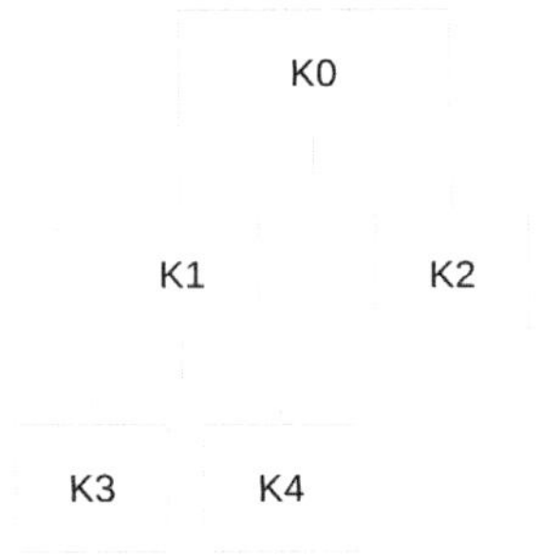

Figure 2. Contour tree defining the structure from figure 1.

To test whether a contour is located within another one, a point is taken from the list of points which defines the contour, and then it is investigated whether this point is inside the other contour using a ray casting algorithm [6]. This algorithm calculates a parameter representation of a line which crosses the investigated area. The points of interception with the contours are calculated. If there is an odd number of solutions in the positive directions, then the point

Journal of Physics: Conference Series **489** (2014) 012045 doi:10.1088/1742-6596/489/1/012045

is within the structure, and vice versa. Once the tree is constructed, contours are coalesced recursively. In the example shown in figure 1 this means that K3 and K4 are connected first, and the resulting contour is connected with K1, which then again is connected to K2. Finally, K0 is representing the entire contour. If a contour has more than 2 child contours attached, a child contour and its closest neighbour are joined first. This way it is ensured that no line will cross a third contour.

The inverse process, converting VOXELPLAN structures to DICOM is straightforward, since there are no additional requirements how contours in a structure are organized.

2.2. DVH and LVH calculation
Once the structures are established, dose-volume histograms and also LET-volume histograms can be calculated in PyTRiP, thereby reducing the need to interact directly with TRiP. LET-volume histograms are a new feature which illustrate the LET-coverage within a tumour in a similar way as dose coverage is shown in a regular dose-volume histogram. This gives a feeling for e.g. the presence of LET hot-spots or cold spots within the volume of interest, which is of relevance for the recent focus on LET-optimized particle therapy methods [7–11].

2.3. TRiP execution and platform considerations
One aim with PyTRiP is to simplify parametric studies, where many treatment plans are calculated while varying one or more parameters. PyTRiP supplies classes for TRiP execution, following an object oriented design paradigm. The treatment plan is thus an object which can be executed and contains the results after execution. A client-server model is applied, since TRiP itself only is compiled for a limited number of platforms. PyTRiP in turn, is functional on Linux, Mac and Windows based operating systems. The client-server model enables users to use PyTRiP on client PCs, even where TRiP is not installed, as long as an encrypted connection is possible to the server where TRiP is installed. Multiple users can use TRiP simultaneously from several different computers.

2.4. Graphical user interface
A graphical user interface to PyTRiP is provided using the platform independent wxWidgets set [12]. The GUI purports to minimize (and ideally obsolete) direct user interaction with TRiP. A demonstration of an early version of the graphical user interface can be seen on youtube [13] - See more at: `http://youtu.be/6ZqcJ60Z598`.

3. Discussion and Conclusion
PyTRiP has already demonstrated to be a versatile package for generating ion treatment plans and perform parametric studies. As an example, PyTRiP was recently used for calculating maps of robust angles for treating patients with protons and heavier ions [5]. Here, a robustness factor is defined which expresses the sensitivity of a given irradiation field towards setup errors. The field is then applied iteratively for a fine meshed grid of couch and gantry angles, generating a robustness map. These maps show islands of robustness, which suggest preferred treatment angles.

PyTRiP is also used for demonstrating the potential of LET-painting [7]. Here, high-LET components of heavy ion beams are targeted to hypoxic tumour compartments in order to achieve a better therapeutic outcome. Extensive studies to quantify the alleged benefit was recently carried out, where radiobiologic models are included in order to determine the tumour control probability [5]. PyTRiP along with TRiP and the Monte Carlo particle transport code SHIELD-HIT12A [14–16] was used to also investigate whether tumour hypoxia could be overcome using ions heavier than carbon ions, such as oxygen-16 ions, which are available at Heidelberg Ionentherapiezentrum (HIT), Universitätsklinikum Heidelberg, Germany [17].

Using the connectivity between PyTRiP, TRiP and SHIELD-HIT12A it is possible to prepare and calculate depth-dose data sets for any primary ion and facility specific beam lines [18], and apply these for treatment plan studies. The same tool set was also used recently to investigate the impact of secondary ions in clinical settings [19, 20].

Next, we foresee to use this combination of tools to generate antiproton treatment planning studies [21, 22], carefully benchmarked against the radiobiologic data acquired at CERN [21, 23]. We also anticipate PyTRiP will be useful to compare radiobiology models as these evolve, and more data from existing and future radiation laboratories become available [24].

Acknowledgments

Bjarne Thomsen, Dept. of Physics and Astronomy, Aarhus University is thanked for providing cluster computing time. This project was supported by CIRRO - The Lundbeck Foundation Center for Interventional Research in Radiation Oncology, The Danish Council for Strategic Research (`http://www.cirro.dk`), Novo Nordisk Fonden and by the Danish Cancer Society (`http://www.cancer.dk`) project ID# DP08023.

References

[1] Krämer M, Jäkel O, Haberer T, Kraft G, Schardt D and Weber U 2000 *Phys. Med. Biol.* **45** 3299–3317
[2] Krämer M and Scholz M 2000 *Phys. Med. Biol.* **45** 3319–3330
[3] Bassler N and Toftegaard J 2013 PyTRiP [Online] latest status: https://svn.nfit.au.dk/trac/pytrip [28 October 2013]
[4] Höss A, Debus J, Bendl R, Engenhart-Cabillic R and Schlegel W 1995 *Der Radiologe* **35** 583 (in German)
[5] Toftegaard J 2013 *Udvikling af et Dosisplanlægningsystem og Nye Dosisplanlægningsstrategier til Ionterapi* Master's thesis Department of Physics and Astronomy, Aarhus University, Aarhus, Denmark (In Danish)
[6] Roth S D 1982 *Computer graphics and image processing* **18** 109–144
[7] Bassler N, Jäkel O, Søndergaard C S and Petersen J B 2010 *Acta Oncol.* **49** 1170–1176
[8] Bassler N, Toftegaard J, Lühr A, Sørensen B S, Scifoni E, Krämer M, Jäkel O, Mortensen L S, Overgaard J and Petersen J B 2013 *Acta Oncol.* In press
[9] Grassberger C, Trofimov A, Lomax A and Paganetti H 2011 *Int. J. Radiat. Oncol., Biol., Phys.* **80** 1559–1566
[10] Antonovic L, Brahme A, Furusawa Y and Toma-Dasu I 2013 *J. Radiat. Res.* **54** 18–26
[11] Böhlen T T, Brons S, Dosanjh M, Ferrari A, Fossati P, Haberer T, Patera V and Mairani A 2012 *Phys. Med. Biol.* **57** 7983
[12] The wxWidgets Team 2013 wxWidgets [online] Available at: http://www.wxwidgets.org [Accessed 28 October 2013]
[13] Toftegaard J 2013 PyTRiPGUI [video online] Available at: http://youtu.be/6ZqcJ6OZ598 [Accessed 28 October 2013]
[14] Bassler N, Hansen D C, Lühr A and Sobolevsky N 2012 SHIELD-HIT12A [online] Available at: http://www.shieldhit.org [Accessed 28 October 2013]
[15] Hansen D C, Lühr A, Herrmann R, Sobolevsky N and Bassler N 2012 *Int. J. Radiat. Biol.* **88** 195–199
[16] Hansen D C, Lühr A, Sobolevsky N and Bassler N 2012 *Phys. Med. Biol.* **57** 2393–2409
[17] Kurz C, Mairani A and Parodi K 2012 *Phys. Med. Biol.* **57** 5017
[18] Ringbaek T P 2013 *Monte Carlo Simulations of New 2D Ripple Filters for Particle Therapy Facilities* Master's thesis Dept. of Physics and Astronomy, Aarhus University, Aarhus, Denmark
[19] Lühr A, Hansen D C, Teiwes R, Sobolevsky N, Jäkel O and Bassler N 2012 *Phys. Med. Biol.* **57** 5169
[20] Lühr A, Priegnitz M, Fiedler F, Sobolevsky N and Bassler N 2013 *Appl. Radiat. Isotopes* In press
[21] Holzscheiter M H, Agazarayan N, Bassler N, Beyer G, DeMarco J J, Doser M, Ichioka T, Iwamoto K S, Knudsen H V, Landua R, Maggiore C, McBride W H, Møller S P, Petersen J, Smathers J B, Skarsgard L D, Solberg T D, Uggerhøj U I, Withers H R, Vranjes S, Wong M and Wouters B G 2004 *Nucl. Instrum. Methods Phys. Res. B* **221** 210–214
[22] Bassler N, Holzscheiter M H, Jäkel O, Kovacevic S, Knudsen H V and the AD-4/ACE Collaboration 2008 *Phys. Med. Biol.* **53** 793–805
[23] Kavanagh J N, Currell F J, Timson D J, Savage K I, Richard D J, McMahon S J, Hartley O, Cirrone G A P, Romano F, Prise K M *et al.* 2013 *Scientific reports* **3** online, ahead of print
[24] Holzscheiter M H, Bassler N, Dosanjh M, Sørensen B S and Overgaard J 2012 *Radiother. Oncol.* **105** 1–3

A case study of IMRT planning (Plan B) subsequent to a previously treated IMRT plan (Plan A)

F Cao[1], C Leong[2], J Schroeder[2] and B Lee[2]

[1]Department of Medical Physics and [2]Department of Radiation Oncology, Fraser Valley Centre, BC Cancer Agency, Surrey, British Columbia, Canada,

Email: fcao@bccancer.bc.ca

Abstract

Background and purpose: Treatment of the contralateral neck after previous ipsilateral intensity modulated radiation therapy (IMRT) for head and neck cancer is a challenging problem. We have developed a technique that limits the cumulative dose to the spinal cord and brainstem while maximizing coverage of a planning target volume (PTV) in the contralateral neck. Our case involves a patient with right tonsil carcinoma who was given ipsilateral IMRT with 70Gy in 35 fractions (Plan A). A left neck recurrence was detected 14 months later. The patient underwent a neck dissection followed by postoperative left neck radiation to a dose of 66 Gy in 33 fractions (Plan B).

Materials and Methods: The spinal cord-brainstem margin (SCBM) was defined as the spinal cord and brainstem with a 1.0 cm margin. Plan A was recalculated on the postoperative CT scan but the fluence outside of SCBM was deleted. A further modification of Plan A resulted in a base plan that was summed with Plan B to evaluate the cumulative dose received by the spinal cord and brainstem. Plan B alone was used to evaluate for coverage of the contralateral neck PTV.

Results: The maximum cumulative doses to the spinal cord with 0.5cm margin and brainstem with 0.5cm margin were 51.96 Gy and 45.60 Gy respectively. For Plan B, 100% of the prescribed dose covered 95% of PTVb1.

Conclusion: The use of a modified ipsilateral IMRT plan as a base plan is an effective way to limit the cumulative dose to the spinal cord and brainstem while enabling coverage of a PTV in the contralateral neck.

1. Introduction

Intensity Modulated Radiation Therapy (IMRT) is often used for planning head and neck (H&N) cancers due to its ability to deliver a highly conformal dose distribution to a target while limiting dose to nearby critical structures such as spinal cord, brainstem, parotid gland and optic chiasm. The use of conventional 3D planning techniques for either a newly diagnosed or recurrent tumour would result in greater doses being delivered to normal tissues (1,2,3). The challenge of irradiation of the contralateral neck for recurrent disease is to respect the tissue tolerance of critical structures such as spinal cord and brainstem while adequately covering a new PTV.

Journal of Physics: Conference Series **489** (2014) 012046 doi:10.1088/1742-6596/489/1/012046

The Eclipse™ planning system from Varian (Varian Medical Systems Inc., Palo Alto, CA) enables IMRT optimization to be performed by summing a previous base plan with a subsequent plan. This allows one to account for the effects of previous radiation when planning a retreatment.

At our cancer centre, we have developed a technique that sufficiently limits the cumulative dose delivered to the spinal cord and brainstem while enabling coverage of a new PTV.

2. Materials and Methods

2.1. Patient Information

A patient was given ipsilateral radiation treatment with 70Gy to PTVa1 and 56Gy to PTVa2 in 35 fractions for right tonsil carcinoma. A left neck recurrence was detected 14 months later. The patient underwent a neck dissection showing 1/54 nodes positive for carcinoma. The tumour deposit was 3 cm in size with extra nodal extension and a close margin.Postoperative left neck radiation was then given to a dose of 66 Gy to PTVb1 and 54Gy to PTVb2 in 33 fractions.

2.2. CT Simulation

CT data sets were obtained from a Phillips™ (Philips Healthcare, 3000 Minuteman Road, Andover, MA) Brilliance Big Bore scanner using 3mm slices with the patient in a supine position and immobilized using a headrest, Type-S™ thermoplastic S-Frame head shell, Type-S™ head extension and Vac-lock bag from CIVCO™ (1401 8th Street SE, Orange City, Iowa 51041).

2.3. Contouring

A Radiation Oncologist (RO) contoured the gross target volume (GTV) for the initial tumor. A PTV (70 Gy) was generated by expanding the GTV contour with a 10 mm margin in all directions except for bony interfaces. A PTV (56 Gy) was contoured for nodal regions at risk for subclinical disease. After the neck dissection for recurrent disease, a postoperative plan was generated. A PTV (66 Gy) was contoured for the recurrent tumour bed while a PTV (54 Gy) was contoured for nodal regions at risk for subclinical disease. All healthy organs including central nervous system (CNS) organs were contoured by a Radiation Therapist (RT) and verified by the RO.

2.4. Planning

The EclipseTM treatment planning system (version 10.1.20) was used for treatment planning. A seven field IMRT plan (Plan A) with sliding window multileaf collimator (MLC) was used. The beam angles were 25°, 180°, 215°, 245°, 285°, 310° and 340°. There were no collimator rotations and no couch rotations for any of the seven beams. Fig.1(A) shows the seven beam arrangements. 99.01% of PTVa1 and 97.85% PTVa2 were covered by 100% of the prescribed doses with the maximum dose being 107.4% of the prescribed dose. The maximum doses received by spinal cord with 0.5cm margin and brainstem with 0.5cm margin were 46.70 Gy and 41.31 Gy respectively. The 0.5cm margin is the standard margin for the PRV used at our centre. It is also the recommended margin used to account for interfraction patient and organ motion in the NCIC CTG HN.6 protocol [4].

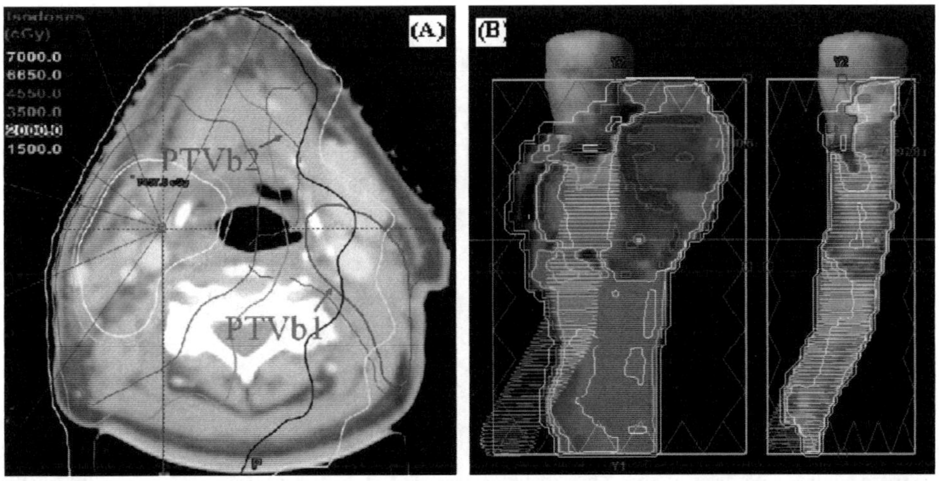

Fig.1 (A)70 Gy in 35 fractions was given for gross disease. The 7 red lines indicate the entry directions of the7 IMRT beams. After neck dissection for recurrent disease, 66 Gy was given to PTVb1(in red) for the nodal bed and 54 Gy was given to PTVb2(in purple) for possible subclinical disease in 33 fractions. (B) Beam's eye view showing one of the seven beam's fluence from plan A before(left) and after(right) the fluence outside SCBM was deleted.*

A structure named SCBM was defined as the spinal cord and brainstem with a 1.0 cm margin. Plan A was recalculated on the postoperative CT scan and this was named Plan A*. The fluence outside of SCBM from Plan A* was deleted and this modified plan was called Plan A**. Plan A** was then used as a base plan to generate the final plan (Plan B) with beam angle of 25°, 50°, 115°, 145°, 180°, 335° that treated the PTVs in the postoperative neck.

2.5. Plan evaluation
The total cumulative dose received by the spinal cord and brainstem from the summation of Plan A* and Plan B were evaluated. The isodose coverage of PTVb1 and PTVb2 was evaluated using Plan B alone.

3. Results
The maximum cumulative doses from the summation of Plan A* and Plan B to the spinal cord with 0.5cm margin and brainstem with 0.5cm margin were 51.96 Gy and 45.60 Gy respectively. 95% of PTVb1 and 81.84% of PTVb2 were covered by 100% of the prescribed doses. 89.98% of the PTVb2 volume was covered by the 95% of the prescribed dose. Since PTVb2 was a complex target spanning the entire contralateral neck, this degree of coverage was deemed acceptable by the RO. The maximum dose from Plan B alone was 112.4%.

4. Discussion
Spinal cord and brainstem toxicity are major concerns when reirradiation of the head and neck is contemplated. Various approaches have been taken to account for previous irradiation. Some centres assume that irradiated CNS tissues will recover by 50% of the dose received if there is an interval of more than one year from the initial treatment (1, 2). Other centres assume no recovery and simply add the physical doses of the initial and subsequent treatment courses when considering tissue tolerances (3). At our centre, we have taken the latter, more conservative approach.

To keep the dose to CNS structures as low as possible, our centre's policy is to avoid using beams that enter CNS structures before reaching the PTV regardless if the plan is an initial one or a retreatment. This approach is illustrated in Fig.1. For patients receiving ipsilateral irradiation, we have designed standardized beam arrangements to facilitate this policy. For a given case, the dosimetrist will load the standardized template to see if the beam arrangement is suitable or not. The beam and collimator angles can be adjusted as needed for each case. The standard beam angles for left sided tumours are 25°, 180°, 215°, 245°, 285°, 310° and 340°. The standard beam angles for right sided tumours are 25°, 50°, 75°, 115°, 145°, 180° and 335°. For most ipsilateral tumours, the standard beam angles do not need to be changed.

At our centre, we have adopted dose constraints outlined in the National Cancer Institute of Canada Clinical Trials Group head and neck clinical protocol (NCIC CTG HN.6)(4). Dose constraints for the spinal cord with 0.5cm margin and brainstem with 0.5cm margin are 52 Gy and 60 Gy respectively. Assuming these structures received a tolerance dose from an initial treatment and that the spinal cord and brainstem recover by 50% of the dose, the dose constraints for re-irradiation would be 28.65 Gy and 39.35 Gy respectively. Although such constraints could be met in some cases, they would not be met in other situations and there remains uncertainty about the validity and amount of dose recovery. For these reasons, we felt that it would be better to have a technique that does not rely on somewhat arbitrary assumptions of dose recovery.

For the case presented above, the maximum dose to the spinal cord with 0.5cm margin was 46.70 Gy from Plan A. Without using a base plan, we would not know the precise location of this maximum dose relative to the new PTVs in the contralateral neck. In order to limit the cumulative dose to the spinal cord with 0.5 cm margin to 52 Gy we would need to give no more than 5.3 Gy to this entire region. The shortest distance between the PTVb2 (54 Gy) and the spinal cord with 0.5 cm margin is about 1 cm. A dose fall-off gradient of 9.0% per millimetre would be required to meet the spinal cord with 0.5 cm margin constraint. As a "rule of thumb", however, the maximum dose fall-off gradient should be less than 7% per millimetre for an IMRT plan. If the dose gradient is greater than the 7% per millimetre, PTV dose conformity and homogeneity tends to be compromised.

The simplest approach for planning treatment of the contralateral neck is to recalculate Plan A on the postoperative scan without any modifications. Assuming that there has been no significant change in the size and shape of the neck, this approach would account for the entire dose previously delivered to the neck. The problem, however, is that such an approach would limit the dose one could deliver to the regions at risk in the contralateral neck since some dose would have been delivered there from the initial ispilateral treatment. Our compromise solution is to delete the fluence outside of SCBM and use this as our base plan. This allows us to fully account for the dose previously delivered to the spinal cord and brainstem while not limiting our ability to deliver an adequate dose to the PTVs in the postoperative neck.

The selection of the margin used for SCBM was made after several analyses. If no margins were chosen on the spinal cord and brainstem, the dose to these structures would be underestimated because the effects of scatter would be ignored. Conversely, if too large of a margin were chosen, you would approach using Plan A* without any modification. Our choice of a 1.0 cm margin for SCBM was felt to be a suitable compromise.

Although Plan A** substantially reduces the dose in the contralateral neck the effects of exit dose through the spinal cord and brainstem into the contralateral neck cannot be deleted by Eclipse. We have adjusted the dose constraints to account for this issue as follows. First, the mean doses in the postoperative PTVs were calculated for Plan A**. These doses were then added to all of the optimization constraints so we could have "room" to deliver the equivalent of the exit dose discussed above. Finally, Plan A** with the adjusted optimization constraints was used as a base plan to generate Plan B.

5. Conclusion

We have developed a novel planning technique for treatment of a contralateral neck recurrence after previous ipsilateral irradiation. The original IMRT plan is recalculated on the CT scan done after the recurrence. Modifications are then made to the plan so it can be used as base plan. This allows us to account for the cumulative dose delivered to the spinal cord and brainstem while enabling coverage of new PTVs in the contralateral neck.

6. References

[1] Sulman E, Scjwartz D, Leang T *et al* 2009 IMRT reradiation of head and neck cancer-disease control and morbidity outcomes *Int J Radiat Oncol Biol Phys* **73** 399–409
[2] Ang KK, Jiang GL, Feng Y *et al* 2001 Extent and kinetics of recovery of occult spinal cord injury *Int J Radiat Oncol Biol Phys* **50** 1013–1020
[3] Zwicker F, Roeder F, Hauswald H *et al* 2011 Reirradiation with intensity-modulated radiotherapy in recurrent head and neck cancer *Wiley Online Library* DOI: 10.1002/hed.21663
[4] Siu L, Waldron J *et al* 2008 A phase III study of standard fractionation radiotherapy with concurrent high-dose cisplatin versus accelerated fractionation radiotherapy with panitumumab in patients with locally advanced stage III And IV squamous cell carcinoma of the head and neck *National Cancer Institute Of Canada Clinical Trials Group (NCIC CTG)*, NCIC CTG TRIAL: HN.6

Journal of Physics: Conference Series **489** (2014) 012047 doi:10.1088/1742-6596/489/1/012047

Radiation treatment dose optimisation using Poisson tumour control probability parameters

G A Cho[1,2], M A Ebert[3,4], L Holloway[2,5], Z Kuncic[2], C Baldock[6] and D I Thwaites[2]

[1]Department of Radiation Oncology, Royal Prince Alfred Hospital, Sydney Local Health District, Camperdown, NSW, Australia
[2]Institute of Medical Physics, School of Physics, University of Sydney, Sydney, NSW, Australia
[3]Radiation Oncology Department, Sir Charles Gairdner Hospital, Hospital Avenue, Nedlands, WA, Australia
[4]School of Physics, University of Western Australia, Crawley, WA, Australia
[5]Liverpool and Macarthur Cancer Therapy Centres, Sydney South Local Health District, Liverpool, NSW, Australia
[6] Faculty of Science, Macquarie University, Sydney, NSW, Australia

E-mail: gwi.cho@sswahs.nsw.gov.au

Abstract. This study examines the Poisson tumour control probability (TCP) γ_{37} and D_{37} parameters of a uniformly irradiated numerical tumour model using changes in tumour burden as a surrogate for treatment response information. An optimum dose D_i for a tumour sub-volume element V_i is described that maximizes TCP as a function of fixed tumour integral dose ξ. TCP was calculated for spatially-varying clonogen density for a total 10^8 cells and radiosensitivity α with mean radiosensitivity in the range 0.4 – 1.0 Gy^{-1}. A bivariate normal distribution is used to describe the radiosensitivity α and the linear term of the linear-quadratic (LQ) cell kill governed the changes in the regional tumour burden within sub-volumes V_i. The optimum dose distribution, D_i, for V_i is obtained as a function of fixed tumour integral dose ξ. For a uniform dose delivery and for TCP = 37%, γ_{37} and D_{37} are described by the effective radiosensitivity $\alpha_{\rm eff}$ and the effective clonogen number $N_{0,\rm eff}$, respectively. $\alpha_{\rm eff}$ is equivalent to differential dose changes in the number of clonogenic cells (tumour burden). The γ_{37} values were found to be inversely correlated with variance of the probability density function of the α distribution. For the biologically optimum dose distribution, γ_{37} was found to converge to the theoretical maximum limit and D_{37} was found to reduce relative to that obtained for the uniform dose case. The TCP parameters γ_{37} and D_{37} could thus be useful in optimising individual radiation treatment doses even when tumour heterogeneity is taken into account.

1. Introduction

With the availability of functional and molecular imaging modalities such as Positron Emission Tomography (PET), the acquisition and application of tumour specific, *in vivo* biological information to routine radiation treatment has become feasible. Biologically guided radiation treatment aims to achieve biological conformity of the dose distribution within tumour regions of higher recurrence risks to maximize the probability of local tumour control [1, 2].

A validated biological response model can be used as an objective function for dose optimization of intensity modulated radiation treatment (IMRT), whereby the search for the optimum dose distribution can be directly based on treatment response to radiotherapy [3]. While experimental biological data are lacking, the effect on treatment response of intra-tumour biological heterogeneity has nevertheless been investigated in several theoretical modeling studies using tumour control probability (TCP) [4, 5, 6]. These studies showed that the variation of the radiosensitivity parameter α has a dominant influence on the TCP and dose response parameters.

In this study, the biological consequence of intra-tumour biological heterogeneity was studied under both uniform and biologically optimum (i.e. maximal TCP) dose distributions using the standard linear-quadratic (LQ) cell survival and mechanistic Poisson TCP models [3, 7]. These models were chosen for the current investigation due to their extensive application to radiation treatment response modeling and dose optimization studies. For uniform dose distribution, the steepness and the position of TCP curves were evaluated quantitatively with normalized dose gradient γ_{37} and the dose required to achieve 37% of tumour control, D_{37}. The normalized dose gradient γ is a relative measure of increase in TCP per given increment in dose. For Poisson TCP model, the steepest point of dose response curve is located at the 37% response level and, γ_{37} is a function of number of (effective) clonogenic cells. Using the steepest part of the Poisson TCP curves and γ_{37} and D_{37} parameters, the aim was to evaluate quantitatively, the effect on the dose response curve of the biologically optimum dose distribution, in relation to the optimization criterion applied (TCP maximization) and also against the uniform dose cases as the reference. The relationship between tumour burden and γ_{37} and D_{37} parameters was investigated. The optimum dose distribution that maximizes TCP for a given fixed tumour integral dose ξ, was described and the gain in TCP from a biologically optimized dose distribution was quantified in terms of γ_{37} and D_{37} parameters.

2. Methods and materials

2.1. Poisson TCP and LQ cell survival models

It was assumed that a tumour is a parallel structure and is curable only if there is no surviving clonogen remaining after treatment [8]. For a discretized tumour model consisting of a group of M independent sub-volume elements (with the i-th element having volume V_i), the TCP is described by

$$TCP(V_1+V_2,\ldots,V_M)=\prod_{M} TCP_M \tag{1}$$

where TCP_M = exp($-N_M$) is the probability of finding zero surviving clonogenic cells when the expected mean number of clonogenic cells is $N_M \neq 0$. The radiation response of tumour cells is described by the linear term of the LQ cell survival model [5]; for element M with local radiosensitivity, α_M, and surviving numberof cells, N_M, after dose D_M, is $N_M = N_{M,0}\cdot\exp(-\alpha_M D_M)$, where $N_{M,0}$ was the initial number of clonogenic cells. The TCP curve was parameterized as a function of γ_{37} and D_{37} parameters [9].

2.2. Optimum dose distribution

The optimum dose distribution that maximizes TCP was investigated using the Lagrange multiplier method. The objective was to find the maximum TCP under the constraints of fixed integral dose to a tumour. The objective function F was defined as the log of the TCP:

$$F(D_1,\ldots,D_M)=\sum_{i=1}^{M}-\left(N_{i,0}\cdot e^{-\alpha_i D_i}\right) \tag{2}$$

where $N_{i,0}$ is the initial number of clonogenic cells, α_i is the radiosensitivity and D_i is the biologically optimum dose calculated for the i-th tumour element. The constraint of the fixed integral dose to a tumour is met if $\sum_{i=1}^{M} D_i - \xi = 0$, where ξ is the tumour integral dose. The constraint function ϕ is defined as:

$$\phi(D_1,\ldots,D_M)=\sum_{i=1}^{M} D_i - \xi = 0 \tag{3}$$

We define the auxiliary function G as:

$$G(D_i,\ldots,D_M,\lambda)=\sum_{i=1}^{M}-\left(N_{i,0}\cdot e^{-\alpha_i D_i}\right)+\lambda\left(\sum_{i=1}^{M}D_i-\xi\right) \tag{4}$$

where λ is the Lagrange multiplier. Taking the partial derivative of G with respect to D_i, we can apply the multiplier condition:

$$\frac{\partial G}{\partial D_i}=\alpha_i N_{i,0}e^{-\alpha_i D_i}+\lambda=0 \tag{5}$$

2.3. Tumour model

A numerical tumour model consisting of 2^{10} elements was generated in MATLAB®. Each element of the tumour model was described by the number of clonogenic cells N and the mean radiosensitivity parameter α. The clonogenic cells were distributed across four consecutive transverse planes with each plane consisting of 2^8 tumour elements. The clonogenic cell distribution was set to have a uniform higher clonogenic cell concentration at the core and exponentially decreasing density towards the edge of the tumour model:

$$\begin{aligned}\rho(r)&=N_n & r\le r_{0,n}\\ &=N_n\cdot\exp(-r/3.2) & r>r_{0,n}\\ n&=1,2,3,4\end{aligned} \tag{6}$$

with $r_{0,1}=r_{0,4}=0.45$ cm and $r_{0,2}=r_{0,3}=0.72$ cm, respectively. For each transverse plane, N_n was set to obtain 2×10^7, 3×10^7, 3×10^7 and 2×10^7 cells, resulting in total 10^8 clonogenic cells.

The radiosensitivity of human tumours is difficult to determine *in vivo* [10]. Currently no known molecular imaging modality can directly provide spatial intra-tumour radiosensitivity distribution *in vivo* [11]. With limited information on the true distribution of radiosensitivity, α was described by a randomly distributed bivariate normal distribution [12], as follows:

$$\alpha_{\mathrm{mn}}=\alpha(x_{\mathrm{m}},y_{\mathrm{n}})=\alpha_{\mathrm{mean}}\cdot c_{\mathrm{normal}}\cdot p_{\mathrm{normal}}(x_i,y_j) \tag{7}$$

$$p_{\mathrm{normal}}(x_i,y_j)=\frac{1}{2\pi\sigma_x\sigma_y}\exp\left[-\frac{(x_i-\bar{x})^2}{2\sigma_x^2}-\frac{(y_j-\bar{y})^2}{2\sigma_y^2}\right] \tag{8}$$

where $c_{\mathrm{normal}}\cdot\langle p_{\mathrm{normal}}(x_i,y_j)\rangle=1$. With equal variance and zero co-variance terms, the probability density function provided a basic form of normally distributed α distributions that can be mapped on the clonogenic cell distribution. A 1-D Gaussian alpha distribution was applied to the LQ cell survival and Poisson TCP models of the same mathematical forms [6]. The variance parameters $\sigma_x^2=\sigma_y^2=10.0$ and 11.0 were chosen to yield the coefficient of variation in surviving fraction at 2 Gy to be between 10 - 15% [13] for $\alpha_{\mathrm{mean}}=0.4$ and less than 30% at $\alpha_{\mathrm{mean}}=0.8$ and were used with mean radiosensitivity $\alpha_{\mathrm{mean}}=0.4-1.0$ Gy^{-1}. Ten series of $p_{\mathrm{normal}}(x_i, y_j)$ were generated for each variance parameter to calculate the corresponding TCPs.

3. Results and discussion

3.1. Biologically optimum dose distribution

The equation for biologically optimum D_i distribution was obtained as a function of fixed tumour integral dose ξ from equation (5).

$$D_i = \frac{\beta + \ln(\alpha_i N_{i,0})}{\alpha_i} \tag{9}$$

$$\beta = -\ln(-\lambda) = \left[\xi - \sum_{i=1}^{M} \ln\left(\alpha_i N_{i,0}\right) \cdot \alpha_i^{-1}\right]\left(\sum_{i=1}^{M} \alpha_i^{-1}\right)^{-1} \tag{10}$$

3.2. TCP γ_{37} and D_{37} parameters for uniform and biologically optimum dose distribution

For a uniform dose delivered to the non-uniform model tumour, the TCP γ_{37} and D_{37} parameters were defined by $\gamma_{37}=\ln N_{0,\mathrm{eff}} \cdot e^{-1}$ and $D_{37}=e \cdot \gamma_{37} \cdot (\alpha_{\mathrm{mean}})^{-1}$ respectively. $N_{0,\mathrm{eff}}$ is the effective clonogenic cell number and α_{eff} is the effective radiosensitivity, which is the differential dose change in the number of clonogenic cells at TCP = 37%. $N_{0,\mathrm{eff}}$ is a function of α_{eff} and D_{37} at TCP = 37% . Figure 1 shows the variation of mean $\alpha_{\mathrm{eff}} \cdot (\alpha_{\mathrm{mean}})^{-1}$ for the uniform dose delivery. Tables 1 and 2 show averages of ten TCP γ_{37} and D_{37} parameters, $\gamma_{37,\mathrm{mean}} = \sum_{k=1}^{10} \gamma_{37,k}/10$ and $D_{37,\mathrm{mean}} = \sum_{k=1}^{10} D_{37,k}/10$ for the uniform and biologically optimum dose distributions for a tumour mean dose of 90 Gy (and with fixed tumour integral dose).

Tumours with higher heterogeneity in radiosensitivity (lower variance parameter, $\sigma_x^2 = \sigma_y^2$), exhibit lower $\gamma_{37,\mathrm{mean}}$ values, which are also below the theoretical limit of 6.777 = $\ln 10^8 \cdot e^{-1}$. $\gamma_{37,\mathrm{mean}}$ values are largely insensitive to variations in α_{mean} and correlate inversely with the coefficient of variation of the α distribution (results not shown). $D_{37,\mathrm{mean}}$ decreases with higher α_{mean} and $\sigma_x^2 = \sigma_y^2$. For the biologically optimum dose distribution, $\gamma_{37,\mathrm{mean}}$ converges to the maximum theoretical limit. Negligible differences in $\gamma_{37,\mathrm{mean}}$ were found between $\sigma_x^2 = \sigma_y^2$ = 10.0 and 11.0. A similar observation was made for $D_{37,\mathrm{mean}}$ (0.5%) (cf. 4.5% for uniform dose case).

In this study, a range of α_{mean} values have been considered up to 1.0 Gy^{-1} as an upper limit. For a tumour with a majority of clonogens consisting of a highly radiosensitive population (e.g. α_{mean} = 1.0 Gy^{-1}), the tumour control could be dominated by the small number of radioresistant sub-population of cells. If this were the primary cause of radiation treatment failure for a given tumour, the current dose optimization (dose redistribution) approach may enable more efficient use of radiation dose to control the regions of high resistance to achieve the optimum tumour control.

An earlier study by Yang and Xing [14] explored an IMRT dose optimization framework incorporating spatial biology distribution using LQ cell survival and Poisson TCP models. The current study investigated a similar question but with a different approach in that the dose prescription was derived as an implicit function of tumour integral dose. In the current approach, the intermediate step that was needed to correlate the reference voxel dose with a voxel dose in question is no longer required. This study also evaluated the effect of uniform and biologically optimised dose distributions using a plan quality indicator, the Poisson TCP at 37% of tumour control, instead of dose volume criteria [14].

Further work is required to incorporate the effect on TCP of uncertainty in the radiosensitivity parameter estimation and to use the realistic (deliverable) dose model that results from biologically optimum dose prescription D_i.

4. Conclusion

For a heterogeneous tumour treated with a uniform radiation dose, TCP = 37% can be adequately described with parameters γ_{37} and D_{37} analogous to those employed for a homogeneous tumour, by introducing an effective radiosensitivity α_{eff} and an effective clonogen number $N_{0,\mathrm{eff}}$. The parameter α_{eff} is equivalent to differential dose changes in the number of clonogenic cells, which could potentially be determined experimentally. For a heterogeneous tumour treated with a biologically optimized dose distribution, on the other hand, γ_{37} is maximized, while D_{37} is reduced. These results suggest that the Poisson TCP parameters γ_{37} and D_{37} could be useful in optimizing individual radiation treatment doses.

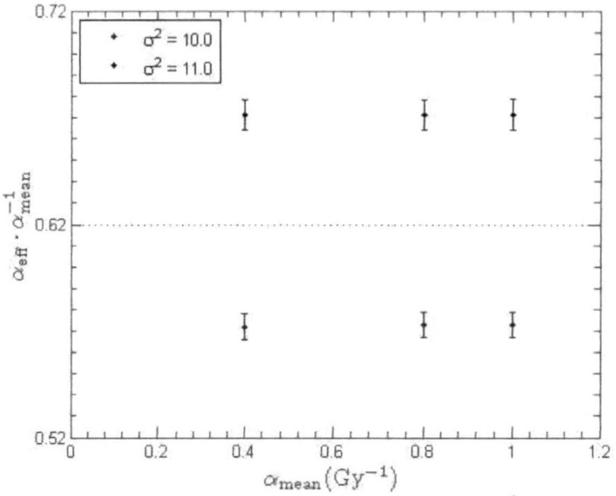

Figure 1. Variation of effective radiosensitivity $\alpha_{eff} \cdot (\alpha_{mean})^{-1}$. Error bars indicate the standard error of the mean (SEM).

Table 1. TCP $\gamma_{37,mean}$ and $D_{37,mean}$ parameters for uniform dose distribution.

α_{mean}	$\sigma_x^2 = \sigma_y^2 = 10.0$		$\sigma_x^2 = \sigma_y^2 = 11.0$	
	$\gamma_{37,\,mean} \pm$ SEM	$D_{37,\,mean} \pm$ SEM	$\gamma_{37,\,mean} \pm$ SEM	$D_{37,\,mean} \pm$ SEM
0.4	5.436±0.038	64.76 ±0.35	5.544±0.019	61.95±0.23
0.8	5.436±0.038	32.38±0.17	5.544±0.019	30.98±0.11
1.0	5.436±0.038	25.90 ±0.14	5.543±0.019	24.78±0.09

Table 2. TCP $\gamma_{37,\,mean}$ and $D_{37,\,mean}$ parameters for biologically optimum dose distribution.

α_{mean}	$\sigma_x^2 = \sigma_y^2 = 10.0$		$\sigma_x^2 = \sigma_y^2 = 11.0$	
	$\gamma_{37,\,mean} \pm$ SEM	$D_{37,\,mean} \pm$ SEM	$\gamma_{37,\,mean} \pm$ SEM	$D_{37,\,mean} \pm$ SEM
0.4	6.770±0.000	47.65±0.03	6.771±0.000	47.42±0.02
0.8	6.775±0.000	23.85±0.01	6.775±0.000	23.73±0.01
1.0	6.776±0.000	19.08±0.01	6.776±0.000	19.01±0.01

References

[1] Ling C, Humm J, Larson S, Amols H, Fuks Z, Leibel S and Koutcher J 2000 *Int. J. Radiat. Oncol. Biol. Phys.* **47** 551
[2] Bentzen S 2005 *Lancet Oncol.* **6** 112
[3] Journal of the ICRU Report 83, Vol 10 2010 *Prescribing, Recording, and Reporting Photon-Beam Intensity-Modulated Radiation Therapy (IMRT)* (Oxford: Oxford University Press)
[4] Brahme A and Agren A 1987 *Acta. Oncol.* **26** 377
[5] Webb S and Nahum A 1993 *Phys. Med. Biol.* **38** 653
[6] Ebert M and Hoban P 1996 *Phys. Med. Biol.* **41** 2125
[7] Li *et al.* 2012 *Med. Phys.* **39** 1386
[8] Munro T and Gilbert C 1961 *Br. J. Radiol.* **34** 246
[9] Bentzen S and Tucker S 1997 *Int. J. Radiat. Biol.* **71** 531
[10] Wigg D 2001 *Applied radiobiology and bioeffect planning.* (Madison:Medical Physics Publishing) p 236 - 238
[11] Gregoire V, Jeraj R, Aldo Lee J and O'Sullivan B 2012 *Lancet Oncol.* **13** e292
[12] Schultheiss T, Zagars G and Peters L 1987 *Radiother. Oncol.* **9** 241

[13] Suit H, Skates S, Taghian A, Okunieff P and Efird T 1992 *Radiother. Oncol.* **25** 251
[14] Yang Y and Xing L 2005 *Med. Phys.* **32** 1473

Journal of Physics: Conference Series **489** (2014) 012048 doi:10.1088/1742-6596/489/1/012048

Automatic Atlas Based Electron Density and Structure Contouring for MRI-based Prostate Radiation Therapy on the Cloud

J.A. Dowling*[1], N. Burdett[1], P.B. Greer[2,3], J. Sun[2,3], J. Parker[2], P. Pichler[2], P. Stanwell[2,3], S. Chandra[1], D. Rivest-Hénault[1], S. Ghose[1], O. Salvado[1] and J. Fripp[1]
[1]CSIRO Australian e-Health Research Centre, Royal Brisbane and Women's Hospital, Queensland, Australia
[2]Calvary Mater Newcastle Hospital, New South Wales, Australia
[3]University of Newcastle, New South Wales, Australia

E-mail: jason.dowling@csiro.au

Abstract. Our group have been developing methods for MRI-alone prostate cancer radiation therapy treatment planning. To assist with clinical validation of the workflow we are investigating a cloud platform solution for research purposes. Benefits of cloud computing can include increased scalability, performance and extensibility while reducing total cost of ownership. In this paper we demonstrate the generation of DICOM-RT directories containing an automatic average atlas based electron density image and fast pelvic organ contouring from whole pelvis MR scans.

1. Introduction

There are a number of benefits of the use of MRI for soft tissue visualisation in prostate cancer radiation therapy. However treatment planning requires the use of computed tomography (CT) scans for simulation. This is necessary as there are currently limited methods to convert MRI intensities to electron density (which is a requirement for dose calculations to be performed directly on MRI scans). Advantages of planning directly on MRI scans would include a reduction in costs and the uncertainties introduced by the registration process. A further motivation is the recent developments in MRI simulation and MRI accelerators (1–3) which bring the potential for MRI based adaptive treatment.

Our research group has been investigating the feasibility of MRI-alone treatment planning for prostate cancer radiation therapy. We have previously published methods for electron density generation from MRI with <2% difference between pseudo-CT and planning CT in point dose at isocentre (4), fast automatic organ contouring (5,6) and proposed a revised MRI-alone workflow (7). To assist with clinical validation of the MR-alone workflow this paper concentrates on a remote cloud platform solution.

* To whom any correspondence should be addressed.

2. Method

2.1.1. Image data

As described in Dowling, et al. (4), a whole pelvis MRI atlas was generated based on 39 manually delineated 1.5T T2w MRI scans from patients prior to external beam radiation therapy for localised prostate cancer. In addition a conjugate electron-density atlas was generated from the co-registered CT-MRI scans (Figure 1). These atlases contain contours for the pelvic bones, bladder, rectum and prostate.

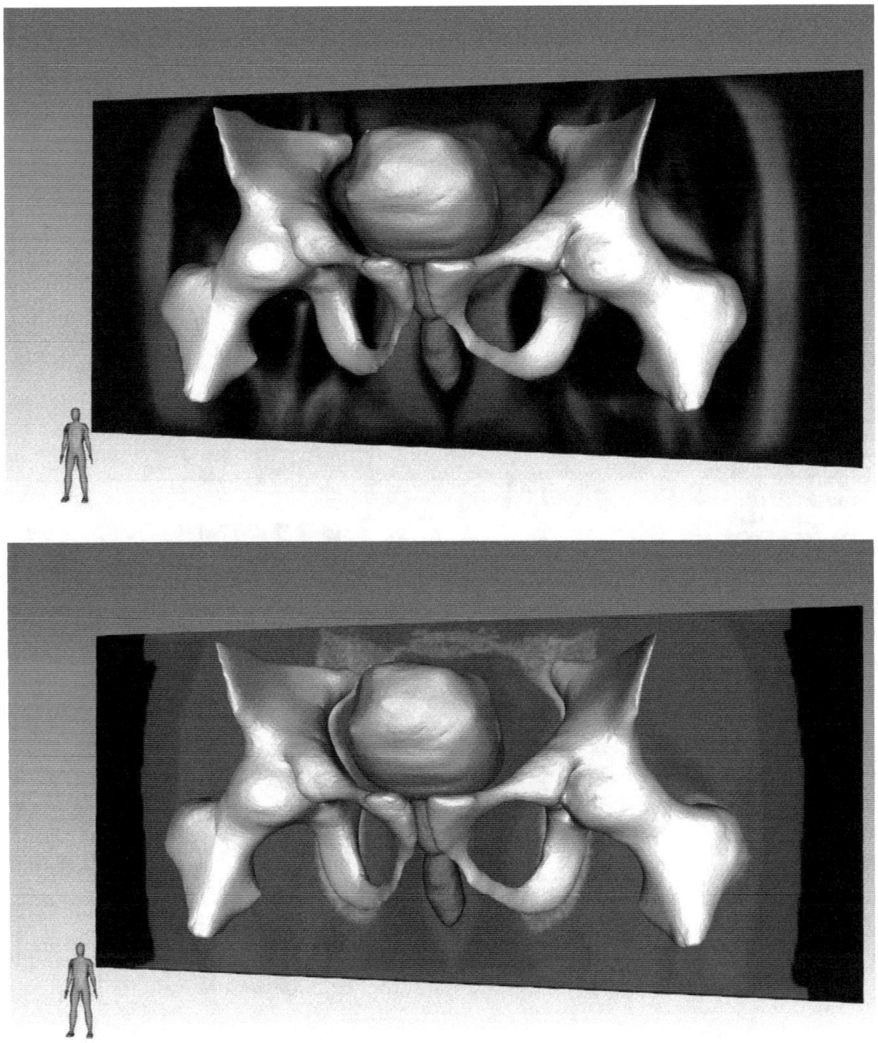

Figure 1. MRI (top) and matching pseudo CT (bottom) pelvic atlas with surface models of the main pelvic organs (bones, bladder, rectum and prostate).

2.1.2. Processing software

The current cloud pipeline for pseudo CT generation and automatic contouring is shown in Figure 2. The interface requires a DICOM series containing a whole pelvis (male) large field of view T2w MRI. Once received the DICOM series is converted into a single 3D volume (NifTI file format, http://nifti.nimh.nih.gov/nifti-1) and is pre-processed prior to average atlas based electron density estimation and initial organ contouring. Atlas based segmentation usually involves an atlas image (generally an average of a set of images) with a matching set of organ labels (eg. Figure 1). To segment a new image, the MRI atlas is registered to a patient's whole pelvis T2w MR image to obtain a good correspondence between structurally equivalent regions in the two images, and then the labels defined on the atlas are propagated to the image.

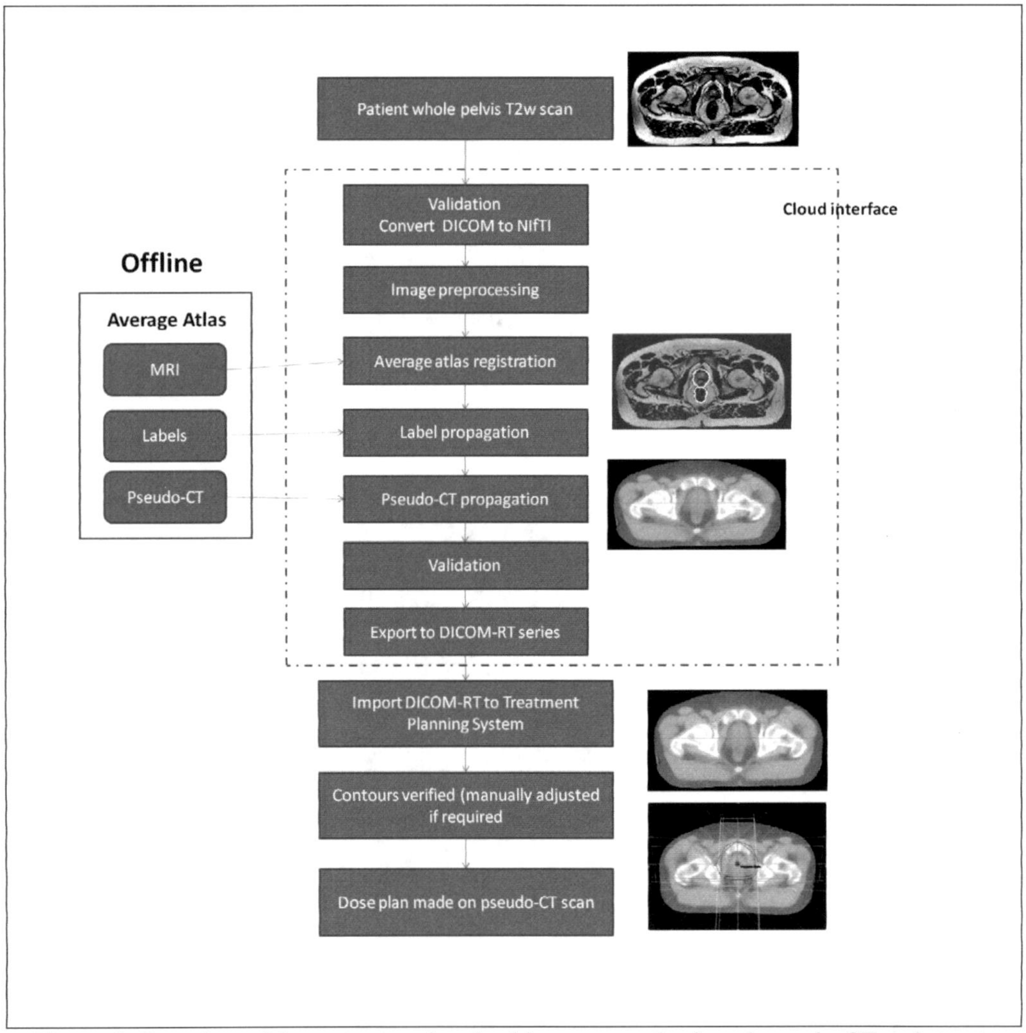

Figure 2. Overview of the main steps involved in average atlas based pseudo-CT and structure generation on the cloud

The technical details of the atlas generation has been reported previously (7). To generate a pseudo-CT from a patient's MRI the same transformation and deformations are then applied to the CT atlas (Figure 1, right) to create a pseudo-CT corresponding to the patient MRI scan anatomy. This pseudo-CT is then used for dose planning and digitally reconstructed radiograph (DRR) generation (Figure 2). In the final step, the resulting volume and structures are converted to a DICOM-RT series for the clinical site to download, QA and modify if necessary.

2.1.3. Cloud interface and environment

A medical imaging analysis and visualization platform has been developed at the CSIRO Australian e-Health Research Centre called MILXView, incorporating a number of advanced algorithms that provide improved automated image analysis and interpretation of medical imaging data. MILXView is designed to process data on a desktop install however the aim is to enable access and execute these algorithms in a cloud environment. The cloud offers several distinct advantages over desktop processing; including achieving economies of scale, reduced spending of technology infrastructure and support, improved accessibility, minimising licensing new software and the ability to deploy software upgrades instantly.

Galaxy (8) has been adapted to execute MILXView binaries in the cloud. Galaxy is a collection of XML and Python scripts that allow users to upload data to a cloud environment and then execute binaries using the uploaded data as arguments. Tools can be quickly added to the user interface by updating a single configuration XML file which states the location of a tool XML file. This tool XML file specifies the binaries or scripts to be executed and their arguments and output parameters.

All data uploaded and produced via Galaxy is stored in a database. Galaxy uses SQLite by default, but can use MySQL or PostgreSQL to overcome concurrency limitations. Galaxy contains a standalone web server. However, some tasks such as serving static content can be offloaded to a dedicated server for efficiency.

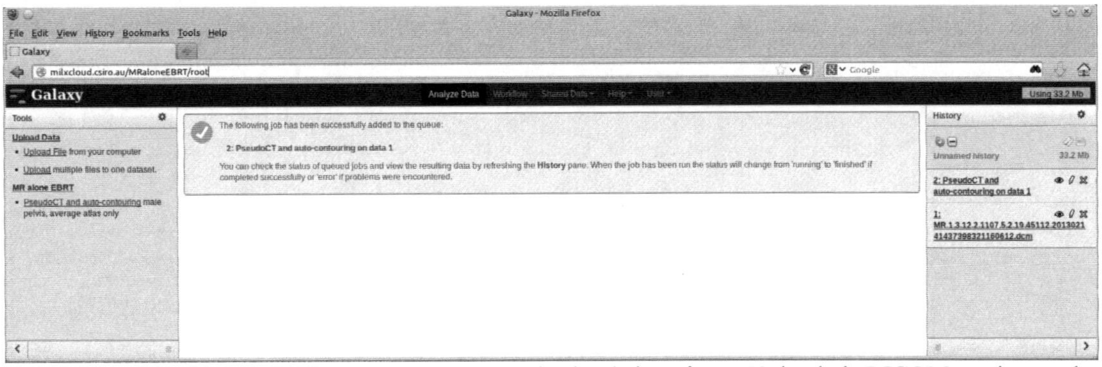

Figure 3. Screen snapshot from the proposed cloud interface. Uploaded DICOM series and automatically generated DICOM-RT directories are maintained in the history panel on the right hand side.

The installation of Galaxy is simple and nearly identical for UNIX/Linux and Mac OS X environments. It is not supported under Windows so users would have to build their own Python Eggs (a method of distributing Python packages). Galaxy can be run on a single desktop, or on a compute cluster. Galaxy currently supports TORQUE, PBS, PBS Pro, Platform LSF, HT Condor, Open grid Engine and Sun Grid Engine clusters and does not require a dedicated or special cluster configuration. Finally, Galaxy can be instantiated on cloud computing infrastructures (primarily Amazon Elastic Compute Cloud (EC2)), or in a user created cloud environment using CloudMan.

3. Results

3.1.1. Pseudo-CT generation

The cloud implementation enables the generation of pseudoCT and automatic bone, bladder, prostate, and rectum contours within approximately 3.5 minutes (using the average atlas shown in Figure 1). Figure 4 shows qualitative results from the automatic conversion of a 3T T2w whole pelvis MRI.

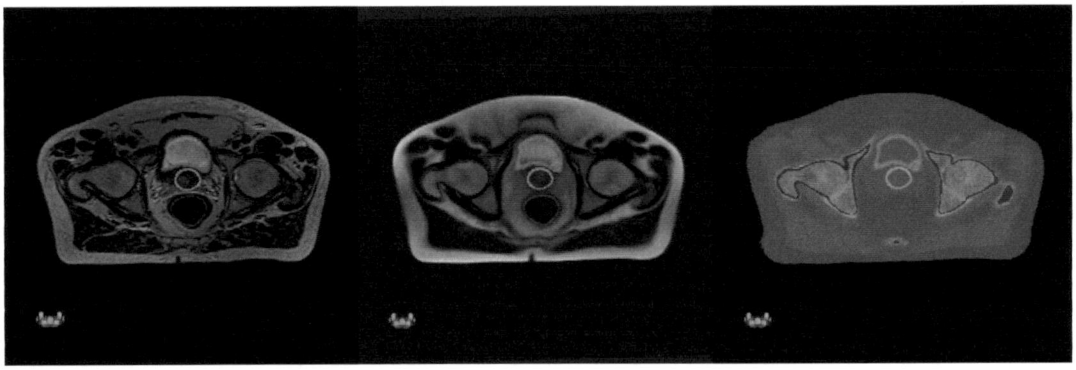

Figure 4. Left: Axial sample slice from a whole pelvis T2w MR scan (not part of the atlas set). The central image shows the result from deformable registration of the whole pelvis MRI atlas (from Figure 1) to the new scan; and the matching pseudo-CT is shown on the right. Automatic organ segmentations for the prostate, bladder, rectum and bones have been overlaid.

4. Discussion

The cloud research in this paper has been motivated by a need to allow increased scalability, performance and extensibility while reducing total cost of ownership and providing access to powerful CPU and GPU clusters for efficient parallel methods in medical image analysis. The cloud architecture has been applied in this work to demonstrate a previously published method working through a cloud interface and producing DICOM-RT files which are ready for import into a commercial treatment planning system for research purposes. The architecture allows us to easily include new methods for pseudo-CT and automatic contouring (for example fast, parallel multi-atlas registration and selection methods), and to interface with both internal and commercial clusters.

Journal of Physics: Conference Series **489** (2014) 012048 doi:10.1088/1742-6596/489/1/012048

Acknowledgments
Jason Dowling acknowledges funding support from the Cancer Council NSW (RG 11-05), Prostate Cancer Foundation of Australia (YI2011), Movember and Cure Cancer Australia.

References

1. Karlsson M, Karlsson MG, Nyholm T, Amies C, Zackrisson B. Dedicated magnetic resonance imaging in the radiotherapy clinic. International Journal of Radiation Oncology, Biology, Physics. Elsevier; 2009 Jun;74(2):644–51.

2. Raaymakers BW, Lagendijk JJW, Overweg J, Kok JGM, Raaijmakers AJE, Kerkhof EM, et al. Integrating a 1.5 T MRI scanner with a 6 MV accelerator: proof of concept. Physics in Medicine and Biology. 2009;54(12):N229–N237.

3. Lee YK, Bollet M, Charles-Edwards G, Flower MA, Leach MO, McNair H, et al. Radiotherapy treatment planning of prostate cancer using magnetic resonance imaging alone. Radiotherapy and Oncology. 2003 Feb;66(2):203–16.

4. Dowling JA, Lambert J, Parker J, Salvado O, Fripp J, Capp A, et al. An atlas-based electron density mapping method for magnetic resonance imaging (MRI)-alone treatment planning and adaptive MRI-based prostate radiation therapy. International Journal of Radiation Oncology, Biology, Physics. 2012 May 1;83(1):e5–11.

5. Dowling JA, Fripp J, Chandra S, Pluim JPW, Lambert J, Parker J, et al. Fast automatic multi-atlas segmentation of the prostate from 3D MR images. Prostate Cancer Imaging. Image Analysis and Image-Guided Interventions. Springer; 2011. p. 10–21.

6. Chandra S, Dowling J, Shen K, Raniga P, Pluim J, Greer P, et al. Patient Specific Prostate Segmentation in 3D Magnetic Resonance Images. IEEE Transactions on Medical Imaging. 2012 Aug 2;31.

7. Greer PB, Dowling JA, Lambert JA, Fripp J, Parker J, Denham JW, et al. A magnetic resonance imaging-based workflow for planning radiation therapy for prostate cancer. The Medical Journal of Australia. 2011 Feb;194(4):S24–S27.

8. Goecks J, Nekrutenko A, Taylor J. Galaxy: a comprehensive approach for supporting accessible, reproducible, and transparent computational research in the life sciences. Genome biology. 2010 Jan;11(8):R86.

Integration of second cancer risk calculations in a radiotherapy treatment planning system

M Hartmann[1,2], U Schneider[1,2]

[1] Radiotherapy Hirslanden AG, Rain 34, Aarau, Switzerland
[2] Physics Institute – Medical Physics, University of Zurich, Zurich, Switzerland

E-mail: matthias.hartmann@hirslanden.ch

Abstract. Second cancer risk in patients, in particular in children, who were treated with radiotherapy is an important side effect. It should be minimized by selecting an appropriate treatment plan for the patient. The objectives of this study were to integrate a risk model for radiation induced cancer into a treatment planning system which allows to judge different treatment plans with regard to second cancer induction and to quantify the potential reduction in predicted risk. A model for radiation induced cancer including fractionation effects which is valid for doses in the radiotherapy range was integrated into a treatment planning system. From the three-dimensional (3D) dose distribution the 3D-risk equivalent dose (RED) was calculated on an organ specific basis. In addition to RED further risk coefficients like OED (organ equivalent dose), EAR (excess absolute risk) and LAR (lifetime attributable risk) are computed. A risk model for radiation induced cancer was successfully integrated in a treatment planning system. Several risk coefficients can be viewed and used to obtain critical situations were a plan can be optimised. Risk-volume-histograms and organ specific risks were calculated for different treatment plans and were used in combination with NTCP estimates for plan evaluation. It is concluded that the integration of second cancer risk estimates in a commercial treatment planning system is feasible. It can be used in addition to NTCP modelling for optimising treatment plans which result in the lowest possible second cancer risk for a patient.

1. Introduction

In several major analyses of the atomic bomb survivor data a dose-response relationship for radiation carcinogenesis up to one or two Gy has been quantified. This low dose range is important for radiation protection purposes. For the estimate of radiation induced cancer risk it is also important to know the shape of the dose-response curve for higher dose ranges. A dose-response model for radiotherapy relevant dose ranges can be achieved by combining the linear-no-threshold model (LNT) derived from the atomic bomb survivors from Hiroshima and Nagasaki with cancer risk data available from about 30,000 patients with Hodgkin's disease who were irradiated with localized doses of up to around 40Gy.

Especially for children and adolescent patients, who were treated with radiotherapy, second cancer risk is one of the important side effect, as other side effects are reduced by using of organ sparing treatment techniques, like IMRT, VMAT or IMPT, developed in the last decade. To minimize radiation induced cancer risk corresponding analyzing tools should be available in the treatment planning process selecting an appropriate treatment plan for the patient. However in recent treatment planning systems this criterion is not subjected.

2. Methods

A dose-response model [1] for carcinoma induction, whose dose ranges are valid for radiotherapy treatment, was integrated into a commercial treatment planning system (Eclipse 11.0, Varian) using the Eclipse Scripting API 11.0 class library. Based on the three-dimensional dose distribution the risk equivalent dose distribution (RED(D)), including cell killing, fractionation and repopulation/repair effects, was calculated. RED is a parameter measured in Gy proportional to second cancer risk. Besides a dose-response model for carcinoma induction,

$$RED(D) = \frac{e^{-\alpha' D}}{\alpha' R}\left(1 - 2R + R^2 e^{\alpha' D} - (1-R)^2 e^{-\frac{\alpha' R}{1-R}D}\right) , \qquad (1)$$

an additional model for sarcoma induction was implemented [2]

$$RED(D) = \frac{e^{-\alpha' D}}{\alpha' R}\left(1 - 2R + R^2 e^{\alpha' D} - (1-R)^2 e^{-\frac{\alpha' R}{1-R}D} - \alpha' R D\right) , \qquad (2)$$

where it assumed that the tissue is irradiated with a fractionated treatment schedule of equal dose fractions d up to a dose D. The number of cells is reduced by cell killing which is proportional to α' and is defined using the linear quadratic model

$$\alpha' = \alpha + \beta d = \alpha + \beta \frac{D}{D_T} d_T , \qquad (3)$$

where D_T and d_T is the prescribed dose to the target volume with the corresponding fractionation dose, respectively. The repopulation/repair parameter R characterizes the repopulation/repair-ability of the tissue between two dose fractions and is 0 if no and 1 if full repopulation/repair occurs. The R values for the different tissues were taken from [2]. It is assumed here an $\alpha/\beta = 3$Gy for all tissues.

Based on the RED-Volume histograms (REDVH) several organ specific risk coefficients are computed. The excess absolute risk (EAR) is factorized into the dose-response function RED(D) and a modifying function μ(agea, agex) that depends on the variables age at exposure (agex) and age attained (agea), including a gender specific factor and the organ specific incidence value β:

$$\mathrm{EAR} = \frac{1}{V_T}\sum_i V(D_i) \cdot \beta \cdot \mathrm{RED}(D_i) \cdot \mu(\mathrm{agex}, \mathrm{agea}) \qquad (4)$$

with the total organ volume V_T, the bins of the dose-volume histogram $V(D_i)$ and dose D_i of the voxel i of the organ.

The organ equivalent dose (OED) [2]:

$$\mathrm{OED} = \frac{1}{V_T}\sum_i V(D_i) \cdot \mathrm{RED}(D_i) \qquad (5)$$

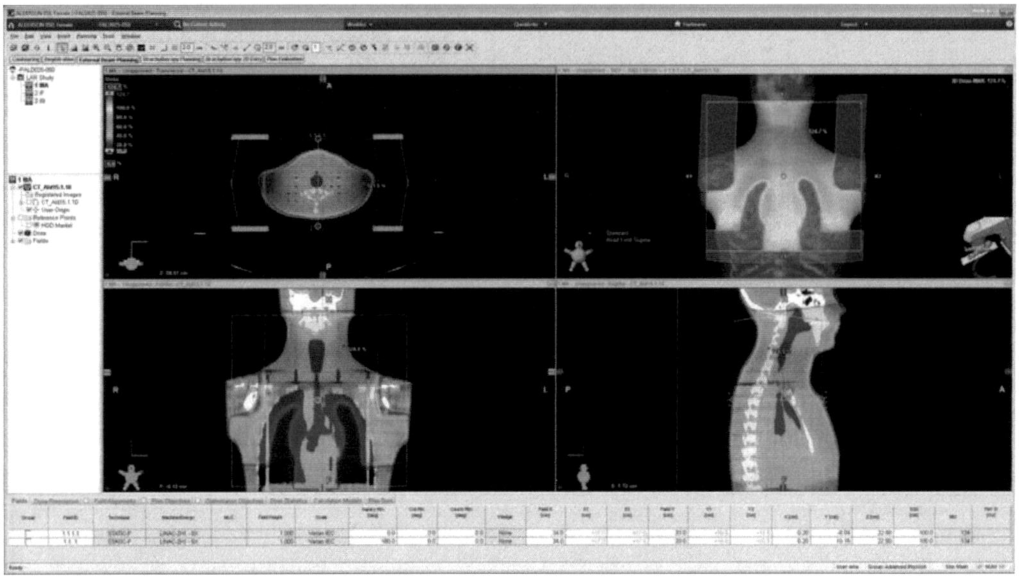

Figure 1. Treatment plan calculated on an Alderson phantom for a typical Hodgkin patient radiation, using a classical mantle field technique.

is a dose-response weighted dose variable averaged over the whole organ volume. OED values are independent of the initial slope β and the modifying function μ. Its corresponding uncertainties are at a minimum.

At last the lifetime attributable risk (LAR) [3] :

$$\mathrm{LAR} = \int_{\mathrm{agex}}^{\infty} \mathrm{EAR}(\mathrm{D}, \mathrm{agex}, \mathrm{agea}) \frac{\mathrm{S}(\mathrm{agea})}{\mathrm{S}(\mathrm{agex})} \, \mathrm{d}(\mathrm{agea}) \tag{6}$$

with the conditional probability S(agea)/S(agex) of a patient alive at age agex to reach at least age agea are calculated for different organs.

In addition to the risk coefficients EAR, OED and LAR of different organs the organ specific RED-volume histograms were computed.

3. **Results**

The TPS-integrated risk model was tested with a typical Hodgkin treatment. Using the Alderson phantom three different Hodgkin plans were prepared. The first plan (MA) is a classical mantle field technique treatment displayed in figure 1, the second one includes involved fields (IF) and for the third an involved node (IN) technique was used. The calculated risk coefficients are showed in figure 2.

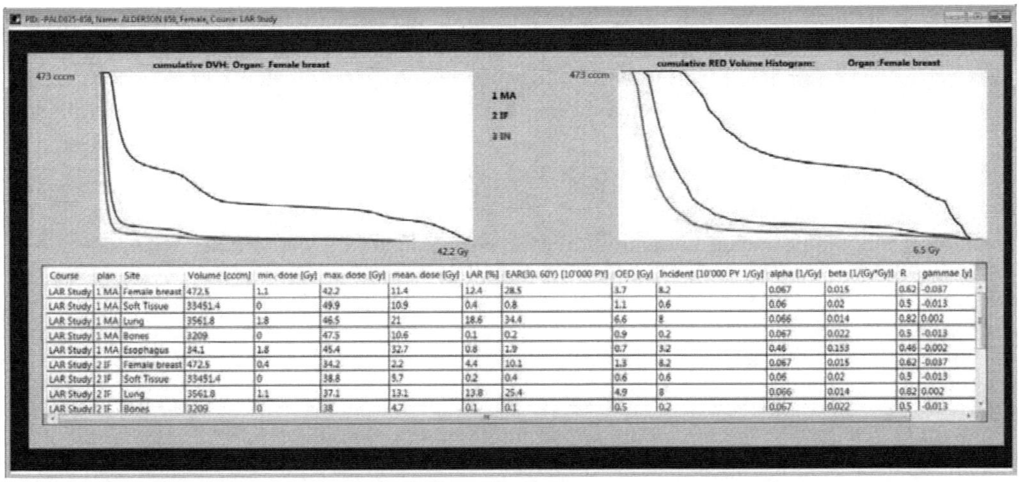

Figure 2. Risk coefficient evaluation (LAR, EAR, OED and RED-volume-histogram) of three different Hodgkin patient plans: 1) mantle field (MF), 2) involved field (IF) and 3) involved node (IN) technique.

For MA the computed LAR for breast and lung cancer was 12.4% and 18.6%, respectively. The calculated risks are in satisfying agreement with literature data of historical mantle field treatments [4, 5]. The expected potential risk improvement for the involved field respectively the involved node techniques are confirmed by LAR values of 4.4% for breast cancer and 13.8% for lung cancer and respectively 2.7% and 8.8% for the IN plan. On the right side of figure 2 the RED-Volume histograms for the breast are shown. EAR is calculated for agex 30 and agea 60 years.

The calculated risks allow to judge different treatment plans with regard to second cancer induction and to choose the best plan with regards to second cancer risk.

4. Discussion

The model described by Equation (1) and (2) include several assumptions to simplify the biological processes leading to cancer induction. This includes the design of tissues, the repopulation process and processes which result in the formation of a tumor cell. This was done to keep the number of model parameters at a minimum. However, this is associated with uncertainties.

One assumption of the presented model is that single dose fractions of a radiotherapy treatment are treated independently. Therefore the linear-no-threshold theory for cancer induction could be applied to each single dose fraction. Although this may be valid for a single exposure lower than 3Gy, it is not clear whether it can be applied for dose fractions which are separated by days when in fact not all cells are fully repaired.

It is also not clear whether the same cell kill parameter α' applies for normal cells and transformed cells as it is applied in Equation (1) and (2). It may be necessary to use a different α'–values.

Another assumption made here is that all biological processes occur instantaneously. It is known that in contrast to the physical processes involved, the biological changes act on a longer time scale. To account for such time delays, in particular for repopulation, a feed-back term in the differential equations would be required.

Many problems and uncertainties are involved in modelling the underlying biology of radiation induced cancer. However, since the models are used for a relative comparison of different treatment plans regarding cancer induction the uncertainties are smaller than for if the models are used for absolute cancer risk estimates.

5. Conclusion

A risk model for radiation induced cancer was successfully integrated in a treatment planning system. Quantitative risk values like OED, EAR and LAR were calculated. It is concluded that the integration of second cancer risk estimates in a treatment planning system is feasible. The risk coefficients can be easily attained during the planning process by one additional click. The plan evaluation tool can be used in addition to NTCP modeling for optimizing treatment plans which result in the lowest possible second cancer risk for a patient.

References

[1] Schneider U. 2009 *Med. Phys.* **36** 1138
[2] Schneider U, Sumila M and Robotka J. 2011 *Theor Biol Med Model.* **26** 27
[3] Kellerer A., Nekolla E. and Walsh L.. 2001 *Radiat. Environ. Biophys.* **40** 249
[4] De Bruin ML, Sparidans J, van't Veer MB, et al. 2009 *J Clin Oncol.* **27** 4239
[5] Hodgson DC, Gilbert ES, Dores GM, et al. 2007 *J Clin Oncol.* **25** 1489

Journal of Physics: Conference Series **489** (2014) 012050 doi:10.1088/1742-6596/489/1/012050

GPU-accelerated ray-tracing for real-time treatment planning

H Heinrich[1], P Ziegenhein[1], C P Kamerling[1], H Froening[2] and U Oelfke[1]

[1] German Cancer Research Center (DKFZ), Heidelberg 69120 Germany
[2] Institute of Computer Engineering, University of Heidelberg, Heidelberg 69120 Germany

E-mail: h.heinrich@dkfz.de

Abstract. Dose calculation methods in radiotherapy treatment planning require the radiological depth information of the voxels that represent the patient volume to correct for tissue inhomogeneities. This information is acquired by time consuming ray-tracing-based calculations. For treatment planning scenarios with changing geometries and real-time constraints this is a severe bottleneck. We implemented an algorithm for the graphics processing unit (GPU) which implements a ray-matrix approach to reduce the number of rays to trace. Furthermore, we investigated the impact of different strategies of accessing memory in kernel implementations as well as strategies for rapid data transfers between main memory and memory of the graphics device. Our study included the overlapping of computations and memory transfers to reduce the overall runtime using Hyper-Q. We tested our approach on a prostate case (9 beams, coplanar). The measured execution times for a complete ray-tracing range from 28 msec for the computations on the GPU to 99 msec when considering data transfers to and from the graphics device. Our GPU-based algorithm performed the ray-tracing in real-time. The strategies efficiently reduce the time consumption of memory accesses and data transfer overhead. The achieved runtimes demonstrate the viability of this approach and allow improved real-time performance for dose calculation methods in clinical routine.

1. Introduction

The computation of the radiological depth data is a vital preprocessing step for radiotherapy treatment planning dose calculations. Since the depth data represents the patient geometry, dose algorithms can use it to correct for tissue inhomogeneities along incident rays. The computation of the radiological depth data relies on time consuming ray-tracing operations typically carried out for every voxel of the patient volume. In adaptive radiotherapy (ART) [1][2][3] treatment planning scenarios with changing patient geometries, this is a severe bottleneck.

Our group investigates a new IMRT treatment planning paradigm (interactive dose shaping) which requires rapid access to radiological depth information for changing patient geometries [4][5].

GPUs are powerful high core count devices that are no longer solely used for graphics applications, but also for general-purpose computing tasks. Compared to CPUs, their computing cores are weaker but the massive amount of these cores results in a significant performance increase. Also, GPUs offer memory bandwidth that is about one order of magnitude higher compared to single-socket CPUs. However, GPUs only excel in performance if (1) their vast amount of computing cores can be kept busy, and (2) if they can perform the computations in-core, i.e. if the input and output data associated with the computation is stored in the on-device memory. Otherwise, performance can degrade due to data movements over the bandwidth-limited PCIe interface. In this work we implemented a GPU-based ray-tracing algorithm using CUDA [6]. We investigated the impact of different strategies to increase the memory access efficiency with respect to the runtime. This includes the potentially inefficient memory access patterns of the ray-tracing kernel functions to the on-device graphics memory as well as the overhead resulting from data transfers to and from the host's main memory. In addition we investigated the benefits of using GPU capabilities for concurrent kernel execution and data movement. Results are obtained for a prostate patient data set.

2. Materials & Methods

2.1. The ray-tracing algorithm

Our ray-tracing implementation is based on the algorithm proposed by Siddon [7], which was adapted for the GPU by utilizing the stepping approach in [8] and the ray-matrix approach in [9]. A ray-matrix is defined at the iso-center of the treatment field perpendicular to each incident beam. The size of the matrix is defined by the extension of the treatment field plus a scattering margin on each side of a beam. Each matrix element represents a point in 3D space. Rays are defined from the beam source, through each matrix element and tracing is performed until a ray exits the patient. The number of points is determined so that the distance of two neighboring rays at the patient's exit plane is equal to the voxel size. The ray-matrix approach can greatly reduce the number of ray-tracing processes in comparison to tracing every voxel of the patient volume for each beam. In addition, every ray can be traced independently from every other ray, which fits well to the CUDA programming model and allowed us to decomposed the problem so that one thread processes the ray-tracing along one ray. For each beam we execute one kernel which spawns a thread for each point in the corresponding ray-matrix.

2.2. Time-to-solution

We use time-to-solution (TTS) as metric to evaluate the performance of our implementation. We define TTS as the accumulated runtimes of the ray-tracing kernel functions on the GPU, the required input data transfer to the graphics device and the transfer of the computed radiological depth data cubes back to the host's main memory as illustrated in Figure 1.

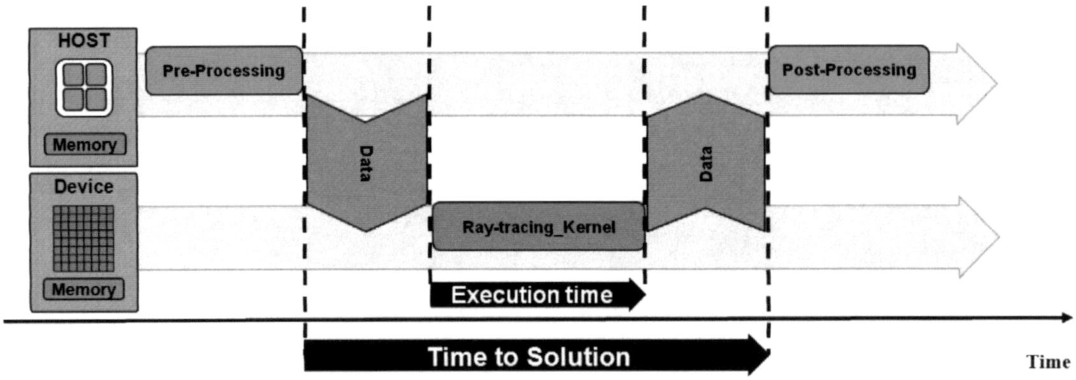

Figure 1. Definition of time-to-solution.

2.3. Acceleration strategies

GPUs employ an explicit memory hierarchy that is exposed to the user using NVidia's CUDA framework. Accessing data from e.g. global or texture memory has an impact on the utilization of the computational resource of the graphics hardware. The global memory is the main memory of NVidia graphics devices. It is cached (although not to reduce access latency, but to reduce contention) and reaches best performance for coalesced access patterns, i.e. when threads access consecutive memory addresses. In contrast, texture memory is optimized for accessing grid data. It reaches best performance for access patterns based on spatial locality.

Our initial implementation of the ray-tracing kernel, the global memory kernel (GMK), accesses the input data stored in global memory. As acceleration strategy (1) we implemented a second version, the texture memory kernel (TMK), which accesses the input data provided as a 3D texture. This strategy focuses on the data transfer from graphics memory and computational units.

The input and output data for the both the GMK and TMK kernel implementations is transferred to and from paged host memory. This can be a potential bottleneck as demand-paging might swap pages out, decreasing access times. To avoid this, the CUDA driver has to copy the data from a paged memory pointer to pinned memory pointer before it can invoke the Host to device (H2D) and device to host (D2H) data transfers using direct memory access (DMA). Allocating the host memory as pinned or page-lock memory permits demand-paging and therefore enables the GPU to access the host's main memory directly using DMA without copy overhead. Strategy (2) focuses on the data transfer between the host's main memory and the device memory. We enhanced our implementation for both, GMK and TMK kernel by using page-lock memory on the host side. The expected benefit is an approximately 2x higher memory bandwidth for the input/output data transfers and thus a reduction of the TTS.

Table 1. Overview of strategies to reduce the TTS.

Strategy (1)	Enhance data transfer between graphics memory and computational units	GMK vs. TMK both implemented using paged host memory
Strategy (2)	Enhance data transfer between host's main memory and graphics memory	Paged vs. Page-lock memory implemented for both GMK and TMK
Strategy (3)	Enhance concurrency of data transfers and computation	Hyper-Q implemented for both GMK and TMK

Strategy (3) focuses on overlapping computational workloads and data transfers. This can be accomplished using Hyper-Q, a novel feature introduced with CUDA 5.0, available on NVidia's Tesla K20 graphics hardware of the current Kepler architecture [10]. To utilize Hyper-Q, data transfers as well as kernel invocation are required to be implemented for asynchronous execution, i.e. that the control flow returns immediately to the CPU after invocation. Dependencies between asynchronous invocations can be expressed with streams, which are first-in-first-out queues of work packets. Separate streams can be processed concurrently by the graphics hardware if sufficient resources are available. The actual scheduling of the different computational tasks and data transfers is carried out transparently to the user. We enhanced our implementation with asynchronous data transfers and kernel invocations and organized independent tasks in separate streams to exploit Hyper-Q. It is expected that computation and memory transfers of multiple beams overlap which can contribute to reduction in TTS. Table 1 provides a quick overview of the different strategies and versions of the ray-tracing implementation.

2.4. Development & test environment

The proposed acceleration strategies were implemented in CUDA 5.0 and C/C++ using MS VisualStudio 2010. We tested the implementations on a Win7 workstation equipped with an Intel Xeon E5 CPU with 64 GByte RAM and an NVidia Tesla K20c with 5 GByte graphics memory.
We used a IMRT treatment plan for a prostate patient with 9 coplanar beams and 256 x 256 x 234 voxels of size $(2.62\ \mathrm{mm})^3$ as test data set. The ray-matrix size for each of the 9 beams includes the beam extension plus an 8 cm scattering margin on each side. The ray-matrix dimensions range 140 x 140 to 186 x 186 points.
To test the accuracy of our GPU implementations used the serial CPU implementation of the ray-tracing algorithm used in [9] as reference.

3. Results

The results of our GPU implementations were compared to the reference implementation. Both produce equal results within single precision accuracy.
The runtimes measured for the different versions of our ray-tracing implementation are depicted in Table 2. The columns 3 to 5 present the runtimes in msec for the Kernel execution on the GPU, the data transfer times H2D and D2H and the TTS. The runtimes are accumulated over 9 kernel invocations, a single data transfer of the input volume and the 9 data transfers of output volumes. For the Hyper-Q versions though, the kernel execution time represents the time from the first kernel start until the last kernel finished.

Table 2. The runtimes for the ray-tracing implementations in msec for a 9 beam prostate case. The presented runtimes are accumulated runtimes over 9 beams.

Rutimes [msec] 9 Beams	Kernel	Kernel Execution	Data transfer H2D / D2H	Time to Solution
Paged Memory	GMK	50	16 / 234	300
	TMK	32	16 / 213	261
Page-lock Memory	GMK	50	10 / 84	144
	TMK	32	10 / 84	126
Hyper-Q	GMK	48	10 / 84	99
	TMK	28	10 / 84	99

A snippet of the output of NVidia's Visual Profiler normalized to an identical time line is shown in Figure 2. It allows an intuitive visual access to the work flow on the GPU and the impact of the implemented acceleration strategies. The golden boxes represent the time consumption of data transfers between host and graphics hardware. The aqua blue boxes show the runtime of the GMK and the purple boxes represent the TMK. The first golden box depicts the single input data transfer to the graphics device and the remaining ones account for the transfers of the output cubes to the host's main memory, one cube for each beam respectively.

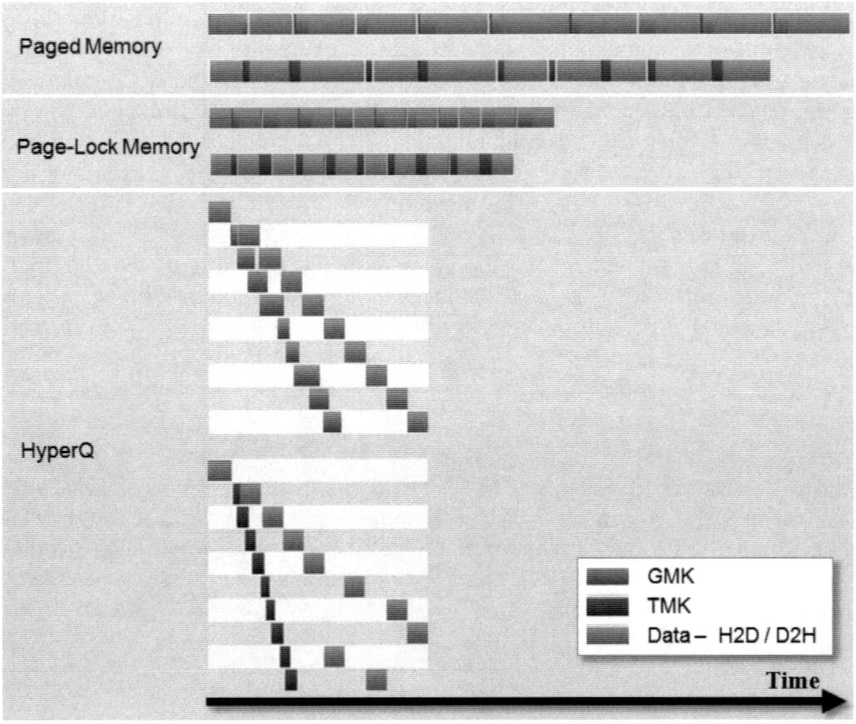

Figure 2. NVidia Visual Profiler output for a 9 beam prostate case. The colored boxes represent runtimes for the different version of the ray-tracing implementations: aqua blue GMK, purple TMK, the respective first golden box the input data transfers and the remaining golden boxes the output data transfers to the host's main memory.

4. Discussion

We investigated three acceleration strategies for enhancing ray-tracing performance for the radiological depth computation for real-time treatment planning. The application of the strategies can enhance the efficiency of the IMRT treatment planning process in particular when changing patient geometries are considered.

We defined the TTS metric to assess the benefit of offloading the ray-tracing to the graphic device for our in-house TPS framework [4][5]. However, many other applications require the results back in the host's memory and can benefit from the proposed strategies. Our first strategy (1) focuses on the impact of using texture memory as opposed to global memory for the ray-tracing kernel implementations. The TMK showed a better performance, since the ray-tracing memory access patterns are based rather on spatial locality than temporal locality. This difference is especially prominent for beam angles which require data access patterns not mapping well to caching strategies based on locality of reference. The profiler output shows reduced sizes of the purple boxes in comparison to the aqua blue boxes and thus indicates the TMK performance to be rather independent from the beam angle.

For strategy (2), we enhanced the implementations by using page-lock host memory with both kernels. The performance gain is shown in the profiler output by reduced sizes of the golden boxes. The memory bandwidth observed was stable above 6 Gbyte/sec across the PCI-e interface which amounts to approximately a factor of 2 versus the implementations using paged host memory.

The final strategy (3) leads to the largest reduction in TTS. It includes overlapping kernel executions and output data transfers back to the host using Hyper-Q. To find that both concurrent implementations for GMK and TMK achieved the same TTS was not expected. It indicates that the rather inefficient memory accesses using global memory implemented in the GMK were efficiently hidden using Hyper-Q.
Combining the strategies, we were able to compute the ray-tracing related tasks in 28 msec for 9 treatment beams in clinical resolution using graphics hardware of NVidia's current Kepler architecture. We exploited its capability of concurrent kernel execution and data movement. We obtained runtimes of 99 msec which corresponds to a speed-up of 3 in TTS in comparison to the initial implementation of GMK and TMK using paged host memory. Future work will include the integration of multi GPU support to study the scaling behavior of our implementation. Furthermore, we will investigate the application of the GPU-based ray-tracing in adaptive treatment planning scenarios for rapid response to changing patient geometries.

5. Conclusion

We have shown that the computation of radiological depth data for each beam and each voxel of a patient volume can be carried out in real-time using graphics hardware. The memory bandwidth between the host and the device is the limiting factor for ray-tracing on the GPU. The presented strategies minimize the time consumption of memory accesses and data movements using novel features of NVidia's Kepler GPU architecture. We found an optimal time-to-solution by combining our texture memory kernel with Hyper-Q.

References

[1] Yan D, Vicini F, Wong J and Martinez A 1997 Adaptive radiation therapy *Phys. Med. Biol.* **42** 123-32
[2] Wu C, Jeraj R, Olivera G H and Mackie T R 2002 Re-optimization in adaptive radiotherapy *Phys. Med. Biol.* **47** 3181-95
[3] de la Zerda A, Armbruster B and Xing L 2007 Formulating adaptive radiation therapy (ART) treatment planning into a closed-loop control framework *Phys. Med. Biol.* **52** 4137-53
[4] Ziegenhein P, Kamerling C P, Oelfke U 2013 Interactive Dose Shaping Efficient Strategies for CPU-based Real-Time Treatment Planning *J. Phys.: Conf. Ser.* presented at ICCR 2013, not yet published
[5] Kamerling C P, Ziegenhein P, Heinrich H, Oelfke U 2013 A 3D isodose manipulation tool for interactive dose shaping *J. Phys.: Conf. Ser.* presented at ICCR 2013, not yet published
[6] NVIDIA CUDA Compute Unified Device Architecture Programming Guide 5.0 edition 2012 *NVIDIA Corp., Santa Clara, CA*
[7] Siddon R L 1985 Fast calculation of the exact radiological path for a three-dimensional CT array. *Med. Phys.* **12** 252-5
[8] de Greef M, Crezee J, van Eijk J C, Pool R, Bel A 2009 Accelerated ray tracing for radiotherapy dose calculations on a GPU *Med. Phys.* **36(9)** 4095-102
[9] Siggel M, Ziegenhein P, Nill S, Oelfke U 2012 Boosting runtime-performance of photon pencil beam algorithms for radiotherapy treatment planning *Physica Medica* **28(4)** 273-80
[10] Whitepaper 2012 NVIDIA's Next Generation CUDA Compute Architecture: Kepler GK110 *NVIDIA Corp., Santa Clara, CA*

Predicting the likelihood of QA failure using treatment plan accuracy metrics

T Kairn[1,2,5], S B Crowe[2], J Kenny[3,4], R T Knight[1] and J V Trapp[2]
[1] Premion Cancer Care, Brisbane, Australia
[2] Science and Engineering Faculty, Queensland University of Technology, Brisbane, Australia
[3] Australian Clinical Dosimetry Service, Australian Radiation Protection and Nuclear Safety Agency, Melbourne, Australia
[4] Radiation Oncology Queensland, Toowoomba, Australia

E-mail: t.kairn@gmail.com

Abstract. This study used automated data processing techniques to calculate a set of novel treatment plan accuracy metrics, and investigate their usefulness as predictors of quality assurance (QA) success and failure. A small sample of 151 beams from 23 prostate and cranial IMRT treatment plans were used in this study. These plans had been evaluated before treatment using measurements with a diode array system. The TADA software suite was adapted to allow automatic batch calculation of several proposed plan accuracy metrics, including mean field area, small-aperture, off-axis and closed-leaf factors. All of these results were compared to the gamma pass rates from the QA measurements and correlations were investigated. The mean field area factor provided a threshold field size (5 cm^2, equivalent to a 2.2 x 2.2 cm^2 square field), below which all beams failed the QA tests. The small aperture score provided a useful predictor of plan failure, when averaged over all beams, despite being weakly correlated with gamma pass rates for individual beams. By contrast, the closed leaf and off-axis factors provided information about the geometric arrangement of the beam segments but were not useful for distinguishing between plans that passed and failed QA. This study has provided some simple tests for plan accuracy, which may help minimise time spent on QA assessments of treatments that are unlikely to pass.

1. Introduction

This study aimed to identify one or more intensity modulated radiotherapy (IMRT) treatment plan dosimetric accuracy metrics, calculable from treatment plan parameters, that could be used to identify plans that were likely to fail routine quality assurance (QA) tests. The use of such metrics to identify failing treatments at the plan completion stage could reduce the linac time required for IMRT QA testing and contribute to the continuous quality improvement of the treatment planning process.

Existing IMRT complexity and deliverability metrics are based on the heterogeneity of the fluence map produced by each beam [1, 2, 3] or on the variations between MLC positions and beam segment aperture geometries [4, 5, 6]. Most of these metrics have been shown to successfully distinguish between IMRT treatment plans with different levels of complexity (for example, distinguishing

[5] To whom any correspondence should be addressed.

between prostate IMRT treatment plans and more-complex head and neck IMRT treatment plans), while showing no clear correlation with increasing dosimetric error for individual treatment sites [3, 6, 7, 8].

This study therefore investigated a set of IMRT treatment plan accuracy metrics that were designed to be sensitive to the treatment plan parameters that are most likely to compromise accurate dose calculations; small field and small segment aperture sizes [9, 10], closed MLC leaves below open linac jaws [11], and small field segments delivered from off-axis positions [12, 13].

2. Method

This study used the results of pre-treatment QA measurements made using a MapCheck2 diode array (Sun Nuclear Corporation, Melbourne, USA), for a small sample of 151 beams from 20 prostate and 3 cranial IMRT treatments, planned and delivered using the BrainLab iPlan treatment planning system (with pencil-beam algorithm) and Brainlab m3 microMLC (Brainlab, Feldkirchen, Germany), with a Varian iX linac (Varian Medical Systems, Palo Alto, USA). The small sample size was determined by the infrequency of microMLC-based IMRT treatments, especially of cranial targets, over the 4 month testing period. 17 of the prostate plans and 2 of the cranial plans were defined as "passed" and approved for treatment, with either 90% of measurement points (above a threshold of 10% of the maximum dose) resulting in $\gamma(2\%, 2mm) < 1.0$, or 95% of measurement points resulting in $\gamma(3\%, 3mm) < 1.0$, when compared with dose to the diode array calculated by the treatment planning system. The remaining plans, which did not meet these criteria, were defined as "failed". This gamma evaluation was performed using MapCheck software (Sun Nuclear Corporation, Melbourne, USA), using the Van Dyk percentage difference evaluation method [14], in absolute mode with no beam smoothing and with a 10% dose threshold.

The treatment plans were exported from the planning system in DICOM format for analysis. Automatic batch calculation of a set of novel treatment accuracy metrics was achieved using the TADA (Treatment And Dose Assessor) code [15], an extension to the MCDTK software suite [16, 17]. The TADA code was adapted to allow automatic batch calculation of metrics including:

- Mean field area (MFA), the average open area of field segments in each beam
- Mean aperture displacement (MAD), the average distance between the midway point between each pair of leaves and the central axis
- Cross-axis score (CAS), the proportion of open leaf pairs where one leaf crossed the central axis
- Closed leaf score (CLS), the proportion of leaf pairs that were closed, downstream of open jaws
- Small aperture score (SAS), the proportion of open leaf pairs that were separated by less than a given threshold distance.

These metrics were calculated separately for each individual beam and then combined in monitor unit weighted averages for each treatment plan.

The treatment plan analysis results obtained using TADA were compared with the pass rates from the QA tests, in order to identify which, if any, of the metrics could be related to the MapCheck2 pass rate or the likelihood of QA failure. The relationship between beam accuracy metrics and individual gamma pass rates was analysed using Excel 2007 (Microsoft, Redmond, USA), with linear regression used for trend estimation and p-value calculation, using two-tailed T-tests. Two-tailed Welch's T-tests were also used to obtain p values expressing the statistical significance of observed differences between the mean values of the accuracy metrics for the prostate treatment beams that passed their QA tests, the prostate treatment beams that failed their QA tests, the cranial treatment beams that passed their QA tests and the cranial treatment beams that failed their QA tests.

3. Results and Discussion

Figures 1(a)-(d) show the results of evaluating the metrics relating to the mean area of the IMRT beam segment apertures and the displacement of those apertures across the linac's central axis. The MFA results shown in figure 1(a) are correlated with gamma pass rates ($p \approx 0.02$) for both the prostate and the cranial treatment fields. Figure 1(a) also suggests that there is a threshold mean aperture size of 5

cm^2 (equivalent to a 2.2 x 2.2 cm^2 square) below which all beams fail their QA tests. Above this threshold, the MFA metric fails to distinguish between plans that pass and fail their QA tests, with 17% of beams with mean field areas between 5 cm^2 and 20 cm^2 failing their QA tests. Results for the MAD, CAS and CLS metrics, shown in figures 1(b), (c) and (d), respectively, provide no useful relationships with gamma pass rates, although they provide information about the geometric arrangement of the IMRT beam segments. The results for MAD and CAS are qualitatively similar, since the number of segments per beam that are centred away from the central axis is correlated with the number of MLC leaves per beam that cross the central axis. The CLS results in figure 1(d) are inversely related to the MFA results in figure 1(a) for similar, geometric reasons.

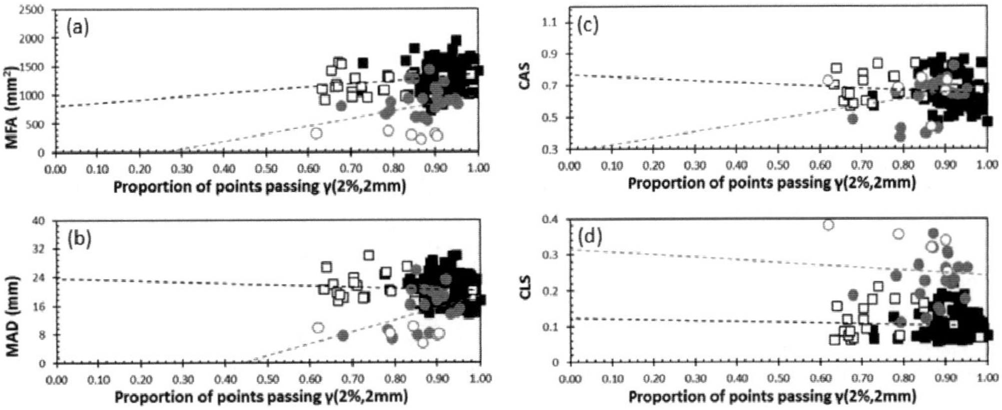

Figure 1. (a) Mean field area, (b) mean aperture displacement (c) cross-axis score and (d) closed leaf score, plotted against the proportion of measurement points in each field that passed a (2%, 2mm) gamma comparison with the planned dose. Filled data points represent fields from plans that were defined as "passed". Open data points represent fields from plans that were defined as "failed". Black squares represent beams from prostate treatments and grey circles represent beams from cranial treatments. Linear trend lines are included as a visual guide only.

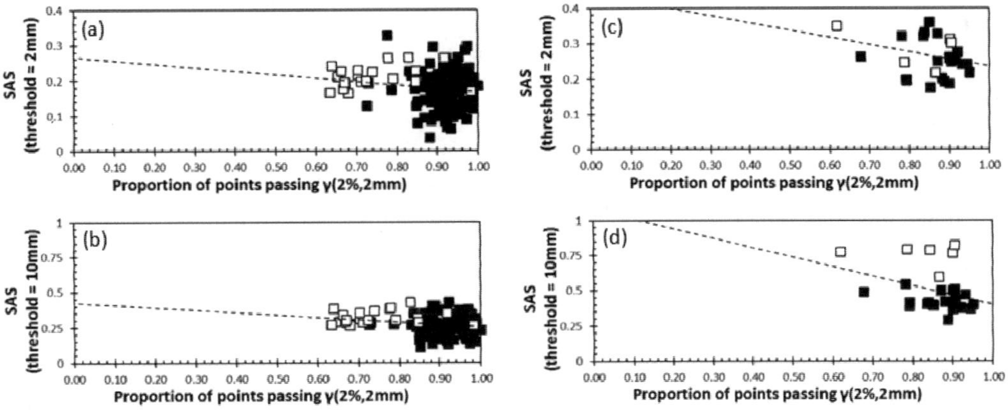

Figure 2. (a) Small aperture scores for prostate treatment beams, calculated using a threshold value of 2 mm, (b) small aperture scores for prostate treatment beams, calculated using a threshold value of 10 mm, (c) small aperture scores for cranial treatment beams, calculated using a threshold value of 2 mm, and (d) small aperture scores for cranial treatment beams, calculated using a threshold value of 10 mm, plotted against the proportion of measurement points in each

field that passed a (2%, 2mm) gamma comparison with the planned dose. Filled data points represent fields from plans that were defined as "passed". Open data points represent fields from plans that were defined as "failed". Linear trend lines are included as a visual guide only.

Figures 2(a) and (b) show the SAS values for the prostate IMRT treatment beams, respectively evaluated using thresholds of 2 mm and 10 mm to define the "small" aperture. Correlations between these metrics and the gamma pass rates for these individual beams are either weak ($p \approx 0.05$ when using the 10 mm threshold) or insignificant ($p \approx 0.2$ when using the 2 mm threshold), however it is apparent from figures 2(a) and (b) that when fewer than 18% of open MLC leaf pairs are open by less than 2 mm, or when fewer than 25% of open MLC leaf pairs are open by less than 10 mm, all beams pass their QA tests. When these small aperture scores are evaluated as monitor unit weighted averages over all beams in each treatment, the small aperture score (defined at 10 mm) has a useful threshold value at 0.32, above which all of the prostate plans that failed their QA tests, plus two of the plans that passed their QA tests, would be identified as likely to fail. All prostate plans with mean SAS values less than 0.32 passed their QA tests.

Results shown in figures 2(c) and (d) suggest that the SAS is similarly able to distinguish between cranial treatment beams that passed and failed their QA tests, when the "small" aperture is defined as 10 mm. Given that a small sample of beams from only one failing and two passing cranial treatments were available for use in this analysis, it is important that this result be confirmed by further study.

Tables 1 and 2 summarise the suitability of the metrics defined in this study for predicting the likely QA outcomes of the IMRT treatment plans examined here. Table 1 lists the mean treatment plan accuracy metrics for the prostate and cranial plans that passed and failed their QA tests. Table 2 provides an indication of the significance of differences between the calculated metrics for the different plan cohorts.

Table 1. Summary of plan accuracy metrics (mean and one standard deviation) for prostate and cranial plans that passed and failed their QA tests. The small aperture thresholds used in the SAS calculations are indicated in parentheses.

Metric	Prostate - pass	Prostate - fail	Cranial - pass	Cranial – fail
MFA (cm^2)	13 ± 2	12 ± 2	9 ± 2	3.0 ± 0.5
MAD (cm)	20 ± 4	21 ± 4	16 ± 6	8 ± 2
CAS	0.65 ± 0.09	0.68 ± 0.08	0.6 ± 0.1	0.7 ± 0.1
CLS	0.10 ± 0.04	0.12 ± 0.05	0.22 ± 0.07	0.35 ± 0.07
SAS (2mm)	0.17 ± 0.06	0.21 ± 0.03	0.25 ± 0.05	0.31 ± 0.08
SAS (10mm)	0.26 ± 0.07	0.32 ± 0.05	0.42 ± 0.06	0.75 ± 0.08

Table 2. Calculated p values (Welch's T-test), for comparisons between metrics listed in Table 1, for prostate plans that passed (PP) and failed (PF) their QA tests and cranial plans that passed (CP) and failed (CF) their QA tests. Small p-values (< 0.001) suggest that the differences between the datasets are unlikely to have arisen by random fluctuations.

PP vs PF	CP vs CF	PP vs CP	PF vs CF
0.02	< 0.001	< 0.001	< 0.001
0.39	< 0.001	< 0.001	< 0.001
0.21	0.52	0.43	0.81
0.13	0.005	< 0.001	< 0.001
< 0.001	0.14	< 0.001	0.03
< 0.001	< 0.001	< 0.001	< 0.001

4. Conclusion

This study evaluated a set of metrics designed to be sensitive to the treatment plan parameters that are most likely to compromise accurate dose calculations. Based upon our small sample our results suggest that some of these metrics may be used to identify plans that are unlikely to pass routine pre-treatment QA tests.The MFA factor provided a threshold field size (5 cm^2), below which all beams failed their QA tests. The SAS provided a useful predictor of plan failure, when averaged over all beams, despite being weakly correlated with gamma pass rates for individual beams. By contrast, the MAD, CAS and CLS factors provided information about the geometric arrangement of the beam segments but were not useful for distinguishing between plans that passed and failed QA. This study has provided some simple tests for plan accuracy, which may help minimise time spent on QA assessments of treatments that are unlikely to pass.

Acknowledgements

This study was supported by the Australian Research Council, the Wesley Research Institute, Premion and the Queensland University of Technology (QUT), through linkage grant number LP110100401.

References

[1] Webb S 2003 *Phys. Med. Biol.* **48**(14) 2051–2062
[2] Llacer J, Solberg T D and Promberger C 2001 *Phys. Med. Biol.* **46**(10) 2637–2663
[3] Nicolini G, Vanetti E, Clivio A, Fogliata A, Korreman S, Bocanek J and Cozzi L 2008 *Radiat. Oncol.* **3** 24
[4] McNiven A L, Sharpe M B and Purdie T G 2010 *Med. Phys.* **37** 505–515
[5] Lee M T, Purdie T G, Eccles C L, Sharpe M B, Dawson L A et al. 2010 *Radiat. Oncol.***5** 115–115.
[6] McGarry C K, Chinneck C D, O'Toole M M, O'Sullivan J M, Prise K M and Hounsell A R 2011 *Med. Phys.* **38** 2027–2034.
[7] Tonigan J, Kry S, Dong L, Purdie T, White R, Ibbott G and Followill D 2011 *Med. Phys.* **38** 3804.
[8] Tonigan J R 2011 Evaluation *of intensity modulated radiation therapy (IMRT) delivery error due to IMRT treatment plan complexity and improperly matched dosimetry data* Master's thesis University of Texas
[9] Kairn T, Crowe S, Kenny J and Trapp J V 2011 *Radiat. Meas.* **46**(12) 1985–1988
[10] Kairn T, Hardcastle N, Kenny J, Meldrum R, Tome W A and Aland T 2011 *Australas. Phys. Eng. Sci. Med.* **34**(3) 333–343
[11] Fenoglietto P, Lalibert´e B, Ailleres N, Riou O, Dubois J B and Azria D 2011 *Radiat. Oncol.* **6** 85
[12] Ahnesjo A and Aspradakis M M 1999 *Phys. Med. Biol.* **44**(11) R99–R155
[13] Brainlab AG 2010 *Brainlab Physics Technical Reference Guide* Revision 1.2
[14] Van Dyk J, Barnett R B, Cygler J E and Shragge P C 1993 *Int. J. Radiat. Oncol. Phys.* **26** 261–273
[15] Crowe S B, Kairn T, Middlebrook N, Hill B, Knight R T, Kenny J, Langton C M and Trapp J V 2013 *Australas. Phys. .Eng. Sci. Med.* **36**(1): 74-74
[16] Crowe S, Kairn T and Fielding A L 2009 *Radiother. Oncol.* **92** S71
[17] Crowe S B, Kairn T, Trapp J V and Fielding A L 2013 Experimental evaluation of MCDTK, the Monte Carlo DICOM ToolKit *World Congress on Medical Physics and Biomedical Engineering May 26-31, 2012, Beijing, China* (Berlin: Springer) pp. 1807–1810

Journal of Physics: Conference Series **489** (2014) 012052 doi:10.1088/1742-6596/489/1/012052

A 3D isodose manipulation tool for interactive dose shaping

C P Kamerling, P Ziegenhein, H Heinrich and U Oelfke

German Cancer Research Center (DKFZ), Department of Medical Physics in Radiation Oncology, Im Neuenheimer Feld 280, 69120 Heidelberg, Germany

E-mail: c.kamerling@dkfz.de

Abstract. The interactive dose shaping (IDS) planning paradigm aims to perform interactive local dose adaptations of an IMRT plan without compromising already established valuable dose features in real-time. In this work we introduce an interactive 3D isodose manipulation tool which enables local modifications of a dose distribution intuitively by direct manipulation of an isodose surface.

We developed an in-house IMRT TPS framework employing an IDS engine as well as a 3D GUI for dose manipulation and visualization. In our software an initial dose distribution can be interactively modified through an isodose surface manipulation tool by intuitively clicking on an isodose surface. To guide the user interaction, the position of the modification is indicated by a sphere while the mouse cursor hovers the isodose surface. The sphere's radius controls the locality of the modification. The tool induces a dose modification as a direct change of dose in one or more voxels, which is incrementally obtained by fluence adjustments. A subsequent recovery step identifies voxels with violated dose features and aims to recover their original dose.

We showed a proof of concept study for the proposed tool by adapting the dose distribution of a prostate case (9 beams, coplanar). Single dose modifications take less than 2 seconds on an actual desktop PC.

1. Introduction

Fluence modulation allows intensity modulated radiation therapy (IMRT) to provide a high spatial conformity of the prescribed dose to the tumor volume while minimizing dose in adjacent healthy tissue. Finding a clinically optimal set of fluence intensities typically relies on the iterative optimization of an objective function. This objective function includes pre-segmented volumes of interest and dose constraints which are weighted by penalty factors [1]. The indirect approach of finding the clinically optimal dose distribution suffers from various inherent shortcomings. First, the control of local dose features is limited to segmented volumes of interest, for example making it difficult to remove cold or hot spots. Second, there is no direct mapping between the parameters of the objective function and the resulting dose distribution. Generating a treatment plan may require a tedious loop of manual constraint adaptation and re-optimization. Third, when patient geometry changes between fractions, it is difficult to adapt the initial treatment plan accordingly. In the worst-case the whole optimization process has to be repeated, which depending on the optimization method involves the recalculation of dose-influence data or plan databases.

To overcome these shortcomings we propose a new planning paradigm: interactive dose shaping (IDS). It aims to perform interactive local dose adaptations of a plan without

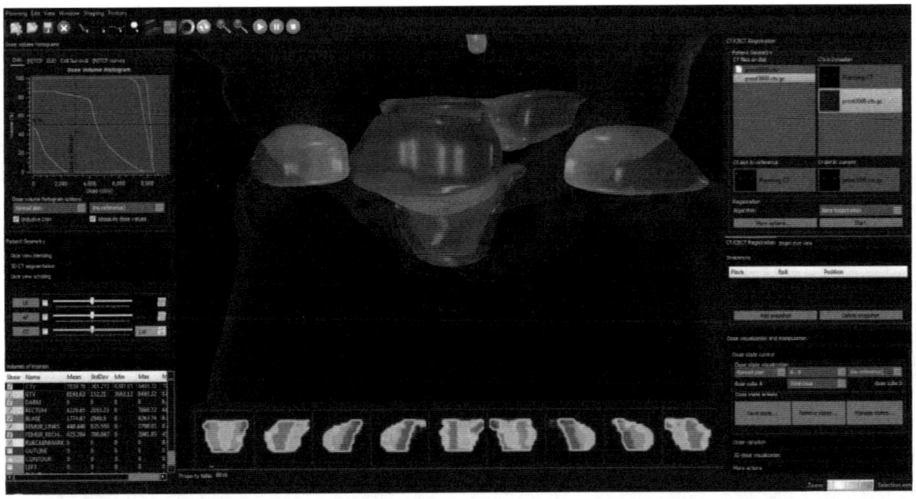

Figure 1. A screenshot from Dynaplan's main view.

compromising already established valuable dose features in real-time. The key operation that facilitates dose shaping is a two-step dose modification and recovery (DMR) strategy. The DMR strategy is triggered by direct interaction with a 3D dose distribution through a graphical user interface (GUI). In this work we introduce an interactive 3D isodose surface manipulation tool which enables local modifications of a dose distribution by sculpting an isodose surface. We show a proof of concept study for a prostate case and compare a plan obtained with our tool to a conventionally optimized treatment plan.

2. Materials & Methods

2.1. Software framework

We developed an in-house IMRT treatment planning system (TPS) framework consisting of a 3D GUI for dose manipulation and visualization (Figure 1). To provide an interactive GUI we required both an intuitive and a responsive graphical representation of patient geometry, beam and fluence setup, dose distribution and quality indicators. To guarantee responsiveness, all our algorithms are optimized for speed and low latency by utilizing various degrees of parallelization.

These considerations led to the following technological choices: For 3D graphics we utilized the 3D graphics engine Ogre3D[1]. All windows and dialogs were implemented using Qt[2], 2D plots using Qwt[3]. The Qt concurrency system was used to ensure a responsive interface, while the computation intensive algorithms run in parallel through OpenMP [2] and AVX[4]. All code was programmed in C/C++. Our software supports 3D Vision[5] shutter glasses, which provide the user with optimal depth information. To enable convenient user navigation and manipulation simultaneously, the SpaceNavigator[6] 3D mouse is supported.

[1] Ogre3D is an open-source multi-platform 3D graphics engine which abstracts over OpenGL (Khronos Group, Beaverton, OR, USA) and Direct3D (Microsoft, Redmond, WA, USA). (www.ogre3d.org)

[2] Qt is a cross-platform application and UI framework by Digia (Helsinki, Finland). (www.qt-project.org)

[3] Qwt is an open-source GUI library (mainly focussing on 2D plotting) based on Qt. (qwt.sourceforge.net)

[4] Advanced Vector Extensions (AVX) is an extension to the x86 instruction set architecture for microprocessors by Intel, Santa Clara, CA, USA. (http://software.intel.com/en-us/avx/)

[5] 3D Vision is a product from NVIDIA, Santa Clara, CA, USA. (www.nvidia.com/object/3d-vision-main.html)

[6] SpaceNavigator is a product from 3Dconnexion, Waltham, MA, USA. (www.3dconnexion.com)

2.2. Interactive dose shaping

The key operation driving IDS is the two-step DMR strategy. The first step, a dose modification, is a direct change of dose in a voxel, selected either directly by the user by GUI interaction or a quality indicator specific algorithm. The fluence adaption will naturally lead to unintended and unwanted dose deviations outside of the selected, local area of modified dose. The subsequent recovery step identifies these voxels and aims to recover their original dose by selecting and modifying a set of fluence amplitudes that does not or only minimally alter the initially achieved dose modification. More information on the DMR strategy can be found in [3].

The local dose modification can be directly imposed by a therapist, by selecting a voxel in our 3D-graphical user interface. The described elemental DMR operation can also be used as building block to achieve more general hierarchical planning goals, such as hot spot removal and whole organ at risk (OAR) sparing. The generalized algorithm for incorporating such planning goals is schematically shown in Figure 2. The first step (1) is the candidate selection of the respective voxels for which the dose will be modified. Subsequently a voxel is chosen (2) from the selection and it's dose is modified (3). The recovery step (4) reduces the deviations introduced by the modification. When the resulting dose distribution does not satisfy the planning quality (5), the algorithm returns to (1) and the loop is repeated. The DMR operation requires several dose computations, which form the computational bottle neck for IDS. Dose calculation is performed with an adapted ultra-fast pencil algorithm and relies on the inherent locality of the fluence

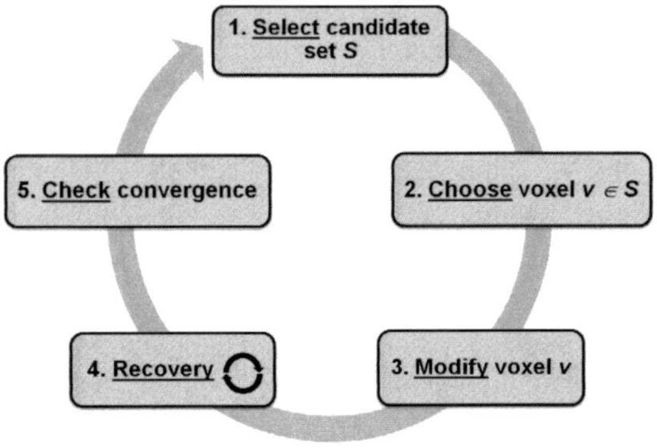

Figure 2. A schematic overview of the IDS algorithm

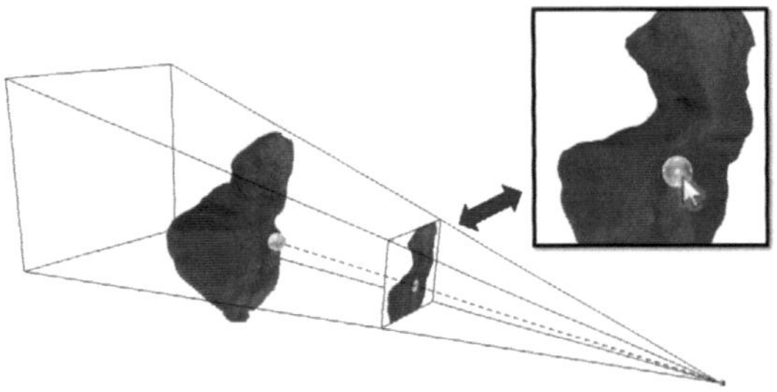

Figure 3. Screen coordinates are projected on the 3D isodose surface by casting a ray from the view frustum's camera origin through the mouse cursor.

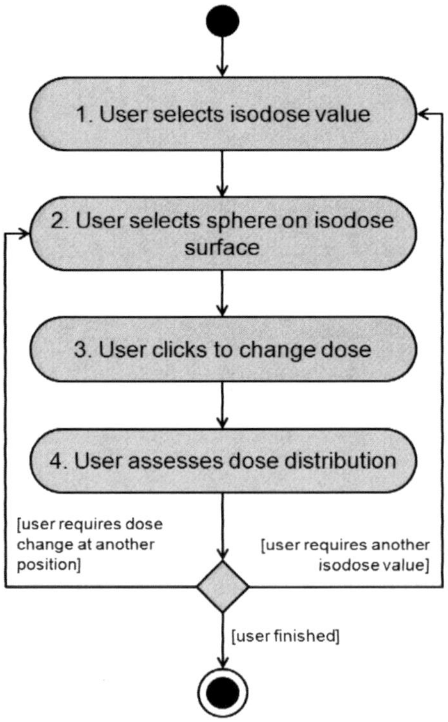

Figure 4. Activity diagram for user interaction with the 3D isodose surface manipulation tool.

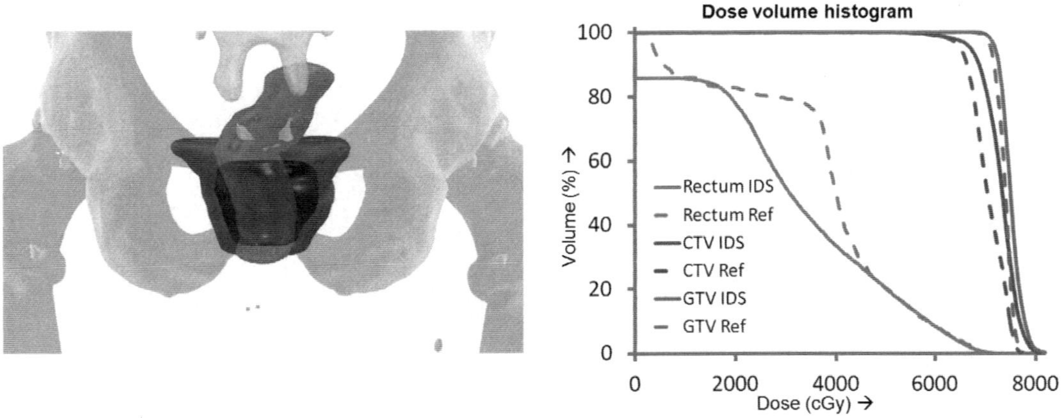

Figure 5. 3D patient geometry of the prostate patient and the dose volume histogram. The IDS plan is represented by the solid curves and the reference plan by the dashed curves.

amplitude changes [4]. The algorithm doesn't require pre-calculated dose influence data, hence it quickly adapts to changing patient geometries.

2.3. Isodose surface manipulation tool

This section describes the isodose surface manipulation tool we propose. The workflow is shown in the activity diagram in Figure 4. As a first step, the user has to select an isodose value (1) for which he wants to observe an isodose surface. This surface is obtained by a real-time volume extraction algorithm which is based on the original Marching Cubes [5] algorithm. It takes a 3D dose distribution as input and generates a triangulated mesh for a specific isodose value. As second step the user positions a sphere on the isodose surface (2) to alter dose at their intersection. Interaction with a mouse is naturally 2D, however the isodose surface manipulation tool requires 3D interaction. Figure 3 shows the view frustum from the imaginary camera which projects the 3D scene on a 2D screen. Now a ray (indicated by the blue dotted line) is cast from the camera source through the mouse cursor. The first intersecting triangle from the isodose surface is the location where the sphere will be positioned. The radius of the sphere can be set manually to alter the locality of the modification. When the user has found the right sculpting position, he clicks to change the dose (3). This input is then translated to a set of candidate voxels to modify and the IDS algorithm starts (Figure 2). The user is provided with immediate feedback; he can now assess the dose distribution (4) using the 3D dose visualization and quality indicators.

2.4. Proof of concept study: clinical prostate case

We provide a proof of concept study for our isodose surface manipulation tool by applying it to a prostate patient and comparing it to a clinically approved conventionally optimized IMRT plan. Both plans consist of 9 coplanar equidistant beams. The reference plan was planned using the KonRad optimization tool (MRC Systems, Heidelberg, Germany). Starting point of our IDS planning approach was a homogeneous, conformal dose of the target volume at the prescribed dose level that does not account for any dose sparing of OARs. This initial dose distribution was interactively modified using the isodose surface manipulation tool to resemble the reference plan. Before plan comparison, the plan was sequenced.

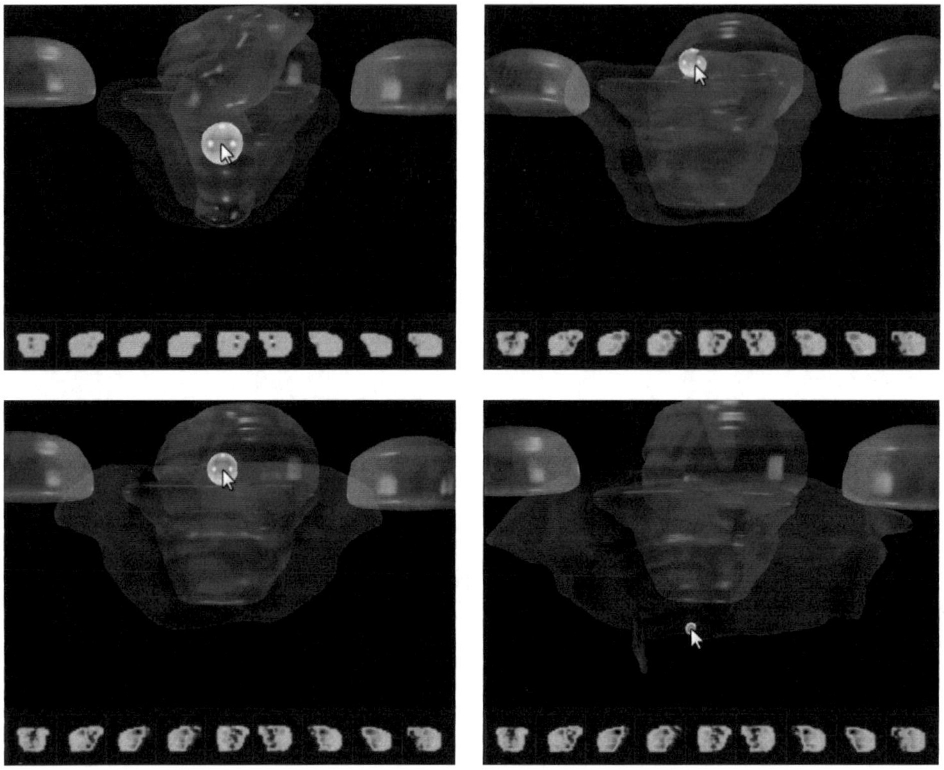

Figure 6. Four screen-shots from successive planning steps. The therapist used the isodose surface manipulation tool to sculpt the dose in and around the rectum. The fluence maps are shown at the bottom of each screen-shot. The mouse cursor was enlarged for better visibility. A video from the planning process can be found in the Supplementary Data.

3. Results

The dose prescription for the prostate case was 76 Gy to the GTV and 70 Gy to the CTV. The planning goal was rectum sparing while maintaining target coverage. The therapist used the isodose surface manipulation tool to sculpt the dose in and around the rectum. Large radii were chosen for coarse dose changes, small radii for fine dose changes. The whole planning process took 5 minutes on a desktop PC configured with a single Intel i7 processor. Dose was calculated on a clinical voxel resolution of $(2\ \text{mm})^3$. A typical dose shaping request by the tool took less than 2 seconds on the desktop PC and less than 1 second on a workstation configured with two Xeon processors. The generation and visualization of the isodose surface after each dose change took typically 30 ms on the desktop PC and 15 ms on the workstation. Figure 5 shows the dose volume histograms. The IDS plan is represented by the solid curves and the reference plan by the dashed curves. The planning process using our isodose surface manipulation tool was captured on video and can be found in the Supplementary Data. Figure 6 shows four screen-shots from successive planning steps.

4. Discussion

The proof of concept study demonstrates that the use of IDS makes it possible to obtain a plan quality at least similar to conventionally optimized IMRT treatment plans. The high dose region for the rectum is comparable, however the volume receiving less than 40 Gy is clearly smaller.

The target volumes from our IDS plan are slightly better, although the difference between GTV and CTV is not very pronounced. Whereas the reference plan can only be adapted by changing penalty weighting factors, our isodose surface manipulation tool makes it more easy to locally change dose features.

As future work we want to evaluate both the interface and the actual dose shaping tools for a patient collective and incorporate the feedback from clinicians. We want to increase the speed of the dose calculation even further. This enables a larger set of voxels to recover and paves the way for guiding strategies: indicating the user which dose changes are still feasible.

The DMR strategy operates in the fluence domain. Before the IDS dose distribution can be delivered, it has to be sequenced, which may deteriorate the local dose manipulations. Currently, we use an external sequencing algorithm, which is separated from the IDS algorithms. We plan to implement a guiding algorithm, which frequently checks whether the currently shaped dose distribution is still deliverable.

Other groups also make efforts to deal with the disadvantages of conventional IMRT treatment plan optimization as stated in the introduction. One example is automatic planning using a heuristic search based on constraint wish lists [6] or plan databases [7]. Multi-criteria optimization (MCO) [8] makes the exploration of the penalty space feasible, but still suffers from the inability to quickly adapt a plan based on an updated patient geometry. These methods are all computationally intensive and require many conventional optimization loops. Therefore research is done on ultra-fast conventional IMRT plan optimization using various parallelization strategies [9] [10]. The IDS algorithm has the potential to cope with all these drawbacks. Future work will include the implementation of strategies to modify voxels based on changing patient geometries.

5. Conclusion

Technically, interactive dose shaping can be done intuitively and in real-time by 3D isodose surface manipulation, as implemented in our in-house TPS Dynaplan. The quality of the obtained plans still has to be evaluated.

References

[1] Bortfeld T 2006 IMRT: a review and preview *Phys. Med. Biol.* **51** R363

[2] Dagum L and Menon R 1998 Openmp: an industry standard api for shared-memory programming. *Comp. Sci. & Eng., IEEE,* **5** 46-55

[3] Ziegenhein P, Kamerling C P, Oelfke U 2013 Interactive Dose Shaping Efficient Strategies for CPU-based Real-Time Treatment Planning *J. Phys.: Conf. Ser.* presented at ICCR 2013, not yet published

[4] Ziegenhein P, Kamerling C P, Heinrich H and Oelfke U 2012 SU-C-213AB-01: Application of Ultra-Fast Dose Calculation in Real-Time Interactive Treatment Panning *Med. Phys.* **39** 3598

[5] Lorensen W E and Cline H E 1987 Marching cubes: A high resolution 3D surface construction algorithm *SIGGRAPH Comp. Graph.* **21** 163-169

[6] Voet P W J, Dirkx M L P, Breedveld S, Fransen D, Levendag, P C and Heijmen B J M 2013 Toward Fully Automated Multicriterial Plan Generation: A Prospective Clinical Study *Int. J. Radiat. Oncol. Biol. Phys.* **51** 866 - 872

[7] Wu B, Ricchetti F, Sanguineti G, Kazhdan M, Simari P, Jacques R, Taylor R and McNutt T 2011 Data-Driven Approach to Generating Achievable DoseVolume Histogram Objectives in Intensity-Modulated Radiotherapy Planning *Int. J. Radiat. Oncol. Biol. Phys.* **79** 1241 - 47

[8] Craft D L, Hong T S, Shih H A and Bortfeld T R 2012 Improved Planning Time and Plan Quality Through Multicriteria Optimization for Intensity-Modulated Radiotherapy *Int. J. Radiat. Oncol. Biol. Phys.* **82** e83 - e90

[9] Ziegenhein P, Kamerling C P, Bangert M, Kunkel J and Oelfke U 2013 Performance-optimized clinical IMRT planning on modern CPUs *Phys. Med. Biol.* **58** 3705-15

[10] Men C, Gu X, Choi D, Majumdar A, Zheng Z, Mueller K and Jiang S 2009 GPU-based ultrafast IMRT plan optimization *Phys. Med. Biol.* **54** 6565-73

Journal of Physics: Conference Series **489** (2014) 012053 doi:10.1088/1742-6596/489/1/012053

Bayesian Decision Support for Adaptive Lung Treatments

Daniel McShan, Yi Luo, Matt Schipper, and Randall TenHaken
Dept. of Radiation Oncology, The University of Michigan, Ann Arbor, MI 48103
E-mail: dlmcshan@umich.edu

Abstract. Purpose: A Bayesian Decision Network will be demonstrated to provide clinical decision support for adaptive lung response-driven treatment management based on evidence that physiologic metrics may correlate better with individual patient response than traditional (population-based) dose and volume-based metrics. Further, there is evidence that information obtained during the course of radiation therapy may further improve response predictions.
Methods: Clinical factors were gathered for 58 patients including planned mean lung dose, and the bio-markers IL-8 and TGF-β1 obtained prior to treatment and two weeks into treatment along with complication outcomes for these patients. A Bayesian Decision Network was constructed using Netica 5.0.2 from Norsys linking these clinical factors to obtain a prediction of radiation induced lung disese (RILD) complication. A decision node was added to the network to provide a plan adaption recommendation based on the trade-off between the RILD prediction and complexity of replanning. A utility node provides the weighting cost between the competing factors.
Results:. The decision node predictions were optimized against the data for the 58 cases. With this decision network solution, one can consider the decision result for a new patient with specific findings to obtain a recommendation to adaptively modify the originally planned treatment course.
Conclusions: A Bayesian approach allows handling and propagating probabilistic data in a logical and principled manner. Decision networks provide the further ability to provide utility-based trade-offs, reflecting non-medical but practical cost/benefit analysis. The network demonstrated illustrates the basic concept, but many other factors may affect these decisions and work on building better models are being designed and tested.
Acknowledgement: Supported by NIH-P01-CA59827

1. Introduction

Adaptive radiation therapy is an approach to improve treatment outcome by refining treatment techniques based on patient specific clinical findings prior to and during treatment. An increasing number of indicators are now available that can be correlated with the effectiveness of treatment. However this increase means that clinical decisions that utilize these factors become increasingly difficult. Further complicating decision making, these indicators are statistically uncertain in terms of their measurements and predictive value. The Bayesian decision support approach described in this paper provides a principled way to deal with statistical uncertainties and to combine them in order to provide clinicians with a better appreciation of the key factors and of their potential impact on treatment outcome.

This paper will discuss the role of the clinical indicators being used for adaptive decisions. In addition a brief overview of Bayesian statistics and how these statistics with graphical modelling and decision support networks can be used in clinical decision support. Examples will focus on treatment of lung cancer and in particular on predictions of radiation toxicity which limits the effectiveness of radiation treatments.

2. Clinical evidence for individualized radio-biology based adaptive corrections

The most common radiation toxicity in the treatment of lung cancer is radiation pneumonitis, which comes from the irradiation of an excessive lung volume. This is a particularly acute issue because patients are receiving chemotherapy with radiation, which dramatically increases the risk of radiation pneumonitis. To aid in our ability to predict lung injury before it occurs, we have determined if plasma cytokine measurements during treatment can improve our radiation induced lung toxicity (RILT) model over the physical DVH alone. In initial studies, we found that radiation-induced elevation in plasma TGF-ß1(transforming growth factor beta 1) during-RT was significantly correlated with RILT. The increase in the ratio of TGF-ß1 at 4-weeks during-RT compared to pre-treatment was also associated with worse diffusing capacity of the lung for carbon monoxide (DLCO) 3 months post-RT (p=0.05). The significance of the TGF-ß1 ratio was supported through a multicenter analysis of Duke University, Peking Union Medical College, and University of Michigan [2]. Combining TGF-ß1 ratio with mean lung dose had a better predictive accuracy than models using TGF-ß1 ratio or mean lung dose alone.

To attempt to build a more comprehensive model, the correlation with the levels of a second plasma cytokine Interleukin 8 (IL-8) and RILT has been assessed. Lower pre-treatment IL-8 levels were significantly correlated with the development of RILT. CT lung dosimetric parameters were borderline significant for its association with RILT. Using receiver operator characteristic curves for predictive risk assessment modelling, the combination of IL-8, TGF-ß1, and mean lung dose into a single model yielded an improved predictive ability (AUC 0.80, 95% CI 0.66-0.94, p <0.001) compared to our original model using mean lung dose alone.

While logistic regression models and Cox proportional hazards models are commonly used methodology for these assessments, we propose instead to use a Bayesian Network approach which has some advantages for uncertain (probabilistic) data, smaller data sets, for handling missing data, and to include more "speculative or knowledge-based" belief information into a decision making strategies.

3. Data management

Adaptive decisions require data management beyond current usage. Clinical observations need to be gathered along with derived information (ex: ratios) along with relevant treatment planning results (ex: mean lung dose). The timing of these observations with respect to decision points is also critical to capture. The data needed is generally simple measurements but can also be more extensive in particular functional imaging studies and correlations to radiation plan dose distributions. With the variety of data required for adaptive decision making and the goal of incorporating new and different finding that have a correlation with outcomes, we are developing an ontology (semantic knowledge driven) approach using standard terminologies and coding when available.

4. Bayesian statistics

Bayesian statistics began with a posthumous publication in 1763 by Thomas Bayes. In essence Bayes described a subjective way to infer the likelihood (probability or belief) of a particular hypothesis given an initial estimate of that likelihood (prior to new evidence being added) (i.e. the prior likelihood) times the likelihood of changes to the likelihood of that hypothesis based on new evidence. I.E.

Posterior likelihood = Prior likelihood x Likelihood of change based on evidence

This Bayesian inference provides a principled way of combining new evidence with prior beliefs. This statistical approach is contrasted with those based solely on the evidence as a whole, with no reference to prior beliefs. Bayesian inference has many applications in science, engineering, medicine, and law.

Bayes' statistics can be applied iteratively. That is, after observing some evidence, the resulting posterior probability can then be treated as a prior probability, and a new posterior probability computed from new evidence. This allows for Bayesian principles to be applied to various kinds of evidence, whether viewed all at once or over time. This procedure is termed *Bayesian updating*.

5. Bayesian belief networks

In describing a complex system with multiple variables (observations or indicators), a useful approach is to build a graphical model. Using the lung toxicity example introduced earlier, Figure 1 shows "nodes" (boxes) for mean lung dose (MLD), IL-8 at week 2(il8_wk2), and TGF-ß1 at week 2 ratio (compared to pre-treatment value) (tgfbeta-wk2_ratio). These nodes are connected (directed lines) to a lung complication probability node. Although this example is simple, Bayesian networks can become extremely complex with hundreds of nodes and connecting relationships.

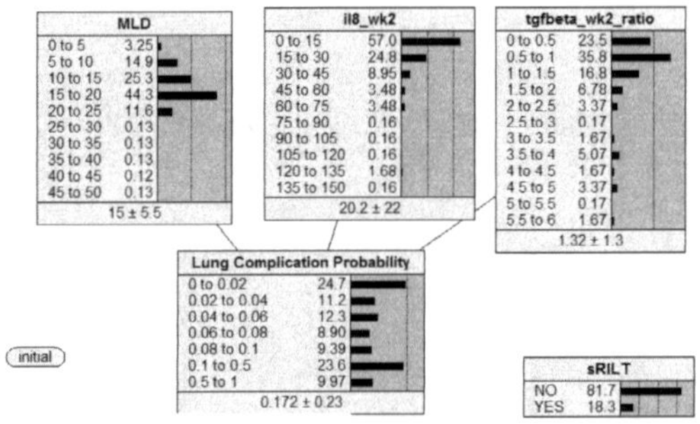

Figure 1 Bayesian belief network for prediction of lung complication probabilitiy

6. Bayesian decision support

Finally we come to Bayesian decision support. What we really want for adaptive therapy decisions are not just a prediction of probabilities but a way to assess the "utility" between no change or making a change (adapting). Clinically, the expected utility would increase with an improvement of tumor control and with a decrease in normal tissue complications. In addition to clinical utility, a decision to make a change in the course of treatment is often tempered by practical factors. Making a change in a course of therapy is generally fairly costly in terms of the need for new treatment planning and evaluation, the need for additional quality assurance checks on a revised approach, the addition of changes in delivery procedures, and the potential of associated hazards. Therefore, there is need to assess the trade-offs between potential therapeutic improvement versus cost and risk assessments.

Within a Bayesian network, decision support nodes are added to the network along with utility nodes which link to other parts of the network. In a Bayesian decision network, the decision node outcome values are optimized to maximize the utility function(s).

In Figure 2, we look at the decision network again, this time entering a finding for a single patient who has a predicted mean lung dose of 17.5, IL-8 value of 7.5 at week 2, and a TGF-ß1 ratio of 2.75 at week 2. The utility node "U" provides relative weights for the lung complication probabilities and for a new node called "Complexity" which is used in this example as an additional factor that would affect the "utility" of making a decision to adapt the treatment. In Figure 2a), the complexity is selected as "Low" and the decision node "Adapt?" indicates a preference to make a change in treatment. However, in Figure 2 b), the complexity is selected as "High" and this changes the decision node towards the preference of not making a treatment change.

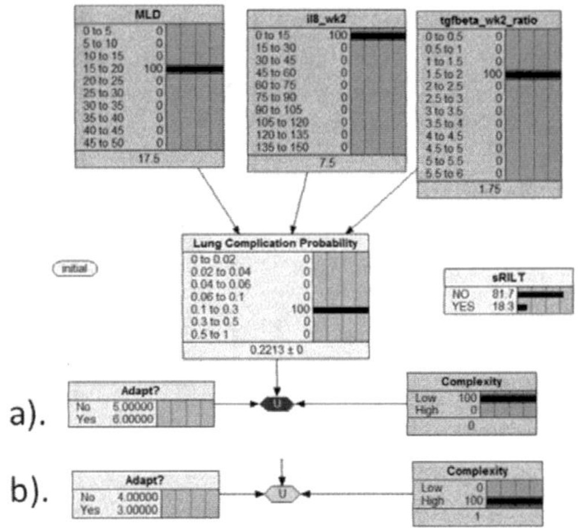

Figure 2 - Decision network for specific case

This decision network is, of course, overly simple in that there will be other clinical factors that affect lung complication predictions and there are other organs such as the heart and oesophagus that may also affect a decision to adaptively change the treatment. Furthermore, there are other factors that would affect the utility of making a change in treatment.

7. Discussion

Computerized clinical decision support has found acceptance in a large number of areas in medicine such as for clinical practice skill improvements, in guideline recommendations, and for checks to reduce preventable medical errors. In general, we are using a computer program to sort through all of the pertinent findings to date to aid the clinician in making the most appropriate decision regarding the on going management of a patient's treatment based on expert knowledge and prior experience. Few of these programs are used in a fully autonomous manner and instead provide decision recommendations along with feedback of the data and the knowledge base factors on which a particular decision recommendation is based.

8. Conclusion

This paper introduced the concepts of Bayesian Decision Networks to provide clinical decision support for adaptive (individualized) radiation therapy. The issues affecting the ability of radiation to be used as an effective treatment are complex and require an understanding of biological processes involved in cancer growth and in cellular repair mechanisms. Utilizing this knowledge and associated clinical tests along with techniques for adjusting the radiation delivery provides radiation oncology with a unique ability to improve treatments over what would be given using a conventional population –based approach.

9. References

[1] Zhao L, Chen M, Ten Haken R, Chetty I, Chapet O, Hayman JA, Kong FM. Three-dimensional conformal radiation may deliver considerable dose of incidental nodal irradiation in patients with early stage nodenegative non-small cell lung cancer when the tumor is large and centrally located. Radiother Oncol. 2007 Feb;82(2):153-9. PubMed PMID: 17287040.

[2] Zhao L, Wang L, Ji W, Wang X, Zhu X, Hayman JA, Kalemkerian G, P, Yang W, Brenner D, Lawrence TS,Kong FM*: Elevation of plasma TGF-beta1 during radiation therapy predicts radiation-induced lung toxicity in patients with non-small-cell lung cancer: combined analysis from Beijing and Michigan. Int.J. Radiat. Oncol. Biol. Phys. 74(5): 1385-90, 2009. PMID: 19231104

Automatic treatment planning implementation using a database of previously treated patients

J A Moore, K Evans, W Yang, J Herman and T McNutt

Department of Radiation Oncology and Molecular Radiation Sciences, Johns Hopkins University, Baltimore, Maryland USA

E-mail: jamoore@jhmi.edu

Abstract. Purpose: Using a database of prior treated patients, it is possible to predict the dose to critical structures for future patients. Automatic treatment planning speeds the planning process by generating a good initial plan from predicted dose values. **Methods:** A SQL relational database of previously approved treatment plans is populated via an automated export from Pinnacle[3]. This script outputs dose and machine information and selected Regions of Interests as well as its associated Dose-Volume Histogram (DVH) and Overlap Volume Histograms (OVHs) with respect to the target structures. Toxicity information is exported from Mosaiq and added to the database for each patient. The SQL query is designed to ask the system for the lowest achievable dose for a specified region of interest (ROI) for each patient with a given volume of that ROI being as close or closer to the target than the current patient. **Results:** The additional time needed to calculate OVHs is approximately 1.5 minutes for a typical patient. Database lookup of planning objectives takes approximately 4 seconds. The combined additional time is less than that of a typical single plan optimization (2.5 mins). **Conclusions:** An automatic treatment planning interface has been successfully used by dosimetrists to quickly produce a number of SBRT pancreas treatment plans. The database can be used to compare dose to individual structures with the toxicity experienced and predict toxicities before planning for future patients.

1. Introduction

Treatment planning involves taking a set of patient specific structures and designing a dose distribution which meets clinical objectives. Unfortunately, beyond the clinical objectives, it can be difficult to estimate the minimum achievable normal tissue dose or even which planning objectives are achievable. By using a database approach, this information can be captured from prior patients to give a good estimate of the achievable dose for a future patient. The use of a database in treatment planning improves plan quality by suggesting the lowest known achievable dose from prior patients whose Organs at Risk (OARs) are closer to the target volumes, which are harder to spare. This system also improves safety by indicating plans that do not meet the objectives met by prior similar patients.

2. Methods

2.1. Database and Software

The automatic treatment planning software uses a Microsoft SQL Server 2005 relational database to store dose and structure information from the clinically approved treatment plans for a set of prior patients. The schema [1] of this database additionally allows for the storage of demographic and toxicity information.

The software interface to the database is written in Python using Tk to generate a user interface. To aid in the building of the database, a series of scripts are used to automatically export the information from the Pinnacle[3] treatment planning system. While this software is initially developed to interface with Pinnacle[3], the software and database exist outside of the planning system and can be interfaced with any other system that permits a scripting interface, or can be used to manually enter the results to the planning system.

2.2. Plan generation and naming Consistency

Unfortunately, using a database to store the vast amount of information used in treatment planning requires consistency in the naming of structures and other plan information. To make the process quicker and easier to use, a renaming interface (**Figure 1**) is developed to automatically map known alternate names to a common set of names. The general naming scheme is based upon published naming recommendations [2]. The naming interface automatically converts all names to lowercase and removes any non-standard characters. The interface also prompts the user if any expected structures are missing. In the case of missing structures, the user can manually rename structures if they were not matched in the automatic process, otherwise the missing structure may need to be contoured.

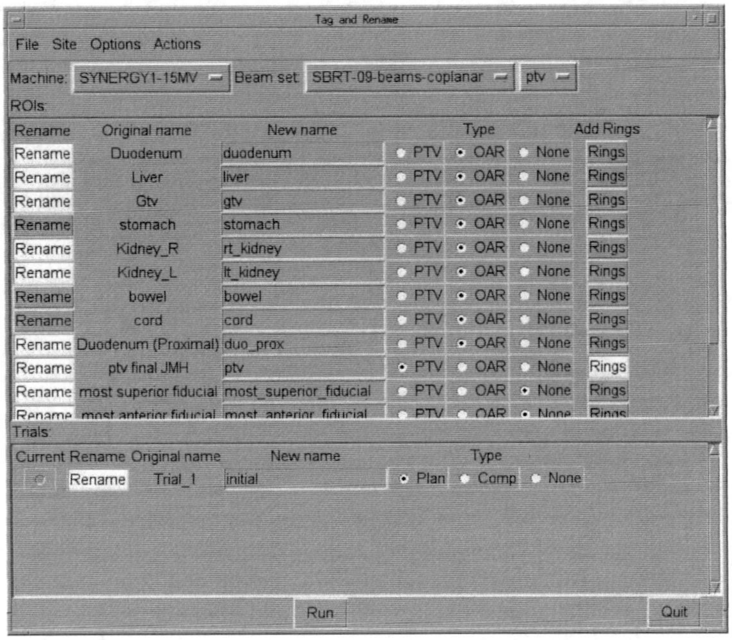

Figure 1. The tagging and renaming interface allows for selection of relevant planning structures and renaming of structures to a common naming scheme. The interface also allows for automatic specification of plan parameters.

For use in automatic planning, each structure must also be tagged as a target or OAR. Structures may also be tagged as none, which excludes them from the database. The renaming interface is also used to generate the basic components of the plan. The contours are cleaned to remove *small (<0.1 cm^2) non-contiguous curves* outside of the structure, ring structures are expanded outside of the target structure to reduce dose outside of the target in planning and combined structures are such as "combined_kidney" are joined from their smaller components (rt_kidney, lt_kidney). The user can select a machine and energy as well as a predefined beam arrangement. The isocenter is set based upon the target selected by the user and the dose grid is automatically specified to cover all of the selected structures.

After the basic plan is generated, overlap volume histograms (OVHs) [3-7] are generated for each target/OAR combination. OVHs are used to select patients from the database which are harder to plan than the current patient. OVH calculation requires approximately 1.5 minutes for a pancreas SBRT plan, but the time depends on the number of contoured structures – roughly proportional to the number of targets multiplied by the number of OARs.

2.3. Database query and plan optimization

Once the OVHs have been computed for the current patient, an objective query interface (**Figure 2**) is used to select the optimization objectives to be used. The user selects a predefined protocol for optimization which contains the basic objectives to be used. The dose value for each objective is queried from the database by selecting the lowest OAR dose from all patients that have the minimum target coverage required and the same or lower distance to overlap for the specified structure. This effectively takes the lowest achievable OAR dose from all patients that are the same or more difficult to plan.

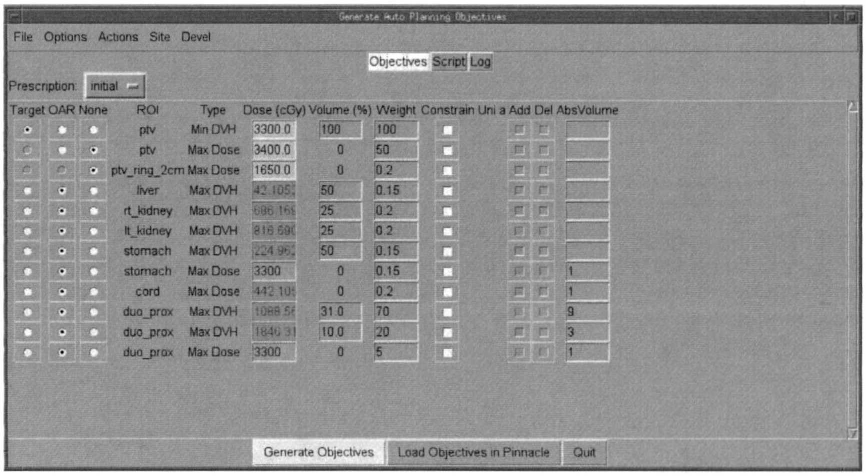

Figure 2. The objective query interface allows for database lookup of the planning objectives. Objectives are color coded to indicate success of the query.

Each planning objective is queried separately from the database. A color coded system is used to indicate the success of the query. Successful queries are colored in green, unsuccessful queries are colored orange. Generally, an unsuccessful query indicates that the current patient has a lower distance to overlap than all other patients in the database. For unsuccessful queries, the user can manually enter a value or accept the tolerance limit which is automatically populated into the field.

New protocols can be developed by creating a new comma separated values file which contains the new objectives and limits for optimization. This allow for easy deployment of the software to new treatment sites and changes of protocol without modifying the software. The software allows for specification of absolute volumes as well as relative volumes for objectives. In the case of absolute volumes for a protocol, the relative volume will be automatically calculated to be exported to the treatment planning system.

Querying the database currently takes approximately 4 seconds for each objective. As the number of patients in the database increases, the time required to query objectives will increase. If the time increases sufficiently to slow planning, pre-computed dose values may be stored in the database for commonly used volumes. This gives up the flexibility of arbitrary volumes in the query but greatly reduces the time required. After the objectives have been queried from the database, a scripting interface allows for the objectives to be directly loaded into the treatment planning system and optimization automatically started.

2.4. Plan evaluation

To aid in evaluating the optimized plans, a protocol check tool (**Figure 3**) is developed which highlights the clinical objectives that are not met by the current plan. This tool automatically exports the dose from the planning system and displays a color coded interface with achieved objectives in highlighted green, acceptable deviations in yellow and unacceptable deviations in red. This tool can be used to determine which planning objectives need to be adjusted to improve the desired trade-offs of the plan and serves as a one click took for quick evaluation.

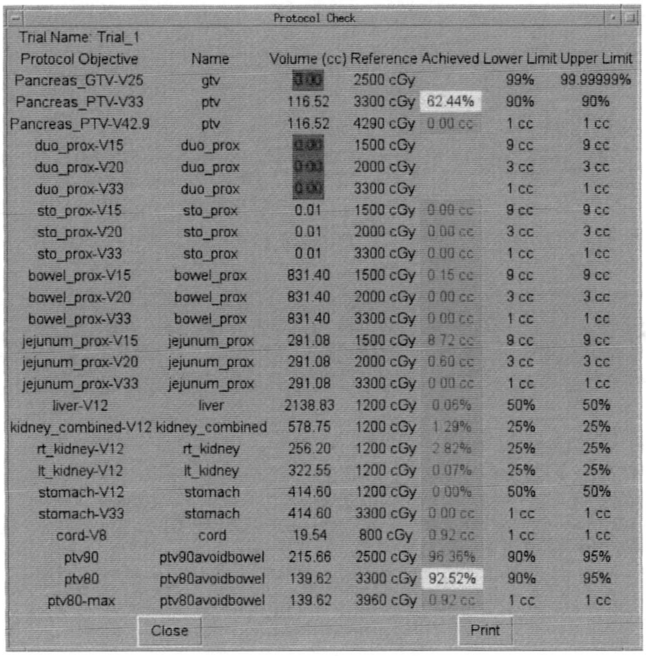

Protocol Check

Trial Name: Trial_1

Protocol Objective	Name	Volume (cc)	Reference	Achieved	Lower Limit	Upper Limit
Pancreas_GTV-V25	gtv	0.00	2500 cGy		99%	99.99999%
Pancreas_PTV-V33	ptv	116.52	3300 cGy	62.44%	90%	90%
Pancreas_PTV-V42.9	ptv	116.52	4290 cGy	0.00 cc	1 cc	1 cc
duo_prox-V15	duo_prox	0.00	1500 cGy		9 cc	9 cc
duo_prox-V20	duo_prox	0.00	2000 cGy		3 cc	3 cc
duo_prox-V33	duo_prox	0.00	3300 cGy		1 cc	1 cc
sto_prox-V15	sto_prox	0.01	1500 cGy	0.00 cc	9 cc	9 cc
sto_prox-V20	sto_prox	0.01	2000 cGy	0.00 cc	3 cc	3 cc
sto_prox-V33	sto_prox	0.01	3300 cGy	0.00 cc	1 cc	1 cc
bowel_prox-V15	bowel_prox	831.40	1500 cGy	0.15 cc	9 cc	9 cc
bowel_prox-V20	bowel_prox	831.40	2000 cGy	0.00 cc	3 cc	3 cc
bowel_prox-V33	bowel_prox	831.40	3300 cGy	0.00 cc	1 cc	1 cc
jejunum_prox-V15	jejunum_prox	291.08	1500 cGy	8.72 cc	9 cc	9 cc
jejunum_prox-V20	jejunum_prox	291.08	2000 cGy	0.60 cc	3 cc	3 cc
jejunum_prox-V33	jejunum_prox	291.08	3300 cGy	0.00 cc	1 cc	1 cc
liver-V12	liver	2138.83	1200 cGy	0.06%	50%	50%
kidney_combined-V12	kidney_combined	578.75	1200 cGy	1.29%	25%	25%
rt_kidney-V12	rt_kidney	256.20	1200 cGy	2.82%	25%	25%
lt_kidney-V12	lt_kidney	322.55	1200 cGy	0.07%	25%	25%
stomach-V12	stomach	414.60	1200 cGy	0.00%	50%	50%
stomach-V33	stomach	414.60	3300 cGy	0.00 cc	1 cc	1 cc
cord-V8	cord	19.54	800 cGy	0.92 cc	1 cc	1 cc
ptv90	ptv90avoidbowel	215.66	2500 cGy	96.36%	90%	95%
ptv80	ptv80avoidbowel	139.62	3300 cGy	92.52%	90%	95%
ptv80-max	ptv80avoidbowel	139.62	3960 cGy	0.92 cc	1 cc	1 cc

Close Print

Figure 3. The plan evaluation interface allows for easy determination of the achieved plan objectives. Color coded values indicate achieved (green), deviation (yellow), or missing values (volumes highlighted red).

The tool uses a file format similar to the protocols for database query to allow for easy generation of new protocols for different sites. The final output of the tool can be included in treatment plan documentation.

2.5. Database Submission
Upon completion of a treatment plan and clinical approval, the plan can be submitted to the database with all structure and dose information. By adding new plans to the database, the quality of future plans is improved. An automated interface is used to export all required data from the planning system and submit it to the appropriate database. Each new plan added to the database improves the quality of future plans queried.

3. Results
A database driven automatic planning tool has been developed and put into clinical use for SBRT pancreas patients. OVH calculations require less than 1.5 minutes for a typical pancreas patient and database lookup time is approximately 4 seconds per objective. We estimate the overall time required for patient to be processed using this tool is less than the time required for one plan optimization step. If the tool results in even one less optimization step to generate an acceptable plan, a net time savings is realized.

4. Conclusions
Automatic planning has the ability to improve quality and safety of treatment plans while reducing the overall planning time required. A database approach allows for similar patients to be selected and has the potential to select similar patient plans based upon dose and toxicity information. Automatic planning tools allow for less experienced planners to generate high quality plans based upon prior patients.

5. References
[1] McNutt T, Wong J, Purdy J, Valicenti R and DeWeese T 2010 OncoSpace: A New Paradigm for Clinical Research and Decision Support in Radiation Oncology *Proceedings of the XVIth Int'l Conf on Computers in Radiotherapy*

[2] Santanam L, Hurkmans C, Mutic S, Vliet-Vroegindeweij C, Brame S, Straube W, Galvin J, Tripuraneni P, Michalski J and Bosch W 2012 Standardizing Naming Conventions in Radiation Oncology *International Journal of Radiation Oncology, Biology, Physics* **83** 1344

[3] Kazhdan M, Simari P, McNutt T, Wu B, Jacques R, Chuang M, and Taylor R 2009 A Shape Relationship Descriptor for RadiationTherapy Planning *Medical Image Computing and Computer-Assisted Intervention* **5762** 100

[4] Wu B, Ricchetti F, Sanguineti G, Kazhdan M, Simari P, Chuang M, Taylor R, Jacques R and McNutt T 2009 Patient Geometry-Driven Information Retrieval for IMRT Treatment Plan Quality Control *Medical Physics* **36** 5497

[5] Wu B, Ricchetti F, Sanguineti G, Kazhdan M, Simari P, Jacques R, Taylor R and McNutt T 2011 Data-driven approach to generating achievable dose-volume histogram objectives in intensity modulated radiotherapy planning *International Journal of Radiation Oncology, Biology, Physics* **79** 1241

[6] Simari P, Wu B, Jacques R, King A, McNutt T, Taylor R and Kazhdan M 2010 A Statistical Approach for Achievable Dose Querying in IMRT Planning *Medical Image Computing and Computer-Assisted Intervention* **6363** 521

[7] Petit S, Wu B, Kazhdan M, Dekker A, Simari P, Kumar R, Taylor R, Herman J and McNutt T 2012 Increased organ sparing using shape-based treatment plan optimization for intensity modulated radiation therapy of pancreatic adenocarcinoma *Radiotherapy and Oncology* **102** 38

6. Acknowledgments
Funding for this research was provided by Philips Radiation Oncology Systems and the Commonwealth Foundation.

Journal of Physics: Conference Series **489** (2014) 012055 doi:10.1088/1742-6596/489/1/012055

Clinical implementation of dose-volume histogram predictions for organs-at-risk in IMRT planning

K L Moore[1,3], L M Appenzoller[2], J Tan[2], J M Michalski[2], W L Thorstad[2] and S Mutic[2]

[1] Department of Radiation Medicine and Applied Sciences, University of California, San Diego, La Jolla, CA 92093, USA
[2] Department of Radiation Oncology, Washington University in St. Louis, St. Louis, MO 63110, USA

E-mail: kevinmoore@ucsd.edu

Abstract. True quality control (QC) of the planning process requires quantitative assessments of treatment plan quality itself, and QC in IMRT has been stymied by intra-patient anatomical variability and inherently complex three-dimensional dose distributions. In this work we describe the development of an automated system to reduce clinical IMRT planning variability and improve plan quality using mathematical models that predict achievable OAR DVHs based on individual patient anatomy. These models rely on the correlation of expected dose to the minimum distance from a voxel to the PTV surface, whereby a three-parameter probability distribution function (PDF) was used to model iso-distance OAR subvolume dose distributions. DVH models were obtained by fitting the evolution of the PDF with distance. Initial validation on clinical cohorts of 40 prostate and 24 head-and-neck plans demonstrated highly accurate model-based predictions for achievable DVHs in rectum, bladder, and parotid glands. By quantifying the integrated difference between candidate DVHs and predicted DVHs, the models correctly identified plans with under-spared OARs, validated by replanning all cases and correlating any realized improvements against the predicted gains. Clinical implementation of these predictive models was demonstrated in the PINNACLE treatment planning system by use of existing margin expansion utilities and the scripting functionality inherent to the system. To maintain independence from specific planning software, a system was developed in MATLAB to directly process DICOM-RT data. Both model training and patient-specific analyses were demonstrated with significant computational accelerations from parallelization.

1. Introduction

Recent studies[1-5] focused on IMRT treatment plan quality have quantified that the use of inverse-planned radiotherapy does not guarantee a truly optimal treatment plan for every patient. IMRT treatment plan quality assessments are currently compromised by their reliance on individual expertise, alertness, and subjective evaluation, with no effective quantitative measures to ensure high plan quality with respect to what is actually achievable. Plan quality variations can be so great that a significant proportion of patients who should have been at low risk of radiation-induced complications

[3] To whom any correspondence should be addressed.

are at much higher risk for poor outcome[6].

To address this problem, we have developed a methodology of model-based algorithms for IMRT treatment plan dose-volume histogram prediction (pDVH) which use patient-specific anatomical features to predict achievable organ at risk (OAR) DVHs. The pDVHs facilitate quantitative assessment of measured vs. expected values for OAR dosimetric parameters, and in this work we describe two separate clinical implementations of this quality control (QC) method.

2. Model-based predictions for OAR DVHs in IMRT

Our framework for model-based predictions of OAR DVHs was described in Ref. [1], and here we summarize the method and primary results. The initial assumption for this work involves the identification of a cohort of *N* site-similar plans that were developed using identical clinical goals and quality assessment criteria. The planning datasets are comprised of structure sets SS_{ij} ($i=1...N$ cases, $j=1...M$ structures with $j=1$ representing the PTV and $j=2...M$ representing M−1 OARs) and dose matrices $D_i = D_i(\vec{x})$, where $\vec{x}$ is the 3-D position vector with arbitrary origin. Operations on D_i, SS_{ij}, and derivative structures result in differential and cumulative DVHs for volumes of interest.

Figure 1 depicts the pDVH process, whereby the boundary distance vector $\vec{r}_i(\vec{x})$ forms the geometric information that will be correlated to the measured DVHs in the training sets. (The use of this quantity has also been explored by other investigators[2, 4].)

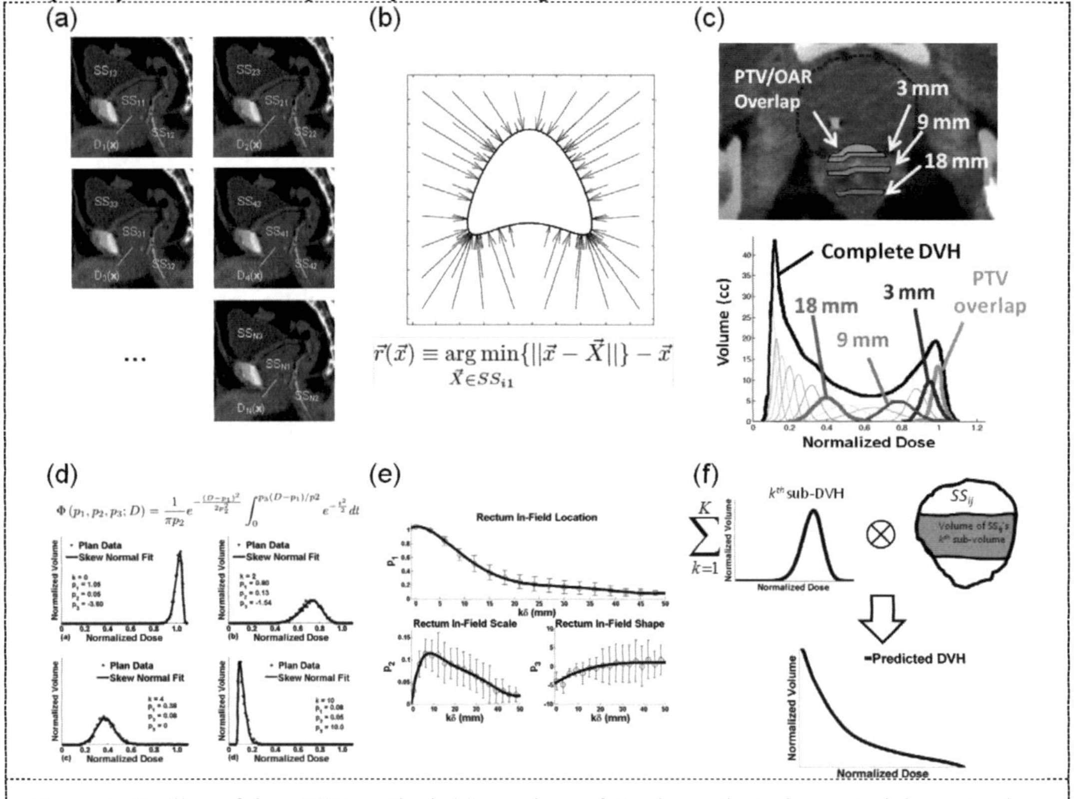

Figure 1. Outline of the pDVH method. (a) A cohort of N prior patients form a training set, where (b) the boundary distance vector field quantifies the geometric arrangement of OAR voxels around PTV(s). (c) OAR voxels are grouped into iso-distance shells, themselves structures with differential DVHs, or "sub-DVHs" of a full OAR. (d) A skew-normal probability function forms a three-parameter basis function that is used to fit the measured iso-distance sub-DVHs , and (e) the evolution of the skew-normal parameters as a function of distance results in (f) the ability to compute pDVHs for any structure set SS_{ij} on new or prior patients, i.e. in the absence of any D_i.

Clinical cohorts of 20 prostate and 24 head-and-neck plans trained pDVH models for rectum, bladder, and parotids. Sums-of-residuals (SR) quantified the integrated difference between clinical DVHs and model-predicted DVHs. The predictive ability for new patients was initially evaluated on a cohort of 20 prostate plans. The models identified plans with large SR values, indicating clinical DVHs that significantly exceeded model-predicted DVHs. All plans were replanned, and the correct identification of outliers was assessed by quantifying the correlation between realized dosimetric gains (SR_{replan}-$SR_{original}$) and the model-predicted gains ($SR_{original}$). The OAR DVHs of the training cohorts were accurately modelled with small mean sums-of-residuals compared to standard deviations. The training and validation sets' sums-of-residuals were statistically indistinguishable, confirming predictive power in new patients. The models exhibited excellent suboptimal plan detection with large correlation coefficients between predicted and realized gains: $r_{rectum}=0.92$, $r_{bladder}=0.88$, and $r_{parotid}=0.84$.

3. Clinical implementation in Philips PINNACLE[3] treatment planning system

From the perspective of clinical implementation, it is advantageous to incorporate quality control elements directly into the treatment planning system[7] as this reduces the need for extraneous data transfers[5] during the planning process. The results of the prostate pDVH models were encoded into a PINNACLE script that can be executed by the planner for any new patient, and has been in use at Washington University in St. Louis for over a year. The script functions by successive use of PINNACLE's contour expansion and Boolean operations to generate the iso-distance shells depicted in Figure 1c. While this is a very computationally inefficient method, it is mathematically equivalent to direct computation of the boundary distance field. The script then computes volume and sub-DVH information for the OAR shells, which can be used to compute a pDVH for the active plan. As shown in Figure 2, multiple pDVH models (derived from different training sets) can be presented as feedback to the user who can compare the current DVH to the pDVH(s). In addition to the full DVH curves, discrete DVH cut points can be encoded and displayed, as well as red flags for situations where the active plan exceeds expected variation. With a simple extension of the PINNACLE scripting routines it was also possible to convert a pDVH into IMRT optimization objectives, a facilitating step for automated treatment planning.

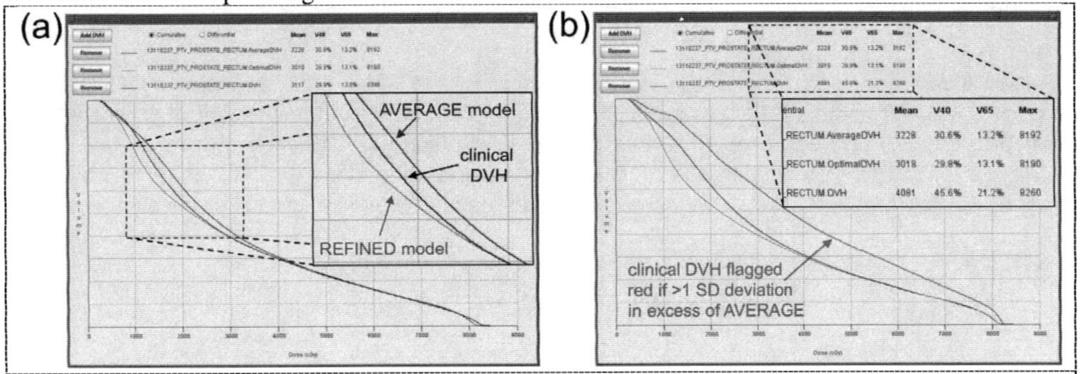

Figure 2. Clinical pDVH module for prostate implemented with PINNACLE scripting. (a) AVERAGE models (trained with uncontrolled prior patient sample) and REFINED models (trained with high plan quality patient sample) for rectum and bladder allow reference pDVHs to be shown aside the current DVHs of an active PINNACLE plan. (b) If the sum-of-residuals analysis shows that the plan is more than one standard deviation outside of the norm, the clinical DVH is flagged in red. Report also shows mean and max dose, as well as V_{40} and V_{65}, for comparison.

4. A universal system for predictive DVH modeling and analysis

To support all treatment planning systems (TPS), we developed an efficient DICOM-based tool to create pDVH models for IMRT QC independent of the TPS. RT dose and structure files exported from

Journal of Physics: Conference Series **489** (2014) 012055 doi:10.1088/1742-6596/489/1/012055

different TPSs including Varian's Eclipse, Philips' PINNACLE[3], and Accuray's Tomotherapy as input to the pDVH DICOM tool (Figure 3).

To date, models have been created for many different treatment sites (intact prostate, prostate bed, head-and-neck, endometrial, cervical, CNS, and lung). MATLAB was used to implement multiple techniques that accelerate the computation of a distance vector field for efficient model building. The program can run on GPUs or, if GPUs are not available, in parallel on multiple CPU cores. The inclusion of GPUs reduces computation time by 200 times, requiring <1 second to compute an OAR vector field and only a few seconds to build a pDVH model. The tool can run in batch mode with minimal user interaction to create a pDVH model for a cohort of patients with a specified treatment site. The tool automatically identifies outliers for treatment plan quality control and pDVH model refinement (Figure 3b).

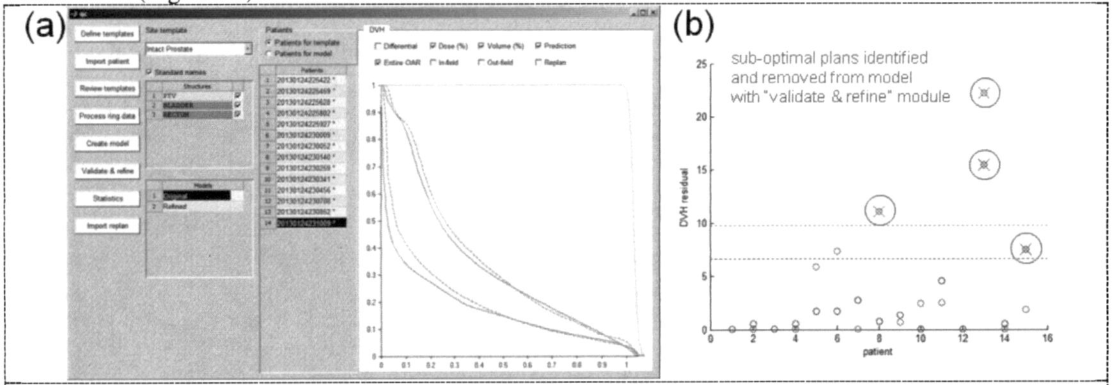

Figure 3. pDVH module in MATLAB. (a) Program interface allows user to import patients and efficiently train new OAR pDVH models. (b) "Validate & refine" module of system allows for the elimination of suspected outliers via residuals analysis, accomplished by removing plans from the training of refined model if their sum-of-residuals score exceeds the standard deviation on any OAR.

5. Summary and conclusions

We have demonstrated that the pDVH methodology can serve as an efficient tool for QC of IMRT treatment plans, providing clinicians a novel and highly efficient plan quality control mechanism by predicting OAR DVHs based on local institutional best practices. This technology can be used to facilitate inter-institutional and inter-TPS quality comparisons based on aggregate clinical data instead of contrived benchmark planning sets, as well as facilitating automated planning routines.

6. References

[1] Appenzoller L M, Michalski J M, Thorstad W L, Mutic S and Moore K L 2012 *Med Phys* **39** (12), 7446

[2] Wu B, Ricchetti F, Sanguineti G, Kazhdan M, Simari P, Chuang M, Taylor R, Jacques R and McNutt T 2009 *Med Phys* **36** (12), 5497-5505

[3] Wu B, Ricchetti F, Sanguineti G, Kazhdan M, Simari P, Jacques R, Taylor R and McNutt T 2011 *Int J Radiat Oncol Biol Phys* **79** (4), 1241-1247

[4] Zhu X, Ge Y, Li T, Thongphiew D, Yin F F and Wu Q J 2011 *Med Phys* **38** (2), 719-726

[5] Siochi R A, Balter P, Bloch C D, Santanam L, Blodgett K, Curran B H, Engelsman M, Feng W, Mechalakos J, Pavord D, Simon T, Sutlieff S and Zhu X R 2011 *J Appl Clin Med Phys* **12** (1), 3479

[6] Moore K L, Brame R S, Low D A and Mutic S 2011 *Int J Radiat Oncol Biol Phys* **81** (2), 545-551

[7] Yang D and Moore K L 2012 *Med Phys* **39** (3), 1542-1551

Dose distribution transfer from CyberKnife to Varian treatment planning system

W Osewski[1], K Ślosarek[2] and B Karaszewska[2]

[1] Radiotherapy Department, Maria Skłodowska-Curie Memorial Cancer Center and Institute of Oncology, Gliwice Branch, Wybrzeże Armii Krajowej 15, 44-101 Gliwice, Poland
[2] Radiotherapy and Brachytherapy Planning Department, Maria Skłodowska-Curie Memorial Cancer Center and Institute of Oncology, Gliwice Branch, Wybrzeże Armii Krajowej 15, 44-101 Gliwice, Poland

E-mail: wosewski@io.gliwice.pl

Abstract. The aim of this paper was to introduce one of the options of the locally developed DDcon.exe which gives the possibility to transfer the dose distribution from CyberKnife (Accuray) treatment planning system (CK TPS) to Varian treatment planning system (Eclipse TPS, Varian). DICOM format is known as a universal format for medical data. The dose distribution is stored as RTdose file in DICOM format, so there should be a possibility to transfer it between different treatment planning systems. Trying to transfer RTdose file from CK TPS to Eclipse TPS the error message occurs. That's because the RTdose file in CK TPS is connected with Structure_Set_Sequence against Eclipse TPS where it's connected with RT_Plan_Sequence. To make it transferable RTdose file from CK TPS have to be 'disconnected' from Structure_Set_Sequence and 'connected' with RT_Plan_Sequence. This is possible thanks DDcon software which creates new RTdose file by changing proper DICOM tags in original RTdose file. New homemade software gives us an opportunity to transfer dose distribution from CyberKnife TPS to TPS Eclipse. This method opens new possibilities to combine or compare different treatment techniques in Varian TPS.

1. Introduction

Nowadays there are a lot of different treatment techniques such as TomoTherapy, CyberKnife, IMRT, VMAT, Brachytherapy etc., so it should be possible to compare or combine all these treatment techniques. To perform the action the dose distribution transfer from other treatment planning systems to Varian TPS Eclipse, might be available. The dose distribution is stored as RTdose file in DICOM format. It is known as an universal format for medical data for all treatment planning systems. Unfortunately it isn't. RTdose file transfer from TomoTherapy or Brahytherapy Masterplan to Varian TPS cause any problem, but it's impossible for CyberKnife RTdose file. To transfer dose distribution from CK TPS to Varian TPS the new option in DDcon software was designed.

Journal of Physics: Conference Series **489** (2014) 012056 doi:10.1088/1742-6596/489/1/012056

2. Material and methods

Accuray - CyberKnife Treatment Planning System version MultiPlan 4.6.0 [2] and Varian Medical Systems: ARIA v.10 – Electronical Medical Records and Information, and Tretment Planning System – Eclipse v.10 was used.

RTdose is a DICOM file, where the calculated in TPS dose distribution is stored. Dose distributions may be represented as: 2D or 3D grids, isodose curves and named or unnamed dose points scattered throughout the volume. It may also contain dose-volume histogram data, single or multi-frame overlays, audio annotations, and application-defined lookup tables [1].

To go through with idea of transfer the dose distribution from CK TPS to Varian TPS a new option in DDcon software was made. It was written in Delphi programming language whereby it is fully transferable between different computers and any special instalation is required. To transfer RTdose file from CK TPS to Varian TPS the error message occurs. This is because the RTdose file in CK TPS is connected with *Structure_Set_Sequence* against Varian TPS with *RT_Plan_Sequence* connectivity. To make RTdose file transferable from CK TPS it must be 'disconnected' from *Structure_Set_Sequence* firstly and 'connected' with *RT_Plan_Sequence*. This is possible by DDcon software usage which creates new RTdose file by changing proper DICOM tags in original RTdose file (figure 1). All operations are based on our own algorithm of reading and converting DICOM files [2,3,4].

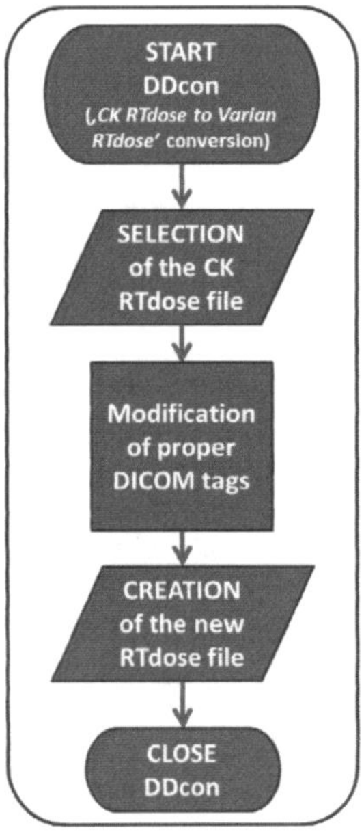

Figure 1. Flowchart of DDcon software showing how to change CK RTdose file to make it transferable to Varian TPS.

3. Results

To transfer RTdose file from CK TPS to Varian TPS firstly the CT scans and structure set from CK TPS have to be transferred to Varian TPS. Secondly treatment plan, based on the CT scans and structure set imported from CK TPS, with minimum one field and without performing the dose distribution calculations should be prepared in Eclipse TPS, Varian. After that it is possible to import

RTdose file, created by DDcon software, by connecting it with previously performed treatment plan. Figure 2 shows CT scans, structures, beams setup and dose distribution from CK TPS. Figure 3 shows imported CT scans with structure set and converted CK dose distribution in Varian TPS.

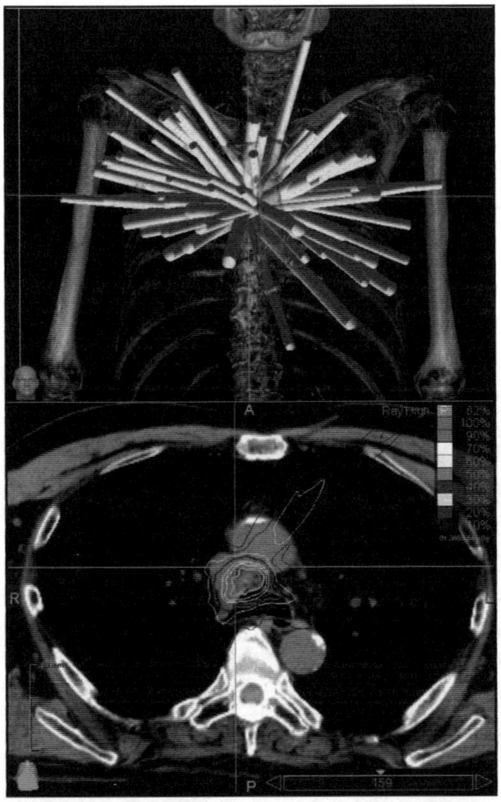

Figure 2. Printscreean from CK TPS showing beams setup and dose distribution which will be transferd to Varian TPS.

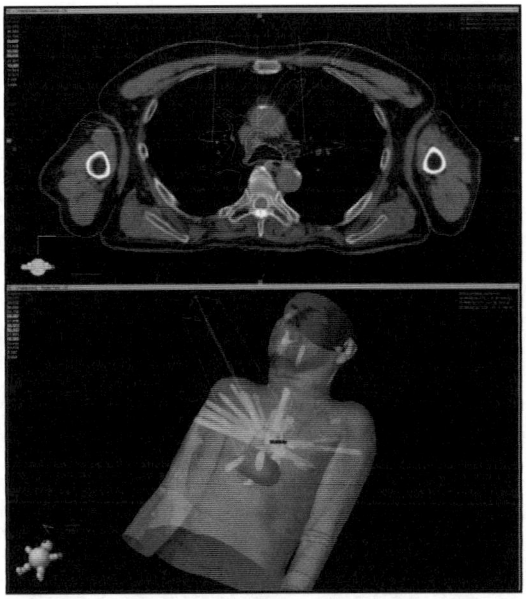

Figure 3. CK dose distribution transferred to Varian TPS.

4. Conclusion

The locally designed software gives the opportunity to transfer the dose distribution from CyberKnife TPS to Varian TPS. This method opens new possibilities to combine or compare different treatment techniques in Varian TPS.

References

[1] *Digital Imaging and Communications in Medicine (DICOM) Supplement 11, Radiotherapy Objects*

[2] Accuray 2007 *DICOM Conformance Statement for Accuray CyberKnife System* (Revision G)

[3] Varian Medical Systems 2011 *System Server – DICOM Conformance Statement* (VA10003D3CS)

[4] Pianykh O S 2008 *Digital Imaging and Communications in Medicine (DICOM). A Practical Introduction and Survival Guide* (Springer)

Journal of Physics: Conference Series **489** (2014) 012057 doi:10.1088/1742-6596/489/1/012057

A comparison of coordinate systems for use in determining a radiotherapy delineation margin for whole breast

EM Pogson[1,2], L Bell[1,2], V Batumalai[2,3], ES Koh[2,3], G DELANEY[2,3], P Metcalfe[1,2], L Holloway[1-4]

1. Centre for Medical Radiation Physics, University of Wollongong, Wollongong, NSW 2522, Australia;
2. Liverpool and Macarthur Cancer Therapy Centres and the Ingham Institute, Liverpool, NSW, Australia
3. SWSCS, University of New South Wales, Sydney, Australia
4. Institute of Medical Physics, University of Sydney, Sydney Australia

E-mail: elisep@uow.edu.au

Abstract. Cartesian co-ordinates, traditionally used for radiotherapy margins, calculated at 6 points, may not adequately represent changes in inter-observer contour variation as necessary to define a delineation margin. As a first step, this study compared the standard deviation (SD) in contour delineation using Polar and Cartesian co-ordinates for whole breast. Whole breast Clinical Target Volumes (CTV) were delineated by eight observers for 9 patients. The SD of contour position was determined for Polar co-ordinates at 1° increments for 5 slices and averaged across all patients. The mean centre of mass (COM) was used as the origin for the right breast, for the left the COM was shifted 1cm superiorly to avoid clipping. The SD was determined for Cartesian co-ordinates for medial-lateral and anterior-posterior positions. At slice Z=0cm considering Polar co-ordinates, the SD peaked medially reaching 3.55cm at 15° for the right breast, and 1.44cm at 171° for the left. The SD of the remaining slices maintained a similar distribution, with variation in the peak occurring within 10° of the Z=0cm positions. By comparison, for Cartesian co-ordinates at slice Z=0cm, the largest SD in the medial-lateral and anterior-posterior directions was 0.54/0.57cm and 1.03/0.67cm respectively for right/left breasts. The SD for inter-observer variation for whole breast varies with anatomical position. The maximum SD determined with Polar co-ordinates was greater than with Cartesian co-ordinates. A delineation margin may thus need to vary with angle over the entire structure and Cartesian co-ordinates may not be the best approach for margin determination for whole breast.

1. Introduction and Background

Radiotherapy treatment techniques have improved significantly with the implementation of; Intensity Modulated Radiotherapy (IMRT), Volumetric Modulated Arc Therapy (VMAT) and Image Guided Radiotherapy (IGRT). Further improvements in radiotherapy are achieved by setting clear delineation protocols, minimising set-up errors and imaging errors, and imaging for internal motion management (1). Delineation uncertainty arising from inter-observer variability has traditionally been ignored as larger systematic errors dominate. Inter-observer variability is still a significant uncertainty, even with clear delineation protocols and uncertainties from delineation, may now play a significant role in influencing the success of these improved techniques.

Although the inclusion of delineation margins in PTV margin recipes is well acknowledged in the literature as necessary, particularly with the introduction of more conformal delivery techniques calculation of these margins is rarely attempted. Studies have been performed for cancer sites such as the Prostate (2, 3) and the Brain (4) to optimise set-up and motion margins ignoring the inter-observer delineation uncertainties. These have shown that margins are necessary to reduce the likelihood of geographical misses, reduce effects from systematic errors (under-dosage) and reduce effects from random errors (blurring) (1). Margin formulae using the combined preparation/systematic Σ standard deviation and execution/random σ standard deviation such as $2.5\Sigma+0.7\sigma - 3mm$ (5), incorporate inter-observer variability within the preparation errors Σ. However this is performed with one value over the whole structure equally or with 3 values in 3 planes. As the inter-observer variation may be accurately determined over the whole structure (multiple angles and points), perhaps this adaptation should be performed before the other margin formula is applied. The appropriate amount or weight that this should have compared to the whole margin formula needs to be investigated further. Introduction of an additional margin, or ensuring PTV margins include the inter-observer variability margin, to account for breast delineation uncertainties would reduce the clinical impact of inter- and intra- observer variation. This has previously been performed using Cartesian co-ordinates for sites such as the prostate and seminal vesicles (6), partial breast (7), whole breast (8) and performed using spherical co-ordinates for sites such prostate (9) and bladder (10). Shape of the target will play a significant role in what co-ordinate system should be used.

2. Method

Datasets from 9 patients from a previous study were utilised (11). Volumes of the CTV were contoured by four breast cancer radiation oncologists (with experience ranging from 6 to over 20 years in oncology) and four radiation therapists (with experience ranging from 3 to 13 years). One sided whole breast Clinical Target Volumes (CTV) were outlined on transverse slices with a standard window level (0) and width (500). Radio-opaque wire was used to aid delineation. Each observer was allocated a unique letter to de-identify their contours. Observers were blind to other observer contours.

The standard deviation (SD) in delineation uncertainty using Polar co-ordinates for 360 degrees in 1 degree increments and in Cartesian co-ordinates (the maximum in each X and Y direction) for whole breast was compared. As the CTV structure may be easily elongated many papers calculate the deviation in the cardinal directions (from the geometric center co-ordinates in 6 directions) and perform this adaption enabling some change over the entire volume. This has been generally undertaken to account for tumour motion, however for interobserver error a more detailed expansion at many directions may be necessary. For this study 360 interpolated points were calculated for each slice Z in Cylindrical coordinates and compared to the Cartesian cardinal points and extensions. The interpolation at 1 degree increments were compared with the original structures to ensure the shape of the structures remained constant.

The standard deviation (SD) of this was calculated for slices as in equation (1) for Cylindrical co-ordinates and equation (4) for Cartesian co-ordinates.

2.1. Cylindrical Co-ordinates

The contours were input into CERR as DICOM RT structure sets and associated Cartesian co-ordinates (X,Y,Z) were determined for each patient and observer at the selected slices. The Cartesian co-ordinates were then converted into Cylindrical co-ordinates for analysis. The SD of contour positions were determined for cylindrical co-ordinates at θ=1° increments for 5 slices *(Z=-2cm, -1cm, 0cm, 1cm and 2cm from the origin)* for each patient using equation (1).

$$SD = \Sigma^{j}_{interobserver,radial}(\theta, Z) = \sqrt{\frac{\sum_{i=1}^{N}\left(r_i^j(\theta,Z)-\bar{r}^j_{obs}(\theta,Z)\right)^2}{N-1}} \quad (1)$$

Where N is the total number of observers per patient and $\bar{r}$ is the radius averaged over all observations, Z is the slice or Height of the structure in cm, j is the inter-observer and θ is the Polar angle ranging from 0-360 degrees. The mean center of mass (COM) was used as the origin for the right breast, for the left the COM was shifted 1cm superiorly to avoid clipping of the structure.

If an observer did not have a contour on a distal slice e.g. Z=2cm the number of observers was reduced to those present. A margin to account for margin delineation uncertainty (MDU) was then calculated as in equation (2). This is not recommended and is simply a preliminary study into the validity of including inter-observer margins in multiple directions in PTV margin recipes.

$$MDU_{\theta,Z}^{i} = 4 \times SD_{\theta,Z}^{i} \quad (2)$$

Where SD is the standard deviation, i the patient number and θ,Z indicate the direction. This margin was selected such that 95% of the contoured union volume would be included for any of the initial contours assuming a guassian distribution of contour variation. The margin to be used on a new patient would be calculated using all 9 previous patient datasets as outlined in equation (3). Where P is the total number of patients.

$$MDU_{\theta,Z}^{general} = \frac{\sum_{j=0}^{P} MDU_{\theta,Z}^{j}}{P} \quad (3)$$

2.2. Cartesian Co-ordinates

The SD and MDU were determined for Cartesian co-ordinates for medial-lateral and anterior-posterior positions, following previously published methodology (8). If inter-observer variability is considered in studies the SD is calculated usually in only 6 directions as treatment planning systems typically allow extension only in the anterior (A), posterior (P), superior (S), inferior (I), lateral (L) and medial (M) directions by one value for all points (from the average COM). The SD is outlined in equation (4). Where x_d is the pixel position, in cm, at the edge of the contour in the appropriate direction (d=A, P, S, I, M, or L). And x_{com} is the average centre of mass co-ordinate also in that direction (d).

$$SD = \sqrt{\frac{\sum_{i=1}^{N}\left[(x_{i,d}-x_{i,d,com})-(\bar{x}_{i,d}-\bar{x}_{i,com})\right]^2}{N-1}} \quad (4)$$

The MDU is found as in equation (2-3), where d is the directions replacing θ and Z(A, P, S, I, L or M). The MDU was added onto the observer contour in all 6 directions to generate a CTV with a margin to account for inter-observer errors. For all methodologies the MDU would be clipped to the Skin, Lung, Heart, Muscle interface and the Medial border if necessary.

3. Results

Figure 1 displays the interpolation points and original points of eight contours on one patient at slice Z=0 for Polar interpolation at 1 degree increments, indicating consistency through the interpolation without clipping or diverging.

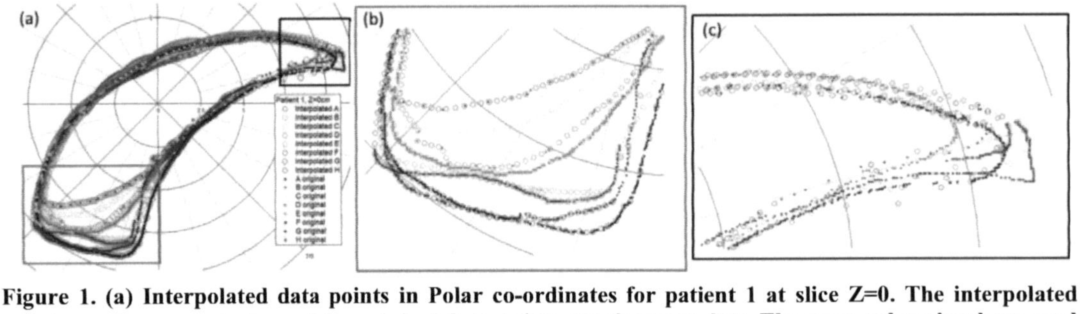

Figure 1. (a) Interpolated data points in Polar co-ordinates for patient 1 at slice Z=0. The interpolated points are shown as circles and the original data-points are shown as dots. The same colour has been used for each observer. Parts (b) and (c) are zoomed in pictures of the whole structure as indicated on part (a) using a blue and black box respectively.

The interpolation used linear interpolation at 1 degree increments. The over-all uncertainty for Z=0 is plotted for patient 1 in Figure 2.

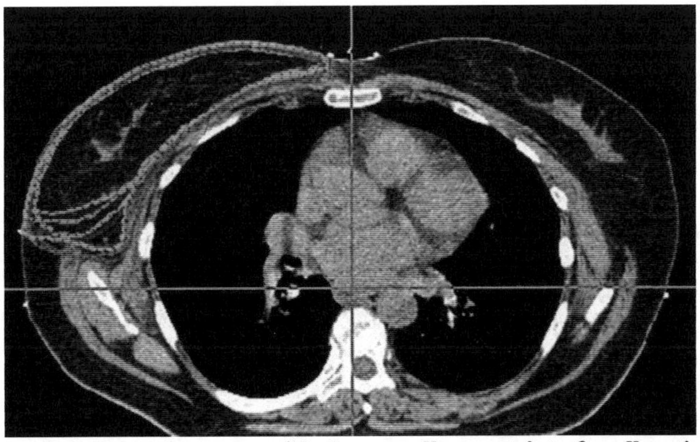

Figure 2. The contours at Z=0 for patient one, with the overall uncertainty for all patients at Z=0 overlaid in red.

Figure 2 highlights the anatomical borders that are most affected by the uncertainty.

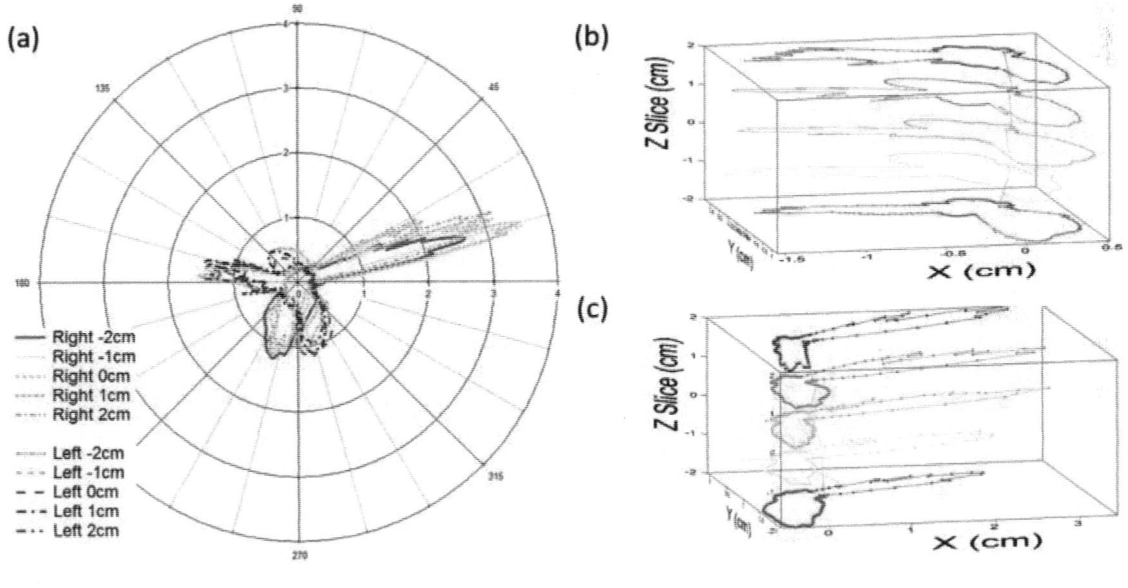

Figure 3. (a) The uncertainty (one standard deviation) derived from 9 patients (5 patients with right breast cancer and 4 with left breast cancer) utilising 8 observers calculated in Polar co-ordinates for Z=-2,-1,0,1,2cm. The R axis is displayed here in cm. This is shown in three dimensions in (b) and (c) for the left and right breast respectively.

At slice Z=0cm considering Polar co-ordinates (lateral=0°), the SD peaked medially reaching 3.55cm at 15° for the right breast, and 1.44cm at 171° for the left (see Figure 3.). The SD of the remaining slices maintained a similar distribution, with variation in the peak occurring within 10° of the Z=0cm positions. In comparison, for Cartesian co-ordinates at slice Z=0cm, the largest SD in the medial-lateral and anterior-posterior directions was 0.54/0.57cm and 1.03/0.67cm respectively for right/left breasts. The average MDU for the cardinal directions in Cartesian co-ordinates is displayed in Table 1 (8).

Table 1. Standard Deviation and Margin to account for delineation uncertainty in Cartesian co-ordinates.

Direction	Anterior	Posterior	Medial	Lateral	Superior	Inferior
Right SD (cm)	0.40	0.22	1.15	1.48	0.52	0.79
Left SD (cm)	0.32	0.46	1.70	1.13	0.46	0.64
Left MDU (cm)	1.28	1.82	1.13	1.70	1.83	2.57
Right MDU (cm)	1.61	0.87	1.15	1.48	2.07	3.17

It may be seen in Figure 4 that for every contour the additional margin of 4 SD is in line with the statistical account of 95% for any observers contour. This was also the case for Z=-2cm,-1cm, 1cm and 2cm as shown in Figure 4 for Patient 1.

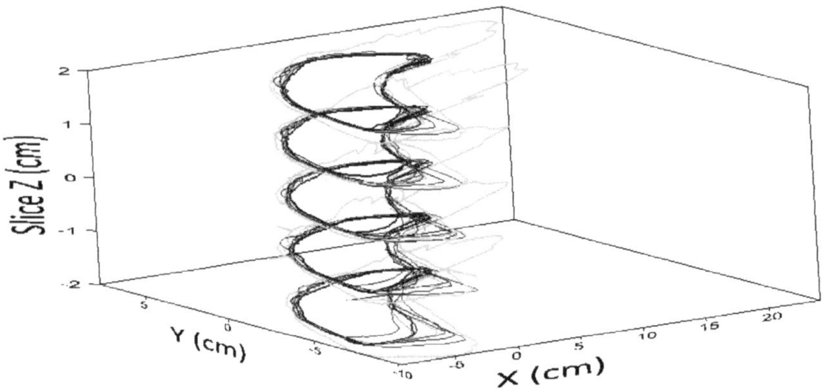

Figure 4. Patient 1 (Right breast) contours are all within the minimum MDU calculated from Cylindrical co-ordinates. All 8 observers lie within the minimum observers contour with the addition of the MDU. The thin lines are observers 1-7, observer 8 contour (minimum contour) is the thick dark blue line. The light blue thick line is the MDU of observer 8. Each observers contour is of a different colour consistent through all slices.

This is the case for all the patients, only one patient is shown as an example for clarity. The average standard deviation was calculated separately for left and right breast patients.

4. Conclusion

It has been shown that there is more accuracy in calculating a standard deviation at multiple points (1 degree increments) to extend the CTV to account for inter-observer error, rather than extending X, Y and Z by their cardinal variances. Margins are larger in some directions using Cylindrical co-ordinates than if the contour was extended by these cardinal variances. This highlights the regions where inter-observer variation is high such as both edges of the contours for both left and right breast patients. At slice Z=0cm considering Polar co-ordinates (lateral=0°), the SD peaked medially reaching 3.55cm at 15° for the right breast, and 1.44cm at 171° for the left. The SD of the remaining slices maintained a similar distribution, with variation in the peak occurring within 10° of the Z=0cm positions. By comparison, for Cartesian co-ordinates at Z=0cm, the largest SD in the medial-lateral and anterior-posterior directions was 0.54/0.57cm and 1.03/0.67cm respectively for right/left breasts. Future work will calculate and compare Cartesian co-ordinates at every point to Cylindrical co-ordinate calculations. Sensitivity to COM variations and a spherical co-ordinate analysis method is also being investigated. Incorporation into a margin recipe that includes organ motion, set-up errors and random errors should be investigated for clinical implementation. A method of extending contours outside six directions (e.g. 360 directions) used in treatment planning systems also requires investigation.

5. Acknowledgement
This research is funded by a Cancer Australia and National Breast Cancer Foundation grant project number 1033237 and from the Liverpool Hospital radiation oncology trust fund.

References

[1] van Herk M 2004 'Errors and margins in radiotherapy,' *Seminars in Radiation Oncology*. **14** 1 52-64

[2] van Herk M, Remeijer P, Rasch C, Lebesque JV 2000 'The probability of correct target dosage: dose-population histograms for deriving treatment margins in radiotherapy,' *International Journal of Radiation Oncology Biology Physics* **47** 4 1121-35

[3] Mutanga TF, de Boer HCJ, van der Wielen GJ, Hoogeman MS, Incrocci L, Heijmen BJM 2011 'Margin Evaluation in the Presence of Deformation, Rotation, and Translation in Prostate and Entire Seminal Vesicle Irradiation With Daily Marker-Based Setup Corrections,' *International Journal of Radiation Oncology Biology Physics* **81** 4 1160-7

[4] Heesters MA, Wijrdeman HK, Struikmans H, Witkamp T, Moerland MA 1993 'Brain tumor delineation based on CT and MR imaging. Implications for radiotherapy treatment planning,' *Strahlentherapie und Onkologie : Organ der Deutschen Rontgengesellschaft [et al]*. **169** 12 729-33

[5] van Herk M, Remeijer P, Lebesque JV 2002 'Inclusion of geometric uncertainties in treatment plan evaluation,' *International Journal of Radiation Oncology Biology Physics* **52** 5 1407-22

[6] Fiorino C, Reni M, Bolognesi A, Cattaneo GM, Calandrino R 1998 'Intra- and inter-observer variability in contouring prostate and seminal vesicles: implications for conformal treatment planning,' *Radiotherapy and Oncology* **47** 3 285-92

[7] Petersen RP, Truong PT, Kader HA, Berthelet E, Lee JC, Hilts ML, et al. 2007 'Target Volume Delineation for Partial Breast Radiotherapy Planning: Clinical Characteristics Associated with Low Interobserver Concordance,' *International Journal of Radiation Oncology Biology Physics* **69** 1 41-8

[8] Holloway LC, Jameson MG, Batumalai V, Koh E, Papadatos G, Lonergan D, et al. 2010 'Estimating a Delineation Uncertainty Margin to Account for Inter-observer Variability in Breast Cancer Radiotherapy,' *International Journal of Radiation Oncology Biology Physics* **78** 3 S741

[9] Rasch C, Barillot I, Remeijer P, Touw A, van Herk M, Lebesque JV 1999 'Definition of the prostate in CT and MRI: a multi-observer study,' *International Journal of Radiation Oncology Biology Physics* **43** 1 57-66

[10] Meijer GJ, Rasch C, Remeijer P, Lebesque JV 2003 'Three-dimensional analysis of delineation errors, setup errors, and organ motion during radiotherapy of bladder cancer,' *International Journal of Radiation Oncology Biology Physics* **55** 5 1277-87

[11] Batumalai V, Koh ES, Delaney GP, Holloway LC, Jameson MG, Papadatos G, et al. 2011 'Interobserver Variability in Clinical Target Volume Delineation in Tangential Breast Irradiation: a Comparison between Radiation Oncologists and Radiation Therapists,' *Clinical Oncology* **23** 2 108-13

Journal of Physics: Conference Series **489** (2014) 012058 doi:10.1088/1742-6596/489/1/012058

Automatic Treatment Planning with Convex Imputing

G. A. Sayre, D. Ruan
University of California – Los Angeles, Department of Radiation Oncology

E-mail: druan@mednet.ucla.edu

Abstract. Current inverse optimization-based treatment planning for radiotherapy requires a set of complex DVH objectives to be simultaneously minimized. This process, known as multi-objective optimization, is challenging due to non-convexity in individual objectives and insufficient knowledge in the tradeoffs among the objective set. As such, clinical practice involves numerous iterations of human intervention that is costly and often inconsistent. In this work, we propose to address treatment planning with convex imputing, a new-data mining technique that explores the existence of a latent convex objective whose optimizer reflects the DVH and dose-shaping properties of previously optimized cases. Using ten clinical prostate cases as the basis for comparison, we imputed a simple least-squares problem from the optimized solutions of the prostate cases, and show that the imputed plans are more consistent than their clinical counterparts in achieving planning goals.

1. Introduction

Radiation therapy planning is a challenging process that relies heavily on dose planner expertise to achieve clinical planning objectives. Specifically, the dose planner must balance the conflicting objectives of target coverage and organ-sparing – both of which depend on patient geometry – by iteratively modifying dose-volume goals. Moreover, the dose planner must ensure that the optimal plan does not suffer from unwanted spatial dose phenomena, such as hot spots. This heavy dependence on operator judgment risks solution sub-optimality and plan inconsistency, in addition to the high cost of human operator time and effort. Therefore, it is important to design techniques to streamline the planning process to improve plan quality and consistency and to increase patient throughput.

In this study, we propose a novel method of convex imputing [1] for radiotherapy treatment planning to overcome these limitations. Specifically, we aim to obtain a latent imputing function whose M-estimator best approximates the behaviour of previously optimized clinical plans. Since previously validated and approved clinical plans reflect both the quantitative goals of target coverage and organ-sparing and the qualitative dose-shaping goals of the physician, the imputing operator would "inherit" such properties implicitly and impose such on new plans. In doing so, we establish a method that: 1) automates the treatment planning process; 2) enables learning of planning priorities; and 3) consistently produces high quality plans.

To establish the feasibility and efficacy of this novel method, we tested our imputing framework on a set of ten clinical IMRT prostate cases that were previously optimized. In Section II, the fundamental theory of convex imputing will be introduced and our specific formulation will be developed. In

Section III, important results will be reported and discussed. In Section IV, we will conclude with a summary of our work.

2. Methods

To begin, we consider an optimization problem where solution x is obtained by optimizing an objective function OF subject to constraints, where both OF and the constraints depend on a parameter set p. The goal of convex imputing is to learn the form of OF, i.e. weights w, given paired observations of previously optimized solutions x_k and their associated parameters p_k. It is assumed that the constraint functions are known and that OF is both convex and affine (Eq. 1).

$$minimize \underbrace{\sum_{n=1}^{N} w_n f_n(x,p)}_{OF} \quad s.t. \quad g_i(x,p) \le 0 \; for \; i = 1,\dots,m; \; A(p)x = b(p) \qquad (1)$$

For a given p, solution x is optimal if the dual variables λ and ν satisfy the KKT conditions. By contrast, for a given p, solution x is approximately optimal if the KKT conditions approximately hold. Thus, approximate optimality interprets the KKT conditions are small residuals, as shown below in Equations 2a-d.

$$r_{ineq} = \left(g_i(x,p)\right)_+, \qquad i = 1,\dots,m \qquad (2a)$$

$$r_{eq} = A(p)x - b(p) \qquad (2b)$$

$$r_{stat} = \nabla_x\left(\mathrm{OF}(x,p)\right) + \sum_{i=1}^{m} \lambda_i \nabla_x g_i(x,p) + A(p)^T \nu \qquad (2c)$$

$$r_{comp} = \lambda_i g_i(x,p), \qquad i = 1,\dots,m \qquad (2d)$$

Since the primal residuals, $r_{ineq}$and r_{eq}, are fixed for each $\{x_k, p_k\}$ and do not depend on weights w, they cannot be used to impute weights w. Therefore, to impute weights w we solve the following problem:

$$w^{imputed} = \underset{w}{argmin} \sum_{k=1}^{K} \underbrace{\phi\left(r_{stat}^k, r_{comp}^k\right)}_{\substack{non-neg. \\ convex\ pen.}} \quad s.t. \quad \underbrace{\lambda^k \ge 0}_{low\ bnd.}, \quad \underbrace{w \ge 0}_{pos.}, \quad \underbrace{\sum_{n=1}^{N} w_n \ge 1}_{norm.} \qquad (3)$$

The first term in Equation 4 penalizes deviation of the stationarity and complimentary slackness residuals from 0, thereby making the proposed forward problem (Eq. 1) approximately optimal with respect to the paired observations $\{x_k, p_k\}$. For our formulation, we selected the L_2 norm to be used for this purpose, i.e. $\phi\left(r_{stat}^k, r_{comp}^k\right) = \left\|r_{stat}^k\right\|_2^2 + \left\|r_{comp}^k\right\|_2^2$. The inequalities in Equation 3 assure that: 1) the dual function of the imputed OF gives the lower bound of the forward problem for observations $\{x_k, p_k\}$; 2) the imputed OF is convex; and 3) the imputed solution is not trivial, e.g. $w = 0$. Since both r_{stat}^k and r_{comp}^k are linear, the imputing problem is a simple least-squares formulation and may be efficiently solved to its global minimum. The imputed objective function $\sum_{n=1}^{N} w_n^{imputed} f_n(x,A)$ can then be used to automatically optimize future instances of the forward problem that will be consistent with previously optimized solutions x_k Thus, by applying this method to radiotherapy, the imputed forward problem may be used to automatically optimize new cases such that the resulting optimal plans reflect the quantitative dosimetric goals of the institution and the qualitative dose-shaping goals of the physician present in training cases.

In designing an appropriate forward problem to be imputed, we utilized three main premises. First, forward problems with complex/exotic penalties are less efficient to optimize than forward problems with simple penalties. Thus, it was desirable to design the forward problem to have simple penalties, so we restricted the *OF* to have quadratic penalties only. Second, while problem geometry largely limits achievable organ-sparing [2], planning consistency is more indicative of organ priority. For example, the rectum absorbs more dose than the femoral heads due to its geometric proximity with the PTV, but rectum-sparing is prioritized higher than femoral-head sparing. Thus, it was desirable to minimize this effect by normalizing structure penalties with an average measure of prescription violation, so we normalized penalties for each structure s with an inter-case average of dose variance $\overline{\sigma_s^2} = {}^{1}/_{C} \sum_c \sigma_s^2(case\ c)$. Third, penalizing the PTV and organs alone would not be sufficient to capture dose-shaping behaviours present in the clinical plans. Thus, several ring structures, which were created from successive volume expansions of the PTV in 0.5 cm radial increments, were used to help shape dose falloff from the PTV.

3. Results and Discussion

Consistently delivering high-quality treatment plans is crucial for the efficacy of radiotherapy. Excepting machine failures and inter-operator contouring differences, treatment planning is the largest obstacle to achieving this goal since dose planners work under time constraints and submit plans of variable quality. In this section, we will demonstrate how convex imputing may be used to improve plan quality and consistency.

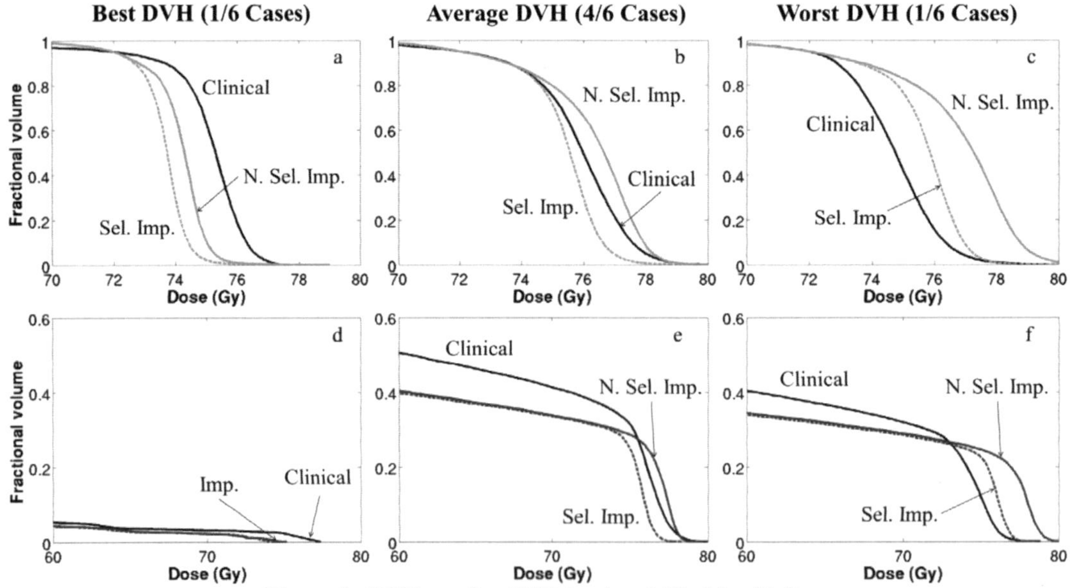

Figure 1. DVH results: PTV (a-c) and Bladder (d-f).

Two sets of imputed objectives were learned for six test cases from separate training sets: 1) the non-selectively imputed objective for each test case was learned from the other nine cases; and 2) the selectively imputed objective for each test case was learned from four cases possessing excellent PTV homogeneity and high-dose bladder-sparing. In Figure 1, the imputed plans were clustered into three categories according to how well they performed w.r.t. the clinical plans in terms of PTV homogeneity and high-dose bladder-sparing. The selectively imputed plans possessed greater PTV homogeneity and high-dose bladder-sparing than: 1) the clinical plans in the majority of cases and 2) the non-selectively imputed plans in all cases. In addition, the selectively imputed plans improved upon the non-selectively imputed plans and clinical plans in terms of rectum-sparing and bulb-sparing, as shown by the characteristic cases in Figure 2.

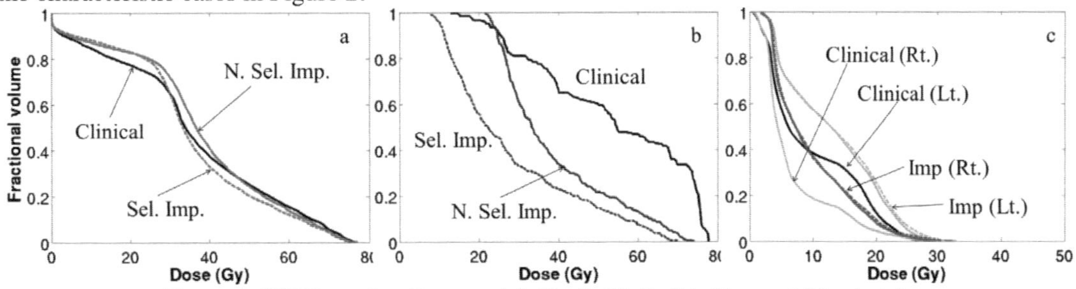

Figure 2. DVH results: Rectum (a), Penile Bulb (b), Femoral Heads (c).

In Figure 3, the non-selectively imputed plan is shown to possess relatively large hot spots whereas the selectively imputed plan does not, further demonstrating that imputing may be improved via proper case selection. In addition, both imputed plans possessed dose falloff from the PTV that was more conformal than that of the clinical plan, and the selectively imputed plan possessed dose falloff that was more conformal than the non-selectively imputed plan, as evidenced by the 50% isodose line.

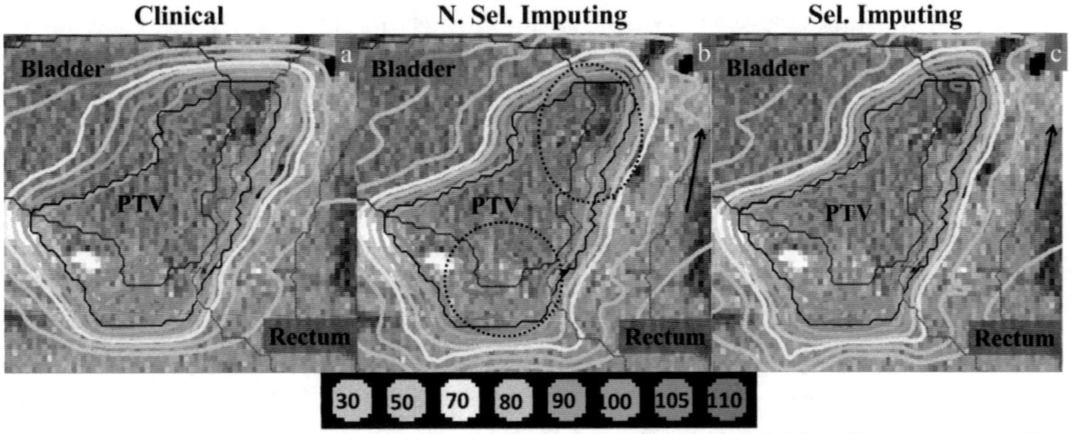

Figure 3. Dose maps. Isodose lines shown. PTV (black), Bladder (dark blue), Rectum (dark red).

4. Conclusions

We have developed a novel method that uses convex imputing to automate the treatment planning process by learning the form of a convex objective function from a training set of previously optimized cases. We demonstrated that only a small number of training cases are needed to learn the DVH and dose-shaping preferences of the physician, and showed that the imputed plans are more consistent than their clinical counterparts in achieving planning goals.

References

1. Keshavarz, A., Wang Y., Boyd, S., *Imputing a Convex Objective*, in *2011 IEEE International Symposium on Intelligent Control (ISIC)*2011, IEEE: Denver, CO, USA. p. 613-619.
2. Sayre, G.A. and D. Ruan, *Dose-shaping using targeted sparse optimization*. Medical physics, 2013. **40**(7): p. 071711.

Journal of Physics: Conference Series **489** (2014) 012059 doi:10.1088/1742-6596/489/1/012059

Modeling Clinical Radiation Responses in the IMRT Era

J L Schwartz[1], D Murray[2], R D Stewart[1] and M H Phillips[1]

[1]University of Washington Medical Center, Seattle, Washington, USA

[2]Cross Cancer Institute, Edmonton, Alberta, Canada

Abstract. The purpose of this review is to highlight the critical issues of radiobiological models, particularly as they apply to clinical radiation therapy. Developing models of radiation responses has a long history that continues to the present time. Many different models have been proposed, but in the field of radiation oncology, the linear-quadratic (LQ) model has had the most impact on the design of treatment protocols. Questions have been raised as to the value of the LQ model given that the biological assumption underlying it has been challenged by molecular analyses of cell and tissue responses to radiation. There are also questions as to use of the LQ model for hypofractionation, especially for high dose treatments using a single fraction. While the LQ model might over-estimate the effects of large radiation dose fractions, there is insufficient information to fully justify the adoption of alternative models. However, there is increasing evidence in the literature that non-targeted and other indirect effects of radiation sometimes produce substantial deviations from LQ-like dose-response curves. As preclinical and clinical hypofractionation studies accumulate, new or refined dose-response models that incorporate high-dose/fraction non-targeted and indirect effects may be required, but for now the LQ model remains a simple, useful tool to guide the design of treatment protocols.

1. Review

Radiation-induced toxicity in tumors and normal tissues arises from the interplay of multiple integrated pathways. Traditionally, our understanding of these pathways came from reductive methods where individual pathways were dissected and analyzed. The creation and testing of alternative and competing models that integrate multiple pathways is an important adjunct to mechanistic laboratory and preclinical studies of radiation response. The formulation of useful and parsimonious dose-response models at the multi-cellular or tissue level are also needed to guide the treatment planning process and for the retrospective analysis of clinical data. In practical terms, a parsimonious dose-response model is often the one that minimizes the number of adjustable (fitted) parameters.

Modeling radiation effects has a long history going back to target theory, developed by Lea in 1946 and published in 1955 [1], the linear-quadratic (LQ) model of Chadwick and Leenhouts [2] in 1974, and the dual action theory of Kellerer and Rossi [3] in 1978. These models were originally developed to help understand the nature of the initial radiation-induced molecular or cellular damage that leads to cytotoxicity, but none of these models fully account for the complexity and range of possible biological responses initiated by radiation-induced DNA damage [4, 5]. Neither do they consider the different modes of radiation-induced cell death [5]. New models of radiation responses that take into account some of our growing understanding about radiation-induced effects continue to be proposed (see for example [6-8]). Most of these alternative and competing models often reduce to the LQ for low doses or low dose rates [9], which is surprising given the complexity of cellular radiation responses.

The LQ model often captures key aspects of how reproductive cell death changes with dose and only has two adjustable parameters (α and α/β). It continues to be a popular choice to design treatments and evaluate clinical outcomes. Perhaps the most often used LQ-based approach to estimate clinical radiation effectiveness is the calculation of Biologically Equivalent (or Effective) Dose (BED) [10]. BED is defined as $-\ln(S)/\alpha$, where S denotes the fraction of the cells that remain viable in the reproductive sense after treatment and α is the LQ model parameter related to the induction of lethal damage by individual radiation tracks. For a radiation treatment delivered as n daily fractions of size d (in Gy), the BED is:

$$\text{BED} = nd[1+ d/(\alpha/\beta)]$$

Here, the LQ parameter α/β quantifies the overall sensitivity of a cell to changes in fraction size. Although estimates of α/β are sometimes derived from *in vitro* or *in vivo* studies, the most appropriate way to estimate α/β is to derive an estimate from clinical dose-response data. Regardless, estimates of α/β are specific to the normal tissue or tumor of interest. Historically, BED has mainly been used for designing and comparing treatments that vary in fraction number and/or the dose per fraction [10]. The advent and increasing use of hypofractionation with its short treatment courses, reduced fraction number and large dose per fraction (hypofractionation) has raised questions about the applicability of the LQ model and BED concept.

The hypothesized basis for the successful use of fractionation in radiation therapy is that fractionation provides time for repair of sublethal damage and tissue repopulation, thus reducing the impact of radiation dose in normal tissues while allowing for reoxygenation and reassortment in the cell cycle in tumor cells, both of which tend to increase tumor radiation sensitivity. Improvements in radiation targeting such as the use of SBRT (Stereotactic Body Radiation Therapy) now provide the opportunity to reduce the volume of normal tissue exposed to radiation, increase dose to tumors, and reduce fraction number. The ability to avoid critical normal tissue structures allows for the delivery of very high radiation doses to tumors in 1 or just a handful of fractions. A technical question that has arisen is whether traditional modeling with BED accurately predicts the clinical response to high dose/fraction and reduced fraction number.

It has been suggested that the LQ model tends to over-estimate BED when high dose fractions are used [11, 12]. The LQ model predicts a continuous bending down of the survival curve at high doses, while in vitro measurements indicate a linear-quadratic-linear (LQL) behavior, i.e., the cell killing is exponential for small and large doses and non-linear for intermediate doses. To address the issue of over-estimation of biological effects, modifications to traditional radiation models have been proposed. Guerrero and Li [11] used the lethal-potentially lethal (LPL) model proposed by Curtis [7] to motivate the derivation of a high-dose correction to the standard LQ model. Park et al. [12] proposed a hybridization of the LQ and target theory models to create a Universal Survival Curve (USC) model. Mehta et al. [13] did a comparative analysis of stage I non-small cell lung cancer treatments using BED and USC. The studies of Mehta et al. [13] included clinical data for both SBRT and 3-

dimensional conformal RT (3D-CRT), and fraction number ranged from 1 to more than 10. The authors reported that while BED was larger than the USC model for equivalent tumor control, the clinical significance of this difference for tumor control probabilities greater than 90% was minimal. Thus, while it is possible to account for the problem of BED over-estimation in the LQ model at high doses, the need for such corrections is so far not compelling, especially when one considers the additional fitted parameters introduced by other models. Future studies might lead to a change in this conclusion..

Another question that has been raised by hypofractionation studies is whether the excellent clinical outcomes associated with hypofractionation are due to the delivery of larger (biological) doses with SBRT or to new biology that becomes especially significant when the dose per fraction is large. Brown and colleagues [14] addressed this question in a discussion of the data presented in Mehta et al. [13]. They argued that the observed monotonic relationship between the tumor control probability and BED is consistent with dose escalation as the primary factor underlying the observed clinical responses; they found no difference in tumor control for single fraction SBRT, multifraction SBRT or 3D-CRT for treatments that deliver the same BED.

In contrast to the observations of Mehta et al. [13] and Brown et al. [14], Zelefsky et al. [15] reported on tumor local progression-free outcomes after treatment with single-dose or hypofractionated regimens for extracranial metastases from renal cell primary tumors. The authors reported 3-year progression-free survival rates of 88% for single doses greater than 24 Gy and 17% for hypofractionated doses (total doses ranged from 20 Gy to 60 Gy). In this example, BED for the single-dose treatment was much less than BED for hypofractionation. Single doses less than 24 Gy (18-22 Gy) had much lower 3-year survival rates (21%) than that observed for the 24 Gy single dose (88%), suggesting a dose threshold for some radiation effect that influences this clinical endpoint. The work from Zelefsky, Fuks, Kolesnick, and colleagues [16-18, 15] suggests that, at least in their clinical model, the biological mechanisms underpinning low dose radiation effects, i.e., the dose range in which target theory and the LQ model are most appropriate (accurate), may differ in substantial ways from high dose biology mechanisms.

Target theory and the LQ model are largely premised on the hypothesis that only those cells that are directly irradiated (damaged) by radiation manifest a response. However, there is growing evidence that non-targeted and indirect effects may play an important role in high dose tumor biology [19-23]. Moreover, the non-targeted and indirect effects of radiation do not necessarily show the same dose response characteristics traditionally observed in cell survival experiments.

Effects on tumor microvasculature and immune responses following radiation exposure are two examples of indirect effects that might not be explained by BED. Fuks, Kolesnick, and colleagues [16-18] emphasize the importance of ceramide-mediated endothelial cell apoptosis of tumor microvasculature as an indirect regulator of radiation-induced tumor cell toxicity that might account for the apparent dose threshold reported in Zelefsky et al. [15].

Abscopal effects, where significant tissue responses are observed at sites far from the radiation-targeted tumor, are another type of indirect radiation effect. Abscopal effects have been associated with immune response stimulation [24-28]. In fact, immune response is important for local control following radiation exposure. Stone et al. [29] reported many years ago that the dose needed to control tumors (TCD50) was twice as high in mouse models that lack T cells. Demaria, Formenti and their colleagues have published extensively on the role of radiation as an immunological adjuvant [30, 24, 31, 5]. In their studies, combining radiation treatment of one tumor with anti-CTLA-4 antibody to overcome T cell tolerance led to significant growth inhibition both in the irradiated tumor and a distant unirradiated tumor [24]. In a review of the importance of dose and fractionation on immune response, Demaria and Formenti [30] reported that single doses of radiation can induce an antitumor immune response in a dose-dependent manner, but there is no consensus yet on the shape of the dose response or type of cell inactivation mechanism required for that stimulation. As for fractionation, while Dewan et al. [24] reported that fractionation was more effective than single doses in inducing the immune-mediated abscopal effect, in their review on this subject, Demaria and Formenti [30] note that other investigators find that single doses are more effective than fractionated doses. Use of different tumor models might explain the discrepancies between different publications in terms of dose and fractionation dependence of immune stimulation. What is important for this discussion is that the immune stimulation is a non-targeted, indirect effect of radiation that is unlikely to show an LQ dose response.

2. Conclusion

It is important to remember that target theory and the LQ model ultimately provide descriptions of phenomena rather than an explicit model of specific biological mechanisms or pathways. As operational tools, they are useful models to guide and compare the effects of fractionated radiation treatments. Because of technological improvements in our ability to deliver highly conformal doses to tumor targets while minimizing the dose to normal tissue, it appears likely that radiation oncology is rapidly moving towards the use of treatments delivered in a single or a few fractions. Although the LQ is likely to remain a useful model for the analysis of clinical outcomes and for guiding the refinement of treatment protocols, caution needs to be exercised in the application of the LQ to large dose per fraction treatments. In particular, there is increasing evidence that the success of high dose per fraction treatments may be due more to non-targeted or indirect radiation effects than to classic radiobiological mechanisms motivating the LQ and target theory. It can be argued that, when applied to clinical data, the LQ implicitly includes some of the indirect effects of radiation. If the LQ model fits the clinical data as suggested [14], it is possible to claim that all of the relevant mechanisms underlying the clinical observation are included (implicitly) in the model. Additional laboratory and preclinical studies are needed to test this hypothesis.

3. Acknowledgement

This work supported by grant CA118130 from the US National Institutes of Health (JLS).

4. References

[1] Lea D E 1955 *Actions of Radiations on Living Cells* (New York, NY, USA: Cambridge University Press)
[2] Chadwick K H and Leenhouts H P 1973 *Phys Med Biol* **18** 78-87
[3] Kellerer A M and Rossi H H 1978 *Radiat Res* **18** 78-87
[4] Thompson L H 2012 *Mutat Res* **751** 158-246
[5] Golden E B, Pellicciotta I, Demaria S, Barcellos-Hoff M H and Formenti S C 2012 *Front Oncol* **2** 88
[6] Carlson D J, Stewart R D, Semenenko V A and Sandison G A 2008 *Radiat Res* **169** 447-59

[7] Curtis S B 1986 *Radiat Res* **106** 252-70
[8] Ballarini F 2012 *J Nucleic Acids* **2010** 350608
[9] Brenner D J, Hlatky L R, Hahnfeldt P J, Huang Y and Sachs R K 1998 *Radiat Res* **150** 83-91
[10] Hall E J and Giaccia A 2011 *Radiobiology for the Radiologist* (Philadelphia, PA: Lippincott, Williams & Wilkins)
[11] Guerrero M and Li X A 2004 *Phys Med Biol* **49** 4825-35
[12] Park C, Papiez L, Zhang S, Story M and Timmerman R D 2008 *Int J Radiat Oncol Biol Phys* **70** 847-52
[13] Mehta N, King C R, Agazaryan N, Steinberg M, Hua A and Lee P 2012 *Practical Radiation Oncology* **2** 288-95
[14] Brown J M, Brenner D J and Carlson D J 2013 *Int J Radiat Oncol Biol Phys* **85** 1159-60
[15] Zelefsky M J, Greco C, Metzer R, Magsanoc J M, Pei X, Lovelock M, Mechalakos J, Zarchy J, Fuks Z and Yamada Y 2012 *Int J Radiat Oncol Biol Phys* **82** 1744-8
[16] Fuks Z and Kolesnick R 2005 *Cancer Cell* **8** 89-91
[17] Garcia-Barros M, Lacorazza D, Petrie H, Haimovitz-Friedman A, Cardon-Cardo C, Nimer S, Fuks Z and Kolesnick R 2004 *Cancer Res* **64** 8285-91
[18] Garcia-Barros M, Paris F, Cordon-Cardo C, Lyden D, Rafii S, Haimovitz-Friedman A, Fuks Z and Kolesnick R 2003 *Science* **300** 1155-9
[19] Bouchet A, Lemasson B, Le Duc G, Maisin C, Brauer-Krisch E, Siegbahn E A, Renaud L, Khalil E, Remy C, Poillot C, Bravin A, Laissue J A, Barbier E L and Serduc R 2010 *Int J Radiat Oncol Biol Phys* **78** 1503-12
[20] Burdak-Rothkamm S and Prise K M 2009 *Eur J Pharmacol* **625** 151-5
[21] Crosbie J C, Anderson R L, Rothkamm K, Restall C M, Cann L, Ruwanpura S, Meachem S, Yagi N, Svalbe I, Lewis R A, Williams B R and Rogers P A 2010 *Int J Radiat Oncol Biol Phys* **77** 886-94
[22] Schmid T E and Multhoff G 2012 *Front Oncol* **2** 80
[23] Sun Y and Nelson P S 2012 *Clin Cancer Res* **18** 4019-25
[24] Dewan M Z, Galloway A E, Kawashima N, Dewyngaert J K, Babb J S, Formenti S C and Demaria S 2009 *Clin Cancer Res* **15** 5379-88
[25] Frey B, Rubner Y, Wunderlich R, Weiss E M, Pockley A G, Fietkau R and Gaipl U S 2012 *Curr Med Chem* **19** 1751-64
[26] Ludgate C M 2012 *Clin Cancer Res* **18** 4522-5
[27] Postow M A, Callahan M K, Barker C A, Yamada Y, Yuan J, Kitano S, Mu Z, Rasalan T, Adamow M, Ritter E, Sedrak C, Jungbluth A A, Chua R, Yang A S, Roman R A, Rosner S, Benson B, Allison J P, Lesokhin A M, Gnjatic S and Wolchok J D 2012 *N Engl J Med* **366** 925-31
[28] Levy A, Chargari C, Cheminant M, Simon N, Bourgier C and Deutsch E 2013 *Crit Rev Oncol Hematol* **85** 278-87
[29] Stone H B, Peters L J and Milas L 1979 *J Natl Cancer Inst* **63** 1229-35
[30] Demaria S and Formenti S C 2012 *Front Oncol* **2** 153
[31] Formenti S C and Demaria S 2009 *Lancet Oncol* **10** 718-26

XVII International Conference on the Use of Computers in Radiation Therapy (ICCR 2013) IOP Publishing
Journal of Physics: Conference Series **489** (2014) 012060 doi:10.1088/1742-6596/489/1/012060

Evaluation of atlas based auto-segmentation for head and neck target volume delineation in adaptive/replan IMRT.

R Speight[1], E Karakaya[2], R Prestwich[3], M Sen[3], R Lindsay[1], R Harding[1] and J Sykes[1,4]

[1] Department of Medical Physics, St. James's Institute of Oncology, Leeds, UK.
[2] Dr. A.Y. Ankara Oncology Education and Research Hospital, Ankara, Turkey
[3] Department of Clinical Oncology, St. James's Institute of Oncology, Leeds, UK
4 Institute of Medical Physics, University of Sydney, Crown Princess Mary Cancer Centre, and Nepean Cancer Care Centre, Sydney, Australia

Richard.Speight@leedsth.nhs.uk

Abstract IMRT for head and neck patients requires clinicians to delineate clinical target volumes (CTV) on a planning-CT (>2hrs/patient). When patients require a replan-CT, CTVs must be re-delineated. This work assesses the performance of atlas-based autosegmentation (ABAS), which uses deformable image registration between planning and replan-CTs to auto-segment CTVs on the replan-CT, based on the planning contours. Fifteen patients with planning-CT and replan-CTs were selected. One clinician delineated CTVs on the planning-CTs and up to three clinicians delineated CTVs on the replan-CTs. Replan-CT volumes were auto-segmented using ABAS using the manual CTVs from the planning-CT as an atlas. ABAS CTVs were edited manually to make them clinically acceptable. Clinicians were timed to estimate savings using ABAS. CTVs were compared using dice similarity coefficient (DSC) and mean distance to agreement (MDA). Mean inter-observer variability (DSC>0.79 and MDA<2.1mm) was found to be greater than intra-observer variability (DSC>0.91 and MDA<1.5mm). Comparing ABAS to manual CTVs gave DSC=0.86 and MDA=2.07mm. Once edited, ABAS volumes agreed more closely with the manual CTVs (DSC=0.87 and MDA=1.87mm). The mean clinician time required to produce CTVs reduced from 169min to 57min when using ABAS. ABAS segments volumes with accuracy close to inter-observer variability however the volumes require some editing before clinical use. Using ABAS reduces contouring time by a factor of three.

1. Introduction

Manual delineation of CTV volumes for head and neck (H&N) IMRT patients on a planning-CT (pCT) can take more than two hours per patient and is subject to large inter and intra-observer error [1-2]. During treatment H&N patients sometimes undergo significant anatomical changes during their treatment often due to weight loss [3]. This can lead to inadequacies in their immobilisation or rotations that cannot be accounted for through rigid body translations. In these cases, it is often necessary to carry out a replan-CT (rCT) so that the dosimetric implications of anatomical changes on the original plan can be assessed and to decide if a full re-plan is required. This involves the clinician re-delineating CTV volumes on rCT and re-calculating the dose from the original plan to highlight either regions of under dosage of the target, or over dosage of surrounding normal tissue. The re-delineation of CTV volumes on rCT again can take more than two hours per patient and is subject to large inter- and intra-observer error [1-2]. Atlas Based Autosegmentation (ABAS) is an example of commercially available software which facilitates the transfer of structures between different CT data sets using deformable image registration, so that contours can be generated automatically. This has the potential benefits of reducing contouring time and both inter- and intra-observer variability. The performance of ABAS has been demonstrated for the lung [4], gynaecology [5], breast and anorectal [6] and H&N [7-8].

The aim of this work was to assess the accuracy of automatically delineated CTV volumes produced by ABAS on rCT and compare them to volumes manually delineated by multiple observers at multiple time points, thus allowing inter- and intra-observer errors to be estimated. A further aim of this work was to estimate the potential time that ABAS could save in the adaptive H&N pathway.

2. Method

Fifteen patients, previously treated with IMRT for H&N carcinoma, were included in this retrospective study. Each patient had a pCT and rCT acquired with a Siemens Somatom Sensation wide bore CT scanner. Patients had a range of prescription doses and tumour locations (five nasopharynx, two oral cavity, two tongue, two larynx, one tonsil, one glottis, one oropharynx and one nasopharynx). Four groups of volumes were delineated for each patient: manual delineation on the pCT; automatic segmentation on rCT; manual delineation on the rCT by up to three clinical oncologists (clin) at multiple time points; and manual editing of the automatic segmentations on rCT. A schematic of all groups of volumes produced is shown in figure 1. For inter-observer results three clinicians contoured a subset of five patients and for intra-observer results one clinician contoured a subset of two patients on three occasions separated by at least one week.

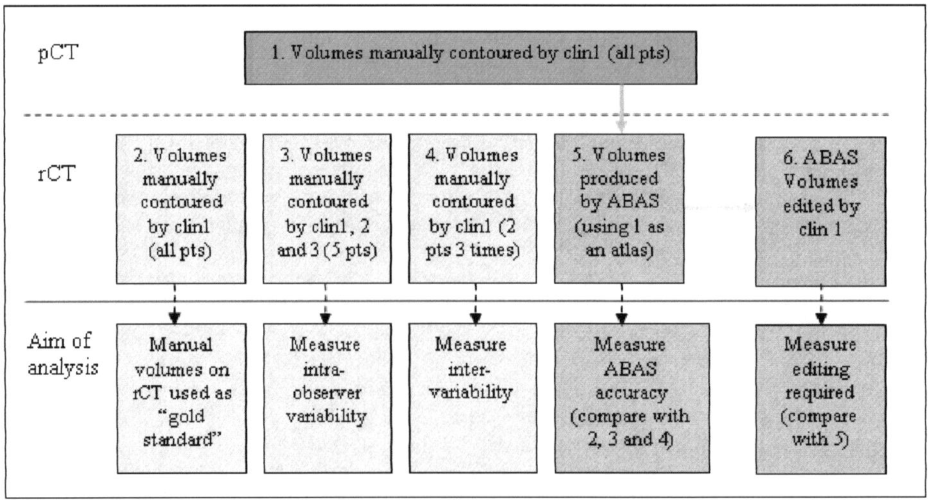

Figure 1. Schematic of the contours delineated on both the pCT and rCT and the aims of the analysis from each step. Green = Manual contours, Orange = ABAS contours.

Manual CTV volumes were delineated according to local protocols. Volumes were automatically segmented on rCT with the H&N algorithm in a research version of ABAS proto8 v0.53 (Elekta AB, Stockholm, Sweden) using the manual delineation from clinician 1 from pCT as an atlas. All volumes were observed in Focal 4D v4.8 (Elekta AB, Stockholm, Sweden).

Different volume sets were compared using the; dice similarity coefficient (DSC) [9], which ranges from zero to one indicating no overlap and a perfect agreement, respectively; and mean distance to agreement (MDA), which reduces for improved agreement. The final analysis was to time observer 1 while manually delineating CTV volumes in steps 2 and 6 in figure 1. This allowed the potential time savings due to ABAS to be measured. Contouring was performed by a research fellow without routine clinical time pressures and therefore may be longer than in the clinical setting.

3. Results

CTV volumes were produced by (1) manual contouring by up to three observers; (2) automatic segmentation by ABAS and (3) manual editing of the ABAS volumes. Differences in volumes were observed between the three manual delineations and the ABAS volume which needed altering in the anterior aspect to make it clinically acceptable (figure 2).

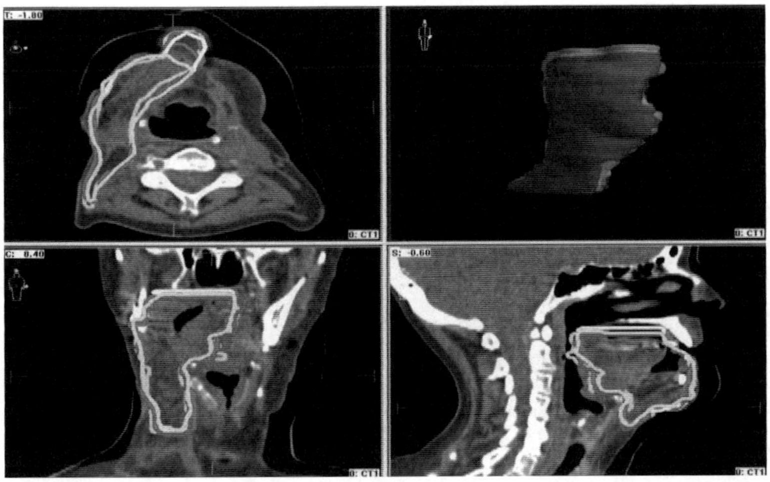

Figure 2. Example CTV volumes produced for patient 1. The volumes shown are manual delineations from observer 1 (red), observer 2 (magenta), observer 3 (green), automatic segmentations by ABAS (cyan) and ABAS volumes edited by observer 1 (yellow).

Mean inter and intra-observer errors were assessed along with ABAS data against each manual observer (figure 3a,b). The mean inter-observer variability (DSC>0.79 and MDA<2.1mm) was greater than the intra-observer variability (DSC>0.91 and MDA<1.5mm). Also the variability between ABAS and manual delineation was different for all observers (DSC 0.82-0.88 and MDA 1.63-2.61mm) and comparable to the inter-observer variability. Finally, although ABAS volumes produced acceptable results (i.e. observer 1 vs. ABAS for all 15 patients gave mean DSC=0.86 and MDA=2.07mm). While ABAS volumes agreed more closely with the manual CTVs, after being edited by observer 1, (mean DSC=0.87 and MDA=1.87mm) the edited volumes remained closer to the volumes auto-segmented by ABAS.

ABAS reduced the delineation time for all patients. The mean clinician delineation and editing times were 169min (σ=25min) and 57min (σ=11min) respectively (figure 4).

4. Conclusion

Inter-observer errors in delineation are higher than intra-observer errors for the H&N volumes considered in this work. ABAS segments volumes with accuracy close to inter-observer variability. However, small regions with significant discrepancies may require editing before clinical use. Use of ABAS reduces contouring time by a factor of three which helps facilitate adaptive radiotherapy for the H&N IMRT patient group.

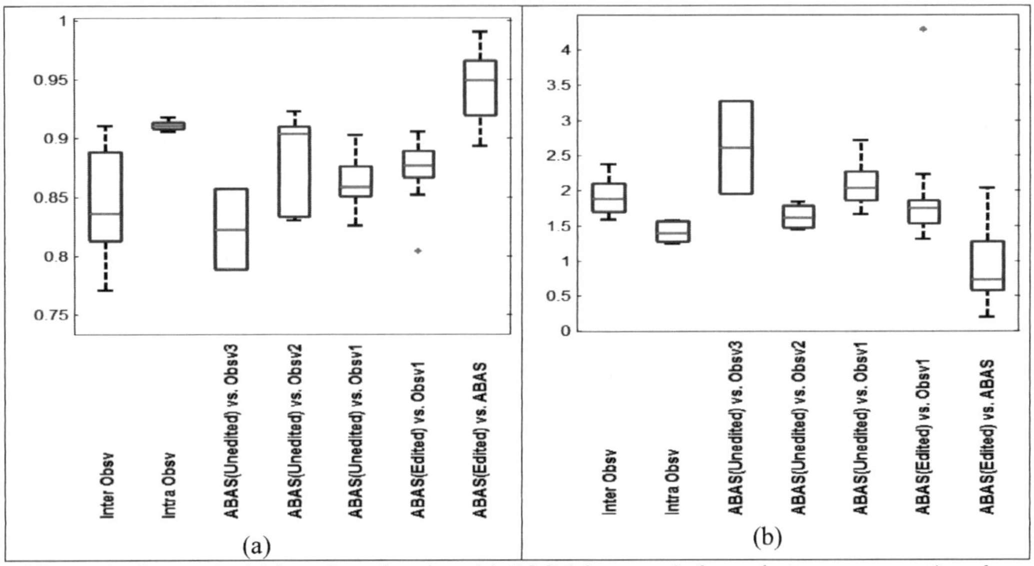

Figure 3. Box and whisker plots showing (a) DSC (closer to 1 shows better agreement) and (b) MDA (reduced number shows better agreement) for CTV volumes produced in different ways.

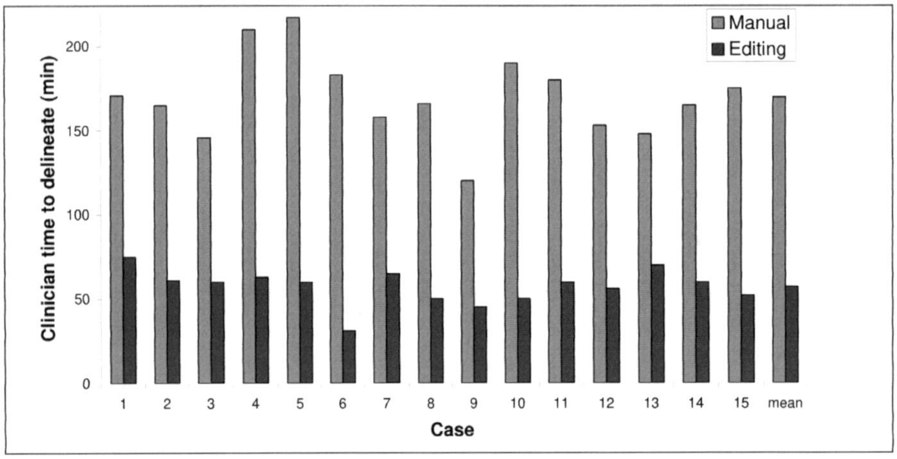

Figure 4. Time taken by observer 1 to both manually delineate and edit ABAS CTV volumes for all 15 patients and the mean.

Acknowledgement: The authors would like to thank Elekta for supplying ABAS as part of a research agreement.

References

[1] Van de Steene J *et al* 2002 *Radiother Oncol* **62** 37–49
[2] Louie AV *et al* 2010 *Radiother Oncol* **95** 166–71
[3] Schwartzl D l and Dong L 2011 *Journal of Oncology* **690595** 1-13
[4] Speight R *et al* 2011 *Radiother Oncol* **98(2)** 277-283
[5] Comsia ND *et al* 2009 *Int J Radiat Oncol Biol Phys* **75** S369–70
[6] Anders L C *et al* 2012 *Radiother Oncol* **102(1)** 68-73
[7] Voet P W J *et al* 2011 *Radiother Oncol* **98(3)** 373-377
[8] Teguh D N *et al* 2011 *Int J Rad Onc Bio Phys* **81(4)** 950-957
[9] Dice L R. 1945 *Ecology* **26** 297–302

Journal of Physics: Conference Series **489** (2014) 012061 doi:10.1088/1742-6596/489/1/012061

Two-dimensional inverse planning and delivery for precision preclinical radiobiological investigations

J M P Stewart[1,2], **P E Lindsay**[2,3] **and D A Jaffray**[1,2,3,4,5]

[1]Institute of Biomaterials and Biomedical Engineering, University of Toronto, Toronto, Ontario, CANADA
[2]Radiation Medicine Program, Princess Margaret Cancer Centre, University Health Network, Toronto, Ontario, CANADA
[3]Department of Radiation Oncology, Faculty of Medicine, University of Toronto, Toronto, Ontario, CANADA
[4]Department of Medical Biophysics, University of Toronto, Toronto, Ontario, CANADA
[5]The Techna Institute for the Advancement of Technology for Health, Toronto, Ontario, CANADA

E-mail: james.stewart@rmp.uhn.on.ca

Abstract. Advances in preclinical radiotherapy systems have provided the technical foundations for delivering highly heterogeneous dose distributions for unique radiobiological experiments, but methods to deliver arbitrary dose distributions are in their infancy. This study developed a method to optimize and automatically deliver planar dose distributions on a recently developed preclinical radiotherapy platform. The method was based on empirically determined dose kernel distributions from radiochromic film measurements. These kernels were used to determine optimal animal stage positions and beam weights to deliver a desired dose distribution at a given depth using a sequential quadratic programming optimization algorithm. The method was validated by end-to-end delivery of two dosimetric challenges designed to quantify targeting and dosimetric accuracy. The results revelead an overall targeting accuracy of 112 μm and a dosimetric delivery error, calculated along four line profiles in radiochromic film measurements, of 6.8%. Mean absolute delivery error across a linear dose gradient between 0 and 1 Gy over 7.5 mm was 0.03 Gy. These results confirm the optimization framework is an effective platform for delivery of millimetre scale heterogeneous dose distributions with sub-millimetre accuracy.

1. Introduction

Continual technological advances in clinical radiotherapy practice have conferred the ability to accurately and conformally modulate dose at millimetre scale resolutions. By and large, the sophisticated beam arrangements permitted by these advances were not validated on animal models, partially due to a corresponding innovation lag in preclinical radiotherapy tools. This gap has recently been mitigated by the introduction of preclinical radiotherapy systems which support sub-millimetre animal positioning setup based on fluoroscopic, cone-beam computed tomography (CBCT), or CT imaging with interchangeable or variable aperture collimated dose delivery at rates from 2-4 Gy/min [1, 2, 3]. A missing element for preclinical dose delivery is to exploit this ability to optimally deliver arbitrary dose distributions. The goal of this work was to develop a general framework for two-dimensional dose optimization on a recently developed preclinical irradiator [1].

2. Methods

The overall method, illustrated in Figure 1, was designed on the basis of empirically measured dose kernel primitives of the small animal radiotherapy system. Dose kernel measurements were made in a 95 × 45 mm solid water phantom at a depth of 10 mm with 29 mm of backscatter. The irradiation isocenter was set under CBCT guidance (to mimic a common small animal setup protocol) to a depth of 10 mm. Phantom and films were irradiated using a 1 mm circular collimator for 60 s at a tube output of 225 kVp / 13 mA.

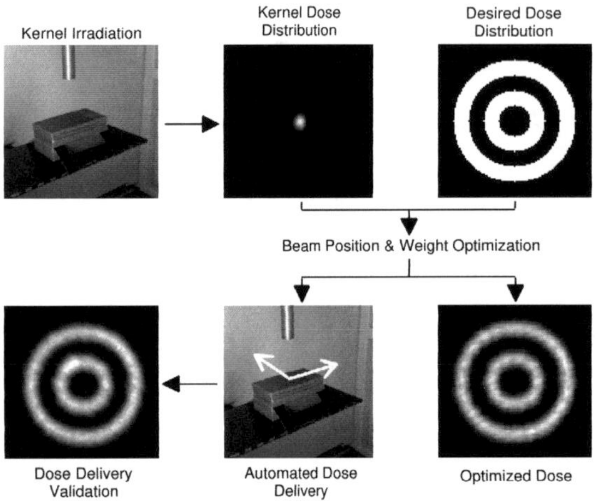

Figure 1. Overview of the two-dimensional dose optimization framework. Clockwise from top left: setup for 1 mm diameter collimator kernel irradiation with radiochromic film placed at 1 cm depth in a solid water phantom. The optimization framework used the kernel dose distribution and the desired dose distribution to determine a set of beam positions and weights to deliver a dose that minimized the difference between the desired and optimized dose distributions. Beam positions and weights were mapped to a set of stage motions for automated delivery with the delivered dose validated with radiochromic film.

This dose kernel primitive, together with an arbitrary desired dose distribution, were used in a sequential quadratic programming optimization method [4] to define a set of beam positions and weights to minimize the difference between the desired and delivered doses. The initial beam positions were defined in a hexagonal arrangement based on maximizing density in a circle packing arrangement [5]. Initial beam weights were defined, for each beam, to deliver the mean of the dose within the half maximum area of that beam. These beam positions and weights were used as the initial conditions in the optimization method. Denoting the desired two-dimensional dose by D_d, the optimization problem sought to minimize

$$\|D_d - D_c(p, \omega)\|_2, \ \omega \geq 0, \tag{1}$$

where $D_c(p, \omega)$ is the calculated dose for beam positions p and weights ω, and the norm is calculated across all pixels of the dose error. Dose was calculated based on a convolution of the beam positions, multiplied by their respective weights, with the dose kernel. Optimized beam positions were converted to stage positions using an in-house developed program which automated dose delivery by iterating through the set of stage positions, pausing at each position according to the beam weight of the respective beam.

End-to-end efficacy of the proposed approach was quantified for two desired dose distributions designed to test the overall targeting and dosimetric accuracy. The first distribution, shown in the top left panel of Figure 2, was a bullseye pattern consisting of 1 mm wide rings with the rings alternating between desired doses of 0 and 2 Gy. A 27 gauge (0.4 mm diameter) needle tip fiducial was embedded in the radiochromic film for the bullseye distribution to define the desired delivery isocenter and quantify targeting accuracy. The second distribution (top left panel of Figure 3) was a 'sock' pattern consisting of a 1 Gy region exponentially decaying 1.5 mm horizontally over 3 mm vertically (labelled A in the figure), a homogeneous constant 1 Gy 1.5 × 4 mm rectangular region (B), a linear dose gradient rectangle of 7.5 mm × 4 mm (C), and a 2 mm radius semicircle with constant 1 Gy dose (D).

3. Results

The dosimetric results for the bullseye distribution are outlined in Figure 2. The effects of the embedded fiducial marker have been removed by masking out a 5 mm radius circle at the center of the relevant distributions. Dose along the vertical and horizontal line profiles, illustrated in the desired dose distribution, are shown along the bottom row. Mean absolute dose differences between the delivered and desired distribution were 0.41 Gy and 0.38 Gy along the horizontal and vertical line profiles, respectively. Between the delivered and optimized distributions, the respective differences were 0.12 Gy for both line profiles. Targeting accuracy, assessed by the difference between the position of the embedded needle fiducial (denoting the desired delivery isocenter) and the actual delivered isocenter was 112 μm.

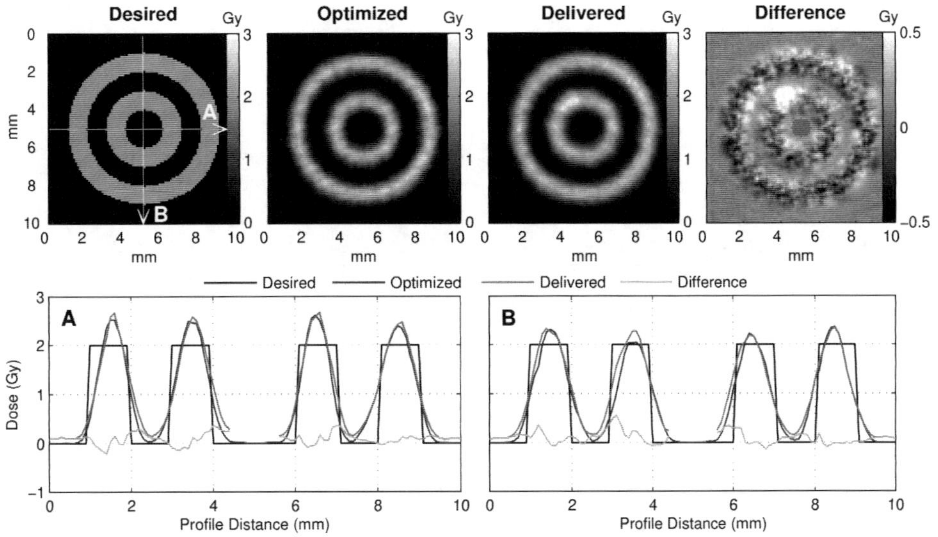

Figure 2. Dosimetric results for the bullseye dose distribution optimization and delivery. Top row: desired, optimized, delivered, and difference (delivered minus optimized) dose distributions. A 0.5 mm radius circle was masked from the center of the delivered and difference dose distributions to eliminate the effects of the embedded fiducial marker. Bottom row: one dimensional dose profiles along horizontal and vertical paths indicated in the desired dose distribution.

The results of the sock dose delivery are illustrated in Figure 3. As with the analysis of the bullseye distribution, the top row shows the desired, optimized, delivered, and difference (delivered minus optimized) dose distributions. The dose along four profiles, visually indicated in the desired dose profile, are shown along the bottom row. The mean absolute difference along the decaying exponential in panel A between the delivered and desired dose is 0.20 Gy and the dose falls below the desired 1 Gy for the last 1.9 mm of the 1 Gy region. The delivered dose of the linear gradient in panel C closely corresponds to the desired dose with a mean absolute difference of 0.03 Gy between 2.5 and 10 mm. Across all four panels, the mean absolute difference between the delivered and optimized doses is 0.04 Gy (relative difference of 6.8% for all profile portions greater than 0.05 Gy). The same difference between the delivered and desired doses is 0.11 Gy (relative difference of 13.0% for all profile portions greater than 0.05 Gy).

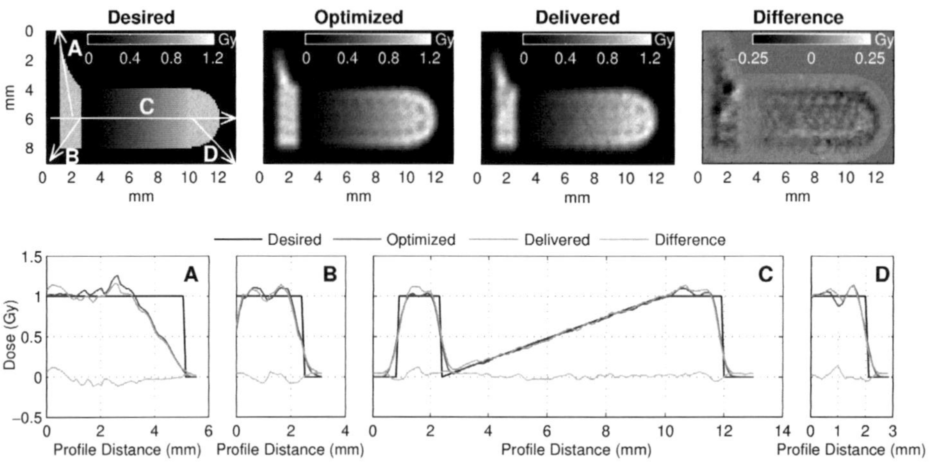

Figure 3. Dosimetric results for the 'sock' dose distribution optimization and delivery. Top row: desired, optimized, delivered, and difference (delivered minus optimized) dose distributions. Bottom row: one dimensional dose profiles along paths indicated in the desired dose distribution. Profiles shown through exponentially decaying region (A), through corner of homogeneous rectangular region (B), across profile width (C), and from center to exterior of semicircle at 45° angle (D). Profile plot dimensions are scaled to have equal abscissa spacing.

4. Conclusions

The optimized preclinical treatment planning and delivery framework developed in this work provides delivery of arbitrary two-dimensional dose distributions to mm scale dosimetric and geometric constructs with a targeting accuracy of approximately 0.1 mm.

Acknowledgments

The authors thank Steve Ansell and Paul DeJean for valuable discussions on the X-RAD 255Cx small animal irradiator. This work is supported by the Natural Sciences and Engineering Research Council of Canada and the Orey and Mary Fidani Family Chair in Radiation Physics.

Conflict of Interest

P. E. Lindsay and D. A. Jaffray are listed as inventors of the system described herein. This system has been licensed to Precision X-Ray Inc. for commercial development.

References

[1] Clarkson R, Lindsay P, Ansell S, Wilson G, Jelveh S, Hill R and Jaffray D 2011 *Med Phys* **38** 845–856
[2] Wong J, Armour E, Kazanzides P, Iordachita I, Tryggestad E, Deng H, Matinfar M, Kennedy C, Liu Z, Chan T, Gray O, Verhaegen F, McNutt T, Ford E and DeWeese T 2008 *Int J Radiat Oncol Biol Phys* **71** 1591–1599
[3] Graves E, Zhou H, Chatterjee R, Keall P, Gambhir S, Contag C and Boyer A 2007 *Med Phys* **34** 4359–4367
[4] Boyd S and Vandenberghe L 2004 *Convex optimization* (Cambridge university press)
[5] Stephenson K 2005 *Introduction to circle packing: The theory of discrete analytic functions* (Cambridge University Press)

Journal of Physics: Conference Series **489** (2014) 012062 doi:10.1088/1742-6596/489/1/012062

PlanJury: probabilistic plan evaluation revisited

M Witte, J-J Sonke and M van Herk

The Netherlands Cancer Institute, Amsterdam, The Netherlands

E-mail: m.witte@nki.nl

Abstract. *Purpose:* Over a decade ago, the 'Van Herk margin recipe paper' introduced plan evaluation through DVH statistics based on population distributions of systematic and random errors. We extended this work for structures with correlated uncertainties (e.g. lymph nodes or parotid glands), and considered treatment plans containing multiple (overlapping) dose distributions (e.g. conventional lymph node and hypo-fractionated tumor doses) for which different image guidance protocols may lead to correlated errors.
Methods: A command-line software tool 'PlanJury' was developed which reads 3D dose and structure data exported from a treatment planning system. Uncertainties are specified by standard deviations and correlation coefficients. Parameters control the DVH statistics to be computed: e.g. the probability of reaching a DVH constraint, or the dose absorbed at given confidence in a (combined) volume. Code was written in C++ and parallelized using OpenMP. Testing geometries were constructed using idealized spherical volumes and dose distributions.
Results: Negligible stochastic noise could be attained within two minutes computation time for a single target. The confidence to properly cover both of two targets was 90% for two synchronously moving targets, but decreased by 7% if the targets moved independently. For two partially covered organs at risk the confidence of at least one organ below the mean dose threshold was 40% for synchronous motion, 36% for uncorrelated motion, but only 20% for either of the organs separately. Two abutting dose distributions ensuring 91% confidence of proper target dose for correlated motions led to 28% lower confidence for uncorrelated motions as relative displacements between the doses resulted in cold spots near the target.
Conclusions: Probabilistic plan evaluation can efficiently be performed for complicated treatment planning situations, thus providing important plan quality information unavailable in conventional PTV based evaluations.

1. Introduction

Compensation for geometric uncertainties when planning external beam radiation treatment is traditionally accomplished through the application of margins, expanding the Clinical Target Volume (CTV) to a Planning Target Volume (PTV), or an Organ At Risk (OAR) to a Planning organ-at-Risk Volume (PRV). These surrogate volumes are then used both for plan optimization and plan evaluation purposes. However, such surrogate volumes offer only limited information about the dose that will likely be absorbed in the various tissues. Probability based plan optimization techniques which do not use PTV margins are under development [1]. Furthermore, Monte Carlo plan evaluation techniques based on population statistics of geometric uncertainties have been proposed [2–4]. Recently, ICRU report 83 stated that confidence intervals on absorbed dose should be considered part of Level 2 (*state-of-the-art techniques*) or Level 3 (*optional research-and-development*) reporting whenever possible [5].

Monte Carlo plan evaluation relies on the estimation of dose absorbed in tissues for many possible geometrical instances. This process is much simplified if the 3D dose distribution is assumed not to change as a result of the geometrical variations, usually valid for deep seated tumors in homogeneous tissue. Then, geometrical instances are approximated as displacements of delineated volumes with respect to the dose distribution. This assumption also implies that patient setup errors and organ motion have an identical effect and can be grouped together.

Uncertainties of shape invariant Regions Of Interest (ROIs) have previously [3] been described using Standard Deviations (SDs) of systematic and random translations and rotations. In this work we extend this description to the case of multiple ROIs [6] and/or multiple dose distributions [7], for which correlated uncertainties may exist. For example, patient alignment based on bony structures could result in correlated systematic errors in multiple tumors, while correction based on a variable tumor position could induce correlated random errors in lymph node positions. Like the SDs, such correlations should be measured from patient series for each imaging and set-up protocol.

2. Methods

2.1. PlanJury

A standalone command-line software tool 'PlanJury' was developed for the Microsoft Windows, Oracle Solaris and GNU/Linux operating systems. PlanJury is configured from a MS Windows ini-style file, and will read 3D planned dose and structure data from files exported using the scripting capability of Philips Pinnacle, or from an in-house developed file format. Structures may be read in as contour sets, CT masks or surface triangulations.

A cloud of random points is generated representing a specified ROI. Systematic translational and optional rotational errors are sampled from Gaussian distributions, and for each systematic error N_{frac} random errors. Every volume point is transformed using a systematic error and each of the associated random errors consecutively, dose values are acquired at the transformed locations using trilinear interpolation, and accumulated dose is computed with and/or without α/β correction. The accumulated dose values are summarized in a dose-volume histogram per systematic instance. An efficient computational structure was established by first preparing all necessary transformation matrices, which can then be used by all random volume points. The loop over all ($\approx 10^4$) random points is parallelized using an OpenMP pragma in the C++ language, using a single atomic (i.e. non-parallel) statement to accumulate the DVH values.

From the sampled DVHs per systematic instance, any number of dosimetric parameters may be extracted. For example, by specifying the lines `GTV.DVH.1.Dose = 95%` and `GTV.DVH.1.Volume = 99%` in the ini-file, PlanJury will return the probability of acquiring at least 95% of dose prescription to 99% of the GTV. The lines `GTV.DVH.2.Prob = 90%` and `GTV.DVH.2.Volume = 99%` could be added to also return the dose which is absorbed at 90% confidence in 99% of the GTV. Similarly, confidence levels and/or expectation values of EUD, TCP and NTCP can be computed.

2.1.1. Multiple ROIs PlanJury may combine multiple ROIs into a compound volume. Error standard deviations can be set per sub-volume, and correlation coefficients for similar errors (e.g. systematic translations in the left-right direction) between sub-volumes can be specified. Gaussian errors generated using the Box-Muller transform are converted into correlated samples using a linear transformation based on the spectral decomposition of the covariance matrices (e.g. `http://en.wikipedia.org/wiki/Multivariate_normal_distribution`).

The way in which dose in the compound volume needs to be scored depends on the clinical relation between the sub-volumes. Two modes of operation are available to work with a compound volume:

Combined. Transformations on the random dots are performed per sub-volume, and dots are combined into a single DVH per systematic instance. This mode can e.g. be used to compute

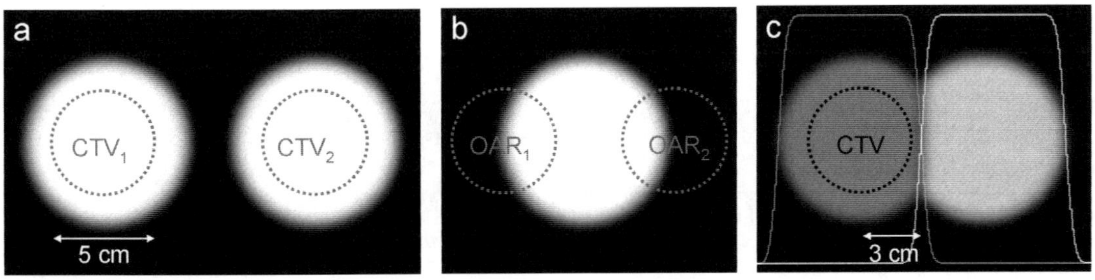

Figure 1. Testing geometries for plan evaluation computations. *a)* Two spherical targets each lie within a perfectly conforming dose distribution with Gaussian penumbra. *b)* Two organs at risk are partially covered with high dose. *c)* Two spherical dose distributions were intersected by a plane and joined to form a continuous dose distribution. Relative displacements between the two doses may result in cold spots near a CTV inside one of the composing distributions. Dose profiles through the center line are indicated.

the minimum dose to either of a number of delineated targets.
Separately. The sampling process results in DVHs per systematic instance for every sub-volume separately. When extracting the dosimetric parameters, a qualifier (e.g. Minimum) specifies which DVH to select for any given systematic instance. This mode can e.g. be used to handle the situation in which one out of two (or more) functioning organs ensures uncomplicated outcome, and one is interested in the maximum dose to the remaining organ (receiving lowest dose).

2.1.2. Multiple dose distributions Treatment plans containing multiple dose distributions can be handled by defining a number of sets. Each of these sets contain a number of dose distributions to be delivered on the same days. Accumulation is always performed without α/β correction for doses within a set, whereas accumulation over sets (i.e. different days) may include fractionation effects. Systematic and random SDs can be specified per dose distribution, and correlation coefficients between doses can be specified for similar random errors within a set, and for systematic errors both within and between sets. (It is assumed random errors may correlate during a day but not from one day to another.) Multiple dose distributions and multiple targets may be combined.

2.2. Testing geometries

Testing geometries were prepared based on a ⌀5 cm spherical target (modeled as a geodesic sphere) and perfectly conforming dose with a Gaussian penumbra ($SD_{pen} = 0.32$ cm). A 1 mm dose grid resolution was used, and the sphere was sampled with a density of 250 random points per cc. Dose prescription was set at 66 Gy in 24 fractions. Systematic and random errors were assumed to be SD = 0.3 cm, and the margin between target and the 95% isodose was fine-tuned such that PlanJury would estimate a 90.0% probability of receiving 95% dose prescription to 99.9% of the spherical volume. Geometries were created for the case of two targets, two OARs, and two overlapping dose distributions (fig. 1a–c).

3. Results

Computations were performed on a workstation with dual Intel Xeon X5650 hexa-core processors running at 2.67GHz and hyper-threading enabled, thus capable of computing 24 threads simultaneously. For a given number of systematic error samples, stochastic residuals were estimated by repeating the determination of each result based on the

Table 1. Confidence levels on minimum DVH parameters for the combined CTV_1 and CTV_2 depicted in fig. 1a. Computed variables are printed boldface, variables in regular typeface were fixed. Stochastic noise intervals are quoted between square brackets. Three correlation coefficients for CTV_1 and CTV_2 were chosen between fully synchronous (= 1) and fully independent (= 0), and apply to systematic and random translations in all directions.

Correlation coefficients		Dose (%) (at least)	Probability (%)	Volume (%) (at least)
	D	**95.0** [94.9, 95.0]	90	99.9
1	P	95	**89.9** [89.8, 90.0]	99.9
	V	95	90	**99.9** [99.9, 99.9]
	D	**93.5** [93.4, 93.5]	90	99.9
0.5	P	95	**84.3** [84.2, 84.5]	99.9
	V	95	90	**99.6** [99.6, 99.6]
	D	**93.2** [93.2, 93.3]	90	99.9
0	P	95	**83.2** [83.1, 83.3]	99.9
	V	95	90	**99.6** [99.6, 99.6]

first and last half of the samples only. The resulting intervals will be quoted between square brackets. Computation times for the probability of receiving at least 95% dose in 99.9% of a single CTV were recorded for various numbers of systematic samples:
5000 samples took 11.6 seconds with resulting $P = 89.4\%$ [88.5%, 90.3%];
10000 samples took 22.6 seconds with resulting $P = 90.1\%$ [89.8%, 90.4%];
50000 samples took 111.5 seconds with resulting $P = 90.0\%$ [90.0%, 90.0%].
Subsequent computations were done using 50000 systematic instances.

3.1. Two targets

DVH results for the case of two CTVs inside their respective high dose regions (fig. 1a) are listed in table 1. For true minimum dose computations (volume = 100%) the results for fully correlated motion would be expected to be identical to the case of a single CTV. However, the computation of true minimum dose in the sampled CTV would depend on a single random point, resulting in considerable stochastic fluctuations. As a compromise, near-minimum dose computations for 99.9% of the volume were performed. For the near-minimum dose the computed probability of $P = 89.9\%$ still corresponds well with the fine-tuned single CTV probability of $P = 90.0\%$. For fully uncorrelated motion the confidence is less at $P = 83.2\%$

3.2. Two OARs

Results for the two OARs partially covered by high dose (fig. 1b) are listed in table 2. It is seen that the probability of receiving no higher than 20 Gy to whichever OAR happens to receive the lowest mean dose is highest for fully correlated motion and decreases with the correlation coefficients, but is always higher than the corresponding probability in either of the OARs separately (results for OAR_2 are the same as for OAR_1 and are omitted).

3.3. Two dose distributions

In the case of two abutting dose distributions (fig. 1c) the DVH analysis for fully correlated motion is similar to the case of a perfectly conformal dose distribution (table 3); the slight increase to $P = 90.8\%$ can be contributed to the effectively larger margin behind the joining

Table 2. Confidence levels on maximum mean ROI dose for the two OARs depicted in fig. 1b. Results for OAR_{Min} are based on whichever OAR receives lowest mean dose for each systematic instance. Correlation coefficients apply to systematic and random translations in all directions.

ROI	Correlation coefficients		Mean dose (Gy) (at most)	Probability (%)
OAR_{Min}	1	*D*	**24.0** [24.0, 24.0]	95
		P	20	**40.0** [39.7, 40.3]
	0.5	*D*	**26.4** [26.4, 26.4]	95
		P	20	**38.9** [38.8, 39.1]
	0	*D*	**28.2** [28.2, 28.3]	95
		P	20	**36.2** [36.1, 36.2]
OAR_1	–	*D*	**33.3** [33.3, 33.3]	95
		P	20	**19.9** [19.7, 20.0]

Table 3. Confidence levels on minimum DVH parameters for a CTV irradiated using two dose distributions (fig. 1c). The correlation coefficients govern the relative systematic and random shifts between the doses in all directions.

Correlation coefficients		Dose (%) (at least)	Probability (%)	Volume (%) (at least)
1	*D*	**95.3** [95.2, 95.3]	90	99.9
	P	95	**90.8** [90.7, 90.9]	99.9
	V	95	90	**99.9** [99.9, 99.9]
0.5	*D*	**84.9** [84.8, 85.0]	90	99.9
	P	95	**67.3** [67.1, 67.5]	99.9
	V	95	90	**97.3** [97.3, 97.4]
0	*D*	**79.4** [79.4, 79.5]	90	99.9
	P	95	**61.7** [61.6, 61.8]	99.9
	V	95	90	**96.6** [96.6, 96.7]

interface. For uncorrelated uncertainties cold spots can occur near the joining interface, and as a result the probability of delivering adequate dose to the CTV decreases to 61.7%.

4. Discussion

By concentrating on a single spherical target with an idealized perfectly conformal dose distribution, the original work [3] on DVH sampling could lead to the (in)famous margin recipe $M = 2.5\Sigma + 0.7\sigma$. In the current work we have extended this sampling approach and considered more complicated clinical circumstances, involving multiple ROIs or multiple dose distributions. To describe such situations, correlations between the errors which occur need to be addressed. As there may be many different ways in which complications could present, we did not strive for parametrized simplifications (recipes). Instead, we developed a software tool which may be clinically deployed to perform probability based plan evaluation under various clinically challenging circumstances. By applying this tool to idealized test situations we have connected our results to those published before, and plotted out the ways the results may change for various challenging clinical scenarios.

The rationale to combine volumes when computations are performed for two or more targets is that the clonogenic cells present in all target sub-volumes should receive adequate dose to ensure global tumor control. Likewise, we have shown how to handle two OARs with a common physiological function, and for which one of the organs can suffice to ensure uncomplicated treatment outcome. On the other hand, we did not consider correlations between dose deposition to e.g. tumors and OARs, or to OARs with independent functionality. Coherent motion patterns and patient setup errors can certainly result in correlated dose deposition in such volumes. However, correlations between the associated clinical effects are typically not relevant. The reason is that dose-effects are assumed to be local, and the clinical effects are assumed not to influence one another. For this reason, clinical trials are usually not separately analyzed for toxicity in patient subgroups with and without treatment failure, nor for failure depending on the occurrence of toxicity.

For treatment sites involving tissue inhomogeneities and/or air-tissue interfaces (such as lung and head-and-neck) the 3D dose distribution may itself be sensitive to the occurrence of geometrical variations. Especially for charged particle therapies this sensitivity may be high. Then, Monte Carlo plan evaluation should re-evaluate the dose as geometries are sampled, and distinguish rigid patient setup errors from organ motion, which should be described as patient deformations. Population based methods to generate patient geometries using principal component analysis have been investigated [8, 9], but full 3D dose reconstruction for the numerous sampled geometries may be impractical.

While the tools described here enable advanced treatment plan evaluation, they do not directly suggest how to improve a plan which does not meet clinical requirements. A solution would be to incorporate the techniques involving multiple ROIs and/or dose distributions also in plan optimization software [1], a major effort which was not yet endeavored.

5. Conclusions

An efficient and flexible tool for probabilistic plan evaluation was developed, able to handle challenging clinical planning circumstances. PlanJury can provide important plan quality information unavailable in conventional PTV or PRV based evaluations, thus aiding the creation of treatment plans with an improved balance between the probability of proper target dose on the one hand and the risk of treatment induced complications on the other.

References

[1] Bohoslavsky R, Witte M, Janssen T and van Herk M 2013 *Phys. Med. Biol.* **58** 3563–3580

[2] Killoran J, Kooy H, Gladstone D, Welte F and Beard C 1997 *Int. J. Radiat. Oncol. Biol. Phys.* **37** 213–221

[3] van Herk M, Remeijer P, Rasch C and Lebesque J V 2000 *Int. J. Radiat. Oncol. Biol. Phys.* **47** 1121–1135

[4] Baum C, Alber M, Birkner M and Nusslin F 2004 *Phys. Med. Biol.* **49** 5475–5488

[5] ICRU 2010 *ICRU Report 83*

[6] Schaake E, Belderbos J, Buikhuisen W, Rossi M, Burgers J and Sonke J J 2012 *Radiother. Oncol.* **105** 150–154

[7] Franssen G, Peulen H, Glaser L, Belderbos J, Sonke J, Rossi M, Heres Diddens F and Damen E 2013 *Radiother. Oncol.* **106S** 109

[8] Budiarto E, Keijzer M, Storchi P, Hoogeman M, Bondar L, Mutanga T, de Boer H and Heemink A 2011 *Phys. Med. Biol.* **56** 1045–1061

[9] Sohn M, Sobotta B and Alber M 2012 *Phys. Med. Biol.* **57** 3693–3709

PyCMSXiO: an external interface to script treatment plans for the Elekta® CMS XiO treatment planning system

Aitang XING[1], Sankar ARUMUGAM[1], Lois HOLLOWAY[1,2,3], Gary GOOZEE[1,4]

[1]Liverpool and Macarthur cancer therapy centres and Ingham Institute, Liverpool Hospital, NSW, Sydney, Australia.
[2]Institute of Medical Physics, School of Physics, University of Sydney, Sydney, NSW, Australia.
[3]Centre for Medical Radiation Physics, University of Wollongong, Wollongong, NSW, Australia.
[4]South Western Sydney Clinical School, University of New South Wales, Sydney, NSW, Australia

aitang.xing@sswahs.nsw.gov.au

Abstract. Scripting in radiotherapy treatment planning systems not only simplifies routine planning tasks but can also be used for clinical research. Treatment planning scripting can only be utilized in a system that has a built-in scripting interface. Among the commercially available treatment planning systems, Pinnacle (Philips) and Raystation (Raysearch Lab.) have inherent scripting functionality. CMS XiO (Elekta) is a widely used treatment planning system in radiotherapy centres around the world, but it does not have an interface that allows the user to script radiotherapy plans. In this study an external scripting interface, PyCMSXiO, was developed for XiO using the Python programming language. The interface was implemented as a python package/library using a modern object-oriented programming methodology. The package was organized as a hierarchy of different classes (objects). Each class (object) corresponds to a plan object such as the beam of a clinical radiotherapy plan. The interface of classes was implemented as object functions. Scripting in XiO using PyCMSXiO is comparable with Pinnacle scripting. This scripting package has been used in several research projects including commissioning of a beam model, independent three-dimensional dose verification for IMRT plans and a setup-uncertainty study. Ease of use and high-level functions provided in the package achieve a useful research tool. It was released as an open-source tool that may benefit the medical physics community.

1. Introduction

Computer technology has dramatically reshaped radiotherapy in the last three decades. Dedicated software and hardware have been developed and extensively used in every aspect starting from CT image acquisition and reconstruction, radiotherapy plan simulation and dose calculations and extending to delivery incorporating record and verification systems and computer-controlled radiotherapy equipment such as linear accelerators (Linacs). For external beam radiotherapy, simulation of radiotherapy plans and dose calculation is performed with a specialized software tool called a radiotherapy treatment planning system (RTP). The Radiotherapist/Medical physicist uses the

RTP to find the optimal plan that delivers the prescribed dose to the target while sparing organs at risk. For each patient the user performs contouring of the tumor and other organs on a CT slice, adjusts the treatment and dose calculation parameters calculates dose and evaluates plan quality and then further adjusts parameters as necessary. Some or all of these tasks can be recorded or written as a series of commands in a file. Once this file is loaded into the TPS, the tasks recorded in the file can be performed automatically. Performing the tasks in the radiotherapy planning system in this way is known as treatment plan scripting [1].

Scripting can only be used in a TPS that provides an internal scripting interface. The Pinnacle (Philips) treatment planning system is one system that has inherent scripting capabilities. Its versatile scripting functionality was explored in a notable online document that contains entire collections of Pinnacle script files for various clinical tasks with a comprehensive user guide [1]. Scripting in Pinnacle has been used routinely in many centers through simple record-and-then-replay hot scripts. Advanced use of its scripting capability has also been reported for checking the integrity of treatment plans [2] as well as automation of plan generation and comparison for intensity modulated radiotherapy (IMRT) and volume modulated radiotherapy (VMAT) [3]. Another treatment planning system, Raystation from RaySearach Laboratories, also provides a built-in research interface for the user to directly interact with the core algorithms used in the system. Elekta CMS XiO is a treatment planning system that is widely used in radiotherapy centres around the world and in our centre along with the Pinnacle system. Although XiO has a non-advertised command line control scripting process, it is a non-flexible scripting interface and not recommended to be used by customers. In this work, an external interface, PyCMSXiO, was developed to enable scripting of radiotherapy plans within XiO.

2. Materials and Methods

2.1. Development of the package

The external scripting interface, PyCMSXiO, was developed using the Python programming language. Python has gained growing popularity in medical physics communities. In comparison with other programming languages, it is easy to use and has built-in advanced data structures such as list and dictionary. Because of these features, it has been used to develop several open-source packages and applications for medical physics, such as pydicom, a python package for reading and writing radiotherapy DICOM files [4], dicompyler [5], an extensible open source radiation therapy research platform and QATrack+, a radiotherapy quality assurance management system [6].

The interface was implemented as a Python package using an object-oriented programming (OOP) technique. The syntax for OOP programming in Python is simple and understandable so that one can rapidly develop a complicated software system or package that realizes design goals and can enable easy maintenance once developed. The package was well structured and contains several well-organized maintainable classes (objects).

2.2. Use of this package

As a reusable python package/library, PyCMSXiO is a tool created to enable the user to write an application to manipulate radiotherapy plans in XiO; this enables a broad range of common functions to be reused to save time and effort. The library contains common routines at a high level that allow the user to directly access and modify plan related objects. In the current version of the package, the dose recalculation within XiO for scripted plans can be triggered via the XiO user interface.

The principle and methodology for using PyCMSXiO to script XiO are exactly the same as those used for scripting in the Pinnacle system. The only difference between them is that Pinnacle scripting uses the internal interface built into the system by the manufacturer while XiO scripting utilizes an interface external to the system. In order to script the plan objects in Pinnacle and XiO, one has to first establish a class of plan object. The plan objects can then be accessed through either the internal or external interface. This OOP scripting method is not only used for simple scripting tasks but also for complex ones usually required for a research project. The use of this package was validated for a number of research projects in our centre. For all of these projects initial plans were validated by comparing intended and resulting plan parameters.

3. Results and Discussion

3.1. *Structure of the package*

The PyCMSXiO was organized as a family of classes (objects). The object-oriented hierarchy of the library class is shown in figure 1 as a Unified Modelling Language (UML) diagram. There are four logical groups of classes and each is indicated by the shaded rectangular area at the top of diagram. In terms of the class relationship, Beam (object), XIOTel (object), XIOPlan (object) and Segment (object) are independent entries in the library, parallel to each other. IMRTBeam (object) and CRTBeam (object) are the children of Beam (object). These two classes inherit the properties and actions from their parents. Each class corresponds to an object in a radiotherapy plan.

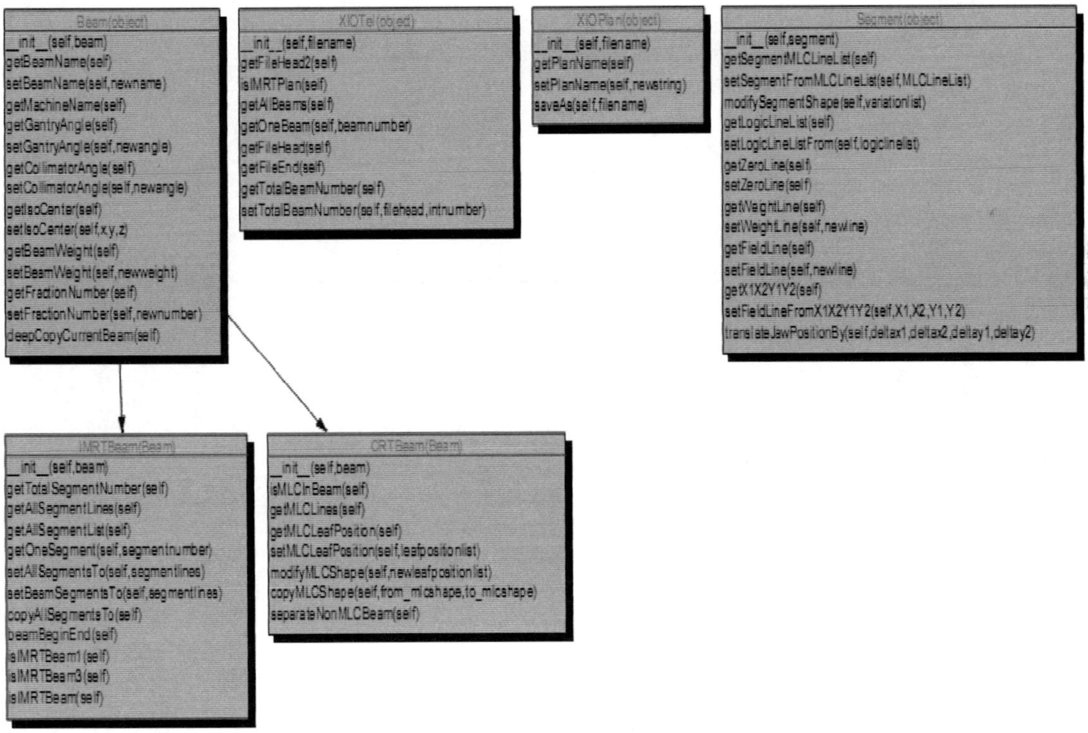

Figure 1. The hierarchy of classes (objects) in the PyCMSXiO python package.

The role of each class is to provide an interface that allows the user to interact with the corresponding object. Beam is a common object in the external radiotherapy plan. Through beam class, one can set or get beam parameters via a class function or class property. As shown in figure 1, there are thirteen object functions in its interface and these can be used to copy the beam, set or get the machine name, collimator and gantry angle, isocenter and weight point. As IMRTBeam (object) and CRTBeam (object) are the subclasses of Beam (object), they also can use the interfaces from beam object except for their own interfaces.

3.2. Comparison between XiO and Pinnacle scripting

Figure 2 demonstrates how to use PyCMSXiO for scripting in XiO and equivalent Pinnacle scripting using its internal interface. In this example, the task was to set the name, collimator and gantry angle for the current chosen beam. This is presented for scripting for both XiO and Pinnacle systems. The Beam object was established using the first- line command and the remaining lines are used to manipulate the beam parameters via the interfaces of the Beam class. The scripting syntax is slightly different between XiO and Pinnacle. This is because the Pinnacle built-in scripting interfaces are implemented as the property of class while PyCMSXiO uses object functions as its class interface. Except for this small syntax difference, XiO scripting with PyCMSXiO is comparable with Pinnacle scripting using its internal scripting interface in term of ease of use and scripting capability.

```
Beam=TrialList.Current.BeamList.Current
Beam.Name=" Sup"
Beam.Gatry=270
Beam.Collimator=180
```

(a)

```
Beam=Beam(BeamList[0])
Beam.SetBeamName("Sup")
Beam.SetGantryAngle(270)
Beam.SetCollimatorAngle(180)
```

(b)

Figure 2. Comparison between (b) XiO scripting with PyCMSXiO and (a) Pinnacle scripting using its internal interface.

3.3. Application of the package

The initial motivation for developing PyCMSXiO came from a research project that required manipulation of treatment plan objects in the XiO system. Since development, the package has been used in several research projects in our department including beam model commissioning [7], independent 3D dose verification for IMRT plan [8] and patient setup studies [9]. Verification of plans generated in the initial stages of each of these projects using the PyCMSXIO package demonstrated accurate generation of treatment plan changes as expected. The use of this package is not limited to the aspects explored in our centre it could be utilised for many other research areas.

Current experience with the PyCMSXIO package has demonstrated that the ease of use and high-level functions encapsulated in this package can help users to quickly develop a tool to meet their specific needs for different research projects. Another advantage of using PyCMSXiO for scripting in XiO is that there is limited risk of harming the core XiO system as the package is an external interface and the scripting is through the Python programming language.

3.4. Package availability and limitations

PyCMSXiO was developed using free programming languages and standard libraries developed by expert volunteers from the Python community. The authors would like to distribute it as an open source tool that may be useful and benefit the medical physics community. The source code and a simple user guide are available on request. The package was only tested and used in XiO version 4.4 and is released under GNU GPL licences without any warranty.

4. Conclusion

An open-source Python package was developed for the Elekta CMS XiO treatment planning system. The package adds scripting capability to XiO through an external interface and enables the user to script radiotherapy plans generated in XiO. It has been used in several research projects and proven to be a useful research tool.

References

[1] Geoghegan S, Gooze G et.al. *Scripting on Pinnalce 3 treatment planning system.* 2007 (www.medphysfiles.com).
[2] Yang D and Moore K. *Med. Phys.* 2012 **39** 1542
[3] Janssen T, Kesteren Z V et.al. *Int J Radiation Oncol Biol Phys.* 2013 85 873.
[4] Mason D. *Pydicom user guide.* 2013 http://code.google.com/p/pydicom/.
[5] Panchal A and Keyes R. *Med. Phys.* 2010 **37** 3245.
[6] Taylor R. *QATrack+: a tool for managing QA program of a radiotherapy department.* 2013 (http://randlet.com/projects/qatrack/).
[7] Xing A, Arumugam S, Holloway L and Gooze G. *Validation of an IMRT beam model using a secondary treatment planning system as a 3D dosimeter. J. Phys.: Conf. Ser.* **444** 012051
[8] Arumugam S, Xing A , Gooze G and Lolloway L. *Independent calculation based verification of IMRT plans using a 3D dose calculation engine. Medical dosimetry.* 2013 **13** 00058-7.
[9] Xing A, Holloway L, Kumar S et.al. *The effect of set-up error on delivered dose in curative radiotherapy for Non-Small Cell Lung Cancer*. 14th World Conference on Lung Cancer (WCLC 2011) Amsterdam RAI, Amsterdam, The Netherlands, July 3 - 7, 2011.

Journal of Physics: Conference Series **489** (2014) 012064 doi:10.1088/1742-6596/489/1/012064

Commissioning of SharePlan: The Liverpool Experience

Aitang XING[1], Shrikant DESHPANDE[1,3], Sankar ARUMUGAM[1], Armia GEORGE [1], Lois HOLLOWAY[1,2,3,4], Gary GOOZEE[1,4]

[1]Liverpool and Macarthur cancer therapy centres and Ingham Institute, Liverpool Hospital, NSW, Sydney, Australia.
[2]Institute of Medical Physics, School of Physics, University of Sydney, Sydney, NSW, Australia.
[3]Centre for Medical Radiation Physics, University of Wollongong, Wollongong, NSW, Australia.
[4]South Western Sydney Clinical School, University of New South Wales, Sydney, NSW, Australia

aitang.xing@sswahs.nsw.gov.au

Abstract. SharePlan is a treatment planning system developed by Raysearch Laboratories AB to enable creation of a linear accelerator intensity modulated radiotherapy (IMRT) plan as a backup for a Tomotherapy plan. A 6MV Elekta Synergy Linear accelerator photon beam was modelled in SharePlan. The beam model was validated using Matrix Evolution, a 2D ion chamber array, for two head-neck and three prostate plans using 3%/3mm Gamma criteria. For 39 IMRT beams, the minimum and maximum Gamma pass rates are 95.4% and 98.7%. SharePlan is able to generate backup IMRT plans which are deliverable on a traditional linear accelerator and accurate in terms of clinical criteria. During use of SharePlan, however, an out-of-memory error frequently occurred and SharePlan was forced to be closed. This error occurred occasionally at any of these steps: loading the Tomotherapy plan into SharePlan, generating the IMRT plan, selecting the optimal plan, approving the plan and setting up a QA plan. The out-of-memory error was caused by memory leakage in one or more of the C/C++ functions implemented in SharePlan fluence engine, dose engine or optimizer, as acknowledged by the manufacturer. Because of the interruption caused by out-of-memory errors, SharePlan has not been implemented in our clinic although accuracy has been verified. A new software program is now being provided to our centre to replace SharePlan.

1. Introduction

SharePlan is a treatment planning system (TPS) developed by Raysearch Laboratories AB (Stockholm, Sweden) for TomoTherapy (Tomo) [1]. The system uses a Tomotherapy plan generated with the TomoTherapy Hi ART system as input and produces a step-and-shoot IMRT plan that is deliverable on a traditional linear accelerator (Linac). SharePlan was designed to replace the procedure of manually creating a backup IMRT plan for treatment on a traditional Linac for each patient being treated on Tomotherapy. Clinical use of SharePlan is expected to result in more efficient generation of backup-plan Tomotherapy plans for centres where only one Tomo unit is installed.

The first step towards introducing SharePlan into routine clinical practice is to commission the system. Commissioning and verification of a beam model in SharePlan was reported by Kristoffer Petersson and his colleagues [2]. Their investigation proved that SharePlan was a useful and time-saving complement for single Tomo unit clinics. In June 2012, one TomoTherapy Hi ART system was installed and commissioned in our centre. One step-and-shoot IMRT plan was created as a back-up plan in Pinnacle (Philips) treatment planning system for each Tomotherapy patient. As a high clinical priority, SharePlan was evaluated with the intention to integrate it into our clinical TomoTherapy procedure. The purpose of this paper is to report our experiences of commissioning SharePlan for clinical use.

2. Materials and Methods

2.1. The virtual linear accelerator in SharePlan

SharePlan includes a generic virtual linear accelerator model that can be adjusted to match the mechanic and dosimetric characteristics of a real Linac. The real Linac either manufactured by Elekta, Varian or Siemens is simulated using a fluence and dose engine. The fluence engine is used to generate the energy fluence map in air at an isocenter plane of a Linac for a given external beam, while the dose engine calculates the dose in patient or phantom for this beam. The fluence engine in SharePlan is a four-source model [3] that takes into account the photons and electrons coming from the target and flattening filter, whereas the dose engine is a collapsed cone (CC) type of algorithm based on the work by Ahnesjo [4].

The dose and fluence engine were implemented using a low-level programming language C/C++ as calculation of dose and fluence is computationally intensive [3]. The implementation details of the dose and fluence engine was hidden from the end-user, but the adjustable parameters of the fluence model and dose model were available to users through different user interfaces. These parameters have to be adjusted to make the virtual Linac model match a specific type of physical linear accelerator.

2.2. Establishment of a 6MV beam model

A virtual Linear accelerator for an Elekta Synergy Linac 6MV photon beam was established as a treatment unit model based on an existing template model in the machine data base. The virtual Linac head was configured to be the same as a Synergy Linac head using the geometric parameters provided by Elekta. The energy spectrum for the 6MV photon beams emitted from the virtual Linac were matched to the Elekta Linac by adjusting dose and fluence parameters to ensure the dose-engine calculated percentage depth dose data (PDDs) matched the PDDs measured on the Synergy Linac for the following fields: 2x2,3x3, 5x5,10x10,15x15,20x20 and 30x30 cm. Similarly the photon fluence calculated by the fluence engine and emitted from the Synergy Linac was matched at the isocenter plane by matching the in-plane and cross-plane dose profiles calculated with the measured profiles in water at depths of 1.5, 5, 10 and 20 cm. The matching criteria for PDDs were 2% for the depth beyond dmax and 2mm distance-to-agreement (DTA) for the build-up region. Criteria of 2% were used for matching the central part of dose profiles and 2 mm DTA for penumbra region [5].

Matching the energy spectrum and energy fluence emitted from the virtual Linac model in SharePlan with ones from a physical Synergy Linac is an iterative procedure. The adjustable parameters for the dose and fluence engines were editable via simple user interfaces. There are two types of adjustable parameters for the energy-spectrum model: the relative weight of the energy spectrum and the output correction factor. SharePlan has a built-in automation tool to aid beam modelling. The auto-modelling tool was found to only be useful as a starting point for manual adjustment. After auto-beam modelling, PDDs and profiles were not matched in all regions for all field sizes. Time-consuming manual adjustment was still required to commission a beam model in SharePlan. The same procedure was used for adjusting the fluence engine for the beam model. The adjustable parameters for fluence engine and how to adjust them are described in detail in the SharePlan physics manual [1].

2.3. Validation of the beam model

The commissioned beam model was used to generate step-and-shoot IMRT plans that were deliverable on an Elekta Synergy Linac. The optimizer implemented in SharePlan uses objective functions that take into account the difference between the dose volume histogram (DVH) of the plan under optimization and the DVH of the Tomotherapy plan to establish an optimal Linac plan. The accuracy of the optimization engine along with the dose and fluence engine can only be verified using IMRT beams. The accuracy of the commissioned beam model for square non-IMRT fields was validated by matching calculated PDD and dose profile with measured ones for these fields during beam modelling. End-to-end tests were performed to verify the beam model using two head neck and three prostate Tomotherapy plans.

For each patient plan, the Tomotherapy plan used for treatment was loaded into SharePlan. A corresponding step-and-shoot IMRT plan was generated for the Elekta Synergy linear accelerator. Seven beams were used for prostate while nine beams for head-neck cases. A quality-assurance (QA) plan was created using Matrix Evolution. Matrix is an array of chambers routinely used for patient delivery quality assurance for our Tomo plans. The QA plan was delivered to Matrix Evolution on an Elekta Synergy Linac. The measured dose distribution was compared with calculated ones utilising a two dimensional (2D) Gamma map with 3%/3 mm criteria [7].

3. Results and Discussion

3.1. Beam model verification results

Figure 2 shows the overlap between calculated PDDs and dose profiles for the simple square fields used for beam model commissioning. For PDDs, a good agreement between the model calculated and measured dose was achieved, indicating that the energy spectrum used by the commissioned 6MV virtual Linac model matched the physical energy spectrum of the 6MV photon beam emitted from the Elekta Synergy Linear accelerator. The accuracy of the fluence engine used for calculating the energy fluence at the isocenter plane in air was shown in figure 2(a). The calculated and measured dose profiles were in good agreement for the central part and penumbra regions. A trade-off had to be made ensuring a good match in the shoulder regions for small fields less than 15x15 cm while slightly sacrificing the shoulder regions for large fields.

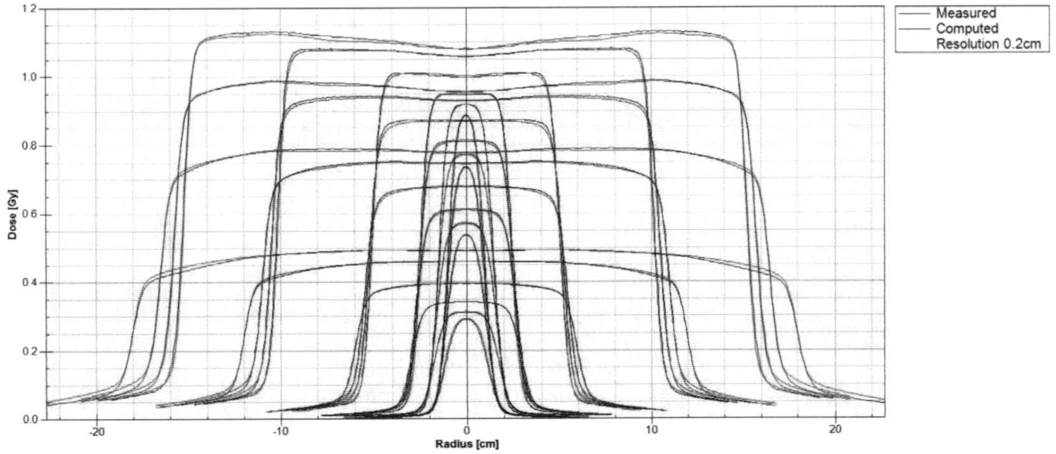

(a)

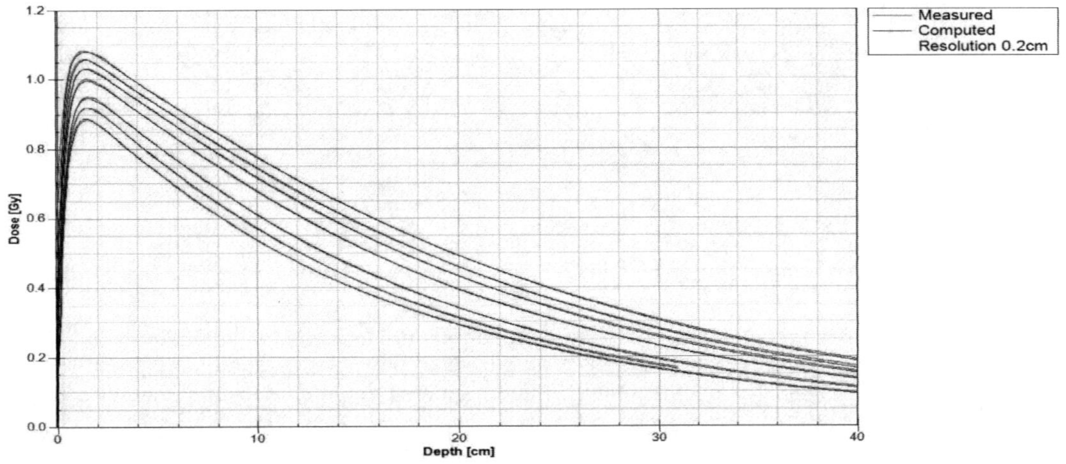

(b)

Figure 2. Comparison between measured dose in water and the dose calculated by the commissioned beam model for (a) cross-plane profiles and (b) PDDs for the following fields: 2x2,3x3, 5x5,10x10,15x15,20x20 and 30x30 cm. The measured data is represented by the redline while the blue line stands for the calculated data.

For end-to-end testing, a 2D dose distribution was measured for each of a total of 39 IMRT beams and compared with those calculated with SharePlan. The minimum and maximum pass rate was 95.4% and 98.7% with a mean value of 97.05%. The clinically acceptable criteria in our centre are over 95% [7]. The average gamma pass rate for each patient plan is shown in table 1 and figure 3 shows the gamma map for one IMRT beam from a prostate patient plan.

Table 1. Gamma pass rates averaged over patient IMRT beams for five patients

Patient #	Averaged Gamma pass rate (%)
1	97.72
2	95.17
3	98.40
4	97.23
5	97.75

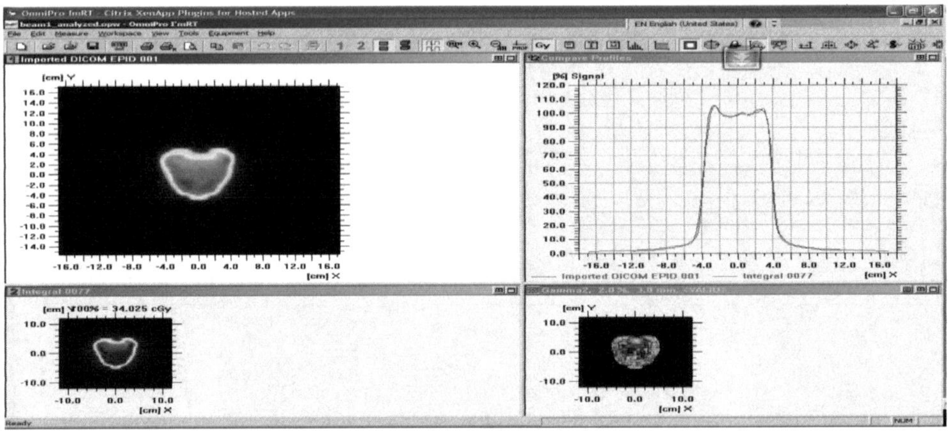

Figure 3. The Gamma map calculated for one IMRT beam from one prostate patient plan. The 2D dose map in the upper left corner shows the Matrix measured dose map while the bottom left corner shows the dose map calculated by SharePlan. The calculated Gamma map is shown at the right bottom panel. The profiles along x axis passing through the center of dose image are plotted at upper right panel.

The end-to-end results demonstrate that the generic linear accelerator implemented in SharePlan can be customized to match a physical linear accelerator via a commissioning procedure. After the beam model was commissioned, the dose and fluence engine were demonstrated to accurately calculate the dose and energy fluence in patients. The IMRT plan generated by the SharePlan optimization engine is deliverable and accurate in term of clinically acceptable criteria, which was also demonstrated by Petersons [2].

3.2. Out-of-memory errors in the SharePlan workflow

To generate a step-and-shoot IMRT plan from a Tomotherapy plan, there are four simple steps to follow after importing the patient Tomotherapy plan into the system: generate plans, plan selection, plan approval and QA setup. The patient IMRT plan and QA plan used for verification were generated following these steps. During our commissioning as shown in figure 4, an error message window frequently popped up saying the system is running out of memory. It occurred on occasion during all of these steps, requiring SharePlan to be closed and restarted again. To avoid this restart requiring the entire workflow to be reinitiated, progressive work had to be regularly saved by the user. The calculation grid was reduced to reduce memory requirements however this did not improve the situation. For each plan, in order to generate the patient IMRT plan and QA plan, SharePlan or the computer had to be restarted many times.

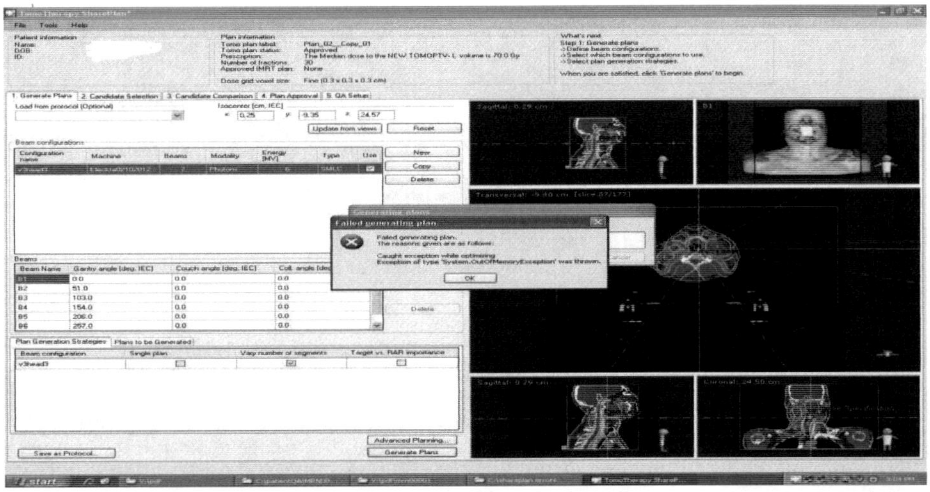

Figure 4. An example of the out-of-memory error thrown out during the use of SharePlan. Following this error SharePlan must be closed and restarted.

The out-of-memory issue is caused by one or more C/C++ functions implemented in the SharePlan fluence engine, dose engine or optimizer. In these functions, some variables are allocated memory, but the memory assigned is not freed at the end of the lifetime of these variables. For each occasion that the functions were called during the dose calculation or optimization increased memory is required resulting in the system running out of available memory. This is a well-know memory-leakage issue in C/C++ programming [6]. As a result of the out-of-memory issue, clinical workflow using this software was not considered feasible. The manufacturer has acknowledged these problems and has now released another software product to replace SharePlan to generate TomoTherapy back-up plans.

4. Conclusion

SharePlan was commissioned with the intention of clinical use. SharePlan was shown to generate backup plans which are accurate and deliverable on a traditional Linac. During the use of SharePlan, however out-of-memory errors resulted in workflow issues resulting in SharePlan not being used clinically. The manufacturer has acknowledged the problem and new software will shortly be installed in our centre to replace SharePlan.

References

[1] RaySearch Laboratoirs AB. 2011 *Physics reference mannual SharePlan 1.1.*

[2] Petersson K , Ceberg C et.al. *Med. Phys.* 2011 **38** 40

[3] Borglund N. *Physics reference manual Ray dose..* 2009 RaySearch Laboratories AB.

[4] Ahnesjo A. *Med. Phys.* 1989 **16** 577.

[5] Vensellar J, Welleweerd H et.al. *Radiother Oncol.* 2001 **60** 191.

[6] Richard R. 2013 *Understanding and using C pointers.* (O'Reilly Media, Inc., 1005 Gravenstein Highway North, Sebastopol, CA 95472).

[7] Vial P Gray A et al. *Australas. Phys. Eng. Sci. Med.* 2013 **36** 138

Journal of Physics: Conference Series **489** (2014) 012065 doi:10.1088/1742-6596/489/1/012065

Inverse planning in the age of digital LINACs: station parameter optimized radiation therapy (SPORT)

Lei Xing and Ruijiang Li
Department of Radiation Oncology, Stanford University, Stanford, CA 94305-5847

lei@stanford.edu & rli2@stanford.edu

Abstract. The last few years have seen a number of technical and clinical advances which give rise to a need for innovations in dose optimization and delivery strategies. Technically, a new generation of digital linac has become available which offers features such as programmable motion between station parameters and high dose-rate Flattening Filter Free (FFF) beams. Current inverse planning methods are designed for traditional machines and cannot accommodate these features of new generation linacs without compromising either dose conformality and/or delivery efficiency. Furthermore, SBRT is becoming increasingly important, which elevates the need for more efficient delivery, improved dose distribution. Here we will give an overview of our recent work in SPORT designed to harness the digital linacs and highlight the essential components of SPORT. We will summarize the pros and cons of traditional beamlet-based optimization (BBO) and direct aperture optimization (DAO) and introduce a new type of algorithm, compressed sensing (CS)-based inverse planning, that is capable of automatically removing the redundant segments during optimization and providing a plan with high deliverability in the presence of a large number of station control points (potentially non-coplanar, non-isocentric, and even multi-isocenters). We show that CS-approach takes the interplay between planning and delivery into account and allows us to balance the dose optimality and delivery efficiency in a controlled way and, providing a viable framework to address various unmet demands of the new generation linacs. A few specific implementation strategies of SPORT in the forms of fixed-gantry and rotational arc delivery are also presented.

1. Introduction

Digital linac with a few other important features has emerged for clinical use[1-6]. A distinct feature of digital linac is that parameters characterizing radiation delivery such as the gantry angle, collimator angle, couch angle, and dose rate are discretized and can be controlled easily in a programmable fashion. The "fundamental unit" is a "station" point (a "station" can also be called a "node" in CyberKnife™ system). In the emerging digital era of radiation therapy (RT), a treatment will be done through the modulation of station parameters as the delivery is done "station by station", instead of "beam by beam". Here we emphasize that a station describes the state of delivery system (including LINAC configurations such as beam energy, aperture shape and weight, gantry/collimator/couch angle). When the auxiliary equipment is stationary, a station is no different from an MLC or jaw-shaped beam. Thus a conventional intensity-modulated beam consists of a collection of stations with the same gantry angle but different MLC segments. Existing radiation therapy scheme does not fully take the digital features of new RT scheme into consideration and how to leverage these new developments to overcome many of the limitations of current RT techniques becomes an urgent issue. Given the fact that the fundamental unit in the new era is "station", it is fair to say that next-generation of RT will be all about the optimization of station parameters, which we call station parameter optimized radiation therapy (SPORT)[4, 7]. VMAT and IMRT are simply two special, and often non-optimal, cases of SPORT. The premise of SPORT is that it will enable us to realize the enormous potential of digital linacs through optimal weighting and spatial distribution of the station points (including non-coplanar, non-isocentric distribution, and even multiple isocenters). In the following, some logistic issues related to SPORT implementation are discussed.

2. Implementation strategies of SPORT

Inverse planning and delivery are generally intertwined. While the station points can be, in principle, distributed arbitrarily in physically realizable parameter space, adding restriction to limit the size of solution space is often necessary for practical reason. The actual delivery of SPORT can be realized in different ways. To familiarize readers with the novel SPORT scheme, here we describe three categories of SPORT delivery: rotational arc SPORT, fixed-station SPORT, and a combination of the two.

2.1. Rotational arc SPORT

In rotational arc delivery, analogous to VMAT and/or sliding window IMRT beam delivery, radiation is on when going from one station to the next. We have implemented a novel scheme of a rotational arc SPORT and shown that the need for multiple arcs in conventional VMAT can be eliminated through more intelligent distribution of the station points in the same coplanar sampling space. Conventional VMAT[8-10] such as Varian RapidArc™ discretizes the angular space into equally spaced station points (178 stations) during planning and then optimizes the apertures and weights of the stations. The aperture at an angle in between two stations is obtained through interpolation. This wisdom tacitly ignores the differential need for intensity modulation of different angles. Unless the discretization is infinitely fine (which is computationally prohibitive and avoidably oversamples some angles), the approach is incapable of providing sufficient modulation for some or all directions for many disease sites. As thus, multiple arcs are often required to produce a clinically acceptable treatment plan. In our one-arc SPORT implementation, the angular sampling rate (i.e. the number of stations per unit angle) is modulated according to the need of individual angles for intensity modulation[7] (see Fig. 1 for a sketch). The essence here is how to identify the need of an individual angle for intensity modulation and provide the necessary intensity modulation for those angles that need it. Obviously, treatment plan for the single arc SPORT can be obtained by a direct optimization of weights and aperture shapes of the station points and their coplanar distribution. Development of such algorithm is to be published elsewhere. In our original study, for the ease of computation, we took a heuristic approach and planned the treatment in an adaptive fashion by starting from a conventional 1-arc VMAT plan with equally spaced stations. The angles that need higher intensity modulation are identified with the help of a "demand metric" defined as[7]:

$$DI(s)=\sum_{\substack{k=-K\\k\neq 0}}^{k=K}\left[\sum_{i=1}^{60}\left(|x_i^A(s)-x_i^A(s+k)|+|x_i^B(s)-x_i^B(s+k)|\right)\right]\left|\left|\frac{MU(s)-MU(s+k)}{\alpha(s)-\alpha(s+k)}\right|\right| \quad (1)$$

where, $x_i^A(s)$ and $x_i^B(s)$ are the i-th MLC leaf positions of banks A and B at the s-th station, MU is the cumulative monitor units, α is the gantry angle. Intuitively, the metric is the geometric modulation weighted by the segmental MU per gantry angle. The neighbouring 2K stations (K=10 in our study) are used to calculate the demand metric of a point. To boost an angle with a high "demand metric" value, we add a number of stations around the point. The rationale here is that, if the aperture or MU changes a lot at an angle in the original one-arc VMAT, it suggests that more apertures or stations are needed in the vicinity of the angle to meet the angular demand for more intensity modulation. The original and the added stations are then optimized again. A single arc sequence is then constructed for delivery using the TrueBeam™.

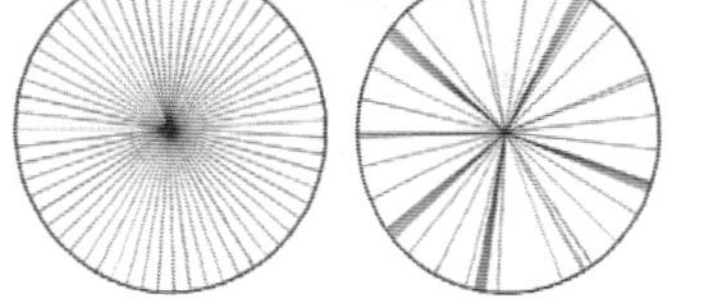

Fig. 1. Schematic plot of the beam angular distributions for VMAT and our rotational SPORT method (Li R & Xing L, Ref. 7). Our method provides angular modulation (modulation per unit angle) through optimal angular sampling of station points.

We have applied the technique to a few clinical cases and all demonstrated the success of the proposed scheme. For a clinically complicated head and neck case, for example, we found that the SPORT approach achieves much better target dose uniformity and sparing of right parotid (~20% EUD reduction) as compared to the 2-arc VMAT. The EUDs of the targets and critical structures for different plans reveal the same for other cases. In general, the MUs required for our approach lie between those for 1- and 2-arc VMAT, yet it leads to the best dose distributions.

As compared to the 2-arc VMAT, the single arc SPORT decreases the MUs by 20%-30%, depending on the case. This reduction will not only speed up dose delivery and improve clinical workflow, but also decrease the total-body radiation dose and the risk of secondary cancers[11-13]. The SPORT plans for these cases have been successfully delivered on a TrueBeam™ Linac.

2.2 Fixed-station delivery

SPORT can also be delivered station by station, much like fixed-gantry IMRT delivery. It is important to note that SPORT is the most general scheme and more conformal and efficiently deliverable dose distribution can be realized. Intuitively, one may consider that SPORT is a scheme with increased angular beam sampling while eliminating dispensable segments of the incident fields. In digital linacs, SPORT delivery is made efficient by (1) removing the redundant segments in the incident beams; (2) concatenating the stations so that they can be delivered in sequence automatically (i.e., auto-field sequencing); and (3) using high dose-rate beams[5, 6, 14]. The technique, which we termed as DASSIM-RT (dense angularly sampled and sparse intensity modulated radiation therapy) (Fig. 2)[1], presents a truly optimal RT scheme with optimal angular sampling (including non-coplanar beams), beam intensity modulation, and possible energy and collimator modulation when going from one gantry angle to another[15].

In passing, we note that the redundancy/inefficiency in intensity modulation can be realized using different techniques. Along this line, a compressed sensing (CS)-based strategy has been proposed by Zhu and Xing[16, 17]. For completeness, this is summarized below. With the use of CS-based inverse planning technique, it is straightforward to find fixed-station SPORT solution that most effectively utilizes all the available station parameters. Dramatically different from simply increasing the number of beams for IMRT or increasing the arcs in VMAT, the dosimetric improvement here is achieved by considering the interplay between the angular sampling and intensity modulation. The SPORT framework will enable us to harness the newly available features of digital linacs and provide truly optimal patient treatment.

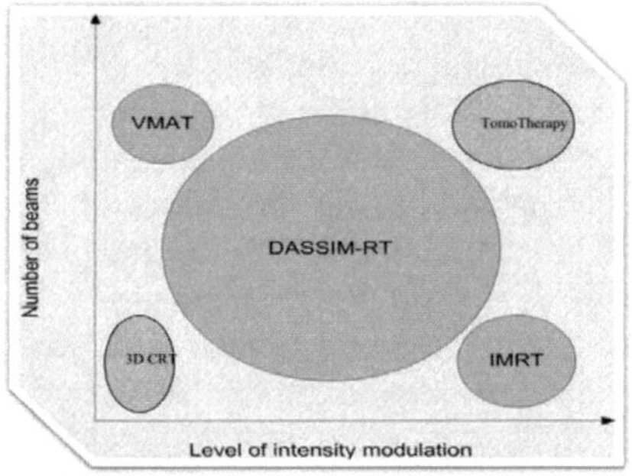

Fig. 2. Schematic plot of DASSIM-RT, conventional 3D CRT, IMRT, VMAT, and TomoTherapy in terms of the number of beams and level of intensity modulation. DASSIM-RT completes the "phase space" extended by inter- and intra-beam modulation for RT.

2.3 CS-based SPORT inverse planning

CS-based method is an effective way to obtain SPORT or IMRT plan with no redundant intensity modulation. General, a treatment plan is selected from a large pool of physically feasible solutions by optimization of an objective function (OF). BBO [18, 19] and DAO[20-22] are two commonly used approaches for inverse planning. DAO attempts to tackle the problem from the delivery aspect by enforcing a pre-chosen (often unjustified) number of segments for each beam and then optimizes the shapes and weights of the apertures. Searching for an optimal solution by using DAO is inherently complicated because of the highly non-convex dependence of the OF on the MLC coordinates and the optimality of final solution is not guaranteed. Furthermore, the optimal number of apertures for a field must be chosen empirically, not by the actual field-specific need. In BBO, on the other hand, the delivery is completely decoupled from optimization and each beamlet is an independent variable - the optimized intensity map is highly complex and entails a large number of segments for delivery, which reduces both delivery efficiency and dosimetric accuracy. Instead of simply smoothing the fluence[23], the maps are encouraged to be piece-wise constant by introducing an L-1 total-variation regularization (TVR) in our approach. The maps so obtained will contain sharp transitions, which are otherwise smoothed if quadratic smoothing is used. The difference between smoothing and TVR can be found in a textbook by Boyd and Vandenberghe [24].

The key to substantially improving the two existing inverse planning methods lies in finding an effective way to seamlessly integrate the machine delivery constraints into the dose optimization process. Instead of directly including the non-convex physical constraints in optimization, we used CS method[16, 17, 25] to accomplish the goal. The technique takes the interplay between planning and delivery into consideration and suppresses the dispensable/redundant intensity modulation of fluence maps naturally. Mathematically, it recasts the conventional DAO defined as[17]

$$\begin{cases} \textit{Minimize} \sum \lambda_i (A_i x - d_i)^T (A_i x - d_i) \\ \text{subject to:} \quad x \geq 0 \quad \text{\& is achievable using apertures (apertureconstraint).} \end{cases} \tag{2}$$

into

$$\begin{cases} \textit{Minimize} \sum \lambda_\sigma (A_i x - d_i)^T (A_i x - d_i) + \kappa \sum_{f=1}^{N_f} \sum_{\mu=1}^{N\mu} \sum_{v=1}^{N_v} \left(\left| x_{\mu,v,f} - x_{\mu-1,v,f} \right| + \left| x_{\mu,v,f} - x_{u,v-1,f} \right| \right) \\ \text{subject to:} \quad x \geq 0 \end{cases} \tag{3}$$

where, the index i denotes different organs, λ_i is the importance factor; A is the dose matrix, and d_i is the prescribed dose. In Eq. (2), the aperture constraint is neither linear nor quadratic. The L-1 norm in Eq. (3) is included to reduce the number of signal levels of the intensity maps by encouraging piece-wise continuity, removing the aperture constraint. Afeature of the above optimization is that it can be reformulated rigorously as a quadratic programming (QP) problem[17, 26], and thus efficiently solved using the standard QP. When applied to IMRT, the algorithm can reduce the total MUs for delivery by ~35% while keep the quality of the resultant IMRT plan unchanged! [26] It is also interesting to point out that the CS-based formalism in the FFF domain can be easily formulated to take the inherent shapes of incident beam profiles into account[27]. By introducing an L-1 objective function specific to the known non-flat beam profile characteristics, the method searches for fluencies that are piece-wise constant in a domain defined by the basis function of the non-uniform beams. We demonstrated that optimization of the system provides IMRT fluences that are piece-wise constant in the selected FFF-domain, thus eliminating the inefficiency of dose delivery caused by non-flatness of the incident beams[28].

2.4 ***Role of prior geometric and dosimetric knowledge in SPORT optimization***

Because of involvement of a large number of variables, a brute-force SPORT optimization can be computationally intractable. Incorporation of partial system knowledge is an effective way to reduce the complexity of the optimization. We have introduced a metric, named beam eye's view dosimetrics (BEVD), for *a priori* ranking of the angular space and facilitating the beam orientation optimization in IMRT [29-31]. The general consideration in constructing the BEVD is that a beam is more preferable if it can deliver more dose to the target without exceeding the dose tolerance of the OARs located on the path of the beam. For computational purpose, a beam portal is divided into a grid of beamlets. Each beamlet crossing the target is assigned the maximum intensity that could be used without exceeding the dose tolerances of the OARs. A forward dose calculation using the "maximum" beam intensity profile is then performed and the score of the given beam direction (indexed by *f*) is calculated according to

$$S_f = \frac{1}{N_T} \sum_{n \in \text{Target}} \left(\frac{d_{nf}}{D_T^P} \right)^2 , \tag{4}$$

where d_{nf} is the maximum dose delivered to the voxel n by the beam from the direction indexed by f, N_T is the number of voxels in the target, and D_T^P is the target prescription. The BEVD score function goes beyond the simple geometric BEV concept in 3D CRT and captures the main feature of a planner's judgment about the quality of a beam. The inclusion of *a priori* BEVD ranking of the angular space should lead to significantly improved convergence and computational speed in SPORT configuration optimization. For a given patient, BEVD reveals that some beam directions are better/worse than others in terms of the dose deliverable to the target volume. This has demonstrated, although indirectly, by the interesting work of Rocha et al[32], who have developed a BEVD-guided pattern search algorithm for beam placement in IMRT.

2.5 Delivery efficiency of fixed station SPORT

The delivery time is described by Eq. (1). For a given SPORT configuration, CS-based fluence optimization will find the irreducible fluence maps, which means that further sparsification of the maps will deteriorate the resultant dose distribution. As exemplified in References 25, 27, the MU can be reduced by 30%-63%, depending on the complexity of the case and the characteristics of the beams (i.e., FFF or flattened beams). In general, the total MLC travel time (the last term in Eq. (1)) is also reduced significantly when the dispensable MLC segments are eliminated. As a result, the delivery time will be reduced greatly as a result of intensity sparsification (the 2nd and 3rd terms in Eq. (1) are generally small compared to the other two terms). The Eq. (1) should include the contributions resulted from the couch motion when non-coplanar points are included in the SPORT treatment.

3. Summary

The emergence of digital linac calls for new strategies for treatment plan optimization and delivery. The concept and implementation of SPORT has been summarized. In general, existing 3D conformal RT (CRT), IMRT and VMAT represent special cases of SPORT. SPORT has a number of unique features and makes it possible to maximally utilize the technical capacity of digital linac. In rotational arc SPORT, for example, it completely eliminates the need for 2 or more arcs, yet providing dose distributions with unprecedented conformality and normal tissue sparing. On a more fundamental level, the benefit of SPORT stems from the better definition of search space and explicit incorporation of parameters characterizing the station points in inverse planning. SPORT enlarges the solution space through improved angular and intensity sampling of the stations[1]. As has indicated in numerous studies, conformal dose distribution is essential to the improved treatment outcome and better patient care. Therefore, the SPORT formalism and tools should have a measurable clinical impact resulting in improved patient care in the future.

Acknowledgement: The work was partially supported by grants from NIH (1R01 CA133474, 1R21 CA153587 and 1K99 166186).

References

1. Li R, Xing L. Bridging the gap between IMRT and VMAT: dense angularly sampled and sparse intensity modulated radiation therapy. Med Phys. 2011;38: 4912-4919.
2. Wang L, Kielar KN, Mok E, Hsu A, Dieterich S, Xing L. An end-to-end examination of geometric accuracy of IGRT using a new digital accelerator equipped with onboard imaging system. Phys Med Biol. 2012;57: 757-769.
3. Fahimian B, Yu V, Xing L, Horst K, Hristov D. Trajectory Modulated Prone Breast Irradiation: A LINAC-based Technique Combining Intensity Modulated Delivery and the Motion of the Couch. Int J Radiat Oncol Biol Phys. 2013: submitted.
4. Xing L, Philips M, Orton C. DASSIM-RT is likely to become the method of choice over conventional IMRT and VMAT for delivery of highly conformal radiotherapy. Med Phys. 2013;40: 020601-020603.
5. Cho W, Bush K, Mok E, Xing L, Suh T-S. Development of a fast and feasible spectrum modeling technique for flattening filter free beams. Medical Physics. 2013;40: in press.
6. Cho W, Kielar KN, Mok E, et al. Multisource modeling of flattening filter free (FFF) beam and the optimization of model parameters. Medical Physics. 2011;38: 1931-1942.

7. Li R, Xing L. An adaptive planning strategy for station parameter optimized radiation therapy (SPORT): Segmentally boosted VMAT. Med Phys. 2013;40: 050701.
8. Yu CX. Intensity-modulated arc therapy with dynamic multileaf collimation: an alternative to tomotherapy. Phys Med Biol. 1995;40: 1435-1449.
9. Otto K. Volumetric modulated arc therapy: IMRT in a single gantry arc. Med Phys. 2008;35: 310-317.
10. Crooks SM, Wu X, Takita C, Matzich M, Xing L. Aperture modulated arc therapy. Phys Med Biol. 2003;48: 1333-1344.
11. Hall EJ. Intensity-modulated radiation therapy, protons, and the risk of second cancers. Int J Radiat Oncol Biol Phys. 2006;65: 1-7.
12. Hardcastle N, Metcalfe P, Ceylan A, Williams MJ. Multileaf collimator end leaf leakage: implications for wide-field IMRT. Phys Med Biol. 2007;52: N493-504.
13. Williams PO, Hounsell AR. X-ray leakage considerations for IMRT. Br J Radiol. 2001;74: 98-100.
14. Georg D, Knoos T, McClean B. Current status and future perspective of flattening filter free photon beams. Med Phys. 2011;38: 1280-1293.
15. Zhang P, Happersett L, Yang Y, Yamada Y, Mageras G, Hunt M. Optimization of collimator trajectory in volumetric modulated arc therapy: development and evaluation for paraspinal SBRT. International journal of radiation oncology, biology, physics. 2010;77: 591-599.
16. Zhu L, Lee L, Ma Y, Ye Y, Mazzeo R, Xing L. Using total-variation regularization for intensity modulated radiation therapy inverse planning with field-specific numbers of segments. Phys Med Biol. 2008;53: 6653-6672.
17. Zhu L, Xing L. Search for IMRT inverse plans with piecewise constant fluence maps using compressed sensing techniques. Med Phys. 2009;36: 1895-1905.
18. Bortfeld T. Optimized planning using physical objectives and constraints. Seminars in Radiation Oncology. 1999;9: 20-34.
19. Xing L, Wu Y, Yang Y, Boyer AL. Physics of intensity modulated radiation therapy In: Mundt AJ, Roeske JC, editors. Intensity Modulated Radiation Therapy: A Clinical Perspective. Hamilton & London: BC Decker Inc., 2005:20-52.
20. Shepard DM, Earl MA, Li XA, Naqvi S, Yu C. Direct aperture optimization: a turnkey solution for step-and-shoot IMRT. Med Phys. 2002;29: 1007-1018.
21. Bednarz G, Michalski D, Houser C, et al. The use of mixed-integer programming for inverse treatment planning with pre-defined field segments. Phys Med Biol. 2002;47: 2235-2245.
22. Cotrutz C, Xing L. Segment-Based Dose Optimization Using a Genetic Algorithm. Physics in Medicine & Biology. 2003;48: 2987-2998.
23. Matuszak MM, Larsen EW, Fraass BA. Reduction of IMRT beam complexity through the use of beam modulation penalties in the objective function. Med Phys. 2007;34: 507-520.
24. Boyd S, Vandenberghe L. Convex Optimization. New York: Cambridge Univ Press, 2004.
25. Zhu L, Choi K, Boyd S, Xing L. Total-Variation Regularization Based Inverse Planning for Volumetric Modulated Arc Therapy. Med Phys. 2009;36: submitted
26. Zhu L, Lee L, Ma Y, Ye Y, Mazzeo R, Xing L. Using total-variation regularization for intensity modulated radiation therapy inverse planning with field-specific numbers of segments. Physics in Medicine & Biology. 2008;53: 6653-6672.
27. Kim T, Zhu L, Suh TS, Geneser S, Meng B, Xing L. Inverse planning for IMRT with nonuniform beam profiles using total-variation regularization (TVR). Med Phys. 2011;38: 57-66.
28. Wang L, Mok E, Xing L. Pros and Cons of Flattening Filter Free IMRT: A Comparison with Conventional IMRT with Flattened Beams. Med. Phys. 2010;37: 3375.
29. Pugachev A, Xing L. Pseudo beam'-eye-view as applied to beam orientation selection in intensity-modulated radiation therapy. Int J Radiat Oncol Biol Phys. 2001;51: 1361-1370.
30. Pugachev A, Xing L. Incorporating prior knowledge into beam orientation optimization. Int J Radiat Oncol Biol Phys. 2002;54: 1565-1574.
31. Schreibmann E, Xing L. Dose-volume based ranking of incident beam direction and its utility in facilitating IMRT beam placement. Int J Radiat Oncol Biol Phys. 2005;63: 584-593.
32. Rocha H, Dias JM, Ferreira BC, Lopes MC. Beam angle optimization for intensity-modulated radiation therapy using a guided pattern search method. Phys Med Biol. 2013;58: 2939-2953.

Journal of Physics: Conference Series **489** (2014) 012066 doi:10.1088/1742-6596/489/1/012066

Interactive Dose Shaping - efficient strategies for CPU-based real-time treatment planning

P Ziegenhein, C P Kamerling and U Oelfke

German Cancer Research Center (DKFZ), Department of Medical Physics in Radiation Oncology, Im Neuenheimer Feld 280, 69120 Heidelberg, Germany

E-mail: p.ziegenhein@dkfz.de

Abstract. Conventional intensity modulated radiation therapy (IMRT) treatment planning is based on the traditional concept of iterative optimization using an objective function specified by dose volume histogram constraints for pre-segmented VOIs. This indirect approach suffers from unavoidable shortcomings: i) The control of local dose features is limited to segmented VOIs. ii) Any objective function is a mathematical measure of the plan quality, i.e., is not able to define the clinically optimal treatment plan. iii) Adapting an existing plan to changed patient anatomy as detected by IGRT procedures is difficult.

To overcome these shortcomings, we introduce the method of Interactive Dose Shaping (IDS) as a new paradigm for IMRT treatment planning. IDS allows for a direct and interactive manipulation of local dose features in real-time. The key element driving the IDS process is a two-step Dose Modification and Recovery (DMR) strategy: A local dose modification is initiated by the user which translates into modified fluence patterns. This also affects existing desired dose features elsewhere which is compensated by a heuristic recovery process.

The IDS paradigm was implemented together with a CPU-based ultra-fast dose calculation and a 3D GUI for dose manipulation and visualization. A local dose feature can be implemented via the DMR strategy within 1-2 seconds. By imposing a series of local dose features, equal plan qualities could be achieved compared to conventional planning for prostate and head and neck cases within 1-2 minutes.

The idea of Interactive Dose Shaping for treatment planning has been introduced and first applications of this concept have been realized.

1. Introduction

The IMRT planning process traditionally consists of an iterative Newton-based optimization of an objective function which is defined from dose volume constraints of pre-segmented volumes of interest [2]. This approach is one of the most popular methods to generate an IMRT plan in clinical environments. However, it suffers from various inherent shortcomings. First, the locality of dose prescription is too coarse. Local dose features can only be restrained to pre-segmented VOIs. Eliminating undesired hotspots in the therapy plan is difficult, controlling the shape of an individual isodose line is almost impossible. Second, the trade-off between planning goals in the objective function is expressed by a set of scalar, relative weighting factors. The meaning of these factors is clinically not intuitive and the planner has to find the right set of parameters through a time intensive trial-and-error approach. Third, using the traditional optimization method, it is difficult to rapidly adapt already optimized treatment plans to a changed patient anatomy as it is required in adaptive radiation therapy (ART). In addition to these clinical drawbacks, the method of the iterative Newton based optimization as used in radiation therapy also has several disadvantages from the computational point of view. In [8] it is claimed that the optimization method is a memory bound problem when a pre-calculated dose influence matrix (D_{ij}) is used. The runtime of the optimization is determined by the transfer speed of the dose influence data between the memory of the planning system to the arithmetic unit (e.g. CPU or GPU). It could be shown that the arithmetic performance potential of modern computer systems cannot be exploited by the optimization method. Since the (parallel) arithmetic computation speed of modern computers develops faster than memory bandwidth, one has to conclude that the Newton-based IMRT plan optimization is not well suited for modern throughput oriented computer systems as used for instance in [5]. Furthermore, the iterative nature of the optimization method prevents an efficient implementation on distributed memory systems.

In this work, an alternative therapy planning paradigm called Interactive Dose Shaping (IDS) is introduced. It is designed to overcome the clinical shortcomings of the traditional IMRT plan optimization. In contrast to [3] the presented interactive planning method does not employ an objective function which is driving a Newton-based optimization but a so called Dose Modification and Recovery strategy which is introduced in this work.

2. Materials & Methods

2.1. The Concept of Interactive Dose Shaping

The development of the IDS paradigm is driven by three main key features. First, the new planning technique involves the therapist more into the planning process. In contrast to the traditional method which consist of an automated optimization block, the therapist should be enabled to interactively control the plan generation process by imposing a series of local dose features into the plan. This paradigm shift facilitates the transition of clinical knowledge from the therapist to the plan generating algorithm. Second, using IDS it should be possible to adapt pre-optimized therapy plans to a changed patient geometry easily and third, the IDS engine has to be implemented by compute-bound algorithms to guarantee proper performance scaling on modern computational architectures.

In order to realize these three features, the IDS concept employs a two-step Dose Modification and Recovery (DMR) strategy. In the dose Modification step, the IDS engine imposes a local, desired dose feature into the treatment plan as requested by the therapist. The request can be formulated by interactive graphical tools via a 3D user interface in the planning framework. The desired local dose feature is implemented into the plan by manipulating the relevant fluence amplitudes of the beam setup. Changing the fluence amplitudes of the beams does not only impose the intended dose feature but also affects other areas of the plan. These unintended dose variations are compensated in a subsequent Recovery step. The Recovery identifies affected areas outside the desired locality and generates a manipulation pattern for the fluence amplitudes which re-installs already established dose features without eradicating the foregoing dose Modification step. Depending on the complexity of the clinical plan, 10 to 30 locations have to be recovered in order to compensate for unwanted, unintentional dose changes outside the Modification area. The initial plan on which the DMR process is applied can either be a pre-optimized plan (in order to impose small alterations due to geometrical changes of the patient) or an unmodulated plan to start from scratch as shown in the example in figure 1.

The Dose Modification and Recovery strategy is explained in detail in section 2.3. As mentioned before, using the D_{ij} matrix approach for the dose calculation is a memory-bound problem and therefore not suited for the interactive planning paradigm. It is replaced by an on-the-fly dose calculation method which is described in section 2.2. The IDS engine is embedded into a new treatment planning framework which provides a powerful three-dimensional graphical representation of the plan together with a set of tools which provide an interface for implementing the interactive dose Modification. Some aspects of the dose Modification interface and parts of this framework is described in [4]

2.2. Ultra-fast, low-latency dose calculation

It is necessary to re-calculate the dose distribution after imposing a dose Modification and after each Recovery step due to two reasons: On the one hand, the effect of the dose Modification must be visible to the subsequent Recovery module. On the other hand, the Recovery locations are chosen sequentially based on the effect of the foregoing Recovery. Thus, 11 to 31 dose calculations (one dose Modification and 10 to 30 Recovery steps) have to be performed for each DMR process within a real-time constraint (1-2 seconds).
To solve this task an ultra-fast CPU-based dose calculation algorithm is introduced based on the pencil beam convolution technique with singular value decomposed kernels according to [1]:

$$D^f(x_p, y_p, d) = \sum_{i=1}^{3} D_i^{'}(d) \int_{-\infty}^{\infty} \int_{-\infty}^{\infty} \Psi^{'}(x,y) f(x,y) \times w_i(x - x_p, y - y_p) dxdy \tag{1}$$

The dose D in the radiological depth d at the point (x_p, y_p) relative to the central axis is calculated as a sum over the depth dependent kernel components $D_i^{'}(d)$ multiplied with the convolution of the fluence $\Psi^{'}(x,y)$, the modulation (transmission) function $f(x,y)$ and the kernel weight $w_i(x - x_p, y - y_p)$. Usually, the convolution is taken out in frequency domain via a fast Hartley transform [1] [6]. For the use in IDS the convolution is calculated in a discretized spatial domain which has three main advantages: i) The DMR strategy imposes local modifications of the fluence amplitudes. In spatial domain these can be considered as relatively small additional terms to the convolution. ii) The discrete convolution in spatial domain consists of simple multiply add operations which are highly optimized in hardware on modern CPUs. iii) By controlling the radius of the finite pencil beam kernel weight $w_i(x - x_p, y - y_p)$ the trade-off accuracy/time can be individually selected for each region of interest.

2.3. The Dose Modification and Recovery strategy

The Dose Modification and Recovery process as a central part of the IDS engine is shown in figure 1a). The IDS process begins with a request from the user who specifies a local dose Modification to the plan (first green box). The request is realized by a set of interactive tools which provide a variety of Modification means. A sophisticated request which can be used for clinical planning is for instance the manipulation of a three-dimensional isodose surface which is explained in [4]. For the scope of this paper which is intended to demonstrate the concepts of IDS, a much simpler user request is considered: The user demands a dose Modification of only one voxel of the plan. An example of the voxel Modification tool is depicted in figure 2 on a phantom setup containing a horse-shoe shaped target and a spherical OAR in close proximity. The figure shows a screen shot from our newly designed planning framework with the voxel selecting tool enabled (the tool proposes groups of 4x4x4 voxels which are shown as color-washed cubes). On the example of requesting a dose Modification on the encircled voxel in figure 2 the DMR process is described in the following while the change of the dose pattern throughout the process is shown in figure 1 b),c) and d) on one transversal slice of the planning setup.

The initial plan on which the user request the dose Modification is shown in figure 1b). The beam fluences are not modulated and the shown isodose line denotes a prescribed dose to the target volume. The location of the Modification is marked by a white dot in figure 1c) and 1d). The user requests a dose reduction of 30% in that voxel $\vec{m}$. The first step of the DMR strategy is to impose the request into the plan. To accomplish that, the dose Modification module (first blue box in Figure 1a) imposes a manipulation of the fluence map Δf_M so that the requested

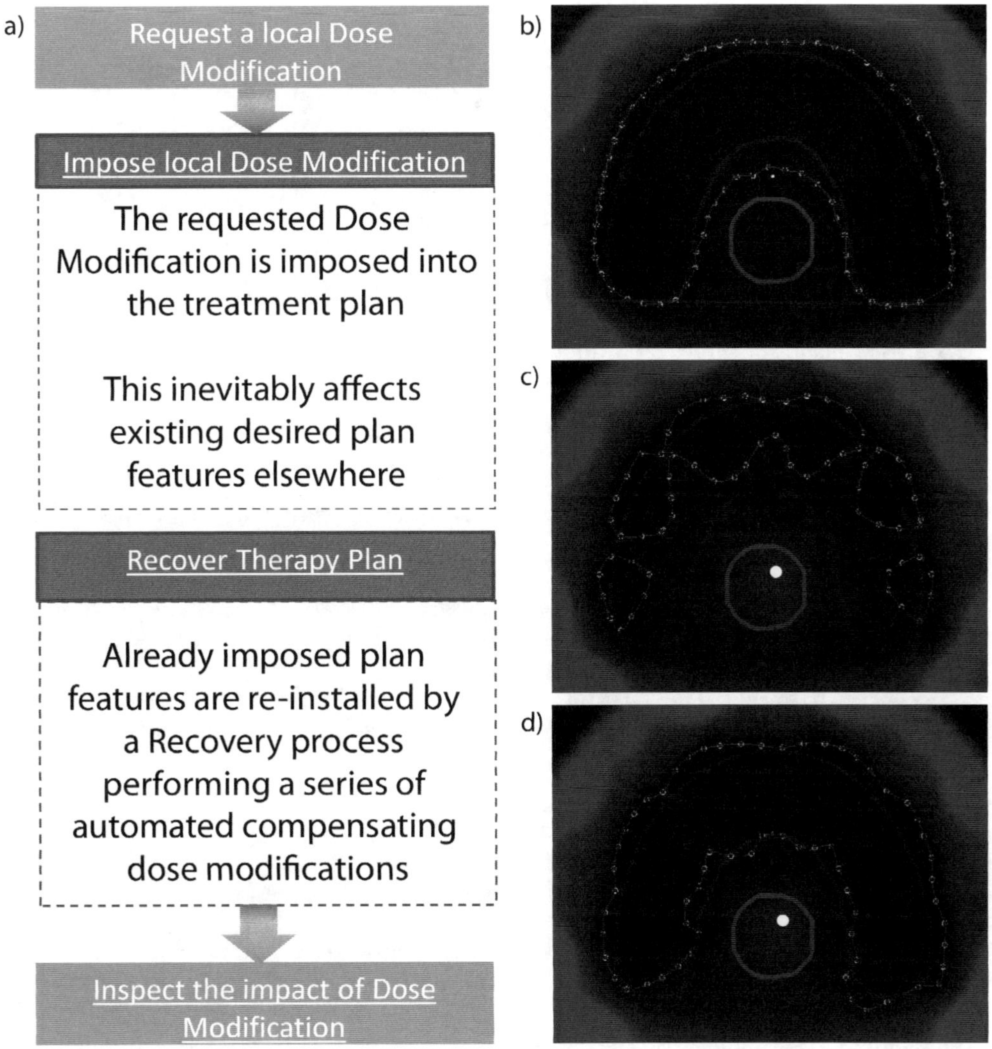

Figure 1. a)The process of the Dose Modification and Recovery strategy (blue boxes) and the relation to the graphical user interface (green boxes). b)The initial planning state of the phantom case c)after the dose Modification of one voxel (white dot) d)after the Recovery process.

feature in $\vec{m}$ is installed. The result is shown in Figure 1c). The dose Modification not only imposes the desired local dose feature $D^{(f+\Delta f_M)}(\vec{m})$ but also affects the global dose distribution, that means there is another voxel $\vec{r}$ so that $D^{(f+\Delta f_M)}(\vec{r}) \neq D^f(\vec{r})$. In this example the effect on the target volume is very distinct: The isodose line in figure 1c) reveals the undesirable cold spots in the target volume as a side effect of the foregoing fluence manipulation.

In a second step the undesired change on the dose distribution outside the area of the Modification to $\vec{m}$ is compensated by the Recovery. The Recovery module (second blue box in figure 1a) identifies locations $\vec{r}$ which have been affected unintentionally by the dose Modification and provides one adaptation Δf_R^k to the fluence map in each recovery step k to re-install the initial

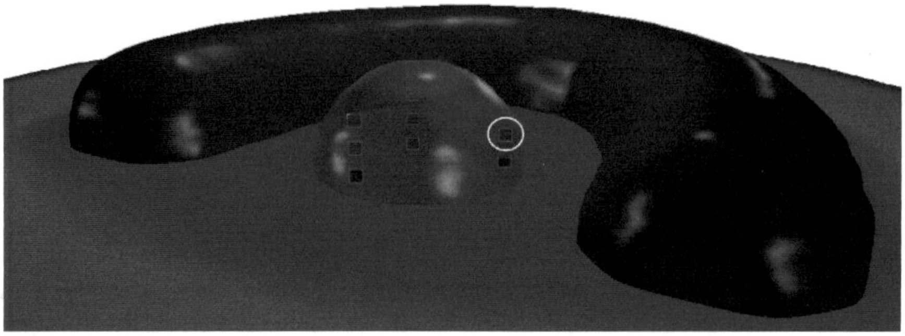

Figure 2. Screenshot of the voxel Modification tool as implemented in the IDS framework.

dose distribution:

$$D^{(f+\Delta f_M+\Delta f_R^k)}(\vec{r}) \approx D^f(\vec{r}) \quad (2)$$

For that, the Recovery strategy prefers fluence amplitudes manipulations Δf_R^k that have a minimal effect on the previously requested dose feature in $\vec{m}$. The effect of the Recovery process is shown in figure 1d). On the one hand, the dose distribution in the target is almost fully recovered while on the other hand, the requested dose Modification in voxel $\vec{m}$ is still in place. After the Recovery process is finished, the resulting 3D dose distribution is presented to the user for evaluation. The whole DMR process can be done in real-time (about 1-2 seconds) on a single processor desktop PC. This guarantees the interactive character of the IDS concept and provides the therapist with an immediate feedback on the requested dose Modification.

3. Results

The concepts of the Interactive Dose Shaping paradigm containing the DMR strategy, the ultra-fast dose calculation module, several dose Modification tools and a three-dimensional graphical representation of the patient geometry/dose have been implemented into a new treatment planning framework.

The implementation of the DMR strategies and the adapted dose calculation method is fast enough to allow for interactive planning of clinical 3D treatment plans within real-time. The runtime of the dose calculation is in the order of 20 ms for typical nine-beam plan.

In [4] the planning result of a prostate case is discussed using the 3D isodose manipulation tool for IDS. A planning result for a more complex intracranial case is shown in figure 3(b). The target dose was prescribed to 6000 cGy. The maximum dose to the left eye was set to 1000 cGy while 2000 cGy was allowed to the right eye due to the close proximity to the tumor. The IDS planning process took less than 2 minutes by imposing various local dose Modifications on a nine-beam photon setup. This planning time includes user interactions and the computational runtime of the DMR.

4. Discussion

The short runtime of the DMR implementation and of the adapted dose calculation method shows that the IDS method is feasible to realize a new interactive planning method from a computational point of view. In contrast to the traditional Newton-based optimization method using pre-calculated dose influence data, IDS is able to exploit the parallel arithmetic capabilities of modern CPU-based hardware which enables scaling to high performance computing and especially future architectures.

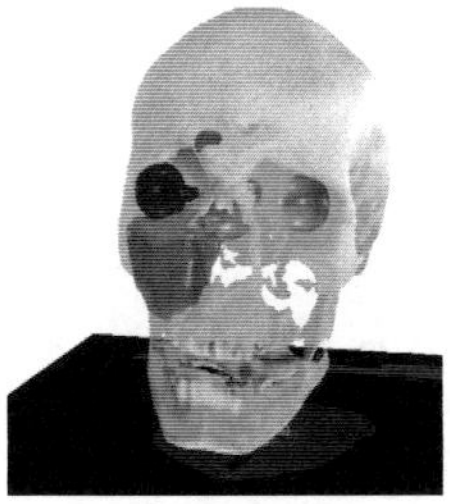

(a)

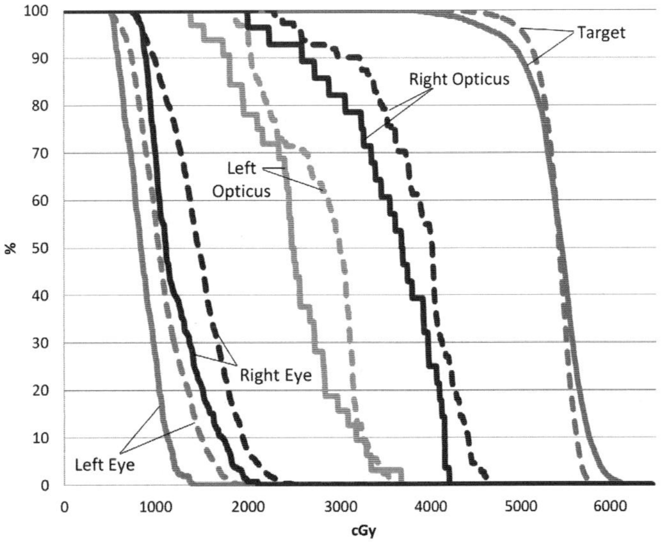

(b)

Figure 3. IDS planning results on a complex intracranial patient case. a) The geometrical patient setup. b) The DVH comparing the traditionally Newton based optimization (dashed curves) and using IDS (solid curves)

The first results presented in this paper and in [4] demonstrate that the IDS planning method is able to derive plans with a similar quality than achieved by traditional methods.

5. Conclusion

The IDS planning paradigm has the potential to become an new interactive planning method which can be used in the clinical routine some day. Although, the concepts are still under development we could demonstrate that the new method is accurate and fast enough to allow for interactive planning which directly includes the therapist into the plan generation process.

References

[1] Bortfeld T 1993 Decomposition of pencil beam kernels for fast dose calculations in three-dimensional treatment planning *Med. Phys.* **20** 311

[2] Bortfeld T 2006 IMRT: a review and preview *Phys. Med. Biol.* **51** R363

[3] Craft D L, Hong T S, Shih H A and Bortfeld T R 2012 Improved Planning Time and Plan Quality Through Multicriteria Optimization for Intensity-Modulated Radiotherapy *I. J. of Rad. Onc. Bio. Phys.* **82** e83 - e90

[4] Kamerling C P, Ziegenhein P, Oelfke U 2013 A 3D isodose manipulation tool for interactive dose shaping *J. Phys.: Conf. Ser.* presented at ICCR 2013, not yet published

[5] Men C, Gu X, Choi D, Majumdar A, Zheng Z, Mueller K and Jiang S 2009 GPU-based ultrafast IMRT plan optimization *Phys. Med. Biol.* **54** 6565-73

[6] Siggel M, Ziegenhein P, Nill S and Oelfke U 2012 Boosting runtime-performance of photon pencil beam algorithms for radiotherapy treatment planning *Physica Medica* **28(4)** 273-80

[7] Ziegenhein P, Kamerling C P, Bangert M, Kunkel J and Oelfke U 2013 Performance-optimized clinical IMRT planning on modern CPUs *Phys. Med. Biol.* **58** 3705-15

[8] Ziegenhein P, Kamerling C P, Heinrich H and Oelfke U 2012 SU-C-213AB-01: Application of Ultra-Fast Dose Calculation in Real-Time Interactive Treatment Panning *Med. Phys.* **39** 3598

Journal of Physics: Conference Series **489** (2014) 012067 doi:10.1088/1742-6596/489/1/012067

A comprehensive EPID-based 3D validation technique for TrueBeam-delivered VMAT plans

W Ansbacher[1,3] I M Gagne[1,3] and C-L Swift[2]

[1]BC Cancer Agency, Vancouver Island Centre, Victoria BC, Canada
[2]BC Cancer Agency, Vancouver Centre, Vancouver BC, Canada
[3]Department of Physics and Astronomy, University of Victoria, Victoria, BC

E-mail: wansbach@bccancer.bc.ca

Abstract.

Purpose: To develop and validate a pre-treatment EPI dosimetry method on Varian TrueBeam linacs using continuous imaging, with reconstruction in a 3D cylindrical phantom geometry.
Methods: Delivery of VMAT plans with continuous imaging is currently possible only in Research Mode on TrueBeam linacs, with images acquired in a proprietary format. An earlier technique was adapted to take advantage of technical improvements in EPID delivery, and was tested under various acquisition conditions. The dosimetry of VMAT plans was evaluated at isocentre and within patient volumes that had been transferred to the virtual phantom.
Results: Approximately 60 portal image projections per arc were found to be adequate for 3D reconstruction in phantom volumes of 28cm diameter. Twelve prostate, CNS and Head & Neck deliveries were evaluated in Research mode relative to the corresponding Eclipse (v.10) treatment plans, and to measurements on an ArcCheck device in Treatment mode. Mean dose differences at isocentre were within 2% for the three-way comparison, and in PTV volumes were within 1% (s.d. 1%). However, some discrepancies were observed in ArcCheck results that may be related to the small dimensions of certain VMAT apertures.
Conclusions: EPI dosimetry with 3D dose reconstruction is an accurate, comprehensive and efficient pre-treatment validation technique for VMAT delivery. Although currently limited to a research mode on TrueBeam, it has the potential to be implemented for clinical use.

1. Introduction

Electronic Portal Image Dosimetry (EPID) has become a standard tool for evaluation of intensity-modulated radiation treatments [1]. The EPID has normally been used in integrating mode, which is ideal for acquisition of IMRT fields, as the image data can be used for 2D validation per field as well as 3D volumetric reconstruction of a delivered dose distribution. Integrated acquisition [2] is less useful for Volumetric Modulated Arc Therapy (VMAT), where the gantry rotates during acquisition, because there is no spatial correlation between the so-called "collapsed" 2D portal image and the 3D dose matrix, except for points along the axis of rotation.

3D dose reconstruction for VMAT requires cine (continuous) imaging where each image projection represents the dose averaged over a relatively small gantry angle increment [3]. However, cine mode was developed for imaging applications rather than dosimetry so has proved to be less than ideal. McCurdy and Greer [4] investigated the IAS3 system on Varian Trilogy linacs and reported some dosimetric nonlinearity. Inaccuracy in the gantry angle recorded for each image acquisition has also been reported but despite this a workable 3D reconstruction model has been proposed [4].

The recent introduction of Varian's TrueBeam accelerator line has complicated and inhibited further attempts at 3D reconstruction because the only cine acquisition mode available in Treatment, ("movie" mode), includes dynamic windowing and signal clipping in order to optimize image contrast, and is therefore quite unsuitable for dosimetry. However, Varian has provided an optional Research (Developer) application on TrueBeam which gives the user direct control of the machine axes and most aspects of the image acquisition, and which can be programmed though the native xml scripting language on which the accelerator platform is built.

In this work we investigated the potential of Research mode for dosimetry-quality cine acquisition, and characterized the dose linearity and gantry angle accuracy, the two elements that had previously been determined as critical. We then adapted a 3D virtual-phantom reconstruction technique that has proven very successful in IMRT QA [6] to make use of these cine images. This technique, known as "3dPD", was used to compare clinical plan reconstructions to both ionization chamber measurements in a cylindrical phantom, and to a widely-accepted measurement technique using an ArcCheck device. In order to perform a full 3D comparison, the ArcCheck measurements were performed using the "3DVH" module rather than simply evaluating the diode signals around the periphery of the device. Twelve plans representing several clinical sites were investigated.

2. Methods and Materials

In Research mode, the creation of the xml code is facilitated by importing a clinical Dicom plan which can then be saved as an xml script. Imaging code was then added with a standard xml editor. All radiation deliveries were performed on Varian TrueBeam linear accelerators at the Vancouver Island Centre in Victoria, using matched 6MV photon beams.

Initially we had expected that acquisition starts and stops would be determined by control points (CPs), as are the motions of the machine axes. However, dosimetry-quality frames are acquired at a fixed rate of approximately 10 f.p.s, and are not synchronized to beam pulses. This proved to be incompatible with CP programming, which is linked to the accumulation of beam pulses, because there is no way to determine CP timing directly. As a result, frames were occasionally dropped, causing as much as a 10% signal loss for a typical 10-frame acquisition. Instead, we discovered that the xml coding allowed the user to access the individual frames accumulated during a single dosimetry acquisition. Ideally it would have been possible to save every N^{th} frame; however, the current version of scripting provides only for the final frame, or all frames, to be saved. Projection images P for small gantry increments were then formed by calculating differences between cumulative frames F

$P_k = F_{(k+1)N} - F_{kN}$ *i.e.,* between every N^{th} frame.

Determining the gantry angle recorded in the image header for each frame or each acquisition is critical for accurate 3D reconstruction. Earlier investigations relied on inclinometers to image some identifiable arrangement of wires or ball bearings from which to determine gantry angle. For this study, we took advantage of the xml programming to design VMAT plans where the collimator angle was equal to the gantry angle at each control point. On TrueBeam, the collimator rotates much faster than the gantry, and we had independently determined from analysis of trajectory log files that the collimator angle was accurate to a small fraction of a degree; the collimator orientation, which could easily be measured on each image, was thus a surrogate for the imaged gantry angle.

A third component of 3dPD was the incorporation of backscatter from the portal image support, which had been unnecessary in the original IMRT code. An empirical model developed by Berry *et.al.*[7] was adapted to deal with the irregular apertures and modulation of VMAT delivery, by redefining Berry's "Y1" collimator parameter to be the effective length of any imaged subfield in the gun-target direction.

Clinical validation consisted of 3-way comparisons of 3dPD, ArcCheck and PTW 30010 ionization chamber measurements at isocentre with VMAT plans calculated with the AAA algorithm (Eclipse v10.0, Varian Medical Systems) for 11 prostate, oropharynx and base of tongue cases (One prostate case resulted in two plans). These plans were converted to verification plans on a 26.6 cm diameter phantom. The phantom density was 1.2 for the ArcCheck and ionization chamber cases, and the treatment couch was included. 3dPD phantom density was 1.0 with no couch. Although both 3dPD and ArcCheck's 3DVH can evaluate clinical PTV and OAR volumes, for consistency in this study, dose differences were evaluated within volumes defined by 80% and 40% of the maximum dose. Pass statistics at the 3%/3mm DTA level for both volumes were also recorded but were not discussed, or reported here due to space limitations.

3. Results

Several difficulties were initially encountered with Research mode. As mentioned above, the acquisition mode chosen resulted in the saving of every frame. These images must be saved to a network drive, which can take several minutes for a typical data transfer of more than 1GB, and paralyzes the application for the duration. It was also found that in the current TrueBeam version (1.6), the export fails if a large number of Dicom-formatted images are to be saved. In addition, a bug in the Dicom dosimetry format causes a data overrun, making the image data invalid for more than 7 frames. As a result, a decoder for Varian's proprietary .xim format had to be developed. (These bugs are apparently resolved in a future TrueBeam release)

The gantry angle written to the image header was determined to be within 0.6° of the angle at the mid-point of any acquisition, whether that was a single or multiple (up to 12) frame acquisition. Linearity with MU was determined to be the same as for normal dosimetry acquisition. This was expected of course, as the acquisition was in fact identical to that used for a normal single-integrated image, but with inclusion of the intermediate frames.

The number of projections for 3D reconstruction was fixed at 60, corresponding to between 10 and 15 frames per projection, depending on the instantaneous dose rate and gantry speed. Although a larger number of projections does smooth the dose reconstruction, the dose differences between 3dPD and Eclipse (evaluated with a 3%/3mm DTA tolerance), did not show any change when more than 45 projections were used. The empirical backscatter correction reduced, but did not eliminate, the over-response at the target (inferior) end of the EPID images. A small decrease in agreement was observed on most 3dPD-Eclipse comparisons within 2cm of the inferior extent of the dose distributions.

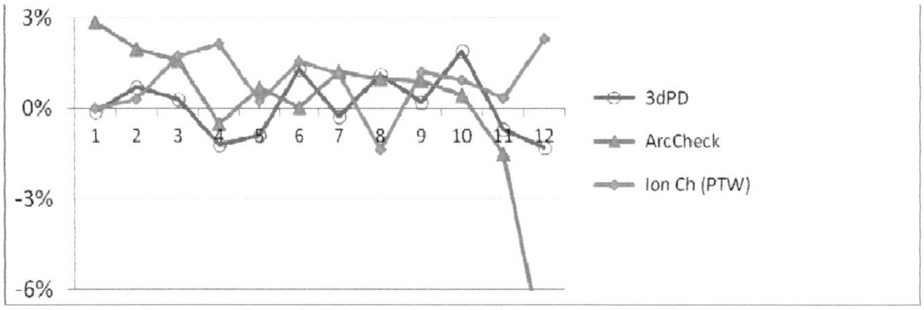

Figure 1. Isocentre dose differences from Eclipse for twelve Head & Neck (1-5) and Prostate (6-12) VMAT plans. In case 12, the ArcCheck difference was -8%

With one exception, all clinical case comparisons were well within expected levels of consistency for all three measurement techniques. The mean dose differences at isocentre were within 1%, with a maximum discrepancy of less than 3% (Figure 1), and both mean high dose (80%) and low dose (40%) ROI volume differences were within 0.5% (Figure 2). However, the results for one plan (case 11) were problematic because the "Global Correction" parameter computed by ArcCheck was -5.4%. According to the vendor's documentation, this correction should be within ±3% for a reliable interpretation, and for all other cases the correction was within 1%. The correction, when applied to the ROI difference of -3% put the ArcCheck validation out of tolerance. As a result, the case was re-planned as #12, where the Global Correction was 1% but the dose differences were -8% and -7% for the isocentre and ROI respectively. The two plans, 11 and 12, did appear to have more than the average number of small apertures, but were not otherwise judged to be problematic. Results using 3dPD did not show any dose discrepancy.

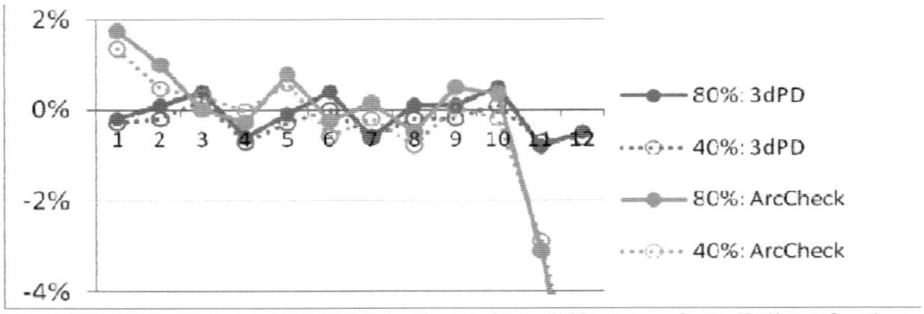

Figure 2 80% and 40% ROI volume dose differences from Eclipse for the same clinical cases. In case 12, the ArcCheck differences were -7% for both ROIs

4. Discussion and Conclusion

Research mode on Varian TrueBeam accelerators has been used to implement a pre-treatment 3D dose reconstruction technique, 3dPD, for efficient VMAT plan validation. The technique employs a user-determined number of image projections, and 60 projections were found to be adequate. The accuracy of gantry angle reconstruction and dosimetric linearity is equal to similar IMRT validation techniques. 3dPD can be used to measure absolute dose at a point with a precision equal to the dosimetric accuracy of the EPID device, and to measure volumetric differences using % dose and DTA metrics within PTV and OAR structures transferred from a CT plan.

A preliminary study has indicated that 3dPD should perform at least as well as a widely-used measurement device, ArcCheck, but requires minimal setup time. One case (consisting of two plans) passed 3dPD validation but failed ArcCheck for reasons that are not clear at this time, but may be related to the spatial resolution of ArcCheck's diode matrix. The 1cm grid is comparable to the dimensions of the small and multiply-connected apertures commonly found in VMAT delivery.

One drawback of the current 3dPD implementation is that it can be used only for *pre*-treatment QA because the necessary images cannot be acquired in normal Treatment mode. In addition, some knowledge of Varian's proprietary .xim image format is required. It is to be hoped that these limitations will be resolved in a future TrueBeam software release.

Acknowledgement

Partial funding for this work was provided by Varian Medical Systems.

We would like to thank Vince LaPointe, BCCA Vancouver Centre, for initial work on Research mode, and Thanos Etmektzoglu, Varian Medical Systems, for assistance with xml scripting.

References

[1] van Elmpt W, McDermott L, Nijsten S, Wendling M, Lambin P and Mijnheer B, 2008 *Radiother. Oncol.* **88,** 289-309

[2] Nicolini G, Vanetti E, Clivio A, Fogliata A, Korreman S, Bocanek J, Cozzi L, 2008 *Radiat. Oncol.* **3,** 24

[3] Mans A, Remeijer P, Olaciregui-Ruiz I, Wendling M, Sonke J, Mijnheer B, van Herk M and Stroom J C, 2010 *Radiother. Oncol.* **94,** 181-187

[4] Ansbacher W, Swift C-L and Greer P B, 2010 J. Phys. Conf. Ser. 250, 012022

[5] McCurdy B M and Greer P B, 2009 *Med.Phys.* **36,** 3028-3039

[6] Ansbacher W, 2006 *Med. Phys.* **33,** 3369-3382

[7] Berry S L, Polvorosa C S and Wuu C-S, 2010 *Med. Phys.* **37,** 2425-34

Journal of Physics: Conference Series **489** (2014) 012068 doi:10.1088/1742-6596/489/1/012068

A comprehensive tool to analyse dynamic log files from an Elekta-Synergy accelerator

Sankar Arumugam[1], Aitang Xing[1], Claire Pagulayan[2], Lois Holloway[1,2,3,4]

[1]Liverpool and Macarthur cancer therapy centres and Ingham institute, Sydney, New South Wales, Australia.
[2]Centre for medical radiation physics, University of Wollongong, Wollongong, New South Wales, Australia.
[3]Uinversity of New South Wales, Sydney, New South Wales, Australia.
[4]Institute of Medical Physics, University of Sydney, Sydney, New South Wales, Australia

Sankar.Arumugam@sswahs.nsw.gov.au

Abstract. This study presents the development of a software tool 'Treat Check' to analyse the dynamic log files from an Elekta – Synergy accelerator. The software generates formatted output in the form of a plot presenting errors in various treatment delivery parameters such as gantry angle, Multi Leaf Collimator (MLC) leaf position, jaw position and Monitor Units (MU) for each of the control-points (CP) of the treatment beam. The plots are automatically saved in Portable Document Format (pdf). The software also has the functionality to introduce these treatment delivery errors into the original plan in the Pinnacle (Philips) treatment planning system (TPS) in order to assess the clinical impact of treatment delivery errors on delivered dose.

1. Introduction

Modern radiotherapy techniques such as Intensity Modulated Radiation Therapy (IMRT) and Volumetric Modulated Arc Therapy (VMAT) offer superior dose conformity to target volumes (TV) with improved control over the dose delivered to critical structures[1]. This advantage is realised through a highly complex process in both planning and delivery. In dynamic IMRT delivery Multi Leaf Collimator (MLC) leaves continuously move during radiation delivery in order to generate the intended intensity modulation. The delivery of VMAT is more complex than IMRT as many of the treatment delivery parameters such as MLC leaves, jaws, gantry, and dose rate continuously vary during the treatment delivery [2]. To ensure the dosimetric accuracy of treatment delivered by these complex techniques pre-treatment dosimetric verification is highly recommend by international governing bodies[3].

Generally pretreatment dose verification is performed using dose measuring systems in well defined and homogeneous phantom geometries. While such dose measurements ensure the accuracy of treatment plans in dosimeter geometry it assures the accuracy of treatment delivery only for that particular treatment delivery. The treatment delivery accuracy of individual fractions of patient treatment can be verified using in vivo dose measurements. Van Herk and Mijnheer demonstrated the possibility of performing in vivo 3D dose verification based on transit images acquired using Electronic Portal Imaging Device (EPID)[4]. Litzenberg et al[5] proposed the use of dynamic log files recorded by the control system of Varian linear accelerator (Varian medical systems, Palo Alto, USA) to verify individual fractions of patient IMRT treatment.

The control system of Elekta-Synergy accelerators (Elekta, Inc. Crawley, UK) records the error between prescribed and delivered treatment parameters for individual control-points (CPs) of treatment plans. However this information can only be reviewed on the service page of the control system. Also some of the parameters such as MLC leaf positions and jaw positions are presented in the system as raw potentiometer readings. This presents a challenge to the end user to assess the significance of errors which occurred during treatment delivery. In this work we present the development of an in-house computer program, Treat Check, which processes the log files from the Elekta –Synergy control system and presents the result in a user friendly manner. The program also contains a module to introduce these errors into the original treatment plan in Pinnacle (Philips, Inc. Fitchburg, USA) TPS to assess the impact of these errors on delivered dose.

2. Materials and Methods

Elekta-Synergy accelerator with Desktop pro, v 7.0, control system software was used in this study. The Synergy accelerator used in this study is capable of producing 6 and 18 MV photon beams and 6, 8, 10, 12, 15 and 18 MeV electron beams. The photon beams can be delivered with one of the following nominal dose rates: 550, 275, 138, 69, 35, 18 and 9 MU/min. The electron beams can be delivered with a nominal dose rate of 400 MU/min. The secondary beam collimating system consists of 40 pairs of MLC leaves and a pair of backup jaws in the X direction and a pair of jaws in the Y direction. Mosaiq, v 2.30.0D1, (Impac Medical Systems, Inc. California, USA) record and verification (R&V) system was used for the transfer of treatment plan parameters from Pinnacle TPS to the accelerator.

2.1. Dynamic log files

The control system of Synergy accelerators records the maximum positional error between prescribed and delivered parameters such as gantry angle, collimator angle, jaw position, MLC leaf position, MU, table lateral, longitudinal, vertical and rotation values for each of the CP of the beam in an individual file for delivered treatment plans. The tolerance values for each of these parameters is also recorded in the same file. However, reviewing this data has many practical limitations as described in the introduction. Figure 1 shows a screen capture of the dynamic log file view page of the Synergy control system. The control system presents the error values for each of the CP in an individual page. Analysing the error values in this manner is time consuming as IMRT and VMAT plans contain large numbers of CPs. The data are stored in the log files in a binary format and it presents a challenge to the user to interpret the stored data.

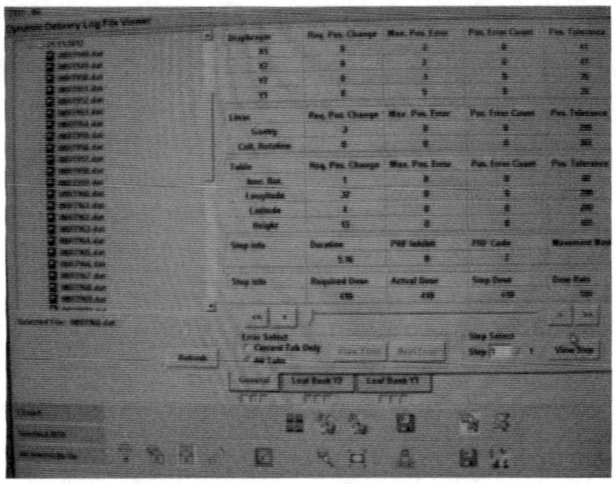

Figure1: Screen capture of the dynamic log file view page in the Elekta-Synergy control system.

2.2. Treat Check - In-house computer program

A computer program, Treat Check, was developed in-house using MATLAB, v 7.11.0.584 (R2010b), (MathWorks, MA, USA) to process the log files derived from the Synergy control system. The overall function of the program is shown in figure 2. Treat Check reads the log file and processes the data to identify the parameter type. The positional error values of jaws and MLC leaf positions are stored as raw potentiometer values. These values were converted in to real positional values using the following formula prescribed by the Synergy manufacturer.

$$\text{ositional error in milimeter} = \text{Rawpositional error vaule} * \text{Scaling Value} \qquad (1)$$

Where, the raw positional error value is determined from the log file and the scaling value is calculated following equation.

$$\text{Scaling Value} = \frac{204.8}{\text{Gain value of the parameter}} \qquad (2)$$

The gain value for each of the jaws and MLC leaf bank is obtained from synergy control system.

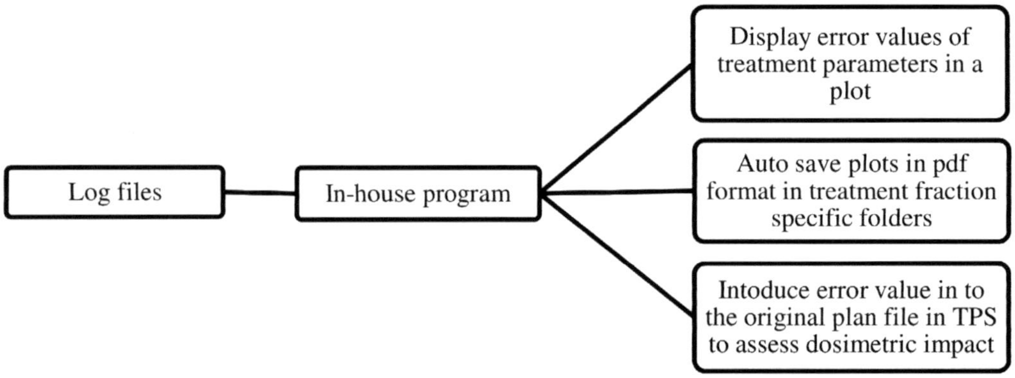

Figure2: Flowchart showing the overall functions of Treat Check.

For the easy interpretation and review of the error values all the CP error values of each of the parameters are plotted individually by Treat Check . The plots are also automatically saved by the Treat Check in a treatment specific folder in pdf for future review. Treat Check also includes a module to introduce these error values into the original plan in the TPS. This module reconstructs a plan file for the Pinnacle TPS using a treatment delivery file of a plan in radiotherapy prescription (RTP) format and errors read from Elekta dynamic log files.[6] This allows the assessment of the impact of these errors on the delivered dose by comparing the dose distribution and Dose Volume Histogram (DVH) parameters. The accuracy of data interpretation of the software was tested by comparing the error values calculated by software and those displayed on the synergy control system for an IMRT and VMAT delivery.

3. Results and Discussion

3.1. Performance of Treat Check

Treat Check successfully reads the dynamic log files from the Synergy control system and performs the tasks as described in the figure2.The comparison of error values of the various parameters calculated by Treat Check and the same displayed on the control system showed no discrepancies, demonstrating accurate performance of the program.

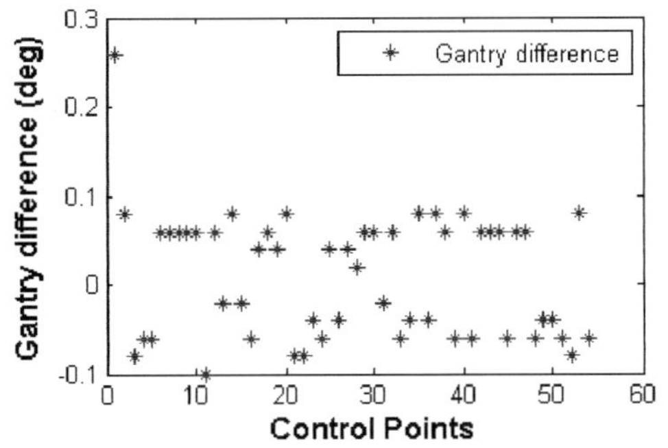

Figure 3: Gantry positional error plot generated by Treat Check for a prostate VMAT plan.

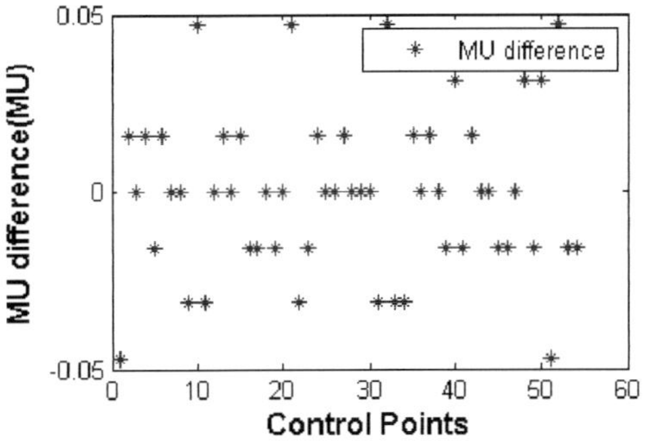

Figure 4: Monitor Units error plot generated by Treat Check for a prostate VMAT plan.

Figures 3 and 4 show the Treat Check output for the error in gantry position and MU delivery for each of the CPs of a prostate VMAT plan with 60 CPs. The gantry error for most of the CPs except the first were within $\pm 0.1^0$. The relatively larger error in the first CP delivery could be due to inertia of the gantry at the start of the VMAT delivery. The MU delivery error for each of the CPs was within ± 0.05 for the studied VMAT plan.

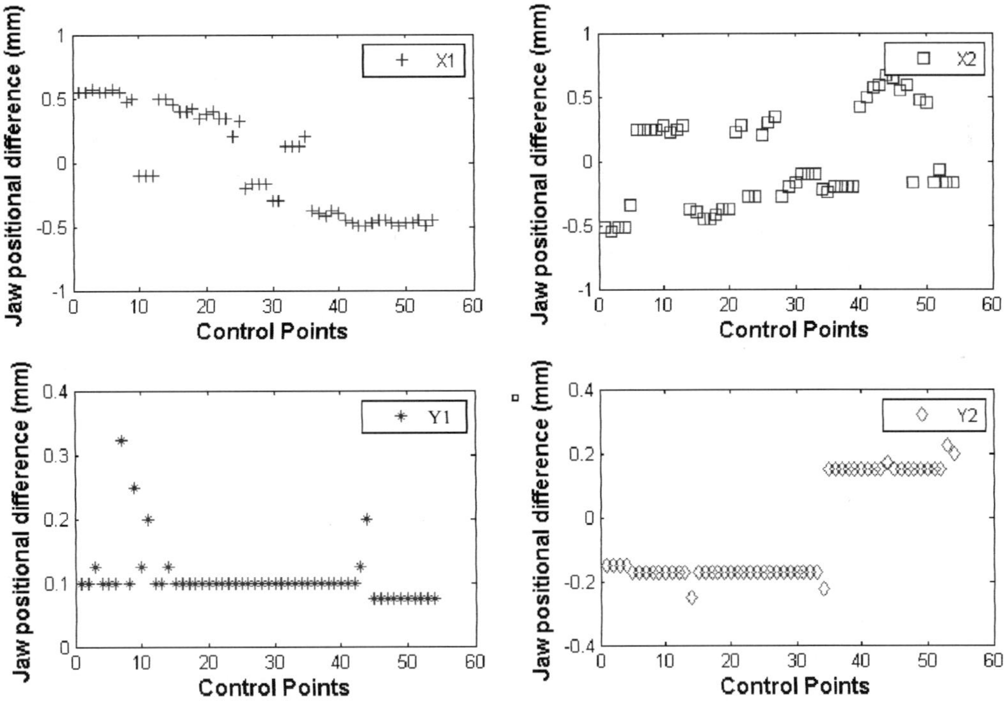

Figure 5: Jaw positional error plot generated by Treat Check for a) X1, b) X2, c) Y1 and d) Y2 jaws.

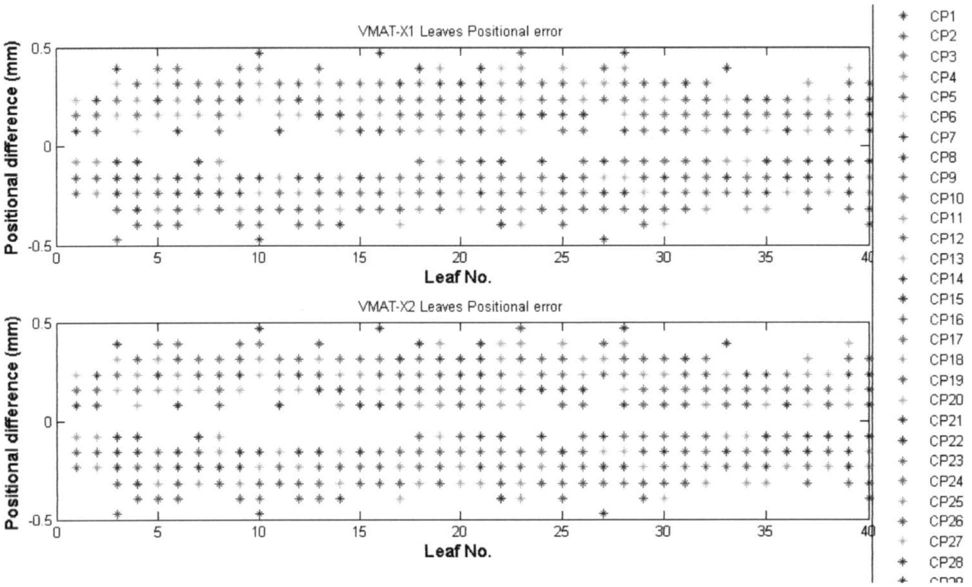

Figure 6: Leaf positional error plot generated by Treat Check for X1 and X2 banks of MLC leaves for all CPs of the studied VMAT plan.

Figures 5 a, b, c and d show the positional error for X1, X2, Y1 and Y2 jaws for all the CPs of the studied VMAT plan. The positional errors for most of the CPs were within ±0.5 mm for X1 and X2 jaws. However for some CPs the error for X2 jaw was up to 0.8 mm. The error values observed for the Y jaws did not vary between CPs greatly because in VMAT delivery Y jaws, which define the length of the field do not move whereas X jaws track the outer most leaf position of their respective bank in order to reduce leaf leakage. Figure 6 shows the positional error for all the leaves of X1 and X2 MLC banks for all the CPs of the studied VMAT plan. The positional error values for the leaves were within ±0.5 mm for the studied VMAT delivery.

3.2. Assessment of delivered dose

Figure 7 shows the DVH of target volumes (PTV-Planning Target Volume, Prostate and SV-Seminal vesicle) and critical structures (Bladder, Rectum, Right Head of Femur and Left Head of Femur) of the original and delivery error introduced plans for the studied prostate VMAT plan. Planned and delivered doses of various structures do not demonstrate any notable differences for the considered fraction of the treatment due to relatively small errors in the delivery parameters (Figures 3, 4, 5 and 6). This aspect of the program, the ability to introduce delivery errors into the planning system, provides the potential to assess delivered plans in clinically meaningful manner by evaluating delivered dose distributions and DVHs of structures.

In the present study the dose calculations of a delivered plan was performed on the planning CT data set of a studied patient. The dose calculation of the delivered treatment on the 3D image data set acquired on the day of treatment would contain the delivered dose information including treatment delivery errors and patient anatomical changes. This information will be of interest in the context of dose accumulation for adaptive radiation treatment. Future work will investigate dose calculations on daily 3D verification images.

3.3. Important considerations

It is important to acknowledge that the error values recorded in the log files of the control system are only true with respect to the accuracy of the calibration of a given parameter. If there are any systematic errors in the calibration these will not be reflected in the error values recorded in the log files. Hence the absolute accuracy of the treatment delivery parameters should be verified as recommended by AAPM TG142[7] guidelines. Further verifying patient treatment using log files does not replace the in-vivo dose measurements, because log files do not provide systematic errors in the treatment delivery parameters. However the analysis of the error values recorded in the log files can be used as supplementary data ensure the accurate delivery of patient treatment.

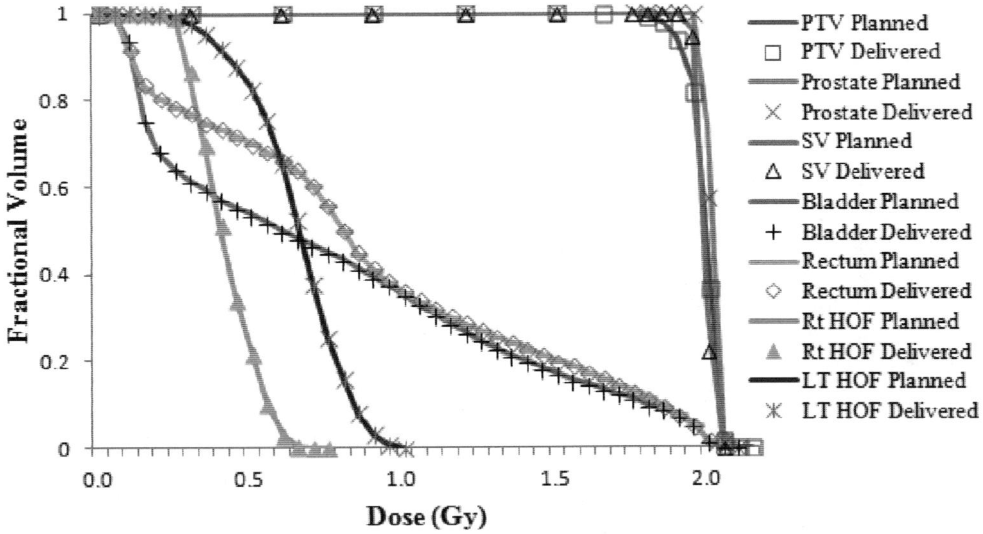

Figure 7: Planned and delivered dose for a treatment fraction of studied prostate VMAT treatment.

4. Conclusion

An in-house computer program was developed and demonstrated to successfully read the dynamic log files from an Elekta-Synergy accelerator. The program successfully generates comprehensive formatted output of the error values that can be easily reviewed by the user. The error values can also be transferred to the TPS allowing assessment of delivered dose in a clinically meaningful manner.

References

[1] D. Wolff, F. Stieler, G. Welzel, F. Lorenz, Y. Abo-Madyan, S. Mai, C. Herskind, M. Polednik, V. Steil and F. Wenz 2009 *Radiother Oncol.* **93** (2) 226
[2] K. Otto 2008 *Med. Phys.* **35** 310
[3] G. A. Ezzell, J. M. Galvin, D. Low, J. R. Palta, I. Rosen, M. B. Sharpe, P. Xia, Y. Xiao, L. Xing and X. Y. Cedric 2003 *Med. Phys.* **30** 2089
[4] M. VAN HERK and B. MIJNHEER 2007 *Int. J. Radiation Oncology Biol. Phys* **67** (5) 1568-1577
[5] D. W. Litzenberg, J. M. Moran and B. A. Fraass 2002 *J Appl Clin Med Phys.* **3** (2) 63-72
[6] S. Arumugam, A. Xing, G. Goozee and L. Holloway 2013 *Med. Dosim(in print).*
[7] E. E. Klein, J. Hanley, J. Bayouth, F.-F. Yin, W. Simon, S. Dresser, C. Serago, F. Aguirre, L. Ma and B. Arjomandy 2009 *Med Phys* **36** 4197

A decision support tool to optimize IMRT QA workflow in a multi-vendor equipment environment

Sankar Arumugam[1], Aitang Xing[1], Philip Vial[1,2], Lois Holloway[1,2,3,4]

[1]Liverpool and Macarthur cancer therapy centres and Ingham institute, Sydney, New South Wales, Australia.
[2]Institute of Medical Physics, University of Sydney, Sydney, New South Wales, Australia
[3]Centre for medical radiation physics, University of Wollongong, Wollongong, New South Wales, Australia.
[4]Uinversity of New South Wales, Sydney, New South Wales, Australia.

Sankar.Arumugam@sswahs.nsw.gov.au

Abstract. Development of a software tool to ease the Intensity Modulated Radiation Therapy (IMRT) pre-treatment Quality Assurance process is presented in this study. The delivery of IMRT involves equipment from multiple vendors. The limitations of the equipment involved in this chain will impact on the best choice of equipment. This often results in the user needing to use multiple pieces of equipment before determining the most appropriate choices to optimise the QA work flow. This is a time consuming process and potentially delays the start of patient treatment. Software was developed in-house to assist the decision making process, validating deliverability of beam delivery parameters and selecting appropriate detector systems and configuration for QA of IMRT plans. The software has been demonstrated to be accurate and improves efficiency of IMRT pre-treatment QA.

1. Introduction

Intensity Modulated Radiation Therapy (IMRT) involves equipment from multiple vendors at various stages of the treatment course such as treatment planning, Quality Assurance (QA) and treatment delivery. Due to the complex nature of this treatment technique plan specific pre-treatment dosimetric verification is highly recommended to ensure the accurate and safe delivery of treatment[1]. Dose verification of IMRT plans is performed using a combination of 1D, 2D or 3D detector systems. In this process the patient plan is recalculated on the phantom image dataset and the phantom and detector position is optimised to cover the entire treatment field. The phantom position on treatment machine is derived during this process. Extensive dosimetric verification of IMRT plans is an essential component of IMRT to validate the accuracy of dose calculations performed by the planning system and treatment delivery by the Linear accelerator (linac). In the clinic routine pre-treatment verification is optimised to simple measurement geometry for efficient workflow. This optimisation process should be supported with sufficient experience and confidence, backed by detailed dose verification, on the Treatment planning System (TPS) and delivery system[2].

IMRT pre-treatment QA on a per beam basis using an Electronic Portal Imaging Device (EPID) has been shown to be more efficient in routine clinical practice[3]. Many approaches are in practice to convert an EPID image into dose matrix[4]. In simpler approaches EPID images are converted in to dose images at the isocentre plane using appropriate calibration factors and the corresponding dose matrices are calculated in the TPS on a hypothetical phantom[5]. The EPID mount system on different vendor linacs offers different degrees of movement in-order to accommodate a range of field sizes and field asymmetries. If the field size exceeds the EPID measurement area the system triggers an interlock to avoid irradiation of the electronics section of the EPID. In practice the appropriate EPID position is identified at the treatment machine for each field. This consumes time for each patient dataset due to possible different EPID positional configurations available on different linacs. Further occasionally TPSs calculated IMRT field segments violate MLC motion limitations. Often this is not identified until the pre-treatment verification process causing delay in both QA and the initiation of patient treatment.

In this study we present an in-house software tool that identifies the presence of undeliverable segments in an IMRT beam. The developed software also predicts the appropriate EPID position configuration by considering the IMRT field size and vendor specific position limitations.

2. Materials and Methods

2.1. Multi-vendor equipment environment

South Western Sydney Local Health District (SWSLHD) provides a cancer treatment service to the south west region of Sydney, Australia through Liverpool Cancer Therapy Centre (LCTC) and Macarthur Cancer Therapy Centre (MCTC). The radiation oncology department in these centres includes radiation treatmentequipment from a range of vendors. The distribution of TPSs and linacs used in these centres for our IMRT programme is shown in figure 1. Routine pre-treatment IMRT verification in our centres is performed using the Electronic Portal Imaging Device (EPID) associated with the linacs. If the field size of the IMRT beam exceeds the active area of the EPID verification is performed using the I'mRT MatriXX ion chamber 2D array (IBA Dosimetry GmbH, Germany) to avoid irradiation of the electronics sections of the EPID device. Some of the important characteristics of the Siemens- Oncor (Siemens AG, Inc. Erlangen, Germany) and Elekta – Synergy (Elekta, Inc. Crawley, UK) EPIDs used for IMRT verification are shown in Table 1.

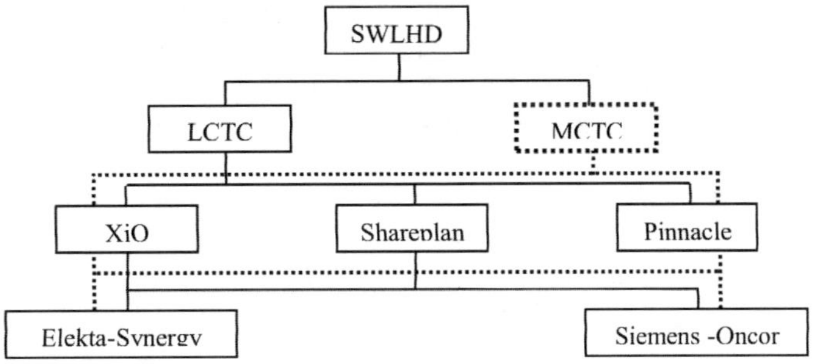

Figure1: Distribution of equipment used for IMRT in SWLHD

Table 1: Key characteristics of Synergy and Oncor linac EPIDs

EPID physical characteristics	Siemens-Oncor	Elekta-Synergy
Source to detector distance	115cm and 145 cm	160 cm
Maximum field width that can be measured	31.6 cm @ 115cm 25.0cm @ 145 cm	26 cm
Lateral movement of EPID	Not Available	±11.8 cm
Longitudinal movement of EPID	Not Available	±11.8 cm

The beam segments calculated by XiO, version 4.6 (Elekta CMS software,Inc, MO,USA) TPS occasionally include an MLC segment that is not deliverable by the Elekta-Synergy linac. Figure 2 shows a segment of a prostate IMRT plan beam generated by XiO where two MLC leaves are interdigitized in the plan but the Elekta-Synergy MLC does not have this capability. Similarly the segments calculated by Pinnacle version 9.0 (Philips, WI, USA) for the Siemens-Oncor linac occasionally come with Y jaw over travel that is more than the actual Y jaw travel. These issues lead to IMRT beams containing invalid segments which are undeliverable. Usually the presence of invalid segments is identified at the time of pre-treatment QA and this causes a delay in the start of patient treatment.

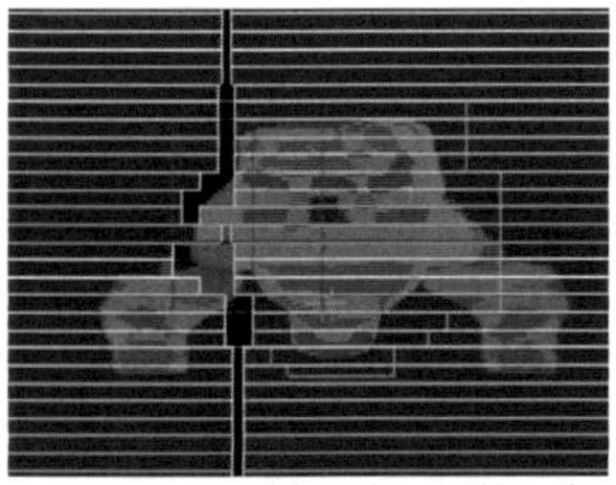

Figure2: Undeliverable IMRT segment generated by the XiO planning system for an Elekta-Synergy linac.

2.2. In-house computer program

In order to identify the presence of undeliverable segments in IMRT fields and to predict suitable QA devices and physical measurement configuration a software tool was developed in-house using the Python language, version 2.6.5.The software reads the treatment delivery file in Radiation Therapy Prescription (RTP) format from any TPS and searches for the presence of segments that violate the position limitations of MLC leaves and jaws in the Elekta-Synergy and Siemens-Oncor accelerators and records the segment number if they are present.

The direct relation of field size defined by X1,X2,Y1 and Y2 jaw positions in IEC coordinate systems to the detectors active area holds good only for collimator(coll) angle 0^0. For coll angles other than 0^0 the maximum extent of the radiation field on the detector changes due to the rotation of the hypotenuse of the field. Figure 3 shows the change in field vertex co-ordinates due to coll rotation θ^0. The co-ordinates of the field vertex due to coll rotation θ^0 from co-ordinates derived from the IEC system is calculated using equations 1 and 2. From the calculated field vertices the effective extent of the radiation field on the detector is calculated.

$$X^{|} = X\cos\theta - Y\sin\theta \quad (1)$$

$$Y^{|} = X\sin\theta + Y\cos\theta \quad (2)$$

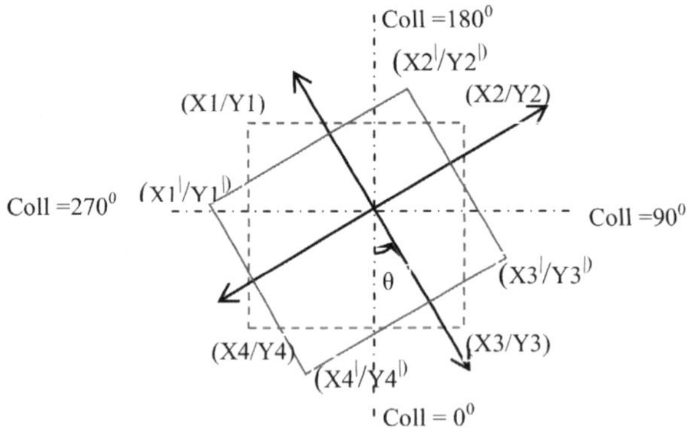

Figure3: Change in co-ordinates of field vertices due to collimator rotation.

The decision tree implemented in the in-house software for the selection of appropriate detectors and the position configuration of the EPID is shown in figure 4. In the decision making process the possible EPID position configurations in Elekta and Siemens linacs (Table 1) and the field vertices calculated using equation 1 and 2 are considered. If the field extent exceeds the active area of the EPID the ion chamber 2D array will be recommended for the measurement. In Elekta linacs the following four EPID configurations are possible based on IMRT field size and collimator angle:

1. A centred position of the EPID and planned collimator and gantry angle for the measurement will be recommended if,

$$|X'|_i \; and |Y'|_i \leq \; 13.0 \; cm \tag{3}$$

Where, $X_i^|$ and $Y_i^|$ are the co-ordinates of the field vertices and i ∈ {1,2,3,4}.

2. A centred position of the EPID and planned gantry angle with the collimator angle set to 0^0 will be recommended for measurement if,

$$(|X'|_i \; or |Y'|_i > \; 13.0 \; cm \;) \text{ and } (\; |X|_i \; and |Y|_i \leq 13.0 \; cm) \tag{4}$$

3. An off centred position of the EPID and planned collimator and gantry angle will be recommended if

$$(|X'|_i \; or |Y'|_i > \; 13.0 \; cm \;) \text{ and } (|X|_i or |Y|_i > 13.0 \; cm \;) \text{ and } (FEx \; or \; FEy \leq 26.0 \; cm) \tag{5}$$

Where, FEx and FEy are the field extents on the detector in X and Y directions. The required EPID offset in lateral and longitudinal directions is calculated as follows

$$\text{Lateral offset} = (X_1^| + X_3^|)/2 \tag{6}$$

$$\text{Longitudinal offset} = (Y_2^| + Y_4^|)/2 \tag{7}$$

4. An off centred position of the EPID and planned gantry angle but collimator set to 0 will be recommended for measurement if

$$(|X'|_i \; or |Y'|_i > \; 13.0 \; cm \;) \text{ and } (|X|_i or |Y|_i > 13.0 \; cm \;) \text{ and } (\; Fx \; or \; Fy \leq 26.0 \; cm) \tag{8}$$

Where, Fx and Fy are the field width and height in X and Y directions. The required EPID offset in lateral and longitudinal directions is calculated using equations 6 and 7 by using field vertices at collimator angle 0^0.

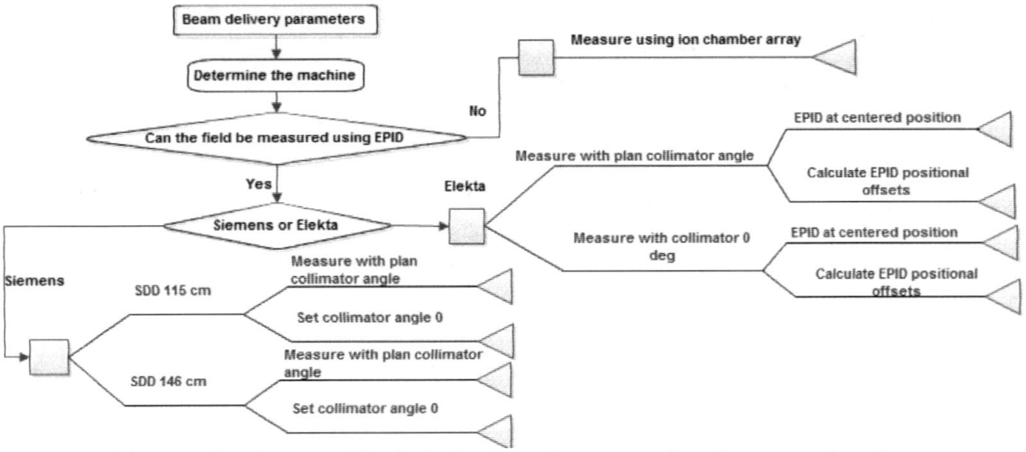

Figure4: Decision tree implemented in the in-house software to select the appropriate detector system and measurement setup considering limitations of equipment and IMRT field size.

In Siemens linacs the following four EPID configurations are possible based on IMRT field size and collimator angle:

1. EPID at 146 cm distance and planned collimator and gantry angle for the measurement will be recommended if,

$$|X'|_i \, and |Y'|_i \leq \; 12.5 \; cm \tag{9}$$

2. EPID at 146 cm distance and planned gantry angle but collimator angle set to 0^0 will be recommended for measurement if,

$$(|X'|_i \, or |Y'|_i > \; 12.5 \, cm\,) \text{ and } (\, |X|_i \, and |Y|_i \leq 12.5 \; cm) \tag{10}$$

3. EPID at 115 cm distance and planned gantry and collimator angle will be recommended for measurement if,

$$12.5 \; cm \geq (|X'|_i \, and |Y'|_i) \leq \; 15.8 \; cm \tag{11}$$

4. EPID at 115 cm distance and planned gantry angle but collimator angle set to 0^0 will be recommended for measurement if,

$$[12.5 \; cm \geq (|X'|_i \, or |Y'|_i > \; 15.8 \; cm\,)] \text{ and } [12.5 \; cm \geq (\, |X|_i \, and |Y|_i \leq 15.8 \; cm)] \tag{12}$$

Based on the individual field specific detectors choice and EPID setup the overall action will be presented by the software for efficient measurement of IMRT fields.

3. Results and Discussion

The user interface of the software and its sections provide various options such as treatment delivery file selection, clinic selection, date, planner and physicist name, output report file (in PDF) name and destination selection, display of analysis results and the option to enter additional comments (Figure 5).

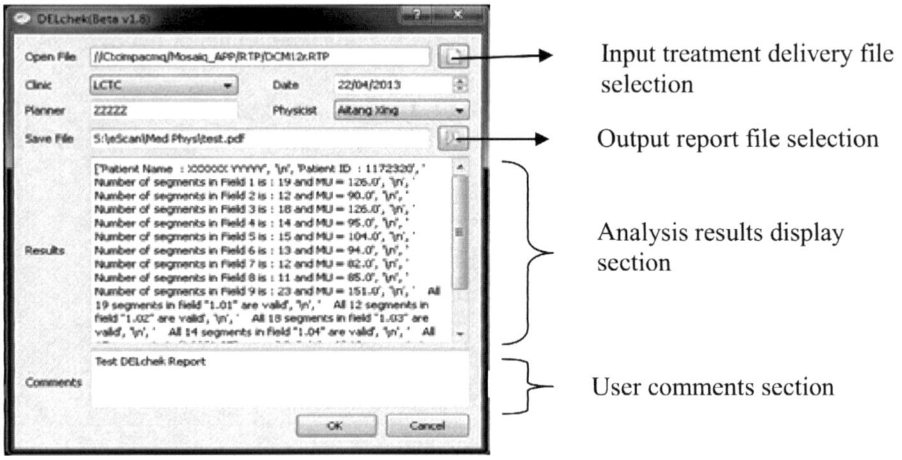

Figure 5: Annotated figure showing various sections of the user interface of in-house software

The software successfully generates the intended report from the treatment delivery file. Figure 6 shows a sample report generated by the software from the treatment delivery file derived from the Pinnacle TPS for a Siemens linac. The patient details, plan details, presence of total number of IMRT and setup fields and presence of total number of prescription and identification of correct treatment delivery machine within the selected clinic have been accurately identified by the software (Plan details section of Figure 6). Similarly the treatment parameter details such as field ID, jaw positions, gantry and collimator angles, beam Monitor Units (MU) and minimum and maximum MU of segments in an IMRT beam also accurately identified by the software from treatment delivery file(Beam parameters section of Figure 6). The total number of segments per beam and deliverability is correctly predicted by the software. In the sample case the presence of an invalid segment (segment no 11) in field ID 1.16 due to violation of the over travel limit of the Y jaw has been accurately detected by the software. The selection of the appropriate detector system (ion chamber array or EPID) and setup of EPID, is accurately determined by the software considering possible positional options available for EPID of Elekta and Siemens linacs (Segment validity and detector selection section of figure 6). Based on these results the overall recommended action is also presented by the software to escalate the identified issues for corrective action.

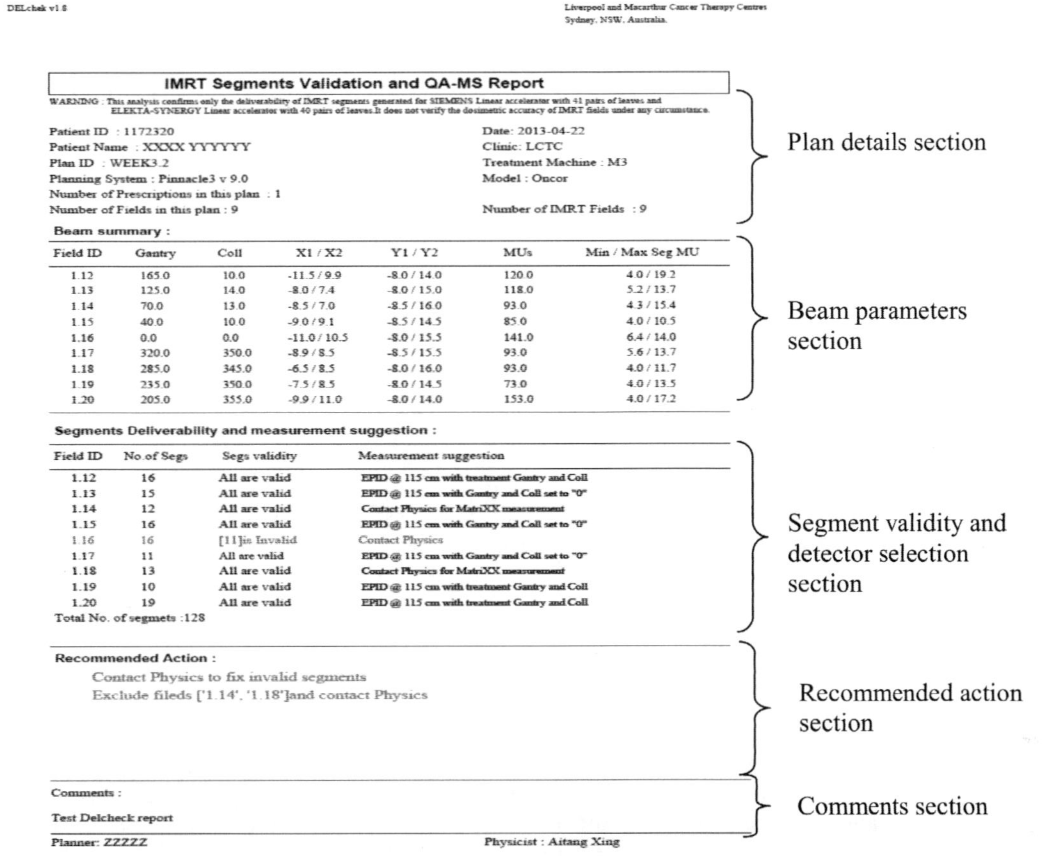
DELchek v1.8

Liverpool and Macarthur Cancer Therapy Centres
Sydney, NSW, Australia.

IMRT Segments Validation and QA-MS Report

WARNING : This analysis confirms only the deliverability of IMRT segments generated for SIEMENS Linear accelerator with 41 pairs of leaves and ELEKTA-SYNERGY Linear accelerator with 40 pairs of leaves.It does not verify the dosimetric accuracy of IMRT fields under any circumstance.

Patient ID : 1172320
Patient Name : XXXX YYYYYY
Plan ID : WEEK3.2
Planning System : Pinnacle3 v 9.0
Number of Prescriptions in this plan : 1
Number of Fields in this plan : 9

Date: 2013-04-22
Clinic: LCTC
Treatment Machine : M3
Model : Oncor
Number of IMRT Fields : 9

Beam summary :

Field ID	Gantry	Coll	X1 / X2	Y1 / Y2	MUs	Min / Max Seg MU
1.12	165.0	10.0	-11.5 / 9.9	-8.0 / 14.0	120.0	4.0 / 19.2
1.13	125.0	14.0	-8.0 / 7.4	-8.0 / 15.0	118.0	5.2 / 13.7
1.14	70.0	13.0	-8.5 / 7.0	-8.5 / 16.0	93.0	4.3 / 15.4
1.15	40.0	10.0	-9.0 / 9.1	-8.5 / 14.5	85.0	4.0 / 10.5
1.16	0.0	0.0	-11.0 / 10.5	-8.0 / 15.5	141.0	6.4 / 14.0
1.17	320.0	350.0	-8.9 / 8.5	-8.5 / 15.5	93.0	5.6 / 13.7
1.18	285.0	345.0	-6.5 / 8.5	-8.0 / 16.0	93.0	4.0 / 11.7
1.19	235.0	350.0	-7.5 / 8.5	-8.0 / 14.5	73.0	4.0 / 13.5
1.20	205.0	355.0	-9.9 / 11.0	-8.0 / 14.0	153.0	4.0 / 17.2

Segments Deliverability and measurement suggestion :

Field ID	No.of Segs	Segs validity	Measurement suggestion
1.12	16	All are valid	EPID @ 115 cm with treatment Gantry and Coll
1.13	15	All are valid	EPID @ 115 cm with Gantry and Coll set to "0"
1.14	12	All are valid	Contact Physics for MatriXX measurement
1.15	16	All are valid	EPID @ 115 cm with Gantry and Coll set to "0"
1.16	16	[11]is Invalid	Contact Physics
1.17	11	All are valid	EPID @ 115 cm with Gantry and Coll set to "0"
1.18	13	All are valid	Contact Physics for MatriXX measurement
1.19	10	All are valid	EPID @ 115 cm with treatment Gantry and Coll
1.20	19	All are valid	EPID @ 115 cm with treatment Gantry and Coll

Total No. of segmets :128

Recommended Action :

Contact Physics to fix invalid segments
Exclude fileds ['1.14', '1.18']and contact Physics

Comments :
Test Delcheck report

Planner: ZZZZZ
Physicist : Aitang Xing

Figure 6: Annotated figure showing various sections of the report generated by in-house software

4. Conclusion

The software successfully generates a comprehensive report that includes a summary of prescription and beam parameters and identification of undeliverable beam segments if present. It provides a seamless workflow to validate the deliverability of segments and chose appropriate detector system and measurement setup. The introduction of this software tool has increased the efficiency of our pre-treatment IMRT QA process in a multi-vendor environment.

References

[1] G. A. Ezzell, J. M. Galvin, D. Low, J. R. Palta, I. Rosen, M. B. Sharpe, P. Xia, Y. Xiao, L. Xing and X. Y. Cedric 2003 *Med. Phys.* **30** 2089.

[2] J. M. Moran, M. Dempsey, A. Eisbruch, B. A. Fraass, J. M. Galvin, G. S. Ibbott and L. B. Marks 2011 *Med Phys* **38** 5067.

[3] G. J. Budgell, Q. Zhang, R. J. Trouncer and R. I. Mackay 2005 *Med Phys* **32** 3267.

[4] W. van Elmpt, L. McDermott, S. Nijsten, M. Wendling, P. Lambin and B. Mijnheer 2008 *Radiother Oncol.* **88** (3) 289.

[5] C. Lee, F. Menk, P. Cadman and P. B. Greer 2009 *Med Phys* **36** 984.

Automatic image registration performance for two different CBCT systems; variation with imaging dose

J BARBER[1,3], J R SYKES[2], L HOLLOWAY[3], D I THWAITES[3]

[1] Nepean Cancer Care Centre, Penrith, NSW, Australia
[2] St James's Institute of Oncology, Leeds, UK
[3] Institute of Medical Physics, University of Sydney, Sydney, NSW, Australia

jeffrey.barber@swahs.health.nsw.gov.au

Abstract. The performance of an automatic image registration algorithm was compared on image sets collected with two commercial CBCT systems, and the relationship with imaging dose was explored. CBCT images of a CIRS Virtually Human Male Pelvis phantom (VHMP) were collected on Varian TrueBeam/OBI and Elekta Synergy/XVI linear accelerators, across a range of mAs settings. Each CBCT image was registered 100 times, with random initial offsets introduced. Image registration was performed using the grey value correlation ratio algorithm in the Elekta XVI software, to a mask of the prostate volume with 5 mm expansion. Residual registration errors were calculated after correcting for the initial introduced phantom set-up error. Registration performance with the OBI images was similar to that of XVI. There was a clear dependence on imaging dose for the XVI images with residual errors increasing below 4mGy. It was not possible to acquire images with doses lower than ~5mGy with the OBI system and no evidence of reduced performance was observed at this dose. Registration failures (maximum target registration error > 3.6 mm on the surface of a 30mm sphere) occurred in 5% to 9% of registrations except for the lowest dose XVI scan (31%). The uncertainty in automatic image registration with both OBI and XVI images was found to be adequate for clinical use within a normal range of acquisition settings.

1. Introduction

The high uptake of kV imaging systems mounted on medical linear accelerators in recent years has placed tools for automatic image registration at the treatment console for routine use. These automatic matching algorithms have been shown to be accurate for matching 3D cone beam computed tomography (CBCT) scans to planning CT scans, on both the Elekta XVI system [1,2] and Varian OBI system [3,4]. There are some significant differences in how the two systems reconstruct 3D image sets, such as the handling of scatter and the sharpness of image reconstruction filters used. They also use different algorithms in their respective automatic registration software, ie – OBI uses a mutual information algorithm, whilst XVI uses a Chamfer algorithm for bone matching and a correlation ratio algorithm for the grey value match.

In a previous study [1], the accuracy of XVI automatic image registration is shown to have a dependence with imaging dose. This study looks at whether there is a similar relationship in the OBI system, and the impact of using images from both systems in a single automatic algorithm, in order to determine any differences in performance between the two systems.

2. Method

A CIRS model 801-P-B Virtually Human Male Pelvis phantom (Computerized Imaging Reference Systems, Inc., Norfolk, Virginia, USA) was used for all images. Cone beam images were acquired on both an Elekta Synergy (Elekta AB, , Stockholm, Sweden) and a Varian TrueBeam (Varian Medical Systems, Palo Alto, CA). For each, the phantom was set up and positioned using the CBCT system using a reference scan acquired on a GE Lightspeed RT scanner (GE Medical Systems, Waukesha, WI) with 1.25mm slice thickness. The initial phantom position was corrected in 6 degrees of freedom (DoF) on the Synergy using a Hexapod couch and 3 DoF on the TrueBeam.

Once positioned to within 0.5mm and 0.5° of the reference scan, a series of CBCT images were acquired on each system with mAs settings chosen to cover a range both above and below that used typically in clinical practice. The settings chosen were matched to be as similar as possible between systems. The exposure was varied from 68 mAs to 1360 mAs with XVI (5 settings), and from 132 mAs to 680 mAs with OBI (4 settings). The effective dose for each scan was calculated using published in-air dose/mAs factors for XVI and OBI [5].

The OBI images were imported into the XVI software, and then all 9 image sets were analysed against the initial reference CT scan. For each image set, an initial random translational offset was applied. Then the correlation ratio algorithm (grey value match) was run to register a mask region in 6 DoF. This mask encompassed the phantom prostate volume with an additional 5mm expansion. The registration result was subtracted from the initial known offset to provide the residual error of the registration algorithm. This process was repeated 100 times for each image set to simulate inter-fraction differences, using random initial offsets sampled from a 3D Gaussian distribution based on variations, observed clinically, in daily setup of prostate patients [6].

The mean of each image set's residual error was subtracted as it was assumed to be the systematic error in phantom positioning, accounting for initial positioning differences. The Target Registration Error (TRE) was determined as the maximum distance of a point on the surface of a sphere with residual translation/rotations to the initial sphere position [2]. A sphere of 30 mm radius was used to approximate the surface of the prostate. This was computed using dual quaternions to combine both translational and rotational components of the residual error and the distance denoted TRE_{30} [2].

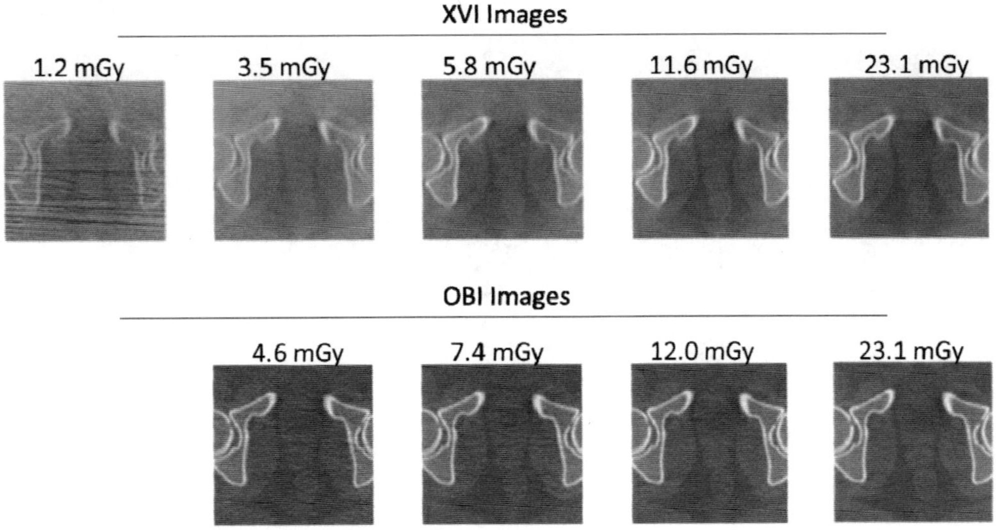

Figure 1 – Series of phantom CBCT collected for varying exposure. The top row shows XVI images and the bottom row shows OBI images. Exposure increases going from left to right.

3. Results

The image quality for the XVI and OBI images both deteriorated with decreasing dose, showing increased noise. For the lowest dose XVI image artifacts are observed which are likely due to photon starvation (Figure 1). Note that very low dose images could not be obtained with the OBI system to compare to the XVI system. Image quality comparisons between the systems are rather subjective due to the differing sharpness and noise present in each system's images. However, for images in the typical range of clinical doses (10-30 mGy), similar image quality was observed based on the ability to resolve anatomical landmarks in soft tissue such as the rectal wall. 5 mm diameter plugs in the phantom were only visible in scans of doses more than 23 mGy for both systems.

a)

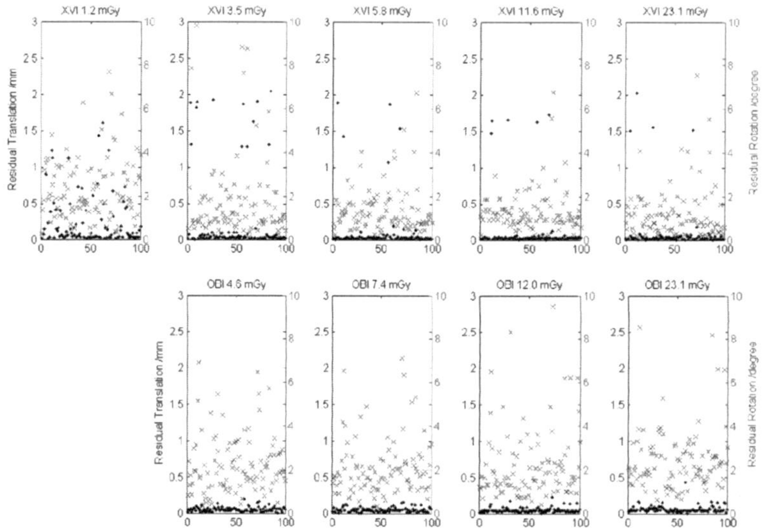

b)

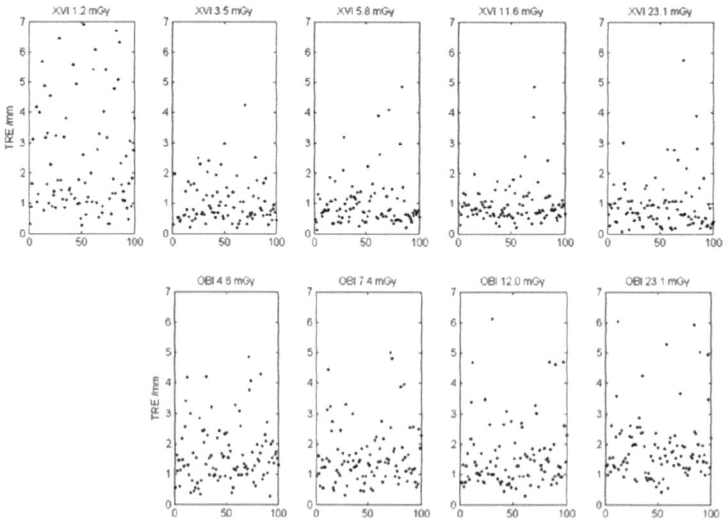

Figure 2 – a) Plots of residual translational error (dots) and residual rotational error (crosses) as 3D vector magnitude in the automatic image registration process for each image-set. b) Plots of TRE_{30}. Each plot shows the results of 100 repeat registrations

Registration performance with the OBI images was similar to that of XVI with residual translations <0.5mm (1σ) (Figure 2a). OBI residual rotations were typically 1.2° (1σ) compared to 0.8° for XVI (Figure 2a). Over the clinical dose range, automatic registration provided similar residual error for both imaging systems. There was a dose dependence on residual translation and rotation errors for the XVI images, with both increasing for images with dose below 4 mGy, which correlated with the work of Sykes et al [2]. A similar dose dependence could not be investigated in the OBI results, because scans below 4 mGy could not be obtained.

The TRE_{30} in the clinical range was similar between the two imaging systems (Figure 2b). There was little difference in the range of TRE_{30} results across image sets except for XVI scans with dose less than 4 mGy where values increase. Using a pass/fail tolerance for any given automatic registration of $TRE_{30} = 3.6$mm, failures of the algorithm occurred in 5 - 9% of registrations except for the lowest dose XVI scan (31%). This indicates the algorithms can provide clinically suitable registrations more than 90% of the time, and also agrees with other results in the literature.

4. Conclusions

The uncertainty in automatic image registration with both OBI and XVI images was found to be adequate for clinical use, and similar between the systems, within the normal range of acquisition settings.

Acknowledgements

The authors would like to thank Adrian Bailey, Radiation Oncology Orange Base Hospital NSW, Australia for the loan of the phantom used in this study.

References

[1] Sykes J R, Brettle D S, Magee D R and Thwaites D I 2009 Investigation of uncertainties in image registration of cone beam CT to CT on an image-guided radiotherapy system. *Phys. Med. Biol.* **54** 7263–83

[2] Sykes J R 2010 Quantification of geometric uncertainties in image guided radiotherapy (University of Leeds) http://etheses.whiterose.ac.uk/3864/

[3] Fox T, Huntzinger C, Johnstone P, Ogunleye T and Elder E 2006 Performance evaluation of an automated image registration algorithm using an integrated kilovoltage imaging and guidance system. *J. App. Clin. Med. Phys.* **7** 97–104

[4] Grams M P, Brown L C, Brinkmann D H, Pafundi D H, Mundy D W, Garces Y I, Park S S, Olivier K R and De los Santos L E F 2013 Analysis of automatic match results for cone-beam computed tomography localization of conventionally fractionated lung tumors *Prac. Radiat. Oncol.* (Article in Press DOI: 10.1016/j.prro.2013.02.008)

[5] Sykes J *et al.* 2010 *CEP10071: X-ray tomographic image guidance radiotherapy systems* http://nhscep.useconnect.co.uk/CEPProducts/Catalogue.aspx?reportId=450

[6] Smitsmans M H P, Wolthaus J W H, Artignan X, De Bois J, Jaffray D a, Lebesque J V and Van Herk M 2004 Automatic localization of the prostate for on-line or off-line image-guided radiotherapy. *Int. J. Radiat. Oncol. Biol. Phys.* **60** 623–35

Evaluation of the ability of a 2D ionisation chamber array and an EPID to detect systematic delivery errors in IMRT plans

Omemh Bawazeer[1], Alison Gray[2], Sankar Arumugam[2], Philip Vial[1, 2], David Thwaites[1], Joseph Descallar[2] and Lois Holloway[1, 2, 3, 4]

[1] Institute of Medical Physics, School of Physics, University of Sydney, NSW Australia
[2] Liverpool and Macarthur Cancer Therapy Center and Ingham Institute, NSW Australia
[3] Centre for Medical Radiation Physics, University of Wollongong, NSW Australia
[4] South West Clinical School, School of Medicine, University of New South Wales, Sydney, NSW Australia.

Obaw8273@uni.sydney.edu.au

Abstract. Two clinical intensity modulated radiotherapy plans were selected. Eleven plan variations were created with systematic errors introduced: Multi-Leaf Collimator (MLC) positional errors with all leaf pairs shifted in the same or the opposite direction, and collimator rotation offsets. Plans were measured using an Electronic Portal Imaging Device (EPID) and an ionisation chamber array. The plans were evaluated using gamma analysis with different criteria. The gamma pass rates remained around 95% or higher for most cases with MLC positional errors of 1 mm and 2 mm with 3%/3mm criteria. The ability of both devices to detect delivery errors was similar.

1. Introduction

Intensity modulated radiotherapy (IMRT) utilises complex motion of the Multi-Leaf Collimator (MLC) to achieve highly conformal dose distributions. The position and movement of the leaves as well as other delivery parameters are susceptible to errors in IMRT delivery [1] so pre-treatment patient quality assurance (QA) is recommended. This QA is commonly performed with 2D or 3D dosimeters and the gamma evaluation technique which combines both dose difference and distance to agreement criteria and has become the current standard to provide a quantitative comparison [2]. However, there is limited evidence in the literature about the ability of such dosimeter systems to detect errors in IMRT. For example, pass rates of ArcCHECK measurements made with deliberately introduced systematic gap width errors of up to 2 mm were reported to be similar to error-free plans, using a 3%/3mm or 3%/2mm gamma criteria [3], whilst the MapCHECK device was reported to be more sensitive to MLC positioning errors than radiochromic film [4].

Recent recommendations indicate that each department should investigate the ability of their QA processes to detect errors [5]. The aim of this work was to investigate the ability of an EPID and a 2D ionisation chamber array to detect systematic MLC leaf position and collimator errors. A number of hypotheses were proposed to allow the evaluation of gamma pass rate results. The first hypothesis was that the detector systems can detect the smallest clinically significant error. For this to be true the gamma analysis pass rate will reduce by a statistically significant value when these errors are introduced. Secondly, that the sensitivity to errors varies between detector systems. For this to be true the slope describing the reduction in pass rate with introduced error will be steeper for one of the detector systems. Thirdly, that a detector system becomes more sensitive to errors when the gamma tolerances are tightened. For this to be true the slope describing the reduction in pass rate with error will be steeper when tighter gamma criteria are used for the analysis.

2. Methods

Two clinical step-and-shoot IMRT plans, one prostate (PR) and one head and neck (H&N) plan, were selected. Both plans utilised 7 fields, with an average number of segments per field of 7.1 and 13.7 for the PR and H&N plans respectively. The plans were created using the Pinnacle[3] treatment planning system (TPS), V 9.0 (Philips Healthcare, USA) and calculated with a grid size of 0.25 cm^3. An in-house computer programme written in Python was used to introduce errors to all segments for each field for both plans. Two MLC position errors were introduced. The first of these shifted all leaf pairs in opposite directions with offsets of 1, 2, and 3 mm applied, resulting in a larger leaf gap.The second error introduced shifted all leaf pairs with offsets of 1, 2, 3, and 4 mm in the same direction, resulting in the field aperture remaining the same size but being shifted in the X direction.

The plans were copied and modified in the MOSAIQ record and verify system, V2.30.04D1 (Elekta Ltd, UK), to allow field by field measurements with the gantry angle set to zero degrees. Collimator errors were also introduced with the initial collimator angle position of all beams increased by 1°, 2°, 3° and 5° respectively in the clockwise direction by manually editing the plan in the record and verify system. The original and edited plans were delivered using a Synergy linear accelerator (linac) (Elekta Ltd, UK). The delivered dose was measured with the EPID attached to the linac and a 2D ionisation array, MatriXX Evolution used with OmniPro-I'mRT V1.4.0.1 (IBA Dosimetry, Germany).

The measured dose for each field was compared to the calculated dose for the 'no-error' plan, using gamma analysis with 3%/3mm, 3%/2mm, and 2%/2mm criteria with a 10% threshold using OmniPro-I'mRT. It is assumed that the no-error plan was delivered without any errors. To allow import into OmniPro-I'mRT, the EPID images were converted to dose via a pixel calibration factor, which links the measured pixel value (under reference EPID conditions) of each element detector to the dose in water (under reference IMRT QA conditions). This correction was applied using an in-house program written in Matlab (Version 7.11.0.584 R20106). A renormalisation was applied to both calculated and measured (EPID, MatriXX) dose distribution to allow a global percentage difference (relative to the maximum dose in the field) to be used in the gamma analysis. The maximum dose value of the TPS calculated dose distribution was determined and then the inverse of this value was applied to result in a situation where 100% value was equivalent to the maximum dose in the TPS dose distribution. The renormalisation value applied to the TPS dose distribution was then applied to the measured dose distribution. A correction for the linac output factor at the time of measurement was applied to the measured data if its value was greater than 1%. The grid size of the raw MatriXX data was also converted to match that of the TPS calculated data using a cubic spline interpolation, due to its inherent resolution of 7.6 mm. OmniPro-I'mRT provides ability to shift the measured and planned dose distributions relative to each other prior to the analysis. This feature may be used to correct for setup errors or uncertainties. For the purpose of reproducibility this facility was not utilised in the methodology of the current study. The assumption was made that the detector was in all cases aligned correctly so the central pixel of the image created from the measurement was aligned automatically with the central pixel of the image created from the planning system when the data was imported into OmniPro-I'mRT.

To evaluate the first hypothesis, a Z-test was used to determine if for the given detector, a different pass rate was observed between the no-error plan and a plan with the smallest clinically significant error introduced. The smallest clinically significant error was defined as 1 mm for MLC position and 2° for collimator rotation, as errors of this magnitude are reported in the literature to have a clinically significant impact on IMRT plans [6, 7]. The p-value represents the probability that the null hypothesis is correct. If a resulting value was less than 0.05, then there is a difference between the pass rate for the no-error plan and the pass rate for the plans with errors introduced. For the second and third hypotheses, a multiple linear regression test was used to compare the sensitivity of each detector for a given error type and plan. Detector, error size (magnitudes of error type) and the interaction between detector and error size were used as variables in each of the multiple linear regression models. Also, the 0.05 p-value was used to represent if the null hypothesis is correct. If a resulting value was less than 0.05, then the slopes of the gamma pass rate vs error size between the detectors were considered to be different.

To determine the magnitude of the uncertainty, the prostate plan was measured three times for each detector on the same day to assess the short term reproducibility, and the plan was delivered on three different following weeks to evaluate the longer term reproducibility. The short and longer term reproducibility values were then combined [8].

3. Results

For the no-error plans, the gamma pass rates were 99.3% and 94.4% for the prostate plan and 91.9% and 93.3% for the H&N plan using EPI and MatriXX respectively. The results for the no-error plan and for the plans with MLC position and collimator rotation errors introduced are presented in Figure 1 and Figure 2. In order to allow comparison between detectors, the plotted values represent the average pass rate of all the fields in the plan. The error bars represent the one standard deviation, uncertainty values, for each device, which were ±0.67 and ±1.33 for EPID and MatriXX respectively. Pass rates were expected to reduce as the magnitude of the errors increased. However, pass rates increased when 1mm and 2 mm MLC position errors were introduced.

The expected trend of pass rates reducing with increasing errors was seen with the collimator errors for both plans, e.g. the pass rate dropped from 90.7% to 68.6% when errors were introduced of 1° and 5° in the head and neck plan using the EPID for measurement. Similarly, the pass rate using the MatriXX reduced from 91.2% to 62.5%.

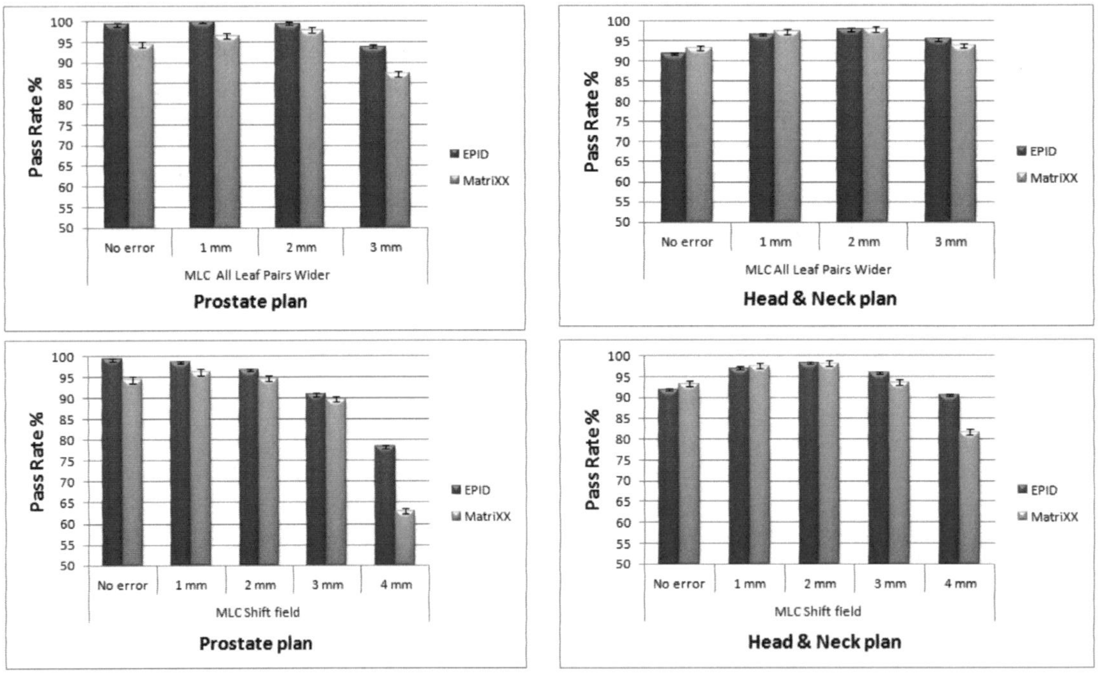

Figure 1: The gamma analysis pass rate measured using EPID and MatriXX detectors with 3%/3 mm criteria for prostate and head and neck IMRT plans without errors and with introduced MLC position errors.

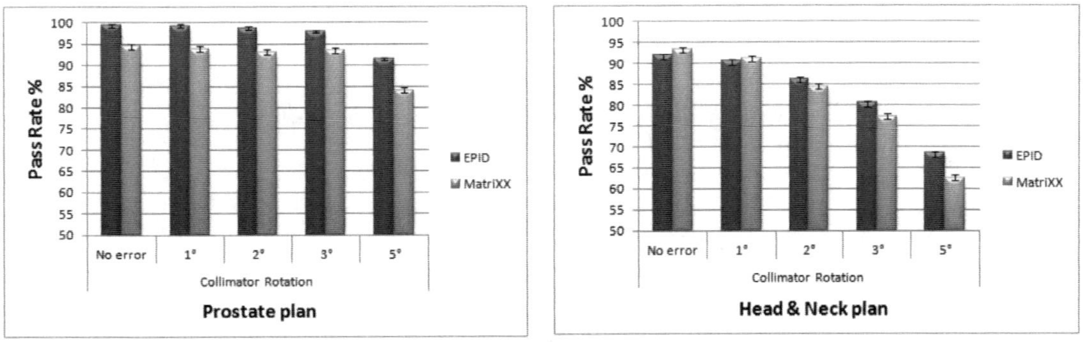

Figure 2: The gamma analysis pass rate measured using EPID and MatriXX detectors with a 3%/3mm criteria for prostate and head and neck IMRT plans without errors and with introduced collimator rotation errors.

The results for the first hypothesis are presented in Table 1. These results indicate that both detector systems had a similar lack of ability to detect the smallest clinically significant errors. Neither detector system considered was able to detect 1 mm MLC position errors or a 2° collimator error for the prostate plan. However, both detector systems were able to detect the 2° collimator error for the H&N plan.

The results for the second hypothesis are presented in Table 2. All of the comparisons produced a value greater than 0.05 which indicates that both detector systems (using the 3%/3mm gamma evaluation criteria) had a similar sensitivity for each plan and all error types with the exception of the collimator rotation errors in the H&N plan, in this instance the MatriXX was more sensitive for this type of error compared to the EPID.

The results for the third hypothesis are presented in Table 3. All of the results produced a value greater than 0.05, indicating that the rate of reduction in pass rates with increasing error magnitude did not change when the gamma criteria was tightened, i.e. the slopes were similar.

Table 1: The p-values from the Z-test to evaluate the ability of detector systems to detect the smallest clinically significant errors (1 mm or 2°) introduced, using 3%/3mm gamma evaluation criteria.

Plan	Error Type	EPID	MatriXX
PR	MLC all leaf pairs wider	0.822	0.938
PR	MLC shift field	0.158	0.916
PR	Collimator rotation	0.201	0.156
H&N	MLC all leaf pairs wider	1.000	0.998
H&N	MLC shift field	1.000	0.999
H&N	Collimator rotation	0.000	0.000

Table 2: The p-values from the linear regression test to evaluate the significance of differences between the detector systems (EPID-MatriXX), using 3%/3mm gamma evaluation criteria.

Plan	MLC all leaf pairs wider	MLC shift field	Collimator rotation
PR	0.878	0.307	0.699
H&N	0.723	0.350	0.005

Table 3: The p-values from the linear regression test to evaluate the significance of the difference between different gamma criteria for a given detector.

	EPID			MatriXX		
Plan	MLC all leaf pairs wider	MLC shift field	Collimator rotation	MLC all leaf pairs wider	MLC shift field	Collimator rotation
PR	0.869	0.286	0.442	0.938	0.937	0.830
H&N	0.356	0.756	0.306	0.878	0.232	0.757

4. Discussion

In general, the pass rates were higher for the prostate plan compared to the head and neck plan. This was attributed to the increased modulation and irregular field shapes in the head and neck plan. In addition, due to the large field sizes in the head and neck plan, for some fields part of the beam extended outside the detector area, resulting in missing data. In cases where the introduced MLC shift moved the delivered field further away from the detector area, a reduction in gamma pass rates would be expected. However, the magnitude of the effect of the missing data on the gamma analysis pass rates was not investigated in this study. Based on these results, the choice of acceptable pass rate may need to be reduced to less than 95% for complex plans. Although it was assumed for the purpose of this study that the (no-error) plans were delivered accurately in reality all measurements were subject to delivery uncertainties. To assess the increased pass rate when small MLC position errors were introduced, the monthly QA results were reviewed. The MLC position accuracy was found to be within the 1 mm tolerance, however, the field size was around 0.5 mm smaller than expected for each MLC bank. This may have contributed to the increased pass rates observed when a 1 mm error was introduced, however does not explain the pass rate increases when a 2 mm error was introduced. Further investigation to identify the cause of this result is planned. For the collimator rotation errors, the prostate plan was less sensitive than the head and neck plan, which was attributed to the relatively small field sizes used for the prostate plan while the head and neck plan has elongated fields.

The results indicate that none of the detector systems have the ability to consistently detect a 1 mm MLC position error when using the gamma analysis method. A 3%/3mm gamma criteria as presented here for hypothesis 1 and 2 is large when aiming to detect a 1 mm MLC position error. However, the analysis comparing tighter gamma criteria (from 3%/3mm to 3%/2mm to 2%/2mm) demonstrated reduced pass rates (for error-introduced, but also for no-error plan deliveries), but no significant differences in the change in pass rate with increasing error magnitude. Considering that the tightest gamma criteria used was 2%/2mm it was not expected that a 1 mm error would be identified. It may be possible to identify 1 mm errors using alternative methods such as leaf end detection or pixel intensity deviation or by using a tighter distance to agreement criterion in the gamma analysis.

5. Conclusion

Neither the EPID nor the 2D ion chamber array investigated in this study when utilised with a gamma analysis criteria between 2%/2mm and 3%/3mm were able to detect clinically significant MLC position errors of 1 mm, whereas both were able to detect 2° rotation errors for a head and neck plan. The sensitivity of both devices was similar in most cases investigated. Moreover, no significant improvement in the ability of the devices to detect errors was observed when changing gamma criteria as the pass rates reduced in a consistent manner for no-error and for error-introduced plan deliveries as the criteria became tighter. This preliminary work has highlighted the need to establish detector systems and assessment criteria that can detect clinically significant errors as well as the need to adapt QA systems based on particular treatment plan types.

References

[1] Mu G. et al., 2008, Phys. Med. Biol. **53** p. 77-88.
[2] Low D. et al., 2011, Med. Phys. **38** p. 1313-38.
[3] García-Vicente F. et al., 2011, J. Appl. Clin. Med. Phys. **13** p. 111-23.
[4] Yan G. et al., 2009, J. Appl. Clin. Med. Phys. **10** p. 120-8.
[5] Moran J. et al., 2011, Pract. Radiat. Oncol. **1** p. 190-5.
[6] Sastre-Padro M. et al., 2007, Phys. Med. Biol. **52** p. 1147-56.
[7] Low D. et al., 1997, Med. Phys. **24** p. 1123-39.
[8] Gregory K. et al., 2005, Australas. Phys. Eng. Sci. Med. **28** p. 131-9.

XVII International Conference on the Use of Computers in Radiation Therapy (ICCR 2013) IOP Publishing
Journal of Physics: Conference Series **489** (2014) 012072 doi:10.1088/1742-6596/489/1/012072

Impact of Image Noise on Gamma Index Calculation

M Chen, X Mo, D Parnell, G Olivera, D Galmarini, and W Lu

21st Century Oncology, USA

mchen@rtsx.com

Abstract. Purpose: The Gamma Index defines an *asymmetric* metric between the evaluated image and the reference image. It provides a quantitative comparison that can be used to indicate sample-wised pass/fail on the agreement of the two images. The Gamma passing/failing rate has become an important clinical evaluation tool. However, the presence of noise in the evaluated and/or reference images may change the Gamma Index, hence the passing/failing rate, and further, clinical decisions. In this work, we systematically studied the impact of the image noise on the Gamma Index calculation. Methods: We used both analytic formulation and numerical calculations in our study. The numerical calculations included simulations and clinical images. Three different noise scenarios were studied in simulations: noise in reference images only, in evaluated images only, and in both. Both white and spatially correlated noises of various magnitudes were simulated. For clinical images of various noise levels, the Gamma Index of measurement against calculation, calculation against measurement, and measurement against measurement, were evaluated. Results: Numerical calculations for both the simulation and clinical data agreed with the analytic formulations, and the clinical data agreed with the simulations. For the Gamma Index of measurement against calculation, its distribution has an increased mean and an increased standard deviation as the noise increases. On the contrary, for the Gamma index of calculation against measurement, its distribution has a decreased mean and stabilized standard deviation as the noise increases. White noise has greater impact on the Gamma Index than spatially correlated noise. Conclusions: The noise has significant impact on the Gamma Index calculation and the impact is asymmetric. The Gamma Index should be reported along with the noise levels in both reference and evaluated images. Reporting of the Gamma Index with switched roles of the images as reference and evaluated images or some composite metrics would be a good practice.

1. Introduction

The Gamma Index defines an asymmetric metric between two images for quantitative image comparison [1, 2]. The dosimetric tolerances, e.g. the dose difference (DD) of 3% and the distance-to-agreement (DTA) of 3 mm, are formulated as weighting factors in the metric, so that the Gamma Index can be used as an indicator for dosimetric pass (<=1) or fail (>1). However, image noise can change the Gamma Index value and thus change the passing/failing rate, and further, clinical decision. For example, the flat panel imager (portal imager) has a certain level of electronic noise. By using different image resolutions, it results in different noise statistics and different Gamma passing rates.

The noise effect in the Gamma Index was studied in [3, 4] for simulated cases and Monte Carlo (MC) based dose calculation, respectively. During the conference, we were informed of the study [3], which was similar to ours but with different emphasis: their emphasis was on the MC dose calculation, while ours was more on the portal comparisons. Due to the asymmetry of the Gamma Index, noise in the two images, the evaluated (searching) image and the reference (searched) image, could cause opposite effects. Here we refer to the image in which the samples are looking for the closest match, as the evaluated (searching) image, and the other image, in which the samples are being searched, as the reference (searched) image. Intuitively, the presence of noise in the reference image makes the search more likely to find a good match between the two compared images, so that the Gamma Index becomes lower and the passing rate higher. On the other hand, assuming the difference of the evaluated and the reference image is caused by noise, the presence of noise in the evaluated image moves the searching samples away from the reference and makes it harder to find a good match, so that the Gamma Index becomes higher and the passing rate lower. For simple cases, such as a constant reference image, we could establish analytic formulation to describe the statistics of the Gamma Index, assuming a certain statistics of noise. For other cases, the analytic formulations become very complicated, so numerical simulations were used instead to obtain the statistics of the Gamma Index. We also look at the effect of noise in portal image comparison.

2. Methods

The Gamma Index of an image g against another image f is defined as

$$\Gamma\left(g(x),f\right)=\min_{y}\left(\sqrt{\left(\frac{f(y)-g(x)}{D}\right)^2+\left(\frac{y-x}{d}\right)^2}\right) \tag{1}$$

where D and d are DD and DTA criteria, respectively.

For simple distributions, such as:

$$f(x)=a \tag{2}$$

and

$$g(x)=b+n(x) \tag{3}$$

where a and b are constant, and $n(x)$ is the noise at x depicted by a Gaussian distribution, the Gamma Index of $f(x)$ against g can be calculated as

$$\Gamma\left(f(x),g\right)=\min_{y}\left(\sqrt{\left(\frac{b-a+n(y)}{D}\right)^2+\left(\frac{x-y}{d}\right)^2}\right) \tag{4}$$

and the Gamma Index of $g(x)$ against f can be calculated as

$$\Gamma\left(g(x),f\right)=\min_{y}\left(\sqrt{\left(\frac{b-a+n(x)}{D}\right)^2+\left(\frac{x-y}{d}\right)^2}\right)=\left|\frac{b-a+n(x)}{D}\right| \tag{5}$$

Note that the difference $b-a$ can be absorbed into the mean μ of noise $n(x)$, i.e. $b-a$ can be added into μ. Therefore, without loss of generality, we can assume $b=a$. If the noise $n(x)$ has a mean μ and standard deviation σ, then it can be derived that $\Gamma(g(x),f)$ has a mean of

$$E\left(\left|\frac{X}{D}\right|\right)=\frac{\mu}{D}\cdot\mathbf{erf}\left(\frac{\mu}{\sqrt{2}\sigma}\right)+\sqrt{\frac{2}{\pi}}\frac{\sigma}{D}e^{\frac{-\mu^2}{2\sigma^2}} \tag{6}$$

where X denotes a Gaussian random variable of mean μ and standard deviation σ and E denotes the expectation. On the other hand, the statistics for $\Gamma\left(f(x),g\right)$ are much more complicated to derive. Therefore, we only simulated its statistics by varying the mean and standard deviation.

For simulations, we defined the image $f(x,y)=100$, for $|x|,|y|\leq 5$ cm, and 0 otherwise, in a 20×20 cm^2 field, and the image $g(x,y)=f(x,y)+n(x,y)$, where $n\sim N(\mu,\sigma)$ is Gaussian distributed. We simulated various μ and σ and summarized the results of the Gamma Index in terms of its mean and passing rate. We also simulated the impact of correlated noise, which were generated by convolving the noise n with local averaging kernels of various sizes.

For real data, we compared portal images of the same treatment beam but in different fractions and compared the measured portals with the corresponding calculated portals. Reduction of noise standard deviation (STD) σ was effected by local averaging, and σ was estimated using panel data outside of the treatment field. We then calculated the Gamma Index passing rate with respect to the noise STD. We also calculated DD and plotted its passing rate, based on the 3% criteria, along with the Gamma passing rate.

3. Results

In Figure 1, we show the mean of the Gamma Index calculated using the formula (6) for a noisy evaluated image against a noise free reference image. The mean is displayed as the image intensity and plotted against the noise mean μ (Y-axis) and STD σ (X-axis).

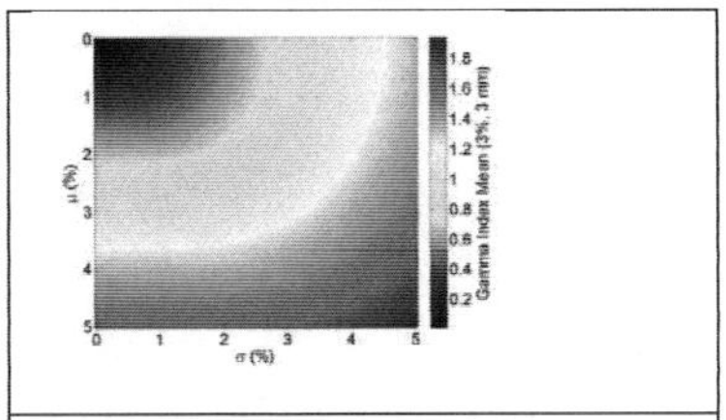
Figure 1 The mean of the Gamma Index calculated using the formula (6).

Figure 2 are the results of numerical simulations for the Gamma Index mean and the passing rate of Γ(*f*(*x*),*g*) and Γ(*g*(*x*),*f*). Because of the symmetry with respect to $\mu=0$, we only plot the results for $\mu\geq 0$. Note that the result of simulation (Figure 2 (a)) agrees with that of calculation based on the formula (Figure 1). We can see that the mean of the Gamma Index increases and the Gamma passing rate decreases, as the mean and the STD of noise in the searching image increase (Figure 2 (a)), and the mean of the Gamma Index decreases, as the STD of noise in the searched image increases.

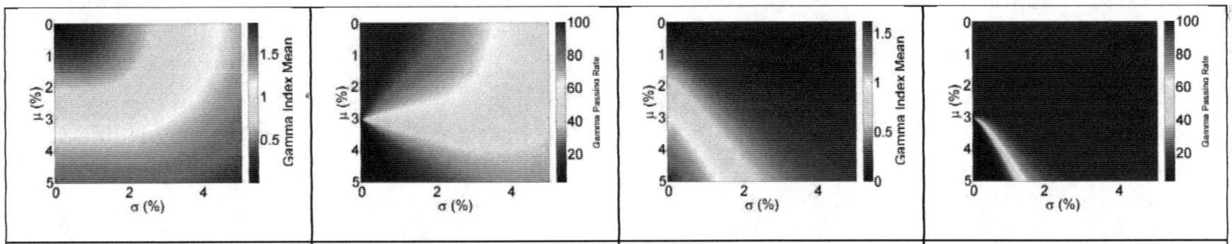

(a) mean of Γ(*g*(*x*),*f*)	(b) pass rate of Γ(*g*(*x*),*f*)	(c) mean of Γ(*f*(*x*),*g*)	(d) pass rate of Γ(*f*(*x*),*g*)

Figure 2 Numerical simulations. The Gamma Index was calculated based on 3%, 3 mm criteria. (a) The mean of Γ(*g*(*x*),*f*). (b) The passing rate of Γ(*g*(*x*),*f*). (c) The mean of Γ(*f*(*x*),*g*). (d) The passing rate of Γ(*f*(*x*),*g*).

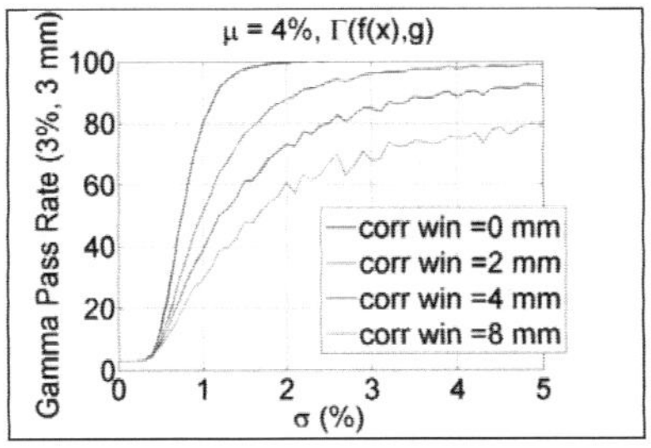

Figure 3 Impact of noise correlation. The legend describes the window sizes of the convolution kernels.

Figure 3 indicates that the noise correlation increase the mean of the Gamma Index and thus decreases the passing rate. The 0 mm correlation window indicates white noise. In Figure 4 and Figure 5, we show the results of portal measurement to measurement and measurement to calculation, respectively. Figure 4 shows (a) the reference portal, (b) the difference of two portals of the same beam in different fractions, and (c) the Gamma and DD passing rate with respect to the noise STD. Figure 5 (a) shows the calculated portal for the same delivery setup of the measurement in Figure 4 (a). It can be seen that the impact of noise for clinical data is similar to that for simulations: noise in the reference image increases the Gamma passing rate, as shown in Figure 4 (c) and Figure 5 (c), and noise in the evaluated image decreases the Gamma passing rate, as shown in Figure 5 (b). These results agreed with intuition.

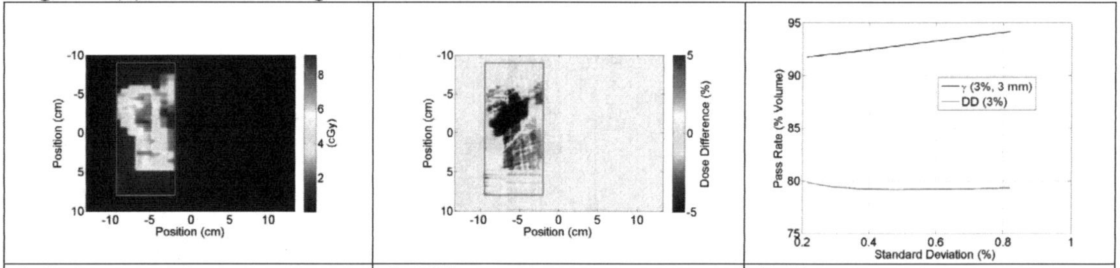

(a) Reference portal | (b) Difference of two measurements. | (c) The gamma and DD pass rate.

Figure 4 Clinical data. (a) The reference portal image. (b) The difference image (test-ref). (c) The Gamma passing rate and dose difference passing rate against noise STD σ.

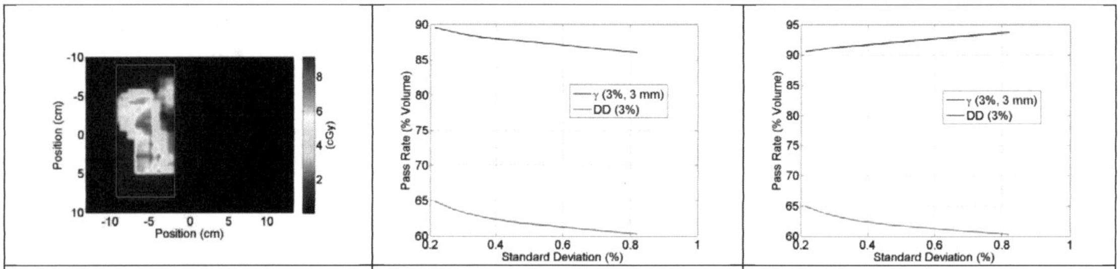

(a) Calculated portal. | (b) Meas. against calc. | (c) Calc. against meas.

Figure 5 Measurement compared with calculation. (a) Calculated portal. The Gamma passing rate and DD passing rate w.r.t. noise STD for (b) calculation against measurement and (c) measurement against calculation.

4. Conclusions

Noise could have significant impact on the Gamma Index calculation. For example, the impact of noise can be seen when using different image resolution: lowering image resolution reduces noise and thus results in an increased Gamma Index and a decreased Gamma Index passing rate. Hence, it is necessary to include the STD of image noise when reporting the Gamma Index, as it can be referenced as uncertainty for the Gamma Index evaluation.

References

[1] Low, D. A., Harms, W. B., Mutic, S. and Purdy, J. A. A technique for the quantitative evaluation of dose distributions 1998 *Med Phys* **25** 5 656-61

[2] Chen, M., Lu, W., Chen, Q., Ruchala, K. and Olivera, G. Efficient gamma index calculation using fast euclidean distance transform 2009 *Phys Med Biol* **54** 7 2037-47

[3] Graves, Y. J., Jia, X. and Jiang, S. B. Effect of statistical fluctuation in monte carlo based photon beam dose calculation on gamma index evaluation 2013 *Phys Med Biol* **58** 6 1839-53

[4] Low, D. A. and Dempsey, J. F. Evaluation of the gamma dose distribution comparison method 2003 *Med Phys* **30** 9 2455-64

Assessing local outcomes in heterogeneous gliomas

Nicholas Dowson[1], Paul Thomas[2,3], Yaniv Gal[3], Michael Fay[2,3,4], Rosalind L. Jeffree[2], Craig Winter[2], Alan Coulthard[2,3], Jye Smith[2,3], Pierrick Bourgeat[1], Olivier Salvado[1,3], Stuart Crozier[3], Stephen Rose[1,3]
[1]Australian e-Health Research Centre (CSIRO), Level 5, UQ Health Sciences Building, Royal Brisbane and Women's Hospital, Herston, QLD 4053, Australia.
[2]Departments of Nuclear Medicine, Radiology, Radiation Oncology, and Neurosurgery, Royal Brisbane and Women's Hospital, Herston, QLD 4053, Australia.
[3]School of IT & Elec. Eng. University of Queensland, St Lucia, QLD, 4072, Australia.
[4]Eberhard Karls Universität Tübingen, Germany (Note: ND & YG contributed equally)
Email: nicholas.dowson@csiro.au

Abstract. Tumours are known to be heterogeneous, yet typical treatment plans consider them as a single unit. This may influence treatment outcomes. However, treatment cannot be customised to intra-tumour variation without a method to establish outcomes at an intra-tumour scale. This work proposes a method to both assess and measure outcomes locally within tumours. **Methods**: Four patients were scanned at two post-surgery time points using contrast enhanced MRI and 3,4-dihydroxy-6-[18F]-fluoro-L-phenylalanine (18F-DOPA) PET. The shell of active tumour tissue is divided into a set of small subregions at both time points. Local outcome is measured from changes in subregion volume over time. The utility of the proposed approach is evaluated by measuring the correlation between PET uptake and documented growth. Correlation with overall survival time was also examined. **Results**: Local outcomes were heterogeneous and evidence of a positive correlation between local 18F-DOPA uptake and local progression was observed. **Conclusions**: Given that intra-tumour outcomes are heterogeneous the consistently positive correlation between FDOPA uptake and progression, local analysis of tumours could prove useful for treatment planning.

1. Introduction

It is known that even within individual tumours there is variation in biological behaviour [1]. Individual cell lines are highly adapted to their immediate environment, e.g. tumour cells in a hypoxic environment have different adaptations to those in normoxic conditions [2]. Tumour heterogeneity is important and is a determinant of patient outcome. However, most imaging studies of malignancy and therapeutic monitoring appear to examine tumours as a whole [3], [4], [5], [6] possibly because of the difficulty of establishing outcomes at an intra-tumour scale. The issue of heterogeneity is particularly appropriate to advanced glioma, which is an aggressive malignancy to which most patients succumb in less than a year [7]. Treatment involves maximal, safe surgical resection to debulk the tumour followed by radiotherapy and adjuvant chemotherapy [8] to control residual tumour. Even so, tumours typically, recur within months, because chemo-therapy and radiotherapy are limited by toxicity to normal tissues [8]; the blood brain barrier can limit access by chemotherapy agents[9]; and the tumour is microscopically infiltrative, which may result in geographic misses during radiotherapy.

This work attempts to address the above issues by introducing a polar spatial model for tumours similar to the polar map images used for myocardial perfusion imaging [10], to allow measurement of biological heterogeneity and to provide a method for comparison of serial follow up measurements.

The method is applied to clinical examples of PET scans in patients with advanced glioma, by using it to examine the predictive capabilities of PET one month after a course of radio- and chemo-therapy.

2. Materials and Methods

Four patients were scanned at two post-surgery time-points using contrast enhanced MRI and 3,4-dihydroxy-6-[18F]=fluoro-L-phenylalanine (FDOPA) PET. Baseline scan was 4 weeks after a 6-week course of chemo-radio-therapy (12-weeks post-surgery). Follow up was 6 months later post-surgery. Scans at each time point were taken within a two day window. The study was approved by the Royal Brisbane and Women's Hospital ethics committee. Informed consent was provided for study inclusion.

MRI data was acquired using a 3T Siemens Trio employing a Magnetization-prepared Rapid Acquisition Gradient-echo (MPRAGE) sequence with the following parameters (FOV 24x25.6x17.6 cm, TR/TE/TI 2300/2.26/900 ms, flip angle of 90, 1 mm isotropic resolution). Images were acquired before and after administration of contrast agent Gadovist, (Bayer HealthCare Pharmaceuticals). FDOPA was synthesised according to the method in [11]. An FDOPA activity of 151MBq on average was administered intravenously (range 138 to 164MBq). PET imaging was performed using a Philips Gemini GXL scanner. A transmission CT scan was acquired first. Subsequently, FDOPA was administered intravenously and a 75-minute acquisition initiated. The images were reconstructed using ordered subset expectation maximisation with corrections for attenuation and scatter. The final volume has a matrix size of 128x128, consisting of 90 planes of 2x2x2mm3 voxels.

The follow-up MRI was registered to baseline using standard multi-scale mutual information rigid registration [12], and the remaining images within each time-point were registered to their respective pre-contrast MRI. Afterwards, a standard non-rigid registration algorithm [13] was applied at a coarse level (40mm node spacing) to account for brain shift.

An experienced nuclear medicine physician (PT) manually contoured each tumour at each time point using fused FDOPA PET and MR images, to define the outer extent of the tumour. The contour of the tumour cavity was obtained by thresholding away the central regions below the median PET intensity on the outer surface. A tumour centre point was defined by calculating the centre of gravity in the baseline study. The tumour centre was reused in the follow-up images. Rays corresponding to each combination of longitude and latitude are projected from the tumour centre to its outer extent, and used to divide the tumour up into a set of small subregions at both timepoints as shown in Fig. 1.

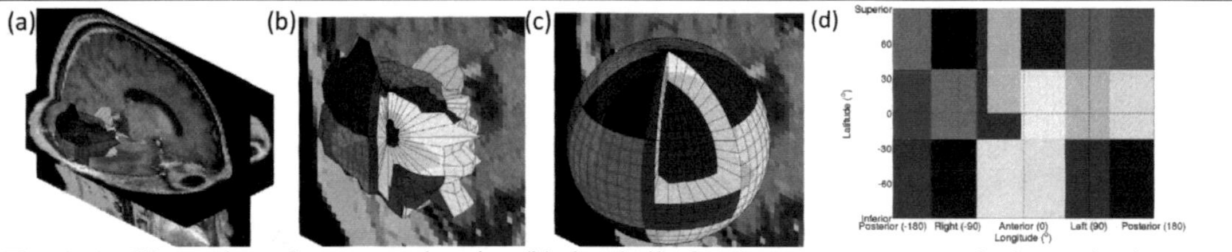

Fig. 1: (a) Illustration of a tumour showing (b) the shell of active tumour surrounds the excised core shown in cutaway. The shell is split into segments (coloured quadrilaterals), which may be projected onto a (c) sphere or (d) Mercator projected surface, to visualize statistics within each segment.

Local outcome is measured in terms of the changes in the volume of each subregion between the baseline and follow-up images. Volume changes are reported in percentage terms relative to the regional volume at baseline. The mean PET uptake within each subregion was also computed. In order to make inter-patient comparisons, PET uptake was measured as Standardized Uptake Value (SUV) ratio, where all PET intensities are divided by the mean PET intensity within the ipsilateral cerebellum. A value of one indicates uptake identical to normal tissue.

3. Results and discussion

Several angular resolutions were applied to find the optimal number of latitudes and longitudes using Receiver Operator Curve (ROC) analysis. The Area of each Curve (AoC) was measured and the results are shown in Fig. 2a. Three latitudes and six longitudes gave the largest area of curve of 0.83. Variation in performance occurs because errors arise from grouping heterogeneous biological regions. Also, changes in both metabolism and location occur, which registration cannot disambiguate. A trade-off between these errors exists: larger segments risk grouping more heterogeneous tissues but reduce correspondence error, and *vice-versa*. Unbiased segment correspondences rely on movements in the tumour centre point, caused by brain shift, remaining small (below 10% of diameter). Although such shifts did not occur in this dataset, this should be checked for and corrected if necessary.

Cox hazard analysis was used to examine correlation between survival time, max segment intensity, and whole-tumour mean for the four patients. Despite the low sample size, an r^2 of 0.23 was obtained (versus 0.59 for the whole-tumour mean). More samples are needed to properly evaluate segments' correlation with survival.

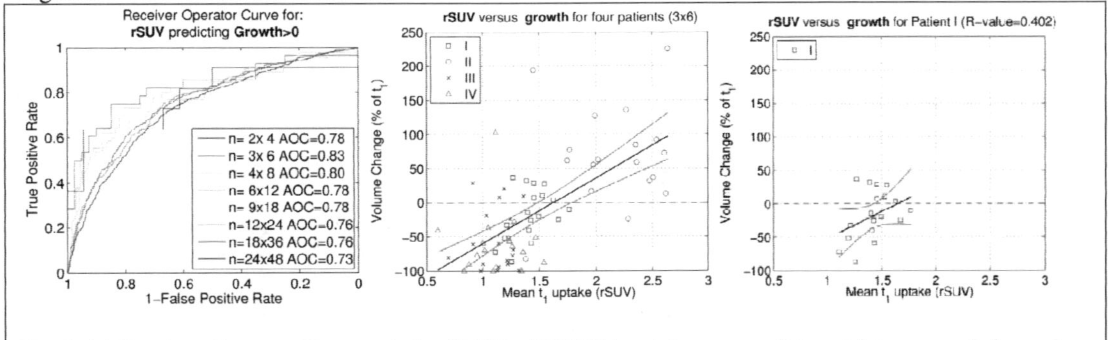

Fig. 2. (a) Receiver Operator Characteristics (ROC) of FDOPA uptake as a predictor of future growth for various angular resolutions of the spatial models, using all patient data. The Area of Curve (AOC) of each curve is also shown. The best result is obtained for 3 latitudes and 6 longitudes.
(b) Scatter plot of FDOPA uptake versus growth for all patients, with linear regression. A growth of: 0% implies no change in volume, -100% implies local disappearance of tumour and +100% implies a doubling in size. Volumes at baseline were respectively 54, 89, 28, and 44ml and at follow-up were 33, 120, 12 and 19ml.
(c) Scatter plot of patient I with locally heterogeneous outcomes. The positive correlation between PET uptake and growth holds, again with a significant R-value of 0.4.

The linear regression applied to the SUVR and growth values of the four patients is shown in Fig. 2b. The Pearson correlation coefficient (or R-value), was 0.656, indicating that uptake is a predictor of future growth. The 95% confidence interval was also relatively narrow. Inter-patient variation had substantial influence on the results. Patients III and IV had lower uptake and general shrinkage. Considering patients individually, positive correlations between SUVR and growth occurred in all cases with respective R-values for patients I to IV of 0.402, 0.135, 0.048 and 0.016. The correlation indicates significant correspondence between FDOPA uptake and future growth. This is a useful result, because it means FDOPA could potentially be used as an early marker of treatment response. In the remaining patient with a mixture of growth and shrinkage, the correlation between uptake and volume change was more significant, implying that local analyses are not only possible but useful, and provide an advantage over analyses of tumours as a whole.

Although positive correlations between uptake and local volume changes were found, a limitation of this study is the low number of patients. In particular, more patients with heterogeneous growth are needed. Even so, the positive correlations in patients with relatively uniform behaviour globally imply that local analysis has potential utility, because it allows a better interrogation of the data, with reduced tissue heterogeneity within each segment relative to the tumour as a whole.

4. Conclusion

A method for dividing tumours into subregions and evaluating outcomes locally in longitudinal studies has been introduced. This may be useful in removing some of the confounding effects of tumour heterogeneity. It may also facilitate research into risk adapted local therapies (such as radiotherapy boost doses to local tumour areas at risk) by identifying and localizing risk biomarkers.

In testing on four longitudinal FDOPA PET studies of patients with advanced glioma, heterogeneity was consistently evidenced both in local outcomes and local PET uptake. Although two patients showed general tumour regression everywhere in all cases positive (if weak) correlations between FDOPA and growth were obtained. Notably, in the single patient with a mixture of tumour growth and shrinkage had the highest R-value (0.402) signifying that the FDOPA has some ability to predict future growth locally within individual patients. Local analysis of tumours could help realize ambitions for tailoring treatment not only to the individual, but at an intra-tumour level.

Acknowledgements
Much assistance was provided by staff in several departments at the Royal Brisbane and Women's Hospital. This work was supported by a Queensland Government Smart State NIRAP grant (MedTeQ) and the Australian National Health & Medical Research Council (NHMRC) grant number 631567.

References

[1] G. H. Heppner, "Tumour Heterogeneity," *Cancer Res.*, vol. 44, no. 6, pp. 2259 –2265, Jun. 1984.
[2] A. R. A. Anderson, A. M. Weaver, P. T. Cummings, and V. Quaranta, "Tumour Morphology and Phenotypic Evolution Driven by Selective Pressure from the Microenvironment," *Cell*, vol. 127, no. 5, pp. 905–915, Dec. 2006.
[3] R. L. Wahl, H. Jacene, Y. Kasamon, and M. A. Lodge, "From RECIST to PERCIST: Evolving Considerations for PET response criteria in solid tumours," *J. Nucl. Med. Off. Publ. Soc. Nucl. Med.*, vol. 50 Suppl 1, p. 122S–50S, May 2009.
[4] C. J. Ledezma, W. Chen, V. Sai, B. Freitas, T. Cloughesy, J. Czernin, and W. Pope, "18F-FDOPA PET/MRI fusion in patients with primary/recurrent gliomas: initial experience," *Eur. J. Radiol.*, vol. 71, no. 2, pp. 242–248, Aug. 2009.
[5] Schiepers, W. Chen, T. Cloughesy, M. Dahlbom, and S.-C. Huang, "18F-FDOPA Kinetics in Brain Tumours," *J Nucl Med*, vol. 48, no. 10, pp. 1651–1661, Oct. 2007.
[6] K. R. Swanson, G. Chakraborty, C. H. Wang, R. Rockne, H. L. P. Harpold, M. Muzi, T. C. H. Adamsen, K. A. Krohn, and A. M. Spence, "Complementary but distinct roles for MRI and 18F-fluoromisonidazole PET in the assessment of human glioblastomas," *J. Nucl. Med. Off. Publ. Soc. Nucl. Med.*, vol. 50, no. 1, pp. 36–44, Jan. 2009.
[7] F. L. Greene, Ed., "AJCC Cancer Staging Manual," 6th ed., Springer, 2002, pp. 387–390.
[8] R. Stupp, P.-Y. Dietrich, S. O. Kraljevic, A. Pica, I. Maillard, P. Maeder, et al., "Promising Survival for Patients With Newly Diagnosed Glioblastoma Multiforme Treated With Concomitant Radiation Plus Temozolomide Followed by Adjuvant Temozolomide," *J. Clin. Oncol.*, vol. 20, no. 5, pp. 1375 –1382, Mar. 2002.
[9] N. J. Abbott and I. A. Romero, "Transporting therapeutics across the blood-brain barrier," *Mol. Med. Today*, vol. 2, no. 3, pp. 106–113, Mar. 1996.
[10] S. G. Nekolla, C. Miethaner, N. Nguyen, S. I. Ziegler, and M. Schwaiger, "Reproducibility of polar map generation and assessment of defect severity and extent assessment in myocardial perfusion imaging using positron emission tomography," *Eur. J. Nucl. Med. Mol. Imaging*, vol. 25, pp. 1313–1321, Sep. 1998.
[11] F. Füchtner, J. Zessin, P. Mäding, and F. Wüst, "Aspects of 6-[18F]fluoro-L-DOPA preparation. Deuterochloroform as a substitute solvent for Freon 11," *Nukl. Nucl. Med.*, vol. 47, no. 1, pp. 62–64, 2008.
[12] M. Jenkinson and S. Smith, "A global optimisation method for robust affine registration of brain images," *Med. Image Anal.*, vol. 5, no. 2, pp. 143–156, Jun. 2001.
[13] D. Rueckert, C. Hayes, C. Studholme, P. Summers, M. Leach, and D. J. Hawkes, "Non-rigid registration of breast MR images using mutual information," in *Medical Image Computing and Computer-Assisted Interventation — MICCAI'98*, vol. 1496, W. M. Wells, A. Colchester, and S. Delp, Eds. Berlin/Heidelberg: Springer-Verlag, 1998, pp. 1144–1152.

XVII International Conference on the Use of Computers in Radiation Therapy (ICCR 2013) IOP Publishing
Journal of Physics: Conference Series **489** (2014) 012074 doi:10.1088/1742-6596/489/1/012074

Constructing a clinical decision-making framework for image-guided radiotherapy using a Bayesian Network

C Hargrave[1,2,3], M Moores[2,3], T Deegan[1], A Gibbs[1], M Poulsen[1], F Harden[2,3] and K Mengersen[2,3]

[1] Radiation Oncology Mater Centre, Princess Alexandra Hospital, QLD Health, Brisbane, QLD, 4101, Australia
[2] Queensland University of Technology, Brisbane, QLD, 4001, Australia
[3] Institute of Health and Biomedical Innovation, Queensland University of Technology, Kelvin Grove, QLD, 4059, Australia

E-mail: c.hargrave@qut.edu.au

Abstract. A decision-making framework for image-guided radiotherapy (IGRT) is being developed using a Bayesian Network (BN) to graphically describe, and probabilistically quantify, the many interacting factors that are involved in this complex clinical process. Outputs of the BN will provide decision-support for radiation therapists to assist them to make correct inferences relating to the likelihood of treatment delivery accuracy for a given image-guided set-up correction. The framework is being developed as a dynamic object-oriented BN, allowing for complex modelling with specific sub-regions, as well as representation of the sequential decision-making and belief updating associated with IGRT. A prototype graphic structure for the BN was developed by analysing IGRT practices at a local radiotherapy department and incorporating results obtained from a literature review. Clinical stakeholders reviewed the BN to validate its structure. The BN consists of a sub-network for evaluating the accuracy of IGRT practices and technology. The directed acyclic graph (DAG) contains nodes and directional arcs representing the causal relationship between the many interacting factors such as tumour site and its associated critical organs, technology and technique, and inter-user variability. The BN was extended to support on-line and off-line decision-making with respect to treatment plan compliance. Following conceptualisation of the framework, the BN will be quantified. It is anticipated that the finalised decision-making framework will provide a foundation to develop better decision-support strategies and automated correction algorithms for IGRT.

1. Introduction

Decision-making in image-guided radiotherapy (IGRT) requires radiation therapists (RTs) to consider and weigh up many factors within tight clinical time constraints [1]. Choosing the most optimal set-up correction to apply in the presence of residual patient positioning errors adds further uncertainty to the decision-making process. The clinical decision-making processes associated with IGRT have become more complex as on-board imaging and treatment delivery technologies have advanced [2].

We aim to use a systems approach to develop a decision-support framework for IGRT to assist RTs to perform IGRT efficiently whilst ensuring that any applied set-up correction, based on their analysis of the images, will maximise treatment delivery accuracy. A Bayesian network (BN) is a tool ideally suited for developing such a framework and has been previously applied to radiotherapy treatment plan selection [3]. BNs are able to model complex systems such as IGRT decision-making, providing both a graphical map of the process via a directed acyclic graph (DAG) as well as probabilistic quantification of the system knowledge and uncertainty using conditional probability tables (CPTs)

[4]. There are various core and iterative processes in the development of a BN [5]. This paper presents the work to date to conceptualise the BN graphical structure and the outcomes of a review workshop undertaken to begin the process of validating the network structure.

2. Materials and Methods

It is crucial to define the purpose of any BN at the outset in order to determine the outcome of interest requiring decision-support. This outcome is defined as the BN's target node. Key factors that influence the outcome of interest and their causal relationships are then identified and mapped as nodes and directional arcs to create the DAG. Clinical protocols and data from published studies are ideal sources for clarifying important outcomes and their influencing factors for the IGRT decision-making process. The construction of the DAG is an iterative process, where a model is proposed, reviewed and modified until structure confidence is achieved. RTs, radiation oncologists (ROs) and physicists with extensive clinical IGRT experience are able to contribute valuable expertise towards the BN review process.

2.1. Review of IGRT clinical practices and the literature

The starting point for conceptualising the BN was to examine the IGRT practices at a local radiotherapy department. IGRT protocols for prostate and head and neck cancer patients were evaluated as planar and volumetric on-board imaging technologies are routinely used in both the online and offline environments. Decision flowcharts were constructed to map the actions that are taken for typical image matching scenarios. A systematic literature review was then performed to ensure a comprehensive analysis of IGRT practices and technologies. Internet searches were conducted to access texts, technical reports and guidelines published by relevant national and international professional bodies, working parties and regulatory authorities. Searches were also conducted on online databases (PubMed, Embase, ScienceDirect, and the Cochrane Library) for peer reviewed publications. 580 publications met the inclusion criteria after abstract review. However for the purpose of conceptualising the BN, 160 articles, either published in 2012 or those frequently cited, were deemed an adequate sample of current knowledge and expertise of the IGRT technologies and practices. Quantitative and qualitative data was extracted from these publications and tabulated according to technology, technique, tumour, OAR, and study results.

2.2. BN structure development

The decision flowcharts and the data extracted from the literature review identified that there are multiple outcomes of interest in the IGRT decision-making process, requiring multiple interacting target nodes. An object oriented BN (OOBN) format was used to create sub-networks relating to the various target nodes. A dynamic component using several OOBN time slices was used to model the sequential decision-making required to assess accurate delivery of the treatment plan as the patient's treatment progresses [4]. A prototype DAG for the BN was then manually constructed by mapping the key factors associated with the different target nodes. Input nodes were used to allow evidence relating to the IGRT process, such as IGRT modality (e.g. planar 2D kilo-voltage (kV) imaging with fiducial marker matching), tumour site and its associated organs at risk (OAR), to be specified in the network.

2.3. BN review

A workshop was conducted where an RT, RO and physicist reviewed the prototype IGRT decision-making BN. Prior to the workshop a list of the network nodes (50 in total), their function and definitions, was provided to the participants. These were reviewed at the beginning of the session to determine whether the outcomes of interest and key influencing factors of the network had been correctly identified. Yellow sticky notes with the target nodes and their key factors written on them were then used to independently construct the BN. The prototype BN was then examined so that missing or redundant nodes, or incorrect directional arcs, could be identified.

3. Results

3.1. Conceptualisation of a prototype BN structure

The multiple outcomes of interest identified for the IGRT decision-making process were measurement or assessment uncertainty, determining the probability of accurately delivering the treatment plan in the presence of residual errors in the online environment and, on-going assessment of treatment delivery compliance with the treatment plan, given the treatments delivered to date. Figure 1 demonstrates the prototype dynamic OOBN DAG created to model the relationship between the target nodes used to represent these various outcomes of interest. An IGRT technology and technique assessment sub-network, and separate online and offline IGRT decision-making sub-networks were linked by their target nodes.

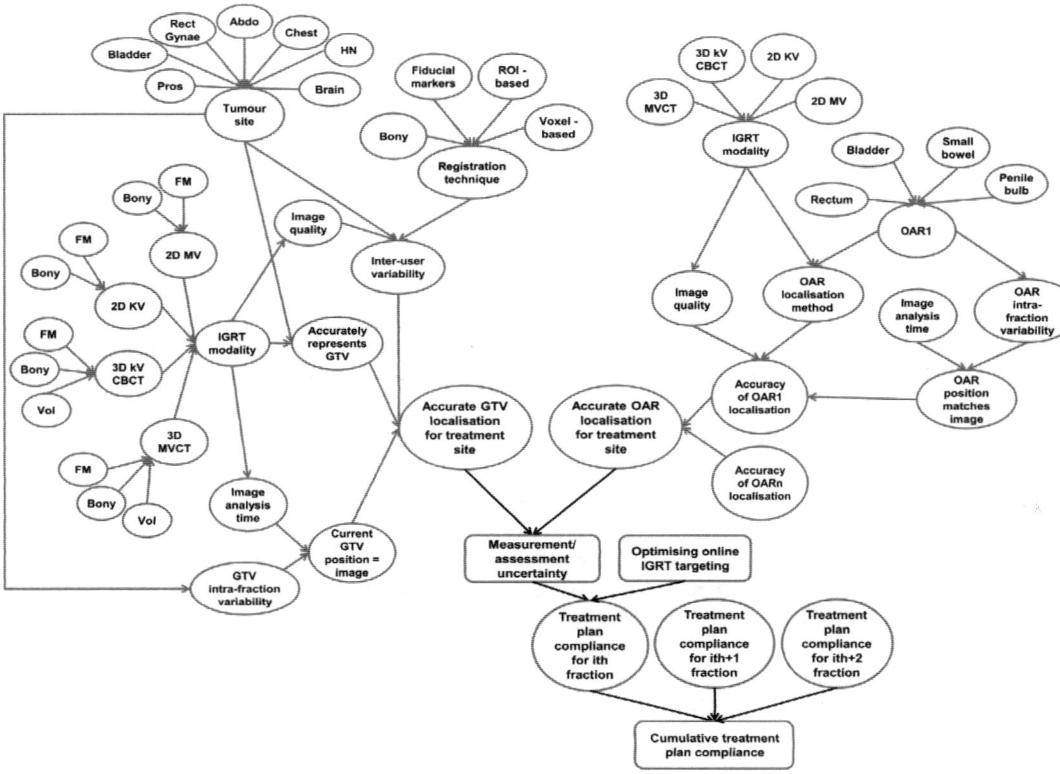

Figure 1. The prototype DAG of the BN. The rectangular target nodes represent the various outcomes of interest in the IGRT decision-making process. The IGRT technology and technique assessment sub-network (target node=Measurement/assessment uncertainty) has been expanded to demonstrate its division into tumour and OAR sub-nets. The treatment plan compliance nodes for individual fractions demonstrate the time slicing component required to assess cumulative treatment plan compliance.

3.2. BN review

The prototype DAG structure was generally accepted by the review process conducted in the workshop. However the role of image quality, radiation dose and patient compliance (including voluntary and involuntary patient-specific issues) in the IGRT decision-making process and their place in the BN structure were key discussion points in the workshop. These factors were considered highly influential in the decision to re-position and re-image the patient in the online IGRT environment. Patient compliance, treatment intent and delivery technique were considered to be key factors to apply permanent changes to the patient set-up or to adapt the patient's plan in the offline IGRT environment. Consensus was reached that additional weighted nodes could be linked with the target nodes to decision nodes to provide advice different IGRT related interventions.

4. Discussion

The various sub-networks of the BN in effect model the various aspects of uncertainty associated with delivering IGRT treatments, including technology and technique error, and random and systematic setup errors. It is important to note that the BN design has been limited to x-ray based on-board imaging devices as these are the most commonly utilised IGRT technologies [2]. Real-time tracking systems were also excluded as these technologies are relatively new and remain developmental in the broader clinical context.

Further work is required to finalise the conceptualisation of the network structure. While the systematic literature review yielded a high number of studies examining technology and accuracy precision, there are a number of key factors and their role in the IGRT decision-making process that are less widely reported. Examples include image matching criteria to optimise tumour targeting, residual error thresholds, image quality, radiation dose and patient compliance. The necessary inclusion of these factors in the BN is supported by a limited number of publications [6], documents such as the National Health Service IGRT guidelines [2], the departmental IGRT decision flowcharts and the outcomes of the review workshop. Future elicitation workshops and interviews have been planned to utilise the expertise of RTs, ROs and physicists from multiple radiotherapy departments to further refine the placement and function of the factors in question in the BN.

An interesting outcome of the workshop was a discussion on how useful visualising the key factors and their relationships in graphical form was for the individual participants in evaluating both their departmental IGRT practices and their own knowledge-base in this field. The primary aim of a BN is that its target nodes predict an outcome of interest or update beliefs when new information about the system is added to the network. However BNs have also been shown to be beneficial in the evaluation of current standards of practice as well as informing individual practice when they have been developed and applied in other fields [7].

5. Conclusion

The graphic structure of a BN to support clinical IGRT decision-making has been developed and reviewed. Further workshops are required to finalise the BN structure. The BN will then be quantified, using both expert clinical knowledge and data sourced from peer-reviewed clinical studies, to generate the conditional probability tables for the network nodes. It is anticipated that the finalised decision-making framework will provide a foundation to develop better decision-support strategies and automated correction algorithms for IGRT.

Acknowledgement

This research is part of a collaborative project between the Radiation Oncology Mater Centre and Queensland University of Technology

References

[1] Kron T 2011 *J Radiother Pract* **10** 71
[2] National Radiotherapy Implementation Group Report 2012 Image Guided Radiotherapy (IGRT): Guidance for implementation and use. *NHS: National Cancer Action Team* http://ncat.nhs.uk/our-work/ensuring-better-treatment/radiotherapy?#tab-igrt
[3] Smith W P, Doctor J, Meyer J, Kalet I J and Phillips M H 2009 *Artif Intell Med* **46** 119
[4] Korb K B and Nicholson A E 2011 *Bayesian artificial intelligence*: (Chapman & Hall/CRC Press)
[5] Johnson S 2009 *Integrated Bayesian network frameworks for modelling complex ecological issues*. School of Mathematical Sciences, (Brisbane: Queensland University of Technology) pp 148
[6] van Beek S, van Kranen S, Mencarelli A, Remeijer P, Rasch C, van Herk M and Sonke J-J 2010 *Radiother Oncol* **94** 213
[7] Marcot B G, Holthausen R S, Raphael M G, Rowland M M and Wisdom M J 2001 *Forest Ecol Manag* **153** 29

An automatic dose verification system for adaptive radiotherapy for helical tomotherapy

Xiaohu Mo, Mingli Chen, Donald Parnell, Gustavo Olivera, Daniel Galmarini and Weiguo Lu

21st Century Oncology, Madison, WI, USA

xmo@rtsx.com

Abstract. Purpose: During a typical 5-7 week treatment of external beam radiotherapy, there are potential differences between planned patient's anatomy and positioning, such as patient weight loss, or treatment setup. The discrepancies between planned and delivered doses resulting from these differences could be significant, especially in IMRT where dose distributions tightly conforms to target volumes while avoiding organs-at-risk. We developed an automatic system to monitor delivered dose using daily imaging. Methods: For each treatment, a merged image is generated by registering the daily pre-treatment setup image and planning CT using treatment position information extracted from the Tomotherapy archive. The treatment dose is then computed on this merged image using our in-house convolution-superposition based dose calculator implemented on GPU. The deformation field between merged and planning CT is computed using the Morphon algorithm. The planning structures and treatment doses are subsequently warped for analysis and dose accumulation. All results are saved in DICOM format with private tags and organized in a database. Due to the overwhelming amount of information generated, a customizable tolerance system is used to flag potential treatment errors or significant anatomical changes. A web-based system and a DICOM-RT viewer were developed for reporting and reviewing the results. Results: More than 30 patients were analysed retrospectively. Our in-house dose calculator passed 97% gamma test evaluated with 2% dose difference and 2mm distance-to-agreement compared with Tomotherapy calculated dose, which is considered sufficient for adaptive radiotherapy purposes. Evaluation of the deformable registration through visual inspection showed acceptable and consistent results, except for cases with large or unrealistic deformation. Our automatic flagging system was able to catch significant patient setup errors or anatomical changes. Conclusions: We developed an automatic dose verification system that quantifies treatment doses, and provides necessary information for adaptive planning without impeding clinical workflows.

1. Introduction

As modern external beam radiotherapy techniques become more complex and allow for highly conformal treatment plans, accurate verification of the dose delivered to patients is also becoming a necessity. Patients are never treated exactly as planned, which could be due to setup error, anatomical change or machine behavior. Without a comprehensive delivery verification program, these uncertainties in treatment may go undetected. In vivo dosimetric verification is especially important for hypo-fractionated treatments, dose escalation schemes, and stereotactic body radiation therapy, where errors in a single fraction could have a significant dosimetric effect on the treatment outcome.

Therefore, in this work, we present an automatic dosimetric verification system for helical tomotherapy (HT) in order to monitor the aforementioned potential treatment deviations, and provide triggers and guides to clinic workflow such as adaptive planning. The system re-computes treatment doses on daily images and provide cumulative dose during the course of treatments using deformable registration. We anticipate the system will lead to automatic adaptive planning, which greatly facilitate adaptive radiotherapy.

2. Methods

The diagram in Figure 1 illustrates the workflow for each patient fraction treated. 1) The input is naturally the planning information, including the treatment plan and CT; as well as daily imaging, machine output etc. In order to recalculate dose on the same image space as the plan CT, a merged image is generated by filling the missing information, both radially and in the superior/inferior direction, of the daily CT (approximately 40 cm field of view) with plan CT as described in Langen et al. [1]. This method assumes anatomy outside of field of view to be the same with plan CT, and users are warned if regions of interest are not imaged. 2) The planned beams are then re-computed on the merged image using our in house dose calculation engine which employs collapsed cone convolution/superposition algorithm [2] and implemented on graphic processors. Note that to allow for correct density mapping, we have also incorporated density calibration in routine quality assurance of imaging devices. 3) The deformation vector field (DVF) between the merged and planning CT is then computed with Morphons algorithm [3], and used to warp the planning contours to the merged image. The analytically computed inverse DVF [4] is used to warp recomputed dose to planning CT for dose accumulation. 4) Subsequently max dose, mean dose, dose volume histogram (DVH), volume, center of mass shift, and dice similarity indices are evaluated to collect trending information for data analysis and reporting. The automatic triggering system would notify the corresponding staffs based on a user-customizable tolerance table of these parameters. 5) And lastly, a reviewing system consists of a web interface and DICOM viewer is developed to enable users to quickly review and check any abnormalities throughout the course of patient treatments. All data are stored in DICOM formats as standardized as possible. Although this framework could be applicable to linac as well, but in this work the results are for HT only.

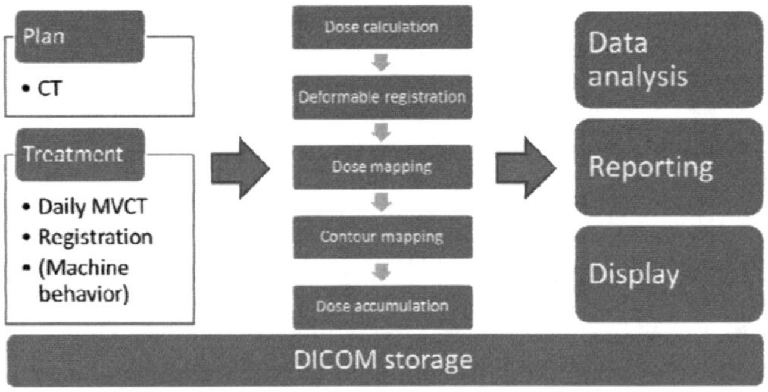

Figure 1. Basic workflow of the system

3. Results

The dose recalculation was validated with two independent methods. 1) To verify the dose recalculation implementation, we compared the plan dose exported from HT against recomputed plan dose using three dimensional Gamma Index [5]. Among 122 randomly selected plans, 118 cases passed the Gamma test at 2%/2mm of 97% voxels whose doses are greater than 10% of maximum plan dose. 2) The whole chain of events from the initial planning to the final dose recalculation on mega voltage CT was tested using a phantom plan. For each HT system, a treatment plan was created on the HT planning station using the kVCT scan of the HT cheese phantom. The phantom was then scanned, aligned and treated on the HT unit and the target dose was measured with an ion chamber. For all 14 HT units tested, the recalculated doses are within 2% compared with the ion chamber measurements.

The deformed contours and doses were visually verified by experts on all fractions for 50 patients, including head and neck, lung, and prostate cases etc. The deformation results were considered acceptable for delivery verification purposes in 47 of the 50 patients. We will be incorporating more metrics to facilitate a more objective evaluation of the deformation in the future.

We demonstrate two retrospective cases where this system can be used clinically. The first case is a patient with chest wall target who was treated with setup error on fraction 3. As shown in Figure 2, this fraction was highlighted on the dashboard of the web report since PTV mean dose is out of tolerance. Figure 3 shows the plan summary page where the under-dosage can be easily visualized and compared to the results in other fractions. Further investigation is available through our DICOM viewer which allows users to visualize the treatment registration, daily and cumulative dose and contours for all fractions as shown in Figure 4.

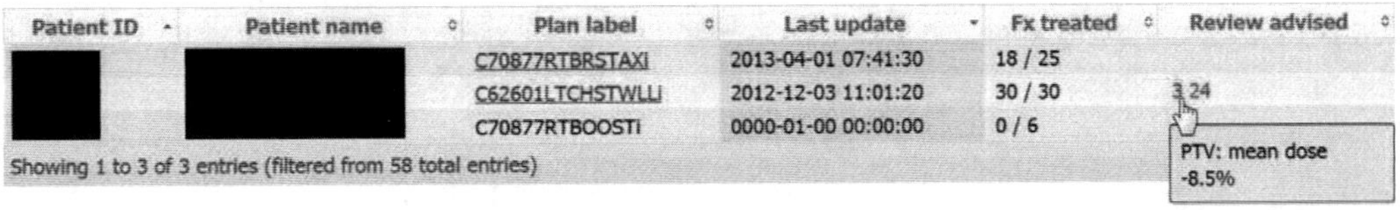

Patient ID	Patient name	Plan label	Last update	Fx treated	Review advised
[redacted]	[redacted]	C70877RTBRSTAXI	2013-04-01 07:41:30	18 / 25	
		C62601LTCHSTWLLI	2012-12-03 11:01:20	30 / 30	3,24
		C70877RTBOOSTI	0000-01-00 00:00:00	0 / 6	

Showing 1 to 3 of 3 entries (filtered from 58 total entries)

PTV: mean dose -8.5%

)ortal of the review system showing review suggestions

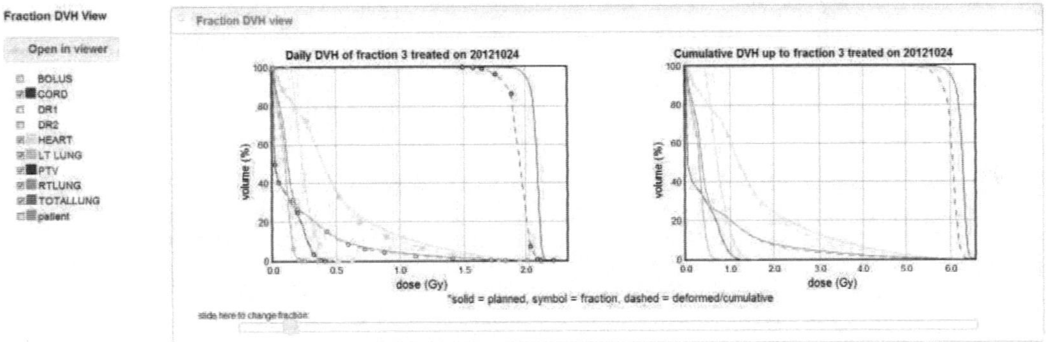

Figure 3. DVH comparison of plan dose (solid) and delivered dose (dashed). The legend on the left side indicates the correspondence of color and the region of interest. The left DVH panel displays fraction dose, and the right panel displays cumulative dose. The displays proceed with respect to fraction as the scroll bar on the bottom slides from left to right.

The second case is a head and neck patient who had been losing weight during the course of treatments. The over-dosage to the right parotid can be clearly identified in our DVH point trends as shown in Figure 5. Since our system is automatic, every patient will be monitored and review or adaptation will be triggered when the metrics are outside of their tolerances.

4. Conclusions

An automatic dose verification system was developed. The system is capable of verifying and tracking treatment doses, as well as providing necessary information for adaptive planning without impedes clinical workflows. Since this system uses pre-treatment images for re-computation, it works in combination with our existing exit detector based in vivo dosimetry program [6]. Both systems are implemented in 14 HT clinics in our network and more than 10,000 fractions have been calculated with this system.

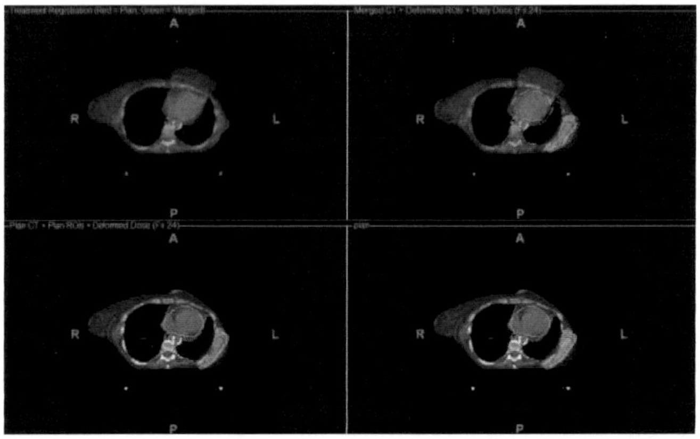

Figure 4. DICOM viewer showing the discrepency of patient breast setup induced PTV underdose

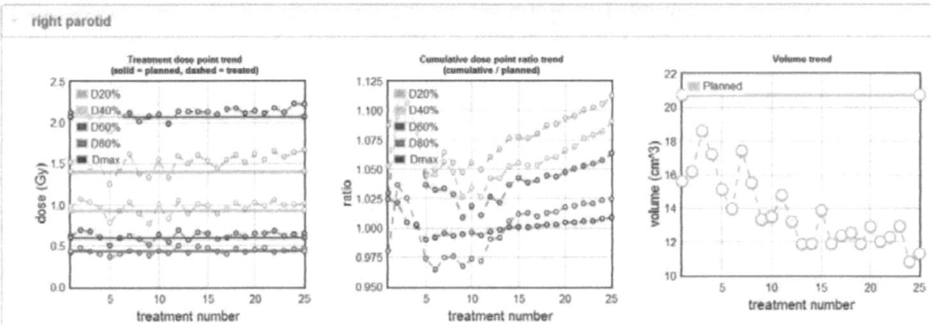

Figure 5. Panels of metrics used to trigger the review. The metrics include DVH points and D_{max}. The cumulative DVH shows an increasing trend of parotid dose.

References

[1] Langen KM, Meeks SL, Poole DO, Wagner TH, Willoughby TR, Kupelian PA, et al. The use of megavoltage CT (MVCT) images for dose recomputations. Phys Med Biol. 2005 Sep 21;50(18):4259–76.

[2] Lu W, Olivera GH, Chen M-L, Reckwerdt PJ, Mackie TR. Accurate convolution/superposition for multi-resolution dose calculation using cumulative tabulated kernels. Phys Med Biol. 2005 Feb 21;50(4):655–80.

[3] Janssens G, Xivry JO de, Fekkes S, Dekker A, Macq B, Lambin P, et al. Evaluation of nonrigid registration models for interfraction dose accumulation in radiotherapy. Med Phys. 2009;36(9):4268–76.

[4] Chen M, Lu W, Chen Q, Ruchala KJ, Olivera GH. A simple fixed-point approach to invert a deformation field. Med Phys. 2008 Jan;35(1):81–8.

[5] Low DA, Harms WB, Mutic S, Purdy JA. A technique for the quantitative evaluation of dose distributions. Med Phys. 1998 May;25(5):656–61.

[6] Lu W, Parnell D, Olivera G, Galmarini D. SU-E-T-59: A Clinical Workflow for TomoTherapy Delivery Verification. Med Phys. 2011;38(6):3499.

Segmentation of cone-beam CT using a hidden Markov random field with informative priors

M Moores[1,3], C Hargrave[1,2,3], F Harden[1,3] and K Mengersen[1,3]

[1] Queensland University of Technology, Brisbane QLD 4000 Australia
[2] Radiation Oncology Mater Centre, Queensland Health, South Brisbane QLD 4101 Australia
[3] Institute of Health and Biomedical Innovation, Kelvin Grove QLD 4059 Australia

E-mail: matthew.moores@qut.edu.au

Abstract. Cone-beam computed tomography (CBCT) has enormous potential to improve the accuracy of treatment delivery in image-guided radiotherapy (IGRT). To assist radiotherapists in interpreting these images, we use a Bayesian statistical model to label each voxel according to its tissue type. The rich sources of prior information in IGRT are incorporated into a hidden Markov random field model of the 3D image lattice. Tissue densities in the reference CT scan are estimated using inverse regression and then rescaled to approximate the corresponding CBCT intensity values. The treatment planning contours are combined with published studies of physiological variability to produce a spatial prior distribution for changes in the size, shape and position of the tumour volume and organs at risk. The voxel labels are estimated using iterated conditional modes. The accuracy of the method has been evaluated using 27 CBCT scans of an electron density phantom. The mean voxel-wise misclassification rate was 6.2%, with Dice similarity coefficient of 0.73 for liver, muscle, breast and adipose tissue. By incorporating prior information, we are able to successfully segment CBCT images. This could be a viable approach for automated, online image analysis in radiotherapy.

1. Introduction

Due to the level of precision and detail required for sufficient volume variation analysis of daily cone-beam CT (CBCT), coupled with the short timeframe in which the radiotherapy procedure is to be carried out in the clinic, radiation therapists are currently unable to perform a detailed analysis of the CBCT scans prior to irradiation. Automated methods may aid in the decision-making process by labelling the image voxels according to tissue type, estimating the boundaries of the tumour and neighbouring organs, and highlighting any regions where changes in these boundaries might have exceeded tolerance. Multi-thresholding using an independent Gaussian mixture model (GMM) is error-prone due to the low contrast-to-noise ratio (CNR) in CBCT scans of human tissue. The low CNR is exemplified by the axial slice from a CBCT shown in figure 1. For comparison, a fan-beam (FBCT) scan of the same subject is illustrated in figure 2. The voxel intensity values alone are insufficient to accurately segment the image according to tissue type; therefore more complex methods are necessary.

Recent approaches to boundary estimation in CBCT have used deformable models that evolve according to partial differential equations. The two major approaches are 3D mesh models [1] and level sets [2]. These models need to be initialized very close to the boundary of interest, as they are prone to becoming stuck on local minima. Thus, they are more suited as a post-processing

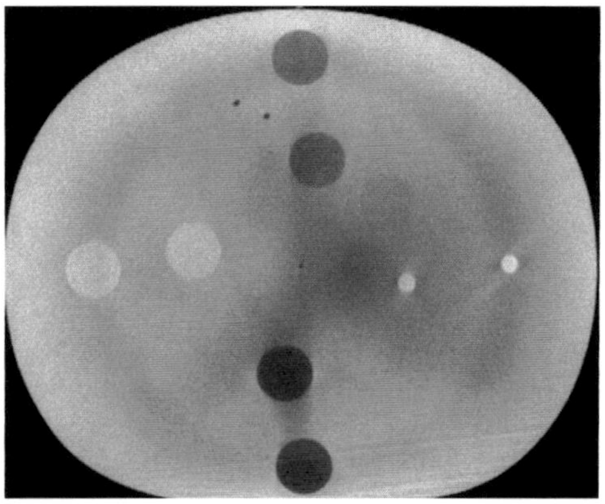

Figure 1. CBCT scan of the ED phantom.

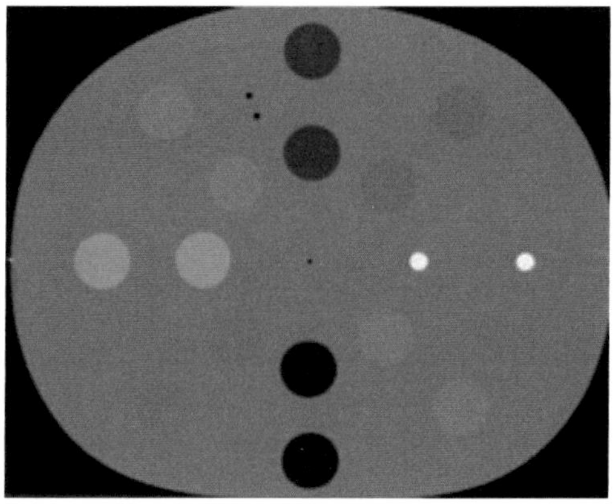

Figure 2. FBCT scan of the ED phantom.

step to refine an approximate solution that was obtained via other means, as in [3, 4, 5]. Other approaches include hidden Markov random field (MRF) models and graph-based segmentation. It has been shown that segmentation by graph cut is equivalent to fitting a discrete MRF [6], so these are essentially two different approaches with the same underlying model.

In this paper we introduce new methods for utilising the rich sources of prior information that are available in IGRT by extending the hidden MRF model. We demonstrate these methods on CBCT scans of an electron density (ED) phantom and show that the external field prior results in a substantial improvement in segmentation accuracy.

2. Materials and Methods

The planning CT along with the contours of the gross tumour volume (GTV) and organs at risk (OAR) constitute a valuable source of prior information for analysing CBCT scans. They can be viewed as a snapshot of the patient at a previous point in time. This patient-specific information can be combined with published studies of organ motion and setup error to produce a probabilistic spatial distribution of voxel labels. The spatial distribution is computed relative to the scan isocentre, avoiding the need for a separate registration step.

2.1. ED phantom experiment

We have evaluated the performance of our method using CBCT scans of a tissue-equivalent ED phantom (CIRS, Inc. model 062). The ED phantom was manufactured from epoxy and contains cylindrical inserts that mimic the X-ray absorption of human tissue: lung (inhale); lung (exhale); adipose; breast (50% fat); water-equivalent solid; muscle; liver; spongy (trabecular) bone; and dense (cortical) bone. The CBCT scans were acquired from a Varian linear accelerator with On-Board Imager (OBI) using a half bow-tie filter to achieve a 450mm field of view. 53 CBCT scans were obtained in total, with 27 held out for testing and 26 incorporated into the Bayesian prior. We also acquired 28 FBCT scans of the same phantom from a Siemens SOMATOM Sensation Open. These scans represent prior information, as will be explained in section 2.2. Example CBCT and FBCT scans of the ED phantom are illustrated in figures 1 and 2.

2.2. Informative priors for CT number

CBCT scanners are not calibrated to the Hounsfield scale. Nevertheless, there is a linear relationship between electron density in a homogeneous region of tissue and the mean intensity of the corresponding voxel values [7]. This relationship has been quantified by fitting a regression model to 26 CBCT scans of the ED phantom. We were thus able to rescale estimates of tissue density from a patient's planning CT scan to predict the distribution of intensity values in a CBCT scan of the same patient. These estimates were incorporated into the hidden MRF model as priors for the mean and variance of each mixture component [8].

2.3. External field prior for voxel labels
We have also included spatial prior information in the MRF model, in the form of an external field. The external field was centred on the planning contours of the GTV and the OARs, relative to the scan isocentre. Geometric uncertainty was incorporated into the model using quantifications of organ motion from published studies. For our ED phantom experiment, we used a mean displacement of 1.2mm with a standard deviation of 7.3mm, which is typical of prostate motion [9]. The external field prior for the ED phantom is illustrated in figure 3.

2.4. Iterated conditional modes
We used the iterated conditional modes (ICM) algorithm [10] to fit our hidden MRF model. This algorithm was implemented by modifying the source code of the R package mritc [11] to accommodate the external field prior and to support a 9 component mixture model.

3. Results and discussion
We first estimated the relationship between electron density ($\times 10^{23}$/cc) and voxel intensity. By fitting the regression model to 26 CBCT scans, we obtained posterior means of 152 for the slope and -761 for the intercept, with standard deviations of 0.97 and 0.30 respectively. Tissue densities were estimated from the 28 FBCT scans and then rescaled using the regression equation to obtain prior distributions for the voxel intensities. This regression relationship is specific to the CBCT scanner in question, but it would be straightforward to fit the same regression model to each on-board imaging device in the radiotherapy department.

The remaining 27 CBCT scans were segmented using between 34 and 100 iterations of the ICM algorithm. This took an average of 9 minutes per scan, running in a single thread. We anticipate that a parallel implementation of the ICM algorithm could reduce this execution time substantially. Overall, the mean misclassification rate was 6.4%. An example image segmentation is shown in figure 4. Table 1 shows a breakdown of segmentation accuracy by tissue type, measured using Dice similarity coefficients [12].

Table 1. Average segmentation accuracy for 27 CBCT scans.

Tissue Type	Dice ($\pm\sigma$)
Lung (inhale)	0.886 ± 0.010
Lung (exhale)	0.805 ± 0.024
Adipose	0.741 ± 0.039
Breast	0.721 ± 0.036
Water	0.964 ± 0.004
Muscle	0.728 ± 0.052
Liver	0.737 ± 0.035
Spongy bone	0.755 ± 0.022
Dense bone	0.723 ± 0.066

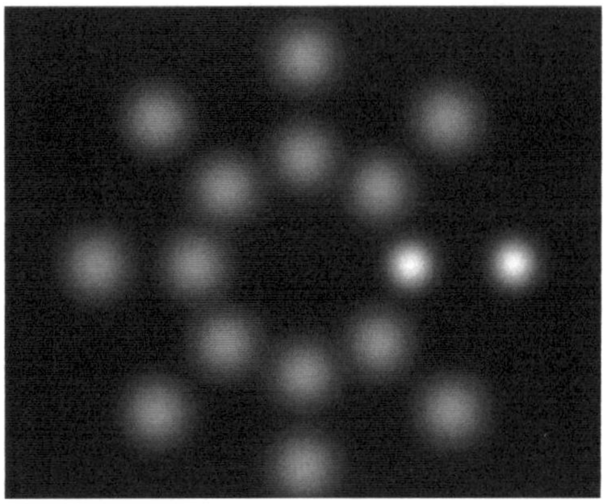

Figure 3. External field prior.

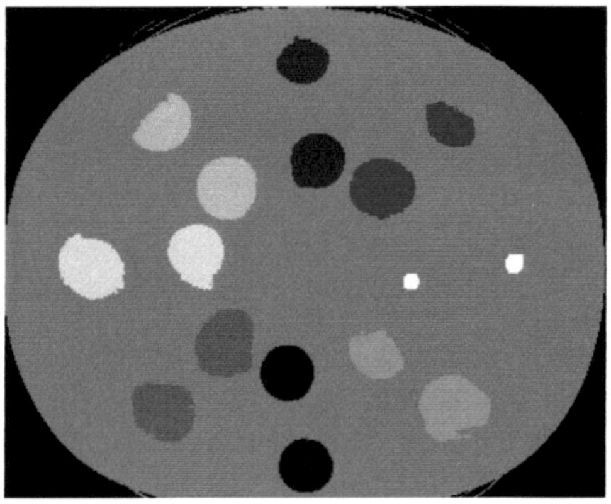

Figure 4. Segmentation result.

4. Conclusion
The external field prior substantially improves segmentation accuracy by incorporating additional spatial information into the model. We have demonstrated this method by application to CBCT scans of an ED phantom so that we are able to evaluate segmentation accuracy against the ground truth. The next step will be to apply the method to radiotherapy patient data.

Acknowledgments
This research is part of a collaborative project between Queensland University of Technology (QUT) and the Radiation Oncology Mater Centre (ROMC), Queensland Health. Computational resources used in this work were provided by the HPC and Research Support Group, QUT.

References
[1] Costa M J, Delingette H, Novellas S and Ayache N 2007 *Proc. 10th Int. Conf. on Medical Image Computing and Computer Assisted Intervention (Brisbane)* part I LNCS **4791** (Berlin: Springer) 252–60
[2] Chen S and Radke R J 2009 *Proc. 12th Int. Conf. on Computer Vision (Kyoto)* (IEEE) 763–70
[3] Chen T, Kim S, Zhou J, Metaxas D, Rajagopal G and Yue N 2009 *Proc. 12th Int. Conf. on Medical Image Computing and Computer Assisted Intervention (London)* part I LNCS **5761** (Berlin: Springer) 43–50
[4] Zhou J, Kim S, Jabbour S, Goyal S, Haffty B, Chen T, Levinson L, Metaxas D and Yue N J 2010 *Med. Phys.* **37** 1298–1308
[5] Lu C, Chelikani S, Papademetris X, Knisely J P, Milosevic M F, Chen Z, Jaffray D A, Staib L H and Duncan J S 2011 *Med. Image Anal.* **15** 772–85
[6] Winkler G 2003 *Image Analysis, Random Fields and Markov Chain Monte Carlo Methods: A Mathematical Introduction* 2nd ed. (Heidelberg: Springer-Verlag)
[7] Kalender W A 2011 *Computed Tomography: Fundamentals, System Technology, Image Quality, Applications* 3rd ed. (Erlangen: Publicis)
[8] Alston C L, Mengersen K L, Robert C P, Thompson J M, Littlefield P J, Perry D and Ball A J 2007 *Comp. Stat. Data Anal.* **51** 4282–96
[9] Frank S J, Dong L, Kudchadker R J, de Crevoisier R, Lee A K, Cheung R, Choi S, O'Daniel J, Tucker S L, Wang H and Kuban D 2008 *Int. J. Radiat. Oncol. Biol. Phys.* **71** 813–20
[10] Besag J 1986 *J. R. Stat. Soc. Ser. B* **48** 259–302
[11] Feng D and Tierney L 2011 *J. Stat. Soft.* **44**(7)
[12] Dice L R 1945 *Ecology* **26** 297–302

Deformable image registration for geometrical evaluation of DIBH radiotherapy treatment of lung cancer patients

W Ottosson[1,2,3], JA Lykkegaard Andersen[2], S Borrisova[2], A Mellemgaard[2] and CF Behrens[2]

[1]Center for Nuclear Technologies, Technical University of Denmark, DTU Risø Campus, DK-4000 Roskilde, Denmark
[2]Department of Oncology, Radiotherapy Research Unit, Herlev Hospital, University of Copenhagen, DK-2730 Herlev, Denmark

E-mail: wiot@dtu.dk

Abstract. Respiration and anatomical variation during radiotherapy (RT) of lung cancer yield dosimetric uncertainties of the delivered dose, possibly affecting the clinical outcome if not corrected for. Adaptive radiotherapy (ART), based on deformable image registration (DIR) and Deep-Inspiration-Breath-Hold (DIBH) gating can potentially improve the accuracy of RT. *Purpose:* The objective was to investigate the performance of contour propagation on repeated CT and Cone Beam CT (CBCT) images in DIBH compared to images acquired in free breathing (FB), using a recently released DIR software. *Method:* Three locally advanced non-small cell lung cancer patients were included, each with a planning-, midterm- and final CT (pCT, mCT, fCT) and 7 CBCTs acquired weekly and on the same day as the mCT and fCT. All imaging were performed in both FB and DIBH, using Varian RPM system for respiratory tracking. Delineations of anatomical structures were performed on each image set. The CT images were retrospective rigidly and deformable registered to all obtained images using the Varian Smart Adapt v. 11.0. The registered images were analysed for volume change and Dice Similarity Coefficient (DSC). *Result:* Geometrical similarities were found between propagated and manually delineated structures, with a slightly favour of FB imaging. Special notice should be taken to registrations where image artefacts or low tissue contrast are present. *Conclusion:* This study does not support the hypothesis that DIBH images perform better image registration than FB images. However DIR is a feasible tool for ART of lung cancer.

1. Introduction

Anatomical changes and variations due to respiration influence the accuracy of imaging, treatment planning and treatment delivery, and may affect the outcome of the planned treatment if not corrected for [1]. In ART, the treatment plan is adjusted during the course of treatment to minimize the divergence from the planned treatment, in terms of target dose coverage and spearing of dose to adjacent healthy organs at risk. Thus, ART has the potential to account for major anatomical changes not accounted for by applied margins [1]. Conventionally, lung cancer patients are treated in FB. However, breathing adapted radiotherapy (BART) by means of DIBH may suppress the geometric and

[3] Author to whom any correspondence should be addressed

dosimetric uncertainties related to respiration. Furthermore BART makes it possible to safely reduce the margins to the targets, yielding a reduction of irradiated volume [3,4]. The advantages of BART and ART illustrates that more individualized treatments are called for to improve the quality of RT. ART is, however, a time consuming process since anatomical structure delineations are needed on each new image set. DIR may be a promising tool in assisting with the delineation process, by deforming the reference contours from the planning CT into the anatomy of a second CT or CBCT. Our hypothesis is that image registrations performed based on DIBH images will result in improved image registrations, with enhanced correlation of volumes and higher scoring of DSC, since they often have visually better image quality compared to images in FB. The objective of this study was to investigate the performance of contour propagation on repeated CT and CBCT images of the thorax over the course of lung cancer treatment, imaged in DIBH, compared to conventional FB, using a recently released DIR software, based on a modified demons algorithm [5].

2. Material and methods

2.1. Patient data

Three locally advanced non-small cell lung cancer patients treated in 33 fractions (fx) using volumetric modulated arc therapy with a prescribed dose of 66 Gy (2 Gy/fx, 5 fx/week) at Herlev Hospital, between December 2012 and May 2013, were included in this study. They were treated on Varian Clinac iX 2300 linear accelerators [6,7] (Varian Medical Systems, Palo Alto, CA) equipped with On-Board Imagers (OBI) capable of performing FB and DIBH CBCT, using version 1.4 of the OBI software.

2.2. Image acquisition

Each patient was dual-CT scanned (acquiring a 4DCT in FB and a DIBH CT) before the start of, in the middle of, and after completion of the course of treatment, (pCT, mCT, fCT, respectively). All imaging were performed in treatment position [6]. They were scanned in a 16 slice Philips Brilliance CT Big Bore, version 3.5.17001 (Philips Medical Systems, Cleveland, OH) integrated with a Varian real-time position management (RPM) system, version 1.7 (Varian Medical Systems), for monitoring the patients' respiration during CT scanning. Intra venous (IV) contrast was administered to the patients during both 4DCT and DIBH imaging, for better contrast of nodal anatomy in the mediastinum. During DIBH imaging, the patients were audio-visually guided to hold their breath within a predefined amplitude level and gating window of 2-3 mm width. Additionally, all patients had 6-7 dual-CBCTs (in FB and DIBH) acquired on the treatment unit using the OBI. The CBCTs were acquired weekly and on the same day as the mCT and fCT. The Varian RPM system was once more used for monitoring the respiration to acquire DIBH gated imaging.

2.3. Definition of target and organs at risk

Delineations of anatomical structures for each patient were performed according to standard protocol by only one experienced oncologist (JLA or SB) on all image sets for that patient [6]. Delineations were carried out in the treatment planning system Eclipse v. 10 (Varian Medical Systems). Contouring of FB and DIBH Gross Tumor Volumes (GTVs) were performed by the oncologist in collaboration with an experienced radiologist. Residual FB and DIBH structures such as Clinical Target Volume (CTV), Planning Target Volume (PTV), medulla, heart, oesophagus, lung and body were additionally delineated by the oncologist. The heart, oesophagus, and lung were not delineated in the CBCT images, since they extended the CBCT scanning range.

2.4. Deformable registration and contour propagation

All CT images were retrospective pre-aligned by semi-automatic rigid registration and subsequently automatic deformed registered to all obtained images (both CBCT and secondary CT images) using Smart Adapt v. 11.0 (Varian Medical Systems). This resulted in 54 rigid and 54 deformable registrations per patient (corresponding numbers were 48 for the one patient with only 6 CBCT scans), which in total resulted in 312 registrations for all patients. The rigid pre-alignment increased the accuracy of the subsequent DIR, and prevented large unrealistic deformations [5]. The initial rigid registrations were based on the bony anatomy of the columna (50-3000 Hounsfield units), where the rest of the settings were predefined from the manufacturer. The rigid registration was done in three steps, each step with a higher image resolution, which improved the efficiency of the algorithm. The default DIR algorithm used in Smart Adapt was derived from a modified demons algorithm [5] based on a diffusion model. In the algorithm additional interaction forces (demons) are added to the original demons algorithm [8] and the voxel resolution of the images gradually increases during the optimization process. The floating image is warped to form a deformed image that match the reference image as closely as possible voxel-by-voxel [5,8]. The driving forces are based on the intensity differences between the two images, as well as the gradient of the image object. A symmetric force is also built into the solver to fulfill the consistency requirement for DIR, i.e. the transformation that maps the reference image to the floating image should be consistent with the inverse transformation that maps the floating image to the reference image. The DIR in this study was restricted to the field of view of the smallest image (typically the CBCT scan), resulting in a volume of interest (VOI) which was further analyzed. All structures extending the VOI (typically the heart, oesophagus, medulla, and body) were cropped to be comprised by the VOI, such that a common ground for comparison was created. For each DIR, the contours on the floating CT image were propagated onto the resulting deformed image (dCT) using the obtained image transformation.

2.5. Geometrical comparison

Rigidly and deformed registered structure volumes ($V_{RIG,DIR}$) were analyzed relative to the corresponding structure volumes (V_{REF}) on the reference image for volume change and DSC, where DSC was calculated according to equation 1 [9].

$$DSC = 2 \cdot \frac{V_{\text{RIG,DIR}} \cap V_{\text{REF}}}{V_{\text{RIG,DIR}} + V_{\text{REF}}} \tag{1}$$

For statistical analysis MATLAB Statistics Toolbox version 8.3 (R2013b) (The MathWorks, Natick, MA) was used. Paired t-tests to evaluate the median DSC measures of the various structures were performed, where differences were considered significant for $p < 0.05$.

3. Results/Discussion

Figure 1 (open triangles) illustrates that the manually delineated reference CT and CBCT GTVs overall are decreasing over the course of treatment. The DIBH volumes (upward pointing triangles) are in general smaller compared to FB (downward pointing triangles) which may be due to the lesser image artefacts (e.g motion blurring) in the DIBH images resulting in smaller targets. The dCT GTVs do not correlate well with the manually delineated reference CBCT (figure 1 (a)) or CT (figure 1 (b)) volumes. If a perfect volume correlation was achieved between the dCT structures and the reference volumes, the filled triangles (figure 1) would follow the corresponding curves with open markers (figure 1 (a)) or be the same volume as the open markers (figure 1 (b)). The dCTs of the pCT at fx 2 and 7 of patient 1 correlate well with the reference CBCTs, but for the remaining fractions the dCT of the mCT and fCT have better correlation with the reference CBCTs. This illustrates the difficulty to perform DIR registration when the anatomy changes over time. In this case, the patient would benefit from ART. It is clearly seen in figure 1 (a), at fx 15-20 for patient 2, that appearing image ring artefacts (figure 2) in the reference CBCT image affect the DIR process, and thus the propagation of structures. The effect was equally seen for DIBH and FB images, and for all CT-CBCT registrations at fx 15-20 for patient 2. The CT-CBCT and CT-CT registrations of patient 3 (figure 1) systematically underestimate the GTVs, both for FB and DIBH images. This may be due to DIR issues in regions with low level of tissue contrast (since the demons algorithm uses intensity values for registration) as the tumor for this patient was closely situated to the mediastinum.

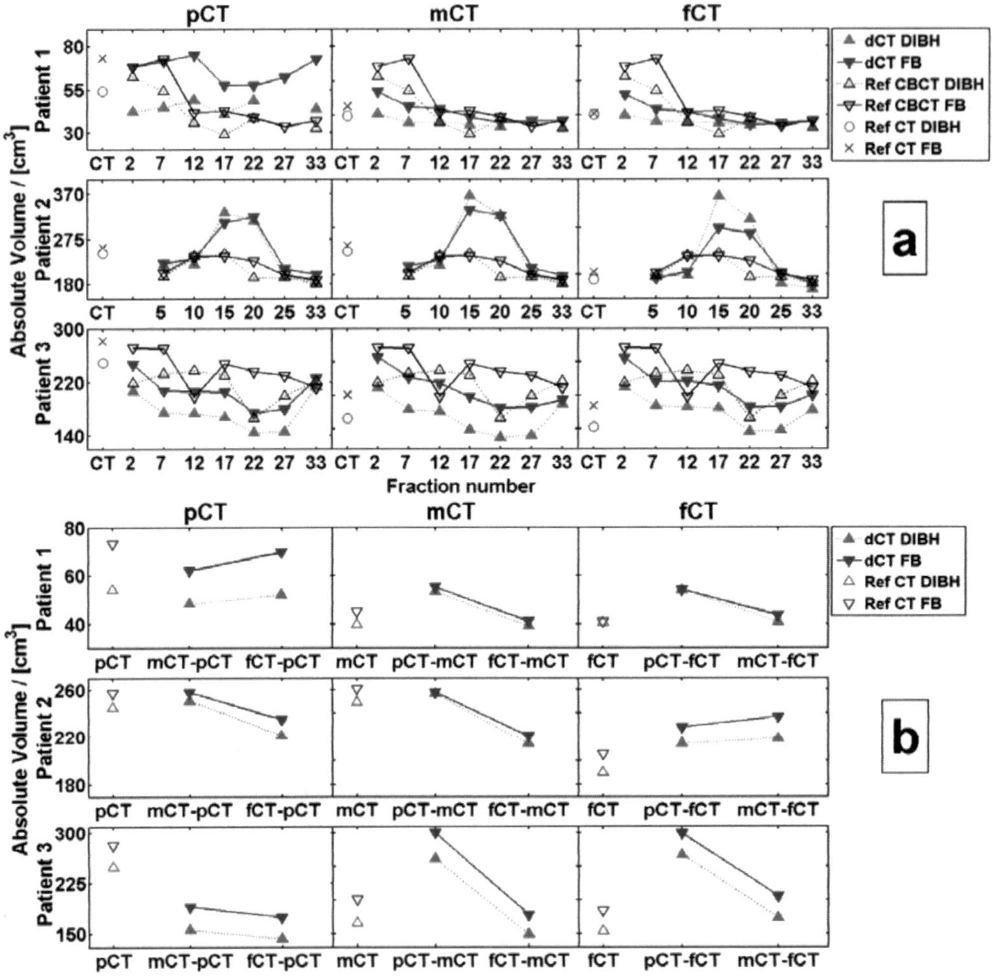

Figure 1. **(a)** Absolute volume per each fraction of the dCT GTVs (filled triangles) in relation to the manually delineated reference CBCT GTVs (open triangles) along with the manually delineated GTVs of pCT, mCT and fCT (x and circles) utilized when performing CT-CBCT DIR. Upward pointing triangles correspond to DIBH, and downward pointing triangles represent FB. In the pCT column it is the pCT that is deformed to match the CBCTs and similar for the columns labelled mCT and fCT. **(b)** Corresponding absolute volume of dCT GTVs (filled triangles) in relation to manually delineated reference CT GTVs (open triangles) when performing CT-CT DIR between the pCT, mCT and fCT. In the pCT column it is the mCT and fCT that is deformed to match the pCT and similar for the columns labelled mCT and fCT.

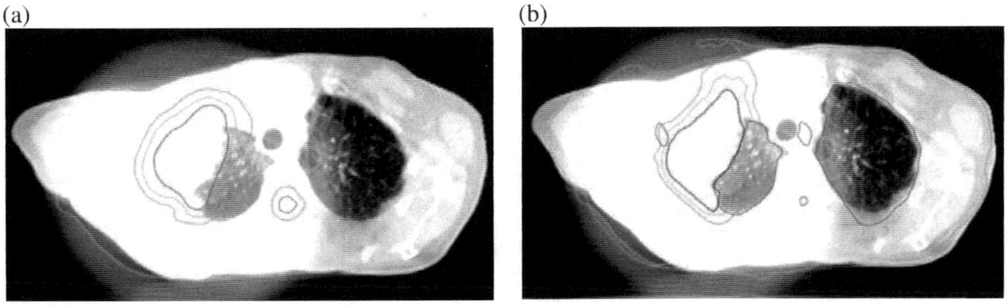

Figure 2. (a) The reference CBCT (fx 15) of patient 2 with manually delineated CBCT structures. **(b)** The same CBCT (fx 15) overlaid with the dCT structures, i.e. the propagated deformed pCT structures. The red, pink, blue and the outer light green delineations represent GTV, CTV, PTV and body contour, respectively. Colors are available in the online version.

No significant difference among DSC measures of the various structures were observed between DIBH and FB CT-CBCT and CT-CT image registrations over the course of treatment (table 1). However, the FB images had slightly higher DSC median value compared to DIBH images (both for CT-CBCT registrations and CT-CT registrations). This observation conflicts with our hypothesis that DIBH imaging resulting in visually less image artefacts compared to FB imaging, yielding better image registration. The applied DIR improved the median DSC measure, compared to rigid registrations, except for some CT-CBCT registrations (indicated by bold font), though still within one standard deviation, and no significant difference was observed.

Table 1. Median DSC measures (one standard deviation) of the propagated deformed structures and rigidly registered structures for all patients during CT-CBCT and CT-CT registration. Bold values indicate that rigidly registered structures have a higher DSC score compared to the propagated structures of the dCT.

Structures	CT-CBCT				CT-CT			
	FB		DIBH		FB		DIBH	
	RIG[a]	DIR[b]	RIG[a]	DIR[b]	RIG[a]	DIR[b]	RIG[a]	DIR[b]
GTV	0.74(0.10)	0.78(0.13)	0.73(0.13)	0.74(0.15)	0.76(0.13)	0.84(0.16)	0.75(0.13)	0.83(0.20)
CTV	0.80(0.07)	0.81(0.09)	**0.79(0.10)**	0.78(0.11)	0.82(0.09)	0.88(0.12)	0.79(0.11)	0.86(0.17)
PTV	0.83(0.07)	0.84(0.07)	**0.82(0.08)**	0.81(0.08)	0.86(0.07)	0.89(0.09)	0.83(0.10)	0.87(0.14)
Body	0.98(0.01)	0.98(0.01)	0.97(0.01)	0.98(0.04)	0.97(0.01)	0.99(0.01)	0.97(0.01)	0.99(0.01)
Medulla	**0.75(0.10)**	0.72(0.13)	**0.77(0.05)**	0.74(0.14)	0.74(0.05)	0.76(0.05)	0.75(0.08)	0.78(0.07)
Oesophagus	NaN[c]	NaN[c]	NaN[c]	NaN[c]	0.58(0.10)	0.73(0.08)	0.59(0.09)	0.70(0.08)
Heart	NaN[c]	NaN[c]	NaN[c]	NaN[c]	0.85(0.05)	0.90(0.06)	0.87(0.13)	0.93(0.08)
Total lung	NaN[c]	NaN[c]	NaN[c]	NaN[c]	0.89(0.01)	0.95(0.01)	0.93(0.04)	0.97(0.02)

[a] Rigid image registration
[b] Deformable image registration
[c] Not available number

4. Conclusion

This study does not support the hypothesis that DIBH images result in better image registrations (both rigid and deformed) when using a modified demons type algorithm for deformable image registration. Large variation was observed for organs without sharp contrast boundaries (including tumors close to mediastinum, atelectasis and thorax wall). Notice should also be taken to image artefacts and bad image quality that can affect the outcome of DIR (e.g. fx 15-20 for patient 2, figure 2). Our observations indicate that ART planning may be necessary during the course of treatment for optimal lung cancer RT, since the pCT not always can be registered correctly to subsequent CBCT and CT images over the course of treatment due to large and rapid volume changes (eg. fx 12 for patient 1, figure 1 (a)). Based on this small patient dataset, it turned out that DIR sometime resulted in a worse median DSC measure compared to conventional rigid registration, though still within one standard deviation. The median DSC of the DIR for these three patients was however mostly higher compared to rigid registrations. Unrealistic deformation vector fields were also sometimes created during DIR. These unrealistic deformation vector fields should carefully be employed if used for other applications, e.g. such as dose deformation. DIR analysis provides a feasible and promising tool for indicating if adaptive re-planning is necessary based on geometrical variations throughout the course of lung cancer treatment, with slightly better correlation for FB than DIBH imaging.

Acknowledgment
The authors greatly acknowledge Susanne Lind for help with coordinating the clinical workflow and Henriette Klitgaard Mortensen for coordinating the scanning personnel.

References
[1] Cho B C J, Bezjak A and Dawson L A 2010 Image guidance in non-small cell lung cancer. *Semin Radiat Oncol* **20** 164–70
[2] Sonke J-J and Belderbos J 2010 Adaptive radiotherapy for lung cancer. *Semin Radiat Oncol* **20** 94–106
[3] Barnes E A, Murray B R, Robinson D M, Underwood L J, Hanson J and Roa W H Y 2001 Dosimetric evaluation of lung tumor immobilization using breath hold at deep inspiration. *Int. J. Radiation Oncology Biol. Phys.* **50** 1091–8
[4] Mageras G S and Yorke E 2004 Deep inspiration breath hold and respiratory gating strategies for reducing organ motion in radiation treatment. *Semin Radiat Oncol* **14** 65–75
[5] Wang H, Dong L, O'Daniel J, Mohan R, Garden A S, Ang K K, Kuban D A, Bonnen M, Chang J Y and Cheung R 2005 Validation of an accelerated "demons" algorithm for deformable image registration in radiation therapy. *Phys. Med. Biol.* **50** 2887–905
[6] Ottosson W, Baker M, Hedman M, Behrens C F and Sjöström D 2010 Evaluation of setup accuracy for NSCLC patients; studying the impact of different types of cone-beam CT matches based on whole thorax, columna vertebralis, and GTV. *Acta oncologica* **49** 1184–91
[7] Sjöström D, Bjelkengren U, Ottosson W and Behrens C F 2009 A beam-matching concept for medical linear accelerators. *Acta oncologica* **48** 192–200
[8] Thirion J P 1998 Image matching as a diffusion process: an analogy with Maxwell's demons *Medical Image Analysis* **2** 243–60
[9] Dice L R 1945 Measures of the amount of ecologic association between species *Ecology* **26** 297–302

Journal of Physics: Conference Series **489** (2014) 012078 doi:10.1088/1742-6596/489/1/012078

Development of A Novel Image Guidance Alternative for Patient Localization using Topographic Images for TomoTherapy

X. Sharon Qi[1], Benjamin White and Daniel A. Low

Department of Radiation Oncology, University of California Los Angeles, Los Angeles, CA, USA

Email:xqi@mednet.ucla.edu

Abstract: To develop a faster and lower dose topogram based image registration for TomoTherapy as an alternative image guidance tool to volumetric megavoltage computed tomography (MVCT). Topogram procedures were performed for an anthropomorphic thorax phantom on a TomoTherapy HD unit (Accuray Inc., Sunnyvale, CA) using couch speeds from 1-4 cm/s and gantry angles of 0 and 90 degrees, other scanning parameters are: 1 mm imaging jaw, compression factor of 1, 30 seconds scanning duration with all multileaf collimators (MLCs) open. The raw exit detector data was exported after each scan. The topogram was reconstructed from a fan beam source for TomoTherapy beam and detector geometry at a SSD of 85 cm. A reference image, so called Digitally Reconstructed Topogram (DRT) was created by integrating the trajectories through the kVCT simulation with the topogram geometry. Image registration was performed by visually aligning the bony structure in topogram to the DRT. Image resolution was determined by the radius of curvature for the detector array, source to axis distance, source to detector distance, detector spacing, and number of detectors. The localization errors were 1.5, 2.5 mm in medio-lateral and anterior-posterior direction, larger errors in cranial-caudal direction was observed for faster couch speeds (i.e., ≥3cm/s). The topographic imaging time was 30 sec (versus 3-5 minutes for MVCT thorax scan) with imaging dose less than 1% of MVCT scan. Topograms with appropriate couch speed provide reliable patient localization images while significantly reducing pre-treatment imaging time. Topogram can be used as an alternative and/or additional patient alignment tool to MVCT on TomoTherapy.

1. Introduction

The unique geometric design and integrated on-board imaging system [1-5] allow for acquisition of megavoltage CT (MVCT) image but also topographic images on Tomotherapy Hi·ART system. The linear accelerator mounted on CT-like gantry, treatments are delivered with continuous and simultaneous gantry and couch rotation [1-3].

In this work, a fast and low-dose alternative (Topogram) to current 3D megavoltage CT (MVCT) localization tool for patient alignment on Tomo was presented. The Topogram could be employed as a precursor or post-treatment alignment verification tool for patient localization. In addition, Topogram offers large longitudinal field of view, which is desirable for cranial-spinal irradiation (CSI), total morrow irradiation (TMI), etc.

Journal of Physics: Conference Series **489** (2014) 012078 doi:10.1088/1742-6596/489/1/012078

The purposes of this work are 1) to investigate patient localization using topographic images on a TomoTherapy unit using various phantoms; and 2) to study the impact of different couch moving speeds on MV topogram localization.

2. Methods and Materials

Topogram procedures were created on TomoTherapy treatment console and for three anthropomorphic phantoms, including head (20 cm length), thorax (45 cm length) and spine (17 cm length) phantoms. The topogram procedures consisted of four couch speeds, 1 cm/s, 2 cm/s, 3 cm/s, and 4 cm/s (maximum couch speed) with a scanning duration of 30 seconds. All MLC leaves were opened to provide the largest field of view. A data compression factor of 1 was used for the topographic scans. An air scan was acquired to normalize the raw detector output. For patient localization, two topograms were acquired for each phantom and four different couch speeds at gantry angles of 0◦ and 90◦.

To obtain the correct reference image from the simulation CT, a digitally reconstructed topogram (DRT) was generated in the TomoTherapy beam and detector geometry [4-6]. To assess daily setup errors, known shifts were applied to the phantom for the topographic images. Image registration was performed by rigidly aligning the visible anatomy of the phantom in the topogram with the anatomy of the created DRT. MVCT scans were performed for the same phantoms using the normal imaging jaw width (4 mm). For MVCT imaging registration, we used bony and soft tissue to align the MVCT to the kVCT. The MVCT was set up with known shifts from the kVCT position. The shifts derived from topogram were compared to the MVCT image.

3. Results

3.1 The reference image and the MV Topograms on Tomo

A digitally reconstructed topogram (DRT), the reference image, was generated based on the CT simulation (KVCT) in the TomoTherapy beam and detector geometry. Figure 1 shows the reconstructed DRT for a thorax phantom. Figure 2 illustrates the reconstructed MV topogram at couch speeds of 1 cm/s and 4 cm/s (the fastest couch moving speed on Tomo).

Figure 1 The digitally reconstructed topogram for a thorax phantom.

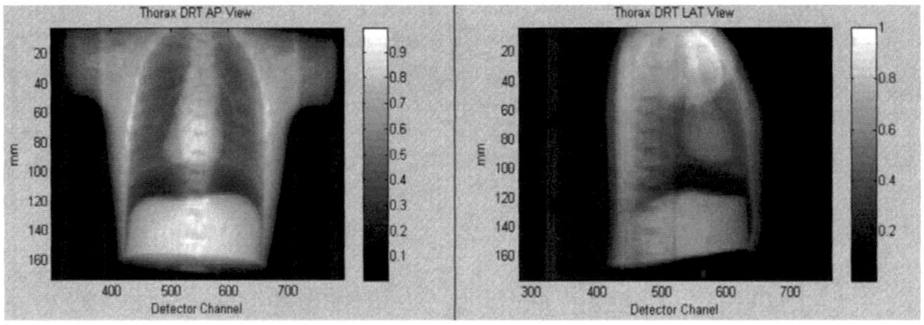

Figure 2 Reconstructed MV Topograms on Tomotherapy at couch speeds of 1 cm/s versus 4.0 cm/s.

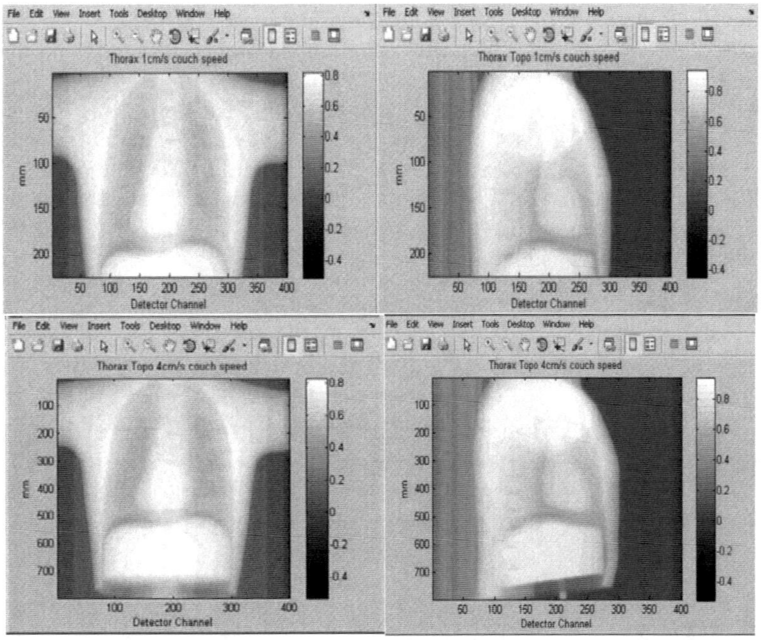

3.2 MV based Topogram registration

Topogram registration was performed by rigid aligning the visible anatomy of the phantom with the anatomy of the created DRT. Figure 3 illustrated MV based Topogram.

Figure 3 MV based Topogram registration for the thorax phantom.

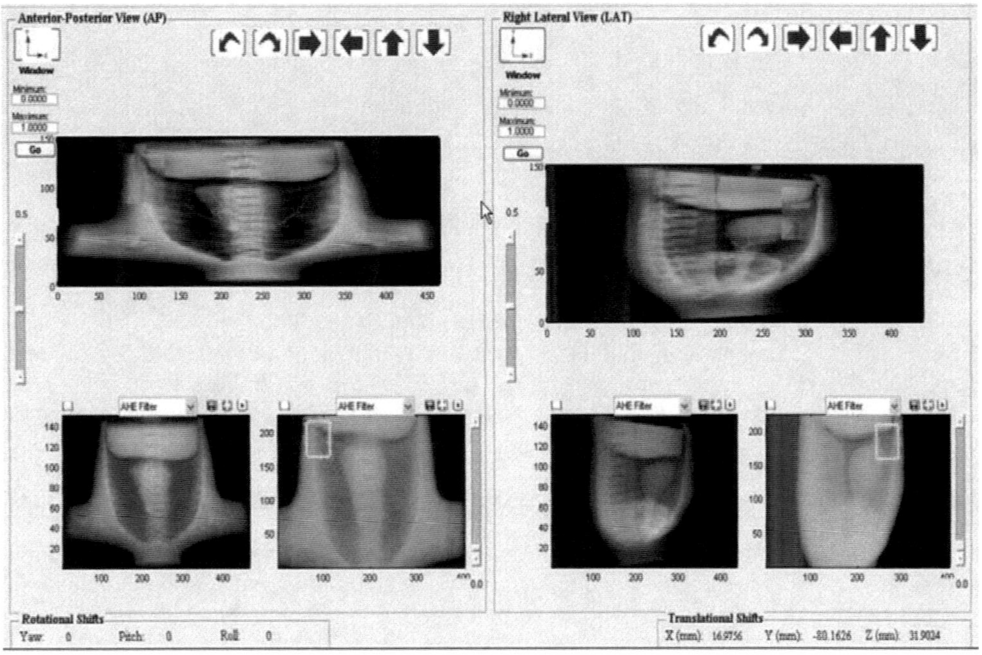

Table 1 shows relative positional errors using MVCT versus Topogram scans at different couch speeds. At maximum couch speed (4 cm/s), the topogram was markedly blurred in the cranial-caudal (CC) direction, and the blurred voxels gave a high positional error of 8.1 mm in the CC direction. The errors in the remaining directions, such as medio-lateral (ML), cranial-caudal (CC), were on the order of the slice thickness of the MVCT so are at an acceptable level. Based on these our preliminary measurements, a couch speed of 3 cm/s or slower might be the couch speed achievable with reasonable positional errors due to voxel deformation in the topogram. At couch speeds of 1 cm/s, 2 cm/s and 3 cm/s, the relative shifts were within 5 mm in all ML, CC and AP directions; while for faster couch speed of 4 cm/s, the relative shifts between different couch speeds might be larger in AP direction.

Table 1 Relative positional errors using Topogram versus MVCT Tomotherapy.

Direction	MVCT (mm)	Topo (mm)
ML	2	1.5
CC	2	8.1
AP	2	2.5

3.3 Scanning image time

Compared with the imaging acquisition time of 2-5 min for MVCT scans, the time required to acquire clinically acceptable topograms for the selected phantom were significantly shorter using topograms. Based on these results, a couch speed of 3 cm/s is the fastest couch speed achievable with reasonable positional errors due to voxel deformation in the topogram as shown in Table 1.

4. Conclusion

Topograms can be a fast alternative and/or additional patient alignment tool to megavoltage CT guidance on Tomotherapy. The implement of topogram on Tomotherapy may significantly reduce pre-treatment MVCT imaging time while provides reliable patient localization information compared to current MVCT guidance.

References

[1] Mackie TR, Holmes T., Swerdloff S., Reckwerdt P., Deasy JO, Yang J, Paliwal B., Kinsella T. Tomotherapy: a new concept for the delivery of dynamic conformal radiotherapy. Med. Phys., 20(6):1709-1719 (1993).

[2] Mackie TR, Kapatoes J., Ruchala K., Lu WG, Wu C., Olivera G., Forrest L., Tome W., Welsh J., Jeral R., Harari P., et al. Image guidance for precise conformal radiotherapy. Int. J. Radiat. Oncol. Biol. Phys., 56(1): 89-106 (2003).

[3] Mackie TR. History of tomotherapy. Phys. Med. Biol. 51:R427-453 (2006).

[4] Meeks SL, Harmon JF, Langen KM, et al. Performance characterization of megavoltage computed tomography imaging on a helical tomotherapy. Med. Phys. 32: 2673-81 (2005).

[5] Keller H., Glass M., Hinderer R, et al. Monte Carlo study of a highly efficient gas ionization detector for megavoltage imaging and image-guided radiotherapy. Med. Phys. 29(2): 165-175 (2002).

[6] Moore KL, Palaniswaamy G., White B., Goddu AM and Low DL. Fast, low-dose patient localization on Tomotherapy via topogram registration. Med Phys. 37(8): 4068-4077 (2010).

XVII International Conference on the Use of Computers in Radiation Therapy (ICCR 2013) IOP Publishing
Journal of Physics: Conference Series **489** (2014) 012079 doi:10.1088/1742-6596/489/1/012079

The Reconstruction Toolkit (RTK), an open-source cone-beam CT reconstruction toolkit based on the Insight Toolkit (ITK)

S Rit[1], M Vila Oliva[1,2], S Brousmiche[2], R Labarbe[2,3], D Sarrut[1], G C Sharp[4]

[1]Université de Lyon, CREATIS; CNRS UMR5220; Inserm U1044; INSA-Lyon; Université Lyon 1; Centre Léon Bérard, France

[2]iMagX project, ICTEAM Institute, Université catholique de Louvain, Louvain-la-Neuve, Belgium

[3]IBA, B-1348, Louvain-la-Neuve, Belgium,

[4]Department of Radiation Oncology, Massachusetts General Hospital, Boston, MA 02114, USA

E-mail: simon.rit@creatis.insa-lyon.fr

Abstract. We propose the Reconstruction Toolkit (RTK, http://www.openrtk.org), an open-source toolkit for fast cone-beam CT reconstruction, based on the Insight Toolkit (ITK) and using GPU code extracted from Plastimatch. RTK is developed by an open consortium (see affiliations) under the non-contaminating Apache 2.0 license. The quality of the platform is daily checked with regression tests in partnership with Kitware, the company supporting ITK. Several features are already available: Elekta, Varian and IBA inputs, multi-threaded Feldkamp-David-Kress reconstruction on CPU and GPU, Parker short scan weighting, multi-threaded CPU and GPU forward projectors, etc. Each feature is either accessible through command line tools or C++ classes that can be included in independent software. A MIDAS community has been opened to share CatPhan datasets of several vendors (Elekta, Varian and IBA). RTK will be used in the upcoming cone-beam CT scanner developed by IBA for proton therapy rooms. Many features are under development: new input format support, iterative reconstruction, hybrid Monte Carlo / deterministic CBCT simulation, etc. RTK has been built to freely share tomographic reconstruction developments between researchers and is open for new contributions.

Cone-beam CT scanners are used worlwide for guiding radiation therapy since it has been proposed to mount them on the gantry of linear accelerators [1]. Several commercial solutions are available today based on this principle and it is currently being extended to gantries for proton therapy. In contrast to diagnostic CT scanners, these scanners allow access to raw data, i.e., the set of 2D projection images and the corresponding acquisition geometry, and many research groups have been applying their new reconstruction algorithms to patient data acquired for image-guided radiotherapy (IGRT).

A large part of these developments are internal to each research group and are not publically available. An exception is Plastimatch (http://www.plastimatch.org/) which provides fast cone-beam CT reconstruction. Extending Plastimatch to new CT reconstruction algorithms is not straightforward because the development model is not modular. This observation has lead to the launch of the RTK initiative in collaboration with the Plastimatch team.

1. Method

RTK (`http://www.openrtk.org/`) is an open-source and cross-platform toolkit for fast tomographic reconstruction based on the Insight ToolKit (ITK). RTK is developed by the RTK consortium under the same Apache 2.0 license as ITK v4, which is a non contaminating BSD-like license approved by the Open Source Initiative. The current members of the RTK consortium are CREATIS, the Massachusset General Hospital, the Université catholique de Louvain and IBA. The members are bound by a gentleman's agreement available from the website.

1.1. Development tools

RTK is developed in collaboration with Kitware and uses the software process suite: CMake, CDash, CTest, etc. The source repository is hosted on GitHub. The user's documentation is available on the RTK wiki and the developer's documentation uses Doxygen. Tests and examples datasets are hosted on MIDAS, a Kitware open-source server for managing data. The quality of the platform is daily checked with regression tests.

1.2. Dependencies

Insight Toolkit (ITK) RTK is based on ITK v3.20 or v4.3 which is its only mandatory dependency. Many ITK features are essential to the functioning of RTK: input-output factories, filters pipelining, streaming, mini-pipelines, multi-threading, smart pointers, FFTW wrapping, etc. Every RTK filter is meant to be ITK compliant and RTK can be thought of as an ITK module.

Gengetopt (`http://www.gnu.org/software/gengetopt/gengetopt.html`) RTK also provides a set of command-line applications that use Gengetopt to define their options. Gengetopt is a tool to generate command line option parsing code for C programs.

GPU: CUDA and OpenCL RTK proposes CUDA implementation of a few time-consuming filters. Part of the code has been taken and adapted from Plastimatch and NiftyRec (`http://sourceforge.net/projects/niftyrec/`), two other open-source tomographic packages. RTK also proposes OpenCL implementation of one filter but less efforts have been put so far in using OpenCL.

1.3. Features

3D circular geometry RTK has yet been developed for the 3D cone-beam CT geometry only, i.e., a source and a flat panel rotating along a circle. Deviations from this strict geometry can be described for each projection image by means of nine degrees of freedom in total: 3 coordinates for the position of the punctual source, 3 coordinates for the flat panel position and 3 angles for the flat panel orientation. RTK provides classes to read and write the geometry in its own XML file format as well as readers for the geometry of several commercial scanners.

Input and Output file formats RTK uses the ITK factories for the input and output of projection images. As for the geometry, we also provide new image readers for several commercial scanners. Simple conversion from raw data to linear attenuation is provided which we aim at improving in the future with calibration, beam hardening correction and scatter correction to reach quantitative CT, i.e., pixel values in Hounsfield units.

Forward- and back-projectors At the core of tomographic reconstruction are forward- and back-projectors. They have been designed in a typical object-oriented fashion to allow testing of various implementations. GPU versions are also provided for faster reconstruction.

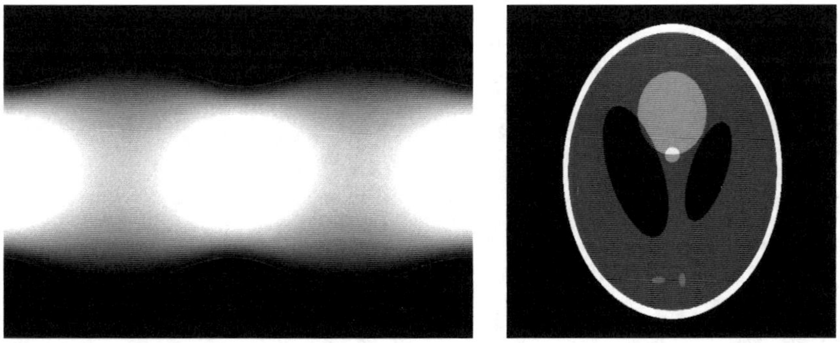

Figure 1. Slices of the 3D simulated sinogram (left) and the corresponding CT image (right) of the Shepp-Logan phantom obtained with RTK.

Reconstruction algorithms The main reconstruction algorithm that is currently implemented is the Feldkamp-David-Kress (FDK) algorithm [2]. Weighting of the projection images prior to FDK reconstruction is available to handle short scans [3] and displaced detector scans for the reconstruction of larger field-of-views [4]. There is also an implementation of the Simultaneous Algebraic Reconstruction Technique (SART) [5] and other iterative algorithms should follow.

Numerical phantoms RTK can simulate projection images and reference CT images of geometric numerical phantoms such as the Shepp-Logan phantom (Figure 1). These phantoms are the most commonly used to validate tomographic reconstruction algorithms. They are also used in RTK functional tests that are nightly run on the dashboard.

2. Results

2.1. Numerical phantom

Figure 1 illustrates RTK ability to simulate sinograms and reference CT images of the Shepp Logan phantom. Examples of reconstruction times are provided in Table 1.

Table 1. Multi-threaded CPU vs. CUDA GPU timing of several RTK filters. The volume size was 512^3 voxels and 512 projection images with 512^2 pixels were used. The measurement excludes hard drive input/output but includes the required transfers between the RAM and the memory of the GPU. The tests have been peformed with 48 threads on a cluster node equipped with two Intel Xeon E5-2630L CPUs and one Nvidia Tesla M2090 GPU.

	CPU	GPU
FDK ramp filtering	83 s (Vnl) 17 s (FFTW)	3 s (CuFFT)
FDK backprojection	110 s	22 s
Forward projection	84 s	17 s

2.2. Catphan acquisitions

The axial slice of the CT image of a quality assurance phantom is provided for real projection images acquired on 3 different cone-beam CT scanners: Elekta Synergy, Varian OBI and IBA

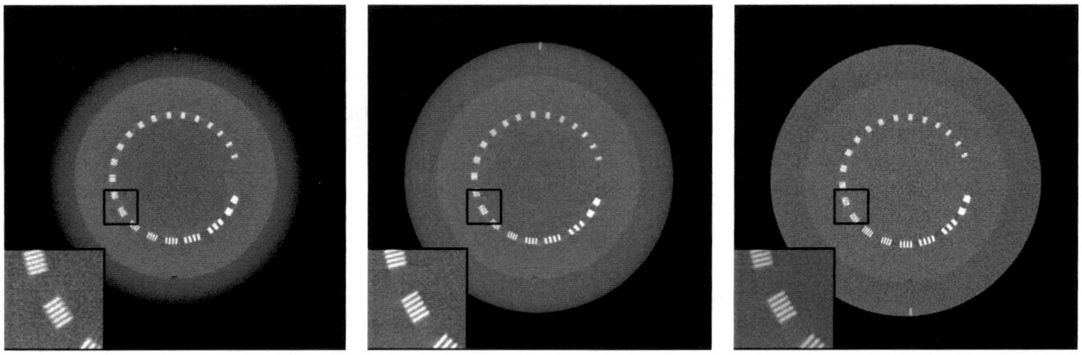

Figure 2. Axial slice with a zoom of cone-beam CT images of the CTP 528 resolution module of Catphan phantoms, reconstructed with the RTK implementation of the FDK algorithm from raw data acquired on an Elekta Synergy in France (left), on a Varian OBI in the USA (middle), and on an IBA prototype scanner in Belgium (right). Raw datasets are publicly available.

ImagX testbench for their prototype scanner. The phantom was the Catphan CTP 503 for the first two and the Catphan 600 for the later.

3. Discussion and Conclusion

RTK is a software toolkit and its code is meant to be used in other platforms. We have already used it to develop proton CT [6] and 2D/3D registration in Elastix [7, 8]. We will also use RTK in Gate [9] to enhance its X-ray simulation capabilities. Others have used it via the command line tools [10]. RTK will also be used in the new cone-beam CT scanner for proton therapy that will be sold by IBA.

The development of RTK is on-going and we intend to add many more features, e.g., new forward- and back-projectors, iterative reconstruction algorithms, etc. Other geometries could also be added, for example the 3D parallel geometry that is currently being developed. RTK is open to new users and developers and those who have made a significant contribution to the platform could join the RTK consortium.

Acknowledgments

Funding for this project has been provided by iMagX, a private-public partnership between IBA and the UCLouvain funded by the convention #1017266 with the Walloon Region of Belgium.

References

[1] Jaffray D, Siewerdsen J, Wong J and Martinez A 2002 *Int J Radiat Oncol Biol Phys* **53** 1337–1349
[2] Feldkamp L, Davis L and Kress J 1984 *J Opt Soc Am A* **1** 612–619
[3] Parker D 1982 *Med Phys* **9** 254–257
[4] Wang G 2002 *Med Phys* **29** 1634–1636
[5] Andersen A and Kak A 1984 *Ultrason Imaging* **6** 81–94
[6] Rit S, Dedes G, Freud N, Sarrut D and Létang J 2013 *Med Phys* **40** 031103 (pages 9)
[7] Klein S, Staring M, Murphy K, Viergever M and Pluim J 2010 *IEEE Trans Med Imaging* **29** 196–205
[8] Delmon V, Vandemeulebroucke J, Pinho R, Vila Oliva M, Sarrut D and Rit S 2013 *International Conference on the Use of Computers in Radiation Therapy (ICCR)* (Melbourne, Australia)
[9] Jan S, Benoit D, Becheva E, Carlier T, Cassol F, Descourt P, Frisson T, Grevillot L, Guigues L, Maigne L, Morel C, Perrot Y, Rehfeld N, Sarrut D, Schaart D, Stute S, Pietrzyk U, Visvikis D, Zahra N and Buvat I 2011 *Phys Med Biol* **56** 881–901
[10] Martin J, McClelland J, Yip C, Thomas C, Hartill C, Ahmad S, O'Brien R, Meir I, Landau D and Hawkes D 2013 *Phys Med Biol* **58** 1809–1822

Journal of Physics: Conference Series **489** (2014) 012080 doi:10.1088/1742-6596/489/1/012080

Potential interoperability problems facing multi-site radiation oncology centers in The Netherlands

J Scheurleer[1], Ph Koken[2] and R Wessel[2]

[1] *Inholland University of Applied Sciences, Bijdorplaan 15, 2015 CE Haarlem, The Netherlands*
[2] *Dept. of Radiation Oncology, VU University Medical Center (VUmc), Amsterdam, The Netherlands*

E-mail: jelle.scheurleer@inholland.nl

Abstract Aim: To identify potential interoperability problems facing multi-site Radiation Oncology (RO) departments in the Netherlands and solutions for unambiguous multi-system workflows. Specific challenges confronting the RO department of VUmc (RO-VUmc), which is soon to open a satellite department, were characterized. Methods: A nationwide questionnaire survey was conducted to identify possible interoperability problems and solutions. Further detailed information was obtained by in-depth interviews at 3 Dutch RO institutes that already operate in more than one site. Results: The survey had a 100% response rate (n=21). Altogether 95 interoperability problems were described. Most reported problems were on a strategic and semantic level. The majority were DICOM(-RT) and HL7 related (n=65), primarily between treatment planning and verification systems or between departmental and hospital systems. Seven were identified as being relevant for RO-VUmc. Departments have overcome interoperability problems with their own, or with tailor-made vendor solutions. There was little knowledge about or utilization of solutions developed by Integrating the Healthcare Enterprise Radiation Oncology (IHE-RO). Conclusions: Although interoperability problems are still common, solutions have been identified. Awareness of IHE-RO needs to be raised. No major new interoperability problems are predicted as RO-VUmc develops into a multi-site department.

1. Introduction

A general definition of interoperability is "The ability of systems and organizations to effectively collaborate". Systems can interact with each other at different levels, each of which may have its own specific requirements [1]. Interoperability problems can occur in three areas:
(1) Technical interoperability - required so that systems can exchange data.
(2) Tactical interoperability - requires an unambiguous message structure.
(3) Semantic interoperability - requires correct interpretation of the delivered information.

In the past most Radiation Oncology (RO) departments used equipment from a single vendor, and were based at one location with a limited, mostly non-electronic, data stream. Due to the increasing use of medical imaging techniques and the rise of interdisciplinary cancer care, the information flow has increased significantly [2] and is no longer limited to the department itself. A substantial data stream comes from other departments in the hospital (intramural) and from departments outside the hospital (extramural). In addition, communication has become not only faster, but also more complex with less human intervention.

Despite an increasing number of medical standards such as DICOM and HL7, successful operational integration of medical information systems still cannot be guaranteed nor can interoperability be ensured [1-6]. It is suggested [7] that interoperability problems in health care are not specific for the present time, but their impact has increased. To improve interoperability the HIMSS (Healthcare Information and Management Systems Society) and the RSNA (Radiological Society of North America) started a joint initiative called 'Integrating the Healthcare Enterprise' (IHE) in 1998. This is a partnership between providers and professionals. IHE promotes the use of established standards rather than developing them [8, 9] and defines a technical framework for each field of work. The framework consists of integration profiles that accurately describe the processes and tasks in that field. In addition, an information model is developed which describes the messages that need to be exchanged between different systems in order to successfully complete a given process [8]. Initially, IHE focused on diagnostic radiology alone but nowadays 9 work fields are covered, including radiation oncology (IHE-RO).

In December 2012 the VU University medical center (VUmc) in Amsterdam, The Netherlands opened its first satellite RO facility located about 50 km away in a general hospital. In January 2011 we began a study to identify potential interoperability issues in the RO domain associated with the new configuration of the department. The aims of this study were (1) to identify interoperability problems at multi-site RO departments in the Netherlands, and (2) to identify possible interoperability problems at the VUmc RO department specifically in order to enable unambiguous workflows in the multi-center situation.

2. Methods

An online nationwide questionnaire survey was conducted to identify possible interoperability problems and solutions that had been applied. The questionnaire consisted of 52 questions in three parts. The first part focused on the equipment and the information systems used in any given department. The second part of the questionnaire addressed interoperability issues that the departments had experienced. The strategies used to tackle interoperability problems were subject of the third part of the questionnaire. In the case of initial non-response, this was followed up by mails and telephone calls. A pilot was conducted at two departments before the optimized questionnaire was sent to all Dutch departments (n=21)

Further detailed information was obtained by in-depth semi-structured interviews in 3 Dutch RO institutes already operating as a multi-site department (n=2) or with advanced plans for a satellite department (n=1).

3. Results

The final response to the survey was 100% (n=21). The first part of the survey, which describes the complexity of the department, confirmed that departments use a variety of information systems. None of the departments exclusively used systems from a single vendor. Almost all departments (19) also used software that had been developed in-house (e.g. e-chart, independent dose-verification) and 17 RO departments shared imaging modalities with other departments in their facility, e.g. departments of radiology and nuclear medicine.

Pooled results from the second part of the survey identified 95 interoperability issues, of which 65 (69%) were DICOM or HL7 related (Table 1). A substantial number of these (n=28, 29%) were DICOM-RT related and frequently hindered communication between the treatment planning system (TPS) and the record-and-verify (R&V) system at the radiotherapy treatment machines. Typical

Journal of Physics: Conference Series **489** (2014) 012080 doi:10.1088/1742-6596/489/1/012080

examples included different interpretation of DICOM tags and the incomplete transfer of treatment data between the TPS and R&V system. To solve such issues four departments had developed their own applications. Intramural issues (n=74) were mainly related to DICOM (n=47) and HL7 (n=12) interoperability problems and predominantly hampered workflow applications and the Hospital Information System (HIS). To overcome these issues data were often manually copied from one system to another. Two departments indicated that these issues were caused by not permitting the import of patient data from the PACS or HIS of other hospitals for safety or legal reasons. Consequently, these data need to be transferred on CD or manually. Similar problems, causes and consequences were observed at the extramural level.

Table 1. The interoperability problems reported by the 21 Dutch Radiation Oncology departments categorized to the standard they were related to.

	Within Dept	Intramural	Extramural	**Total**
DICOM	12	7	5	**24**
DICOM-RT	26	2	0	**28**
HL-7	2	10	2	**14**
Remaining	7	8	14	**29**
Total	**47**	**27**	**21**	**95**

The third and last part of the survey identified which strategies were being used to prevent interoperability problems and demonstrated a department's acquaintance with the work and role of IHE-RO (Figure 1). Only slightly more than a quarter of the departments (n=6) negotiated solutions with the vendor or with third parties; solutions were always dedicated to the local specific situation rather than generic ones, i.e. applicable for similar situations. About 50 % of the departments solved problems on their own (n=10), some of them even developed their own applications if necessary (n=4). Only 2 departments prevented problems by reducing data-transfer between systems. The remaining three departments opted for a one-vendor solution (n=1) to prevent interoperability problems or accepted the existence of interoperability problems (n=2). Knowledge of IHE-RO was limited in all departments; Although 15 of the respondents were aware of its existence only 1 department had implemented a single IHE-RO profile; the reason was that own solutions were favored over the IHE-RO solution.

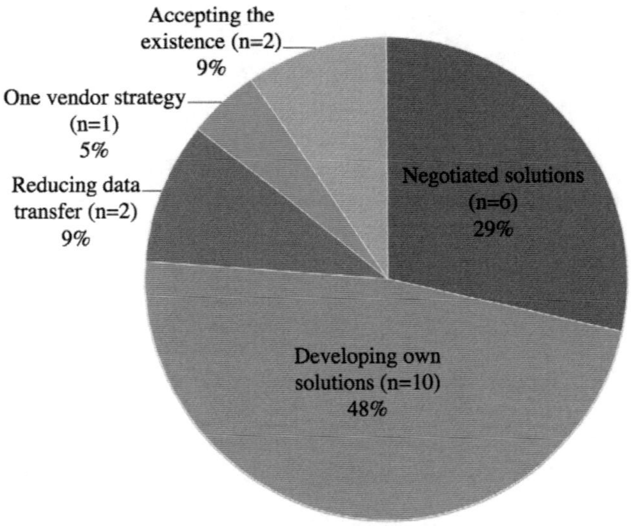

Figure 1: The strategies used by the 21 Dutch Radiation Oncology departments to prevent or overcome interoperability issues.

The further in-depth interviews confirmed the results described above, and with regard to multi-center institutions, three points of interest were highlighted. First, the use of paper charts can sometimes lead to urgent and immediate problems (e.g. chart is not present on the right location or the chart is missing); for this reason 2 out of the 3 departments worked in a paperless environment and the third one had advanced plans to become paperless. Second, concerning the availability of various specialists: all 3 departments had one or more physicians at the satellite location. Only one department also had a physicist available on location and the other two used 'cockpit' approaches, meaning that information on the satellite unit's consoles could be viewed and (if necessary) controlled from the home-location. Third, all 3 departments had enlarged their computer-network to the satellite location as a bulge and mirrored networks were installed to ensure continuity in case of failures, e.g. network loss or unplanned downtime.

4. Discussion

This study investigated interoperability problems in RO departments in The Netherlands and how they were being addressed. Specifically, the situation in our own department was investigated in order to prevent possible interoperability problems in a multi-site environment.

Although technical interoperability problems are rare, the real challenges are to acquire and maintain interoperability on both a tactical and semantic level. Interoperability problems within a department predominantly occurred between the treatment planning and the record-and-verify systems. Intramural interoperability problems mainly took place between the workflow applications and the HIS, and could significantly increase the workload if data had been manually copied. Most of these problems were DICOM and HL7-related, some were due to institutional safety policies. Similar extramural interoperability problems occurred mainly because of financial and legal reasons.

The current freedom of choice in implementing standards is a weak point because it can lead to interoperability problems. This will only be solved by consensus. Naturally, a balance between flexibility and rigidity is essential for consensus; on the one hand, a standard which leaves too much room for interpretation still risks producing interoperability problems . On the other hand, the use of new technological developments may be hindered if these are only allowed after the adoption of standards. Therefore interoperability is a long-term goal, requiring the continuous effort of all parties to reach a detailed consensus on how standards should be applied.

The IHE could play a role in radiation oncology in The Netherlands. Their workgroups include both users and vendors, which is necessary for wider acceptance. However, users in The Netherlands are insufficiently aware of IHE procedures and how they can enhance interoperability. A shared initiative from the already existing Dutch board "Information Technology (IT) in RO" could possibly give new input to the IHE-RO group in the development of Integration Profiles. Additionally the government could contribute by creating the proper preconditions for a necessary legal framework concerning safety and privacy at an extramural level.

5. Conclusions

An interoperability study, based on both a national survey and in-depth interviews with selected departments, was conducted with the participation of all Dutch radiation oncology departments. We showed that

(1) Interoperability problems on a technical level are rare and mainly occur between the TPS and the R&V systems within a department, or between RO-dedicated systems and the HIS on an intramural level;

(2) Only a few of the identified problems in a national survey are relevant for our own department's growth towards a multi-site center;

(3) The majority of the identified interoperability issues could be solved by already existing IHE-RO profiles, or profiles that are being developed.

References

[1] Curran B 2007 IHE - Radiation Oncology. The International Health Enterprise

[2] Swerdloff S J 2007 Data handling in radiation therapy in the age of image-guided radiation therapy *Semin Radiat Oncol* **17** 287-92

[3] Channin D S, Siegel E L, Carr C and Sensmeier J 2001 Integrating the healthcare enterprise: a primer. Part 5. The future of IHE *Radiographics* **21** 1605-8

[4] Bosch W, Curran B and Swedloff S 2010 IHE-RO: Interoperable Data Standards for Radiation Oncology. In: *ICRR 2010,* (Amsterdam

[5] Ciottl V 2010 Generation HL7. Of the 15 founding fathers of HL7, just one remains today, leaving many to question whether interoperability is attainable *Healthc Inform* **27** 48

[6] Siegel E L and Channin D S 2001 Integrating the Healthcare Enterprise: a primer. Part 1. Introduction *Radiographics* **21** 1339-41

[7] Knaup P and Dickhaus H 2009 Perspectives of medical informatics: advancing health care requires interdisciplinarity and interoperability. Special topic on the occasion of the 35th anniversary of the Heidelberg/Heilbronn curriculum of medical informatics *Methods Inf Med* **48** 1-3

[8] Channin D S 2001 Integrating the Healthcare Enterprise: a primer. Part 2. Seven brides for seven brothers: the IHE integration profiles *Radiographics* **21** 1343-50

[9] Channin D S, Parisot C, Wanchoo V, Leontiev A and Siegel E L 2001 Integrating the Healthcare Enterprise: a primer: Part 3. What does IHE do for ME? *Radiographics* **21** 1351-8

Journal of Physics: Conference Series **489** (2014) 012081 doi:10.1088/1742-6596/489/1/012081

Target point correction optimized based on the dose distribution of each fraction in daily IGRT

Markus STOLL[1], Kristina GISKE[1], Eva M STOIBER[1,2], Michael SCHWARZ[1] and Rolf BENDL[1,3]

[1]Division of Medical Physics in Radiation Oncology, German Cancer Research Center (DKFZ), Heidelberg, Germany;

[2]Department of Radiation Oncology, University Hospital Heidelberg, Heidelberg, Germany;

[3]Faculty of Informatics, Heilbronn University, Heilbronn, Germany;

E-Mail: m.stoll@dkfz.de

Abstract. **Purpose**: To use daily re-calculated dose distributions for optimization of target point corrections (TPCs) in image guided radiation therapy (IGRT). This aims to adapt fractioned intensity modulated radiation therapy (IMRT) to changes in the dose distribution induced by anatomical changes. **Methods**: Daily control images from an in-room on-rail spiral CT-Scanner of three head-and-neck cancer patients were analyzed. The dose distribution was re-calculated on each control CT after an initial TPC, found by a rigid image registration method. The clinical target volumes (CTVs) were transformed from the planning CT to the rigidly aligned control CTs using a deformable image registration method. If at least 95% of each transformed CTV was covered by the initially planned D95 value, the TPC was considered acceptable. Otherwise the TPC was iteratively altered to maximize the dose coverage of the CTVs. **Results**: In 14 (out of 59) fractions the criterion was already fulfilled after the initial TPC. In 10 fractions the TPC can be optimized to fulfill the coverage criterion. In 31 fractions the coverage can be increased but the criterion is not fulfilled. In another 4 fractions the coverage cannot be increased by the TPC optimization. **Conclusions**: The dose coverage criterion allows selection of patients who would benefit from replanning. Using the criterion to include daily re-calculated dose distributions in the TPC reduces the replanning rate in the analysed three patients from 76% to 59% compared to the rigid image registration TPC.

1. Introduction

Patient positioning during treatment delivery of fractioned IMRT can be realized by a stereotactic setup. Uncertainties in this setup and anatomical changes, caused by different organ filling levels, tumor shrinkage or weight loss, make an additional correction of the patient position necessary. This can be realized by an additional table shift to change the target point – a target point correction (TPC).

In image guided radiation therapy (IGRT) a control CT scan can be used to calculate a TPC with image based methods like rigid image registration in the tumor region. Anatomical changes can be taken into account by using deformable image registration approaches [1]. The effects of anatomical changes to the dose distribution are still a matter of current research. In this work we extend the image-derived TPC calculation by incorporating the dosimetric changes.

2. Methods and Material

We use a deformable image registration to propagate the volumes of interest (VOIs) from planning CT to control CT (see Figure 1), which is acquired with an in-room on-rail spiral CT. The dose is re-calculated on the control CT and a TPC is calculated to maximize dose coverage of the propagated clinical target volumes (CTVs).

We evaluate our dosimetric TPC calculation method by comparing its results to a TPC calculated with rigid image registration based on mutual information in a retrospective study.

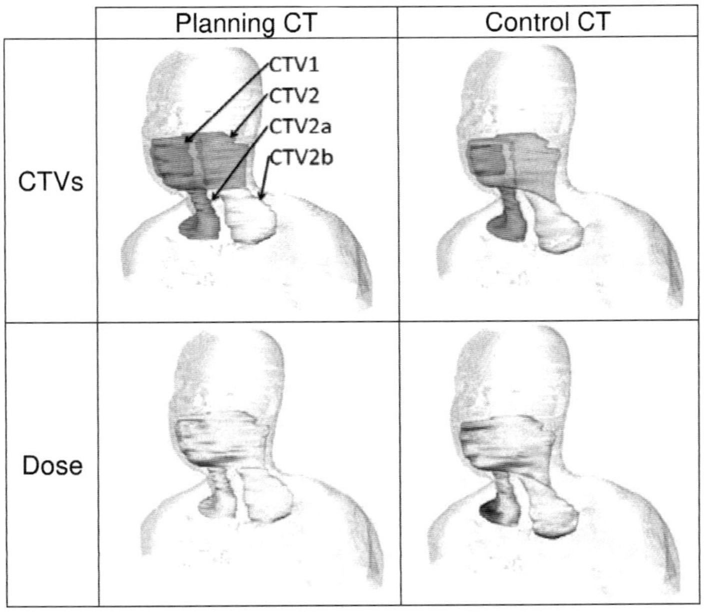

Figure 1: CTVs in planning-CT (top-left) are transformed to the control CT (top-right) using an elastic image registration method. If the dose distribution of the planning CT (shown on CTV surfaces bottom-left) is re-calculated on the control CT, under-dosage may occur (shown as blue areas on the transformed VOI surfaces bottom-right).

2.1. Patients

Three head-and-neck cancer patients, who received postoperative IMRT in 28-32 fractions, were included in the retrospective study. 68 daily images (30 of patient 1, 30 of patient 2 and 8 of patient 3) acquired with an in-room on-rail spiral CT-Scanner were available and enabled dosimetric TPC calculations. All patients had a customized fixation device, which was used for patient immobilization and stereotactic setup.

2.2. IMRT plans

Two CTVs (CTV1, CTV2) were defined for each patient. A dose of 70.4 Gy was prescribed to the CTV1, which contains only the gross tumor region. The CTV2, where a dose of 57.6 Gy was prescribed, includes the cervical and supraclavicular lymph nodes. The CTV2 was split into three parts (see Figure 1, top-left) to allow separated coverage analyses of the three partial volumes, which have different deformability. A 3.0 mm CTV-to-PTV margin was applied. The maximum dose for the spinal cord was limited to 40 Gy.

2.3. TPC optimization

The optimization of the TPC is performed in four steps (see Figure 2). Steps 2-4 are repeated until an acceptable coverage was found or a maximum number of iterations was reached.

2.3.1. Registration and VOI propagation

The control CT is rigidly aligned to the planning CT using stereotaxy. A displacement vector field is calculated using a fully automated deformable image registration method to propagate the contours of all CTVs from the planning CT to the control CT [2]. The propagated contours are reviewed by a radiation oncologist to exclude fractions with registration errors.

2.3.2. Dose re-calculation on control CT after TPC
The target point of the plan is modified according to the TPC vector and the dose distribution is re-calculated on the control CT. In the first iteration the TPC vector is initialized using rigid image registration. The result of the coverage optimization is used in the following iterations.

2.3.3. Dose coverage calculation
The D95 values of the PTVs are extracted from the DVHs of the initial plan and used as a threshold dose value for the corresponding CTV. Coverage for each CTV of more than 95% was chosen as acceptance criterion. If this criterion is fulfilled or a maximum number of iterations is reached, the optimization is stopped.

The coverage is calculated using a mask for each CTV marking the dose region having dose values above the corresponding threshold. The overlap of each region with the corresponding voxel grid mask of the VOI, normalized by the total volume of the VOI, is the coverage.

2.3.4. Coverage optimization
A simplex optimizer alters the TPC translation vector to reach coverage of 95% or more for all CTVs. During this step the dose distribution is assumed to be invariant in respect to the TPC. Thus the dose distribution is re-calculated in the next step (2.3.2) to allow an exact calculation of the coverage.

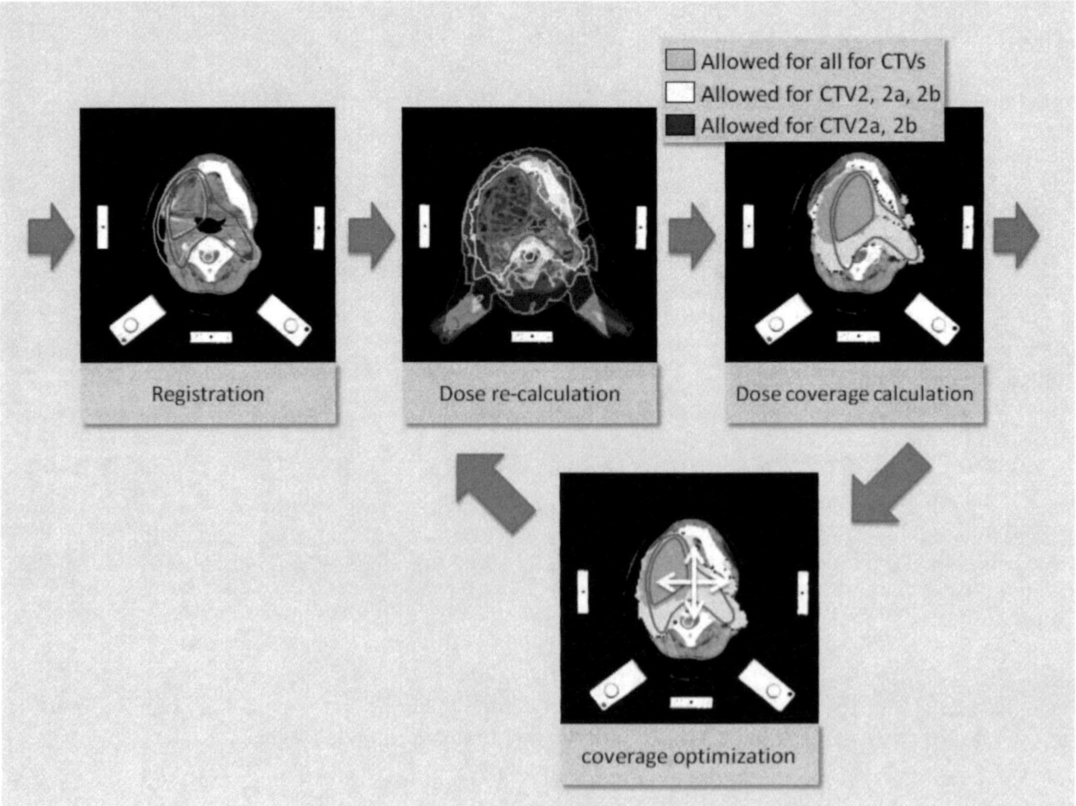

Figure 2 optimization workflow: The VOIs are propagated using a deformable image registration, the dose is re-calculated on the control CT, areas having dose values above CTV specific threshold are extracted to calculate the coverage and the coverage is optimized by altering the TPC vector.

3. Results

59 out of 68 fractions of all three patients passed the review of the propagated contours. In 24% (14 out of 59) of these fractions the coverage criterion was already fulfilled after the initial image based TPC without any coverage optimization steps. In 17% of these fractions (10 out of 59) the TPC was successfully optimized to fulfill the coverage criterion. In 53% of these fractions (31 out of 59) the coverage could be increased but the criterion could not be fulfilled after of 10 iterations. In 7% of these fractions (4 out of 59) the coverage could not be increased by the TPC optimization. Figure 3 shows the volumes having a lower dose value than the corresponding dose threshold. The 5% tolerance of the coverage criterion was subtracted.

To conform to the initially planned dose distribution a new plan optimization is needed. Our method supplies an objective measure to select candidates for replanning

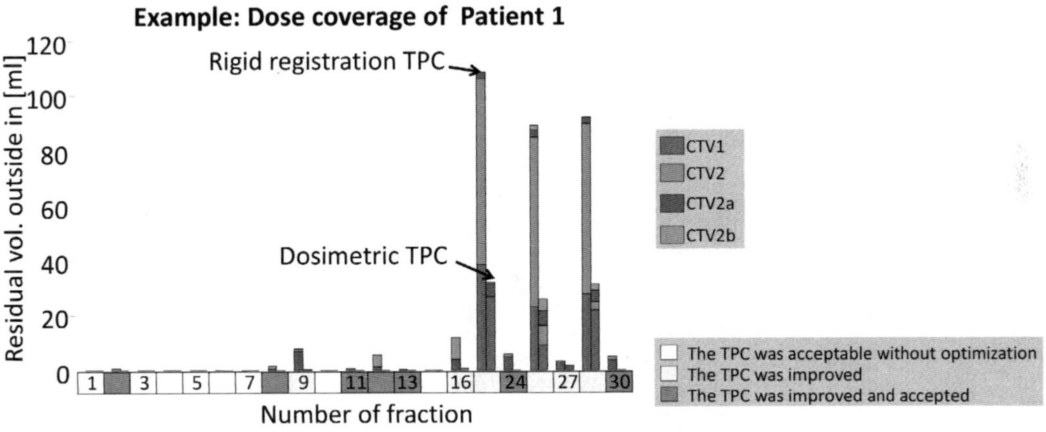

Figure 3: Volume outside corresponding dose threshold for all CTVs

4. Conclusions

The proposed dose coverage criterion can help to identify fractions with under-dosages, in which case replanning is suggested. Using the criterion to optimize the TPC reduces the replanning rate in the analyzed three patients from 76% to 59% compared to the rigid image registration TPC.

References

[1] Stoiber E M, Schwarz M, Debus J, Bendl R, Giske K 2013 *An optimised IGRT correction vector determined from a displacement vector field: A proof of principle of a decision-making aid for re-planning* (Acta Oncologica) accepted for publication

[2] Malsch U, Thieke C, Huber P E, Bendl R 2006 *An enhanced block matching algorithm for fast elastic registration in adaptive radiotherap* (Physics in medicine and biology vol. 51) p 4789

Acknowledgement:

This research was partly funded by the German Research Foundation (DFG) as part of project C02, SFB/TRR 125 Cognition-Guided Surgery.

Journal of Physics: Conference Series **489** (2014) 012082 doi:10.1088/1742-6596/489/1/012082

Improving accuracy of markerless tracking of lung tumours in fluoroscopic video by incorporating diaphragm motion

M Schwarz[1], H Teske[1,2], M Stoll[1] and Rolf Bendl[1,2]

[1] Department of Medical Physics in Radiation Oncology, German Cancer Research Center (DKFZ), Heidelberg, Germany

[2] Department of Medical Informatics, Heilbronn University, Heilbronn, Germany

E-mail: m.stoll@dkfz.de

Abstract. Purpose: Conformal radiation of moving tumours is a challenging task in radiotherapy. Tumour motion induced by respiration can be visualized in fluoroscopic images recorded during patients breathing. Markerless methods making use of registration techniques can be used to estimate tumour motion. However, registration methods might fail when the tumour is hidden by ribs. Using motion of anatomical surrogates, like the diaphragm, is promising to model tumour motion. Methods: A sequence of 116 fluoroscopic images was analyzed and the tumour positions were manually defined by three experts. A block matching (BM) technique is used to calculate the displacement vector relatively to a selected reference image of the first breathing cycle. An enhanced method was developed: Positions, when the tumour is not located behind a rib, are taken as valid estimations of the tumour position. Furthermore, these valid estimations are used to establish a linear model of tumour position and diaphragm motion. For invalid estimations the calculated tumour positions are not taken into consideration, and instead the model is used to determine tumour motion. Results: Enhancing BM with a model of tumour motion from diaphragm motion improves the tracking accuracy when the tumour moves behind a rib. The error (mean $\pm$ SD) in longitudinal dimension was 2.0 $\pm$ 1.5 mm using only BM and 1.0 $\pm$ 1.1 mm when the enhanced approach was used. Conclusion: The enhanced tracking technique is capable to improve tracking accuracy compared to BM in the case that the tumour is occluded by ribs.

1. Introduction

Precise lung tumor irradiation requires not only paying attention to interfractional position variations of the target but also gaining information about the target trajectory throughout the treatment. The 2D motion induced by respiration can be visualized with fluoroscopic images recorded during patients breathing. Since implanting fiducial markers is not widely accepted, we prefer a marker-less method making use of registration techniques to extract the tumor motion.

Directly tracking for tumor was assesed by a number of research articles coping with fluoroscopic images or 2D-plane MR images [1, 2].

The projective character of fluoroscopic images becomes apparent in the varying image intensities throughout a breathing cycle. Moreover a local structure, which is distinguishable in one image, may disappear in another image.

To overcome these intensity changes, we have developed an approach using multiple reference images to track for tumor motion in fluoroscopic video with present illumination changes [3]. The multiple template approach was first mentioned in terms of tumor tracking in fluoroscopic images by Cui *et al* [4].

Here, we present an alternative method to cope with these issues. It enhances the direct tracking method based on block matching by making use of the image information available in the rest of the image, e.g. when the tumor lies behind the rib. In this work, we track the tumor and an internal surrogate motion simultaneously. Whenever the tumor is hidden by another structure, we fall back to the surrogate tracking. This way the limitations of a direct tracking techniques can be compensated.

2. Methods

2.1. Fluoroscopic image data

A sequence of 116 fluoroscopic images of a single patient recorded during 10 breathing cycles from identical view angles was analyzed to determine the induced tumor motion. Distances of 1540 mm from the radiation source to flatpanel and 1000 mm from the radiation source to the isocenter were used. The pixel size was 0.8 mm x 0.8 mm on the panel with a matrix size of 512 x 512 pixels.

To generate ground truth data, the tumor positions were manually defined by one radiation oncolocist and two computer scientists with experience in reading fluoroscopic images. The range of motion was 14-19 mm in superior-inferior (SI) direction. In this direction the averaged interobserver standard deviation was found to be around 0.8 mm. The motion in left-right (LR) direction was found to be only 0.8-1.6 mm. Since the averaged inter-observer standard deviation of the ground truth in LR direction was in the same range, results for the LR direction were discarded in this study.

2.2. Direct tracking technique with block matching

The reference image for the direct tracking approach was selected out of the images recorded during the first exhalation. In this study, the selected reference image represents the position at maximum inhalation. To estimate the tumor motion, a block matching technique is used to calculated the motion vector relatively to the selected reference image. The tumor motion is assumed to be rigid. The region of interest (ROI) is limited to a rectangle that encompasses the whole tumor contour.

2.3. Diaphragm motion and linear correlation model

The diaphragm is sharp in contrast to the neighboring lung tissue and therefore is more visible in the fluoroscopic images than the tumor. Hence, the calculation of diaphragm motion supplies accurate results regardless of the respiratory phase. If there is a psoitive correlation between the movements of the tumor and diaphragm, the diapgragm can be used as an internal anatomic surrogate for the tumor. A linear correlation model between the tumor and the diaphragm motion is built to access the partial occlusion of the tumor with ribs (see Fig. 1).

First, a template matching for the tumor and the diaphragm is performed. A collision detection for the contour of the tumor and the static contours of the ribs, created manually from the first image of the sequence, determines whether the tumor is located behind a rib. If this is not the case, the tumor position is taken directly from the template matching. Additionally, the positions of the diaphragm and tumor are added to the data set for the linear correlation model. Otherwise, the calculated diaphragm movement is used to determine the tumor's position based on the established correlation model.

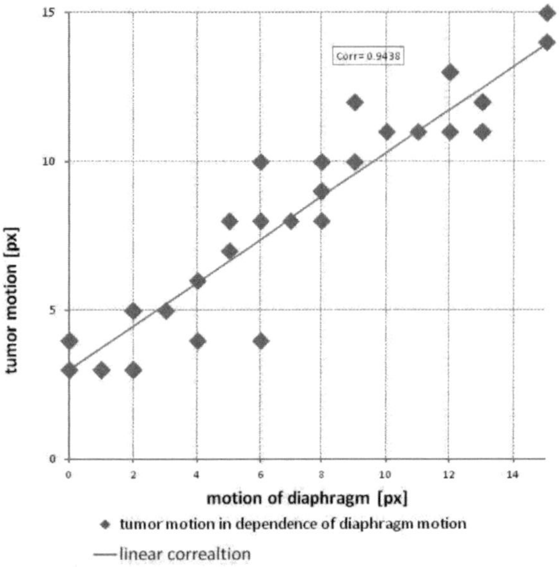

Figure 1. The estimated tumor positions based on direct tracking by use of a block matching technique- for the images without overlapping of tumor and ribs - show a positive correlation to the corresponding diaphragm movements.

3. Results

The error (mean ± SD) in longitudinal dimension was 2.0 ± 1.5 mm (2.5 ± 1.9 pixels) using only BM and 1.0 ± 1.1 (1.3 ± 1.4 pixels) when the enhanced approach was used. Regarding the interobserver variability of about 1 pixel in the ground truth data, the enhanced approach shows no significant difference in its error rate (see Fig. 2). The inter-observer variability is shown as a band surrounding the dotted blue line.

4. Conclusion

In difficult conditions, such as superposition of rib and tumor, advanced tracking techniques can contribute to a significant reduction of the tracking error. The described extension of the template matching technique permits coping with partial occlusions of the tumor by a rib. The incorporation of the modeling component improves the tracking result compared to direct estimation of the tumor's position by template matching.

The decision whether the tumor is located behind a rib is conducted by comparing the calculated position of the ribs and the estimated tumor position. As a preferred alternative, a threshold based on the correlation coefficient after matching could have been used to make the decision. The threshold method was successfully used by Cervino *et al* to determine, if contrasting structures in 2D MR sequences were moving away from the selected layer [2]. For the present data, a positive correlation between occlusion and change in the maximum correlation coefficient has been established. This knowledge will be incorporated into future work.

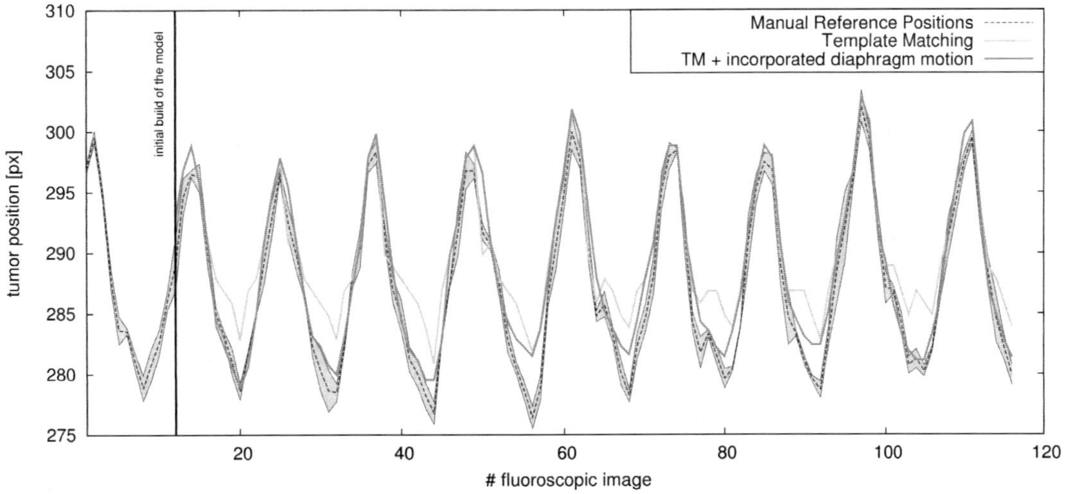

Figure 2. Realtime tracking algorithm based on cross correlation (green). Enhanced approach (direct tracking plus incorporated diaphragm motion) to cope with the situation when the tumor is hidden by a rib (pink). For comparison, the manually determined reference positions are plotted (dotted-blue).

References

[1] Moser T *et al* 2008 Detection of respiratory motion in fluoroscopic images for adaptive radiotherapy. *Phys Med Biol* **53** 3129–3145

[2] Cervino LI *et al* 2011 MRI-guided tumor tracking in lung cancer radiotherapy. *Phys Med Biol* **56** 3773–3785

[3] Schwarz M *et al* 2011 Comparison of two methods for fluoroscopic tracking of tumor mass. *Proc 11th Biennial ESTRO* 451

[4] Cui Y *et al* 2007 Multiple template-based fluoroscopic tracking of lung tumor mass without implanted fiducial markers. *Phys Med Biol* **52** 6229–6242

Acknowledgement

This research was carried out with the support of the German Research Foundation (DFG) as part of project C01, SFB/TRR 125 Cognition-Guided Surgery. The authors would like to thank Philipp Steininger and all involved colleagues of the Institute for Research and Development on Advanced Radiation Technologies, Salzburg, for providing the sequence of 2D fluoroscopic images.

Journal of Physics: Conference Series **489** (2014) 012083 doi:10.1088/1742-6596/489/1/012083

Towards adaptive radiotherapy for head and neck patients: validation of an in-house deformable registration algorithm

C Veiga[1], J McClelland[1], S Moinuddin[2], K Ricketts[1], M Modat[1], S Ourselin[1], D D'Souza[3] and G Royle[1]

[1] Department of Medical Physics and Bioengineering, University College London, London, UK
[2] Department of Radiotherapy, University College London Hospital, London, UK
[3] Radiotherapy Physics Group, University College London Hospital, London, UK

E-mail: catarina.veiga.11@ucl.ac.uk

Abstract. The purpose of this work is to validate an in-house deformable image registration (DIR) algorithm for adaptive radiotherapy for head and neck patients. We aim to use the registrations to estimate the "dose of the day" and assess the need to replan. NiftyReg is an open-source implementation of the B-splines deformable registration algorithm, developed in our institution. We registered a planning CT to a CBCT acquired midway through treatment for 5 HN patients that required replanning. We investigated 16 different parameter settings that previously showed promising results. To assess the registrations, structures delineated in the CT were warped and compared with contours manually drawn by the same clinical expert on the CBCT. This structure set contained vertebral bodies and soft tissue. Dice similarity coefficient (DSC), overlap index (OI), centroid position and distance between structures' surfaces were calculated for every registration, and a set of parameters that produces good results for all datasets was found. We achieve a median value of 0.845 in DSC, 0.889 in OI, error smaller than 2 mm in centroid position and over 90% of the warped surface pixels are distanced less than 2 mm of the manually drawn ones. By using appropriate DIR parameters, we are able to register the planning geometry (pCT) to the daily geometry (CBCT).

1. Introduction

The concept of adaptive radiation therapy (ART) was introduced by Yan *et al* as a closed-loop feedback process, where the variational positioning of the beams and internal motion of the patients are systematically monitored and characterized, and the plan is modified accordingly [1]. Cone-beam computed tomography (CBCT) is an online imaging method that provides valuable 3D information of the patient in treatment position but with incorrect Hounsfield units (HU) for dose calculations [2]. Deformable image registration (DIR) can help resolve the major challenges in ART: planning computed tomography (pCT) scans warped to match the daily cone-beam CT (CBCT) can be used for reliable dose calculations and to facilitate dose summation and automatic recontouring [3]. A fast, accurate, and robust CT-CBCT DIR algorithm is then a fundamental step in ART implementation. While there are a wide variety of studies assessing the quality of CT-CT deformable registration with patient data [4, 5], for CT-CBCT (a quasi-intermodality case) they are scarcer [6, 7]. Particularly, Castadot *et al* (2008) compare 12 CT-CT registration strategies combining different registration algorithms

[8]. However, in our opinion it is as important to carefully tune the registration parameters as it is to choose an appropriate algorithm, as a bad choice of parameters can cause in sub-optimal or even unacceptable registration results.

In this work we investigate a DIR algorithm implemented at University College London (UCL) to register the planning CT and weekly CBCT images taken from head and neck (HN) patients, a cohort known to benefit from ART [9].

2. Methods and Materials

2.1. Patients data acquisition and registration

Data from 5 HN patients treated in our clinic was used in this study. These patients presented considerable anatomical changes throughout their treatment, and were referred for midtreatment re-planning. The CT was acquired on a GE Widebore 16 slice system, while the VARIAN (Palo Alto, USA) OBI (on-board imaging) system integrated in the linear accelerator was used to acquire weekly CBCT images. The resolution was $0.979 \times 0.979 \times 2.5$ mm^3 and $0.897 \times 0.897 \times 2$ mm^3, for the CT and CBCT images respectively.

NiftyRef is an open-source registration package developed by UCL's Centre for Medical Image Computing, containing a GPU implementation of the B-spline deformable registration algorithm [10]. On a preliminary study we investigated proper registration parameters to be used in CT-CBCT registration in the HN region [11]. The optimal range of parameters found was: bending energy (BE) weight between 2 and 6%, logarithm of the jacobian weight of 1%, a maximum of 1000 iterations, a binning of 32 for the joint histogram calculation, intensities thresholded to the range [-1000 2000] and a control point spacing (CPS) between 5 and 10 voxels. Lower BE allows more liberty for the algorithm to perform larger contractions and expansions, however if its weight is too small the registrations are under-constrained and unrealistic deformations are likely to occur. The logarithm of the jacobian penalty term encourages the registrations to be folding-free. Masking refers to defining the region in the CBCT where the deformation field is optimized, considerably speeding the computation. For a full explanation of the different parameters see [11]. As a follow-up of this study we found evidence that pre-processing the CBCT image with non-linear intensity adjustment (gamma correction) improved the registrations. These conclusions led to the selection of a variety of parameters settings, tested here more extensively. The aim is to find a single set of parameters that performs well over all the datasets available.

2.2. Assessment of the quality of the registrations

Registrations were compared by visual inspection, computation time and similarity of warped structures with a gold standard. The resulting deformation field can be used to map points and regions of interest delineated in the pCT dataset to the CBCT dataset. Structures delineated in the pCT were warped and compared with contours manually drawn by the same physician on the CBCT. Due to the noise and low contrast inherent to CBCT imaging, it is difficult to define points or delineate structures with confidence and consistency. We chose a set of structures that can be seen unequivocally on the CT and CBCT, even though it may not contain some of the typical organs-at-risk (OAR) of the HN region. The set contains a mixture of soft tissue and bones, and was representative of the patient daily positioning and weight loss process. Vertebras C1, C4 and C7 were considered as bone landmarks for different regions, while external body and sternocleidomastoid muscles were chosen as surrogates for soft tissue.

Structures manually drawn in the CBCT and deformed from the pCT were compared using Dice similarity coefficient (DSC) [8], Overlap index (OI) [5], Distance transformation (DT) [4] and centroid position error (CoM) [4]. Dose comparisons, considering dose-differences (DD) and gamma analysis (γ), were performed to assess whether reduced accuracy but faster registrations had a noticeable effect on dosimetry. Dose calculations for the same IMRT plan were performed using VARIAN Eclipse analytical anisotropic algorithm.

Table 1. Mean values of DSC, OI, DT_{2mm} and CoM, grouped by varying parameter.

Parameter	Value	DSC	OI	DT_{2mm} (%)	CoM (mm)
Bending Energy weight (BE)	3%	0.843	0.867	7.7	1.42
	5%	0.837	0.861	8.4	1.43
Control Point Spacing (CPS)	5	0.851	0.877	6.8	1.46
	10	0.828	0.852	9.2	1.39
Masking the reference image	Yes	0.839	0.862	8.3	1.44
	No	0.841	0.867	8.1	1.39
Pre-processing the reference image	Yes	0.842	0.864	7.7	1.39
	No	0.837	0.866	8.8	1.46

3. Results

A total of 16 registrations per dataset were performed by varying 4 different inputs: (i) different BE (3% and 5%), (ii) different CPS (5 and 10 voxels), (iii) pre-processing of the CBCT and, (iv) masking. We compared the results obtained for the different quantities averaged over all structures and datasets (Table 1). The DIR was initialized with a rigid-only alignment. Separating individual structures and datasets we found improvement in 66%, 59%, 70% and 44% of the cases when using lower BE, smaller CPS, pre-processing and masking, respectively. Our findings suggest that combining pre-processing, CPS=5 and BE=3% leads to better registrations. Masking had an unpredictable effect on the registrations accuracy. This assessment considered that the effect of a parameter on the registration was independent of the other parameters, which is not the case. To confirm that the chosen parameters do stand out compared with the rest, we looked at mean DSC, OI, DT_{2mm} and CoM for each registration separately. Figure 1 shows how the registrations in the light blue region (CPS=5 and pre-processing) with odd number (BE=3%) give the best results in general.

Dose distributions were calculated using the same IMRT plan on the registrations that do and do not use a mask. The goal was to assess if accuracy for speed tradeoff has a big impact in the dose calculated. The two dose distributions obtained were very similar (Figure 2). DD was smaller than 2% of the prescribed dose (pD) on 96.2% of the voxels, with a mean value of 0.25 Gy (0.4% pD). The comparison passed a γ-test on 99.4% of the voxels (mean γ of 0.082).

4. Discussion and conclusions

Properly choosing the registration parameters is crucial in using DIR on clinical data as the parameters determine the actual deformations produced. The parameters should be tuned for different sites, imaging modalities, image size and quality, etc. Often default parameters which work well on a wide range of different images will be sub-optimum for a particular task. Non-linear intensity adjustment changes the CBCT contrast map, which had a positive impact in bin size and distribution of the histograms (Figure 1). Masking has an unpredictable effect on the registrations accuracy, depending on the particular dataset. The reduction of computation time is good reasoning for always masking the registrations and we confirmed that this has a small effect on dose calculations, particularly in OARs (Figure 2).

Our study aimed at tuning our B-Spline based deformable registration algorithm for excellent performance at CT-CBCT registrations in the head and neck site, for images obtained in our clinic. The optimal parameters found for the deformable registration were: BE of 3%, logarithm of the jacobian weight of 1%, a maximum number of 1000 iterations, a binning of 32 for the joint histogram calculation, thresholding the intensities to the range [-1000 2000], CPS of 5 voxels, non-linear intensity adjustment of the CBCT and masking. We achieved a median value of 0.845 in DSC, 0.889 in OI, error smaller than 2 mm in centroid position and over 90% of the warped

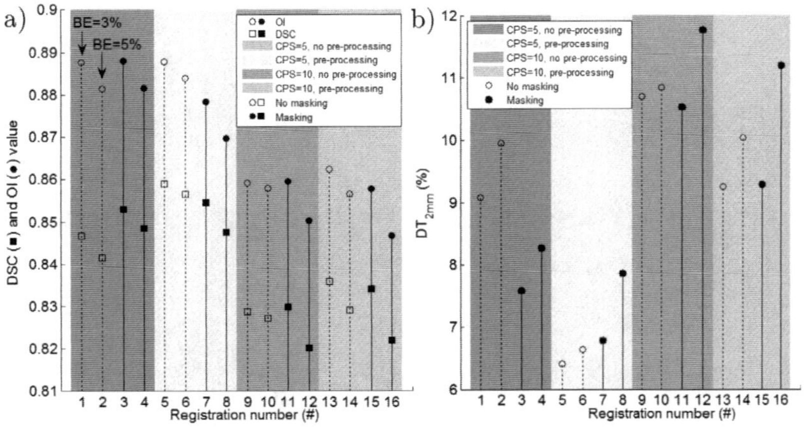

Figure 1. Variation of a) DSC (squares) and OI (circles) and b) DT_{2mm} for different combinations of parameters: BE=3% (#1,3,5,7,9,11,13,15) and BE=5% (#2,4,6,8,10,12,14,16); CPS=5 (#1-8) and CP=10 (#9-16); mask (#3,4,7,8,11,12,15,16) and no mask (#1,2,5,6,9,10,13,14); pre-processing (#5-8,13-16) and no-preprocessing (#1-4,9-12). Smaller control point spacing translated in larger values of DSC and OI. Pre-processing the CBCT improved DSC and DT_{2mm} values.

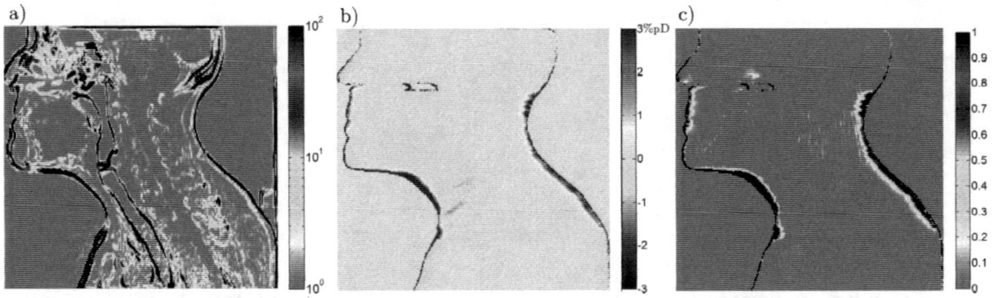

Figure 2. For the two registration approaches: a) difference in pixel intensity, b) DD and c) γ-value. The similarity between the images is high, with most of the differences smaller than 10HU. The regions where the differences in dose are higher are outside the CBCT field of view, in the skin, immobilization mask and airways, and these have little effect on the dosimetry of the OARs.

surface pixels are distanced less than 2 mm of the manually drawn ones. Future work will focus on further validation on a larger cohort to further support the effectiveness and efficiency of the proposed method in clinical applications.

References

[1] Yan D *et al* 1997 *Phys. Med. Biol.* **42** 123-132
[2] Stock M *et al* 2009 *Radiother. Oncol.* **93** 1-7
[3] Lee C *et al* 2008 *Int. J. Radiat. Oncol. Biol. Phys.* **71** 1563-1571
[4] Zhang T *et al* 2007 *Int. J. Radiat. Oncol. Biol. Phys.* **68** 522-530
[5] Tsuji S *et al* 2010 *Int. J. Radiat. Oncol. Biol. Phys.* **77** 707-714
[6] Hou J *et al* 2011 *Med. Phys.* **38** 2088-2094
[7] Peroni M *et al* 2012 *Int. J. Radiat. Oncol. Biol. Phys.* **84** e427-e433
[8] Castadot P *et al* 2008 *Radiother. Oncol.* **89** 1-12
[9] Hansen E *et al* 2006 *Int. J. Radiat. Oncol. Biol. Phys.* **64** 355-362
[10] Modat M *et al* 2010 *Comput. Methods Programs in Biomed.* **98** 278-284
[11] Veiga C *et al* 2012 *Image-Guidance and Multimodal Dose Planning in Radiation Therapy Workshop, MICCAI*

Journal of Physics: Conference Series **489** (2014) 012084 doi:10.1088/1742-6596/489/1/012084

Streamlining EPID-based IMRT quality assurance: auto-analysis and auto-report generation

Aitang XING[1], Sankar ARUMUGAM[1], Shrikant DESHPANDE[1,3], Armia, GEORGE, Lois HOLLOWAY[1, 2, 3,4], Gary GOOZEE[1, 4], Alison GRAY[1], Philip VIAL[1, 2]

[1]Liverpool and Macarthur cancer therapy centres and Ingham Institute, Liverpool Hospital, NSW, Sydney, Australia.
[2]Institute of Medical Physics, School of Physics, University of Sydney, Sydney, NSW, Australia.
[3]Centre for Medical Radiation Physics, University of Wollongong, Wollongong, NSW, Australia.
[4]South Western Sydney Clinical School, University of New South Wales, Sydney, NSW, Australia

aitang.xing@sswahs.nsw.gov.au

Abstract. Using an EPID for patient specific IMRT QA is an efficient way to verify patient plans prior to the treatment. Our centres' current EPID dosimetry method is based on a water equivalent depth approach, where EPID images of each IMRT field are converted to dose images and compared to Treatment planning system calculated dose images using a commercial software tool. Two physicists across two sites perform the analysis for an average of 12 new IMRT patients per week. To speed up this process, an in-house program called AutoEPIDIMRTQA was developed. The program automatically performs the following tasks sequentially: reading and converting raw EPID images acquired on either a Siemens Oncor or Elekta Synergy linear accelerator, registering them with planar dose images calculated by Pinnacle (Philips) or CMS XiO (Elekta) treatment planning systems, analyzing the profiles for registered images and calculating the Gamma map. Finally an IMRT QA report is automatically generated. AutoEPIDIMRTQA was validated against commercial software. The analysis time for a typical 9-beam IMRT head-neck patient decreases from 30 minutes to 4 minutes. The total QA time was reduced by 40% using AutoEPIDIMRTQA. Thus we have demonstrated a significant reduction in the time burden for physics staff performing IMRT QA.

1. Introduction

In modern radiotherapy, emerging new treatment techniques such as intensity-modulated radiotherapy (IMRT), intensity modulated arc therapy (IMAT) and Tomotherapy, provide more choices for oncologists and radiotherapist/dosimetrists to personalize patient treatment for the specific treatment situation. The resulting treatment plans are inherently more complex. To ensure the prescribed treatment dose is accurate, deliverable and within clinically acceptable error tolerance, medical physicists need to conduct quality assurance (QA) for each patient before treatment [1]. This presents a challenge for radiotherapy cancer centres with limited physics resources and a large number of new IMRT patients per week.

The clinical pressures to ensure QA is completed on time and the practical need for improving QA efficiency has resulted in many centres using electronic portal imaging devices (EPIDs) for patient specific quality assurance [2, 3]. EPIDs are a standard component of a large fraction of modern linear accelerators (Linacs). Compared with other dosimetry equipment used for IMRT QA such as 2D ion chambers, diode arrays or film, the use of EPIDs is expected to be more time efficient. High spatial resolution and relatively high sensitivity for detecting delivery errors are among the other advantages for adapting an EPID-based IMRT QA approach [4, 5]. EPID-based IMRT QA was implemented in our department in 2011 to replace the previous phantom and 2D ion chamber array approach. An in-house program was developed to convert the raw EPID images into water equivalent dose images based on a previously reported EPID dosimetry method [6]. In this work, a new software tool, AutoEPIDIMRTQA, was implemented to save physics time spent analyzing IMRT QA results by automating the procedure. The purpose of this report is to present the software implementation and evaluation of initial results.

2. Materials and Methods

2.1. Current IMRT QA system used for EPID-based IMRT QA

The EPID dosimetry method developed by Lee et al was originally tested for Varian aS500 EPID device from a Varian 21 EX Linear accelerator [6]. In our centres, two Elekta Synergy (Elekta Ltd, Crawley, UK) Linacs with EPID devices (iViewGT) and two Siemens Oncor Linacs equipped with EPID devices (OPTIVUE™ 500) are being used to deliver IMRT treatment to patients using 6MV photon beams. IMRT treatments are planned using either Pinnacle V9.0 (Philips) or XIO V 4.6 (Elekta). We implemented an IMRT QA system based on the approach developed by Lee et al [6] and adapted for our mix of equipment.

The IMRT QA system is based on flood-field corrected integrated EPID images acquired either on Elekta or Siemens Linacs as input and outputs DICOM-RT image files. The EPID image is converted to a 'dose image' calibrated to the dose at a reference depth in water, in our case 5.4 cm depth in water at the isocenter plane. The planar dose is calculated at the same depth in a virtual water phantom by the TPS (Pinnacle or XIO). This methodology was validated against a two dimensional (2D) ion chamber array [7] and has been used clinically for more than two years in our centre. In current routine QA practice, radiation therapists (RTs) acquire the EPID images and physicists use an in-house developed software program to convert the EPID image to dose image. The EPID and TPS calculated dose images for each field are manually loaded into OmniPro-I'mRT for profile and gamma map analysis and to enable generation of an IMRT QA report.

2.2. Implementation of AutoEPIDIMRTQA

The new software tool, AutoEPIDIMRTQA, was designed not only to convert the raw EPID images into dose images but also to automate the process of analyzing the EPID QA results and generating a QA report. The software was implemented in Matlab (2010a). Figure 1 shows a flow chart of the program structure. The program reads the EPID and TPS data and processes the dose images as per the current program described above. AutoEPIDIMRTQA automatically aligns and registers EPID and TPS dose images using a sub-pixel image registration algorithm [8]. The profiles passing through the center of the registered images along x and y axis are analyzed and a 2D gamma map is also calculated using user-specified criteria. This procedure is repeated automatically for each IMRT beam. Finally, the program saves the dose image into DICOM-RT files and compiles an IMRT QA report that contains the dose profile results and calculated gamma maps.

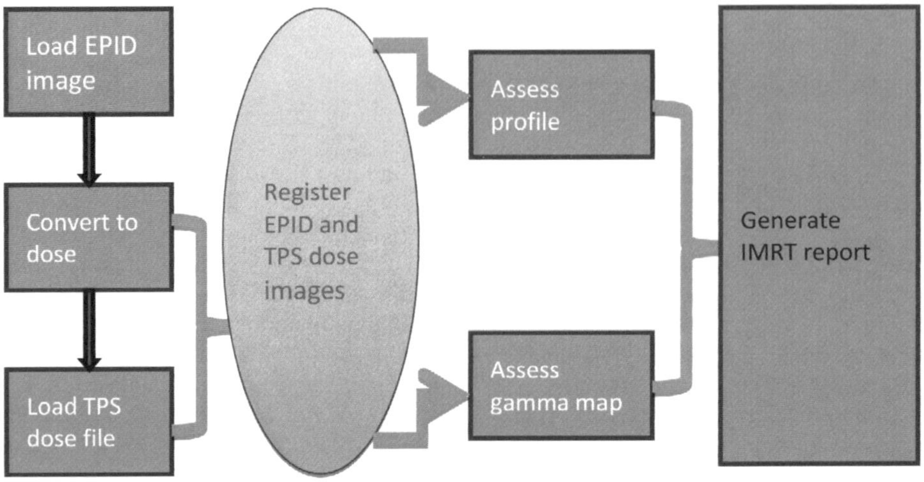

Figure 1. The internal flow chart implemented in AutoEPIDIMRTQA.

2.3. Software validation

There are four sub functions implemented in AutoEPIDIMRTQA: i) conversion of EPID images to dose images, ii) registration of converted EPID dose images and TPS-calculated dose images, iii) the calculation of the gamma map and profiles and iv) report generation. The EPID images are converted to dose using the same pixel-to-dose conversion algorithm used in the current clinical program. This code was validated against a 2D ion chamber array [7] and also by patient-specific QA results of over 100 patients clinically using 2%/3mm gamma criteria (global), 10% low dose threshold with acceptable gamma pass rates of over 95% for each IMRT beam. The image registration algorithm used in the program was verified by comparing the introduced image shifts and algorithm-calculated shifts for given EPID images. The accuracy of the gamma function was verified by comparing it with a commercial software tool. Once the first three sub functions are verified, the sub function of report generation just inserts the automatically generated profile and Gamma image into a word document with customizable format. The efficiency improvement as a result of using AutoEPIDIMRTQA was compared among different IMRT QA schemes used in our centres.

3. Results and Discussion

3.1. The unique features of AutoEPIDIMRTQA

The AutoEPIDIMRTQA program has incorporated several features to address clinical needs for our IMRT QA program: (1) Ease of use. As shown in figure 2, the only user interaction is to select i) the folder containing the raw EPID images and exported TPS dose images, ii) select the TPS system utilised and ii) if reference EPID images for a 10x10cm field was taken for this patient. (2) Real-time animation of gamma map and profile analysis. The program will analyze profiles and calculate the gamma map while compiling the results into a word document report. During this process, two pop-up windows containing the profile images and gamma images are displayed providing a convenient way for the physicist to quickly review results for all beams before the final QA report. (3) Backward compatibility to the program currently being used. Except for generating a QA report, the program also automatically outputs a DICOM-RT image file containing an EPID dose image for each beam. The EPID dose image can be loaded into other software tools for further analysis if necessary.

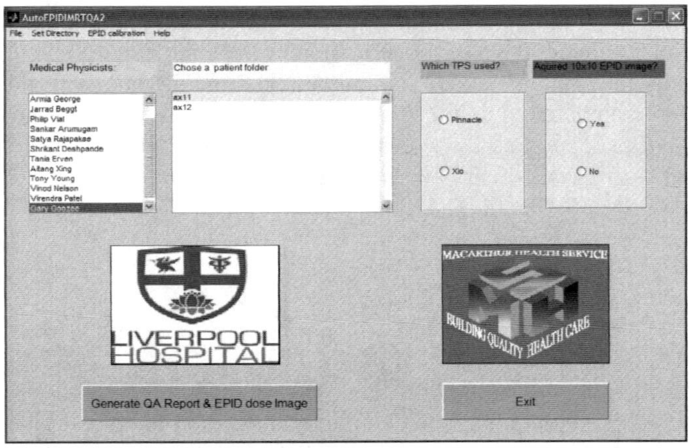

Figure 2. The simple graphic user interface designed for AutoEPIDIMRTQA.

3.2. Image alignment and Gamma function evaluation

The image registration algorithm implemented in the program can achieve sub-pixel accuracy for image alignment [8]. The implementation of this registration algorithm in the AutoEPIDIMRTQA program was validated by introducing known shifts to images along X and Y directions and then registering the offset image with original one. The calculated and introduced shifts in the units of pixels are presented in Table 1.

Table1. Comparison of introduced and registration algorithm calculated shifts

	Shifts (pixels) along X and Y directions (X,Y)					
Introduced shifts (pixels)	(-1, 2)	(1,2)	(2,3)	(4,1)	(3,0)	(6,-9)
Calculated shifts (pixels)	(-0.17,1.86)	(1.2,2.03)	(1.45,2.89)	(3.48,0.16)	(2.72,0)	(5.34,-8.5)
Differences (pixels)	(-0.83,0.14)	(0.2,-0.03)	(0.55,0.11)	(0.52, 0.84)	(0.28,0)	(0.64,-0.5)

Table 2. Comparison of the Gamma pass rate calculated by AutoEPIDIMRTQA and OminiPro for a seven-beam prostate patient.

	Gamma pass rate						
Beam Number	1	2	3	4	5	6	7
AutoEPIDIMRTQA (%)	98.1993	99.6652	98.8681	99.4242	99.2272	98.6741	98.9585
OminiPro-I'mRT (%)	98.66	97.97	99.14	99.47	99.13	99.13	98.76
Difference	-0.4607	1.6952	-0.2719	-0.0458	0.0972	-0.4559	0.1985

The Gamma pass rate calculated by AutoEPIDIMRTQA and OminiPro-I'mRT was compared for 90 IMRT beams using a Gamma criteria of 3%/3mm. The maximum and average difference in gamma pass rate was 2.1 % and 1.3 % respectively. Table 2 shows the Gamma results for one typical seven-beam prostate patient. The differences may be explained by the fact that the manual alignment is not exactly the same as automatic alignment and the different implementation of the Gamma functions in the two programs. Figure 3 shows manual and automatic alignment for one IMRT beam from one head-neck patient. The collimator angle for this beam was 40 degrees. In order to manually achieve a good alignment of these two images shown here in OmniPro-I'mRT, one image had to be rotated by 40 degree clockwise and then translated in both horizontal and vertical directions. The gamma pass rate for this IMRT beam by AutoEPIDIMRTQA and OminiPro was 98.06% and 97.91%, respectively.

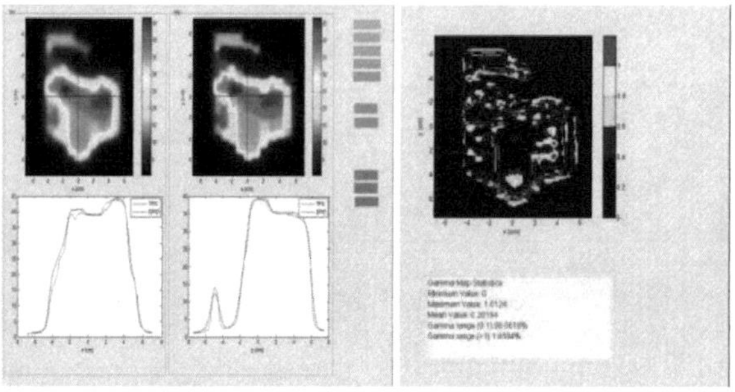

(a)

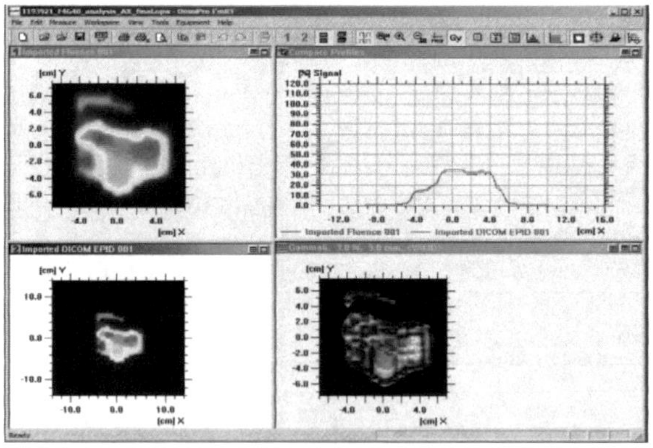

(b)

Figure 3. The profiles analysis and Gamma map calculated by (a) AutoEPIDIMRTQA and (b) by physicist in OmniPro-I'mRT.

3.3. Clinical benefits and future development

The benefit from the implementation of AutoEPIDIMRTQA is that it will further improve the efficiency of our EPID-based IMRT QA. Although the EPID images are acquired by radiation therapists at a scheduled machine time for each patient in our centres, a dedicated physicist was still assigned to perform the following tasks: converting the EPID raw images into dose images, loading them and TPS dose image into OminiPro IMRT, manually aligning them, analysing the profile, calculating Gamma map and compiling a QA report. Figure 4 shows the total time for performing IMRT QA measurement and analysis using different QA devices and software tools. The time shown here for each step was averaged over 8 patients. The times in figure 4 also reflects the evolution of IMRT QA methodology in our clinic. The time for Radiation Therapists to acquire and export EPID image is the same for methods B and C (figure 4). The QA analysis time decreases from an average of 30 minutes for a typical 9-beam head-neck IMRT patient to 4 minutes with AutoEPIDIMRTQA. The physics QA time and the total QA time was reduced by 86% and 40% respectively when changing from method B to method C (figure 4). In comparison with the Matrix-only QA method (method A), IMRT QA with EPID plus AutoEPIDIMRTQA decreased the total QA time by 57%.

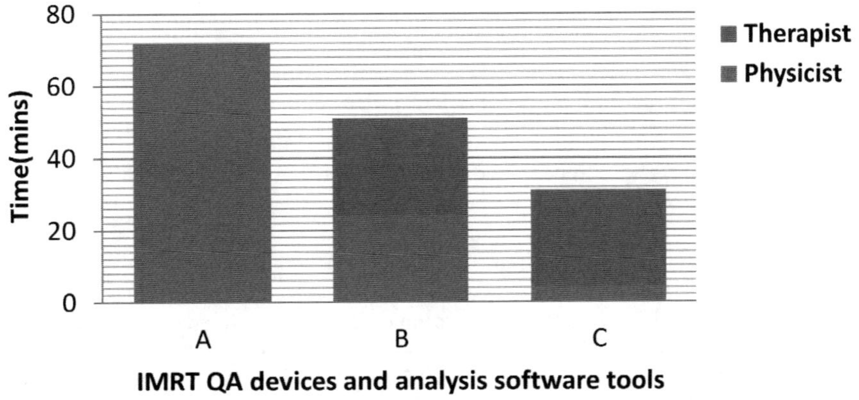

Figure 4. The total time taken to perform IMRT QA and analysis using (A) ion chamber array and commercial software (B) EPID and commercial software (C) EPID and AutoEPIDIMRTQA. The red and blue parts of the columns indicate the time spent by therapists and physicist respectively. For matrix-based QA (A) physicists perform measurement and analysis. For EPID based QA (B and C) therapists perform measurements and physicists analyze the results.

AutoEPIDIMRTQA is still being actively developed. A full automation is planned to be achieved by implementing it as a daemon or a server. In this manner it will be always running in the background and reviewing a folder in a network driver where patient raw EPID image and TPS dose images are stored as a subfolder. The QA report will then be automatically sent to physicists. A database is also going to be included to record the QA results for further statistical analysis.

4. Conclusion

An in-house software tool was developed and validated to aid the analysis of QA results for EPID IMRT QA. This program achieves significant efficiency gains streamlining clinical practice.

References

[1] Low D A, Moran J M, et al. 2011 *Med. Phys.* **38** 1313
[2] Vial P, Hunt P, Greer P B, et al. 2008 *Australas. Phys. Eng. Sci. Med.* **31** 216
[3] Arumugam S, Xing A, Jameson M and Holloway L. 2013 *Med. Phys.*071724-1.
[4] Elmpt W, McDermott L, et al. 2008. *Radiother Oncol.* **88** 289
[5] Fredh A, Bengtsson J et.al. 2013 *Med. Phys.* **40** 03716-1
[6] Lee C, Menk F, Greer P B et. al. 2009 *Med. Phys.* **36** 984.
[7] Vial P Gray A et al. *Australas. Phys. Eng. Sci. Med.* 2013 **36** 138
[8] Tian Q, Huhns N. Computer Vision ,Graphics and Imaging Processing. 1986 **35** 220.

Journal of Physics: Conference Series **489** (2014) 012085 doi:10.1088/1742-6596/489/1/012085

Fast regional readout CMOS Image Sensor for dynamic MLC tracking

H Zin[1,2], E Harris[2], J Osmond[2] and P Evans[2,3]
[1] Advanced Medical and Dental Institute, Universiti Sains Malaysia, 13200, Kepala Batas, Penang, Malaysia
[2] Joint Department of Physics, The Institute of Cancer Research and The Royal Marsden NHS Foundation Trust, Downs Road, Sutton, Surrey, SM2 5PT, UK
[3] Centre for Vision, Speech and Signal Processing, University of Surrey, Guildford, Surrey, GU2 7XH, UK

E-mail: hafiz.zin@gmail.com

Abstract. Advanced radiotherapy techniques such as volumetric modulated arc therapy (VMAT) require verification of the complex beam delivery including tracking of multileaf collimators (MLC) and monitoring the dose rate. This work explores the feasibility of a prototype Complementary metal-oxide semiconductor Image Sensor (CIS) for tracking these complex treatments by utilising fast, region of interest (ROI) read out functionality. An automatic edge tracking algorithm was used to locate the MLC leaves edges moving at various speeds (from a moving triangle field shape) and imaged with various sensor frame rates. The CIS demonstrates successful edge detection of the dynamic MLC motion within accuracy of 1.0 mm. This demonstrates the feasibility of the sensor to verify treatment delivery involving dynamic MLC up to ~400 frames per second (equivalent to the linac pulse rate), which is superior to any current techniques such as using electronic portal imaging devices (EPID). CIS provides the basis to an essential real-time verification tool, useful in accessing accurate delivery of complex high energy radiation to the tumour and ultimately to achieve better cure rates for cancer patients.

1. Introduction

Volumetric modulated arc therapy (VMAT) is an advanced radiotherapy technique which involves arcing the linac around the patient while delivering the radiation beam shaped by dynamic multileaf collimators (MLCs) [1]. VMAT also has variable dose rate (beam fluence rate) and arcing speed during beam delivery. Electronic portal imaging devices (EPIDs) are widely used tool for geometric verification and dose verification of radiotherapy treatment. EPIDs have a large field of view (typically greater than 20 cm x 20 cm) and allow beam's eye view (BEV) imaging. However the EPID has a frame rate limit of about 12 frames per second and suffers from image artefacts such as ghosting and image lag [2], which limit the ability to monitor a dynamic treatment delivery and result in inaccurate sensor response (i.e., dose measured at a specific irradiation period). An imaging modality with very fast sampling is desirable to monitor the MLC movement as well as changes in the dose rate during VMAT. Otherwise the benefit of these techniques can be degraded due to inability to assess the success of treatment delivery. A few protocols to evaluate VMAT delivery have been published [3, 4] but the evaluation made was of the overall quality of the cumulative dose delivered and did not provide any information regarding the performance of the MLC leaf positions.

[1] To whom any correspondence should be addressed.

Complementary metal-oxide semiconductor Image Sensors (CISs) have evolved over the past 20 years into a competitive imaging technology due to technology scaling of CMOS and the feasibility of high numbers of pixels integrated on-chip. The sensor allows fast region-of-interest (ROI) read-out which enables the user to trade array size, with read-out speed. CMOS has also recently been successfully made radiation hard [5] and can be fabricated into a large array format using stitching technology [6]. Therefore, this paper discusses a preliminary study of using a prototype Complementary metal-oxide semiconductor Image Sensor (CIS) for tracking dynamic MLCs (DMLCs) by utilising fast, region of interest (ROI) readout functionality. This leads to application of the prototype CIS in VMAT verification which has been recently published [7].

2. Materials and Methods

The CMOS Image Sensor used in this work was developed by the MI-3 consortium (UK) [8]. The sensor comprised 1350 rows × 1350 columns of 40 μm square pixels, giving a sensing area of 54 mm × 54 mm. Sensor frame rate could be increased from ~20 frames per second (f/s) to ~400 f/s by reducing the read out from 1350 rows to 70 rows of pixels in ROI read out mode. The CMOS Data Acquisition System (DAQ) used a Memec Virtex-II Pro 20FF1152 FPGA development board (Xilinx, Inc., San Jose, CA, USA). The host PC controls the CIS through the FPGA board via a dynamic link library where the operation of the sensor can be customised. The binary image data is transferred back to the PC at GBit speeds and analysed offline using Matlab (MathWorks, Natick, MA, USA).

Assessment of DMLC leaf tracking success in the presence of variations in frame rate from ROI (which affect the visibility of the moving MLC) was performed by imaging a DMLC sequence designed to employ various leaf speeds. The work was carried on an Elekta Synergy accelerator (Elekta, Crawley, UK) with a Beam Modulator head, using 6 MV photon beam. The sensor was placed away from the radiation beam and was optically coupled in camera configuration with a fast terbium doped gadolinium-oxy-sulfide $Gd_2O_2S{:}Tb$ scintillation screen.

The design of the DMLC sequence is shown in Figure 1. The triangular configuration means that the leaves will have to travel at different speeds to ensure all MLCs close in at the end of the delivery. A series of images of the MLC moving across the field was acquired at frame rate 21.0 f/s, 50.4 f/s, 100.9 f/s, 201.6 f/s and 403.6 f/s. Because different frame rates were achieved by reducing the number of rows through ROI read out, the sensor FOV decrease with increasing frame rate as shown in Figure 1. The middle row of imaging area was positioned on the middle leaves of the prescription to make sure all ROIs used, imaged the fastest MLC leaves in the middle.

The leaf edge in the images was located with an automatic edge tracking algorithm shown in Figure 2. The coordinates of the MLC edges on the image were located along the profile plot for the right edges where the MLC leaves were moving. The detection was repeated row-by-row across the triangular field for all frames acquired. The position of the MLC leaves on the captured images was compared with the expected position from the prescription.

3. Results and Discussion

The edges tracked from the triangle field are presented for every 0.1 s time interval in Figure 3 for the central four MLC leaves (MLC number 18 to 21) at frame rates 201.6 f/s and for the central two MLC leaves (MLC number 19 and 20) at 403.6 f/s. The leaves are moving from right to left in the shown orientation for prescribed positions between 15 mm to -25 mm. The smoothing of the jagged shape of the MLC leaves is due to light scattering before reaching the camera in the indirect detection geometry [9]. Most of the edges tracked have a smooth line except for some edges where the lines are noisier. Examining the corresponding images reveals these are due to spiking pixels characterised in [6] which periodically affects certain pixel columns in the prototype CIS.

To exclude the edge smoothing of the tracked MLC, mean pixel value from the region within the middle of each leaf is selected as the leaf position. Edge detected across each leaf is divided into three bins. The middle bin is selected as the MLC position to exclude the smoothed regions. The edge position was calculated for MLC number 19 (the fastest MLC speed) across MLC position of 5 mm to -25 mm, i.e., the range shown in Figure 3. Position error is defined as the difference between the tracked position and the prescribed positions. The existence of spiking pixel in the data and the variation of sensor frame rates has not caused the sensor tracking accuracy to be worse than 1.0 mm. This is better than the 3 mm maximum error set by the AAPM [10]. The smoothing of the MLC leaves' edge tracked due to light scattering from the sensor in camera configuration has not interfered with the accuracy of the tracked MLC leaves positions.

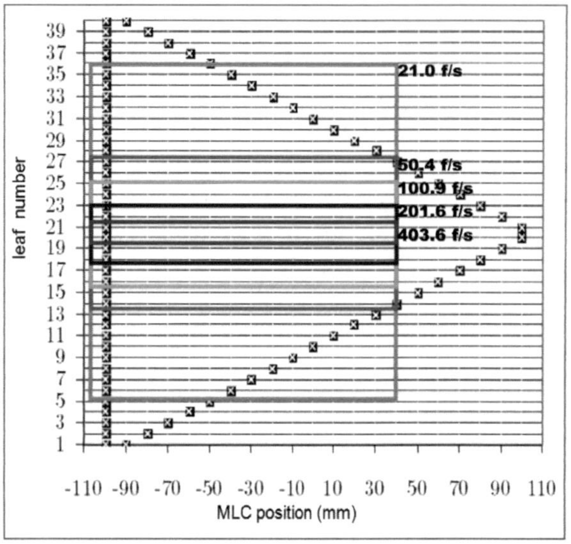

Figure 1: Imaging area for each frame rate used to image the moving triangular field overlaid on the prescription.

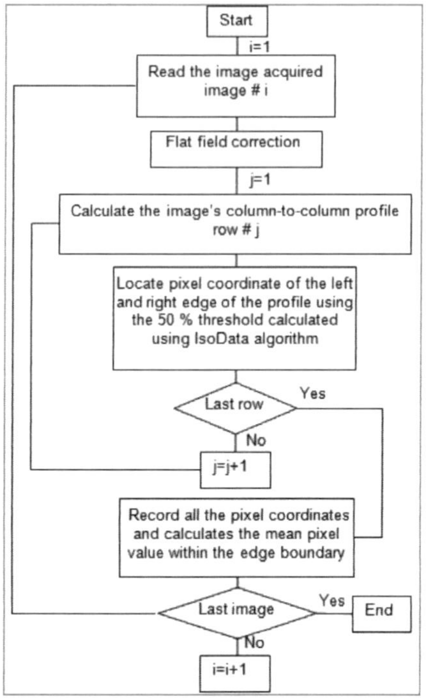

Figure 2: Flowchart of the automatic edge tracking method used to detect MLC edges from acquired CIS images.

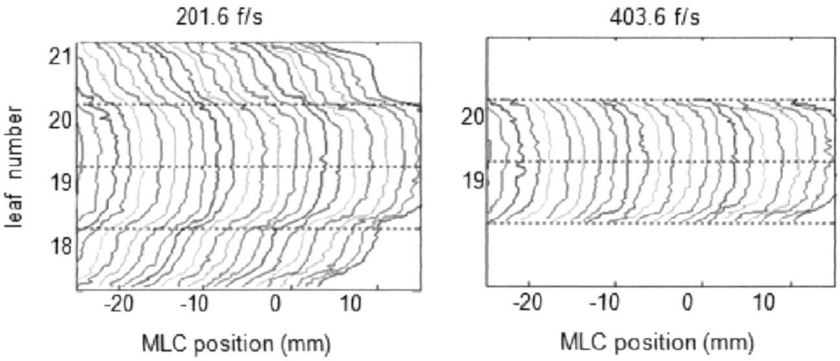

Figure 3: Plot of leaf edge detected for the central four MLCs (MLC number 18 to 21) at frame rates 201.6 f/s and for the central two MLCs (MLC number 19 and 20) at 403.6 f/s. The colour of each leaf edge tracked is shown alternating between different colours at every 0.1 s time interval, moving from right to left.

It has been demonstrated that the prototype CIS is able to track MLCs at a high sampling rate throughout the DMLC delivery. The use of fast ROI read out shows CIS is much faster than currently available alternatives, such as EPIDs. ROI read out, allows the frame rate to increase to ~404 fps using 70 rows regional readout (20 times increase from a full frame readout). Moreover, by using a combination of ROI sizes, speeds and positions, a sequence of ROI read out that changes its size and position across the full frame can be used to selectively image different groups of dynamically moving

MLCs. Imaging only the smaller ROI with the sensor also reduces the image processing workload and allows real-time verification of the MLCs. This has great potentials in commissioning a new treatment technique involving DMLC delivery, where independent assessment of the MLC leaves can be investigated by tracking the leaves' positions across the delivery.

A recent study using this sensor for VMAT verification [7] has demonstrated the novel capability of this sensor for verification of not only dynamic MLC leaves position, but also dose rate at very high frame rate. The MI-3 Plus consortium has recently developed a wafer size CMOS image sensor [11], constructed using the same stitching technique used for the prototype CIS in this study but across the whole wafer. The CIS has a total imaging area of 12.8 cm × 12.8 cm with two sides buttable, which can be doubled to 25.6 cm × 25.6 cm by joining two by two sensors together. The large imaging area is comparable to the size achieved by an EPID, but at the same time has a faster frame rate which is limited in an EPID. The wafer size CIS is expected to allow BEV imaging in direct detection geometry, which eliminates the light scattering problem from camera geometry observed in this work.

4. Conclusion

The use of ROI read out demonstrated that the prototype CMOS Image Sensor is faster and more versatile than currently available alternatives such as EPID. CIS provides the basis to an essential real-time verification tool, useful in accessing accurate delivery of complex high energy radiation in complex radiotherapy delivery such as VMAT.

5. Acknowledgement: This work is supported by EPSRC grant MI-3 plus (EP/G037671). Research at The Institute of Cancer Research is also supported by Cancer Research UK under Programme C46/A10588 and NIHR funding to the Biomedical Research Centre. Travel to ICCR 2013 for HZ was supported by a grant (TPLN) from Universiti Sains Malaysia.

6. References

1. Bedford, J.L. and A.P. Warrington, *Commissioning of Volumetric Modulated Arc Therapy (VMAT).* International Journal of Radiation Oncology Biology Physics, 2009. **73**(2): p. 537-545.
2. Partridge, M., B.M. Hesse, and L. Müller, *A performance comparison of direct- and indirect-detection flat-panel imagers.* Nuclear Instruments and Methods in Physics Research Section A: Accelerators, Spectrometers, Detectors and Associated Equipment, 2002. **484**(1-3): p. 351-363.
3. Ling, C.C., et al., *Commissioning and Quality Assurance of RapidArc Radiotherapy Delivery System.* International Journal of Radiation Oncology Biology Physics, 2008. **72**(2): p. 575-581.
4. Nicolini, G., et al., *The GLAaS algorithm for portal dosimetry and quality assurance of RapidArc, an intensity modulated rotational therapy.* Radiation Oncology, 2008. **3**(1): p. 24.
5. Guerrini, N., et al., *A high frame rate, 16 million pixels, radiation hard CMOS sensor.* Journal of Instrumentation, 2011. **6**(03): p. C03003.
6. Zin, H.M., et al., *Characterisation of regional variations in a stitched CMOS active pixel sensor.* Nuclear Instruments and Methods in Physics Research Section A: Accelerators, Spectrometers, Detectors and Associated Equipment, 2010. **620**(2-3): p. 540-548.
7. Zin, H.M., et al., *Towards real-time VMAT verification using a prototype, high-speed CMOS active pixel sensor.* Phys Med Biol, 2013. **58**(10): p. 3359-75.
8. Allinson, N., et al., *The Multidimensional Integrated Intelligent Imaging Project (M-I3).* Nuclear Instruments and Methods in Physics Research Section A: Accelerators, Spectrometers, Detectors and Associated Equipment, 2009(604): p. 196-198.
9. Heijmen, B.J.M. and et al., *Portal dose measurement in radiotherapy using an electronic portal imaging device (EPID).* Physics in Medicine and Biology, 1995. **40**(11): p. 1943.
10. Gary, A.E., et al., *IMRT commissioning: Multiple institution planning and dosimetry comparisons, a report from AAPM Task Group 119.* Medical Physics, 2009. **36**(11): p. 5359-5373.
11. Esposito, M., et al., *DynAMITe: a wafer scale sensor for biomedical applications.* Journal of Instrumentation, 2012. **6**(12): p. C12064.

Journal of Physics: Conference Series **489** (2014) 012086 doi:10.1088/1742-6596/489/1/012086

The use of a virtual reality simulator to explore and understand the impact of Linac mis-calibrations

Andrew W Beavis[1,2,3,4] and James W Ward[2,3]

[1] Department of Radiation Physics, Queens Centre for Oncology and Haematology, Cottingham, Hull, UK.
[2] Department of Computer Science, University of Hull, Hull, UK
[3] Vertual Ltd, Logistics Institute, University of Hull, Hull, UK
[4] Faculty of Health and Well Being, Sheffield-Hallam University, Sheffield, UK

Andy.beavis@hey.nhs.uk

Abstract.
Purpose: In recent years there has been interest in using Computer Simulation within Medical training. The VERT (Virtual Environment for Radiotherapy Training) system is a Flight Simulator for Radiation Oncology professionals, wherein fundamental concepts, techniques and problematic scenarios can be safely investigated.
Methods: The system provides detailed simulations of several Linacs and the ability to display DICOM treatment plans. Patients can be mis-positioned with 'set-up errors' which can be explored visually, dosimetrically and using IGRT. Similarly, a variety of Linac calibration and configuration parameters can be altered manually or randomly via controlled errors in the simulated 3D Linac and its component parts. The implication of these can be investigated by following through a treatment scenario or using QC devices available within a Physics software module.
Results: One resultant exercise is a systematic mis-calibration of 'lateral laser height' by 2mm. The offset in patient alignment is easily identified using IGRT and once corrected by reference to the 'in-room monitor'. The dosimetric implication is demonstrated to be 0.4% by setting a dosimetry phantom by the lasers (and ignoring TSD information). Finally, the need for re-calibration can be shown by the Laser Alignment Phantom or by reference to the front pointer.
Conclusions: The VERT system provides a realistic environment for training and enhancing understanding of radiotherapy concepts and techniques. Linac error conditions can be explored in this context and valuable experience gained in a controlled manner in a compressed period of time.

1.Introduction.

Recently simulation training has received a lot of attention within medicine and conferences, such as Medicine meets Virtual Reality (MMVR) [1] and Computer Assisted Radiology and Surgery (CARS), offer many examples where computer simulation has been explored. The use of computer simulation or virtual reality applications for Radiotherapy training [2], however have not received the same kind of attention within the radiotherapy community nor typically during ICCR meetings.

Simulation training offers the option to mirror real world equipment and conditions in a virtual reality (VR) environment which is safe for all the participants (trainers, clinical tutors and simulated patients), it also allows the exploration of experiences not normally or easily available in the real world. Visualisation of the isocentre concept is a relatively trivial demonstration to provide, but is extremely effective in discussions with early career professionals, especially for the subset of trainees who find it particularly challenging to understand. For all professionals, no matter what their level of experience, the exploration of treatment errors or Linac mis-calibrations can be extremely informative. These are easily generated in a simulation environment and provide opportunities for staff to learn from unfortunate or errant events.

The VERT system (Virtual Environment for Radiotherapy Training) was developed [3] as an education tool for therapy radiographers/ therapists/ RTTs in order to provide a simulation environment that offered alternative facilities to combat lack of clinical resources for training. Development began in 2002 and continued as a 'garage research' project for a number of years. In 2006 it gained interest from the Society and College of Radiographers and the National Cancer Director (Professor Sir Mike Richards) as a potential solution to a number of identified problems with the training, recruitment and retention of Therapy Radiographers in England. Recommendations within the 2007 NRAG (National Radiotherapy Advisory Group) report and subsequent 2007 Cancer Reform Strategy suggested it be adopted nationally within England. In 2007 a company was created in order to manage activity including the ongoing research and development work of the Radiotherapy Simulation training concept alongside commercial activity.

2.Methods/ materials

VERT comprises of software and hardware that provide a fully interactive and articulated virtual Linac that enables it to be used as a functional surrogate [4] of a real world machine. Various pieces of treatment positioning and QC equipment can be used and imaging devices can be deployed. Patients can be placed on the treatment couch and accurately aligned to the isocentre, having been created from DICOM files generated from any commercial treatment planning system. The graphics are created using large scale Stereoscopic projection and viewed by the users/ participants using 3D glasses.

The virtual Linac is almost fully functional in that it can mimic the motion and operation of its physical counterpart, with the only major exception being the generation of ionising radiation. Contemporary conventional Linac models from Varian, Elekta and Siemens are available. The operation of the machine is controlled via OEM (original equipment manufacturer) hand pendants and/or by GUIs available as part of the main screen or on separate touch screen (or normal) monitors and wireless tablet technology. A high level of attention to detail is provided with features such as set up lasers, field light, TSD indicators all being available and controlled via the appropriate hand pendant to enable the user to set the patient up in the same manner as they would expect to in the treatment room. Features such as couch-side controls (available via the touch screen) and in-room monitors are context specific to the Linac currently displayed. Patient position can be checked using DRR or fluoroscopic mode or using the IGRT module where the current position can be assessed against referenced images. All the images used in the VERT process simulations are generated in real time, including the fluoroscopic images acquired when the user rotates the gantry (manually or during VMAT delivery). In order to achieve the required interactivity these calculations are performed on the GPU.

Patients are represented using plans from treatment planning systems (fig 1), loaded via a DICOM interface. This allows every user to create patients that represent typical planning/ delivery technique appropriate to their practice. These data sets can be sent directly from the planning system, or loaded via memory media, allowing the data to be anonymised. The planning CTs, structure sets, dose and plan descriptions are transferred using the appropriate DICOM objects and are used to construct a 3D representation of that section of the patient they represent. Sagittal and Coronal CT planes are generated from the axial CT planning data set. Structures can be represented as native contours and as volumetric objects created by default when the patient data is first loaded.

Fig 1: VERT simulation of Linac and an IMRT patient loaded from a commercial planning system.

Whereas, by default, the patient is initially placed in the correct treatment position , an error simulation function is available such that random or known positional errors can be introduced with six degrees of freedom. This functionality, along with various dose visualisation tools and MV/ kV image generation tools, allows the implication of patient mis-positioning to be explored.

A number of QC activities or work flows are available within the simulation that allow the output and beam profiles of the Linac to be measured, the laser alignment and light-radiation field coincidence to be checked. For these a suite of QC devices are available and can be loaded in place of the patient. These devices are interactive and 'produce measurements' via 6MV and 15MV beam models built into the simulation which have representative depth dose/ energy characteristics. All measurement simulations made within the virtual system reflect the accuracy, or otherwise, of the equipment alignment and positioning. Similarly as with the patient set up error simulations these allow the trainee or user to explore a range of situations where the equipment has not been correctly set up and be able to appreciate the consequences of such.

The available equipment includes:

- a QC dosimetry phantom which has an ion chamber that can be placed at different depths (dmax for the two beam qualities and 2 to 20 cm depths at 1 cm increments).
- a plotting tank that needs to be levelled correctly, the accuracy of which can be judged using the integrated spirit levels
- a simple radiation field- light field coincidence phantom which fluoresces when irradiated
- a laser alignment phantom
- a water tank that has a reference class chamber and a farmer / departmental chamber mounted side by side to allow the calibration of the departmental chamber

With the exception of the beam models, dosimetric features can be customised to the users own preference such as calibration point/ depth of the Linac, energy response/ calibration factors of the chambers. Additionally the temperature and pressure within the Linac bunker are randomised, within realistic limits, for the last activity listed above and the output of the Linac are subjected to user controlled random fluctuations and offsets. This theme is expanded with the ability to introduce mis-calibrations or variances from the ideal for Linac parameters such as Gantry, Collimator, couch travel parameters, isocentre wobble, calibration of the TSD and the light field projectors, jaw positions and the laser positions. This feature allows the machine to be presented to a trainee in an incorrect state to give them the opportunity or exercise in diagnosing a Quality Control problem and be able to work through the solution. Scenarios can be programmed and captured in order to be repeatedly and exactly reproduced in order to allow (fair) competency testing for individuals or a group/ class of students.

3.User Case scenario: Correct set-up, wrong output?

To illustrate the use of the Physics tools/ activities within VERT and the utilization of the capability to mis-calibrate the virtual Linac the following user case scenario was developed. For this exercise the choice of Linac was arbitrary.

The activity using the output phantom and Farmer ionization chamber was loaded with the chamber set at 5 cm depth and a 10cm x 10cm field was set. This configuration was chosen, being the calibration conditions for the Linac. The vertical height of all the horizontal lasers in the virtual bunker were altered by 2mm, such that they indicated an isocentre position to be further from the target than intended. This was achieved by entering a systematic offset in the *errors and machine variances* menu. The vertical lasers themselves were not altered to ensure the isocentre was still indicated to lie along the CAX of the beam. This 3D scene was captured using the *virtual presenter* feature in order to be able to repeat the exercise easily without having to set up the errors/ offsets following any final recalibration of the virtual Linac.

The couch height and general position was altered in order to set the dosimetry block such that the chamber was centered on the isocentre. This process is performed as if one were using a real Linac using the Lasers and light field; However by making the phantom transparent and zooming in and out of the display additional confirmation could be seen! On accessing the associated virtual electrometer (via iPad display) the output was found to be 0.996 rather than the expected 1.000 Gy.

The set up of the block was then explored further, again following a typical clinical troubleshooting process. For example, in this instance:

- Reconfirmed the position by the lasers
- Switched on the TSD projector / ODI (optical distance indicator)
  - Which indicated that the surface of the block was not precisely at 95 cm as should be expected and therefore raised the question regarding the laser calibration integrity
- Load the linac's *100 cm front pointer* and check the laser projection of the isocentre against the position of its tip
  - Which indicated the position of the isocentre as indicated by the lasers to be below that indicated by the mechanical front pointer
- The surface of the block was set to the lasers
  - This demonstrated the previous observation, given the surface of the block was below the tip of the front pointer.
- Selection the Laser alignment phantom – set the phantom up by aligning its markings to the light field crosshairs (fig 2) with the gantry at 0, 90, 270 degrees. Then assess laser alignment against the phantoms markings.

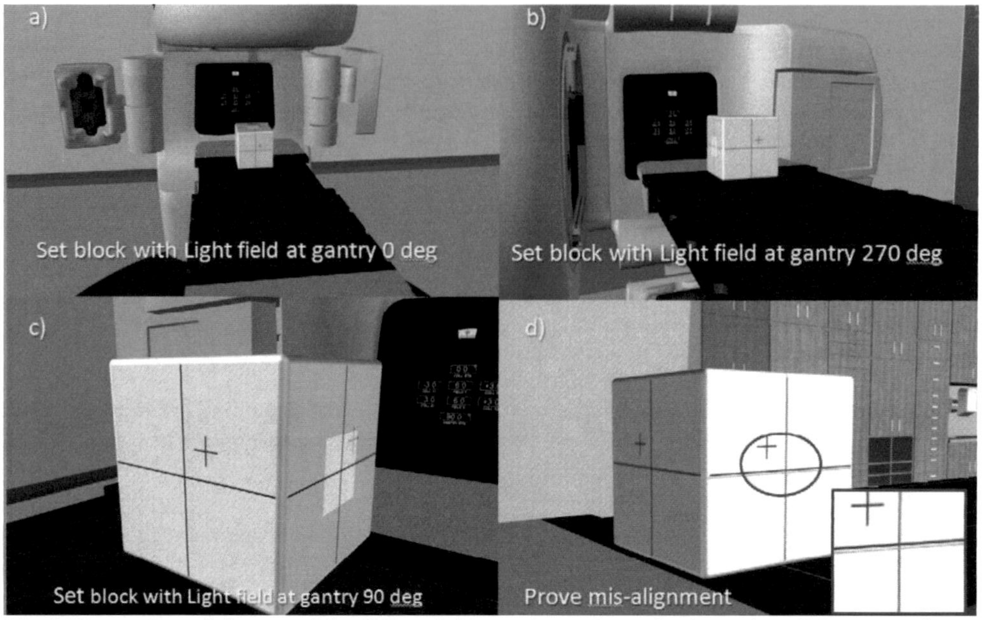

Fig 2 Illustration of the use of the QC activities to investigate the integrity of the set-up laser calibration. Setting the alignment phantom with the light field crosshairs at cardinal angles (a,b,c) and proving the lasers to be 'below' the isocentre plane (d)

Having performed these checks the trainee should conclude that the horizontal lasers were mis-calibrated such that they indicated the isocentre was 2mm further from the target than was really the case and therefore explaining the 0.4% reduction in output of the Linac. Confirmation of this conclusion would be gained by setting the *ion chamber/ output block* up by the ODI or using the *mechanical front pointer* and checking the output at the actual isocentre

4.Discussion

The VERT system was originally created in order to provide a *flight simulator* system for radiographers (therapists, RTTs) in training. It was designed to provide both: 1) the opportunity to gain practical experience away from the busy clinical setting and 2) to give educators a tool for explaining concepts and process issues which was vastly superior to using slides or white-boards to illustrate their presentations. It was immediately apparent that VERT provided a platform to explore treatment errors and patient setup issues and at its outset features were included in the system to facilitate this and simulate such errors using treatment plans from commercial planning systems.

The VERT system was rolled out across all the radiographer teaching schools in England in early 2008 and has been integrated into the teaching of every radiographer trained in England since the beginning (October) of the 2008/9 academic year. Part of the feedback and experience gained was that simulation of Physics concepts and workflow would be helpful for all staff groups. One of our particular interests and continuing areas of research is the use of simulation training and, specifically the VERT system, in the understanding of errors/ process deviations and ultimately treatment integrity and patient safety. The intended use of VERT now expands to all Radiation oncology personnel and the presented scenario is more suited towards, though not exclusively intended for, Physicist trainees and Dosimetrists. In this paper we have explored the use of VERT to provide an (arguably simple) example of how a system mis-calibration might be explored during the routine QC of a machine. We have chosen a scenario that used various existing features of the system in order to illustrate the 'systematic implementation of features throughout the software. In actual fact this extends across all activities and any mis-calibration of the lasers (in this instance) would be apparent and affect the setting up of any loaded patient, until the calibration was corrected.

Computer simulation training is not expected, nor intended to replace practical experiential training, however it most certainly provides the opportunity to understand a wide spectrum of error conditions and their implications in a compressed time frame. It can often take a number of years to be able to experience such things clinically, simply due to their infrequent nature. As described here scenarios can be staged, explained and therefore understood in the classroom, as part of a planned course on quality control, using virtual simulation *on demand* and in a safe and controlled environment. Once the exercises are completed, the virtual Linac can be quickly recalibrated

As stated above the presented example is considered to be fairly simple, however it demonstrates that the process simulation capabilities within the VERT system can reproduce the clinical/ quality control work flow. More complex troubleshooting scenarios can be generated by compounding mis-calibrations or mixing these with set-up errors. In certain activities the output of the Linac can be varied and measurement uncertainties can be assigned to the metrology equipment. In the *calibrate the ion chamber activity* the ambient temperature and pressure are modulated in the treatment bunker in order to demonstrate the importance of making the appropriate corrections to the measurements. In the future we will explore making further systematic fluctuations to simulate calibration drift and further random fluctuations in parameters to simulate, for example a sudden thunder storm.

5.Conclusion

The VERT system is a commercially offered simulation training system that has evolved from the research and education interests of a radiation physics and computer science collaboration. More recently it has been broadened to encompass features that are designed to facilitate training pursuant to a better understanding for patient safety. Simulation training can be used to demonstrate the basic concepts, processes and work-flow of radiotherapy, but it can also be used to simulate errors and process failures and allow participants to examine such scenarios leisurely in the class room setting. Furthermore, experiences typically gained over many years of clinical work can be efficiently demonstrated by experienced professionals to their trainees in the simulation environment where there is zero risk to patients or staff even if a virtual linac is intentionally mis-calibrated. We are continuing our research interest in developing further physics activities and process simulation within our simulation training platform.

References

1. Phillips R, Ward JW, Beavis AW. 2005 Immersive Visualization Training of Radiotherapy Treatment. *Proc of Medicine Meets Virtual Reality 13. Studies in Health Technology and Informatics,* **111**, 390-396.
2. Boejen A, Grau C. 2011 Virtual Reality in radiation therapy training. *Surgical Oncology* **20** 185-188
3. Beavis AW, Phillips R, Ward JW. 2007 A virtual environment for radiotherapy training. *Proc of the international conference on the use of computers in radiation therapy, ICCR Toronto. Journal of Physics, Conference Series, IOP.* 605-609
4. Bridge P, Appleyard RM, Ward JW, Phillips R, Beavis AW. 2007 Development and Evaluation of a Virtual Radiotherapy Treatment Machine using an Immersive Visualization Environment . *Computers and Education,* **49** 481-494.

Journal of Physics: Conference Series **489** (2014) 012087 doi:10.1088/1742-6596/489/1/012087

An MLE method for finding LKB NTCP model parameters using Monte Carlo uncertainty estimates

Martin CAROLAN[1,2], Brad OBORN[1,2], Kerwyn FOO [3], Annette HAWORTH[4], Sarah GULLIFORD[5] and Martin EBERT[6,7]

[1]Illawarra Cancer Care Centre, Wollongong Hospital, Wollongong NSW 2500, Australia

[2]Centre for Medical Radiation Physics, University of Wollongong, Wollongong NSW 2522, Australia

[3]Department of Radiation Oncology, Royal Prince Alfred Hospital, Camperdown, NSW 2050, Australia

[4]Peter McCallum Cancer Centre, Melbourne, Vic, Australia

[5]Joint Dept of Physics, Institute of Cancer Research and Royal Marsden National Health Service Foundation Trust, Sutton, United Kingdom

[6]Sir Charles Gairdner Hospital, Nedlands, WA, 6009, Australia
[7]School of Physics, University of Western Australia, Crawley, WA 6009, Australia.

E-mail: `martin.carolan@sesiahs.health.nsw.gov.au`

Abstract.
The aims of this work were to establish a program to fit NTCP models to clinical data with multiple toxicity endpoints, to test the method using a realistic test dataset, to compare three methods for estimating confidence intervals for the fitted parameters and to characterise the speed and performance of the program.

1. Introduction

The QUANTEC reports [1] have summarised and reported data on radiation induced complication probabilities in multiple organs. The QUANTEC series emphasised the importance of collecting dosimetry and outcome data to enable more accurate modelling of normal tissue complication probability (NTCP). This data has mainly been collected in the context of clinical trials in the past but ideally clinics would record detailed and accurate data for all patients during routine treatment operations allowing a "data mining" approach to development of more accurate NTCP models [2]. Various NTCP models exist but one of the most commonly applied models is the Lyman, Kutcher, Burman (LKB) model [3]. The LKB model describes the NTCP in terms of three model parameters and the dose received by the organ of interest. When the model parameters are known and the dose volume histogram data is available the calculation of the NTCP is trivial and can be performed in a few seconds using a spreadsheet. Determining the appropriate values of the model parameters from a large set of patient dose volume histograms and observed toxicities is significantly more computationally intensive. The maximum likelihood estimation (MLE) method is used to identify the set of parameter values which best describes the observed clinical outcomes. The estimated LKB model parameters are drawn from some distribution of possible values that describe the patient group and it is necessary to determine the confidence interval around these values. There are several methods available to determine the confidence intervals including the profile likelihood method as well as Monte Carlo based techniques such as parametric and non parametric bootstrapping. This paper describes the design and testing of a program to determine LKB model parameters from a number of test data sets and compares parameter confidence intervals derived using parametric and non parametric bootstrapping.

2. Methods

2.1. Lyman - Kutcher - Burman Model

The LKB model [3] is used to generate a normal tissue complication probability based on the dose received by the normal tissue of interest. The dose volume histogram (DVH) for the normal tissue is collapsed into a single dose parameter referred to as the Equivalent Uniform Dose (EUD). Conceptually the EUD may be considered as the uniform dose to a particular organ that would yield the same outcome or toxicity endpoint. The Equivalent Uniform Dose is calculated as follows:

$$EUD = \left[\frac{1}{N_{voxels}} \sum_{i=1}^{N_{voxels}} d_i^a \right]^{\frac{1}{a}} \tag{1}$$

The dose in each voxel of the organ is raised to the power of a (an alternative representation uses $n = 1/a$) and summed over all voxels. The parameter a represents the seriality of the tissue response to radiation. When a is large the overall EUD is dominated by the hottest voxels (serial type response). When a is small the EUD is more influenced by the total volume irradiated (parallel type response). The NTCP is then given by:

$$NTCP = \frac{1}{\sqrt{2\pi}} \int_{-\infty}^{t} e^{\frac{-x^2}{2}} dx \qquad where: t = \frac{EUD - D_{50}}{mD_{50}} \tag{2}$$

Where D_{50} represents the dose to the organ that results in a 50% chance of a particular complication and m defines the slope of the assumed sigmoid dose response curve of the tissue.

2.2. Finding a, m and D_{50} using Maximum Likelihood Estimation

Given a sample of patients where the DVHs for the organ at risk of complication are known and the observed incidence of a particular complication is recorded it is possible to make an estimate of a, m and D_{50}. To achieve this NTCP values for each patient in the population are calculated for a range of a, m and D_{50} values. For each combination of a, m and D_{50} the NTCPs for each patient are compared to the observed toxicities and the log-likelihood (LLH) is calculated [4]. The log likelihood is a measure of agreement between the observed toxicities and the calculated NTCPs and is found as follows:

$$LLH = \sum_{tox=1} \ln(NTCP(a, m, D_{50})) + \sum_{tox=0} \ln(1 - NTCP(a, m, D_{50})) \tag{3}$$

The first term is summed over all patients where a complication is observed and the second is summed over all patients where no complication is observed. The combination of a, m and D_{50} which yields the maximum value of the log-likelihood is the parameter set which best fits the patient data being used. The search for the maximum likelihood can be approached with a smart optimisation algorithm or alternatively by brute force searching through a large grid parameter space. In this work we took the brute force approach.

2.3. Multiple Complication Endpoints

Once suitably large arrays of a, m, D_{50} and NTCP are calculated these are independent of the actual observed endpoint and can be used to find the MLE of the parameters for any observed complications. This calculation efficiency increases with the number of end points fitted.

2.4. Finding the Confidence Intervals

Confidence intervals are commonly determined using the profile likelihood method. This involves finding the maximum log-likelihood and then varying each parameter until the log-likelihood is decreased by an amount equal to half the critical value of the $\chi^2(1)$ distribution at the desired significance level. The search is repeated for each parameter and this yields the respective confidence intervals [5]. A more general but computationally intensive approach also commonly used in NTCP modelling is non-parametric bootstrapping [6]. This involves randomly selecting a population of patients from the data set (with replacement) and then finding the a, m, and D_{50} again. This is repeated many times (e.g. 1000 - 2000) and yields a distribution of a, m, and D_{50} values. The confidence interval is determined by looking at the value of each parameter at the percentile cut-off for the confidence interval required in each distribution. This approach relies on the intrinsic distribution of the patient sample itself. It is robust and the main potential weakness is if the patient group does not adequately cover the whole parameters space. An alternative approach is to use parametric bootstrapping. This assumes that the fitted model is a good representation of the data (and the real population it represents). Therefore a synthesised patient population can be derived from the model rather than the original patient data set. In practice this is achieved by calculating the NTCP for each patient in the original data set using the optimised a, m, and D_{50} values. To generate a new patient population, a random number (r_n) between 0 and 1 is generated for each patient. If $r_n < NTCP$ patients are assigned a complication. If $r_n > NTCP$ the patients are assigned no complication. The values of a, m, and D_{50} for the new synthesised patient population are found using MLE. This process is repeated many times (1000-2000) and the confidence intervals are determined from the resulting distributions. The parametric approach relies on the model being a good fit to the true patient population.

2.5. Implementation

The MLE optimisation and three different confidence interval estimation approaches were programmed using MATLAB (version 7.12.0, The Mathworks Inc., 2011). The code was run on a HP server with 64 cores (four 16 core CPUs) running at 2.6 GHz and with 256 GB of RAM. The array containing the grid of a, m, D_{50} and NTCP for each patient was calculated once. All subsequent MLE searches were performed against this array. Since this array is accessed 1000's of times in the bootstrapping routines using a large precalculated grid is advantageous. The search grid was defined so that a had a range from 10^{-3} to 10^2 in 61 bins equally spaced on a logarithmic scale. The dose parameter D_{50} spanned from 0-150 Gy in 1 Gy increments. The NTCP slope parameter m spanned from 0.01 to 2 in 47 bins. To perform the non parametric bootstrapping 2000 re-sampled populations were broken up into 50 batches of 40 populations thereby using 50 of the 64 cores on the server. The parametric bootstrapping was performed in a similar way with 2000 synthesised patient populations solved across 50 cores.

2.6. Evaluation with a Test Data Set

Three test data sets were based on DVHs and toxicities from 738 patients in the TROG 03.04 RADAR clinical trial [7, 8]. For the purposes of the current work the data set was treated simply as a realistic distribution of DVHs. The DVHs for all patients were used in a single data set without consideration of any other details related to the trial. The test set model parameters derived and used in this work do not have any clinical significance beyond the testing described here and readers are referred to references [7, 8] for more details of the RADAR study itself. Three separate complication endpoints were chosen giving a range of parameter values and numbers of patients with the complication.

3. Results

The test data sets consisted of anal canal and anorectal dose volume histograms from 738 patients in the RADAR data set. All patients for which complete data was available were included. The data set therefore incorporates a reasonably heterogeneous set of patients from all arms of the study with prescribed doses ranging from 66 to 74 Gy. Using various techniques. The complication incidence for the three test data sets was 126/738, 309/738 and 126/138 for test sets 1, 2 and 3 respectively.

3.1. Confidence Intervals

In all cases the confidence interval derived using the likelihood profile method is smaller than either of the bootstrapping methods for deriving confidence intervals. The confidence interval derived from the non parametric bootstrap was always larger than the confidence interval based on the parametric bootstrap. This trend is expected since the likelihood profile confidence interval represents the MLE fit to the specific data set. With the parametric bootstrap the possibility of different outcomes for each member of the patient test population is exercised but with the inherent assumption that the NTCP model accurately describes the distribution of the actual data. Whereas the non parametric model which re-samples the existing patient population does not assume the model accurately describes the patient distributions. Examples of the a, m and D_{50} distributions found for the two different bootstrap methods are shown in figures below.

Table 1. Model parameter values and confidence intervals (68%) for each calculated by three different methods each of the test data sets.

	a	D_{50}	m
Test Set 1	0.6	69	0.6
likelyhood profile CI	0.001-0.7	66-79	0.6-0.65
parametric CI	0.034-0.2	62-81.5	0.525-0.7
non parameteric CI	0.034-3.5	61-86.5	0.47-0.71
Test Set 2	3.2	55	0.55
likelyhood profile CI	2-5	52-59	0.45-0.85
parametric CI	1.46-6.24	50.3-60.6	0.41-0.94
non parameteric CI	0.88-7.24	47.6-61.4	0.37-1.03
Test Set 3	3.2	77	0.4
likelyhood profile CI	2-5	72-88	0.3-0.5
parametric CI	1.17-7.15	70.7-93.5	0.32-0.61
non parameteric CI	1.21-11.73	69.9-94.7	0.26-0.61

3.2. Computation Time

Our test data sets had 738 patients and the MLE search was performed over a very modest grid of a, m, D_{50} values of dimensions $(61 \times 150 \times 47)$. To construct the array of EUD and NTCPs required for each patient with this search grid takes approximately 30 minutes on our hardware. A single MLE solution of a, m, D_{50} for all three of the test data sets took 150 seconds on a single core. Therefore to perform a MLE solution for 2000 patient populations takes approximately 3×10^5 seconds or 83 hours. When this is run on 50 cores simultaneously (without memory sharing) the calculations can be completed in approximately 1 h 40 minutes.

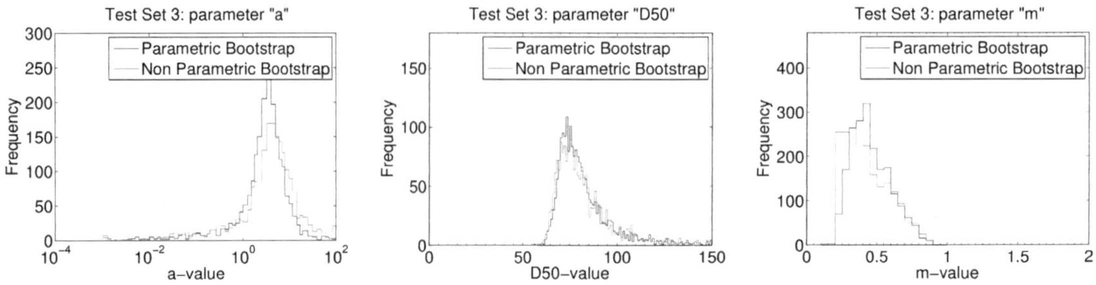

Figure 1. Parametric and Non Parametric bootstrapping for the parameters "a", "D_{50}" and "m".

There was no difference observed in the computing time (150 $\pm$ 5 seconds per iteration) required for the parametric versus the non parametric bootstrapping techniques so in our implementation neither has any advantage over the other in this respect. At these speeds of execution and relying on multiple cores for processing it is possible to analyse a data set with approximately 750 patients and derive NTCP parameters for 20 different complication end points in an overnight batch taking approximately 12 hours.

4. Discussion

4.1. Grid search considerations

Determination of LKB parameters calculating the EUD and NTCP on the fly for a large number of patients and end points is very inefficient. A pre-calculated grid is preferred. The span of the grid is important so that the search space contains any real MLE maxima. Failure to meet this requirement is indicated by overpopulated bins at the extremes of the bootstrap distributions. One shortcoming of our implementation of the "brute force" grid search technique is that it does not easily lend itself to adaptive grid resizing as this would require further NTCP calculations on the fly as the MLE converges on a maximum value. A smarter (no grid) optimisation procedure could give better performance so long as it converges quickly (i.e. the total number of NTCP calculations for all bootstrapping and all endpoints is significantly less than the 4.3×10^5 array entries used for the grid search).

4.2. Parametric or non parametric bootstrap?

This work shows that the choice of either parametric bootstrapping or non-parametric bootstrapping for confidence intervals in NTCP models is probably not critical in most cases. The non parametric approach is used in most publications related to confidence intervals for NTCP models. This assumes the sample of patients in the study group is representative of the wider population to whom the model would be applied. The bootstrapping procedure should give a good indication of the distribution the parameter could have for that population. In contrast to this the parametric approach results in a distribution of parameter values reflecting the model prediction for these values not the actual sample set distribution. This will give narrower confidence intervals than the more pessimistic confidence intervals from the non parametric bootstrap. Derivation of the confidence intervals was based on a simple quantile approach however other approaches to determining the confidence intervals are described in the statistics literature [6]. These are recommended as more rigorous than our simple approach and should be considered for future implementation. A parametric approach is invaluable for predicting distributions of NTCPs for a set of patients and comparing how well different models fit the patient data [9, 10]. Either technique will provide a more conservative estimate of the confidence intervals for an NTCP model than the profile likelihood calculation and in our implementation there was virtually no difference in the computing time.

5. Conclusions

This simple exercise has demonstrated that in a small number of test data sets parametric and non parametric bootstrapping can both be used to generate comparable confidence intervals for NTCP model parameters. Both approaches give more information about parameter distributions than the simple but less intensive likelihood profile method. There is no difference in calculation time in our implementation of the calculations. Further refinement including implementation of an adaptive grid spacing would improve accuracy of the MLE optimisation but this is difficult to incorporate into a simple grid search algorithm without speed penalties. Additional functionalities to allow cross validation of fitted models and development of well characterised data sets as suggested by other authors [11] should also be implemented.

6. Acknowledgements

This work is supported by NHMRC Project Grant 1006447. We thank Dr Anthony Carolan for useful discussions on parametric and non parametric bootstrapping.

References

[1] Marks, L. B., et al. (2010). IJROBP, 76(3 Suppl), S10–9. doi:10.1016/j.ijrobp.2009.07.1754
[2] Deasy, J. O., et al. (2010). . IJROBP, 76(3 Suppl), S151–4. doi:10.1016/j.ijrobp.2009.06.094
[3] Stewart R.D., Li, X.I. (2007), Med Phys 34(10):3739-3751.
[4] Gulliford, S. L., et al, (2012). Radiother and Oncol, 102(3), 347-51. doi:10.1016/j.radonc.2011.10.022
[5] Stryhn, H., & Christensen, J. . In Proceedings of 10th International Symposium Veterinary Epidemiology and Economics, 2003. p208.
[6] Carpenter, J., & Bithell, J. (2000). . Statistics in medicine, (August 1999), 1141-1164. Retrieved from http://onlinelibrary.wiley.com/doi/10.1002/(SICI)1097-0258(20000515)19:9<1141::AID-SIM479>3.0.CO;2-F/abstract
[7] Denham, J. W., et al. (2012). Radiother and Oncol, 105(2), 184–92. doi:10.1016/j.radonc.2012.09.018
[8] Kearvell, R, et al. J Med Imaging Radiat Oncol. 2013 Apr;57(2):247-57. doi: 10.1111/1754-9485.12025. Epub 2013 Jan 7
[9] Seppenwoolde, Y. V. S., et al. (2003) 55(3), 724–735. doi:10.1016/S0360-3016(02)03986-X
[10] Schilstra, C., & Meertens, H. (2001). .IJROBP, 50(1), 147–58. Retrieved from http://www.ncbi.nlm.nih.gov/pubmed/11316558
[11] Xu, C.-J., Van der Schaaf, A., et al (2012). IJROBP, 84(1), e123–9. doi:10.1016/j.ijrobp.2012.02.022

Journal of Physics: Conference Series **489** (2014) 012088 doi:10.1088/1742-6596/489/1/012088

Semi-automated contour recognition using DICOMautomaton

H Clark[1,2], J Wu[2], V Moiseenko[2,3], R Lee[2], B Gill[2], C Duzenli[1,2] and S Thomas[4]

[1] University of British Columbia, Vancouver, B.C., Canada.
[2] British Columbia Cancer Agency - Vancouver Centre, Vancouver, B.C., Canada.
[3] University of California San Diego, San Diego, Ca., USA.
[4] British Columbia Cancer Agency - Fraser Valley Centre, Surrey, B.C., Canada.

E-mail: `hdeanclark@gmail.com`

Abstract. Purpose: A system has been developed which recognizes and classifies Digital Imaging and Communication in Medicine contour data with minimal human intervention. It allows researchers to overcome obstacles which tax analysis and mining systems, including inconsistent naming conventions and differences in data age or resolution.
Methods: Lexicographic and geometric analysis is used for recognition. Well-known lexicographic methods implemented include Levenshtein-Damerau, bag-of-characters, Double Metaphone, Soundex, and (word and character)-N-grams. Geometrical implementations include 3D Fourier Descriptors, probability spheres, boolean overlap, simple feature comparison (e.g. eccentricity, volume) and rule-based techniques. Both analyses implement custom, domain-specific modules (e.g. emphasis differentiating left/right organ variants). Contour labels from 60 head and neck patients are used for cross-validation.
Results: Mixed-lexicographical methods show an effective improvement in more than 10% of recognition attempts compared with a pure Levenshtein-Damerau approach when withholding 70% of the lexicon. Domain-specific and geometrical techniques further boost performance.
Conclusions: `DICOMautomaton` allows users to recognize contours semi-automatically. As usage increases and the lexicon is filled with additional structures, performance improves, increasing the overall utility of the system.

1. Purpose

The ever-growing volume of dosimetric patient data accumulated by medical centers is varied not only in terms of quality, scope, and format, but also in a more subtle way; each center has a unique naming and contouring *dialect*. While the Digital Imaging and Communication in Medicine (DICOM) standard addresses the former (i.e. interoperability between various types of hardware and software), it imposes constraints of the latter type on software. In particular, the flexibility of contour generation, specification, and labeling presents a challenge for systematic identification. For example, 60 patient contour sets from one centre over one year contained more than a dozen labels indicating the left parotid. In some cases both parotids were contoured into a single 'parotids' structure while in others the parotids had not been contoured - sometimes because they had been surgically removed. While an experienced researcher could reliably identify a left parotid, it is not conceivable to perform manual identification in analyses involving hundreds of patients. We present herein a set of practical lexicographical and geometrical techniques, incorporated into `DICOMautomaton`, for performing automated recognition of existing contour data.

2. Methods

We consider the situation whereby a researcher has a collection of existing contour data of mixed origin and a set of structures (e.g. organs) which are to be identified. An example might be locating all left parotids, taking into account that not every structure set is required to contain a left parotid and that unknown aliases may be used.

We define a *lexicon* l to be the collection of exact mappings from a set of raw labels d_i to a unique label c_i such that $l(s) = c_i$ *if and only if* s is a label in d_i. We denote the set of all d_i as d and the set of all c_i as c. The lexicon represents the researcher's domain knowledge and is similar to a thesaurus. For instance a limited, toy lexicon for head and neck cancers might be

$$d_1 = \{\text{l parotid}, \text{lt\_par}, \text{Left Parotid}\} \qquad c_1 = \text{Left Parotid}$$
$$d_2 = \{\text{r partd}, \text{r\_par}, \text{Right Parotid}\} \qquad c_2 = \text{Right Parotid}$$
$$d_3 = \{\text{chiasm}, \text{opt\_chiasm}\} \qquad c_3 = \text{Optic Chiasm}.$$

Using this lexicon, one would be unable to identify a structure named "left par" though it is clear that it most closely relates to d_1. To capture this intuitive notion of similarity we model group membership using string similarity measures, which consider two inputs similar if they share specific features. The similarity between strings s_1 and s_2 for measure J is written as a score $S_J(s_1, s_2) \in [0, 1]$ where $S_J = 0$ denotes no similarity and $S_J = 1$ perfect similarity. Scoring allows us to incorporate the similarity of strings from many measures. Given a lexicon l and input s which refers to an unknown member of c, we choose a combining function F to weight the similarity scores with the elements of d_i into a total score for c_i

$$T_{c_i}(s) = F\left[\left\{S_J(s, \hat{s}) \middle| \hat{s} \in d_i\right\}; l\right] \in [0, 1].$$

Typically, F is chosen to help reduce statistical uncertainty by producing a high $T_{c_i}(s)$ when many measures produce high scores and a low $T_{c_i}(s)$ otherwise. For well behaved general measures, F could be a simple mean. Denoting the set of all $T_{c_i}(s)$ as $T(s)$, the suspected translation $\tilde{c}$ is determined by choosing a mapping function $M(T(s)) = \tilde{c} \;\middle|\; \tilde{c} \in c$. Both F and M are subject to the desirable property that if $s \in d_i$ then $\tilde{c} = c_i$ as one would find using the lexicon without any consideration of similarity. We can judge the validity of $\tilde{c}$ using $T(s)$.

Due to wide diversity of contour labeling encountered, F and M were equipped with tunable minimum thresholds to help reliably control type I and II errors. For this study, F was implemented as a weighted-average while M was taken to be a *maximum* function with additional degeneracy-handling logic.

Well-known lexicographic measures implemented in `DICOMautomaton` include Levenshtein-Damerau [1] and Jaro-Winkler [2] edit measures, Soundex [3], Double Metaphone [4], and Match Rating Approach [5] phonetic measures, statistical measures including longest common substrings/sequences, N-grams of user-specified order N (character or word based in a variety of flavours, including Dice [6], Jaccard [7], and cosine metrics), and a generic bag-of-characters measure. The Levenshtein-Damerau measure is popular due to its speed and simplicity. Alternative lexicographical measures implemented include an artificial neural network-based measure, a self-orthogonalizing measure which ignores all common N-grams by elements of d, a custom measure (which we refer to as DICOMhash) which differentiates labels which are liable to confuse other techniques (e.g. 'CTV+3mm' vs. 'CTV+5mm'), and a domain-specific head and neck measure which hard-codes an individual centre's naming conventions.

Geometrical methods considered are similar to the aforementioned lexicographic methods; instead of a string input, they take a set of ordered two-dimensional contours. They rely on geometrical features, but otherwise serve an identical purpose as the lexicographic measures. Geometric measures implemented include probability spheres and boolean overlap relative-position measures, Fourier descriptor and eccentricity shape-based measures, a domain-specific measure, and simple feature measures involving volume, total perimeter/surface area, and centroids [8]. The most intuitive measures involve the spatial location of objects in $\mathbb{R}^3$, so we describe here probability spheres to demonstrate the basic approach: in normalized coordinates, each target structure is given a centre, an effective radius, and a radially-dependent normalized weighting $W(\vec{r})$. The purpose of weighting is to prescribe inhomogeneous regional similarity. The similarity score between structures is evaluated as $\int W_1(\vec{r})W_2(\vec{r})d\vec{r}$. The more structures overlap in $\mathbb{R}^3$, the more likely we are to think of them as similar.

Cross-Validation Lexicon Folding (CVLF) was used to estimate the overall effectiveness of the system. CVLF involves randomly choosing a fraction f of d and measuring the ability of the system to correctly translate the entirety of the complete lexicon. This is loosely analogous to measuring the abilities of an individual to reconstruct a foreign vocabulary after being given only a sample portion. Random selection of a portion of the lexicon may result in omission of c_i, yielding a system artificially unable to recognize any of the corresponding d_i. Where applicable we correct for this effect. Denoting the number of elements in d as N_d, omission of fN_d elements reduces the maximum CVLF recognition rate, on average, to $1 - \sum_i P(c_i)\, N_{d_i}/N_d$ where

$$P(c_i) = \frac{\Gamma(N_d - N_{d_i} + 1)\,\Gamma(N_d - fN_{d_i} + 1)}{\Gamma(N_d - (f+1)\,N_{d_i} + 1)\,\Gamma(N_d + 1)} \;\Big|\; fN_d \leq N_d - N_{d_i}, \text{ otherwise } 0.$$

To evaluate system performance, contour label data from 60 head and neck cancer patients was used for CVLF, producing a lexicon composed of $N_d = 325$ and $N_c = 18$ elements denoting 16 unique structures and two honey-pots (i.e. for interception of artifacts). The honey-pots comprised 146 strings and the remaining were distributed with an average N_{d_i} of 11.2 ($\sigma = 5.1$).

3. Results

M and F thresholds were found to greatly affect the recognition and error rates. Both recognition and type I error rates increased when decreasing thresholds. Type II error rates increased when increasing thresholds. Reasonable default values were found to be 0.3 for F and 0.5 for M.

Results of CVLF for various retention fractions f are shown in figure 1. We estimate that in typical circumstances a user would possess between $50 - 80\%$ of d and $75 - 100\%$ of c from the complete lexicon. Focusing on an information-deficient situation where $f = 0.3$ (i.e. 30% of d are known), the Levenshtein-Damerau measure performed successful recognition 76% of the time with type I errors occurring 23% of the time. The use of several (mixed) lexicographic measures improved the raw recognition rate 7% and reduced type I errors 6%, representing an effective overall improvement in more than 10% of recognition attempts.

Geometrical techniques performed similar to lexicographic techniques. In general, they produced both lower recognition and error rates compared with lexicographic techniques. Remarkably, the type I error rate did not increase above 7% at any f using mixed geometrical measures (probability spheres, perimeter length, a lateral-position discriminator, and centroid comparison). At $f = 0.3$, mixed geometrical measures performed successful recognition 71% of the time. For probability spheres, it was found that an appropriate effective radii was $r_{eff} = 3\,(3V/4\pi)^{1/3}$ where V was the structure's volume. The preceding factor of 3 was chosen to help handle variations in position due to patient geometry and orientation.

Domain-specific lexicographic measures which hard-coded naming conventions performed dramatically better, achieving an optimal recognition rate of 97% when $f = 0.3$.

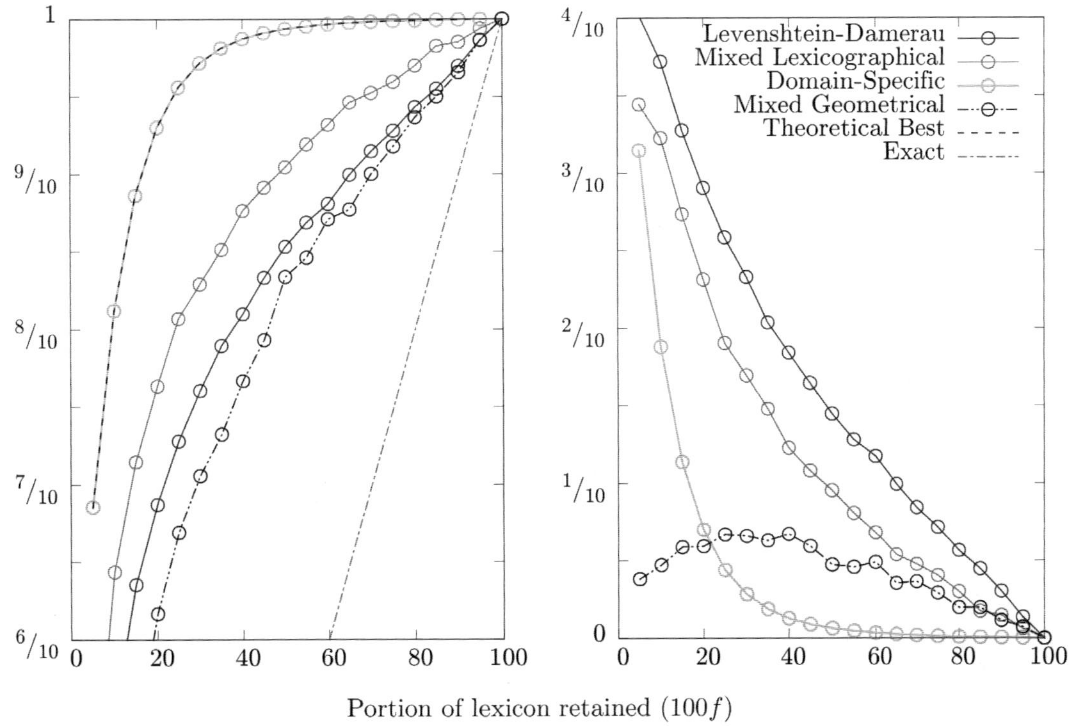

Figure 1. Comparison of select measures. Recognition (left) and type I error (right) rates versus lexicon retention during CVLF. Also shown are the optimal recognition and exact match rates. The latter refers to application of l without the use of similarity measures. The mixed lexicographical measure contains Levenshtein-Damerau, Jaro-Winkler, DICOMhash, N-gram, and bag-of-characters measures. The mixed geometrical measure contains probability spheres, perimeter length, a lateral-position discriminator, and centroid comparison measures. The domain-specific (lexicographic) measure achieves optimal recognition.

4. Discussion

The use of mixed lexicographic measures resulted in an effective overall improvement in more than 10% of recognition attempts compared with the Levenshtein-Damerau approach.

Although geometrical techniques generally showed reduced error rates, they required considerably more processing power, memory, and storage to use. Furthermore, they introduced subtle dependencies on data fidelity and were somewhat unwieldy compared to simpler lexicographic techniques.

Domain-specific measures performed optimally in recognition. Unfortunately, they are by nature often incapable of handling previously-unseen input. It is unclear how to appropriately gauge their performance during real usage. A balance between the uncertainty of domain-specific measures and the less-performant, but more robust, lexicographic and geometrical measures can be achieved with appropriate mixing. However an estimate of f is required for reliable weighting, which may be difficult to assess.

Hybrid techniques could further improve system efficiency by involving meta-information. For example, simultaneous recognition on a collection of mutually exclusive input can ensure that the system does not erroneously detect two spinal cords in a single patient.

Actual usage indicates performance higher than that suggested by CVLF. We believe this is due to CVLF randomly removing elements of d; in reality some labels are encountered frequently (e.g. 'body') while others are rare. This regularity provides increased recognition power.

Finally, we hope to have demonstrated that recognition is, in some cases, a flexible way to handle contour data. It rejects the rigidly-defined database paradigm and encourages data sharing across labeling and contouring conventions.

5. Conclusions

`DICOMautomaton` can be used to perform accurate, semi-autonomous contour recognition. Lexicographic methods are generally suitable to this end when the domain is well-known, while more computationally-burdensome geometrical methods are available for information-deficient situations. Mixing lexicographic measures produces an effective recognition improvement of more than 10% over a pure-Levenshtein-Damerau approach, while domain-specific measures can achieve optimal recognition. Increasing domain knowledge in the lexicon increases performance, and so continued usage will tend to increase the successful recognition rate.

Acknowledgments

The authors acknowledge financial support from The Ride to Conquer Cancer and the BCCA.

References

[1] Levenshtein V 1966 Binary codes capable of correcting deletions, insertions and reversals *Soviet Physics Doklady* vol 10 p 707 2
[2] Winkler W E 1990 String comparator metrics and enhanced decision rules in the fellegi-sunter model of record linkage *Survey Research Methods* (American Statistical Association) pp 354–359 2
[3] Knuth D E 1973 *The art of computer programming, Vol. 3* vol 109 (Addison-Wesley, Reading, MA) 2
[4] Philips L 2000 *CC Plus Plus Users Journal* **18** 38–43 2
[5] Moore G B, Kuhns J L, Trefftzs J L and Montgomery C A 1977 *Accessing individual records from personal data files using non-unique identifiers* (US Department of Commerce, National Bureau of Standards) 2
[6] Dice L R 1945 *Ecology* **26** 297–302 2
[7] Jaccard P 1901 *Etude comparative de la distribution florale dans une portion des Alpes et du Jura* (Impr. Corbaz) 2
[8] Nixon M and Aguado A S 2008 *Feature extraction & image processing* (Academic Press) 3

Journal of Physics: Conference Series **489** (2014) 012089 doi:10.1088/1742-6596/489/1/012089

Development of a Multi-Centre Clinical Trial Data Archiving and Analysis Platform for Functional Imaging

Brandon Driscoll [1], David Jaffray [1,2,3] & Catherine Coolens [1,2,3]

1) Department of Radiation Physics, Princess Margaret Cancer Centre, 610 University Avenue, Toronto, Ontario, M5G 2M9, Canada
2) Department Radiation Oncology, University of Toronto, 150 College St, Toronto, Ontario, M5S 3E2, Canada
3) Techna Institute, University Health Network 124-100 College Street, Toronto, Ontario, M5G 1L5, Canada

Contact Email: Brandon.Driscoll@rmp.uhn.ca

Abstract. Purpose: To provide clinicians & researchers participating in multi-centre clinical trials with a central repository for large volume dynamic imaging data as well as a set of tools for providing end-to-end testing and image analysis standards of practice. Methods: There are three main pieces to the data archiving and analysis system; the PACS server, the data analysis computer(s) and the high-speed networks that connect them. Each clinical trial is anonymized using a customizable anonymizer and is stored on a PACS only accessible by AE title access control. The remote analysis station consists of a single virtual machine per trial running on a powerful PC supporting multiple simultaneous instances. Imaging data management and analysis is performed within ClearCanvas Workstation® using custom designed plug-ins for kinetic modelling (The DCE-Tool®), quality assurance (The DCE-QA Tool) and RECIST. Results: A framework has been set up currently serving seven clinical trials spanning five hospitals with three more trials to be added over the next six months. After initial rapid image transfer (+ 2 MB/s), all data analysis is done server side making it robust and rapid. This has provided the ability to perform computationally expensive operations such as voxel-wise kinetic modelling on very large data archives (+20 GB/50k images/patient) remotely with minimal end-user hardware. Conclusions: This system is currently in its proof of concept stage but has been used successfully to send and analyze data from remote hospitals. Next steps will involve scaling up the system with a more powerful PACS and multiple high powered analysis machines as well as adding real-time review capabilities.

1. Introduction

Quality results from multi-center radiation oncology clinical trials require consistent and robust trial protocols capable of quantifying or eliminating differences across participating institutions. The Quantitative Imaging for Personalized Cancer Medicine (QIPCM) initiative funded by the Ontario Institute for Cancer Research attempts to address these issues through 3 main aims. 1) Establish a stable software platform for image reads/analysis in clinical trials with capabilities for centralized image archival and remote real-time review. 2) Commission a suite of defined imaging procedures and corresponding accreditation tests for use in clinical trials. 3) Establish and operate a subsidized service to implement and support robust, quantitative imaging in clinical trials.

As part of this initiative, the goal of this work was to provide clinicians & researchers participating in multi-center clinical trials with a central repository and cloud-based software tool set for storing/sharing functional imaging data. This platform will facilitate end-to-end testing and image analysis standards of practice as well as a means of performing dynamic contrast enhanced (DCE) QA to enable scanner accreditation.

2. Methods

There are 4 main components to the Clinical Trials involving Functional Imaging (CTFI) data archiving and analysis system: the CTFI Patient Database, the CTFI PACS, the data analysis engine(s) and the high-speed networks that connect them (Figure 1).

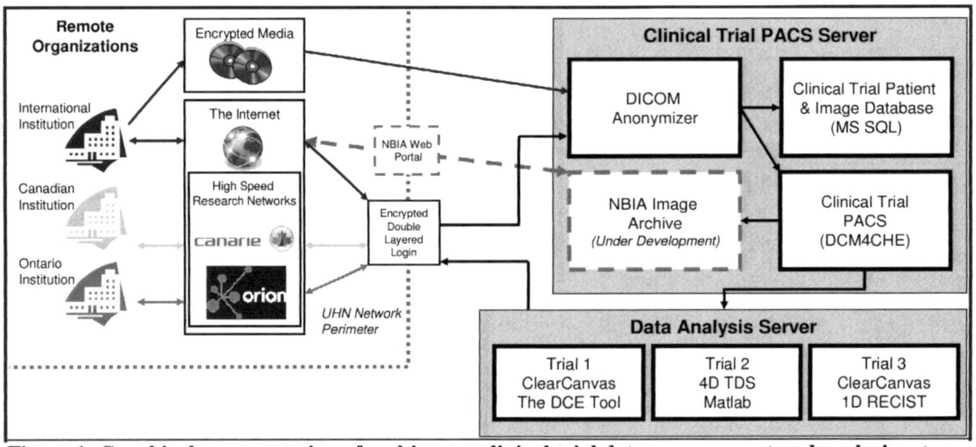

Figure 1: Graphical representation of multicenter clinical trial data management and analysis setup

2.1. Accessing the Network and Transferring Functional Imaging Data

When a clinical trial is registered with the QIPCM initiative, user accounts are generated along with a One Time Pad for remote access. Both the use of the one time pad as well as the users own log in and password are required to access the system. Any computer connected to the internet can gain access to the clinical trial data and management system through a web portal provided they have been assigned the appropriate account information.

Considering the size of image data being transferred the connection between the remote center and the University Health Network (UHN) should be as fast as possible. Many hospitals and research organizations across Ontario are connected to the 100 gigabit per second ORION fibre-optic research network (or CANARIE for the rest of Canada). If the remote PC is connected to one of these organizations, transfer speeds between networks are substantially improved. If the remote PC is not connected to one of the high speed research networks the data can alternatively sent by encrypted hard disk or DVD.

It is important to note that data is only transferred from the remote site to the server once after it is acquired. From then on all data remains within the UHN network and the servers are connected by gigabit connections so transfer between the PACS and analysis PC is very fast.

2.2. Clinical Trial PACS Server

A server has been commissioned which works as a clinical trial data repository. The server currently hosts our data anonymizer, a PACS (picture archiving and communication system) and a Clinical Trial Patient Database running in MS SQL. Figure 2 below outlines how these 3 separate parts interact.

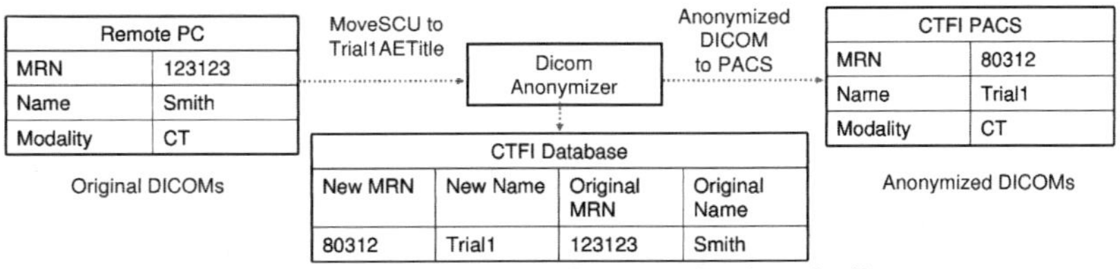

Figure 2: Outline of the CTFI data storage and anonymization pipeline.

2.2.1. Data Anonymization

Since it can never be guaranteed that the data remote sites send to the system will be fully anonymized, the data transfer process includes an anonymizer which receives the original DICOM files and passes them along to the database and the PACS.

A different AE Title is associated with each trial so that different anonymization schemes can be allowed and proper access rights can be enforced. The anonymization works for any modality and has the capability to filter out objects which might contain patient information such as dose reports or other secondary image sets.

2.2.2. Clinical Trial Patient Database

The CTFI database is automatically propagated as DICOM files are received from each clinical trial. The database is designed in MS SQL and queries can be created such that PI's can check the current status of their imaging trial. The patient info that is removed from the DICOM header is stored in a controlled table in the database.

2.2.3. Clinical Trial PACS

The anonymized DICOM data are then forwarded to the Clinical Trial PACS. The PACS runs the DICOM archive and image manager software DCM4Chee (DCM4che.org). The images and study derivatives stored within the PACS are visible only to the appropriate users by AE Title access control.

2.3. Data Analysis Computers

A separate high powered server is used as an access point to the local network. Within this data analysis engine, multiple virtual machines exist (1 per clinical trial). Each virtual machine currently runs a Clear Canvas Workstation and will serve two main purposes; encrypted data transfer from the remote PC to the CTFI server and analysis of the images using one of the custom analysis tools. Two main functional imaging analysis tools were developed for the Clear Canvas platform; one for kinetic modeling (The DCE-Tool®, www.theDCETool.com, section 2.3.1) and one for RECIST (Response Evaluation Criteria in Solid Tumors) (Section 2.3.2). This setup provides the ability to remotely perform computationally expensive operations on very large data archives with minimal end-user hardware. The analysis framework also provides the ability to export the study derivatives back to the Clinical Trial PACS. Other customized tools can also be installed on the clinical trial virtual machine including 4D Functional Image Analysis tools such as 4D TDS [1] or 3D image registration tools.

2.3.1. The DCE Tool

The DCE Tool, within the ClearCanvas framework, is a software plug-in for developing and sharing research results using DCE-CT/MR and PET perfusion studies. Although DCE-CT,MR and PET perfusion studies can provide a direct and quantitative assessment of tumor blood flow, the analysis of these studies are computationally demanding and time consuming.

Some of the important features of the DCE Tool include CT, MR & PET Support, a 4D Time Browser, 2D and 3D ROI Support, MR Intensity to Contrast conversion and a range of different kinetic models including Tofts[2], ATH[3] and Thorwarth[4]. The software is capable of performing

Journal of Physics: Conference Series **489** (2014) 012089 doi:10.1088/1742-6596/489/1/012089

standard 2D/3D ROI analysis or Voxel-Wise analysis and the results can be exported to Microsoft Excel/CSV.

2.3.2. 1D RECIST

1D RECIST is a very common trial endpoint to measure tumour progression[5]. It is not computationally expensive but it is important to have a centralized analysis point for trial results. For this reason the 1D RECIST plug-in was created for ClearCanvas. This tool supports RECIST 1.1 and has automatic long and short axis calculation from a traced ROI. The software can track individual patient response through time via import/export of results and is currently being extended for clinical use.

2.4. Procedures for Tests and Trials: QA and Accreditation of DCE-CT

In order to develop a set of QA procedures for the use of DCE-CT in multi-center clinical trials a previously designed flow phantom[6], was fully validated and then utilized to design a set of standard operating procedures and QA metrics. The developed QA protocol was then applied at three institutions that were part of clinical trial involving DCE-CT and results were compared[7].

An additional ClearCanvas plug-in was designed to support this process in the form of the DCE QA Tool. This tool is capable of performing static and dynamic QA analysis by performing a curve comparison to varied truth protocols in terms of total mass as well as goodness of fit (R^2). These results can be exported to Microsoft Excel for comparison between scanners and protocols.

3. Results

A framework has been set up and currently serves 12 clinical trials spanning 5 hospitals with more trials to be added over the upcoming months. The image store currently holds over 725,000 individual tomographic slices (> 4300 3D volumes) from over 124 functional imaging studies. After initial rapid image transfer, (> 2 MB/s over ORION), all data analysis is done server-side, making it robust and rapid. This has provided the ability to perform computationally expensive operations such as voxel-wise kinetic modeling on very large data archives (+20 GB/50k images/patient) remotely with minimal end-user hardware. This centralized data analysis and storage model also greatly simplifies the addition of new trials and remote sites as very minimal set up is required for the end user.

4. Discussion

The Multi-Centre Clinical Trial Data Analysis and Archival system is fully operational and supporting a number of internal and external trials. The database is constantly growing and the infrastructure needed to support the trials is being upgraded over the next 2 months. At the end of that process the database will have robust backup solutions in place and storage space sufficient to hold 16 million images. Over the next two years the system will also move towards web based viewer platform which is currently in development. This server side rendered viewer will control access to the PACS thus removing the need for one time pads and remote desktop connections.

By having a centralized analysis point using a standardized tool, variations in results can be minimized while allowing multiple observer studies to use the same image files under the same conditions which will increase the reliability of the results[8].

It is important to recognize the work that has been done by the National Cancer Institute (NCI) in attempting to facilitate the archiving and sharing of biomedical imaging through the creation of the National Biomedical Image Archive (NBIA). In order to interact with the many organizations around the world that have begun to use the NBIA standard for sharing and storing images, we are in the process of setting up our own NBIA image repository on the Clinical Trial PACS computer in parallel with our existing setup.

Finally, though the most progress has been made in developing SOPs for DCE-CT QA work has been ongoing on PET and DCE-MR QA protocol development. The phantom has now been validated for DCE-MR using the 3.0 T MR system[9] at the Princess Margaret Cancer Center and the next steps will be to design a robust QA protocol for this modality. A preliminary PET/CT QA protocol has also been developed using the NEMA IEC Body Set phantom based on the general guidelines of the NCI Centers of Quantitative Imaging Excellence PET/CT protocol[10]. This adapted protocol is being utilized to accredit the multiple institutions in a multi-center Melanoma trial involving FAZA imaging.

5. Conclusions

The Multi-Centre Clinical Trial Data Archiving and Analysis Platform is currently in the process of moving from the proof of concept stage to a production system. It has been used successfully to send and analyze data from remote hospitals as well as to support a number of internal high volume functional imaging trials. Next steps will involve scaling up the system with a more powerful PACS and multiple high powered analysis machines as well as adding real-time review capabilities.

References

[1] Coolens C, Wee W, Driscoll B, Jaffray D. "Novel Methodology for Visualization and Analysis of Functional Vasculature for 4D DCE-CT" International Journal of Radiation Oncology Biology Physics, 78(3), S22 (2010)

[2] Tofts, P.S., Brix, G., Buckley, D.L., et al. "Estimating kinetic parameters from dynamic contrast-enhanced T1-weighted MRI of a diffusible tracer: Standardized quantities and symbols." Journal of Magnetic Resonance Imaging, 10:223-232. (1999)

[3] Koh, T.S., Zeman, V., Darko, J., et al. (2001). The inclusion of capillary distribution in adiabatic tissue homogeneity model of blood flow. Physics in Medicine and Biology, 46:1519-1538.

[4] Thorwarth D, Eschmann SM,Paulsen F,Alber M, "A kinetic model for dynamic [18F]-Fmiso PET data to analyse tumour hypoxia." Phys Med Biol. 50(10):2209-24. (2005)

[5] Eisenhauera E.A., Therasseb P. et al. „New response evaluation criteria in solid tumours: Revised RECIST guideline (version 1.1)" European Journal Of Cancer , 45, 228–247, (2009)

[6] Driscoll BD, Keller H, Coolens C, "Development of a dynamic flow imaging phantom for dynamic contrast-enhanced CT." Med Phys. 38(8),4866-80. (2011)

[7] Driscoll BD, Keller H, Jaffray DA, Coolens C, "Development of a dynamic quality assurance testing protocol for multi-site clinical trial DCE-CT accreditation." Med. Phys. 40(8), (2013)

[8] FitzGerald T.J., "What We Have Learned: The Impact of Quality From a Clinical Trials Perspective" Semin. Radiat. Oncol. 22,18-28 (2012)

[9] Lee SL, Foltz W, Driscoll B, Fatemi A, Menard C, Coolens C, Chung C. "Comparison of Arterial Input Functions by Magnitude and Phase Signal Measurement in Dynamic Contrast Enhancement MRI using a Dynamic Flow Phantom". In: Proceedings of the 20th Annual Meeting of ISMRM. Melbourne, Australia; 2012. p. 236

[10] "NCI Centers of Quantitative Imaging Excellence: Manual of Procedures", ACRIN: American College of Radiology Imaging Network Version 3.2 (2013)

Derivation and representation of dose-volume response from large clinical trial data sets: an example from the RADAR prostate radiotherapy trial

MA Ebert[1,2], K Foo[3,4], A Haworth[5,6], SL Gulliford[7], R Kearvall[1], A Kennedy[1], S Richardson[1], M Krawiec[1], N Stewart[1], DJ Joseph[1,8] and JW Denham[9]

[1] Department of Radiation Oncology, Sir Charles Gairdner Hospital, Perth, Austral
[2] School of Physics, University of Western Australia, Perth, Australia
[3] Radiation Oncology, Sydney Cancer Centre, Sydney, Australia
[4] University of Sydney, Sydney, Australia
[5] Department of Physical Sciences, Peter MacCallum Cancer Centre, Melbourne, Australia
[6] University of Melbourne, Melbourne, Australia
[7] Joint Department of Physics, Institute of Cancer Research and Royal Marsden National Health Service Foundation Trust, Sutton, UK
[8] School of Surgery, University of Western Australia, Perth, Australia
[9] University of Newcastle, Newcastle, Australia

E-mail: Martin.Ebert@health.wa.gov.au

Abstract. Large multicentre radiotherapy trials incorporating assessment of multiple outcomes at multiple timepoints can generate extensive datasets. We have investigated graphical techniques for presentation of this data and the associated underlying dose-volume response information, necessary for guiding statistical analyses and translating outcomes to future patient treatments. A relational database was used to archive reviewed plan data for patients accrued to the TROG 03.04 RADAR trial. Viewing software was used to clean and enhance the data. Scripts were developed to export arbitrary dose-histogram data which was combined with clinical toxicity data with a median follow-up of 72 months. Graphical representations of dose-volume response developed include prevalence atlasing, univariate logistic regression and dose-volume-point odds ratios, and continuous cut-point derivation via ROC analysis. These representations indicate variable association of toxicities across structures and time-points.

1. Introduction

We have been investigating the relationship between dose-volume data and clinical outcome for the rectum for patients accrued to the TROG 03.04 'RADAR' prostate radiotherapy trial. In this context, we have developed a series of representations of the data that can provide graphical and statistically informative representations of those relationships. The results are graphical displays to demonstrate the intricate relationships between dose, volume, anatomical definition, specific toxicity and toxicity grade.

2. Description

2.1. *Trial data collection*

The RADAR trial focused on the influence of duration of androgen deprivation on prostate cancer patients and included an external beam radiotherapy (EBRT) dose-escalation component (two-phase treatment to 66, 70, 74 or 78 Gy) and optional HDR brachytherapy. Of 1071 participants accrued from 24 contributing centres, 813 had EBRT alone and of these, 754 complete planning data sets were retrieved from the participating centres' planning systems and archived. Descriptions of the technical aspects of the RADAR protocol, assessed toxicities and principal outcomes have been given elsewhere [1, 2].

A software platform ('SWAN') was developed to facilitate review and processing of participant planning data [3]. SWAN allows visual review and manipulation of most plan objects, automated review of multiple protocol items, reporting back to submitting centres and archive of plan objects into a relational database. SWAN was used to review and archive all data presented here. The RADAR protocol required delineation of rectum from 1 cm below the GTV to where the rectum turns horizontally into the sigmoid colon. This definition was poorly adhered to and most rectums were manually corrected or re-delineated by our team. Rectal contours were also extended inferiorly to the base of the ischial tuberosities to create an 'anorectum' structure.

2.2. *Data processing and extraction*

Histogram calculation, export and link to clinical outcomes. Java scripts were developed to apply SWAN's independent dose-volume histogram (DVH) and dose-surface histogram (DSH) algorithms to archived data. The DVH algorithm calculates the exact area of dose-voxels encompassed by joined delineation points for a structure. DVH and DSH data were automatically calculated at 1 cGy resolution and exported for all archived patient plans for rectum and anorectum. A script was also developed to separate out the lowest 3 cm of the anorectum, defining an 'anal canal', for which DVH/DSH data was exported. Prior to histogram calculation, the multiphase RT data were combined to form an equivalent dose in 2 Gy fractions (EQD_2) based on $\alpha/\beta = 3$ Gy. Dose-volume data was linked (via trial ID) to all patient outcomes data, exported from the clinical database in November 2012. Median follow-up time was 72 months (maximum 108 months). The data were collated as an array (one element per patient) of structures in Matlab 2012b (Mathworks, Natick MA), which was subsequently used to code analysis. Peak grade across the 'late' follow-up period $\geq$ 12 months was determined for each toxicity.

2.3. *Derivation and representation of dose-volume response*

2.3.1. Prevalence – DVH atlasing. The QUANTEC report on normal tissue effects in radiotherapy detailed the advantages of completeness of presentation of data derived from clinical studies [4]. A dose-volume atlas demonstrates, for a given anatomical structure the number of patients whose DVH/DSH passes through a discrete element ('pixel') of dose (or EQD_2)/volume space, and the associated proportion of those patients experiencing an 'event' (a specific toxicity grade definition at a specific time-point or longitudinally across timepoints), and can thus be used to present dose-outcomes data to facilitate meta-analyses. This technique was initially investigated by Jackson et al [5], who previously demonstrated a variety of characteristics of dose-volume data that can be related to patient rectal toxicities [6]. The 'atlas of complication incidence' has since been used for visual association of parameters with trial outcomes [7-9]. Our implementation utilises pixels of dimensions 5 Gy EQD_2 and 5% relative volume/surface area, each coloured according to the fraction of events/patients with a histogram that passes through that pixel. Text values are added to each pixel, being the number of events over the number of histograms passing through the pixel. Alternative representations can be made based on various descriptive statistics [5] or alternative transformations of dose [7].

Figure 1 illustrates atlases for rectal bleeding incidence of grade ≥ 2, as recorded at trial time-point 36 months post-randomisation, and as summarised as the peak grade each patient experiences over the entire follow-up period. With the values tabulated in the plot, a cumulative incidence score as presented by Jackson et al [5] can alternatively be derived.

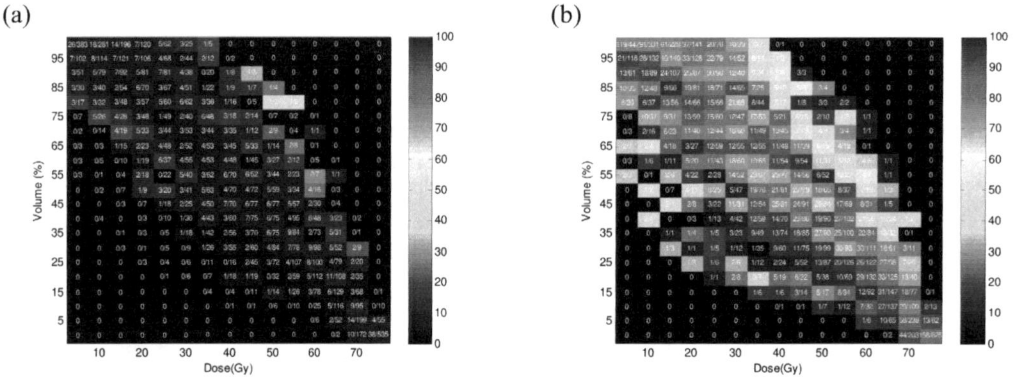

Figure 1. Atlases of prevalence of rectal bleeding of grade ≥ 2 excluding patients with baseline values > 0, at (a) the 36 month time-points, and (b) peak value across whole late ($\geq$ 12 months post radiotherapy) follow-up period. Dose/volume intervals are 5 Gy/5%. Text values indicate event rates in each atlas 'pixel', with the ratio displayed as a colour-scaled percentage, for all patients with follow-up at that time-point and who were not excluded due to baseline toxicity.

2.3.2. Dose-effect – univariate analysis and odds ratios. As pointed out by Jackson *et al* [5], summarising data via an atlas puts it in an excellent form for regression analysis. As a means of illustrating the impact of dose-volume across an atlas, we have combined univariate logistic regression at each EQD_2 interval (arbitrarily selecting 1 Gy intervals), yielding an odds-ratio (OR) for toxicity incidence per % volume increase at each interval, with determination of an OR at each individual dose-volume point for toxicity incidence when the population is dichotomised above and below each volume point at each EQD_2 interval. An alternative atlas summarising this OR-distribution can be generated with OR on a colour scale. The dichotomous OR is considered significant if the lower 95% confidence interval (CI) is greater than 1.0. This can be indicated (in this case, a grey outline) for pixels where the OR is significant. Samples, again for grade > 1 rectal bleeding, are shown in Figure 2. The OR calculation routine developed by Cardillo [10] was used.

2.3.3. Volumetric cutpoints. The ability of specific volumes to discriminate events at each EQD_2 interval can be used to establish volumetric cut-points using the receiver operator characteristic (ROC) curve, derived using volume as the independent predictor [11-13], and assessed using the area under the curve (AUC). A significance level for a cutpoint can be estimated by re-sampling methods (see on-line resources supplementary to this publication). Sample distributions of resulting cutpoints as a function of EQD_2 from the RADAR trial data are shown in Figure 2 for peak late grade of rectal bleeding for grade ≥ 2. Given expected co-linearity of derived cut-points, p-values have been adjusted for multiple testing according to the Holm-Bonferoni step-down method [14]. Derivation of optimal cutpoints using a maximally-selected test statistic has also previously been demonstrated by Buettner et al [15].

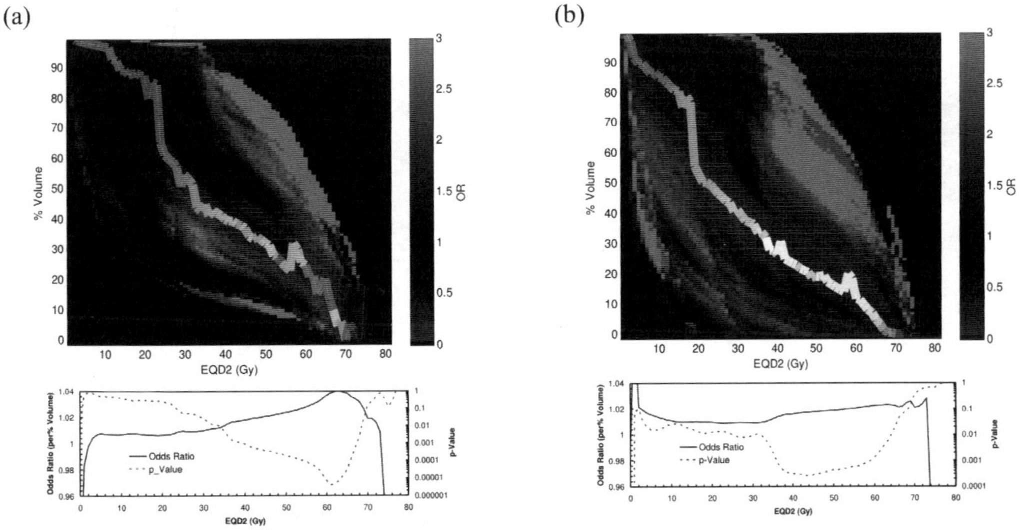

Figure 2. Results of univariate logistic regression, dichotomous OR and optimal cutpoint for rectal bleeding, peak grade across whole late follow-up period of ≥ 2. The bottom plot in each figure shows the OR from logistic regression at each EQD_2 interval plotted with the corresponding p-value. The top plot overlays the ROC-derived cutpoint as a function of dose over the dichotomous OR. The OR is shown as a colour scale at each discrete 1 Gy/1% 'pixel' (as shown), with a grey outline when the lower 95% CI is greater than 1.0. The cutpoint derived from the ROC Youden index is shown as a thick line which is colour-scaled according to the ROC AUC or multiple-testing corrected p-value - from black (AUC ≤ 0.5) to white (AUC at its maximum), and then green for $0.05 < p \leq 0.01$, and red for $p \leq 0.01$. (a) For DVH of whole anorectum, and (b) for DVH of anal-canal.

3. Discussion

Multiple factors can influence the observed associations between dose-volume (or dose-surface area) points derived by these mapping techniques, including the distribution of DVH/DSH data at each EQD_2 value. The absence of significant areas of association when randomly-generated data is used (see supplementary online resources) suggests however, that the significant regions observed on the maps and cutpoint derivations are in fact the result of significant associations between dose-volume points and resulting patient toxicities. The issue then remains of what use can be made of that information.

A significant cutpoint establishes a volumetric index significantly separating patients with and without a specific resulting toxicity. By observing the patterns of this significance, as a function of dose, histogram type and anatomical structure, we can drive hypotheses of the underlying response mechanisms and set constraints for histogram-based optimisation of future patient treatment. By comparing ROC-derived cutpoints with the underlying dichotomous OR values (ie., is shown in the plots of Figure 2) we also establish the precision with which a cutpoint is defined. In Figure 2 for example, we see a wide region of significant dichotomous OR values suggesting the resulting cutpoint values are not well defined, with the potential to take on large confidence intervals. Validation of the cutpoints could be undertaken by comparing them with values derived using alternative techniques, and investigation of such techniques is currently underway.

4. Conclusion

DVH atlasing continues to be established as a standard for the presentation of dose-response data and is ideally suited to the analysis presented above. The variation of significance of association between volume and outcome is apparent when comparing graphical representations spanning anatomical structures (as shown for anorectum and anal canal above) and event definitions. Such representations are now driving derivation of univariate regression and proportional hazards assessment for presentation in more comprehensive publications.

Acknowledgements
We acknowledge funding from the Cancer Council Western Australia (John Nott Fellowship 2012), Cancer Australia and the Diagnostics and Technology Branch of the Australian Government Department of Health and Ageing (grant 501106), the National Health and Medical Research Council (grants 300705, 455521, 1006447), Abbott Laboratories, Novartis Pharmaceuticals. We gratefully acknowledge the support of participating RADAR centers, C Wilcox, the Trans-Tasman Radiation Oncology Group, and past SWAN programmers – J Blight, S Price, B Hooton and E van der Wath.

References

[1] Kearvell R, Haworth A, Ebert MA, Murray J, Hooton B, Richardson S, et al. *JMIRO*. 2013;In press.
[2] Denham JW, Wilcox C, Lamb DS, Spry NA, Duchesne G, Atkinson C, et al. *Radioth Oncol*. 2012;105:184-92.
[3] Ebert MA, Haworth A, Kearvell R, Hooton B, Coleman R, Spry NA, et al. *Radioth Oncol*. 2008;86:200-10.
[4] Jackson A, Marks LB, Bentzen SM, Eisbruch A, Yorke ED, Ten Haken RK, et al. *Int J Radiat Oncol Biol Phys*. 2010;76:S155-S60.
[5] Jackson A, Yorke ED, Rosenzweig KE. *Sem Radiat Oncol*. 2006;16:260-8.
[6] Jackson A, Skwarchuk MW, Zelefsky MJ, Cowen DM, Venkatraman ES, Levegrun S, et al. *Int J Radiat Oncol Biol Phys*. 2001;49:685-98.
[7] Liu F, Yorke ED, Belderbos JSA, Borst GR, Rosenzweig KE, Lebesque JV, et al. *Int J Radiat Oncol Biol Phys*. 2013;85:182-9.
[8] Mutter RW, Liu F, Abreu A, Yorke E, Jackson A, Rosenzweig KE. *Int J Radiat Oncol Biol Phys*. 2012;82:1783-90.
[9] Gulliford SL, Miah AB, Brennan S, McQuaid D, Clark CH, Partridge M, et al. *Radioth Oncol*. 2012;104:205-12.
[10] Cardillo G. Odds: compute odds and risk ratio on a 2x2 matrix. http://www.mathworks.com/matlabcentral/fileexchange/15347; 2007.
[11] Gulliford SL, Partridge M, Sydes MR, Andreyev J, Dearnaley DP. *Radioth Oncol*. 2010;94:241-7.
[12] van der Laan HP, van den Bergh A, Schilstra C, Vlasman R, Meertens H, Langendijk JA. *Int J Radiat Oncol Biol Phys*. 2008;70:1138-45.
[13] Kim B, Park HC, Oh D, Shin EH, Ahn YC, Kim J, et al. *Radiat Oncol J*. 2012;30:43-8.
[14] Westfall PH, Young SS. Resampling-BasedMultiple Testing: Examples and Methods for p-Value Adjustment. New york: Wiley; 1993.
[15] Buettner F, Gulliford SL, Webb S, Sydes MR, Dearnaley DP, Partridge M. *Phys Med Biol*. 2009;54:6535-48.

Journal of Physics: Conference Series **489** (2014) 012091 doi:10.1088/1742-6596/489/1/012091

Combining Dosimetry & Toxicity: Analysis of two UK Phase III Clinical trials

Sarah L Gulliford

Joint Department of Physics, The Institute of Cancer Research & Royal Marsden NHS Foundation Trust, Sutton, United Kingdom

sarahg@icr.ac.uk

Abstract. There are many advantages to performing a clinical trial when implementing a novel radiotherapy technique. The clinical trials framework enables the safety and efficacy of the "experimental arm" to be tested and ensures practical support, rigorous quality control and data monitoring for participating centres. In addition to the clinical and follow-up data collected from patients within the trial, it is also possible to collect 3-D dosimetric information from the corresponding radiotherapy treatment plans. Analysing the combination of dosimetric, clinical and follow-up data enhances the understanding of the relationship between the dose delivered to both the target and normal tissue structures and reported outcomes & toxicity. Aspects of the collection, collation and analysis of data from two UK multicentre Phase III radiotherapy trials are presented here. MRC-RT01 dose-escalation prostate radiotherapy trial ISRCTN47772397 was one of the first UK multi-centre radiotherapy trials to collect 3-D dosimetric data. A number of different analysis methodologies were implemented to investigate the relationship between the dose distribution to the rectum and specific rectal toxicities. More recently data was collected from the PARSPORT trial (Parotid Sparing IMRT vs conventional head and neck radiotherapy) ISRCTN48243537. In addition to the planned analysis, dosimetric analysis was employed to investigate an unexpected finding that acute fatigue was more prevalent in the IMRT arm of the trial. It can be challenging to collect 3-D dosimetric information from multicentre radiotherapy trials. However, analysing the relationship between dosimetric and toxicity data provides invaluable information which can influence the next generation of radiotherapy techniques.

1. Introduction

Radiotherapy in the 21st century allows a carefully shaped dose-distribution to be accurately directed to a tumour with either radical or palliative intent. However it is impossible to completely avoid targeting normal tissue surrounding the tumour. This is partly to ensure that all potential clonogens are included in the irradiated region but largely due to the entrance and exit path of each radiotherapy beam. As the ability to optimise and deliver more complex dose-distributions evolves, it will be possible to be much more prescriptive about the amount and distribution of radiation to normal tissue structures. In order to capitalise on this possibility the dose-response of each normal tissue must be characterised.

The process of characterisation is complex. Analysis requires clinical data including accurate and specific toxicity data and complete dosimetric information on specific organs at risk. Each normal tissue is associated with a number of different toxicities and heterogeneity in the treatment population will influence the side effects each patient will experience. Much of the information currently available on the response of normal tissues is summarised in the QUANTEC report[1]. Each of the organ specific papers is structured to include relevant endpoints, review dose-volume data and summarise recommended dose-volume constraints. The application of mathematical and biological models such as Normal Tissue complication Probability (NTCP)[2] are also considered.

2. Data Collection

The MRC-RT01 dose-escalation prostate radiotherapy trial ISRCTN47772397 [3] was one of the first UK multi-centre radiotherapy trials to collect 3-D dosimetric data. The trial randomised 843 men with localised prostate cancer between a standard radiotherapy dose of 64 Gy or an escalated dose of 74Gy achieved with a 10Gy boost to the prostate only. Although comprehensive follow-up data was available for most patients, it was only possible to combine the toxicity and dosimetric data for 388 patients. The deficiency in dosimetric data was partly attributable to difficulties in retrieving the data from treatment planning systems with non-proprietary export formats. These technical issues have been largely overcome with the wide-scale implementation of DICOM-RT and development of treatment plan analysis software [4,5]. However diligent prospective archiving of data is fundamental to the creation of comprehensive data sets.

3. Analyses

Data sets which combine toxicity and dosimetric data are valuable resources in the effort to further the understanding of the response of normal tissues. They can be used to derive dose-volume constraints using statistical methods [6,7],derive parameters for NTCP models such as the classic Lyman Kutcher Burman model [8,9] or develop novel models [10]. They are also invaluable as independent datasets on which to test potential models.

The initial analysis of the RT01 data took the form of an independent validation of dose-volume constraints for the rectum already proposed in the literature[11]. Two sets of rectal dose-volume constraints were tested; those proposed by Fiorino et al [6] of 40Gy/65%, 50Gy/55%, 60Gy/40%, 70Gy/25% and 75Gy/5% and those proposed for the CHHiP trial [12] of 30Gy/80%, 40Gy/70% 50Gy/60% 60Gy/50% 65Gy/30%, 70Gy/15% and 75 Gy/3%. Each constraint was tested independently on 7 different late rectal toxicity endpoints; Rectal Bleeding (RMH), Proctitis (RTOG) Subjective Sphincter control, Subjective stool frequency and Management of sphincter control (LENT/SOM) and Loose stools and rectal urgency (UCLA PCI).

The effect of retrospectively applying the constraints was tested by calculating Odds ratios defined as the odds of reporting a specified late rectal toxicity as a result of the treatment plan failing to meet a suggested constraint relative to the odds of reporting the specified late toxicity if the constraint had been met. It was demonstrated that some constraints were more effective for some endpoints than others; potentially due to underlying differences in patho-physiology. However, overall there was a clear picture that the more constraints failed (when considering the lowest volume constraint at each dose level) the more likely a patient was to experience toxicity. In addition, it was observed that in the group of patients whose treatment had met all of the constraints, approximately a third of patients reported toxicity graded as either moderate or severe.

4. **Deficiencies in Current Methodology**

There are several explanations for the failure of dose volume constraints and NTCP models to accurately predict toxicity. Firstly it is important to acknowledge that dose-volume constraints, models and model parameters are derived from a dataset which will represent one or more treatment types (e.g. fractionation, beam arrangement). It cannot be presumed that results can be generalised to another treatment type. However there are more fundamental reasons why the incidence of toxicity is not fully predicted by current approaches.

4.1 Deficiencies in Toxicity Reporting/Definitions.

Although standardised scales for toxicity reporting are used there are still challenges in the representation of side effects. Analysis is either performed cross-sectionally (i.e. toxicity reported at 2 years) or cumulatively using the maximum grade reported over the available follow-up period. Both definitions are sensitive to spurious results and neither fully characterises the experience of the patient[13]. An alternative approach is to use an area under the curve approach where the time and severity of toxicity are integrated over the follow-up period[7]. Figure 1 represents 2 idealised examples of follow-up history. Using standard definitions both patients would report grade 2 toxicity. However it is clear that the overall toxicity profile is quite different. Dose-volume constraints for the rectum were derived using both definitions and were shown to provide complimentary information.

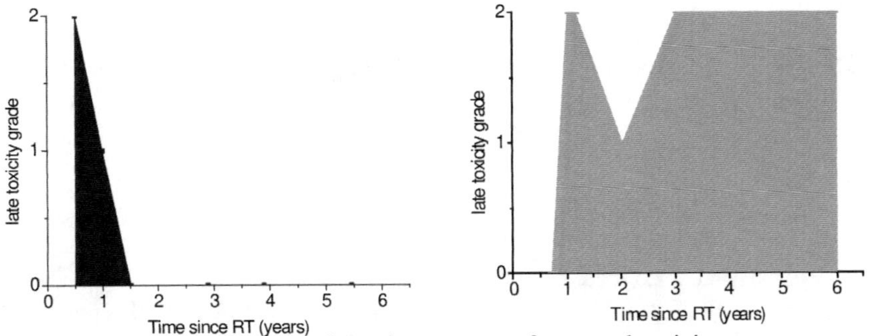

Figure 1 Illustrative examples of the time course of reported toxicity.

4.2 Treatment plan not representative of the dose received by the normal tissue.

Treatment plans are calculated using a CT scan taken prior to treatment and therefore the dose accumulated by normal tissues is unlikely to be accurate. Changes in patient anatomy over the course of radiotherapy treatment are not uncommon. In addition intra- and inter- fraction motion will compound these uncertainties. In the era of IGRT it is possible to improve the estimate of dose received by a specific organ or tissue by recalculating the dose distribution on cone beam CT acquired over the course of treatment. Accumulating this dosimetric information should result in a more accurate representation of the dose distribution received[14].

4.3 Lack of spatial information

The dose distribution to a normal tissue is usually presented as a dose-volume-histogram (dvh). The dvh allows for comparison of the main features of the dose distribution however if the organ or tissue has a heterogeneous dose response then information regarding the location of the dose is also important. An alternative representation of the dose to a tubular structure such as the rectum is a dose surface map (figure 2) where the dose distribution to the rectum surface is unfolded and spatial integrity is preserved. Using dose surface maps to analyse the RT01 trial cohort it was possible to demonstrate that loose stools was most correlated with the longitudinal (superior-inferior) extent of doses in the range of 20-30 Gy, in contrast rectal bleeding was best characterised by the lateral (left-right) extent of doses between 40 and 60Gy[15].

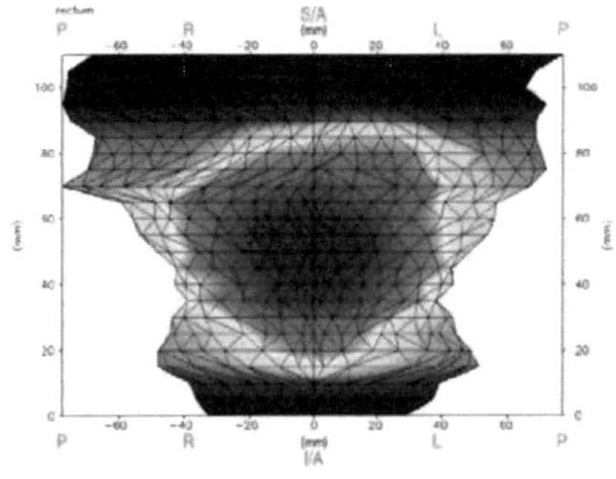

Figure 2. Example of a dose surface map where the dose to the rectal wall has been unfolded along the most posterior point of the contour.

4.4 Patient & Genetic factors

Finally we consider that there are many patient specific factors that contribute to and confound the dose response of normal tissues. These factors may include co-morbidities and concomitant treatments. It is also important to consider inherent genetic differences. Single Nucleotide Polymorphisms (SNPs) are rearrangements of a single pair in a gene. These variations continue to be tested as potential candidates for radiation response although the large number of potential variations has so far led to inconclusive results[16].

5. **The benefits of Clinical trials for the implementation on new radiotherapy techniques.**

In addition to the comprehensive data collected within a clinical trial, there are other advantages to using data collected within this structured framework. The trial protocol ensures that all aspects of the radiotherapy are considered and acceptable techniques are defined. In addition rigorous quality assurance programs ensure that patients recruited to the trial are treated according to protocol. Overall this reduces the heterogeneity of data in the cohort available for analysis.

The importance of using clinical trials to implement new technology was highlighted by the results of the Phase III PARSPORT trial ISRCTN48243537 [17]. This trial was unique in comparing conformal radiotherapy and intensity modulated radiotherapy for head and neck cancer in a fully randomised setting. The use of IMRT allowed for sparing of the contra-lateral parotid gland and trial findings confirmed a decrease in Xerostomia in the group of patients who received IMRT.

An unexpected finding from the trial was that acute fatigue was more prevalent in the IMRT arm of the trial. It was hypothesised that the difference in the incidence of fatigue was related to differences in dosimetry to a number of CNS structures [18]. The dose distributions to the structures were compared to the incidence of fatigue and it was demonstrated that the maximum and mean dose to the posterior fossa, cerebellum and brainstem were correlated with acute fatigue. Subsequent analysis characterised the dose distribution using the dose-volume atlas method proposed by Jackson et al [19]. This method allows the dvh to be summarised in combination with a specific toxicity endpoint. Each box in the atlas represents a small region of dose and volume. The dvh are overlayed and the boxes through which the dvh passes are recorded (denominator). Also recorded in each box are the number of patients whose dvh pass through the box AND reported the toxicity under investigation (numerator). An example atlas generated for the cerebellum is shown in figure 3.

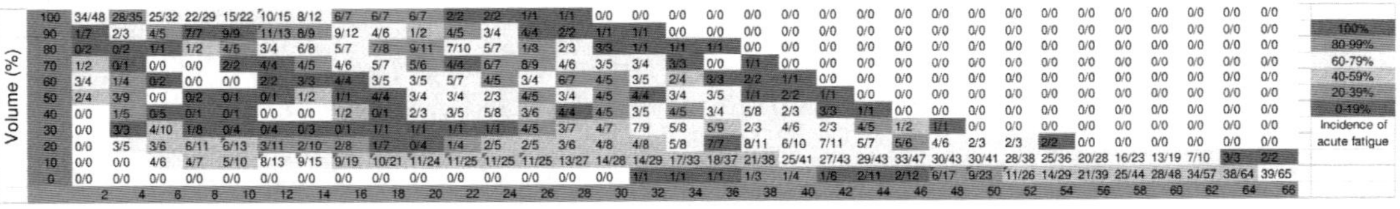

Figure 3. Dose-volume atlas summarising dvh of the cerebellum and incidence of grade 2 fatigue.

Using an atlas approach allowed the analysis of the relationship between the whole dvh and toxicity and has potential for both the derivation of constraints and generation of models. Use of the atlas method is encouraged within the QUANTEC report as it facilitates the combination of data from different patient cohorts. Combining datasets allows for development of generalisable models and independent testing.

6. Summary

Clinical trials are vital to validate and introduce new radiotherapy techniques. The framework of clinical trials facilitates the development of databases of carefully audited information providing valuable information on the response of normal tissues to radiation. Deficiencies in the characterisation of normal tissue response to radiation and confounding factors are acknowledged but ongoing research and feedback will improve future characterisation of the response of normal tissues.

Acknowledgements

The author wishes to gratefully acknowledge collaboration and support from Professor David Dearnaley and the Medical Research Council Clinical trials unit (RT01) and Professor Christopher Nutting and the Institute of Cancer Research Clinical Trials Unit (PARSPORT). Funding was received from Cancer Research UK, Section of Radiotherapy Grant number C46/A2131, Head and Neck Program Grant number C7224/A13407 and NHS funding to the NIHR Biomedical Research Centre.

References

[1] Marks LB, Ten Haken RK Martel MK. Guest editor's introduction to QUANTEC: a users guide. Int.J.Radiat.Oncol.Biol.Phys. 2010;76:S1-S2.

[2] Bentzen SM, Constine LS, Deasy JO, et al. Quantitative Analyses of Normal Tissue Effects in the Clinic (QUANTEC): an introduction to the scientific issues. Int.J.Radiat.Oncol.Biol.Phys. 2010;76:S3-S9.

[3] Dearnaley DP, Sydes MR, Graham JD, et al. Escalated-dose versus standard-dose conformal radiotherapy in prostate cancer: first results from the MRC RT01 randomised controlled trial. Lancet Oncol. 2007;8:475-487.

[4] Deasy JO, Blanco AI Clark VH. CERR: a computational environment for radiotherapy research. Med Phys 2003;30:979-985.

[5] Ebert MA, Haworth A, Kearvell R, et al. Detailed review and analysis of complex radiotherapy clinical trial planning data: evaluation and initial experience with the SWAN software system. Radiother Oncol 2008;86:200-210.

[6] Fiorino C, Fellin G, Rancati T, et al. Clinical and Dosimetric Predictors of Late Rectal Syndrome After 3D-CRT for Localized Prostate Cancer: Preliminary Results of a Multicenter Prospective Study. Int.J.Radiat.Oncol.Biol.Phys. 2008;70:1130-1137.

[7] Gulliford SL, Partridge M, Sydes MR, Andreyev J Dearnaley DP. A comparison of dose-volume constraints derived using peak and longitudinal definitions of late rectal toxicity. Radiother.Oncol. 2010;94:241-247.

[8] Gulliford SL, Partridge M, Sydes MR, Webb S, Evans PM, Dearnaley DP. Parameters for the Lyman Kutcher Burman (LKB) model of Normal Tissue Complication Probability (NTCP) for specific rectal complications observed in clinical practise.Radiother Oncol. 2012 Mar;102(3):347-51

[9] Rancati T, Fiorino C, Gagliardi G. Fitting late rectal bleeding data using different NTCP models: results from an Italian multi-centric study (AIROPROS0101). Radiother Oncol. 2004 Oct;73(1):21-32

[10] Tucker SL, Zhang M, Dong L, Mohan R, Kuban D, Thames HD. Cluster model analysis of late rectal bleeding after IMRT of prostate cancer: a case-control study. Int.J.Radiat.Oncol.Biol.Phys. 2006 Mar 15;64(4):1255-64.

[11] Gulliford SL, Foo K, Morgan RC, et al. Dose-volume constraints to reduce rectal side effects from prostate radiotherapy: evidence from MRC RT01 Trial ISRCTN 47772397. Int.J.Radiat.Oncol.Biol.Phys. 2010;76:747-754.

[12] Khoo VS Dearnaley DP. Question of dose, fractionation and technique: ingredients for testing hypofractionation in prostate cancer--the CHHiP trial. Clin.Oncol.(R.Coll.Radiol.) 2008;20:12-14.

[13] Schmid MP, Potter R, Bombosch V, et al. Late gastrointestinal and urogenital side-effects after radiotherapy--incidence and prevalence. Subgroup-analysis within the prospective Austrian-German phase II multicenter trial for localized prostate cancer. Radiother Oncol 2012;104:114-118.

[14] Jaffray DA, Lindsay PE, Brock KK, Deasy JO Tome WA. Accurate accumulation of dose for improved understanding of radiation effects in normal tissue. Int.J.Radiat.Oncol.Biol.Phys. 2010;76:S135-S139.

[15] Buettner F, Gulliford SL, Webb S, Sydes MR, Dearnaley DP Partridge M. Assessing correlations between the spatial distribution of the dose to the rectal wall and late rectal toxicity after prostate radiotherapy: an analysis of data from the MRC RT01 trial (ISRCTN 47772397). Phys.Med.Biol. 2009;54:6535-6548.

[16] Barnett GC, Coles CE, Elliott RM, et al. Independent validation of genes and polymorphisms reported to be associated with radiation toxicity: a prospective analysis study. Lancet Oncol 2012;13:65-77.

[17] Nutting CM, Morden JP, Harrington KJ, et al. Parotid-sparing intensity modulated versus conventional radiotherapy in head and neck cancer (PARSPORT): a phase 3 multicentre randomised controlled trial. Lancet Oncol. 2011;12:127-136.

[18] Gulliford SL, Miah AB, Brennan S, et al. Dosimetric explanations of fatigue in head and neck radiotherapy: an analysis from the PARSPORT Phase III trial. Radiother Oncol 2012;104:205-212.

[19] Jackson A, Yorke ED Rosenzweig KE. The atlas of complication incidence: a proposal for a new standard for reporting the results of radiotherapy protocols. Semin.Radiat.Oncol. 2006;16:260-268.

Total centralisation and optimisation of an oncology management suite via Citrix®

C James[1], J Frantzis[1], L Ripps[2] and P Fenton[1]

[1] Epworth Radiation Oncology, Richmond, VIC 3121, AUS
[2] Epworth HealthCare IT, Richmond. VIC 3121, AUS

E-mail: chris.james@epworth.org.au

Abstract. The management of patient information and treatment planning is traditionally an intra-departmental requirement of a radiation oncology service. Epworth Radiation Oncology systems must support the transient nature of Visiting Medical Officers (VMOs). This unique work practice created challenges when implementing the vision of a completely paperless solution that allows for a responsive and efficient service delivery. ARIA® and Eclipse™ (Varian Medical Systems, Palo Alto, CA, USA) have been deployed across four dedicated Citrix® (Citrix Systems, Santa Clara, CA, USA) servers allowing VMOs to access these applications remotely. A range of paperless solutions were developed within ARIA® to facilitate clinical and organisational management whilst optimising efficient work practices. The IT infrastructure and paperless workflow has enabled VMOs to securely access the Varian™ (Varian Medical Systems, Palo Alto, CA, USA) oncology software and experience full functionality from any location on multiple devices. This has enhanced access to patient information and improved the responsiveness of the service. Epworth HealthCare has developed a unique solution to enable remote access to a centralised oncology management suite, while maintaining a secure and paperless working environment.

1. Introduction

While computers have become an integral component in the Radiation Oncology department for treatment planning, recording and verification; the composition and storage of patient treatment and medical records have long been paper-based. As such, the management of patient information and treatment planning is traditionally an intra-departmental requirement of a radiation oncology service.

Epworth Radiation Oncology (ERO) is a relatively new private, not-for-profit service in Melbourne, Victoria, Australia, providing fast turn-around for patients between the initial consult appointment with their oncologist, planning CT-simulation and treatment commencement. To achieve this, a streamlined, robust and easy-to-use workflow is essential.

ERO also faces a unique challenge to support the transient nature of Visiting Medical Officers (VMOs) where the referring Radiation Oncologists (ROs) are regularly off-site. The provision of a quick planning to treatment time would be difficult to achieve in a traditional onsite, fat-client application deployed and paper-based service model. The VMOs require quick, functional and secure access to applications and information normally only available if on-site.

2. Methods

2.1. Remote Access

The ARIA® and Eclipse™ software suite is deployed over four dedicated Citrix® servers running Windows® 2008R2 SP1 (Microsoft, Seattle, WA, USA) in a data-centre instead of a traditional workstation based installation.

Staff requiring remote access download Citrix Receiver™ (Citrix Systems, Santa Clara, CA, USA) onto the local device (computer, tablet etc.) to facilitate virtual delivery of applications and desktops. The user navigates to the Epworth XenApp® (Citrix Systems, Santa Clara, CA, USA) web interface URL in their internet browser where the Citrix Access Gateway™ (Citrix Systems, Santa Clara, CA, USA) provides two levels of security. The user is prompted for their Epworth network user name and password and if confirmed, an SMS passcode is generated for the session and delivered to the mobile phone number registered to that particular user. Access is only granted once these two security measures are satisfied.

The web interface displays icons for available applications based on access rights granted to the user's profile by Epworth HealthCare IT. Once an application is selected, it is virtualised and securely delivered to the user by Citrix XenApp®. The application is hosted on the server but displayed to the user via Citrix Receiver™. Mouse and keyboard inputs update the visible screen however raw data does not leave the data-centre. Application delivery is routed through and facilitated by a Citrix Netscaler® (Citrix Systems, Santa Clara, CA, USA) appliance providing secure access to patient data.

2.2. Paperless Environment

To compliment the remote access and enhance its functionality, a range of paperless solutions were developed for all internal ERO processes and forms.

All medical records are maintained within the patient's profile in ARIA®. Patient demographics as well as radiology and pathology results are fed into ARIA® from the hospital patient master index via HL7 standard messaging and constantly updated.

ARIA® provides Microsoft® Word (Microsoft, Seattle, WA, USA) integration allowing for the creation of document templates. Templates were produced for all ERO documents, allowing patient details to insert automatically onto all documents, removing the need for manual data entry or ID stickers.

Integrisign® e-pad-ink® (ePadLink, Simi Valley, CA, USA) electronic signature pads are used to sign any legal documents with signature fields inserted into the templates within ARIA®. Once signed, these fields are locked with a time and date stamp by the Integrisign® software.

3. Results

The paperless workflow has succeeded in providing an efficient and responsive service and has been adopted by all professional disciplines in the department. The use of templates has provided consistency in structure, content and format of each document type used across both ERO sites.

The IT infrastructure has enabled users to securely access Varian™'s oncology software and experience full functionality across a range of devices in any location. ERO VMOs monitor their workload, contour tumours and review plans remotely with access to the patients' full medical record due to the paperless systems in place. In addition, having functional offsite access allowed for the development of a casual remote planning workforce to assist the planning team during heavy planning periods.

4. Discussion

4.1. Paperless Environment

Electronic workflow has been shown to reduce workload and increase efficiency in a Radiation Oncology Department [1,2] with our experience supporting these findings. ERO utilises a lean staff model, yet maintains a high patient throughput while offering advanced treatment techniques. The efficiency is attributed to the streamlined electronic processes and paperless solution built around ARIA™. The time saved in system efficiencies can be spent enhancing patient care [1,2].

ERO found the use of templated documents and forms within the electronic medical record (EMR) has not only reduced workload but also improved the consistency in content and format of our documents, as noted previously [3]. However, the production of these templates required initial investment of time and documents have undergone many variations and updates as input is received from end users or as processes are refined. Templates are easily updated in a paperless environment, promoting responsive process improvement. Version control is ensured, as the end user can only ever access the current document version and general users are unable to edit the template.

Another consideration when creating templates is the degree of freedom to grant the user. There is a balance between restricting customisability to maintain document standards and providing sufficient flexibility in the template to ensure efficient use and adequate detail of the data entered.

One of the final hurdles for implementing a truly paperless workflow is handling signatures for medico-legal documents. ERO's solution follows unpublished work presented by Fenton, Medwell and Thakar and utilises Integrisign® e-pad-ink® electronic signature pads and software.

4.2. Remote Access

Literature regarding the use of Citrix® to implement a truly remote access solution in radiation oncology is scarce. Belard et al. [4] have published their work on providing remote access for Walter Reed Army Medical Centre Radiation Oncologists and Physicists to participate in the planning of proton therapy. Using Citrix®, the group provided secure remote access to proton planning software and enabled the transfer of plans to the University of Pennsylvania patient database for treatment delivery. Colonias et al. [3] discussed deploying an Electronic Health Record across an integrated radiation oncology network using a Citrix® server but did not, however, provide access to a planning system.

The extensive use of Citrix® remote solutions at ERO stemmed initially from the referral pathway of our patients. A majority of our Radiation Oncologists are VMOs who are not regularly on site. Being a private service, ERO is focused on providing a point of difference for all clients through excellent patient care, state-of-the-art equipment with advanced treatment techniques and extremely fast planning to treatment turn-around times. This has been achievable primarily through the provision of fast, efficient and secure access for VMOs to ensure that their tasks in the planning process can be completed in a timely manner from any location.

Certainly, security of data is a concern and has been raised previously in relation to the EMR [5]. These concerns could now be exaggerated by the access of data from remote locations outside of the hospital network; however the infrastructure and systems developed protect patient data on multiple levels. In addition to the hospital-network's security, each individual workstation is protected by user name and password. ARIA®, like most patient information systems, is also protected by its own secure user name and password combination unique to each user and separate from the workstation and hospital network combination.

As outlined, there are additional security-levels applied to remotely-accessed sessions. Remote access must first be enabled on a user's profile by Epworth HealthCare IT. The IT administrators have total control over user access, and are able to monitor and log details of each session and user. The Citrix® web interface is protected by user name and password combination, and then by a session-unique SMS passcode. It has been suggested that security measures such as this result in patient data being far more protected in an electronic environment compared to previous physical paper files [2].

When accessing a Citrix® remote session, no raw data is leaving the datacentre. The applications are hosted on the server within the hospital's network and the Citrix® software and hardware is virtualising the Varian™ applications and publishing them to the user who is only interacting with keyboard and mouse clicks. While this solution offers many levels of security, new concerns may be introduced regarding staff accessing patient information where they can be observed and these will need to first be identified, and then remedied.

4.3. Implementation

Moving toward a truly paperless department is a big undertaking especially for a department with established workflows and processes.

The multidisciplinary nature of a radiation oncology department means that for a paperless workflow to function efficiently and effectively, all professions must "buy in", which has certainly been our experience. Laerum et al. [6] deduce that technology alone is not enough to provide an effective and efficient paperless environment, instead work practices need to be designed around that electronic workflow. This can create difficulties if a department was to transition to being paperless, because a new workflow may disrupt current work roles which has been shown to result in local resistance [7].

The workflow and solutions developed at ERO do not require end users to possess advanced knowledge of information technology or specialist skills, removing one of the challenges of creating an efficient paperless and remotely accessible environment. The user needs to navigate around and interact with the software easily or there will be an increased resistance to adopting electronic processes. While the infrastructure supporting our Oncology Management Suite (individual workstations, hospital network and Citrix® deployment) was developed and is managed by Epworth HealthCare IT, all of our paperless solutions were developed in-house by radiation therapists with no advanced or formal information technology training, further supporting that with a dedicated team and a clear vision, a truly paperless and remote radiation oncology department is achievable.

5. Conclusion

Epworth Radiation Oncology has developed an electronic environment around the intended service model using proven approaches as well as creating new and innovative solutions. A centralised oncology management suite within the data-centre allows secure remote access for users from any location on multiple devices. The unique combination of a truly electronic environment and remote access with full application functionality has enabled ERO to support the VMO referral pathway while maintaining rapid planning times, advanced treatment techniques and excellent patient care.

6. References

[1] Han Y, Huh S J, Ju S G, Ahn Y C, Lim D H, Lee J E and Park W 2005 Impact of an electronic chart on the staff workload in a radiation oncology department *Jpn. J. Clin. Oncol.* **35** 470-74

[2] Middleton M, Bradford C, Frantzis J, Ambler A, Sisson T, Montgomerie D and Martin J 2009 Paperless and paper-based processes in the modern radiotherapy department *Radiography* **15** 300-5

[3] Colonius A, Parda D S, Karlovits S, Fuhrer R, Trombetta M, Strickland S, Luick M, Billy R, Horrigan T and Werts E D 2011 A radiation oncology based electronic health record in an integrated radiation oncology network *J. Radiat. Oncol. Inform.* **3** 3-11

[4] Belard A, Dolney D, Zelig T, McDonough J and O'Connel J 2011 Improving proton therapy accessibility through seamless electronic integration of remote treatment planning sites *Telemed J. E. Health* **17** 370-5

[5] Steward M 2005 Electronic medical records: privacy, confidentiality, liability *J. Leg. Med.* **26** 491-506

[6] Laerum H, Ellingsen G and Faxvaag A 2001 Doctors' use of electronic medical records systems in hospitals: a cross sectional survey *BMJ* **323** 1344-48

[7] Heeks R, Mundy D and Salazar A. 1999 *Why health care information systems succeed or fail* Retrieved 23 August 2013 from http://www.sed.manchester.ac.uk/idpm/research/publications/wp/igovernment/documents/igov_wp09.pdf

Journal of Physics: Conference Series **489** (2014) 012093 doi:10.1088/1742-6596/489/1/012093

An investigation of the impact of variations of DVH calculation algorithms on DVH dependant radiation therapy plan evaluation metrics

A M Kennedy[1], J Lane[1], M A Ebert[1,2]
[1] Department of Radiation Oncology, Sir Charles Gairdner Hospital, Perth, Australia
[2] School of Physics, University of Western Australia, Perth, Australia

E-mail: Angel.kennedy@health.wa.gov.au

Abstract. Plan review systems often allow dose volume histogram (DVH) recalculation as part of a quality assurance process for trials. A review of the algorithms provided by a number of systems indicated that they are often very similar. One notable point of variation between implementations is in the location and frequency of dose sampling. This study explored the impact such variations can have on DVH based plan evaluation metrics (Normal Tissue Complication Probability (NTCP), min, mean and max dose), for a plan with small structures placed over areas of high dose gradient. Dose grids considered were exported from the original planning system at a range of resolutions. We found that for the CT based resolutions used in all but one plan review systems (CT and CT with guaranteed minimum number of sampling voxels in the x and y direction) results were very similar and changed in a similar manner with changes in the dose grid resolution despite the extreme conditions. Differences became noticeable however when resolution was increased in the axial (z) direction. Evaluation metrics also varied differently with changing dose grid for CT based resolutions compared to dose grid based resolutions. This suggests that if DVHs are being compared between systems that use a different basis for selecting sampling resolution it may become important to confirm that a similar resolution was used during calculation.

1. Introduction

Plan review and quality assurance (QA) systems are sometimes used during radiation therapy clinical trials to ensure the quality and consistency of data. This often includes recalculation of dose volume histograms (DVHs). Several studies demonstrated that variation in DVH calculation methods can lead to significant differences in resulting DVHs, which in turn can add unnecessary noise to DVH based plan evaluation [1, 2]. Plan review systems provide convenient access to DVHs, but not necessarily to original plans, with derived information used in presentation of trial outcomes analyses. Compatibility between DVH calculation algorithms is therefore desirable. Variations in DVH calculation algorithms were explored together with the impact of these variations on DVH based plan evaluation metrics.

1.1. Comparison of plan review DVH calculation algorithms

DVH calculation algorithms employed by several plan review/quality assurance systems (see table 1) were reviewed. We found that the algorithms were very similar and in general conformed to the assumptions outlined by the Image Guided Therapy QA Centre (ITC) [2], with VODCA allowing some greater flexibility in user-control that can mean the calculations do not conform to ITC assumptions. These assumptions are described in figure 1. In addition all the systems reviewed were consistent in their selection of 3D linear interpolation. One point on which the ITC assumptions are vague and on which the plan review systems differed is the selection of voxel sampling size. Table 1 presents the range of voxel sampling sizes used by different implementations of the algorithm. Variation in sampling resolution for the DVH calculation described by the ITC has an impact on both volume errors at the borders of the structures and on the points that dose is interpolated to. Previous research has suggested that high dose gradients and small structure sizes increase the sensitivity of DVHs to variations in DVH calculation assumptions [2-4].

ITC DVH Computation
Assumptions:

1. The patient anatomy and target structures are represented by axial, planar, closed loops at the center position of each CT scan.
2. The dose is represented as a 3D dose-grid covering the irradiated volume and sampled at a spatial rate sufficient to represent the continuous dose distribution.
3. Structures are defined as a set of stacked right prisms, whose shapes are defined by structure loops. Each prism is assumed to extend axially from one-half the distance to the next inferior slice location to one-half the distance to the next superior slice location. (The width of prisms at the superior-most and inferior-most slices is assumed to be twice the half-distance to the next neighboring slice location.)
4. The patient volume is broken into cubes of the "spatial sampling size".
5. The dose value at the center of each cube is interpolated in three dimensions from the dose-grid.
6. The DVH bin corresponding to the dose value of each contained cube is incremented by the cube volume. For I = 0,1,...,n-1, the ith bin of the histogram represents the structure volume receiving dose D, such that $i\Delta \leq D < (i+1)\Delta$, where Δ is the bin width in Gy. (DVHs are displayed as cumulative functions where the ith bi represents the structure volume receiving dose $\geq i\Delta$)

Figure 1. ITC assumptions [1] [5]

Table 1. Comparison of DVH algorithms for plan review systems. [6-8]

Software	Structure capping at axial ends	Sampling voxel resolution	Interpolation
Swan[1]	As defined by ITC assumption 3	CT resolution, user selection	Linear 3D
CERR[2]	As defined by ITC assumption 3	CT resolution	Linear 3D
Slicer RT[3]	As defined by ITC assumption 3	2 * dose grid resolution, user selection	Linear 3D
VODCA[4]	User chooses whether or not to extend half a slice beyond end slices in z direction	CT resolution as default but guarantees a minimum of 128*128 voxels in x and y direction of the bounding box of each slice.	Linear 3D

2. Method: examining the impact of sampling resolution

2.1. Algorithmic variations: separating volume and dose differences

[1] Developed by the authors [5].
[2] See [6-7]
[3] Information from Andras Lasso (personal communication).
[4] See [8], Details of assumptions received from Stefano Gianolini (personal communication).

We created a variation on the ITC algorithm that removes the volume error component by adding only the intersecting volume of sampling voxels and structure loops (Exact Voxel Interpolation or EVI). This allowed us to isolate the impact of changes in dose sampling from changes in volume estimation. It also allowed us to include in our comparisons the base case where sampling voxels are lined up with dose grid voxels so that no interpolation occurs but exact dose values are used (Exact Dose or ED). Figure 2 depicts these variations on the original algorithm. Chung et al [4] noted that dose could be varied by several percentage points just by shifting the origin of the dose grid calculated during treatment planning.

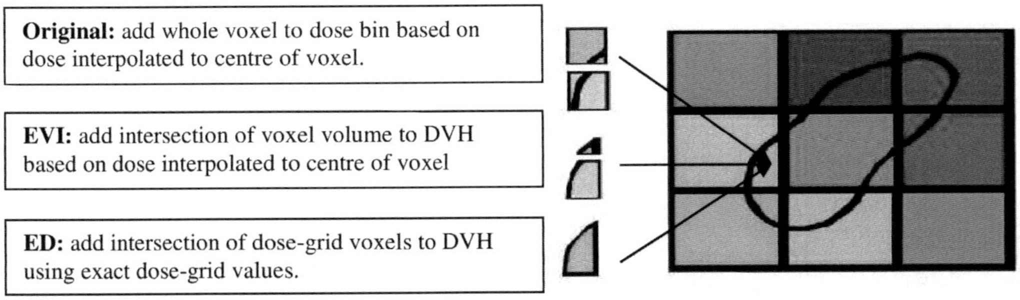

Figure 2. DVH calculation algorithmic variations on sampling and volume calculation strategies.

2.2. Dataset

We selected a stereotactic brain plan created with XiO (Elekta Pty Ltd) with the dose grid exported at resolutions (voxel sizes) from 1mm to 5mm in 1mm increments and at 2.5mm. The plan was selected for having steep dose-gradients (as high as 12Gy on the 0.25cm dose grid) and small structure volumes (ranging from 0.57cc to 26.6cc). Structures were shifted to lie over an area of high dose gradient. The CT slice width was 0.3cm.

2.3. Experimental variations

We compared the EVI variation at a range of resolutions including those used by different plan review systems listed in table 1 as well as the ED base case. We also included some higher resolutions for EVI including 2, 4 and 8 times the CT resolutions and a condition that extended VODCA's variable resolution algorithms to the axial (z) dimension. DVHs were calculated for 5 structures on all 5 dose grids.

2.4. DVH comparison/evaluation

We compared DVHs based on differences in normal tissue complication probability[5] (NTCP), and minimum, mean and maximum dose. We examined the differences between DVHs calculated at different resolutions and how these differences varied with change in dose grid resolution. We also examined how much DVH based metrics at a given sampling resolution changed over different dose grid export resolutions.

3. Results

Figure 3 shows the results of comparing DVH based plan evaluation metrics for each condition across different dose grid resolutions. We found that in general there was a strong impact of dose grid resolution on DVH based NTCP, max dose and to a lesser extent min dose. However differences do not start to get large until dose grid resolutions lower than 0.3cms are used.

[5] The initial NTCP parameters were taken from Luxton et al [9] with TD50 then adjusted to ensure that NTCP values did not go to zero or one for this dataset.

The way that DVH based metrics changed was very similar for EVI using CT based sampling resolutions compared to EVI at twice the dose grid resolution. Interestingly the ED algorithm, which is based on the dose grid resolution, results in differences to mean dose that follow those of EVI at the CT based resolution far more closely than the other dose grid based sampling resolution, EVI at twice the dose grid resolution. It seems that, especially in the case of max dose, interpolation smoothes out variations caused by a change in dose grid but that for NTCP aligning the sampling grid with the dose grid (ED) reduced variation with decreasing dose grid resolution.

(a) (b) (c) (d)

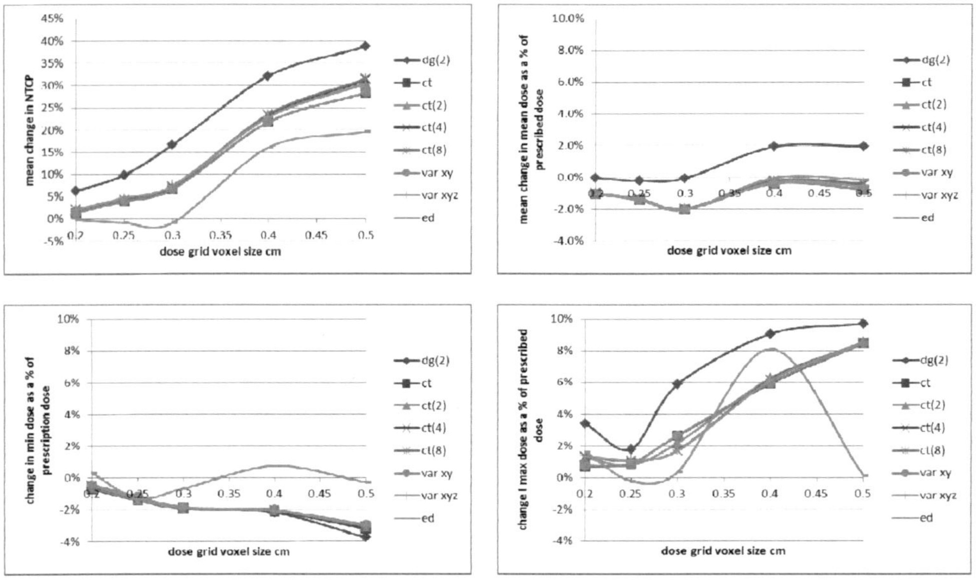

Figure 3. Mean differences between DVH based plan evaluation metrics ((a) NTCP, (b) mean dose, (c) min dose, (d) max dose) from values calculated on finest dose grid (0.1cm) with those calculated at lower dose grid resolution. The value for each algorithmic condition at dose grid resolutions greater than 0.1cm is subtracted from the value for that algorithm on a dose grid of resolution 0.1cm. Min, max and mean dose changes are presented as a percentage of the prescribed dose (60Gy).

(a) (b)

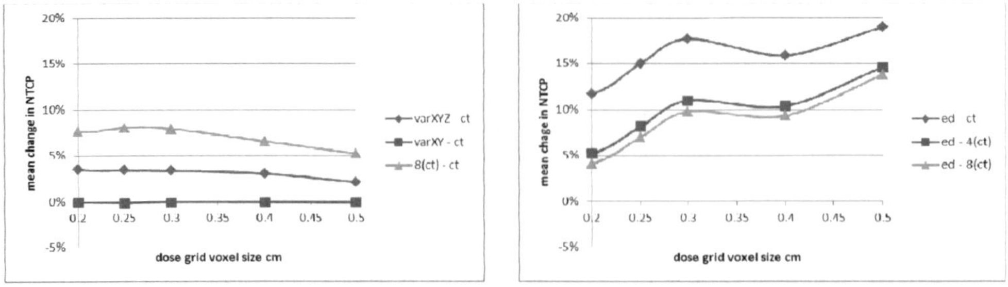

Figure 4. Mean differences in NTCP values between different resolution/algorithm conditions at each dose grid resolution. (a) Differences between NTCP values calculated via EVI at several different CT based resolutions. Increased resolution in z impacted the NTCP far more than increased resolution in x and y. (b) Differences between EVI at CT based resolutions and ED. Increased EVI resolution increases similarity between ED and EVI.

The CT based conditions display similar variations with changing dose grid resolution though there are significant differences in absolute values. Figure 4(a) displays comparisons between NTCP values generated by different algorithmic conditions. It is interesting to note that changes to the resolution in the z direction for EVI had a much greater impact than changes to the resolution in x and y. This could be partly explained by the higher mean dose gradients that existed for these structures in the z (5.37Gy) direction than for either x (1.87Gy) or y (3.71Gy). An examination of the variation for individual structures in comparison to each other suggested a relationship between mean dose gradient for the structure and the magnitude of variation in DVH based plan evaluation metrics between algorithms.

Figure 4(b) depicts differences between NTCP values produced by ED and EVI at different resolutions. Increasing the resolution of CT based EVI increased the similarity of EVI to ED for NTCP significantly.

4. Discussion

In this study, the conditions under which variations were examined were fairly extreme. In cases where dose gradients were less extreme we found minimal variation in DVH based metrics. Most of the plan review systems used or permitted users to select CT based resolutions in any case, which suggests DVH comparability between systems.

The results suggest that in the presence of steep dose gradients, comparable sampling resolutions and dose grid resolutions would facilitate agreement between independent DVH calculation algorithms. They also suggest that under these conditions resolutions finer than the CT resolution or twice the dose grid resolution might be required to achieve high levels of confidence in DVH accuracy (this ignores any inaccuracies that may already exist in the plan data imported into the plan review system including dose calculation inaccuracies).

We attribute the increase in similarity between EVI and ED with increasing resolution of EVI to the fact that for dose voxels entirely inside the structure linear interpolation provides an approximation of the exact dose values. Increasing the resolution should improve this approximation. This would not be the case for voxels on structure borders where, given the dose is assumed to be continuous, interpolation should be more accurate.

Struabe et al [2] noted variations in assumptions with regard to how structures behave between slices in the z direction can have a significant impact on DVHs. In this case we certainly found that sampling resolution in the z direction had a noticeable impact on DVH based plan evaluation metrics. If DVHs need to be calculated on very small structures with high dose gradients it may be worth considering other methods of modelling change in the z direction such as interpolation of structure contours [10].

References

[1] Ebert M A, Haworth A, Kearvell R, et al. 2010 *Phys. Med. Biol.* **55** N337-46
[2] Straube W, Matthews J, Bosch W, et al. 2005 *Med. Phys.* **32** 2021-2
[3] Panitsa E, Rosenwald J C and Kappas C 1998 *Phys. Med. Biol.* **43** 2807-16
[4] Chung H, Jin H, Palta J, et al. 2006 *Phys. Med. Biol.* **51** 4841-56
[5] Haworth A, Kearvell R, Greer P B, et al. 2009 *Radiother. Oncol.* **90** 299-306
[6] Deasy J 2008 Using dose-volume histograms in CERR http://cerr.info/cerrwiki/index.php?title=DVH&oldid=1456, accessed 15/03/2013
[7] J Deasy A A, Y Wang 2003 CERR user guide http://cerr.info/usersguide.php#dvh, accessed 15/03/2013
[8] Visualisation and Organisation of Data for Cancer Analysis (VODCA) www.vodca.ch, accessed 14/03/2013
[9] Luxton G, Keall P J and King C R 2008 *Phys. Med. Biol.* **53** 23-36
[10] Herman G T, Zheng J and Bucholtz C A 1992 *IEEE Comput. Graph.* **12** 69-79

Journal of Physics: Conference Series **489** (2014) 012094 doi:10.1088/1742-6596/489/1/012094

Trust, but verify – Accuracy of clinical commercial radiation Treatment Planning Systems

J Lehmann[1,2], J Kenny[1], J Lye[1], L Dunn[1,2] and I Williams[1,2]

[1] Australian Clinical Dosimetry Service, Australian Radiation Protection and Nuclear Safety Agency, Yallambie, Victoria, Australia
[2] School of Applied Sciences, Royal Melbourne Institute of Technology (RMIT) University, GPO Box 2476, Melbourne, Victoria, Australia

E-mail: Joerg.Lehmann@arpansa.gov.au

Abstract. Computer based Treatment Planning Systems (TPS) are used worldwide to design and calculate treatment plans for treating radiation therapy patients. TPS are generally well designed and thoroughly tested by their developers and local physicists prior to clinical use. However, the wide-reaching impact of their accuracy warrants ongoing vigilance. This work reviews the findings of the Australian national audit system and provides recommendations for checks of TPS. The Australian Clinical Dosimetry Service (ACDS) has designed and implemented a national system of audits, currently in a three year test phase. The Level III audits verify the accuracy of a beam model of a facility's TPS through a comparison of measurements with calculation at selected points in an anthropomorphic phantom. The plans are prescribed by the ACDS and all measurement equipment is brought in for independent onsite measurements. In this first version of audits, plans are comparatively simple, involving asymmetric fields, wedges and inhomogeneities. The ACDS has performed 14 Level III audits to-date. Six audits returned at least one measurement at Action Level, indicating that the measured dose differed more than 3.3% (but less than 5%) from the planned dose. Two audits failed (difference >5%). One fail was caused by a data transmission error coupled with quality assurance (QA) not being performed. The second fail was investigated and reduced to Action Level with the onsite audit team finding phantom setup at treatment a contributing factor. The Action Level results are attributed to small dose calculation deviations within the TPS, which are investigated and corrected by the facilities. Small deviations exist in clinical TPS which can add up and can combine with output variations to result in unacceptable variations. Ongoing checks and independent audits are recommended.

1. Introduction

Radiation Therapy relies heavily on computer based Treatment Planning Systems (TPS) to design and calculate treatment plans for most patients treated. Commercial TPS are generally well designed and thoroughly tested by their developers. Each system undergoes acceptance testing after installation. Medical Physicists commission and verify the calculations of a TPS before the system is used to plan patient treatments at a Radiation Therapy facility. Medical Physicists also perform ongoing quality assurance (QA) of a TPS and additional checks when updates are installed. However, no check can cover all aspects of the system, and the wide-reaching impact of the accuracy of TPS calculations compels the use of additional QA measures.

The Australian Clinical Dosimetry Service (ACDS) has been created by the Australian federal government as a joint initiative between the Department of Health and Ageing and the Australian Radiation Protection and Nuclear Safety Agency. The ACDS is nearing the end of a three year test period during which it designed and implemented a three level national audit system. Audits are provided free of charge to Radiation Oncology facilities throughout Australia. [1] This work reviews findings of the ACDS Level III audit and provides recommendations for checks of TPS.

2. Methods:

The ACDS Level I audit checks the output of the linac for reference conditions. Level II covers part of the treatment chain using a synthetic CT data set as the basis for planning and a planar array as measurement device. The Level III audit represents an end-to-end test that covers the entire chain of procedures a patient experiences at a Radiation Therapy facility from imaging through planning, checks, setup, delivery and record. Radiation Therapists conduct each of the steps in keeping with routine clinical practice so that the audit assesses the actual patient process.

The treatment plans are prescribed by the ACDS and all measurement equipment is brought in for independent onsite measurements by ACDS auditors. In this first version of audits, plans are comparatively simple, involving asymmetric fields, wedges and inhomogeneities. The audit uses an anthropomorphic thorax phantom (IMRT Phantom Model 002LFC CIRS, Norfolk, VA, USA) which contains materials with radiological properties of inhale lung and bone as inhomogeneities (figure 1).

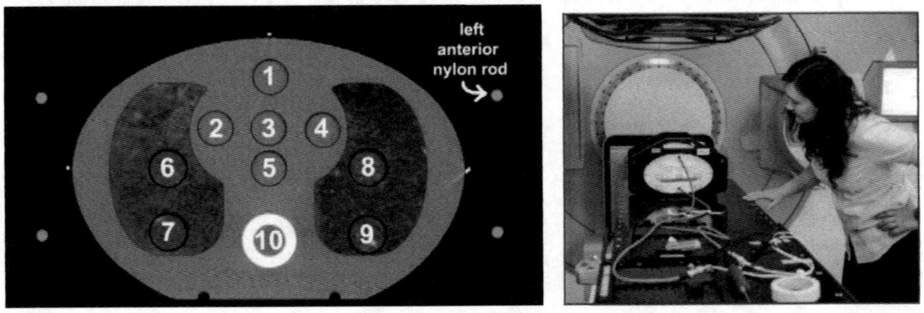

Figure 1. CIRS thorax phantom. Left: CT image of phantom at central location. Right: Phantom setup for measurement at linear accelerator.

The phantom features ten cylindrical ports which, in the default configuration, are filled with solid plugs. For the audit measurements the solid plugs are replaced with plugs that hold Farmer type ionization chambers. Three cases with 6 MV photons are planned and delivered (figure 2). The first case is a measurement near linac reference conditions, which serves as a "sanity" check. A 10 cm x10 cm field is delivered using a 100 cm Source to Surface Distance (SSD) setup with the prescription being in point 1, at 3 cm depth. A second measurement is performed at point 10, which is at 15 cm depth. The chamber is in a water equivalent plug, which is surrounded by a cylindrical shell of bone equivalent material. Cases 2 and 3 have been adopted from an international test case publication [IAEA-TECDOC-1583]. Small changes were made to the field size to allow for a shorter (superior – inferior direction) version of the phantom, which simplified its transport in commercial carriers. Also, in addition to the measurement points suggested in IAEA-TECDOC-1583, additional measurement points were selected. Case 2, which is IAEA-TECDOC-1583 Case 2, uses a lateral wedged field, isocentrically located around point 1, which is also the prescription point. Additional measurement points are in the build-up region behind the lung material (point 4) and at an out of field location in the lung (point 7). Case 3 is based on IAEA-TECDOC-1583 Case 7 with additional measurements taken at points 8 and 10. The measure used to compare plan and measurement is "Variation from ACDS", defined as: (facility planned dose – ACDS measured dose) / ACDS measured dose.

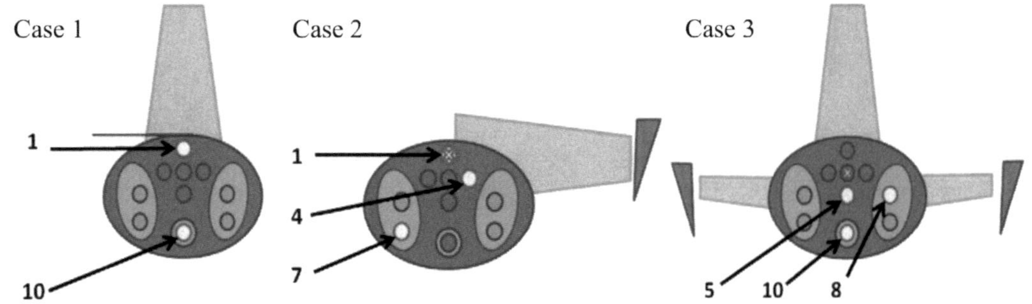

Figure 2. Audit cases for first version of Level III audit. Case 1 is SSD setup, Cases 2 and 3 are isocentric (isocentre marked with cross). Measurement points are indicated with arrows.

Measurement points in low dose areas (outside the field) and points in the lung equivalent material are reported to the facility but not scored (RNS). A measurement point is considered passed at the "Optimal Level" if the "Variation from ACDS" is within 3.3%. It is considered passed at the "Action Level" if the variation is between 3.3% and 5%. The point is considered at the "Outside Tolerance Level" if the variation is outside 5%. The overall audit outcome is equal to the lowest result for an individual measurement point. This approach is in principle based on the ACDS strategy to use 2σ for the limit of the Optimal Level and 3σ for the threshold to Outside Tolerance, with σ being the uncertainty of the measurement. However due to the complexity of the Level III audit, the uncertainty of the measurements could not be assessed with sufficient accuracy. Therefore a clinically acceptable tolerance of 5% [3] was chosen to equal the Outside Tolerance Level (3σ) and the Action Level tolerance was derived from there.

The Level III audit was tested in four field trials between February and April 2012. Radiation Oncology Facilities across Australia with a diverse mix of equipment (TPS, Record and Verify System, Linac) were selected for the field trials. All measurement results in the field trials were at "Pass Optimal Level". Feedback from the facilities was incorporated into the procedures and the Level III audit was deployed clinically in July 2012.

3. Results

The ACDS has performed 14 Level III audits to-date. Six audits returned at least one measurement at Action Level, indicating that the measured dose differed more than 3.3% (but less than 5%) from the planned dose. Two audits failed (difference exceeded 5%).

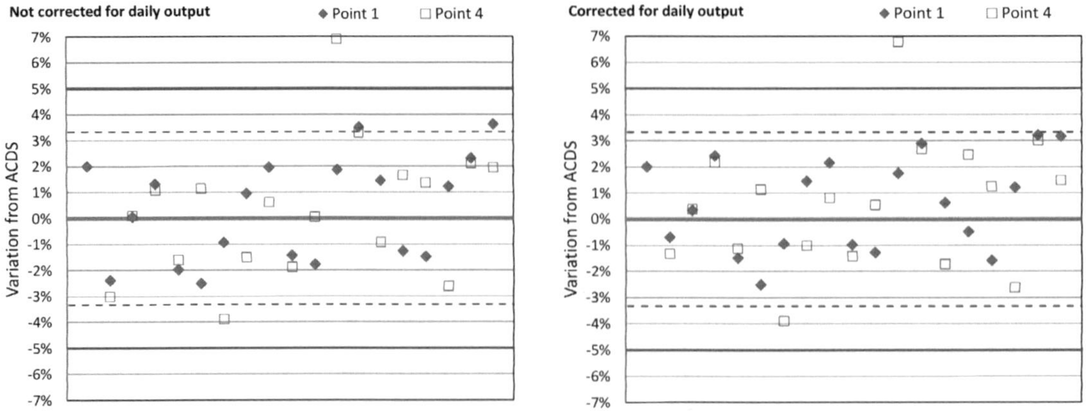

Figure 3. ACDS Level III audit Case 2 results for Points 1 and 4. The left diagram shows the data as measured and scored, the right diagram shows the data corrected for daily output provided by the facility

Figure 3 shows the results for points 1 and 4 of Case 2. The "Variation from ACDS" is plotted for all audited facilities, including the field trials and one international comparison (random order). The ACDS measured dose is not corrected for Linac output for the analysis, because of the principle of the end to end test used for the Level III audit. The left diagram shows this data. Using the output of the Linac can be helpful in understanding the data and in troubleshooting it. Therefore Linac output, measured by the physics staff of the facility, is also recorded during the audit. Results with the ACDS measurements corrected for the Linac output are shown on the right graph.

Both above mentioned audit fails occurred for Case 2. In one of the fails an incorrect field size had been transmitted to the Record and Verify system coupled with QA not being performed. This result corresponds to the first data point of both plots, where the difference for point 4 was > 100% (not shown). The second fail (see data point with a Point 4 variation of above +6.5% in the graphs) was investigated and reduced to Action Level with the onsite audit team finding phantom setup at treatment a contributing factor.

4. Discussion and Conclusions

The Level III audit verifies the accuracy of a beam model of a facility's TPS through comparison of measurements with calculation at selected points in an anthropomorphic phantom. As the audit comprises of several steps, any deviations found could be attributed to several sources. Potentially, errors could also cancel each other out. The treatment plans selected for the audit have been chosen to support troubleshooting of any deviations found. Additionally collected data, such as the complete 3D dose information and radiographic images taken to verify phantom setup at the treatment machine, is available to the audit team in this effort. If further clarification is needed, the ACDS Level II audit can be deployed. The Level II audit uses a synthetic CT data set and a 2D array to verify planar dose delivered to a rectilinear phantom made of water-equivalent material with lung-equivalent inhomogeneities. The cases of the Level II audit represent components of the beams of the Level III audit, helping pinpointing the problem within the TPS or delivery system.

The found Action Level and Fail results are attributed to:

1. Not following provided instructions
2. Failing to observe internal protocols and QA procedures
3. Setup errors
4. Dose calculation deviations in the TPS for wedges, in particular for off axis positions
5. Dose calculation deviations in the TPS for Reference conditions.

All of the above causes can be avoided or rectified in clinical practice. Staff should be trained and supervised to follow instructions and internal protocols and procedures. Tests should be implemented to check TPS and delivery system thoroughly at commissioning and regularly as well as with upgrades and major repairs of any component of the treatment chain.

An independent audit program is warranted.

5. References

[1] Williams I, Kenny J, Lye J, Lehmann J, Dunn L, and Kron T 2012 *Australas Phys Eng Sci Med* **35(4)** 407-11

[2] IAEA-TECDOC-1583 "Commissioning of radiotherapy treatment planning systems: Testing for typical external beam treatment techniques." 2008 International Atomic Energy Agency, Vienna. ISBN:978-92-0-100508-3

[3] American Association of Physicists in Medicine "Report of task group No. 65, Tissue inhomogeneity corrections for megavoltage photon beams" 2004 Medical Physics Publishing, Madison.

Acknowledgement

The Australian Clinical Dosimetry Service is a joint initiative between the Department of Health and Ageing and the Australian Radiation Protection and Nuclear Safety Agency.

XVII International Conference on the Use of Computers in Radiation Therapy (ICCR 2013) IOP Publishing
Journal of Physics: Conference Series **489** (2014) 012095 doi:10.1088/1742-6596/489/1/012095

Browser Based Platform in Maintaining Clinical Activities – Use of The iPads in Head and Neck Clinics

WY Yang[1], J Moore[1], H Quon [1,2,3], K Evans [1], A Sharabi[1], J Herman[1], A Hacker-Prietz[1] and T McNutt[1]

[1] Department of Radiation Oncology and Molecular Radiation Sciences, Johns Hopkins University, Baltimore, Maryland, USA
[2] Department of Otolaryngology-Head and Neck Surgery, Johns Hopkins School of Medicine, Baltimore, Maryland, USA
[3] Department of Oncology, Johns Hopkins School of Medicine, Baltimore, Maryland, USA

Email: tmcnutt1@jhmi.edu

Abstract:
Purpose: Incompatibility between documentation and clinical workflow causes physician resistance in organized data collection, which in turn complicates the use of data in patient care improvement. To resolve the gap, we developed an iPad compatible in situ browser-based platform that integrates clinical activity with data collection and analysis presentation. The ability to perform in-clinic activities and monitor decision making using the iPad was evaluated.
Methods: A browser-based platform that can exchange and present analysed data from the MOSAIQ database was developed in situ, the iPads were distributed in head and neck clinics to present the browser for clinical activities, data collection and assessment monitoring. Performance of the iPads for in-clinic activities was observed.
Results: All in-clinic documentation activities can be performed without workstation computers. Accessing patient record and previous assessments was significantly faster without having to open the MOSAIQ application. Patient assessments can be completed with the physician facing the patient. Graphical presentation of toxicity progression and patient radiation plans to the patient can be performed in single interface without patient leaving the seating area. Updates in patient treatment status and medical history were presented in real time without having to move paper charts around.
Conclusions: The iPad can be used in clinical activities independent of computer workstations. Improvements in clinical workflow can be critical in reducing physician resistance in data maintenance. Using the iPad in providing real-time quality monitoring is intuitive to both providers and patients.

1. Introduction

Clinical workflow, which consists of numerous focus-demanding tasks, has been known to be associated with various inevitable interruptions and distractions which slow down workflow and induce medical errors. While researchers identified clinical documentation to be one of the most vulnerable activities for workflow interruptions [1], efforts to improve the process have been mostly, if not exclusively, devoted to Electronic Health Record (EHR) development. The downside to this EHR reliant approach is that the design of the product is focused on extended generalization instead of customization capability, which resulted in ignorance of institution specific clinical ergonomic factors that affect physicians' ease-of-use (such as documenting while facing patients, fast mobile application start-up solutions, structured data collection, or intuitive data presentation). The compromised fluency of workflow generated physician resistance to adoption [2-4]. In our department of radiation oncology head and neck clinics, the problems were duplicative, and even worse given our excessive amount of patients. To overcome the difficulties, a browser-based platform ("Onco-Browser") as part of our integrative clinical informatics project --"Oncospace" was developed and customized to fit in hand held terminal devices which were proven to be feasible in integrating into clinical workflow [5]. The iPads were deployed in clinics to host the browser for clinical activities management. Performance and clinician acceptance of the solution was observed and evaluated.

2. Methods

A browser-based platform that can exchange and present analyzed data from the MOSAIQ database was developed in situ ("Onco-Browser"), data collection forms that were created in the MOSAIQ system were customized to fit hand-held devices, the browser access was protected using institution specific login portal. The iPads were distributed in radiation oncology head and neck clinics to physicians, nurses and patients. The system was deployed for clinical activities management including medical record documentation, research data collection, patient reported quality-of-life (QOL) assessment collection for patients, radiation machine schedule planning, clinical assessment monitoring and graphical presentation. Data collection by patient alone was maneuvered under "Guided Access" mode in iPads to prevent browsing over to other patients, and all data collection forms are submitted with the provision of clinician username and password. Data presentation was developed under physician requirements to accomplish complex task of patient evaluation and education. Presented data was pulled and analyzed in real time from the MOSAIQ database in the form of graphics and charts for toxicity interpretation and intuitive patient education.

3. Results

3.1. Clinical Activity Performance

All in-clinic activities were able to be performed without workstation computers. Patient demographic data and treatment progress can be viewed on a single page as seen in **Figure 1**. Clinical data collection can be performed on the iPads without the access to mobile workstations, time to access collection form has improved compared to opening the clinical system application on a mobile workstation as the only time lag is the browser opening period. Data collection forms were touch screen friendly with size customizable radio buttons and drop down lists as shown in **Figure 2**, enabling physicians to complete the patient assessment process without facing away from the patient. Patient QOL data collection is achieved via "Guided Access", which prevents the patient from free browsing (**Figure 3**). Medical record documentation is automatically generated by integrating a robust documentation template with data points from in-clinic data collection and/or inter-departmental data (such as phlebotomy etc.). Language processed by the voice recognition tool (Siri) is acceptable by head and neck physicians and has been using the function since deployment. Medical documentation are also occasionally performed by voice input.

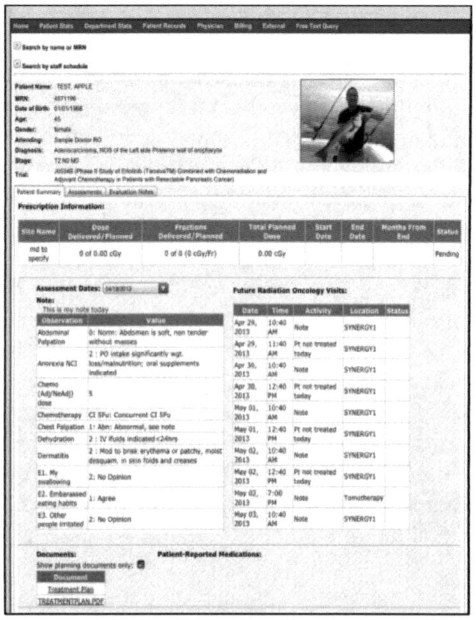

Figure 1. *Patient summary page with test patient, this page was used by physicians for quick overview of patient before entering clinic room.*

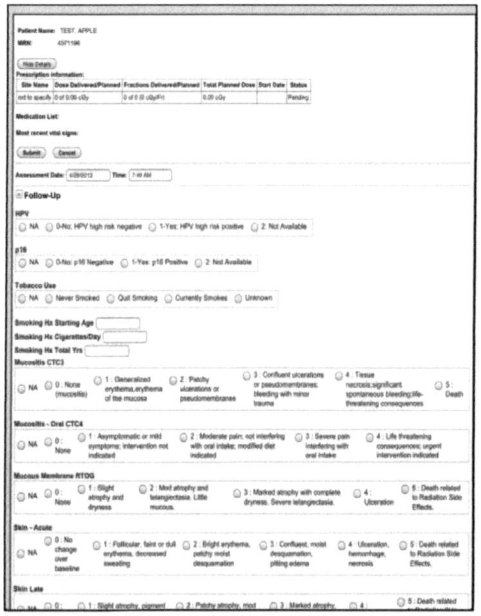

Figure 2. *Clinical data collection page, automatically generated from pre-existing forms in Onco-Browser application.*

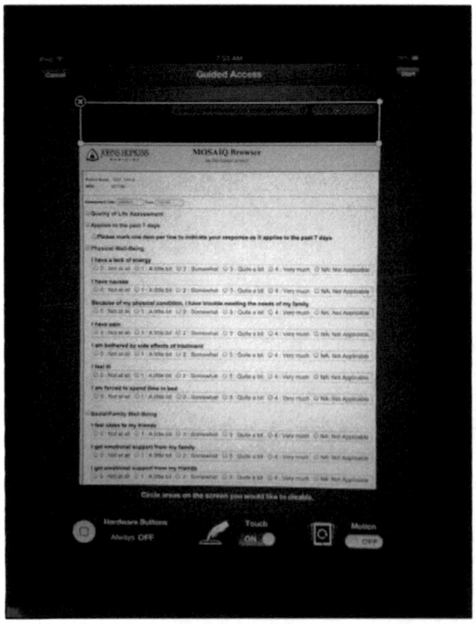

Figure 3. *"Guided Access" mode in the iPad for self-entering QOL data, which was designed to reduce stress for patients and provide more privacy.*

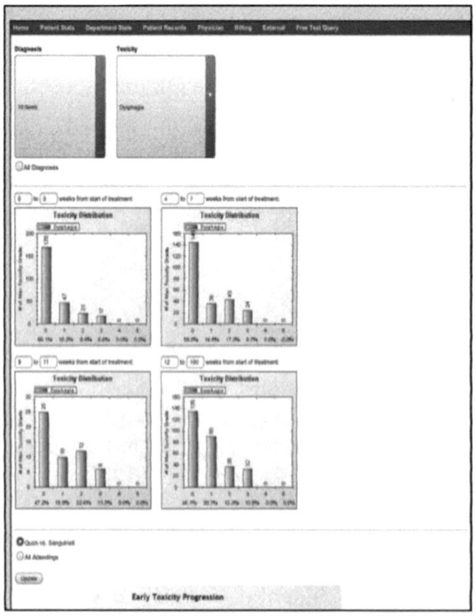

Figure 4. *Toxicity progression comparison in bar plot, four graphs represented different time periods and can be easily customized*

3.2. Data Presentation.

Toxicity progression is presented in numerous graphical means ie. bar graph, star chart, line plot etc. (see **Figure 4**), enabling patient participation on interpretation of toxicity trends. Patient education was significantly improved when real time in-clinic analyzed data was presented via the iPads in intuitive graphical form. Frequent dose prescription bar plot (see **Figure 5**) enabled physician to double check when their prescribed dose to a single diagnosis was deviating from regular prescriptions. Dose-Toxicity plot enabled dosimetrists and physicians to avoid predicted toxicities when planning organ at risk (OAR) dose-volume histogram (DVH) as shown in **Figure 6**.

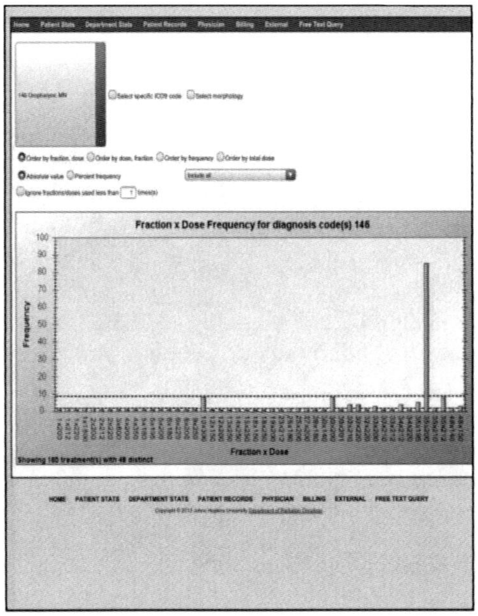

Figure 5. *Aggregated dose prescriptions for a single diagnosis, y axis represent prescription frequency and x axis represents type of prescription*

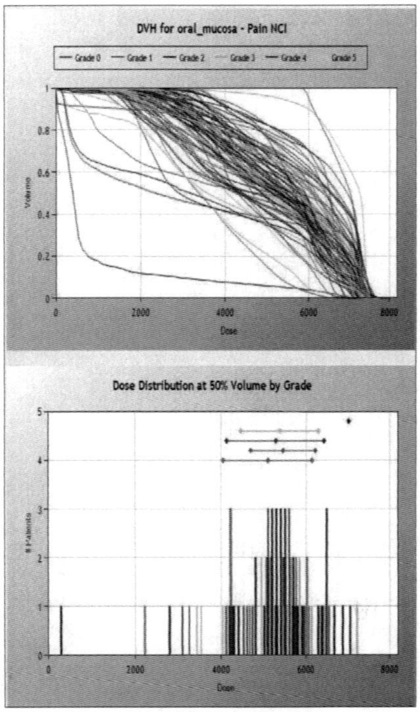

Figure 6. *Dose-Toxicity Plot for a single OAR, different colour represents a toxicity grade.*

4. Discussion

Although the idea of a mobile solution to solve clinical difficulties is not new and has been a topic of research for feasibility in numerous studies [5-8]. General resistance grows over time which results in 'Technology Disadoption' [4] due to the clinical nature of narrow tolerance in workflow breaks. We believe mobile solutions will only be adopted by physicians in clinics if they (and in the order of our assumed priority): 1. Do not exhibit or have minimal buggy behaviours; 2. Maintain intuitive user interfaces; 3. Adopt research capabilities, including fast data collection, clinical/research database data exchange, easy data querying, and data analysing/aggregating tools; 4. Improve workflow of conventional clinical activities; 5. Provide additional functionalities to enhance user experience, ie. Fast access to dosimetric databases for in-clinic dose plan overview, aggregated toxicity progression by diagnosis for patient education and predictive narcotic dose adjustment etc. In addition to above, building relationship with clinical champions who are willing to promote and demonstrate the advantages in adopting mobile solutions is critical to the success of implementation.

This research is the first of its kind to demonstrate performance of an implemented iPad solution, and results shows that the iPads can be used for clinical activities to provide augmented functionalities and enhanced clinical user (physicians, patients, nurses) experiences. Theoretically this platform can be implemented on any terminal device that supports internet browsing, but the iPad was specifically chosen because of its guided access feature provided enhanced security when the patients are operating on the device. Further study is needed to evaluate the efficiency in enhancing clinical workflow of this deployment.

5. Conclusion

The iPad can be used in clinical activities independent of computer workstations. Improvements in clinical workflow can be critical in reducing physician resistance in data maintenance. Using the iPad in providing real-time medication quality monitoring is intuitive to both providers and patients.

6. Acknowledgement

This research is funded by the Commonwealth Foundation.

7. References

[1] Weigl M, Muller A, Zupanc A, Glaser J, Angerer P. Hospital doctors' workflow interruptions and activities: An observation study. *BMJ Qual Saf.* 2011 Jun;**20**(6):491-7.

[2] Tierney WM, Miller ME, Overhage JM, McDonald CJ. Physician inpatient order writing on microcomputer workstations. effects on resource utilization. *JAMA.* 1993 Jan 20;**269**(3):379-83.

[3] Grabenbauer L, Skinner A, Windle J. Electronic Health Record Adoption - Maybe It's not about the Money: Physician Super-Users, Electronic Health Records and Patient Care. *Appl Clin Inform* 2011; Nov 9;**2**(4):460-71.

[4] Lapointe L, Rivard S. Getting physicians to accept new information technology: insights from case studies. *Cmaj* 2006; May 23;**174**(11):1573-8.

[5] Muller M, Rassweiler MC, Klein J, Seitel A, Gondan M, Baumhauer M, et al. Mobile augmented reality for computer-assisted percutaneous nephrolithotomy. *Int J Comput Assist Radiol Surg* 2013; Mar 23;.

[6] Grabenbauer L, Skinner A, Windle J. Electronic Health Record Adoption - Maybe It's not about the Money: Physician Super-Users, Electronic Health Records and Patient Care. *Appl Clin Inform* 2011; Nov 9;**2**(4):460-71.

[7] Gurland B, Alves-Ferreira PC, Sobol T, Kiran RP. Using technology to improve data capture and integration of patient-reported outcomes into clinical care: pilot results in a busy colorectal unit. *Dis Colon Rectum* 2010; Aug;**53**(8):1168-75.

[8] Horng S, Goss FR, Chen RS, Nathanson LA. Prospective pilot study of a tablet computer in an Emergency Department. *Int J Med Inform* 2012; May;81(5):314-9.

Journal of Physics: Conference Series **489** (2014) 012096 doi:10.1088/1742-6596/489/1/012096

Organism-level models: When mechanisms and statistics fail us

M H Phillips[1], J Meyer[1], W P Smith[2] and J K Rockhill[1]

[1] Department of Radiation Oncology, University of Washington, Seattle, WA, USA
[2] Stratton VA Medical Center, Albany, NY, USA

E-mail: markp@u.washington.edu

Abstract. Purpose: To describe the unique characteristics of models that represent the entire course of radiation therapy at the organism level and to highlight the uses to which such models can be put. **Methods:** At the level of an organism, traditional model-building runs into severe difficulties. We do not have sufficient knowledge to devise a complete biochemistry-based model. Statistical model-building fails due to the vast number of variables and the inability to control many of them in any meaningful way. Finally, building surrogate models, such as animal-based models, can result in excluding some of the most critical variables. Bayesian probabilistic models (Bayesian networks) provide a useful alternative that have the advantages of being mathematically rigorous, incorporating the knowledge that we do have, and being practical. **Results:** Bayesian networks representing radiation therapy pathways for prostate cancer and head & neck cancer were used to highlight the important aspects of such models and some techniques of model-building. A more specific model representing the treatment of occult lymph nodes in head & neck cancer were provided as an example of how such a model can inform clinical decisions. A model of the possible role of PET imaging in brain cancer was used to illustrate the means by which clinical trials can be modelled in order to come up with a trial design that will have meaningful outcomes. **Conclusions:** Probabilistic models are currently the most useful approach to representing the entire therapy outcome process.

1. Introduction

Models are of interest in medical science because they promote our understanding of fundamental processes. Just as importantly, however, they allow us to predict–whether it be the type of disease, its future progression or its response to a specific intervention. A more complete model treats more than just the disease or intervention, but also its interaction with all aspects of a person's health. In radiation therapy, we deal with multiple models. One way of categorizing them is the level at which they model the body and its interaction with radiation. We have models at the atomic and molecular levels (dose calculations), the molecular and cellular levels (radiobiology), the cellular and organ levels (TCP, NTCP) and the organism levels.

Another way to categorize our models is the degree to which they encapsulate cause-and-effect pathways (mechanistic models) versus empirical observations (probabilistic models). These categories are associated with the above-mentioned models of scale. The smaller the scale, the more likely the model is to represent specific mechanisms. The larger the scale, the more the models encapsulate clinical observations in the form of probability distributions. Clearly, the complexity of the environment being modeled influences the number of variables and the interactions between them.

Currently, we have a very accurate cause-and-effect model of the deposition of radiation at the atomic and molecular levels based on well-established theories of modern physics. Such exact knowledge fades quickly as we move to the biological effects of radiation absorption, even at the biomolecular and cellular level. Once basic ideas of free-radical-induced DNA strand breaks have been shown to incompletely describe what happens as we have learned more and more about the many molecular pathways of damage and repair. Now our models include very specific mechanisms, but also statistical distributions of measured parameters such as α and β. At the level of organ response, our mechanistic understanding is even less, with suboptimal concepts such as "parallel" and "serial" organs to guide us. Our models even incorporate their uncertain knowledge in their names, such as *tumor control probability (TCP)*, and *normal tissue complication probability (NTCP)*.

Obviously, if we are talking about models of our entire organism, we must rely on probabilistic models–assuming that such models even exist. Clinical exigencies force us to try to make predictions so that we can devise a course of treatment and counsel patients as to the decisions they must make. Recent developments have led to an emphasis on genomics as a source of knowledge that will allow us to develop "patient-specific treatments". The amazing increase in our knowledge of the genetic components of disease and response gives us great hope that we are at the point where we can discard our probabilistic organism-level models for more exact mechanistic models.

In the following sections of this paper, we briefly describe why this may not occur as soon as we would like, and to give some examples of alternatives that we can adopt while we wait while our knowledge catches up to our needs.

2. Genomic models

A fundamental idea that has been the basis of most of theories of cancer development is that one rogue cell somehow escapes the normal bounds of cellular reproduction and growth and through unrestricted division leads to a cancer composed of identical genetic copies of the original cell. Such a simple view of the genetic structure of cancer is being rapidly replaced by a much more complex genetic system. Gerlinger *et al* studied renal cancers and found "intratumor heterogeneity and branched evolution revealed by multiregion sequencing" [1]. Vermaat *et al* looked at hepatic metastases of colorectal cancers and reported that the primary and metastic tumors are genetically different which has many implications for the selection of patients for treatment targeted to a particular genotype [2]. For one, it calls into question the ability to determine the genetic range by means of any type of feasible biopsy. They hypothesize that such differences may lead to tumor adaptation and therapeutic failure through Darwinian selection. Clinicians are not surprised when multiple metastases respond completely differently to identical treatments, in opposition to the conventional wisdom that they are all identical. Another possibility is that different cellular milieus can have profound effects even on genetically identical tumor cells.

The idea that our current genetic knowledge is at a stage where we can develop such accurate models that genetically-targeted treatments are imminent is also proving less robust than some have predicted. For example, a paper by Nebert and Zhang found that most examples of pharmacogenetic traits reflect contributions from innumerable low-effect genes and explain only a small fraction of phenotype variation [3].

Similarly, in a study of monozygotic twins, Roberts *et al* performed whole genome sequencing and compared the incidence of disease of nine common cancers [4]. Their findings cast significant doubt that our current knowledge of genomics is going to be translated into patient-specific models of disease. Summarizing results from all nine cancers, they reported that only 0-40% of true positives would actually test positive; less than 10% of the general population would test positive; and there was only a 0-40% decrease in risk relative to the general population for those testing negative.

Finally, as Liu points out, what is needed is not more lists of mutations, but rather, the ability to detect all relevant components of a system and to describe the complexity in a mathematical model [5]. He further discusses the need to reconstruct this complexity experimentally in order to learn more about their characteristics. Only that way can we make predictive models that explain possible responses of a cancer in terms of its gene networks.

We see that genetics certainly adds to our knowledge but we are still far from a mechanistic model at the organism level. We have no models that incorporates system responses, such as immune response. We also have no models that include environmental and stochastic variables. Even at the cellular level our models use surrogate variables that encapsulate much ignorance.

3. Probabilistic models

Randomized controlled clinical trials (RCT) are our best sources of data for probabilistic models. They also represent our best mechanistic models since it is usually the case that the design of an RCT is based on some prediction of cause and effect. The problem is that our knowledge is so incomplete that the model variables are very imprecise and many key variables and parameters are unknown. For example, three RTOG studies provide data for our models of the responses for the radiation treatment of brain metastases. There are really only three predictive variables (Karnofsky Performance Status, primary tumor status, and age), and the majority of patients actually fall into Class 2 because they don't fall into the other two definitive categories.

Clinical trials are based on our best estimate of cause-effect relationships. The structure of the trial reflects the structure of the model. The results of the trial fill in the values of the model. Unfortunately, since there are so many unaccounted variables, the values are probability distributions over the variables of the trial. RCT's are constrained to vary only one or two variables at a time, so that it takes many trials or other sources of information to fill in the entire model.

As stated before, one of the main uses of a model is to aid in decision making. In the case of medical treatment, these decisions are made under uncertainty. In these circumstances, we can consider a decison model as one that:

- predicts outcomes as a function of a given state and action;
- handles trade-offs between competing outcomes;
- provides some means of ranking outcomes.

Since this model must incorporate uncertainty, further qualities are:

- the use of probabilities as input;
- the ability to combine probabilities;
- the ability to perform inference.

Bayesian networks (BN) and their related structures, influence diagrams (ID), are one form of probabilistic model that are very appropriate for medical decision making [6, 7]. A BN is a directed acyclic graph that represents a joint probability table. It is constructed of variables connected by directed edges that imply causality. Each variable is characterized by more than one state (including continuous ones). There is a probability for an given state of any given variable that is conditioned on the states of each of its parent variables. These conditional probabilities are obtained from (a) randomized controlled trials, (b) medical research literature, (c) expert opinion, and (d) other models. Typically, the initial conditions of a medical situation are used to instantiate some of the variables while the probability calculus of the BN is used to calculate the probabilities of the other variables, usually some form of health outcome. If the model includes the evolution of the health states over time, it is often convenient to couple the

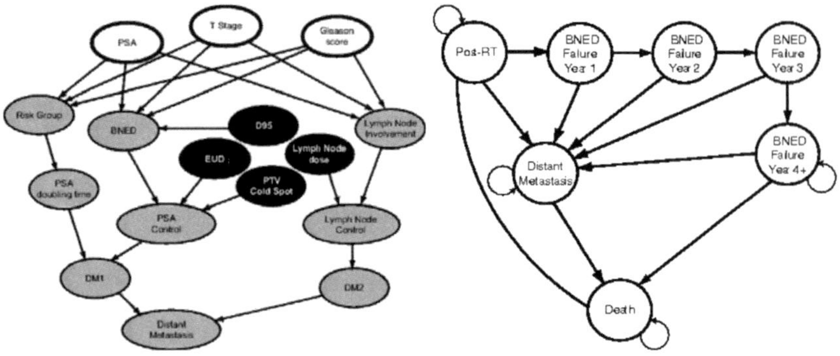

Figure 1. Model of overall outcome after radiation therapy in prostate cancer using a Bayesian network and Markov model.

BN with a Markov model (MM) [8]. If the relative values of all the possible outcome states are measured by means of *utilities*, then the expectation value of the utilities can be calculated by using the probability distributions calculated by the BN, thereby creating an influence diagram [9]. In standard utility theory, the best decision is the one which leads to the maximum expected utility (MEU) [10].

The following sections provide three different influence diagrams that demonstrate different applications of these decision models in radiation therapy.

3.1. Decision model for selection of IMRT plans for prostate cancer

We have constructed an evolutionary multiobjective optimization algorithm to calculate a set of IMRT plans that embody the trade-offs between the competing objectives for prostate cancer [11, 12, 13]. In order to aid in the decision to select one from this set of deliverable plans, we constructed an influence diagram. Since one of the key objectives is to increase survival, and since prostate cancer can take years to develop, we included a Markov model to capture the time evolution [14, 15] .

Figure 1 illustrates part of the BN (the nodes that represent normal tissue responses has been omitted for reasons of space). The probability of biochemical no evidence of disease (BNED) depends on the states of the three clinical variables and the delivered dose, as has been documented in a number of clinical trials. Other nodes and edges represent the decision variables used clinically and the conditional probabilities that they represent have been obtained from the literature.

Measuring patient-specific outcomes for the multiobjective health outcome states, e.g. {tumor control} + {rectal condition} + {bladder condition} + {sexual function}, is difficult. Using average measured utilities for populations of prostate cancer patients, we devised decision models for different classes of patients. The different classes were characterized by age and by their personal preference towards increasing life expectancy (tumor control) vs quality of life (complication-free). The combination of the influence diagram, Markov model and utilities allowed us to calculate the effect of different IMRT plans on the quality adjusted life expectancies, and to rank plans based on that metric.

3.2. Decision model for irradiation of occult lymph nodes in H&N cancer with and without FDG-PET

When designing treatment plans for head and neck cancer in the absence of definitive imaging or clinical evidence of positive lymph nodes, decisions need to be made whether or not to treat these

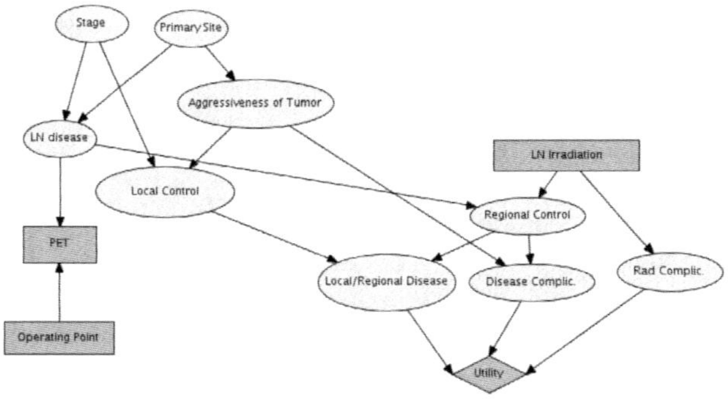

Figure 2. Influence diagram for decisions regarding the efficacy of FDG-PET imaging in the irradiation of occult lymph nodes for H&N cancer. The square nodes indicate action variables.

"occult" nodes. Clinical experience has shown that in some percentage of cases, not treating node-negative regions will result in tumor recurrence. The probability of recurrence depends on the location and stage of the primary tumor, and these conditional probabilities are well-documented. Conventional CT imaging is often unable to visualize small tumors in the lymph nodes. It has been postulated that including FDG-PET imaging could improve the diagnosis of positive nodes. This influence diagram was constructed to model the effect on decision making of this additional information [16]

Possible actions in this model are whether or not to obtain PET imaging, what operating point in the receiver operator curve (ROC) designates a "positive", and what dose to deliver. The ROC describes the accuracy of the information obtained from the PET images. The operating point sets the balance between false positives and false negatives. The selection of this point basically represents whether it is better to treat a tumor that isn't there or to not treat a tumor that is.

The results of our decision modeling provided recommendations for different tumor sites and different stages, representing the varying probabilities of lymph node involvement. In some cases, the model recommends no use of PET since it would not change the clinical decision. This may seem counter-intuitive, but it correctly folds in the fact that there is some probabilitiy that the PET images are wrong. For example, late stage nasopharyngeal tumors result in lymph node metastases over 90% of the time. A negative PET image could represent a false negative and it is usually worth the risk of complications to treat that region. In other words, it is best to ignore the PET information in that setting. Other recommendations included setting of the operating point (usually towards high sensitivity) and the dose level.

An interesting result was the different way in which radiation oncologists view the same health state depending on whether it was the result of disease progression or the result of the radiation therapy. We found that it was necessary to separate out those two situations since clinicians often make conservative treatment decisions in order to avoid the possibility of complications. It is a personal decision as to whether reducing the chance of tumor control falls in the same category of "do no harm".

3.3. Decision model in the design of a clinical trial of the efficacy of FLT-PET in re-irradiation of gliomas

Decisions regarding whether or not to re-treat a patient with a glioma are complicated by a number of factors, such as pseudo-progression and responses to other interventions. The question arose as to whether FLT-PET could be used to distinguish active regions of tumor growth from other, non-malignant metabolic conditions. In other words, what information–what levels of sensitivity and specificity–would be necessary to change the clinical decisions and improve outcomes? A Bayesian network is an appropriate method of constructing a model to set bounds for the performance of PET in order to guide the design of a clinical trial.

Construction of the decision model is relatively straight-forward with respect to the variables and their states as well as in the connections between the variables. The difficulty lies in obtaining the probabilities. Much of the data in the literature has little in the way of a "gold standard" of truth with respect to the condition of the tumor. Thus it is hard to determine the conditional probabilities of the clinical examinations and the MR images.

There are established statistical methods for using a Bayesian approach to the design of clinical trials [17]. This differs from the method described here, although the statistical method can be part of an influence diagram. Basically, the difference has to do with the fact that the statistical approach can deal with a narrow range of relevant variables. The BN can incorporate all of the input variables and possible outcomes into the decision model, thereby providing a broader assessment of the potential influence of PET.

This work is on-going and, in the face of the challenges mentioned, the results are difficult to predict. However, as difficult as it is, it is easier than designing and carrying out trials without such guidance.

4. Conclusion

Hopes for an imminent mechanistic model of organism-level radiation responses based on improved understanding of genetics appears to be unrealistically optimistic. The situation is perhaps best summed up by Topol: "But these predictions will always remain estimates, and DNA's power to predict illness will always be limited. The key to the future of individualized medicine is that such estimates and probabilities lead to actionable steps and prevention of disease." [18] The need to make clinical decisions regarding of the our theoretic understanding compels us to make decision models as best we can. Currently, that means that our models are probabilistic. We have demonstrated that Bayesian networks and the related influence diagram are a productive means for constructing such models and we gave several examples of their use.

References

[1] Gerlinger M, Rowan A J, Horswell S, Larkin J and et al 2012 *N Engl J Med* **366** 883–892
[2] Vermaat J S, Nijman I J, Koudijs M J, L G F and et al 2012 *Clin Cancer Res* **18** 688–699
[3] Nebert D W and Zhang G 2012 *Science* **337** 910
[4] Roberts N J, Vogelstein J T, Parmigiani G, Kinzler K W, vogelstein B and Velculescu V E 2012 *Sci Translat Med* **4** 1–9
[5] Liu E T 2013 *Science* **339** 1493
[6] Nease R F and Owens D K 1997 *Med Decis Making* **17** 263–276
[7] Owens D K, Schachter R D and Nease R F 1997 *Med Decis Making* **17** 241–262
[8] Beck J R and Pauker S G 1983 *Med Decis Making* **3** 419–458
[9] Doctor J N and Miyamoto J M 2003 *J Mathematical Psychology* **47** 557–567
[10] Von Neumann J and Morgenstern O 1944 *Theory of Games and Economic Behavior* (New York: Wiley)
[11] Holdsworth C, Kim M, Liao J J and Phillips M H 2010 *Med Phys* **37** 4986–4997
[12] Holdsworth C, Stewart R, Kim M, Liao J J and Phillips M H 2011 *Med Phys* **38** 2964–2974
[13] Holdsworth C H, Kim M, Liao J and Phillips M 2012 *Med Phys* **39** 2261–2274
[14] Meyer J, Phillips M H, Cho P S, Kalet I and Doctor J N 2004 *Phys Med Biol* **49** 1637–1653
[15] Smith W P, Doctor J, Meyer J, Kalet I J and Phillips M H 2009 *Artificial Intelligence in Medicine* **46** 119–130
[16] Phillips M H, Smith W P, Parvathaneni U and Laramore G 2010 *Int J Radiat Oncol Biol Phys* 1089–1095
[17] Spiegelhalter D J, Abrams K R and Myles J P 2004 *Bayesian Approaches to Clinical Trials and Health-Care Evaluations* (John Wiley and Sons)
[18] Topol E J 2012 *Sci Transl Med* **4** 135

A database paradigm for the management of DICOM-RT structure sets using a geographic information system

Weber Shao[1], Patrick A Kupelian, Jason Wang, Daniel A Low, and Dan Ruan
Department of Radiation Oncology, David Geffen School of Medicine at UCLA

E-mail: wshao@mednet.ucla.edu

Abstract. We devise a paradigm for representing the DICOM-RT structure sets in a database management system, in such way that secondary calculations of geometric information can be performed quickly from the existing contour definitions. The implementation of this paradigm is achieved using the PostgreSQL database system and the PostGIS extension, a geographic information system commonly used for encoding geographical map data. The proposed paradigm eliminates the overhead of retrieving large data records from the database, as well as the need to implement various numerical and data parsing routines, when additional information related to the geometry of the anatomy is desired.

1. Introduction

DICOM-RT, a widely utilized extension to the Digital Imaging and Communications in Medicine (DICOM) imaging standard, represents many aspects of radiotherapy imaging. One of such is the RT structure sets, a group of definitions of anatomical regions of interest. It is desirable to devise a paradigm for representing the RT structure sets in the database management system, in such way that secondary calculations of geometric information can be performed quickly from the existing RT contour definitions.

To achieve this, there are two possible general approaches. One approach is to rely on client-side processing after record retrieval from the database. Using this method, computation can be carried out swiftly if the routine is implemented using a compiled language such as C. A drawback of this approach is the relatively long data retrieval time due to the size of the set of RT contour records.

The alternative approach is to perform computation directly on the database server before the query and retrieval. This method produces shorter data retrieval time than does the client-side processing approach, since only the results, as opposed to the sets of contours, need to be returned by the SQL query. However, the performance would suffer significantly if the mathematical calculations, which would require loops, are implemented purely using a stored procedure language, which is processed through an interpreter.

Therefore, in order for the server-side computation approach to be efficient, it ought to be supplemented by an additional set of SQL functions in the form of pre-compiled binary objects. While it is possible to extend the SQL language by writing and compiling C routines into functions, it is optimal to employ a geographic information system (GIS), which is historically used for storing, managing, and manipulating geographical spatial data. Specifically, we aim to accomplish this goal using PostGIS, an extension to the PostgreSQL database management system (DBMS).

[1] To whom any correspondence should be addressed.

2. Experiment Configuration

An experiment is conducted to measure the performance of the computations of three common quantities pertaining to an anatomical structure set: surface area, volume, and geometric centroid. The experiment encompasses the two aforementioned approaches and is administered in two settings for each approach.

2.1. Settings

One database server and one client program are required for each setting. The database server is hosted on a workstation with 12 Intel Xeon E5645 2.4GHz CPUs and 12GB of RAM. Settings 1 and 2 differ in the location of the client program.

- Setting 1: The client program resides on the same workstation as that of the database server.
- Setting 2: The client program resides on a remote workstation, which has 2 Intel(R) Core(TM) 2 Duo E7400 2.80GHz CPUs and 4GB of RAM, and is connected through a cable modem at a maximum transfer speed of 30Mbps. The remote workstation is 7 networking hops from the database server, and is physically located 23 miles away.

2.2. Approaches

In each approach, the database query and retrieval operations are wrapped inside a C client program supported by the libpq5 library. PostgresSQL 9.1 is used as the database management system.

- Approach 1: The computation is performed on the server side as a SQL query, and the results are returned to the client program as records. PostGIS 2.0.1 is used as the GIS extension to PostgreSQL.
- Approach 2: The contour information is retrieved as a set of records returned by the SQL query, and the computation is performed subsequently on the client side inside a C routine.

3. Database Organization

In this database paradigm, one treatment course contains several structures. Each structure contains a collection of contours, as it is represented in the design of the DICOM-RT standard. Each contour is a closed set of vertices, all of which share a common z-coordinate. We design and populate the database tables *Course*, *Structure*, and *Contour* to model this set of relations. If desired, the *StructureSecondary* table can be used to store the output of the secondary geometric properties that are to be computed.

In Approach 1, the set of vertices describing each contour is modelled by a polygon object stored as *geometry*, a generic PostGIS geometric data type. In Approach 2, each set of vertices is modelled by *real[]*, an array of floating point numbers in PostgreSQL. With the exception of the difference in the data type representation of contours, the database tables and their sets of relations are identical in both approaches.

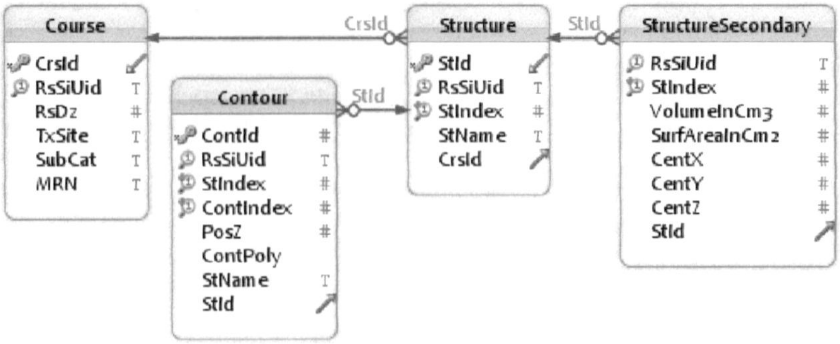

Figure 1. Database organization for RT contours

4. Methods

For Approach 1, we invoke several PostGIS API functions to assist in carrying out the computation of two-dimensional geometric measurements as SQL columns, which are used with aggregate functions to calculate the quantities interest: surface area, volume, and geometric centroid. As of PostGIS version 2.0.1, three-dimensional support is only limited to a few functions related to polygonal surfaces and triangular irregular networks, and therefore is not used by this approach. Approach 2 is implemented by directly coding the mathematics of Equations (1) to (7) inside the C routines.

4.1. Surface Area

For computing the total surface area of an anatomical structure represented in DICOM-RT, the algorithm is to find the product of the total path length and the slice thickness of its z-plane, Δz:

$$A_{surface} = \Delta z \sum_{i=1}^{N-1} \sqrt{(x_{i+1} - x_i)^2 + (y_{i+1} - y_i)^2} \tag{1}$$

The slice thickness in DICOMT-RT is uniform across all contours. In the database representation of the contour, the first vertex is also used as the last vertex to ensure a closed polygon. For this calculation, Approach 1 invokes the *st_boundary()* function to obtain the vertices of the contour, and the *st_length()* function to find the contour path length. Thus the surface areas of all structure sets within a treatment course can be obtained by the following query:

```
SELECT cotr."RsSiUid", cotr."StName",
  SUM(crs."RsDz" * st_length(st_boundary(cotr."ContPoly")))
FROM "Contour" AS cotr, "Course" AS crs
WHERE cotr."RsSiUid" = crs."RsSiUid"
AND crs."MRN" = 'XX-XX-XXX'
GROUP BY cotr."RsSiUid", cotr."StIndex", cotr."StName"
ORDER BY cotr."RsSiUid", cotr."StIndex"
```

4.2. Volume

The volume of a DICOM-RT anatomical structure is given by the product of the slice thickness and the sum of the areas of each contour:

$$V = \Delta z \sum_{j=1}^{M} A_{contour,j} \tag{2}$$

The area of each contour is given by:

$$A_{contour} = \frac{1}{2} \sum_{i=1}^{N-1} (x_i y_{i+1} - x_{i+1} y_i) \tag{3}$$

The SQL query in Approach 2 for the calculation for (2) can be made with the assistance of the invocation of the *st_area()* function, which produces the contour area in lieu of implementing the mathematics of (3) directly. Thus the volumes of all structure sets within a treatment course can be obtained by one query:

```
SELECT cotr."RsSiUid", cotr."StIndex", cotr."StName",
  SUM(crs."RsDz" * st_area(cotr."ContPoly"))
FROM "Contour" AS cotr, "Course" AS crs
WHERE cotr."RsSiUid" = crs."RsSiUid"
AND crs."MRN" = 'XX-XX-XXX'
GROUP BY cotr."RsSiUid", cotr."StIndex", cotr."StName"
ORDER BY cotr."RsSiUid", cotr."StIndex"
```

4.3. Centroid

To calculate the three-dimensional centroid of a structure, first we find the two-dimensional centroid (P_x, P_y) of each contour that belongs to the structure:

$$P_x = \frac{1}{6A_{signed}} \sum_{i=1}^{N-1} (x_i + x_{i+1})(x_i y_{i+1} - x_{i+1} y_i) \tag{4}$$

$$P_y = \frac{1}{6A_{signed}} \sum_{i=1}^{N-1} (y_i + y_{i+1})(x_i y_{i+1} - x_{i+1} y_i) \tag{5}$$

where A_{signed} is the signed area defined by:

$$A_{signed} = \sum_{i=1}^{N-1} (x_i y_{i+1} - x_{i+1} y_i) \tag{6}$$

Thus the three-dimensional centroid (C_x, C_y, C_z) of the structure would be the weighted average of the centroid of each contour, where the weight is the area of each contour, $A_{contour,j}$:

$$\begin{bmatrix} C_x \\ C_y \\ C_z \end{bmatrix} = \frac{1}{\sum_{j=1}^{M} A_{contour,j}} \begin{bmatrix} \sum_{j=1}^{M} A_{contour,j} P_{x,j} \\ \sum_{j=1}^{M} A_{contour,j} P_{y,j} \\ \sum_{j=1}^{M} A_{contour,j} P_{z,j} \end{bmatrix} \tag{7}$$

While P_x and P_y are given by Equations (4) and (5), P_z is simply the z-coordinate value where the contour is defined, which corresponds to the *PosZ* column of the of the *Contour* table in the database structure diagram in Figure 1. For the SQL query in Approach 1, we can use a subquery to compute the reusable quantities of the areas of contours, $A_{contour,j}$. The PostGIS *st_centroid()* function would handle the calculation of the two-dimensional contour centroid, while the *st_x()* and *st_y()* functions would return its x- and y-coordinates, respectively. Consequently, the calculation of the geometric centroids of all the structure sets for one treatment course can be accomplished by one nested query:

```
SELECT bar."RsSiUid", bar."StIndex", bar."StName",
  SUM(bar.area * bar.x) / SUM(bar.area),
  SUM(bar.area * bar.y) / SUM(bar.area),
  SUM(bar.area * bar.z) / SUM(bar.area)
FROM (
  SELECT cotr."RsSiUid",
    cotr."StIndex", cotr."StName", cotr."ContIndex",
    st_area(cotr."ContPoly") AS area,
    st_x(st_centroid(cotr."ContPoly")) AS x,
    st_y(st_centroid(cotr."ContPoly")) AS y,
    cotr."PosZ" AS z
  FROM "Contour" AS cotr, "Course" AS crs
  WHERE cotr."RsSiUid" = crs."RsSiUid"
  and crs."MRN" = 'XX-XX-XXX'
) AS bar
GROUP BY bar."RsSiUid", bar."StIndex", bar."StName"
ORDER BY bar."RsSiUid", bar."StIndex"
```

5. Results and Discussions

After 4 trials each across 24 prostate cases, we obtain the following results of average processing time for all structure sets in one treatment course, under each setting for each approach:

Table 1. Performance comparison of settings and approaches

Setting and Approach	**Surface Area**	**Volume**	**Centroid**
Local Retrieval, PostGIS-Based	148 ms	124 ms	224 ms
Local Retrieval, Client-Based	415 ms	405 ms	405 ms
Remote Retrieval, PostGIS-Based	329 ms	304 ms	406 ms
Remote Retrieval, Client-Based	1296 ms	1300 ms	1290 ms

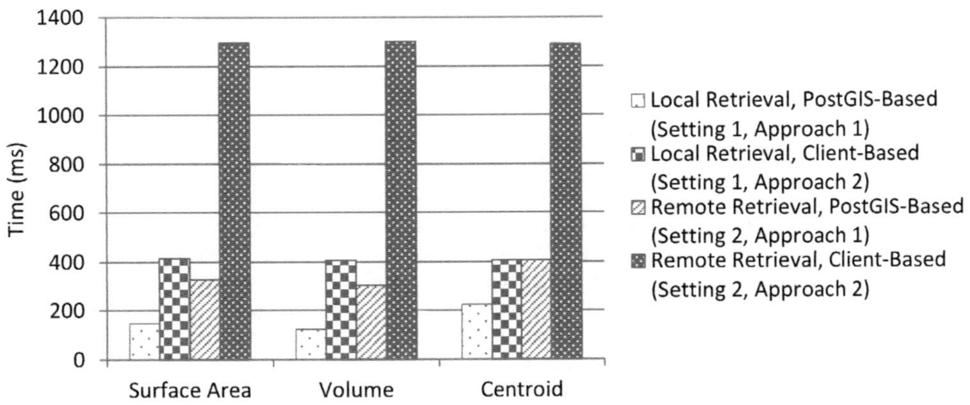

Figure 2. Performance comparison of settings and approaches

The approach of using PostGIS performs noticeably faster than the approach of relying on client-side computation. The difference is more evident in the setting where the workstation is remotely located from the database server. While the result of Setting 2, Approach 2 is dependent upon the hardware processing power of the remote client, as well as the network transfer time, the scenario presented in this experiment is realistic. A database server typically has a more powerful processor and is equipped with more RAM than a client workstation. A server and client could very possibly have a longer distance or a slower connection mechanism in another scenario than in the configuration presented here, and thereby would produce a sharper contrast in the performance of the two approaches.

6. Conclusions

The proposed paradigm for DICOM-RT structure set archival provides several benefits. First, it eliminates the overhead of retrieving large records from the database. Second, it simplifies the implementation of various numerical and data parsing routines, by relying on a solid application framework that is deployed for production worldwide. Additionally, the centralized, server-side design allows its functionalities to be utilized by multiple parties.

7. References

[1] Obe R and Hsu L 2011 *PostGIS in Action* (Shelter Island, NY: Manning)
[2] Gorthi S, Cuadra M, and Thiran, J 2009 "Exporting Contours to DICOM-RT Structure Set" in *The Insight Journal – 2009 January - June*
[3] Zlatanova S 2006 "3D Geometries in Spatial DBMS", Pp. 1-14 in *Innovations in 3D Geo Information Systems* (New York, NY: Springer)
[4] Bourke P 1988 *Calculating The Area And Centroid Of A Polygon*, Retrieved December 12, 2012 from http://paulbourke.net/geometry/polygonmesh/

Has the use of computers in radiation therapy improved the accuracy in radiation dose delivery?

J Van Dyk[1,2] **and J Battista**[1,3]
[1] Western University, London, Ontario, Canada
[2] Formerly, [3] London Regional Cancer Program, London, Ontario, Canada

E-mail: vandyk@uwo.ca

Abstract. Purpose: It is well recognized that computer technology has had a major impact on the practice of radiation oncology. This paper addresses the question as to how these computer advances have specifically impacted the *accuracy of radiation dose delivery to the patient.* **Methods**: A review was undertaken of all the key steps in the radiation treatment process ranging from machine calibration to patient treatment verification and irradiation. Using a semi-quantitative scale, each stage in the process was analysed from the point of view of gains in treatment accuracy. **Results**: Our critical review indicated that computerization related to digital medical imaging (ranging from target volume localization, to treatment planning, to image-guided treatment) has had the most significant impact on the accuracy of radiation treatment. Conversely, the premature adoption of intensity-modulated radiation therapy has actually degraded the accuracy of dose delivery compared to 3-D conformal radiation therapy. While computational power has improved dose calibration accuracy through Monte Carlo simulations of dosimeter response parameters, the overall impact in terms of percent improvement is relatively small compared to the improvements accrued from 3-D/4-D imaging. **Conclusions**: As a result of computer applications, we are better able to see and track the internal anatomy of the patient before, during and after treatment. This has yielded the most significant enhancement to the knowledge of "*in vivo*" dose distributions in the patient. Furthermore, a much richer set of 3-D/4-D co-registered dose-image data is thus becoming available for retrospective analysis of radiobiological and clinical responses.

1. Introduction

While historically, computer applications occurred mainly in the scientific realm, today computer applications are ubiquitous. The question as to whether this has had any societal impact is moot. One way of assessing this would be to see what happens if we turn off any application that involves computer usage; many activities in society today would come to a halt, as evidenced by unexpected failures in telecommunication networks. Because of the highly technical nature of radiation therapy, it was one of the first *medical* specialties to apply computers to routine clinical procedures [1]. This led to the initiation of the International Conferences on the Use of Computers in Radiation Therapy (ICCR), the first of which was held in 1966 in Cambridge, UK. References to each of the proceedings up until 2010 can be found in a book chapter by Van Dyk [1]. In the context of these ICCR conferences, the question being addressed in this paper is whether computer applications have actually impacted the accuracy of dose delivery to the patient in radiation therapy and, if so, how, and to what extent.

2. Methods

A true test of whether accuracy in dose delivery is affected by the use of computers in radiation therapy would be to do a controlled study analyzing dose delivery accuracy with and without computer assistance. However, it is obvious that this is effectively (and ethically) impossible since some of these procedures can simply not be executed 'manually' (e.g., IMRT optimization). The practical alternative is to use a logical, semi-quantitative analysis of the impact.

Since net accuracy is affected by multiple steps in the overall radiation therapy process, it is helpful to review the issues that impact dose delivery accuracy at each stage of that process and then to assess the impact of the role of computers on that stage. In terms of the steps in the radiation treatment process, there are two broad considerations. The first deals with technology-related uncertainties such as radiation beam calibration and the commissioning of the radiation treatment planning system. The second relates to patient-related uncertainties including patient anatomy and its variation. These considerations are summarized in column 1 of Tables 1 and 2, respectively. A brief description of each step is found in column 2 of these Tables along with a description of accuracy and uncertainty issues in column 3.

3. Results

Column 4 of Tables 1 and 2 provides the authors' opinion of the impact of computers in radiation therapy in each stage of the calibration, commissioning or treatment process, based on a literature review. While this is admittedly subjective, it does provide an inventory of items to be considered in assessing gains or losses in dose accuracy. Figure 1 provides a graphical summary of the data in Tables 1 and 2.

4. Discussion

Historically, computer applications in radiation therapy have been most often linked to the treatment planning component of the treatment process. The 1987 ICRU Report 42 [2], *Use of Computers in External Beam Radiotherapy Procedures with High-Energy Photons and Electrons*, specifically produced "… a report on treatment planning and recording and documentation procedures …". Little consideration was given to the other stages of the calibration, commissioning and the downstream treatment process involving the individual patient.

4.1. Calibration and commissioning process

4.1.1. Beam calibration

Basic radiation dosimetry has been, and continues to be the major focus in national standards calibration laboratories, for external quality assurance review agencies, as well as for medical physics professionals involved in developing dosimetry/calibration protocols. In 1976, ICRU Report 24 [3] showed that hospital beam calibrations in photon beams had uncertainties of 1.0 to 2.5% (optimally) or 2.3 to 4.9 % (minimally). In 1984, Svensson [4] indicated that beam calibrations could be performed with an uncertainty of 1.4 to 3.4% depending on whether they were cobalt-60, megavoltage x-ray or electron beams. A 2011 review by Andreo [5] showed that the determination of absorbed dose in a water phantom for therapeutic beams can be performed with an accuracy of 1 to 2% depending on the beam (photons or electrons) and whether the ionization chamber was calibrated in a primary or secondary standards laboratory. Some of the improvements in calibration protocols relate to the use of revised fundamental dosimetric factors generated with Monte Carlo techniques that are computer intensive. Thus, one could argue that computers have aided in the reduction of absolute dose calibration uncertainties by perhaps 1 to 4%.

Table 1. An evaluation of ***the impact of the use of computers on the dose delivery accuracy*** specifically for technology-related aspects of external beam radiation therapy using a five-star scale with one star representing minimal impact and five stars representing major impact.

1. Activity	2. Brief Description	3. Accuracy & Uncertainty Issues	4. Estimated Impact
Calibration protocol	• Data/parameter generation	• Dependent on protocol • Dependent number of particle histories for Monte Carlo	**
Beam calibration	• Use water phantom & detector positioning	• Distance determinations • Detector reading	*
Beam commissioning	• Use of computer controlled detection system	• Detector positioning • Detector size • Detector response	**
Treatment planning system commissioning	• Entry of data • Data fitting to determine model parameters • In phantom analysis	• Data transfer • Data fitting • Algorithm limitations	*****

4.1.2. Treatment planning system commissioning.

Treatment planning system applications are very computer intensive because the 3-D/4-D dose distribution must be determined in a heterogeneous time-dependent absorber – the patient. Modern systems use 3-D scatter integration techniques or Monte Carlo simulations. For many years, even into

Table 2. An evaluation of ***the impact of the use of computers on the dose delivery accuracy*** specifically for the patient-related steps for external beam radiation therapy using a five star scale with one star representing minimal impact and five stars representing major impact. TPS = treatment planning system; OAR = organ-at-risk

1. Activity	2. Brief Description	3. Accuracy & Uncertainty Issues	4. Estimated Impact
Diagnosis/clinical evaluation	• Tumor pathology/staging • Use of imaging	• Imaging limitations (resolution, sensitivity) • False positives/negatives (specificity)	****
Therapeutic decisions	• Cure/palliation • Treatment modalities/techniques	• Physician decision making • Statistical analysis of clinical trials	*
Treatment prescription/ directive	• Dose prescription including normal tissue dose-volume constraints	• Radiobiological outcome data • Limitations on dose-volume data for both tumours & normal tissues • Radiobiological models	*
Patient positioning & immobilization for imaging	• Laser positioning • Patient support, positioning, & immobilization	• Laser accuracy on CT & other imaging units • Congruence with therapy machine isocentre • Stability of immobilization devices • Patient comfort & compliance	*
Imaging for treatment planning	• Set-up patient • Generate images for target & normal tissue localization	• Imaging limitations, e.g., resolution, interpretation of pixel signals • Image fusion accuracy • 3-D/4-D considerations	*****
Contouring (image segmentation) of target volumes & organs at risk	• Use ICRU concepts of GTV, CTV, ITV, PTV, PRV	• Limitations of imaging and segmentation • Limitations of knowledge of microscopic spread not seen in images • Uncertainties associated with CTV to PTV margins • Inter- and intra-observer variations of target/normal tissue delineation	**
Treatment planning (forward or inverse)	• Dose calculations • Treatment optimization	• Quality of beam data measurements • Quality of TPS commissioning • TPS algorithm accuracy in different regions and densities • Degradation in accuracy with IMRT fields (e.g. small segments)	***
Physician approval of treatment plan	• Plan acceptability, esp. regarding tumor dose uniformity or acceptable doses to OARs • Use of DVHs	• Physician knowledge/experience • Clarity of communication between planner and physician • Decision- making influenced by quality of printouts and graphic displays	***
Data transfer & file management	• Done manually in 2-D era • Done through departmental network in 3-D/IMRT/MLC era	• Data transfer has little impact on accuracy/uncertainties unless a gross error occurs (e.g., wrong patient ID, wrong MLC configuration) • More of an impact on reducing significant errors compared to manual copying of data • Computers required for complex 3-D/IMRT plans	*
Plan validation/checking	• For 2-D, would be check of plan & MU • For IMRT, phantom measurement or independent calculation	• Dependent on accuracy of 2nd check • Has little impact on accuracy other than a confirmation and quality assurance to intercept gross errors	**
Treatment machine set-up/immobilization/verification imaging (e.g., IGRT)	• Use same immobilization at imaging for planning and in treatment room	• Patient set-up uncertainties should be determined in each department for each technique • IGRT reduces both systematic and random uncertainties	*****
Treatment dose delivery, possibly with *in vivo* dosimetry	• Dose is delivered with appropriate field sizes, gantry rotations, shielding, MLC settings, etc.	• Machine dosimetry calibration • Reproducibility of patient set-up • Accuracy of *in vivo* dosimetry system • Adaptive therapy protocol • QA audits for IMRT delivery show significant variations at the 7%/4 mm accuracy levels	**

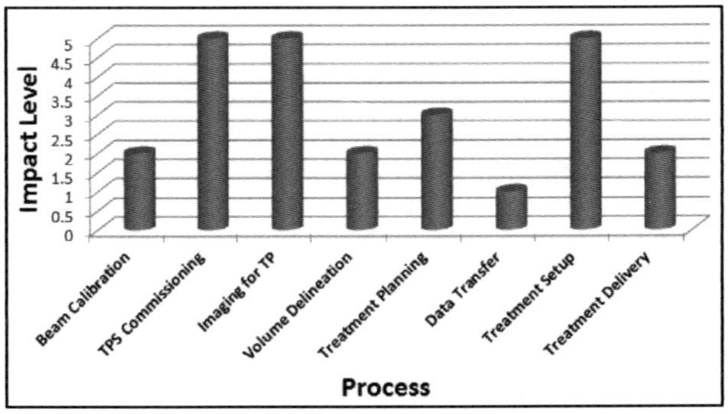

Figure 1. Graphical representation of the impact of computers on dose delivery accuracy for the different stages of the radiation treatment process using a five star scale, with one star representing minimal impact and five stars representing major impact.

the 2000s, tissue inhomogeneity corrections were not performed in many cancer centres. Today such calculations are performed more routinely and have reduced absolute dose delivery uncertainties from as much as 20% without any heterogeneity corrections to a few percent with convolution-superposition or Monte Carlo calculations [6]. Thus improvements in accuracy by as much as 15 to 20% have been noted. These represent potentially large inconsistencies and systematic offsets in the dose delivered in patients during the past decades of clinical experience. In some cases, it invalidates or at least impairs retrospective analysis of clinical dose-response results from the past era in radiation therapy.

4.2. Patient-related process

4.2.1. Imaging for diagnosis, treatment planning and target volume delineation

Imaging technologies have evolved dramatically in the last decades driven by the evolution of computer technologies. The use of 3-D imaging has allowed for very significant improvements in the ability to define target volumes and critical structures so that dramatic improvements can be made in normal tissue sparing with a potential for escalation in tumour doses [7-9]. This has also been extended to 4-D considerations to allow for intra-fraction motion of both tumour and normal tissue zones especially in lung [10]. It is very difficult to quote accuracy improvements in dose delivery; however, it is worth noting that improvements in target volume delineation have significantly decreased the likelihood of a "geographic miss".

4.2.2. Treatment planning and treatment delivery per se (excluding image-guidance)

A major component of dose delivery accuracy to the patient can be assessed by performing end-to-end phantom tests where an anatomical phantom is scanned, planned and treated as if it were a patient. A number of reports have appeared describing such results from multiple institutions [6,11-14]. The conclusions are: (1) simpler dose calculation algorithms (those that do not integrate scatter in 3-D nor handle electron transport) can be inaccurate by as much as 20% in absolute regional dose, (2) 3%/3 mm criteria of acceptability at the k=1 (~1 st. dev.) is reasonable for 3-D conformal radiation therapy (CRT), (3) even with more sophisticated algorithms, as many as 30% of institutions can fail the 7%/4mm criteria of acceptability set by the Radiological Physics Center (RPC) for IMRT dose delivery upon their first attempt [13], and (4) dose delivery accuracy of 5%/4mm should be achievable at the k=1 level (meaning that ~⅓ of situations lay outside these criteria). Furthermore, an examination of the variation in IMRT dose prescription, treatment planning, dose recording, and dose delivery among 803 brain, head-and-neck, and prostate cancer patients who were treated with different treatment planning systems at five different medical institutions indicated significant variations [15]. This raises concerns about the validity of comparing clinical outcomes from IMRT patients and suggests the need for national and/or international guidelines for dose prescription, dose computation, and reporting for meaningful clinical trials in IMRT. *Thus, in summary, dose delivery accuracy has actually been degraded as radiation therapy procedures have moved into the more sophisticated IMRT delivery procedures perhaps prematurely.*

4.2.3. Treatment delivery.
(a) Image-guidance. Image-guided radiation therapy (IGRT) uses imaging in the treatment room and allows for treatment adjustments to account for geometric deviations, both systematic and random. A recent review by Bujold *et al.* [16] demonstrated "that higher-quality dose delivery enabled by IGRT results in higher clinical control rates, reduced toxicity, and new treatment options for patients that previously were without viable options". As with target volume delineation, it is very difficult to quote numerical magnitudes for dose delivery uncertainties associated with geometric displacement uncertainties; however, it is clear that these could be very significant because they are generally in high dose gradient regions, more than 5%/mm, near the tumour or normal tissue structures. IGRT would not be possible without digital technology; hence the "major impact" rating in our evaluation.
(b) Intra-fraction motion. With large CTV to PTV margins, slight tumour motion is relatively insignificant in terms of dose delivery variations. However, with reduced margins, higher dose prescriptions and tumour tracking technologies, the time component has become an extremely relevant consideration in ensuring full and accurate dose delivery to the target while simultaneously keeping normal tissue doses at acceptable levels. A recent review showed that for early staged lung cancers biologically equivalent doses (BED) of 72-80 Gy delivered with conventional techniques have been increased to 100-140 Gy using stereotactic body radiation therapy (SBRT), with tumour tracking or beam-gating techniques [17]. They reported that the average amplitudes of superior-inferior tumour motion were larger than 10 mm in approximately 33% of lung cancers. Clearly this could result in unacceptable dose variations if the treatments were delivered without image guidance and automated beam delivery, all heavily dependent on computer technologies.

4.2.4 Treatment evaluation and follow-up.
In view of the rich set of image data that are generated daily and that can be stored for later evaluation, patient follow-up analysis allows for the generation of much improved dose-response information both for tumour and normal tissues. Dose-volume analysis can now be performed as never before. Two of the 11 research priorities recommend by the Quantitative Analysis of Normal Tissue Effects in the Clinic (QUANTEC) working group included (1) establishment of large continually growing data bases with full access to the 3-D dose matrix and linkage to biomarkers and clinical outcome and (2) development of methods for recording actual delivered dose (i.e., in an individual patient after fractionated radiotherapy) [18]. Both of these priorities are now possible with modern computer technology and are already being implemented [19-21]. Note, however, that proper controls for the accuracy of recorded 3-D/4-D dose distributions must still be applied (e.g., variations in dose algorithm accuracy for different IMRT situations).

5. Conclusions
While dose calibration procedures have yielded relatively small improvements in dose delivery accuracy (perhaps 1-4%) based on computer generated fundamental dosimetry parameters, it is clear that the combination of better dose calculation algorithms and digital imaging have had a more significant impact on dose delivery and geometric accuracy. Thus, the move into 3-D CRT allowed for very significant improvements in accuracy in targeted dose delivery. However, the implementation of IMRT combined with small field dosimetry difficulties has (temporarily) reduced the overall dose delivery accuracy to the point that, on their first attempt, 20 to 30% of institutions were not able to pass the 7%/4 mm criteria set by the RPC.

As a result of computer applications, we are better able to see and track the internal anatomy of the patient before, during and after treatment. This has yielded the most significant enhancement to the overall dose delivery accuracy of radiation therapy of individual patients. Furthermore, a much richer set of 3-D/4-D co-registered dose-image data are becoming available for retrospective analysis of radiobiological and clinical responses paving the way to personalized radiation therapy.

Acknowledgements: This work was influenced by the activities of a consultants' group at the International Atomic Energy Agency (IAEA) which developed a report on *Accuracy Requirements and Uncertainties in Radiation Therapy* as well as research performed at the London Regional Cancer Program through grants provided by the Canadian Institutes of Health Research and the Canadian Cancer Society.

References

[1] Van Dyk J 2012 Commissioning and Quality Assurance. In *Treatment Planning in Radiation Oncology*.(Edited by FM Khan and Gerbi BJ), pp. 144-168. Lippincott, Williams and Wilkins, Philadelphia.
[2] International Commission on Radiation Units and Measurements. ICRU Report 42: Use of computers in external beam radiotherapy procedures with high-energy photons and electrons. 1987 Bethesda, Maryland, International Commission On Radiation Units and Measurements.
[3] International Commission on Radiation Units and Measurements. ICRU Report 24: Determination of Absorbed Dose in a Patient Irradiated by Beams of X or Gamma Rays in Radiotherapy Procedures. 1976 Washington, D.C., International Commission on Radiation Units and Measurements.
[4] Svensson H 1984 Quality assurance in radiation therapy: physical aspects. *Int. J. Radiat. Oncol. Biol. Phys.* **10,** Suppl 1, 59-65.
[5] Andreo P 2011 Accuracy Requirements in Medical Radiation Dosimetry. International Symposium on Standards, Applications and Quality Assurance in Medical Radiation Dosimetry, 1-23. Vienna, Austria, International Atomic Energy Agency (IAEA).
[6] Mijnheer B *et al* (Eds) 2008 *ESTRO Booklet No. 9: Guidelines for the Verification of IMRT*. ESTRO, Brussels, Belgium.
[7] Rasch C, Steenbakkers R and van Herk M 2005 Target definition in prostate, head, and neck. *Semin Radiat Oncol* **15,** 136-145.
[8] Hamilton CS and Ebert MA 2005 Volumetric uncertainty in radiotherapy. *Clin. Oncol. (R. Coll. Radiol.)* **17,** 456-464.
[9] van Herk M 2011 Margins and Margin Recipes. In *Uncertainties in External Beam Radiation Therapy*.(Edited by Palta JR and Mackie TR), pp. 169-190. Medical Physics Publishing, Madison, WI.
[10] Moorees J and Bezak E 2012 Four dimensional CT imaging: a review of current technologies and modalities. *Australas. Phys Eng Sci. Med.* **35,** 9-23.
[11] Gershkevitsh E *et al* 2008 Dosimetric verification of radiotherapy treatment planning systems: results of IAEA pilot study. *Radiother. Oncol.* **89,** 338-346.
[12] Davidson SE *et al* 2008 Technical note: Heterogeneity dose calculation accuracy in IMRT: study of five commercial treatment planning systems using an anthropomorphic thorax phantom. *Med. Phys.* **35,** 5434-5439.
[13] Molineu A *et al* 2005 Design and implementation of an anthropomorphic quality assurance phantom for intensity-modulated radiation therapy for the Radiation Therapy Oncology Group. *Int. J. Radiat. Oncol. Biol. Phys.* **63,** 577-583.
[14] Ibbott GS, Molineu A and Followill DS 2006 Independent evaluations of IMRT through the use of an anthropomorphic phantom. *Technol. Cancer Res. Treat.* **5,** 481-487.
[15] Das IJ *et al* 2008 Intensity-modulated radiation therapy dose prescription, recording, and delivery: patterns of variability among institutions and treatment planning systems. *J. Natl. Cancer Inst.* **100,** 300-307.
[16] Bujold A *et al* 2012 Image-guided radiotherapy: has it influenced patient outcomes? *Semin. Radiat. Oncol.* **22,** 50-61.
[17] Shirato H *et al* 2012 Real-time 4-D radiotherapy for lung cancer. *Cancer Sci.* **103,** 1-6.
[18] Bentzen SM *et al* 2010 Quantitative Analyses of Normal Tissue Effects in the Clinic (QUANTEC): an introduction to the scientific issues. *Int. J. Radiat. Oncol. Biol. Phys.* **76,** S3-S9.
[19] Vinogradskiy Y *et al* 2013 Use of 4-Dimensional Computed Tomography-Based Ventilation Imaging to Correlate Lung Dose and Function With Clinical Outcomes. *Int. J. Radiat. Oncol. Biol. Phys.* **86**, 366-371.
[20] Vinogradskiy Y *et al* 2012 A novel method to incorporate the spatial location of the lung dose distribution into predictive radiation pneumonitis modeling. *Int. J. Radiat. Oncol. Biol. Phys.* **82,** 1549-1555.
[21] Vinogradskiy Y *et al* 2012 Investigation of the relationship between gross tumor volume location and pneumonitis rates using a large clinical database of non-small-cell lung cancer patients. *Int. J. Radiat. Oncol. Biol. Phys.* **82,** 1650-1658.

Journal of Physics: Conference Series **489** (2014) 012099 doi:10.1088/1742-6596/489/1/012099

Integration of a clinical trial database with a PACS

M van Herk[1]
Radiotherapy Department, The Netherlands Cancer Institute / Antoni van Leeuwenhoekhuis, Plesmanlaan 121, 1066 CX Amsterdam, The Netherlands

E-mail: m.v.herk@nki.nl

Abstract. Many clinical trials use Electronic Case Report Forms (ECRF), e.g., from OpenClinica. Trial data is augmented if DICOM scans, dose cubes, etc. from the Picture Archiving and Communication System (PACS) are included for data mining. Unfortunately, there is as yet no structured way to collect DICOM objects in trial databases. In this paper, we obtain a tight integration of ECRF and PACS using open source software. Methods: DICOM identifiers for selected images/series/studies are stored in associated ECRF events (e.g., baseline) as follows: 1) JavaScript added to OpenClinica communicates using HTML with a gateway server inside the hospitals firewall; 2) On this gateway, an open source DICOM server runs scripts to query and select the data, returning anonymized identifiers; 3) The scripts then collects, anonymizes, zips and transmits selected data to a central trial server; 4) Here data is stored in a DICOM archive which allows authorized ECRF users to view and download the anonymous images associated with each event. Results: All integration scripts are open source. The PACS administrator configures the anonymization script and decides to use the gateway in passive (receiving) mode or in an active mode going out to the PACS to gather data. Our ECRF centric approach supports automatic data mining by iterating over the cases in the ECRF database, providing the identifiers to load images and the clinical data to correlate with image analysis results. Conclusions: Using open source software and web technology, a tight integration has been achieved between PACS and ECRF.

1. Introduction

To improve the accuracy and power of a clinical trial, it is important to have full access to treatment planning information and verification images. First, this data allows an independent review of the treatment, and allows analysis of the clinical outcome taking detected deviations into account. Second, when the data is available in digital format, it becomes possible to perform quantitative analysis at the trial bureau, such as calculating dose-volume histograms, biological modelling, or data mining relating images and/or dose distributions to outcome [e.g., 1].
The Radiation Therapy Oncology Group (RTOG) was the first to recognize the need of digital exchange of patient data [2]. RTOG now uses DICOM based solutions [3]. Web-based Electronic Case Report Forms (ECRFs) are becoming the standard way to deal with clinical trials with standardized database formats and dictionaries [4]. However, proper integration of DICOM and e.g., radiotherapy objects in such ECRFs is lacking, other than using unstructured file attachments, screen-

[1] To whom any correspondence should be addressed.

dumps or including limited numeric dose parameters. The aim of this study is to develop a methodology that provides, from within an ECRF environment, access to Hospital PACS to collect DICOM images and objects, anonymize them, provide proper cross-linking of the anonymized objects with the ECRF database and finally submit the data to a central site. The setup should be such that original clinical data and its identifiers are only present in the submitting hospital, anonymized data is only available to the central site, yet that for a data manager; the hospital's internal PACS and the ECRF appear fully integrated.

2. Material and methods

This work connects two pieces of open source software: OpenClinica ECRF builder [5] and the Conquest DICOM server [6] as a PACS gateway. The methodology is, however, applicable to other ECRF's as well.

2.1. OpenClinica

OpenClinica is an open source ECRF system developed by Akaza Research. It conforms to international standards such as the CDISC's (Clinical Data Interchange Standards Consortium) Operational Data Model (ODM). Here we used the free community edition (version 3.1.2) but the company also offers commercial support. OpenClinica depends on modifying a template Excel file to define an event form (figure 1). The Excel file with items 1 and 8 defined are shown in table 1. The integration between OpenClinica and other systems depends on two features of OpenClinica: first that it is possible to use text fields defined in the Excel file to inject JavaScript code into the generated HTML page and second that it is possible to receive data from an external web site. Item 8 is the actual PACS link: JavaScript functions 'link' and 'config' where implemented by the author. The function 'getExternalValue' is part of OpenClinica that allows accessing controlled vocabularies. Here we only use the listener event that it creates to enter information into item 8.

Table 1. Columns of Excel file used to define a typical simple data entry element (1) and PACS link (8) in the ECRF form in OpenClinica shown in figure 1. For the latter item, HTML and JavaScript code is injected to generate the appropriate hyperlinks for PACS coupling. The hyperlink will select a CT scan made on the date entered in item 1 and store the information in item 8. The getExternalValue function is OpenClinica's method to access controlled vocabularies.

Column	**Value (item 1)**	**Value (item 8)**
LEFT_ITEM_TEXT	Date of Physical examination	<A HREF=# onclick = "javascript:link(8,1,'CT'); return false;"> **Baseline CT:** </A>
RIGHT_ITEM_TEXT	2	.
HEADER	Visit information:	language="JavaScript" src=http:includes/marcel.js></script
SUBHEADER		**PACS link:** <A HREF=# onclick="javascript:openconfig(); return false;">Configure</A>
RESPONSE_TYPE		Calculation
RESPONSE_VALUES_OR_CALCULATIONS		func: getExternalValue("http://dum",right,60,40)

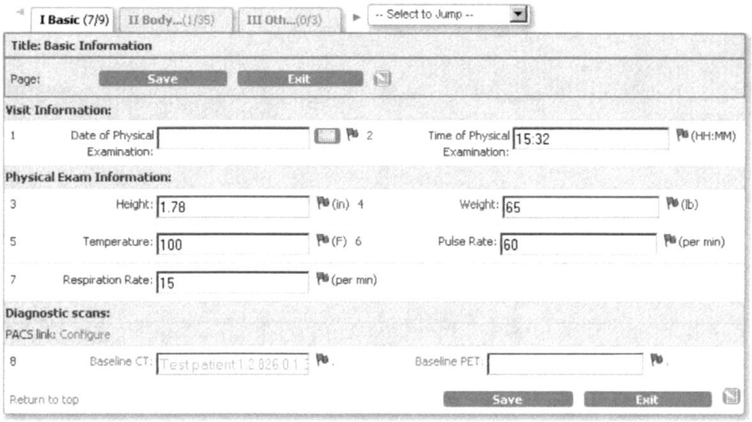

Figure 1. Example OpenClinica data entry form (detail). Items 1 is a standard item, while item 8 is modified to provide the PACS connection.

The PACS link item appears as a standard text item, yet it can only be filled in using the hyperlink left of it.

2.2. Conquest DICOM server

The Conquest DICOM server is based on code from University of California at Davis but it has been greatly extended in the past decade. The system is structured around a server core, with an optional Windows GUI, data storage, a third party SQL server, and an optional web interface. The server core performs all DICOM communication, and provides a scripting system that can interact with almost all tasks of the server. The applied scripting language is Lua which is popular due to its small size, high speed and good versatility in gaming as well as data processing [7]. The script system adds a wide range of operations such as database access, DICOM queries and moves, changing DICOM data, etc. The Conquest DICOM server also provides a web interface in which new pages can be scripted.

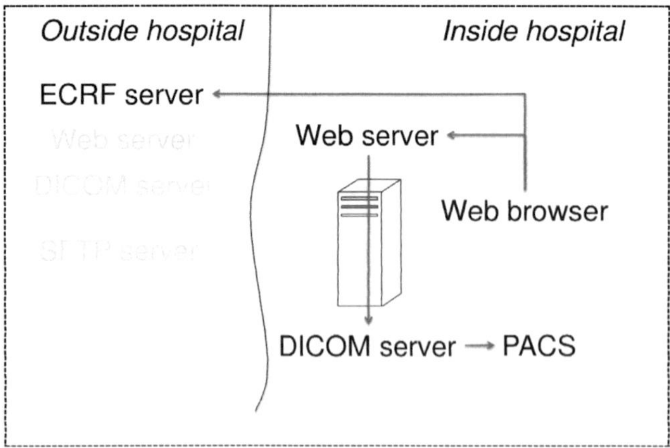

Figure 2. Information flow between the data manager web browser, the central ECRF containing only anonymized data, and the PACS that contains sensitive information. For the data manager, these systems appear fully integrated. The DICOM server gateway keeps anonymisation data.

The integrated view between PACS and ECRF in the data manager's web browser is created by opening specific web pages generated by the Conquest DICOM server web interface from within the ECRF (figure 2). Cross-referencing between ECRF and DICOM occurs by transmission of data from OpenClinica to the Conquest web pages (using a CGI interface) and vice-versa using HTML5's PostMessage function. Two issues exist with this setup: first, the JavaScript code (stored on the outside ECRF server) needs to have access to the (inside) address of the Conquest web server. Then there must be a persistent cross linking between the subject ID of the ECRF and the corresponding Patient ID on the PACS. To solve these issues, site and patient specific data are currently stored in cookies on the web browser. This is acceptable as this browser runs on a system in a protected environment within the hospital.

Screen shots from views presented at time of data entry are shown in figure 3. From the link, a browse window opens that allows users to view the scans that match the selected modality and data using a standard Web Access to DICOM objects (WADO) interface, inspect the original header, inspect the anonymized header, select a study or series, and submit it to the central server. Data to be submitted (e.g., a series with 100 objects) is then passed through an anonymization script integrated in the Conquest DICOM server, collated (zipped) and sent to central storage. The script can be user modified, which is essential as the approval from the PACS administrator to ship data outside the hospital depends on the ability to understand the method, inspect the result (anonymized data can be previewed from the ECRF), and make changes if necessary. Inside the hospital, ECRF data can also be de-anonymised, for instance to verify which PACS data has been submitted previously. Similar screens have been programmed to review and download data for users in other sites than the submitting hospital. They only have access to anonymized data.

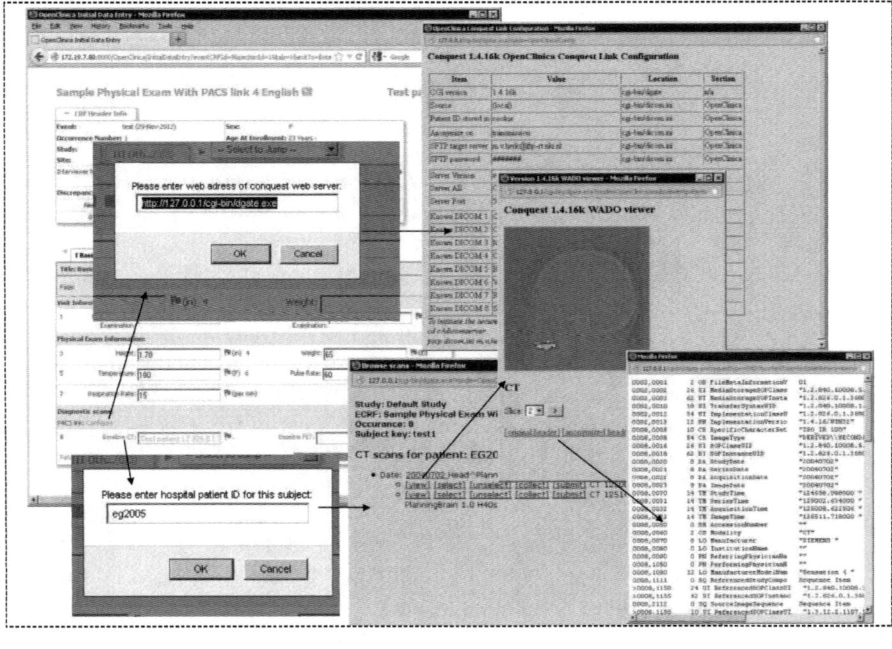

Figure 3. Screens used for cross-linking PACS info (green windows) with ECRF (left-top window). The 'configure' link provides setup of the gateway address and tests its functionality. The 'scan' link asks for the hospital ID and allows inspection and selection of DICOM info.

3. Results and Discussion

An alternative solution to the tight integration sketched here is to send all DICOM data for particular patients or studies to the central office, which leaves the burden of linking the data to events to the central data manager. Because centrally, no access to RIS or patient chart is available this action can be very cumbersome and error-prone. It is therefore much better to provide this linking in the hospital, giving the tools to view and inspect the data to the person with the knowledge or contacts to choose the appropriate scans or other DICOM objects for a certain event. Currently the linking is manual, but if enough information is provided to the ECRF it may be further automated in the future.

OpenClinica is powerful enough to support this integration, but it is far from perfect. Our JavaScript code needs access to variables such as SubjectKey and other essential identifiers. These are, however, not available as JavaScript variables but can only be obtained from the HTML making the integration dependent on the HTML page layout which is OpenClinica version dependent.

The setup described in this paper can easily modified to other ECRF systems. We are currently implementing this approach in the ARTFORCE and AIRFORCE clinical trials, both of which use the TENALEA clinical trial management software [8]. Automated image data mining is possible as follows: First ECRF is exported as ODM XML database - this database contains the UIDs of the anonymized submitted data. We can then loop over the items in this database, load the associated DICOM images and structures using the UIDs in the database, perform image analysis and correlated the results with the clinical clinical findings reported in the same database.

4. Conclusions

A tight connection of PACS and ECRF was demonstrated using open-source software on both ends. The 'hack' depends on JavaScript on the OpenClinica side and Lua scripting on the DICOM server side and is fully user-configurable. The setup allows ECRF-centric collection of DICOM data with all anonymized UIDs stored in the ECRF allowing proper data-mining and is therefore highly preferable to unstructured collection of images. The system will be used be used in the ARTFORCE and AIRFORCE clinical trials.

5. Acknowledgments

This work was supported by the Center for Translational Molecular AIRFORCE project.

6. References

[1] Witte MG, Heemsbergen WD, Bohoslavsky R, Pos FJ, Al-Mamgani A, Lebesque JV, van Herk M 2010 Relating dose outside the prostate with freedom from failure in the Dutch trial 68 Gy vs. 78 Gy. *Int. J. Radiat. Oncol. Biol. Phys.* **77** 131-8

[2] Purdy JA, Harms WB, Michalski J, Cox JD 1996 Multi-institutional clinical trials: 3-D conformal radiotherapy quality assurance. Guidelines in an NCI/RTOG study evaluating dose escalation in prostate cancer radiotherapy. *Front. Radiat. Ther. Oncol.* **29** 255-63

[3] Liu D, Ajlouni M, Jin JY, Ryu S, Siddiqui F, Patel A, Movsas B, Chetty IJ. 2009 Analysis of outcomes in radiation oncology: an integrated computational platform. *Med. Phys.* **36** 1680-9

[4] Ohmann C, Kuchinke W. 2009 Future developments of medical informatics from the viewpoint of networked clinical research. Interoperability and integration. Methods. *Inf. Med.* **48** 45-54

[5] http://www.openclinica.com

[6] http://ingenium.home.xs4all.nl/dicom.html

[7] Roberto Ierusalimschy 2013 *Programming in Lua* (Rio de Janeiro, Brasil: Lua.org)

[8] Wildeman MA, Zandbergen J, Vincent A, Herdini C, Middeldorp JM, Fles R, Dalesio O, van der Donk E, Tan IB. 2011 Can an online clinical data management service help in improving data collection and data quality in a developing country setting? *Trials.* 2011 **12** 190

A DICOM based radiotherapy plan database for research collaboration and reporting

J Westberg[1], S Krogh[1], C Brink[1,2] and I R Vogelius[3]

[1]Laboratory of Radiation Physics, Odense University Hospital, Sdr. Boulevard 29, 5000 Odense, Denmark
[2]Institute of Clinical Research, University of Southern Denmark, Winsløwparken 19, 5000 Odense, Denmark
[3]Department of Radiation Oncology, Rigshospitalet, University of Copenhagen, Copenhagen, Blegdamsvej 9, 2100 København Ø, Denmark

jonas.westberg@rsyd.dk

Abstract. Purpose: To create a central radiotherapy (RT) plan database for dose analysis and reporting, capable of calculating and presenting statistics on user defined patient groups. The goal is to facilitate multi-center research studies with easy and secure access to RT plans and statistics on protocol compliance. Methods: RT institutions are able to send data to the central database using DICOM communications on a secure computer network. The central system is composed of a number of DICOM servers, an SQL database and in-house developed software services to process the incoming data. A web site within the secure network allows the user to manage their submitted data. Results: The RT plan database has been developed in Microsoft .NET and users are able to send DICOM data between RT centers in Denmark. Dose-volume histogram (DVH) calculations performed by the system are comparable to those of conventional RT software. A permission system was implemented to ensure access control and easy, yet secure, data sharing across centers. The reports contain DVH statistics for structures in user defined patient groups. The system currently contains over 2200 patients in 14 collaborations. Conclusions: A central RT plan repository for use in multi-center trials and quality assurance was created. The system provides an attractive alternative to dummy runs by enabling continuous monitoring of protocol conformity and plan metrics in a trial.

1. Introduction

The DICOM protocol and file format was designed in the 1980's to facilitate a standardized way of communicating and interchanging data between medical imaging devices [1]. This made it possible for manufacturers to exchange information in medical images, structured documents, and workflow over a network. The DICOM data model is hierarchical with its patient, study, series and object (modality), and defines a large dictionary of tags to compose the objects and allows vendors to incorporate proprietary information [2, 3]. The DICOM RT supplement includes the RtPlan, RtStruct, RtImage, RtDose and RtRecord modalities [4].

The DICOM standard does not require user credentials to access patient information on a server [5]. A typical PACS only implements a list of clients it is allowed to communicate with, but have no or minimal restrictions on access control after the connection is established.

There exists an increasing demand for tools to interpret and analyze RT plans. With the variety of treatment techniques and treatment planning systems (TPS) it requires technical skills to extract the information from all variations of DICOM datasets. This paper describes a system designed to meet some of the challenges related to dose analysis using DICOM RT data from multiple centers. Similar systems exist [6, 7, 8, 9, 10, 11] with different support of data formats, transmission protocols, data anonymization, distribution model of the computations, review tools, statistics and data aggregation. The goal of this study is to design and implement a system with new features for easy and secure sharing of non-anonymized DICOM RT data.

2. Methods

2.1. Design

The system is designed as a three-tier system as shown in figure 1. Each center has a dedicated DICOM gateway configured on a secure network, which serves as a security entry to protect the central DICOM server. The gateway forwards data to the central DICOM server in the first tier.

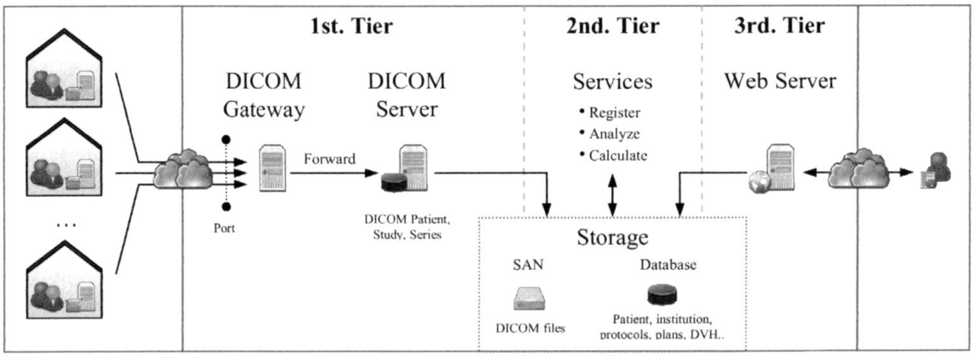

Figure 1. Architecture and design of the system. The centers submit DICOM RT data to the DICOM server (gateway), then data is handled by a three-tier system, and finally the users can access the DICOM RT data and dose reports from a web interface.

The second tier is the core of the system. It contains a series of independent services deployed on computers dedicated for computation. The first service (Register) reads the DICOM file and registers the patient, DICOM series, study and modality in the database. The second service (Analyze) analyzes the content of the file and extracts details depending on the modality (e.g. beams for RT plan). The third service (Calculate) recalculates and stores the DVHs in the database. The third tier implements an ASP.NET website for users to administer and group patient data, configure and retrieve dose reports.

Structure names are not unique and centers can have local naming conventions. To compare DVHs structures names are mapped to a unique identifier. Users can define a structure dictionary to map local structure names to a common name used in the study.

Sharing and grouping data is an important aspect of the system. The database contains a list of data groups and each group can be related to a clinical protocol, a scientific study or created ad-hoc. All new data submitted to the server is by default not included in any data group. A user must log in via the web interface and add the patients data to one or more data groups. The data groups define the basis for collaboration, sharing data and setting up dose reports. Data not included in any group is presumed to be unused and deleted after a certain time.

2.2. Security

A key feature is secure network communication with modern encryption techniques. In our system, the safety of the data transport is provided by The Danish Healthcare Data Network (SDN).

Each data group has a permission matrix with the participating centers as columns and the users as rows as shown in table 1. Three permission rights are available; "group summary" (sum) will allow data from the center to be include in the summary of the dose report, "patient information" (info) allows the user to see patient information too, and "send" gives an option to export the DICOM data from the system to predefined DICOM servers at any center.

User \ Center	A			B			C		
	Sum	Info	Send	Sum	Info	Send	Sum	Info	Send
1				x					
2				x	x				
3							x	x	x

Table 1. Permission matrix example. Each data group has a permission matrix where the collaborating users can be granted access to data from other centers in the group. In this example user 1 only have permission to view summarized data from center B. User 2 is able to do the same and view the plan and patient information too. User 3 has access to all data from center C and able to send the DICOM data to a center. None of the users have access to data in the group from center A.

Special data groups are also provided for transmitting DICOM data ad-hoc between centers.

2.3. DVH Calculation

The DVH calculation depends on the resolution of the dose grid and interpolation method, as the dose grid resolution is often configured to use coarser resolution than the reference CT. This dependency becomes more pronounced for small structures in regions with high dose gradients [12, 13]. To ensure consistency all DVHs are recalculated and stored in the database for comparison and generation of dose statistics. DICOM datasets containing pre-calculated DVHs from TPSs are still imported into the database for visualization purposes but not included in the statistics. DVHs are stored in 1 cGy resolution as binary large objects (BLOB) in the database, and have database fields for the minimum, maximum, and mean doses as well as volumes radiated above specific doses (V_{1Gy}-V_{100Gy} in step of 1 Gy) and dose levels for specific volumes ($D_{2\%}$, $D_{50\%}$ and $D_{98\%}$).

2.4. Data Validation And Verification

Data validation and verification is performed at three levels to comply with the ontology, business logic and ensure quality of data. First the contents of the DICOM files are validated to ensure that it contains a supported modality and verify whether or not the patient, study and series exist in the database. Secondly, the actual RT plan is interpreted to validate the existence of prescription, CT, structures, beams, and dose and verify that all referenced data has been received. Thirdly, it is verified that the submitted patient data in a data group properly describes a treatment course. The third step is the most complicated, and might involve user input for ambiguous cases to determine what action to perform. Examples could be duplicated plans or plan modifications before or during the treatment course. The user is notified in case the automatic process fails.

2.5. Dose Reports

Users can compose dose reports in a data group by selecting the dose or volume to report in the DVH for the structure of interest. The report is generated on-demand from the selected fields in the database when the user views the report in the web interface.

3. Results
All RT centers in Denmark are connected and working in 14 collaboration groups on the server. Currently, the database contains over 2200 patients and almost 300.000 DICOM files. The database is hosted in a SQL hotel, files on a storage area network (SAN), and the DICOM servers, web site, and the DVH calculations run on virtual computers. The NARLAL (National phase II lung trial which is scheduled to include 116 patient [14]) group uses the system to extract the dosimetric information (e.g. dose coverage of tumor and dose to OAR such as lung, spinal cord, heart, and esophagus). This has reduced the costs and workload in reporting dosimetric data and in cases where a patient is referred for re-treatment in another center the previous RT plan is easily exchanged between centers.

4. Discussion
With the transparent setup it is simple to export data from the TPSs and into the system. In the initial setup all the software was running on one computer. As the data traffic increased the core services (from the 2nd tier in figure 1) were moved to another virtual computer to reduce workload. The database was moved to a hosted location and into a more scalable environment.

MedCom handles the SDN in Denmark and provides solutions for foreign countries to connect, with several countries in the EU connected already. Legal issues may arise as countries may have regulations on how data should be handled when transmitting data across borders to an external host.

Computing resources on the internet are moving to cloud based architectures to deliver Software as a Service (SaS). Though cloud computing is an attractive technology to dynamically allocate hosted resources as needed, it has not gained acceptance for storing and handling person related information yet.

Submitting anonymized data could be accommodated. This would require 1) an agreement on how to anonymize DICOM data correctly, 2) ensure that the new patient identifications does not collide, and 3) patients scanned or treated at multiple centers still map to the same identification in all DICOM files. Modifying a DICOM file would require new unique IDs to be generated, which could result in broken references between files. Surprisingly, anonymization software can produce different results and personal information can still be hidden in images as pixels, private DICOM tags, and use of initials or other personal information in text fields.

Missing data or unreported changes is of major concern. E.g. data submitted before end of treatment would describe the intended plan, while the actual treated plan(s) are recorded in the Record and Verify System (RVS). Exporting the treatment records from the RVSs would increase the quality of data. Unfortunately, the export of treatment records in DICOM format is not yet supported by all major RVSs.

5. Conclusion
We have implemented a multi-center collaboration platform for research in radiotherapy based on DICOM RT. Data can be transmitted to the server using DICOM export from a TPS or PACS, and it is not mandatory to anonymize data. Users can set up dose reports from a web interface to ensure protocol compliance.

Acknowledgment
This work was funded by The Lundbeck Foundation Centre for Interventional Research in Radiation Oncology (CIRRO), Work Package 9.

References
[1] NEMA PS3 / ISO 12052, Digital Imaging and Communications in Medicine (DICOM) Standard, National Electrical Manufacturers Association, Rosslyn, VA, USA (available free at http://medical.nema.org/)
[2] NEMA PS3 / ISO 12052, PS3.5: Data Structures and Encoding
[3] NEMA PS3 / ISO 12052, PS3.6: Data Dictionary

[4] NEMA PS3 / ISO 12052, PS3.3: DICOM Information Object Definitions

[5] NEMA PS3 / ISO 12052, PS3.15: DICOM Security and System Management Profiles

[6] Roelofs E et al 2010 Design of and technical challenges involved in a framework for multicentric radiotherapy treatment planning studies *RT Oncol.* **97** 567-71

[7] http://www.elekta.com/healthcare-professionals/products/elekta-software/oncology-informatics-data-alliances/radiation-oncology-data-alliance.html

[8] Hopkins S, Oakes L, Upasani S, Goldwein J 2010 Radiation Oncology Data Alliance as a Vehicle for Radiation Dosing Analysis for Lung, Breast, and Prostate *Am. J. of Clinical Oncol.* **33** 213

[9] McNutt T R, Evans K, Wu B, Kahzdan M, Simari P, Sanguineti G, Herman J, Taylor R, Wong J, DeWeese T 2010 Oncospace: All Patients on Trial for Analysis of Outcomes, Toxicities, and IMRT Plan Quality **78** *IJROBP* S486

[10] Ebert M A, Haworth A, Kearvell R, Hooton B, Coleman R, Spry N, Bydder S, Joseph D 2008 Detailed review and analysis of complex radiotherapy clinical trial planning data: Evaluation and initial experience with the SWAN software system *Radiotherapy Oncol.* **86** 200-10

[11] Palta J R, Efstathiou J A, Bekelman J E, Mutic S, Bogardus C R, McNutt T R, Gabriel P E, Lawton C A, Zietman A L, Rose C M 2012 Developing a national radiation oncology registry: From acorns to oaks *Practical Radiation Oncol.* **2** 10-7

[12] Straube W, Matthews J, Bosch W and Purdy J 2005 SU-FF-T-310: DVH Analysis: Consequences for Quality Assurance of Multi-Institutional Clinical Trials *Med. Phys.* **32** 2021-2

[13] Ebert M A, Haworth A, Kearvell R, Hooton B, Hug B, Spry N A, Bydder S A, Joseph D J 2010 Comparison of DVH data from multiple radiotherapy treatment planning systems *Phys. Med. Biol.* **55** N337–46

[14] http://clinicaltrials.gov/ct2/show/NCT00887783?term=NARLAL&rank=1

AUTHOR INDEX

Ackerly, T 197
Ahmad, S 165
Ajlouni, M........ 32
Akagi, T 1
Altunbas, C........ 169
Ammazzalorso, F 61, 208
Andersen, J A L 373
Ansbacher, W........ 328
Appenzoller, L M 269
Arm, J 161
Arumugam, S 305, 310, 332, 338, 348, 404
Aso, T 1
Ayadi, M 115
Bailey, M........ 95
Baldock, C 226
Bangert, M........ 6, 212
Barber, J 344
Bassler, N........ 12, 17, 217
Battista, J........ 476
Batumalai, V........ 276
Bäuerle, T 143
Bawazeer, O........ 348
Beavis, A W 414
Bednarz, T 208
Behrens, C F........ 373
Bell, L........ 276
Bendl, R 143, 392, 396
Blake, S J 22
Borrisova, S 373
Bouilhol, G........ 115
Bourgeat, P 138, 357
Boydev, C........ 121
Brink, C 487
Brousmiche, S........ 383
Burdett, N........ 232
Cao, F........ 221
Carolan, M........ 95, 420
Champion, C 51
Chandra, S........ 132, 232
Charles, P H 28, 66
Chen, D 32
Chen, J 151, 156
Chen, M 75, 353, 365
Chetty, I J 32
Cho, G A 36, 226
Clark, H 42, 426
Clasie, B 46
Coolens, C........ 431
Coulthard, A........ 138, 357
Crabtree, T 106
Crowe, S B 28, 66, 249
Crozier, S........ 138, 357
Dang, K........ 132
Das, K J M........ 202
Davidson, T N........ 181
Deegan, T 361
Delaney, G........ 276
Delmon, V 127
Denham, J W 436
Depauw, N........ 46
Derraz, F 121
Descallar, J........ 348
Deshpande, S........ 310, 404
Devpura, S........ 32
Dowling, J 132, 187, 232
Dowson, N 138, 187, 357
Driscoll, B........ 431
D'Souza, D........ 400
Dunn, L 456
Duzenli, C........ 42, 426
Dyk, J V........ 476
Ebert, M 226, 420, 436, 451
Evans, K 264, 460
Evans, P........ 147, 410
Fay, M 138, 357
Fenton, P 447
Foo, K........ 420, 436
Foroudi, F 132
Fox, C D 132
Franich, R D 70
Frantzis, J 447
Fränzle, A........ 143
Fripp, J........ 232
Froening, H 243
Fukushima, C........ 173
Gagne, I M 328
Gal, Y 138, 357
Galmarini, D 75, 353, 365
George, A 310, 404
Gholami, Y........ 51
Ghose, S........ 232
Gibbs, A........ 361
Gill, B 42, 426
Gill, S........ 132
Giske, K........ 392
Goozee, G 305, 310, 404
Gordon, J 32
Gray, A........ 348, 404
Greer, P........ 22, 56, 161, 187, 232
Gulliford, S 420, 436, 441
Hacker-Prietz, A........ 460
Hackworth, W 70
Hansen, D C........ 17
Harden, F 361, 369
Harding, R 291
Hargrave, C 361, 369
Harris, E........ 147, 410
Hartill, C 165
Hartmann, M........ 238
Hawkes, D 165

AUTHOR INDEX

Haworth, A 420, 436
Heinrich, H 243, 254
Hennig, P 6
Herman, J 264, 460
Hill, B 28
Hillengass, J 143
Holloway, L 22, 226, 276, 305, 310, 332, 338, 344, 348, 404
Iwai, G 1
Jacques, R 90
Jaffray, D 295, 431
James, C 447
Jeffree, R 138, 357
Jelen, U 61, 208
Jensen, N 151, 156
Jeraj, R 80
Jones, B L 169
Joseph, D J 436
Juneja, P 147
Kairn, T 28, 66, 249
Kamerling, C P 243, 254, 322
Karakaya, E 291
Karaszewska, B 273
Kato, M 111
Kato, S 173
Kavanagh, B 169
Keall, P 161
Kearvall, R 436
Kennedy, A 436, 451
Kenny, J 28, 249, 456
Kim, J-H 36
Kim, T 161
Knight, R 28, 249
Koh, E S 276
Koken, P 387
Kooy, H 46
Krawiec, M 436
Krogh, S 487
Kron, T 70, 132
Kumar, S 32, 202
Kumazaki, Y 173
Kuncic, Z 22, 51, 56, 86, 226
Kupelian, P A 471
Kurosawa, T 111
Labarbe, R 383
Landau, D 165
Lane, J 451
Langton, C M 28
Lausch, A 151, 156
Lee, B 221
Lee, D 161
Lee, R 42, 426
Lee, T Y 151, 156
Lehmann, J 456
Leong, C 221
Li, H 32
Li, R 316
Lindsay, P E 295
Lindsay, R 291
Liu, D 32
Lock, M 151, 156
Lonski, P 70
Low, D A 379, 471
Lu, W 75, 353, 365
Lühr, A 12, 17
Luo, Y 260
Lye, J 456
Mackie, T R 80
Madden, T 46
Maeda, Y 1
Mahou-Lago, S M 106
Martin, D 36
Martin, J 165
McClelland, J 165, 400
McDowall, R 138
McNamara, A L 22, 86
McNutt, T 90, 264, 460
McShan, D 260
Meir, I 165
Mellemgaard, A 373
Mengersen, K 361, 369
Metcalfe, P 276
Meyer, J 465
Michalski, J M 269
Miften, M 169
Millar, J 197
Miyaura, K 173
Mo, X 75, 353, 365
Modat, M 400
Modersitzki, J 181
Moinuddin, S 400
Moiseenko, V 42, 426
Monville, M E 56
Moore, J 264, 460
Moore, K L 269
Moores, M 361, 369
Morishita, Y 111
Moseley, D J 177, 181
Movsas, B 32
Munbodh, R 177
Murray, D 286
Mutic, S 269
Nishio, T 1
Oborn, B 95, 420
Oelfke, U 6, 86, 212, 243, 254, 322
Oliva, M V 127, 383
Olivera, G 75, 353, 365
Osewski, W 273
Osmond, J 410
Ottosson, W 373

AUTHOR INDEX

Ourselin, S 400
Pagulayan, C 332
Parker, J 232
Parnell, D 75, 353, 365
Pasquier, D 121
Perl, J 100
Peshko, O 181
Petersen, J B 17, 217
Peyrodie, L 121
Pham, D 132
Phillips, M H 286, 465
Phipps, A 70
Pichler, P 232
Pickard, S 36
Pinho, R 115, 127
Pistorius, S 193
Pogson, E M 276
Poulsen, M 361
Prajapati, S 80
Prestwich, R 291
Qi, X S 379
Quon, H 460
Radon, M 61
Ralston, A 36
Richardson, S 436
Ricketts, K 400
Ripps, L 447
Rit, S 115, 127, 383
Rivest-Hénault, D 187, 232
Roche, M 106
Rockhill, J K 465
Rose, S 138, 357
Rosenfeld, A 46
Royle, G 400
Ruan, D 282, 471
Ruben, J 197
Saito, N 111
Saitoh, H 173
Salvado, O 138, 232, 357
Santiago, A 61
Sarrut, D 115, 127, 383
Sasaki, T 1
Sayre, G A 282
Schefter, T 169
Scheurleer, J 387
Schipper, M 260
Schneider, U 238
Schroeder, J 221
Schwartz, J L 286
Schwarz, M 392, 396
Sen, M 291
Shao, W 471
Sharabi, A 460
Sharp, G C 383
Shimizu, M 111
Siddiqui, M S 32
Slosarek, K 273
Smith, J 138, 357
Smith, W P 465
Sobolevsky, N 12, 17
Sonke, J-J 299
Speight, R 291
Stanwell, P 232
Stewart, J M P 295
Stewart, N 436
Stewart, R D 286
Stoiber, E M 392
Stoll, M 392, 396
Sumkauskaite, M 143
Sun, J 232
Swift, C-L 328
Sykes, J 291, 344
Takase, W 1
Takata, N 111
Taleb-Ahmed, A 121
Tan, J 269
Tanaka, T 111
Taylor, M L 70
TenHaken, R 260
Teo, P T 193
Terlaky, T 181
Teske, H 396
Thiran, J P 121
Thomas, C 165
Thomas, P 138, 357
Thomas, S 42, 426
Thomsen, B 17
Thorstad, W L 269
Thwaites, D 226, 344, 348
Tin, M M 36
Toftegaard, J 12, 217
Toghyani, M 51
Trapp, J V 28, 66, 249
Tse, R 36
Van Herk, M 299, 482
Vandemeulebroucke, J 127
Veiga, C 400
Vial, P 22, 338, 348, 404
Vogelius, I R 487
Wang, J 471
Ward, J W 414
Watase, Y 1
Wen, C-D 197
Wessel, R 387
Westberg, J 487
White, B 379
Williams, I 456
Williams, M 95
Winter, C 138, 357
Witte, M 299

AUTHOR INDEX

Wittig, A 61
Wong, C 197
Wong, E 151, 156
Wu, J 42, 426
Xing, A 305, 310, 332, 338, 404
Xing, L 316
Yamashita, T 1
Yang, W 264, 460
Yip, C 165
Yoganathan, S A 202
Ziegenhein, P 212, 243, 254, 322
Zin, H 410